Environmental Geochemistry of Sulfide Oxidation

ACS SYMPOSIUM SERIES **550**

Environmental Geochemistry of Sulfide Oxidation

Charles N. Alpers, EDITOR
U.S. Geological Survey

David W. Blowes, EDITOR
University of Waterloo

Developed from a symposium sponsored
by the Division of Geochemistry, Inc.,
at the 204th National Meeting
of the American Chemical Society,
Washington, DC,
August 23–28, 1992

American Chemical Society, Washington, DC 1994

Library of Congress Cataloging-in-Publication Data

Environmental geochemistry of sulfide oxidation : developed from a symposium sponsored by the Division of Geochemistry, Inc., at the 204th National Meeting of the American Chemical Society, Washington, DC, August 23–28, 1992 / Charles N. Alpers, David W. Blowes, [editors].

p.   cm.—(ACS symposium series, ISSN 0097–6156; 550)

Includes bibliographical references and indexes.

ISBN 0–8412–2772–1

1. Sulphides—Oxidation—Congresses.   2. Environmental geochemistry—Congresses.

I. Alpers, Charles N., 1958–   . II. Blowes, David W., 1956–   .
III. American Chemical Society. Division of Geochemistry.
IV. American Chemical Society. Meeting (204th: 1992: Washington, D.C.) V. Series.

QE516.S1E58   1994
628.1'6832—dc20   93–39250
                                                              CIP

The paper used in this publication meets the minimum requirements of American National Standard for Information Sciences—Permanence of Paper for Printed Library Materials, ANSI Z39.48–1984. ∞

Copyright © 1994

American Chemical Society

All Rights Reserved. The appearance of the code at the bottom of the first page of each chapter in this volume indicates the copyright owner's consent that reprographic copies of the chapter may be made for personal or internal use or for the personal or internal use of specific clients. This consent is given on the condition, however, that the copier pay the stated per-copy fee through the Copyright Clearance Center, Inc., 27 Congress Street, Salem, MA 01970, for copying beyond that permitted by Sections 107 or 108 of the U.S. Copyright Law. This consent does not extend to copying or transmission by any means—graphic or electronic—for any other purpose, such as for general distribution, for advertising or promotional purposes, for creating a new collective work, for resale, or for information storage and retrieval systems. The copying fee for each chapter is indicated in the code at the bottom of the first page of the chapter.

The citation of trade names and/or names of manufacturers in this publication is not to be construed as an endorsement or as approval by ACS of the commercial products or services referenced herein; nor should the mere reference herein to any drawing, specification, chemical process, or other data be regarded as a license or as a conveyance of any right or permission to the holder, reader, or any other person or corporation, to manufacture, reproduce, use, or sell any patented invention or copyrighted work that may in any way be related thereto. Registered names, trademarks, etc., used in this publication, even without specific indication thereof, are not to be considered unprotected by law.

PRINTED IN THE UNITED STATES OF AMERICA

1993 Advisory Board

ACS Symposium Series
M. Joan Comstock, *Series Editor*

V. Dean Adams
University of Nevada—
  Reno

Robert J. Alaimo
Procter & Gamble
  Pharmaceuticals, Inc.

Mark Arnold
University of Iowa

David Baker
University of Tennessee

Arindam Bose
Pfizer Central Research

Robert F. Brady, Jr.
Naval Research Laboratory

Margaret A. Cavanaugh
National Science Foundation

Dennis W. Hess
Lehigh University

Hiroshi Ito
IBM Almaden Research Center

Madeleine M. Joullie
University of Pennsylvania

Gretchen S. Kohl
Dow-Corning Corporation

Bonnie Lawlor
Institute for Scientific Information

Douglas R. Lloyd
The University of Texas at Austin

Robert McGorrin
Kraft General Foods

Julius J. Menn
Plant Sciences Institute,
  U.S. Department of Agriculture

Vincent Pecoraro
University of Michigan

Marshall Phillips
Delmont Laboratories

George W. Roberts
North Carolina State University

A. Truman Schwartz
Macalaster College

John R. Shapley
University of Illinois
  at Urbana–Champaign

L. Somasundaram
DuPont

Peter Willett
University of Sheffield (England)

Foreword

THE ACS SYMPOSIUM SERIES was first published in 1974 to provide a mechanism for publishing symposia quickly in book form. The purpose of this series is to publish comprehensive books developed from symposia, which are usually "snapshots in time" of the current research being done on a topic, plus some review material on the topic. For this reason, it is necessary that the papers be published as quickly as possible.

Before a symposium-based book is put under contract, the proposed table of contents is reviewed for appropriateness to the topic and for comprehensiveness of the collection. Some papers are excluded at this point, and others are added to round out the scope of the volume. In addition, a draft of each paper is peer-reviewed prior to final acceptance or rejection. This anonymous review process is supervised by the organizer(s) of the symposium, who become the editor(s) of the book. The authors then revise their papers according to the recommendations of both the reviewers and the editors, prepare camera-ready copy, and submit the final papers to the editors, who check that all necessary revisions have been made.

As a rule, only original research papers and original review papers are included in the volumes. Verbatim reproductions of previously published papers are not accepted.

M. Joan Comstock
Series Editor

Contents

Preface xiii

LABORATORY STUDIES OF SULFIDE-OXIDATION KINETICS

1. **Rates of Reaction of Galena, Sphalerite, Chalcopyrite, and Arsenopyrite with Fe(III) in Acidic Solutions** 2
   J. Donald Rimstidt, John A. Chermak, and Patrick M. Gagen

2. **Laboratory Studies of Pyrrhotite Oxidation Kinetics** 14
   Ronald V. Nicholson and Jeno M. Scharer

3. **Effect of Humidity on Pyrite Oxidation** 31
   Sandra L. Borek

4. **Kinetics of Hydrothermal Enrichment of Chalcopyrite** 45
   Joon H. Jang and Milton E. Wadsworth

MICROBIAL PROCESSES AFFECTING SULFIDE OXIDATION

5. **Oxidation of Inorganic Sulfur Compounds by Thiobacilli** 60
   Isamu Suzuki, C. W. Chan, and T. L. Takeuchi

6. **Microbial Oxidation of Sulfides by *Thiobacillus denitrificans* for Treatment of Sour Water and Sour Gases** 68
   K. L. Sublette, Michael J. McInerney, Anne D. Montgomery, and Vishveshk Bhupathiraju

7. **Solid-Phase Alteration and Iron Transformation in Column Bioleaching of a Complex Sulfide Ore** 79
   Lasse Ahonen and Olli H. Tuovinen

8. Alteration of Mica and Feldspar Associated with the Microbiological Oxidation of Pyrrhotite and Pyrite 90
    Tariq M. Bhatti, Jerry M. Bigham, Antti Vuorinen, and Olli H. Tuovinen

    NUMERICAL MODELING OF SULFIDE OXIDATION IN TAILINGS, WASTE ROCK, AND IN SITU DEPOSITS

9. Rates of Mechanisms That Govern Pollutant Generation from Pyritic Wastes 108
    A. I. M. Ritchie

10. Attempts To Model the Industrial-Scale Leaching of Copper-Bearing Mine Waste 123
    L. M. Cathles

11. A Computer Program To Assess Acid Generation in Pyritic Tailings 132
    Jeno M. Scharer, Ronald V. Nicholson, Bruce Halbert, and William J. Snodgrass

12. Time–Space Continuum Formulation of Supergene Enrichment and Weathering of Sulfide-Bearing Ore Deposits 153
    Peter C. Lichtner

    SOLUBILITY AND SORPTION CONTROL IN FORMATION OF SULFIDE-OXIDATION PRODUCTS

13. Influence of Siderite on the Pore-Water Chemistry of Inactive Mine-Tailings Impoundments 172
    C. J. Ptacek and David W. Blowes

14. Mineralogical Characteristics of Poorly Crystallized Precipitates Formed by Oxidation of Fe^{2+} in Acid Sulfate Waters 190
    E. Murad, U. Schwertmann, Jerry M. Bigham, and L. Carlson

15. Atomic and Electronic Structure of PbS {100} Surfaces and Chemisorption–Oxidation Reactions 201
    Carrick M. Eggleston and Michael F. Hochella, Jr.

TRANSPORT OF SULFIDE-OXIDATION PRODUCTS
IN SURFACE WATERS

16. **Effects of Instream pH Modification on Transport of Sulfide-Oxidation Products**.. 224
    Briant A. Kimball, Robert E. Broshears,
    Diane M. McKnight, and Kenneth E. Bencala

17. **Transport and Natural Attenuation of Cu, Zn, As, and Fe in the Acid Mine Drainage of Leviathan and Bryant Creeks**........ 244
    Jenny G. Webster, D. Kirk Nordstrom, and
    Kathleen S. Smith

18. **Acid Mine Drainage in Wales and Influence of Ochre Precipitation on Water Chemistry**....................................... 261
    Ron Fuge, Fiona M. Pearce, Nicholas J. G. Pearce,
    and William T. Perkins

TRANSPORT AND STORAGE OF SULFIDES AND OXIDATION
PRODUCTS IN SEDIMENTS

19. **Stratigraphy and Chemistry of Sulfidic Flood-Plain Sediments in the Upper Clark Fork Valley, Montana**........................ 276
    David A. Nimick and Johnnie N. Moore

20. **Release of Toxic Metals via Oxidation of Authigenic Pyrite in Resuspended Sediments**.................................... 289
    John W. Morse

21. **Mobilization and Scavenging of Heavy Metals Following Resuspension of Anoxic Sediments from the Elbe River**....... 298
    W. Calmano, U. Förstner, and J. Hong

EFFECTS OF SULFIDE-OXIDATION PROCESSES
ON GROUND-WATER GEOCHEMISTRY

22. **Seasonal Variations of Zn/Cu Ratios in Acid Mine Water from Iron Mountain, California**................................. 324
    Charles N. Alpers, D. Kirk Nordstrom, and
    J. Michael Thompson

23. **Secondary Iron-Sulfate Minerals as Sources of Sulfate and Acidity: Geochemical Evolution of Acidic Ground Water at a Reclaimed Surface Coal Mine in Pennsylvania**.............. 345
    C. A. Cravotta III

24. **Limit to Self-Neutralization in Acid Mine Tailings: The Case of East Sullivan, Quebec, Canada**......... 365
    M. D. Germain, N. Tassé, and M. Bergeron

    SULFIDE-OXIDATION PROCESSES IN WETLANDS
    AND THE OCEANS

25. **Attenuation of Acid Rock Drainage in a Natural Wetland System**......... 382
    Y. T. J. Kwong and D. R. Van Stempvoort

26. **Kinetics of Oxidation of Hydrogen Sulfide in Natural Waters**......... 393
    Jia-Zhong Zhang and Frank J. Millero

    RECENT ADVANCES IN ANALYTICAL METHODS

27. **Determination of Hydrogen Sulfide Oxidation Products by Sulfur K-Edge X-ray Absorption Near-Edge Structure Spectroscopy**......... 412
    Appathurai Vairavamurthy, Bernard Manowitz, Weiqing Zhou, and Yongseog Jeon

28. **Evolved-Gas Analysis: A Method for Determining Pyrite, Marcasite, and Alkaline-Earth Carbonates**......... 431
    Richard W. Hammack

    STABLE ISOTOPE FRACTIONATION AND
    EQUILIBRATION IN OXIDIZING SULFIDE SYSTEMS

29. **Controls of $\delta^{18}O$ in Sulfate: Review of Experimental Data and Application to Specific Environments**......... 446
    D. R. Van Stempvoort and H. R. Krouse

30. **Sulfur- and Oxygen-Isotope Geochemistry of Acid Mine Drainage in the Western United States: Field and Experimental Studies Revisited**......... 481
    B. E. Taylor and Mark C. Wheeler

    SUPERGENE OXIDATION AND ENRICHMENT
    OF SULFIDE ORE DEPOSITS

31. **Applications of Mass-Balance Calculations to Weathered Sulfide Mine Tailings**......... 516
    Edward C. Appleyard and David W. Blowes

32. Oxidation of Massive Sulfide Deposits in the Bathurst Mining Camp, New Brunswick: Natural Analogues for Acid Drainage in Temperate Climates 535
    D. R. Boyle

33. Thiosulfate Complexing of Platinum Group Elements: Implications for Supergene Geochemistry 551
    Elizabeth Y. Anthony and Peter A. Williams

REMEDIATION AND PREVENTION OF THE ENVIRONMENTAL EFFECTS OF SULFIDE OXIDATION

34. Suppression of Pyrite Oxidation Rate by Phosphate Addition 562
    Xiao Huang and V. P. Evangelou

35. Iron Sulfide Oxidation: Impact on Chemistry of Leachates from Natural and Pyrolyzed Organic-Rich Shales 574
    Thomas L. Robl

36. Oxidation of Sulfide Minerals Present in Duluth Complex Rock: A Laboratory Study 593
    Kim A. Lapakko and David A. Antonson

37. Chemical Predictive Modeling of Acid Mine Drainage from Metallic Sulfide-Bearing Waste Rock 608
    W. W. White III and T. H. Jeffers

38. Composition of Interstitial Gases in Wood Chips Deposited on Reactive Mine Tailings: Consequences for Their Use as an Oxygen Barrier 631
    N. Tassé, M. D. Germain, and M. Bergeron

39. Field Research on Thermal Anomalies Indicating Sulfide-Oxidation Reactions in Mine Spoil 645
    Weixing Guo and Richard R. Parizek

INDEXES

Author Index 660

Affiliation Index 661

Subject Index 661

Preface

THE STUDY OF SULFIDE-OXIDATION REACTIONS, the characterization of the products of these reactions, and the prediction of the fate of sulfide-oxidation products in natural environments are increasingly active areas of research. This research is motivated by the locally severe environmental impacts and toxicity associated with the products of sulfide oxidation. Mine wastes are of particular concern because of past and present disposal practices that have resulted in the widespread degradation of surface-water and ground-water quality. Mine wastes have therefore come under increasing regulation.

This book provides an interdisciplinary overview of recent research on geochemical processes of sulfide oxidation. Researchers from a wide variety of disciplines are represented: geochemistry, microbiology, hydrology, mineralogy, physics, and civil engineering. The studies included range from theoretical modeling exercises to laboratory- and field-based studies.

One of our main objectives in organizing the symposium on which this book is based was to foster new collaborations among researchers in related fields. Our hope is that the interdisciplinary nature of this volume will encourage researchers to look beyond the highly specialized literature and to incorporate recent advances in kinetics, microbiology, numerical modeling, surface chemistry, equilibrium solubility relations, analytical methods, and remediation technology in their future research efforts. We understand that the seeds sown at the symposium have already begun to bear fruit: Several such new collaborative efforts have begun.

The structure of this volume is similar to that of the symposium: Twelve different themes related to sulfide oxidation processes are represented as different parts of the book. The order follows a progression of geochemical environments starting at the source area and evolving with increasing distance along both surface-water and ground-water flow paths. Hence, the first two sections of the book deal with laboratory studies of sulfide-oxidation reaction kinetics and microbiological processes.

Subsequent sections of the book include chapters on the solubility and sorption control of aqueous metal concentrations, the transport of sulfide-oxidation products in surface waters, the transport and storage of sulfides and sulfide-oxidation products in sediments, and the effects on ground-water geochemistry. Additional sections are included on analytical methods, on the stable isotopes of sulfur and oxygen in systems undergoing sulfide oxidation, and on "supergene" (weathering-related) oxidation and enrichment of sulfide-bearing ore deposits.

The final section of the book presents chapters on remediation and the prevention of environmental impacts of sulfide oxidation. This topic is of growing concern to mining companies, government regulators, and environmentalists. It is clear that environmental aspects of mining must play an important role in mine design from the initial planning stages. Progress will be necessary in many areas of science to ensure that the mistakes of the past are not repeated and that the environmental impacts of mining will be minimized or eliminated in the decades to come.

Acknowledgments

We thank the authors of the chapters in this book for their commitment to meeting the necessary deadlines and for their patience with our numerous requests. We also thank the many researchers who assisted in the review of the chapters in this book. Each chapter was reviewed by at least one and in most cases two or more external referees as well as by both editors. Many of the reviews were highly constructive and provided new insights that were generously shared with the authors.

We gratefully acknowledge financial assistance from the Waterloo Centre for Groundwater Research and the ACS Division of Geochemistry for the travel costs of foreign speakers at the symposium. The program chairs of the ACS Division of Geochemistry, James A. Davis and Mary Sohn, and the treasurer, Dan Melchior, were helpful and supportive in the planning and execution of the symposium. Chris Hanton-Fong assisted with the final proofreading and typesetting of several of the manuscripts; her efforts are gratefully appreciated. Finally, we thank Rhonda Bitterli, Meg Marshall, and the rest of the ACS Books staff for their assistance during the editing and production of this volume.

CHARLES N. ALPERS
Water Resources Division
U.S. Geological Survey
Room W–2233
2800 Cottage Way
Sacramento, CA 95825

DAVID W. BLOWES
Waterloo Centre for Groundwater Research
University of Waterloo
Waterloo, Ontario N2L 3G1
Canada

RECEIVED August 10, 1993

Laboratory Studies of Sulfide-Oxidation Kinetics

Chapter 1

Rates of Reaction of Galena, Sphalerite, Chalcopyrite, and Arsenopyrite with Fe(III) in Acidic Solutions

J. Donald Rimstidt[1], John A. Chermak[2], and Patrick M. Gagen[3]

[1]Department of Geological Sciences, Virginia Polytechnic Institute and State University, Blacksburg, VA 24061
[2]Mineralogisches Institut, Universität Bern, 3012 Bern, Switzerland
[3]O.H.M. Corporation, 1000 Holcomb Woods Parkway, Roswell, GA 30076

We measured the rates of reaction of Fe(III) with galena, sphalerite, chalcopyrite, and arsenopyrite at 25°C and pH values near two and found:

MINERAL	RATE LAW	ACTIVATION ENERGY
Galena	$r_{Fe^{3+}} = -2.82 \times 10^{-3} (A) (m_{Fe^{3+}})^{0.98}$	40 kJ mol^{-1}
Sphalerite	$r_{Fe^{3+}} = -8.13 \times 10^{-7} (A) (m_{Fe^{3+}})^{0.58}$	27 kJ mol^{-1}
Chalcopyrite	$r_{Fe^{3+}} = -1.78 \times 10^{-7} (A) (m_{Fe^{3+}})^{0.43}$	63 kJ mol^{-1}
Arsenopyrite	$r_{Fe^{3+}} = -1.45 \times 10^{-3} (A) (m_{Fe^{3+}})^{0.98}$	≈18 kJ mol^{-1} (0-25°C) −6 kJ mol^{-1} (26-60°C)

Where $r_{Fe^{3+}}$ is the rate of reduction of Fe(III) to Fe(II) (mol sec^{-1}), $m_{Fe^{3+}}$ is the concentration of Fe(III) (mol kg^{-1}), and A is the surface area of the solid (m^2) exposed to the solution. These results show that the reaction rates, the reaction orders and the activation energies vary substantially from mineral to mineral indicating that the details of the reaction mechanism differ among the various sulfide minerals. This means that pyrite cannot be used as a proxy for these other sulfide minerals when studying the details of sulfide mineral oxidation.

The oxidation of sulfide minerals is a beneficial process in some cases and a detrimental one in others. This process is responsible for the secondary enrichment of many ore deposits making them more valuable; in addition, the natural sulfide mineral oxidation process has been adapted to recover metals from sulfide ores through hydrometallurgy. On the other hand, the oxidation of sulfide minerals exposed naturally, or more often by mining activities, often produces serious pollution problems.

Iron plays a central role in oxidizing sulfide mineral deposits. When pyrite or pyrrhotite oxidize, they release Fe(II) into solution. The dissolution of iron-bearing silicates, oxides, or carbonates in the acidic solutions that are common in this environment also releases Fe(II). At near neutral pH, Fe(II) is rapidly oxidized to Fe(III) by dissolved oxygen (1, 2). At low pH, the rate of this reaction via the inorganic route is quite slow but *Thiobacillus ferrooxidans* and related bacteria accelerate the rate as they derive metabolic energy by using O_2 to oxidize the Fe(II) to Fe(III) (1, 3). Moses (4) suggests that Fe(III) is the primary oxidant of pyrite even at near neutral pH even though it is relatively insoluble under these conditions, and his reaction model suggests that Fe(III) is probably involved in the oxidation of most sulfide minerals. Some of the Fe(III) reacts with other nonferrous sulfides, as well as the iron-bearing ones, to release other metal ions into solution (5) and during this process it is converted to Fe(II). This Fe(II) is reoxidized to Fe(III) to complete the cycle. Finally, the excess Fe(III) precipitates as iron oxyhydroxides whose solubility is controlled by the Fe(III) activity and the pH (6).

The primary goal of this study was to gather information on the rate of reaction of Fe(III) with galena, sphalerite, chalcopyrite, and arsenopyrite in acidic solutions similar to those found in weathering sulfide deposits. These data can be used in models which account for the sources and sinks of Fe(III) in weathering sulfide deposits. Berner (7) reports that dissolved iron concentrations in acidic mine waters are often as high as 100 to 500 mg L^{-1} (2 to 9x10^{-3} m), with pH values of less than two. In our experiments, Fe(III) concentrations ranged from 10^{-2} to 10^{-4} molal (m) and the pH was near two. The predominant galena oxidation reaction under these conditions is

$$PbS + 8\ Fe^{3+} + 4\ H_2O \rightarrow 8\ H^+ + SO_4^{2-} + Pb^{2+} + 8\ Fe^{2+} \qquad (1)$$

where some of the Pb^{2+} and SO_4^{2-} react to precipitate anglesite ($PbSO_4$). The overall sphalerite oxidation reaction is:

$$ZnS + 8\ Fe^{3+} + 4\ H_2O \rightarrow 8\ H^+ + SO_4^{2-} + Zn^{2+} + 8\ Fe^{2+}. \qquad (2)$$

The most important oxidation reaction for arsenopyrite is

$$FeAsS + 13\ Fe^{3+} + 8\ H_2O \rightarrow 14\ Fe^{2+} + SO_4^{2-} + 13\ H^+ + H_3AsO_4(aq) \qquad (3)$$

where some of the Fe(III) can react with the dissolved arsenate to precipitate scorodite ($FeAsO_4 \cdot 2H_2O$). Finally, the most important reaction for chalcopyrite is

$$CuFeS_2 + 16\ Fe^{3+} + 8\ H_2O \rightarrow Cu^{2+} + 17\ Fe^{2+} + 2\ SO_4^{2-} + 16\ H^+. \qquad (4)$$

A second goal of this study was to compare the characteristics of the reaction of galena, arsenopyrite, sphalerite, and chalcopyrite with Fe(III) to the pyrite reaction. There have been several quantitative studies of the reaction of pyrite with Fe(III) both because it is a geochemically important reaction and because pyrite may serve as a model system for all sulfide mineral oxidation. If pyrite is to serve as an effective model system, the characteristics of the pyrite reaction must be substantially the same as for other sulfide minerals.

Little is known about the reaction rates of galena and sphalerite in dilute Fe^{3+} solutions similar to those found in nature as compared to the relatively extensive hydrometallurgical research using very strongly oxidizing solutions. Previous hydrometallurgy studies on galena and sphalerite, tabulated by Chermak (8), used very concentrated (up to 3 molal) $FeCl_3$ solutions which are much more reactive than natural waters. The numerous previous studies of the oxidation rate of chalcopyrite under various conditions were designed to improve the hydrometallurgical recovery of copper, and also were performed using much higher concentrations of oxidants than are normally found in nature; examples are tabulated by Gagen (9). Most of these studies indicated that a significant amount of elemental sulfur is produced by intense oxidation of chalcopyrite; however, our experiments which used much lower concentrations of Fe(III) produced no elemental sulfur. Although arsenic pollution from mine waste dumps is a potentially serious environmental problem, arsenopyrite oxidation has received very little study and there are no reported values of rate constants or activation energy.

Experimental design

All of the sulfide samples used in these experiments were obtained from Ward's Natural Science Establishment, Inc. The galena and sphalerite are from the Mississippi Valley-type deposits of the Tri-state district of Kansas, Missouri, and Oklahoma. The chalcopyrite came from Messina, Transvaal, Republic of South Africa and the arsenopyrite came from Gold Hill, Utah. The samples were hand sorted to remove macroscopic impurities before crushing. Microprobe analysis of the run solids showed that their chemical formulas are $Zn_{0.96}Fe_{0.009}Cd_{0.006}S_{1.02}$, PbS, $CuFeS_2$, and $FeAs_{0.95}S_{1.05}$ (8, 9). The mineral grains were examined in polished section and the chalcopyrite was found to contain minor (<1%) sulfide contaminants (pyrite and covellite) whereas the galena, sphalerite, and arsenopyrite were free of contaminants. The crystals were crushed in a steel mortar and pestle. The resulting powder was passed under a magnet to remove any steel chips and then sieved to isolate the 60-100 mesh size fraction. All samples were rinsed five times in acetone (galena and sphalerite) or ethanol (chalcopyrite and arsenopyrite), to remove the very fine particles which adhere to the grains after crushing. A 2.0000±0.0005 gram sample was used in each experiment. The specific surface area of the unreacted material was determined by three-point N_2 BET analysis to be 0.020± 0.001 $m^2 g^{-1}$ for galena, 0.024± 0.001 $m^2 g^{-1}$ for the sphalerite, 0.049 $m^2 g^{-1}$ for chalcopyrite, and 0.066±0.02 $m^2 g^{-1}$ for arsenopyrite.

Because our goal here was to determine the fundamental rate of reaction of Fe^{3+} with these sulfide minerals, independent of any surface layer diffusion processes, we used $FeCl_3$ solutions instead of Fe_2SO_4 solutions. This avoided the precipitation of $PbSO_4$ on the surface of the galena grains and the lead concentration of the solutions was low enough to avoid precipitation of lead chloride. Run solutions were prepared by adding appropriate amounts of 0.50 molal ferric chloride solution to acidified distilled-deionized water (1.5 mL concentrated HCl per one liter of water). Fe(III) concentrations of the run solutions ranged from 10^{-2} to 10^{-5} molal. The pH was checked prior to each experiment. The initial Fe(III) concentrations were checked by spectrophotometry using the 1,10 phenanthroline procedure of (10).

The mixed flow reactor (MFR) used for most of these measurements was constructed following the design of Rimstidt and Dove (11). The sample was held between two layers of 100 mesh nylon screen. The sample holder was set in the one liter reaction kettle (volume of solution in the operating vessel = 987 mL). The temperature was controlled by immersion in a constant temperature water bath. The feed solution was supplied to the reactor kettle by a peristaltic pump and the Eh of the effluent solution was monitored by an Eh electrode. The feed solution was in equilibrium with oxygen in the ambient air. When the effluent solution reached a constant Eh value the corresponding emf was recorded and approximately 25 mL of effluent solution was sampled; then the reactor kettle was opened and the mineral sample was removed, rinsed in distilled deionized water to arrest the reaction, dried, and stored in a desiccator.

The electrodes used were combination platinum/saturated 4 m KCl/Ag–AgCl Eh electrodes. They were calibrated against a standard Zobell solution (12). The temperature correction for the emf of this Zobell solution is given by Nordstrom(13). The ferric ion content of the effluent was calculated from the following equations:

$$Eh = E° + (RT/nF) \ln (a_{Fe^{3+}}/a_{Fe^{2+}}) \tag{5}$$

$$T_{Fe} = m_{Fe(III)} + m_{Fe(II)} \tag{6}$$

$$a_{Fe^{3+}} = (\gamma_{Fe^{3+}})(m_{Fe^{3+}}) \text{ and } a_{Fe^{2+}} = (\gamma_{Fe^{2+}})(m_{Fe^{2+}}) \tag{7}$$

where the standard state is a hypothetical ideal one molal solution. The electrode readings (*emf*) were adjusted such that

$$Eh = emf (v) + 0.19373 \tag{8}$$

for the galena experiments or

$$Eh = emf(v) + 0.19635 \qquad (9)$$

for the sphalerite, chalcopyrite and arsenopyrite experiments. The corrections specified by equations (8) and (9) are discussed by Wiersma (14). Equations (5) through (9) can be combined to give

$$m_{Fe^{3+}} = \frac{T_{Fe}\left(\gamma_{Fe^{2+}}/\gamma_{Fe^{3+}}\right)\left(e^{(Eh-E°)nF/RT}\right)}{1 + \left(\gamma_{Fe^{2+}}/\gamma_{Fe^{3+}}\right)\left(e^{(Eh-E°)nF/RT}\right)} \qquad (10)$$

This equation was derived by Wiersma and Rimstidt (15). Equation (10) was used to calculate the concentration of Fe(III) in the effluent from the measured emf. The concentration of free Fe(II) and Fe(III) used in equation (6) were adjusted for the formation of chloride, hydroxide, and sulfate complexes.

The apparent rate of consumption of Fe(III), r', was found by multiplying the difference between the Fe(III) concentration (number of moles of Fe(III), $n_{Fe^{3+}}$, per mass of feed solution, M) in the feed solution and the effluent solution by the flow rate (16).

$$r' = (dn/dt)_{rxn} = [(n_{Fe^{3+}}/M)_{in} - (n_{Fe^{3+}}/M)_{out}] [dM/dt] \qquad (11)$$

where dM/dt = the flow rate. The reaction rate for a standard system with 1 m² of surface area is

$$r = r'/A \qquad (12)$$

where A is the amount of mineral surface area exposed to the solution.

The mixed flow reactor and ancillary equipment proved to be generally well suited for this type of experiment. The time for a reaction to reach a steady state varied depending on the sulfide mineral and the flow rate, ranging from 4 to 8 hours for galena with a flow rate of 2.4×10^{-4} kg sec^{-1} to 20 to 32 hours for sphalerite when the flow rate was slowed to 6.55×10^{-6} kg/sec. The slower sphalerite and chalcopyrite reactions produced Fe(II) concentrations very near the lowest detectable limit. In fact, these reaction were so slow at 25°C the rates could not be determined. In order to increase the amount of Fe(II) in the effluent, more solids could be used, the temperature increased, or the flow rate slowed. The minimum rate observable is dictated by the minimum Fe(II) to Fe(III) ratio needed to give an accurate Eh reading. This experiment works best with Fe(III) concentrations ranging from 10^{-2} to 10^{-5} molal. For iron concentrations greater than 10^{-2} molal the extended Debye-Hückel equation does not give an accurate estimate of activity coefficients. At Fe(III) concentrations below 10^{-5} molal the electrode does not respond solely to the ferrous/ferric couple (17). Sample size can probably range from 1 to 10 grams depending on the specific surface area and reaction rate of the mineral. In the sphalerite and chalcopyrite experiments, we increased the reaction rate by raising the temperature to 40° and 60°C and used the Arrhenius equation to extrapolate the rate laws to the lower temperatures expected in acid mine drainage environments.

To determine whether, over the long duration (from 9 to 23 hours) of the mixed flow reactor experiments with arsenopyrite, a scorodite layer might form and inhibit the reaction, a series of batch reactor experiments of short duration (40 minutes) was performed. In these experiments, the arsenopyrite was held in a sample holder similar to the one used in the MFR system. This sample holder was shorter (11.4 cm instead of 16.5 cm) so the entire assembly was completely submerged in the 500 mL of solution used in these experiments. A stirrer with the same type blade used in the MFR circulated the solution downward through the mesh that held the arsenopyrite grains. The Eh electrode was mounted in the kettle outside of the sample holder. This entire apparatus and the FeCl₃ solution were thermally equilibrated in the water bath before 500 mL of FeCl₃ solution was poured into the reactor kettle to initiate the reaction. The emf readings, recorded at five-minute intervals, were used to determine $m_{Fe^{3+}}$ as previously described. The initial rate method described by McKibben and Barnes (18) was used to calculate the apparent rate constant, r', which was then converted to the actual rate, r, using equation (12). Note that this method gives the reaction rate very near $t = 0$, before any scorodite could have formed to inhibit the reaction.

The effect of Fe(III) concentration on the reaction rate can be expressed as

$$dn_{Fe^{3+}}/dt = r = k(m_{Fe^{3+}})^n \qquad (13)$$

where k is the reaction rate constant and n is the reaction order for Fe(III). Taking the logarithm of both sides of equation (13) yields

$$\log r = n \log (m_{Fe^{3+}}) + \log k \qquad (14)$$

so that a graph of log r versus log $m_{Fe^{3+}}$ should be a straight line with a slope of n. The rate data from all experiments are given in Table I; detailed results are listed in Chermak (8) and Gagen (9). Figure 1 (based on data in Table I) shows for example the results from the chalcopyrite oxidation experiments. The rate constants change with temperature as expressed by the Arrhenius equation:

$$k = A_a(e^{-E_a/RT}) \qquad (15)$$

Where A_a is a pre-exponential term, E_a is the activation energy of the reaction, R is the gas constant, and T is the temperature in Kelvins. Taking the logarithm of both sides of this equation gives

$$\log k = -(E_a/2.303R)(1/T) + \log A_a \qquad (16)$$

This can be substituted into equation (14) to give

$$\log r = n \log m_{Fe^{3+}} -(E_a/2.303\ R)(1/T) + \log A_a \qquad (17)$$

This is a linear equation of the form

$$\log r = a (\log m_{Fe^{3+}}) + b (1/T) + c \qquad (18)$$

Results

Coefficients from the regression of the rate data for galena, sphalerite, and chalcopyrite to equation (14) for each temperature and to equation (18) for all temperatures are listed in Table II. Activation energies calculated from the b term of these fits ($b = -E_a/2.303R$) are 40 kJ mol⁻¹ for galena, 27 kJ mol⁻¹ for sphalerite, and 63 kJ mol⁻¹ for chalcopyrite.

The results for the Fe(III)-arsenopyrite reaction are not consistent with the Arrhenius equation. The Arrhenius graph for arsenopyrite (Figure 2) shows a very unusual behavior: from 0 to 25°C the E_a is ≈ 18 kJ mol⁻¹ but from 25 to 60°C the E_a is ≈ – 6 kJ mol⁻¹. Thus, these data could not be fit to equation (18); instead rate laws applicable for each temperature at which experiments were performed are listed in Table II. The significance of the negative activation energy is discussed in the conclusions.

Scanning electron microscope (SEM) examination of run solids proved to be very useful for understanding the reaction processes. Reaction residues of galena experiments show the development of deep etch pits along incipient cleavage traces, a pitted surface composed of 5 to 75 μm etch pits and 1 to 10 μm blobs of elemental sulfur. The deep, well developed etch pits are most likely the result of strain energy introduced into the grains during the crushing process (19, 20). The localization of the elemental sulfur into blobs is good evidence that it was transported to a site of nucleation and deposited as the result of the decomposition of a relatively soluble and stable aqueous species. If it were formed directly from sulfide mineral by a reaction such as

$$MeS = 1/8\ S_8 + 2\ e^- + Me^{2+} \qquad (19)$$

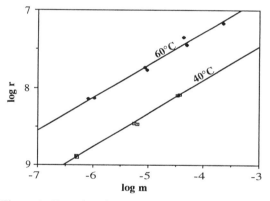

Figure 1. Rate data for the chalcopyrite experiments.

Table I. Rate Data from All Experiments

EXP#	pH	log $m_{Fe^{3+}}$ mol kg^{-1}	log r mol m^{-2} sec^{-1}	EXP#	pH	log $m_{Fe^{3+}}$ mol kg^{-1}	log r mol m^{-2} sec^{-1}
GALENA				**ARSENOPYRITE**			
25°C				**0°C**			
PbS2	1.96	-3.40	-5.99	Aspy 71	1.79	-3.000	-6.19
PbS5	2.00	-4.37	-7.02	Aspy 72	1.82	-3.498	-6.44
PbS7	2.03	-3.45	-5.81	Aspy 73	1.79	-3.002	-6.14
PbS8	1.87	-3.72	-6.02	Aspy 74	1.82	-3.506	-6.65
PbS9	1.50	-2.22	-4.72	Aspy 75	1.79	-4.031	-6.99
40°C				Aspy 76	1.79	-4.043	-7.08
PbS11	1.98	-3.63	-5.79	**15°C**			
PbS13	2.20	-4.62	-6.67	Aspy 68	1.81	-3.072	-6.06
PbS14	1.45	-2.41	-4.62	Aspy 69	1.82	-3.564	-6.47
SPHALERITE				Aspy 70	1.81	-4.067	-6.93
40°C				**25°C**			
ZnS6	2.01	-3.08	-7.64	Aspy 55	1.80	-3.100	-5.88
ZnS7	2.00	-4.21	-8.30	Aspy 56	1.83	-3.569	-6.40
60°C				Aspy 57	1.80	-4.132	-6.83
ZnS1	2.00	-4.34	-8.01	Aspy 58	1.80	-3.098	-5.86
ZnS2	1.98	-3.74	-7.67	Aspy 59	1.83	-3.657	-6.41
ZnS3	2.01	-3.11	-7.51	Aspy 61	1.80	-4.180	-6.98
ZnS4	2.00	-4.29	-8.21	**35°C**			
ZnS5	2.01	-3.20	-7.34	Aspy 62	1.80	-3.171	-5.97
CHALCOPYRITE				Aspy 63	1.83	-3.706	-6.47
40°C				Aspy 64	1.80	-4.291	-7.09
Cpy 30	1.88	-4.42	-8.09	Aspy 65	1.83	-3.741	-6.45
Cpy 31	1.90	-5.20	-8.47	Aspy 66	1.80	-4.278	-7.00
Cpy 32	1.90	-5.25	-8.46	**40°C**			
Cpy 33	1.89	-6.29	-8.90	Aspy 77	1.72	-3.125	-6.00
Cpy 34	1.89	-6.30	-8.89	Aspy 78	1.79	-3.640	-6.58
Cpy 35	1.90	-4.47	-8.10	Aspy 79	1.79	-4.265	-7.14
60°C				**60°C**			
Cpy 16	1.93	-4.30	-7.45	Aspy 81	1.79	-3.782	-6.67
Cpy 17	1.87	-4.36	-7.35	Aspy 82	1.79	-4.231	-7.06
Cpy 20	1.86	-6.07	-8.15	Aspy 83	1.72	-3.398	-6.26
Cpy 21	1.86	-5.96	-8.14				
Cpy 22	1.87	-5.01	-7.77				
Cpy 23	1.87	-5.06	-7.74				
Cpy 29	1.90	-3.64	-7.17				

Table II. Rate Laws Based on Best Fit of Experimental Data [At fixed temperatures the rate law is of the form: $\log r = n \log (m_{Fe^{3+}}) + \log k$; for polythermal data the rate law is of the form: $\log r = n \log m_{Fe^{3+}} - (E_a/2.303R)(1/T) + \log A_a$, where T is the temperature in Kelvins. N is the number of data used in the regression and R is the correlation coefficient]

T, °C	N	R^2	RATE LAW
			GALENA $E_a = 40$ kJ mol^{-1}
25*			$\log r = 0.98(\pm 0.06)\log m_{Fe^{3+}} - 2.55(\pm 3.05)$
25	5	0.97	$\log r = 1.03(\pm 0.11)\log m_{Fe^{3+}} - 2.37(\pm 0.40)$
40	3	1.00	$\log r = 0.93(\pm 0.02)\log m_{Fe^{3+}} - 2.39(\pm 0.07)$
25-40	8	0.98	$\log r = 0.98(\pm 0.06)\log m_{Fe^{3+}} - 2101(\pm 643)/T + 4.50(\pm 2.15)$
			SPHALERITE $E_a = 27$ kJ mol^{-1}
25*			$\log r = 0.58(\pm 0.09)\log m_{Fe^{3+}} - 6.09(\pm 2.45)$
40	2	1.00	$\log r = 0.58(\pm 0.00)\log m_{Fe^{3+}} - 5.84(\pm 0.00)$
60	5	0.88	$\log r = 0.58(\pm 0.12)\log m_{Fe^{3+}} - 5.59(\pm 0.46)$
40-60	7	0.93	$\log r = 0.58(\pm 0.09)\log m_{Fe^{3+}} - 1433(\pm 532)/T - 1.28(\pm 1.68)$
			CHALCOPYRITE $E_a = 63$ kJ mol^{-1}
25*			$\log r = 0.43(\pm 0.10)\log m_{Fe^{3+}} - 6.75(\pm 0.54)$
40	6	1.00	$\log r = 0.43(\pm 0.01)\log m_{Fe^{3+}} - 6.20(\pm 0.07)$
60	7	0.98	$\log r = 0.42(\pm 0.02)\log m_{Fe^{3+}} - 5.60(\pm 0.12)$
40-60	13	1.00	$\log r = 0.43(\pm 0.01)\log m_{Fe^{3+}} - 3287(\pm 120)/T + 4.28(\pm 0.36)$
			ARSENOPYRITE $E_a \approx 18$ kJ mol^{-1} (0-25°C) and ≈ -6 kJ mol^{-1} (25-60°C)
0	6	0.96	$\log r = 0.84(\pm 0.08)\log m_{Fe^{3+}} - 3.62(\pm 0.28)$
15	3	1.00	$\log r = 0.87(\pm 0.02)\log m_{Fe^{3+}} - 3.38(\pm 0.07)$
25	6	0.99	$\log r = 0.98(\pm 0.05)\log m_{Fe^{3+}} - 2.84(\pm 0.17)$
35	5	0.99	$\log r = 0.98(\pm 0.05)\log m_{Fe^{3+}} - 2.85(\pm 0.20)$
40	3	1.00	$\log r = 1.00(\pm 0.06)\log m_{Fe^{3+}} - 2.91(\pm 0.22)$
60	3	1.00	$\log r = 0.96(\pm 0.06)\log m_{Fe^{3+}} - 3.01(\pm 0.23)$

*Estimated from fit of polythermal data.

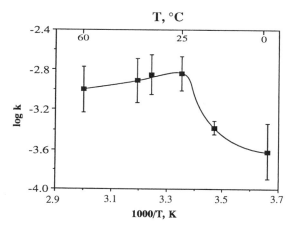

Figure 2. Arrhenius plot for the reaction of Fe(III) with arsenopyrite. Error bars represent one standard error.

where no aqueous transport occurred, the elemental sulfur should appear as a homogeneous coating on the surface. On the other hand, if it were formed from aqueous $S_2O_3^{2-}$ intermediate, as suggested by Rimstidt (21) according to the overall reaction

$$3 \; HS_2O_3^- = 1/2 \; S_8(s) + 2 \; SO_4^{2-} + H^+ + H_2O \qquad (20)$$

(22), the elemental sulfur would preferentially grow upon nucleation sites to form the observed discrete blobs. SEM and X-ray diffraction analysis of the solids showed anglesite and S(s) as reaction products from galena oxidation and S(s) from sphalerite oxidation. SEM photographs of sphalerite after exposure to the ferric chloride solutions showed several deep etch pits of ill-defined shape and arrangement (no such pits were visible on unreacted sphalerite grains). The ill-defined shapes and arrangement of these etch pits suggest that they may not be related to strain in the crystal lattice as was the case for galena. A very few small sulfur globules occurred within some of these pits; sulfur development on the sphalerite was much less extensive than on the galena showing only 1 to 3 μm sized blobs within the etch pits which were 10 to 40 μm in the longest direction. SEM examination of reacted chalcopyrite and arsenopyrite showed no detectable change of the chalcopyrite surfaces and minor scorodite formation on portions of the arsenopyrite grain surfaces from the MFR experiments. The arsenopyrite grains from the initial rate experiments, from which the reported rate data were derived, showed no detectable surface alteration or products. Reflected light examination of the reacted grains showed no traces of covellite or other copper or iron sulfides on the chalcopyrite grains and no visible arsenic or iron sulfides on the arsenopyrite grains. Our experiments showed scorodite as the only solid product of arsenopyrite oxidation by Fe(III). Scorodite deposition was very minor in our experiments because of its relatively high solubility in acidic solutions (23). Elemental sulfur was not observed on either chalcopyrite or arsenopyrite.

Analysis of Pb, Zn, Cu, and As in the effluent solutions from the galena, sphalerite, chalcopyrite, and arsenopyrite experiments (8, 9), respectively, showed that the ratios of these product species to the amount of Fe(III) reduced never matched exactly the stoichiometries predicted by equations (1) through (4). For example, equation (1) predicts that the ratio of Fe(II) to dissolved lead in the effluent solution should be 8 to 1. The measured ratios ranged from 6.2 to 1 to 7.8 to 1 suggesting that the reaction given by equation (1) (i.e. sulfate production) is most important but some sulfur species with a lower oxidation state must have also been produced. Some of these lower oxidation state sulfur species eventually decompose to form the elemental sulfur observed on the grain surfaces. On the other hand, the sphalerite reactions showed Fe(II)/Zn(II) ratios ranging from 2.1 to 4.1 suggesting that lower oxidation state sulfur is an important reaction product. However, SEM observations showed only very small amounts of elemental sulfur on the surfaces of the reacted sphalerite grains. Similar relationships were obtained for chalcopyrite and arsenopyrite. One possible explanation for these observations is that the aqueous sulfur (± arsenic) species with intermediate oxidation states formed in these experiments persist metastably in solution. The lifetimes of such species in the acid, oxidizing solutions used here are unknown but may be quite long. Takano (24) reports that $S_4O_6^{2-}$ and $S_5O_6^{2-}$ persisted at concentrations between 300 and 500 ppm in 0.1 m HCl solutions for more than two years.

Conclusions

The results of this study illustrate the wide range of behaviors for the reaction of Fe(III) with sulfide minerals (Table II). For example, although galena and sphalerite are both simple, cubic monosulfides, they react with Fe(III) at quite different rates. At 25°C and $m_{Fe^{3+}} = 10^{-3}$ the reaction rate for galena is about 300 times faster than that of sphalerite. The reaction order for Fe(III) is 0.98 for galena but only 0.58 for sphalerite.

Arsenopyrite, though commonly considered to be one of the most refractory of the ore minerals, undergoes relatively rapid reaction with acidic ferric chloride solutions. Table II shows that arsenopyrite is among the most reactive of the minerals studied. On the other hand, chalcopyrite and covellite react slowest of the minerals studied and chalcopyrite reacts about 30 times slower than pyrite. At low Fe(III) concentrations used in these experiments, some soluble sulfur anion(s), rather than elemental sulfur, was the dominant sulfur species produced. This is consistent with observations in weathering sulfide deposits where elemental sulfur is rarely found.

The relatively large values of the activation energies for the galena and chalcopyrite reactions (Table II) suggest the reaction rate is controlled by a chemical reaction mechanism rather than rate control by diffusion of ions to or from the mineral surfaces through a layer of reaction products. The activation energy of 27 kJ mol^{-1} for sphalerite is only slightly higher than the 20 kJ mol^{-1} expected for diffusion control. The activation energy of arsenopyrite varies significantly over the temperature range considered. The E_a is $\approx$ 18 kJ mol^{-1} from 0 to 25°C and then becomes $\approx$ –6 kJ mol^{-1} from 25 to 60°C. The slightly negative E_a for arsenopyrite does not appear to be the result of the precipitation of scorodite on the arsenopyrite in the higher temperature experiments because the initial rate method determines the rate at a very small extent of reaction, before significant scorodite could precipitate. The negative activation energy is best explained by a sequence of reactions that contains a branch where one reaction path is dominant at all temperatures while the less vigorous side reaction that produces an inhibitor of the rate-limiting step becomes increasingly important at higher temperatures. Because none of the other iron-bearing sulfide minerals show this behavior, circumstantial evidence points to an arsenic species as the most likely inhibitor. Negative activation energies, have been reported in other complex reactions such as the chemisorption of gases on solids ([25]) and reactions in cool flames ([26], [27]).

Figure 3 compares the relative rates of consumption of Fe(III) by the sulfide minerals studied here to pyrite (data from Rimstidt, *Geochim. Cosmochim. Acta* in press). The figure shows galena, pyrite, and arsenopyrite oxidize relatively quickly whereas sphalerite and chalcopyrite react sluggishly. In nature, the development of anglesite on the exterior of galena and scorodite on the surface of arsenopyrite produces an armoring effect which slows the reaction rate. Given a larger extent of reaction, we would likely have found that the reaction rate was controlled by diffusion of reactants and/or products through such a product layer.

Table III shows a comparison of the reaction rate of Fe(III) with various sulfides. This wide range of behavior together with the negative activation energy for arsenopyrite oxidation demonstrates that the chemistry of sulfide mineral oxidation is quite complex and that simply studying pyrite oxidation as a proxy for the other sulfide minerals is not likely to produce a sufficient understanding of the reaction mechanism. Although this complexity may frustrate our present attempts to understand the reaction mechanism, it offers us the hope that once the mechanism is elucidated the oxidation behavior of the various sulfide minerals can be modified, using catalysts and inhibitors, to improve the technology of hydrometallurgy and to decrease pollution problems arising from mine wastes. This is a particularly important goal in the cases of the minerals studied here. If the oxidation rate of chalcopyrite could be accelerated without producing an armoring layer of elemental sulfur, hydrometallurgical extraction of copper from this mineral would be more economic. On the other hand, the relatively rapid rate of oxidation of arsenopyrite in acidic Fe(III) solutions indicates mine tailings containing modest amounts of pyrite and arsenopyrite could release arsenic into the environment at a significant rate. In this case we must discover a method to inhibit this reaction in order to mitigate the arsenic pollution problem.

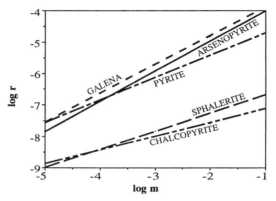

Figure 3. Graph showing the rate of reaction of Fe(III) at 25°C with various sulfide minerals as a function of the Fe(III) concentration in solution at pH = 2. Equations for the lines are given in Table II. Data for pyrite from (9).

Table III. Comparison of the Reaction Rates of Fe(III) with Various Sulfide Minerals [Left side: $m_{Fe^{3+}} = 10^{-3}$ and pH = 2; data for pyrite and marcasite are from (15); data for blaubleibender covellite and covellite are from (29); Right side: $m_{Fe^{3+}} = 10^{-1}$; data from (28)]

MINERAL	RATE mol m^{-2} sec^{-1}	MINERAL	RATE relative rate
Arsenopyrite	1.7×10^{-6}	Pyrrhotite	4.3
Galena	1.6×10^{-6}	Tetrahedrite	3.8
Pyrite	2.7×10^{-7}	Arsenopyrite	1.7
Marcasite	1.5×10^{-7}	Galena	1.5
Blaubleibender Covellite	7.1×10^{-8}	Enargite	1.1
Sphalerite	7.0×10^{-8}	Marcasite	0.9
Chalcopyrite	9.6×10^{-9}	Pyrite	0.8
Covellite	9.1×10^{-9}	Sphalerite	0.5
		Chalcopyrite	0.3

Acknowledgments - This research was supported by National Science Foundation Grants EAR-8318948 and EAR-9004312 and by the Department of Interior's Mining and Mineral Resources Research Institutes program administered by the Bureau of Mines under allotment grant number G1164151. We thank William Newcomb for performing the numerous surface area measurements reported here and for help with many other laboratory tasks. James Light, Don Bodell, and Patricia Dove helped design, build, and maintain the mixed flow reactor system. We thank J. R. Craig, R. J. Bodnar, and Mike McKibben for their comments and review of prior versions of this paper.

Literature Cited

1. Singer, P. C.; Stumm,W. Science 1970, 167, 1121-1123.
2. Millero, F. J.; Sotolongo, S.; Izaguirre, M. Geochim. Cosmochim. Acta 1987, 51, 793-801.
3. Monroe, D. Am. Biotech. Lab. 1985, 10-20.
4. Moses, C. O. Ph.D.Thesis, University of Virginia, 1988.
5. Peters, E. Met. Trans B 1976, 7B, 505-517.
6. Murray, J. W. In Marine Minerals; R. G. Burns, Ed.; Reviews in Mineralogy; Mineralogical Society of America: Washington DC, 1979, 6, pp 47-98.
7. Berner, R. A. Low temperature geochemistry of iron; Handbook of Geochemistry; Springer Verlag: Berlin, 1970; section 26, pp II-1.
8. Chermak, J. A., M.S. Thesis, Virginia Polytechnical Institute and State University 1986.
9. Gagen, P. M. M. S. Thesis, Virginia Polytechnic Institute and State University 1987.
10. Harvey, A. E.; Smart, J. A.; Amis, E. S. Anal. Chemistry 1955, 27, 26-29.
11. Rimstidt, J. D.; Dove, P. M. Geochim. Cosmochim. Acta 1986, 50, 2509-2516.
12. Langmuir, D. In Procedures in Sedimentary Petrology R. E. Carver, Ed.; John Wiley and Sons: New York, 1971; pp 597-635.
13. Nordstrom, D. K. Geochim. Cosmochim. Acta 1977, 41, 1835-1841.
14. Wiersma, C. L. MS Thesis, Virginia Polytechnic Institute and State University 1982.
15. Wiersma, C. L.; Rimstidt, J. D. Geochim. Cosmochim. Acta 1984, 48, 85-92.
16. Levenspiel, O. Chemical Reaction Engineering; John Wiley and Sons: New York: 1972.
17. Natarajan, D. A.; Iwasaki, I. Minerals Science Engineering 1974, 6, 35-44.
18. McKibben, M. A.; Barnes, H. L., Geochim. Cosmochim. Acta 1986, 50, 1509-1520.
19. Lasaga, A. C.; Blum, A. E. Geochim. Cosmochim. Acta 1986, 50, 2363-2379.
20. Brantley, S. L.; Crane, S. R.; Crerar, D. A.; Hellman, R.; Stallard, R. Geochim. Cosmochim. Acta 1986, 50, 2349-2361.
21. Rimstidt, J. D.; Chermak, J. A.; Newcomb, W. D. Fifth International Symposium on Water Rock Interaction, Reykjavik, Iceland, 1986, pp. 471-474.
22. Zaiser, E. M.; Lamer, V. K. J. Colloid Science 1948, 3, 571-585.
23. Dove, P. M.; Rimstidt, J. D. Amer. Min. 1985, 70, 838-844.
24. Takano, B. Science 1987, 235, 1633-1635.
25. Low, M. J. D. Chem. Rev. 1960, 60, 267-312.
26. Gray, B. F. Trans. Faraday Soc. 1968 65 pp 1603-1614.
27. Yang, C. H. J. Phys. Chem. 1969, 73, 3407-3413.
28. Emmons W. H. U.S. Geological Survey Bull. 625, 1917, p. 123.
29. Walsh C. A.; Rimstidt J. D. Can. Min. 1986, 24, 35-44.

RECEIVED April 8, 1993

Chapter 2

Laboratory Studies of Pyrrhotite Oxidation Kinetics

Ronald V. Nicholson[1] and Jeno M. Scharer[2]

[1]Department of Earth Sciences, Waterloo Centre for Groundwater Research, and [2]Department of Chemical Engineering, University of Waterloo, Waterloo, Ontario N2L 3G1, Canada

> A pneumatically mixed flow-through reactor was used to study the oxidation kinetics of pyrite and pyrrhotite. Experiments were conducted at pH values of 2, 3, 4, and 6 with temperatures controlled at 10, 22, and 33°C. A grain-size fraction with an average particle diameter of 105 microns was used in all experiments.
>
> The rates of pyrrhotite oxidation ranged from 6 to 14 x 10^{-9} mol·m^{-2}s^{-1} at 22°C for solution pH values of 2 to 6. Although the rates were not strongly pH-dependent, minimum values were observed for pH values of 3-4. Experimental activation energies were approximately 50 kJ·mol^{-1} at pH values of 2 and 4, but in excess of 100 kJ·mol^{-1} at pH=6. The molar ratio of SO_4^{2-}/Fe in the reactor effluent varied from 0.93 at pH=2 to 0.76 at pH=6, suggesting a preferential retention of sulfur on the pyrrhotite solids with increasing pH. The rates of pyrrhotite oxidation at atmospheric concentrations of O_2 and at 22°C were on the order of 100 times those measured for pyrite. The iron deficiency in pyrrhotite is considered the major factor for these high rates.

Wastes from non-ferrous metal mines often contain environmentally significant quantities of iron sulfide minerals, usually as mixtures of pyrite (FeS_2) and pyrrhotite ($Fe_{1-x}S$). Exposure of these minerals at the ground surface to atmospheric oxygen and water results in oxidation and release of acidity to the environment. The rate of oxidation and the environmental controls on the rates determine, to a great extent, the quality of the water that has passed through the sulfide wastes. The oxidation of pyrite is well understood in comparison to that of pyrrhotite. Although the precise reaction mechanism for the oxidation of pyrite remains under investigation (e.g. ([1])), the macroscopic rates of reaction can be predicted with reasonable certainty and the products of oxidation are well known under most environmental conditions. In contrast, very few investigations of pyrrhotite oxidation have been reported. The rate controls on the reactions are not known and even the oxidation products are poorly understood.

Pyrite oxidation has received a large degree of attention as a focus of research because of the prevalence of that mineral in rocks associated with coal deposits in the United States. A vast amount of research pertaining to pyrite oxidation and acid generation has been reported and many reviews exist (e.g. 2-5). Although pyrrhotite oxidation kinetics have not received much research attention, the proportion of pyrrhotite to pyrite at many base and precious metal mines is quite high and in some instances, pyrrhotite is the only significant waste sulfide present.

Pyrrhotite chemistry is complicated by the deficiency of iron in the crystal structure. The general formula for pyrrhotite is $Fe_{1-x}S$, where x can very from 0.125 (Fe_7S_8) to 0.0 (FeS). The monoclinic form of the iron-deficient Fe_7S_8 is one end member. The intermediate hexagonal or orthorhombic structures are less iron-deficient with stoichiometries of Fe_9S_{10} to $Fe_{11}S_{12}$. The equimolar FeS variety known as troilite is hexagonal (6). The iron deficiency may affect the oxidation of pyrrhotite. It has been speculated that iron vacancies in the crystal structure are compensated by the presence of Fe^{3+} (6). However, this has not been confirmed. It is critical to recognize the iron deficiencies, particularly when writing the oxidation of pyrrhotite by ferric iron as shown below.

When oxygen is the primary oxidant, the overall reaction may be written as;

$$Fe_{1-x}S_{(s)} + \left(2 - \frac{x}{2}\right)O_2 + xH_2O \rightarrow (1-x)Fe^{2+} + SO_4^{2-} + 2xH^+ \qquad (1)$$

This implies that each mole of pyrrhotite can produce up to one-quarter mole of protons for the most iron deficient form ($x=0.125$) or no protons from the stoichiometric FeS ($x=0$). Acid (protons) can also be produced upon oxidation of the dissolved iron and precipitation of ferric hydroxide as illustrated in the following equation:

$$Fe^{2+} + \frac{1}{4}O_2 + \frac{5}{2}H_2O \rightarrow Fe(OH)_{3(s)} + 2H^+ \qquad (2)$$

Alternately, the oxidation reactions may not proceed to completion. If elemental sulfur is produced, the following reaction may apply:

$$Fe_{1-x}S_{(s)} + \left(\frac{1-x}{2}\right)O_2 + 2(1-x)H^+ \rightarrow (1-x)Fe^{2+} + S^0 + (1-x)H_2O \qquad (3)$$

In this case, acid can be consumed, although net consumption will be less if the ferrous iron oxidizes as in equation 2. These reactions and their variable effect on acid formation indicate the importance of defining the precise mechanism of pyrrhotite oxidation.

At low pH, ferric iron concentrations can be significant. Measured concentrations of ferric iron (Fe_T^{III}) in porewater within a pyritic tailings were found to be as high as 2000 mg/L near to the ground surface (7). Oxidation of pyrrhotite by ferric iron may occur as follows:

$$Fe_{1-x}S_{(s)} + (8-2x)Fe^{3+} + 4H_2O \rightarrow (9-3x)Fe^{2+} + SO_4^{2-} + 8H^+ \quad (4)$$

if the sulfide is transformed completely to sulfate or as:

$$Fe_{1-x}S_{(s)} + (2-2x)Fe^{3+} \rightarrow (3-3x)Fe^{2+} + S_{(s)}^0 \quad (5)$$

if elemental sulfur is formed as a reaction intermediate. The oxidation to sulfate can produce almost equimolar quantities of H^+ from Fe^{3+}, whereas sulfur production does not produce acid.

Previous Work

One of the more complete studies involving abiotic pyrrhotite oxidation was reported by Steger and Desjardins (8). The stoichiometry of reaction products was discussed but no data on the rates of reaction were provided. Products of pyrrhotite oxidation have been inferred from reactions for pyrite and field evidence for naturally oxidized tailings containing large fractions of pyrrhotite (9).

Field observations of well oxidized tailings indicate that elemental sulfur exists in quantities up to 2.5% (by weight) as a result of oxidation of pyrrhotite (10). The partial oxidation of pyrrhotite may also lead to the formation of marcasite (FeS_2) coatings on the pyrrhotite. Observations of the mineral zoning of oxidized ore deposits has suggested that pyrite and marcasite are intermediate products of pyrrhotite oxidation. The secondary FeS_2 would later oxidize in the presence of oxygen or aqueous Fe^{3+} (11-14). The formation of pyrite (and/or marcasite) during pyrrhotite oxidation was confirmed experimentally (15-17) by Mössbauer spectroscopy and X-ray photoelectron spectroscopy (XPS) suggesting the following reactions:

$$2Fe_{1-x}S + \left(\frac{1}{2}-x\right)O_2 + (2-4x)H^+ \rightarrow FeS_2 + (1-2x)Fe^{2+} + (1-2x)H_2O \quad (6)$$

followed by the formation of a very iron-deficient sulfide (Fe_yS where $y < 0.5$) and then by the fully oxidized products consisting of Fe^{III} solids and sulfate. Sulfur and marcasite/pyrite are metastable in the presence of oxygen and will also oxidize. In any case, the formation of FeS_2 must either consume acid as in equation 6 or at least result in no net production of protons even if Fe^{2+} is oxidized as in equation 2. This is shown by adding equations 2 and 6 to give:

$$2Fe_{1-x}S + \left(\frac{3}{4}-\frac{3}{2}x\right)O_2 + \left(\frac{3}{2}-3x\right)H_2O \rightarrow FeS_2 + (1-2x)Fe(OH)_{3(s)} \quad (7)$$

However, the presence of these intermediates may have important implications in the determination of the eventual prediction of oxidation rates for pyrrhotite.

Similarly, sulfur oxyanions are known to be produced at high rates of oxidation in pyritic systems (18). It was recognized, however, that the intermediate sulfur species are not stable in the presence of oxygen and neutral to acidic pH and transform to sulfate so readily that analysis needed to be performed very shortly after sampling to avoid loss of intermediates. However, the realization that the intermediates form led to a more thorough understanding of the oxidation mechanism. Similar intermediates may be expected in the oxidation of pyrrhotite when high rates occur.

Present Work

The current laboratory work concerns the abiotic and biological oxidation of pyrrhotite at ambient environmental conditions. There is a dearth of kinetic data on the oxidation of this mineral. The few previous studies on the mechanism and the stoichiometry of pyrrhotite oxidation show pyrrhotite to be more reactive than pyrite (8, 11-13, 19, 20, 21), but kinetic and thermodynamic data lack reproducibility (22). Reliable and reproducible kinetic data are essential for establishing the extent to which short-term laboratory data may be used to predict long-term processes in the field.

Modeling is a useful tool for examining the consequences of oxidation under various physio-chemical conditions. However, the existing uncertainties in the data base for pyrrhotite and the inability to achieve a consistent model for well-controlled scenarios of oxidation diminish the usefulness of modeling studies. The objectives of this work were to develop a kinetic (biological and non-biological) model for pyrrhotite oxidation based on laboratory data. The model should include factors that are known to be important to pyrite oxidation, including, pH, surface area, oxygen concentration, ferric iron concentration and temperature. The stoichiometry or material balance of the oxidation reactions provides insight into the reaction mechanism. The potential significance of interaction between pyrite and pyrrhotite mixtures was also considered because galvanic interaction has been reported in biologically active systems of mixed sulfides (23).

The long-term goal of the investigation is to evaluate the degree of applicability (or translatability) of fundamental laboratory data to observed oxidation rates under conditions similar to those of tailings in a field setting. It is important to establish the extent to which short-term laboratory data on oxidation kinetics and mass-transport properties may be used to predict these coupled processes. For this reason, experimental data on pyrite and pyrrhotite oxidation (both biological and non-biological) should be compared under conditions of "kinetic control" (i.e. controlled laboratory reactors) and "diffusion control" (i.e. dynamic column studies and field scenarios).

The results reported here involve abiotic oxidation by oxygen only under kinetically controlled conditions. Investigation of biologic oxidation and reactions with ferric iron are in progress.

Materials and Methods

Museum-grade pyrrhotite and pyrite were pulverized in a ball mill and passed through a series of standard mesh (Tyler) screens. The XRD method of Arnold (24) was used to determine the iron content (stoichiometry) and crystal structure of the pyrrhotite used. A hexagonal variety with a formula of Fe_9S_{10} was used for these experiments. The mass fraction sized between sieves #120 and #170 mesh (105 micron nominal particle diameter) was used for most kinetic studies. The sample was suspended in methanol, cleaned for 30 to 45 s in an ultrasonic cleaner, washed with 1N HCl followed by distilled water, rinsed several times with fresh aliquots of methanol, and dried before storage. In the mixed-flow kinetic studies, a weighed sample was suspended in 0.001 M EDTA solution and the pH was adjusted to the desired value (pH=2 to pH=6) with either HNO_3 or NaOH. The chelating effects of EDTA on Fe^{3+} was desired to prevent precipitation of ferric oxyhydroxide for experimental pH values above 4.

The mixed-flow kinetic experiments were performed in internal split flow, pneumatically mixed reactors, shown on Figure 1. The reactors had an internal diameter of 5.6 cm and a height to diameter aspect ratio of 7. Each reactor was water jacketed to maintain a constant temperature and experiments were run at 10, 22, and 33°C. A liquid reactor volume of 650 mL containing 3.0 $g \cdot L^{-1}$ to 10.0 $g \cdot L^{-1}$ solid suspension was maintained. Oxygen transfer and mixing were provided by air flow through a sintered glass sparger at the bottom of the vessel at a flow rate of 0.3 $cm^3 cm^{-2} min^{-1}$. The reactors were operated in a continuous flow-through manner. Fresh liquid was delivered into the reactor at a rate of 30 $mL \cdot h^{-1}$ and removed from the headspace by overflow at the same rate, while the solids were retained in the reactor. The duration of the continuous-flow experiments was approximately 50 hours with each fresh batch of pyrrhotite sample.

Fixed-bed column reactors such as those described by Nicholson et al. (25) were used to study the oxidation kinetics of pyrite and pyrrhotite mixtures. The fixed-bed column experiments were performed in 1-cm diameter plastic tubes loaded with 10 g of selected sulfide mixtures. Pyrrhotite fractions were set at 0, 0.25, 0.5, 0.75, and 1.0. Water flow was about 5 $mL \cdot h^{-1}$ and an air flow rate of about 600 $mL \cdot h^{-1}$ was maintained by a slight positive pressure that kept the granular sample partially drained at all times. The experiment was run at room temperature (22±2°C). Effluent samples were collected every day and experiments were run for as long as two weeks. The effluent was analyzed routinely for ferrous ion by colorimetry with bathophenanthroline reagent (26), total iron by atomic absorption spectroscopy and sulfate content by ion exchange chromatography. Usually, steady-state mass flux was achieved after 24 hours of continuous operations.

Measurements of the oxygen mass-transfer characteristics in the mixed-flow reactors were performed using a series-900 galvanic dissolved oxygen electrode (New Brunswick Scientific Co., Edison, NJ). The liquid in the vessel was first deoxygenated with nitrogen. Then, air was introduced through the sparger at various flow rates and the increase of the dissolved oxygen with time was measured in the downcomer (bubble-free) section of the vessel. The values of the mass-transfer coefficients were calculated from log-transformed concentration profiles. The actual

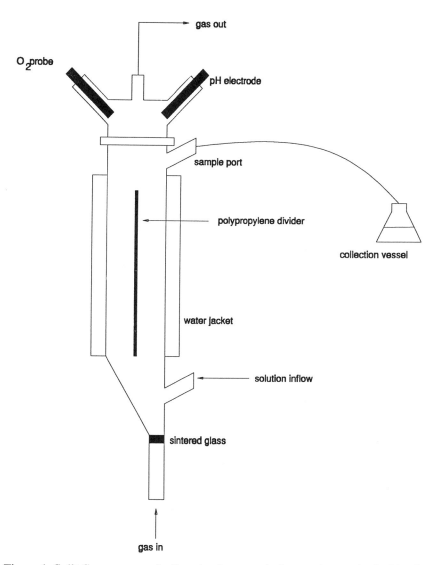

Figure 1. Split-flow, pneumatically mixed reactor design used to study the kinetics of pyrrhotite oxidation.

mass-transfer coefficients at various air velocities were compared with an acceptable minimum value calculated from the stoichiometric oxygen uptake rate expected during pyrrhotite oxidation. It was evident that actual aeration rates employed in this study were adequate to meet the oxygen demand.

Surface area of the sulfides was not measured. Estimates of surface area were made using the empirical relationship demonstrated between grain size and BET measured surface areas for silica grains given by Parks (27). Comparison of pyrite grain size with measured BET values indicates that surface area estimates are well within a factor of two using Parks' relationship for silica grains (data from 1, 28, 29). The regression of BET surface area (A in cm^2/g) on grain diameter (d in cm) is $\log(A)=0.415-\log(d)$ (27) that gives a specific surface area of 248 $cm^2 g^{-1}$ for the 105 micron grains used. This approach allows comparison of rates to other studies on the basis of surface area when grain size is known.

Results

The concentrations in the reactor effluent were typically in the range of 8×10^{-6} to 2×10^{-4} mol·L^{-1} for both iron and sulfate. A typical plot of concentration versus time is shown in Figure 2 for an experiment at 22°C. Initial experiments that were conducted over longer time periods exhibited nearly constant concentrations over periods of days. Most experiments exhibited steady-state concentrations between 500 and 1500 minutes from initiation.

The rates of oxidation derived from iron concentrations ranged from 6.4 to 14.1×10^{-9} mol·m^{-2}s^{-1} for solution pH values in the range of 2 to 6 at 22°C. Varying the temperature from 10 to 33°C resulted in a maximum difference of oxidation rates from 2.5 to 57.6×10^{-9} mol·m^{-2}s^{-1} at pH=6.

Effects of pH. The variation of solution pH from 2 to 6 resulted in mean differences in rates of 40% at 22°C and 60% at 33°C. The effect of pH was not consistent at the three temperatures studied. At the lowest temperature (10°C), the rate of oxidation decreased slightly with increasing pH (Figure 3). At the two higher temperatures (22 and 33°C), the rates appear to be near minima at pH values of 3 to 4. At 22°C, rates were essentially equal at pH values of 2 and 6, whereas, at 33°C, the rate at pH=6 was almost two times the value at pH=2.

Effects of Temperature. The surficial rate constant (k) is defined with units of mol·m^{-2}s^{-1}. Typical Arrhenius behaviour was observed for all pH values investigated (Figure 4). The calculated activation energies (E_a) were not, however, constant across the range of pH values. The lower activation energies were found at pH=4 and at pH=2 but, at pH=6, the value of E_a was almost double the value at pH=4 (Table I).

Effects of Surface Area. With the narrow range of grain size used in the experiments, it was assumed that surface area increased linearly with the mass of pyrrhotite exposed to oxidation. The rate of oxidation (mol·s^{-1}) was therefore compared to the mass of pyrrhotite to determine the dependence of oxidation rates on

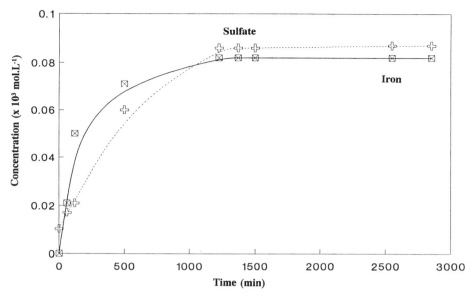

Figure 2. Concentrations of iron and sulfate as a function of time for an experiment at 22°C. Steady-state concentrations usually occur between 500 and 1000 minutes.

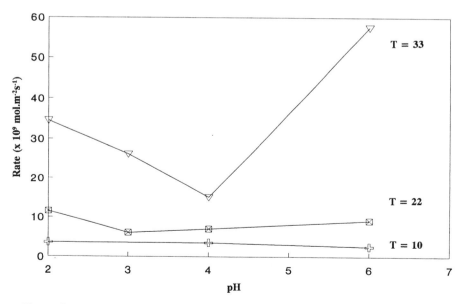

Figure 3. Rate of reaction as a function of pH at temperatures of 10°C, 22°C, and 33°C.

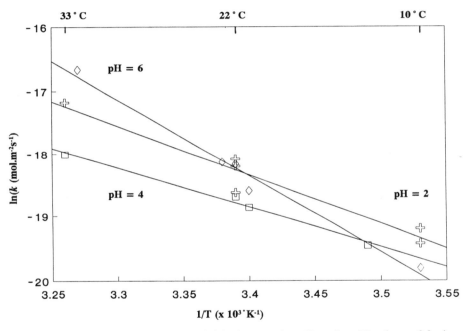

Figure 4. Arrhenius plots with individual regressions lines for pH values of 2, 4, and 6.

Table I. Experimental Activation Energies as a Function of pH

pH	E_a (J/mol)
2	58,100
4	52,400
6	100,400

surface area. The limited data suggest a near-linear dependence of rate on mass and hence surface area (Figure 5). Repeated experiments suggest that one standard deviation in measured rates is about 30% of the measured value. Error bars were placed on the single measurement values in Figure 5 to reflect this uncertainty.

Ratio of Sulfate to Iron. Both iron and sulfate were measured over time in four experiments. The molar ratios of sulfate to iron in the reactor effluent varied with pH. Complete release of oxidation products from Fe_9S_{10} should exhibit a SO_4^{2-}/Fe value of 1.1 in the effluent solution. The value closest to the theoretical was observed at pH=2 ($SO_4^{2-}/Fe = 0.925$). This ratio decreased with increasing pH to a low of 0.76 at pH=6 (Table II). The exception to this trend was exhibited by an experiment at pH=6 with no EDTA that resulted in a ratio of 2.46.

Table II. Molar Ratios of SO_4^{2-}/Fe in Reactor Effluent for the Three Solution pH Values Studied

Exp. No.	pH	EDTA Present	Number of Points	Average SO_4^{2-}/Fe ± Std. Dev.
PO-4	2	YES	5	0.925 ± 0.006
PO-2	4	YES	5	0.83 ± 0.15
PO-8	6	YES	10	0.76 ± 0.13
PO-5	6	NO	7	2.46 ± 0.39

Effects of EDTA. The addition of 0.001 M EDTA to the reactor solution had no significant effect on measured rates for experiments conducted at pH=2 and pH=4

(Table III). At pH=6, however, the rates based on Fe concentrations were as much as 80 times lower when EDTA was absent from solution.

Table III. Values of Rate Constants with and without 0.001 m EDTA in Reactor Solutions at pH Values of 2, 4, and 6

Temp. (°C)	pH	Rate (x 10^{-9} mol·m^{-2}s^{-1})	
		No EDTA	EDTA
10	2	4.6	2.4
22	2	6.5	9.6
22	4	5.0	6.5
30	6	0.7	58.0

Mixtures of Pyrite and Pyrrhotite. Three mixtures and two pure sulfide samples were oxidized in the column (fixed-bed) reactors. The oxidation rate of the pure pyrrhotite was 1.3×10^{-8} mol·m^{-2}s^{-1} and compares well with the oxidation rate measured in the mixed-flow reactor at 22°C and pH=2 (1.2×10^{-8} mol·m^{-2}s^{-1}). The reaction rate of the pure pyrite was 1.13×10^{-10} mol·m^{-2}s^{-1} that compares well with the average rate of five independent studies of $5 \pm 2.1 \times 10^{-10}$ mol·m^{-2}s^{-1} compiled from the literature by Nicholson (unpublished). The rates of the mixtures were slightly higher than calculated values based on the fractions of each sulfide (Figure 6) but the predicted and observed rates are not significantly different considering that standard deviations of average rates are about 30 to 50 percent of the mean.

Discussion

Measured oxidation rates are reproducible (see T=22°C in Figure 4) and relatively small changes in rates caused by changes in pH were easily observed. The most significant effect on the rates of reaction was caused by changes in temperature. There are, however, some interesting trends between rates and pH and between activation energy and pH that deserve further discussion.

The overall effect of pH in the range of 2 to 6 on the rate of oxidation is relatively small, the rates being within about 50% of the mean value. Yet, the trends of oxidation rate with pH were not anticipated. And more surprising was the change in pH trends with temperature (Figure 3). A rate minimum around pH=4 at temperatures of 22 and 33°C has not been reported in other studies of sulfide oxidation. Results of Smith and Shumate ([30]) and those of McKibben and Barnes ([28]) both indicate a very slight increase in oxidation rate for pyrite from pH values less than 2 to pH=4. In a review of experimental oxidation rates for pyrite, Nicholson

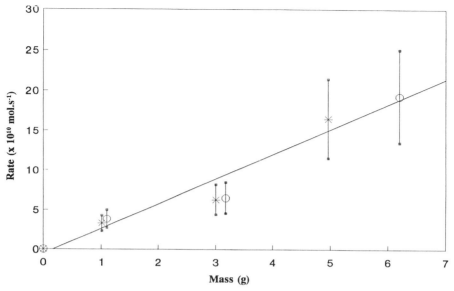

Figure 5. Rate of pyrrhotite oxidation as a function of mass of pyrrhotite (-120 + 170 mesh particles) at 22°C; (o) pH = 2 and (*) pH = 6. Regression line includes the origin.

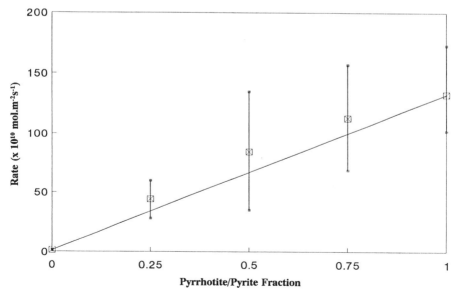

Figure 6. Rates of oxidation (as iron released) as a function of the fraction of pyrrhotite in mixtures with pyrite at pH = 2 and 22°C. The solid line represents calculated rates based on the fractions present and oxidation rates of pure end-members.

(unpublished) found very small differences in rates for pH conditions from 2 to 8. The pyrrhotite oxidation rates observed in this study exhibit decreasing rates from pH=2 to pH=6 at 10°C but show increases by a factor of two to three from pH=4 to pH=6 for the two higher temperatures. This anomaly may be linked to a change in the reaction mechanism that is also suggested by the measured activation energies.

Activation energy also varies as a function of pH with low values around pH=2 to 4 (Table I) and effectively doubling with an increase of pH from 4 to 6. A similar behavior for pyrite oxidation was noted. The data are quite variable, yet E_a for pyrite oxidation by oxygen varies from about 57 kJ·mol^{-1} in acidic solutions (pH from 2 to 4, (28)) to values approaching 90 kJ·mol^{-1} at pH values of 7 to 8 (25).

It is possible that the increase in activation energy represents a change in the rate-limiting step from that of surface controlled or heterogeneous reaction with a lower E_a to a more homogeneous step with a larger E_a, likely involving covalent bond cleavage. However, silicate mineral dissolution also exhibits significant trends of activation energy with pH that have been attributed to other processes. It has been suggested that proton adsorption-desorption can contribute up to 50 kJ·mol^{-1} to the experimental activation energy of dissolution reactions (31). It was further suggested that the difference in E_a values is related to the difference between the experimental pH and the point of zero net proton charge (PZNPC) of the solid surface. Although it would seem logical to attempt to identify the rate-controlling mechanism of the oxidation reaction at various pH levels by the changes in experimental activation energy, the results of such comparisons would be of little value if the large enthalpies of proton adsorption-desorption are not taken into account. It appears that temperature dependence of reaction rates is similar for silicates and sulfides and the latter deserves further attention.

The simplest oxidation reaction (equation 1) indicates that the molar ratios of sulfate to iron should be close to unity. High SO_4^{2-}/Fe ratios imply that iron has either not been released from the sulfide structure or that it has been lost from solution after release. Similar logic applies to the sulfur component for low ratios of SO_4^{2-}/Fe. Previous studies indicate that more iron than sulfate is released under certain conditions of oxidation resulting in formation of secondary marcasite or pyrite are formed (17). A major concern in these experiments is related to the possible oxidation and precipitation of the ferrous iron as ferric hydroxide at pH=6. This process not only interferes with the detection of the released iron in the reactor effluent but also can lead to coating of the pyrrhotite surfaces, resulting in lower rates of oxidation in the same way that has been demonstrated for pyrite (32). The difference in rates by a factor of about 80 with and without EDTA (Table III) suggest that this should be a major concern for experiments above pH values of 4. Although fresh Fe(OH)$_3$ solubility can be on the order of 10^{-7} mol·L^{-1} at pH=4, well below total Fe concentrations observed in the effluent, it is probable that the low abiotic rates of Fe^{2+} oxidation at that pH prevent the formation of significant concentrations of ferric iron in solution. At pH=6, the rate of oxidation of Fe^{2+} by oxygen is orders of magnitude faster (3) and it is therefore necessary to prevent the formation of Fe(OH)$_3$ solids to avoid the loss of iron in solution.

The observed sulfate to iron ratios of less than unity in the effluent indicate a relative depletion of sulfur. There may be several reasons for this. First, the released sulfur could comprise various intermediate sulfoxyanions, particularly sulfite and thiosulfate. Acidic conditions favour the decomposition of thiosulfate to elemental sulfur and sulfite as products (33). Some effluent samples were treated with hydrogen peroxide and reanalysed for sulfate with no significant differences in concentration noted. The missing sulfur, therefore, does not appear to be in the form of dissolved sulfur intermediates. Transformation of the sulfide to elemental sulfur on the surface of the pyrrhotite is the second possibility to account for the lost sulfate. Elemental sulfur has been observed in oxidizing sulfide tailings, up to 2.5% of the bulk weight that had an estimated 25% sulfide content (34). Third, the sulfur may not have been released but may remain in the sulfide structure while iron was released. Secondary marcasite and pyrite may form by oxidation in acidic solutions (17). The results of this study are not conclusive regarding sulfur unaccounted by sulfate. Further analytical work and surface analysis is planned to address this issue.

Mixtures of sulfide minerals have been shown to exhibit galvanic protection of minerals lower in the electromotive series (23). The results of the current study suggest that galvanic enhancement did not occur when pyrite and pyrrhotite were physically mixed. Because the study of the grain surfaces is not yet complete, it is not known whether or not pyrite oxidation was inhibited. However, observed pyrite oxidation rates being a factor of 100 less than those for pyrrhotite, it may not be possible to distinguish such galvanic interaction. These experiments were purposely performed in fixed-bed reactors to insure physical contact of the grains and allow electrical connection that was expected to be necessary for galvanic protection to occur. Lack of a galvanic response suggests that the bulk oxidation rate is linearly dependent on the mole fraction of the sulfides. Additional experiments will be performed under biologic oxidation and ferric iron leaching to verify this response.

Comparisons With Pyrite Oxidation. At atmospheric oxygen saturation and 22°C, pyrrhotite oxidation is a factor of 20 to 100 faster than values reported for pyrite oxidation at 25°C. This is consistent with the observed higher relative reactivity of pyrrhotite that is commonly observed in the field. The deficiency of iron in the crystal structure of pyrrhotite (6) could account for the more rapid oxidation rates because of the lower stability of the crystal lattice. The trend in activation energies with pH for pyrite oxidation is quite comparable to those found for pyrrhotite in this study, with values almost doubling from acidic solutions to near-neutral conditions (Table IV).

Further Work Needed. There are many issues that have not been addressed here that may represent significant roles in the pyrrhotite oxidation process. Oxygen concentration will almost certainly affect reaction rates. It is important to determine the dependence on oxygen for predictive purposes. Oxidation by ferric iron may dominate in acidic environments where rapid oxidation of Fe^{2+} by bacteria can produce significant concentrations of Fe^{3+}, as seen in shallow pyritic tailings (7). It is also known that bacteria such as *Thiobacillus ferrooxidans* play a catalytic role in the oxidative dissolution of sulfide minerals. The biologic system can be investigated to study the enhancement of reaction rates and also to gain insight into the role of sulfur intermediates that may be produced during rapid oxidation of sulfide minerals.

Table IV. Reported Activation Energies of Oxidation of Pyrite by Oxygen at Different Solution pH Values Compared to Activation Energies Found in This Study

pH	Activation Energy (J/mol)	
	Pyrite	Pyrrhotite[c]
2	57,000[a]	58,000
4	57,000[a]	52,000
6		100,000
7-8	88,000[b]	

[a] - McKibben and Barnes (28)
[b] - Nicholson et al. (25)
[c] - this study

Finally, the deficiency of iron in the pyrrhotite crystal structure requires further study to determine what effect this has on rates of reaction. The pyrrhotite used in these experiments was hexagonal with the approximate formula Fe_9S_{10} yet other stoichiometry (e.g. Fe_7S_8) and crystal structures (e.g. monoclinic) are common. The study of the role of the iron deficiency may provide insight into the mechanism for the oxidation reaction. An analysis of stabilization energies of the Fe_7S_8 crystal lattice showed that a mixed valence model [$(Fe_5^{II}Fe_2^{III})S_8^{-II})$] was most reasonable (35). However, Mössbauer studies revealed mixed results. Some early results (36) suggested that Fe^{III} did not exist in the crystal structure as a separate entity for Mössbauer transition times of 10^{-7} to 10^{-8} seconds. A later study (37) indicated a complex Mössbauer spectra for iron that is consistent with the presence of Fe^{III} in the lattice. Although the concept of mixed valence iron in a covalent bonded structure like pyrrhotite is ambiguous, the net deficiency in electrons may account for the increased oxidation rates over those observed for pyrite.

Conclusions

The split-flow, pneumatically mixed reactor is an effective means of studying pyrrhotite oxidation kinetics. Dissolved iron is a good indicator of the oxidation rate but care is required at pH values above about 4 where oxidation of ferrous iron and loss by precipitation can result in erroneously low rates of reaction. A chelating agent such as EDTA can help avoid such iron losses.

At atmospheric oxygen concentrations and at 22°C, the rate of pyrrhotite oxidation is on the order of 1×10^{-8} mol·m^{-2}s^{-1} or about one hundred times higher than that for pyrite under similar conditions. The reaction rates are mildly dependent on pH but the trends are not consistent at all temperatures. For a range of pH from 2 to 6, maximum differences in oxidation rates were about a factor of 2 or within 50% of the mean rate at a specified temperature. The inconsistent trends of rate with pH at different temperatures are related to the differences in activation energy that occur across the range of pH studied. Implications for doubling of activation energy from pH=4 to pH=6 are unclear but may be related to proton adsorption on the sulfide surface, similar to that suggested for silicate minerals. The lack of interaction in pyrite-pyrrhotite mixtures suggests that oxidation rates can be estimated independently for each mineral in mixtures such as sulfide tailings.

The deficiency of sulfate in the effluent solution and the increase of this deficiency with increasing pH indicate that simple oxidative dissolution of the $Fe_{1-x}S$ is not likely. The increased production of non-soluble sulfur products with increasing pH may be linked to a change in reaction mechanism and rate limiting step. Further study of this phenomenon is required.

While some preliminary controls on the surface-based pyrrhotite oxidation reaction have been quantified, many other factors require investigation before a comprehensive kinetic model can be constructed and applied to field conditions.

Acknowledgements

We wish to acknowledge the professional assistance of Mr. Wayne Noble in the laboratory. The authors also wish to thank Charles Alpers for his editorial guidance and appreciate the thoughtful help of two anonymous reviewers. These efforts significantly enhanced the quality of the manuscript. As usual, however, omissions and errors remain the authors' responsibility.

Literature Cited

1. Moses, C. O.; Herman, J. S. Geochim. Cosmochim. Acta 1991, 55, 471-482.
2. Rogowski, A. S.; Pionke, H. B.; Broyan, J. G. J. Environ. Qual. 1977, 6, 237-244.
3. Lowson, R. T. Chemical Reviews, 1982, 82, 461-497.
4. Nordstrom, D. K. Soil Sci. Soc. Amer. Spec. Publ. 1982 10, 37-56.
5. de Haan S. B. Earth Sci. Rev. 1991, 31, 1-10.
6. Vaughan, D. J.; Craig, J. R. Mineral Chemistry of Metal Sulphides; Cambridge University Press: Cambridge, UK, 1978.
7. Dubrovsky N. M.; Cherry, J. A.; Reardon, E. J.; Vivyurka, A. J. Can. Geotech. J. 1984, 22, 110-128.
8. Steger, H. F.; Desjardins, L.E. Chem. Geol. 1978, 23, 225-233.
9. Boorman R. S.; Watson D.M. CIM Bulletin 1976, August, 86-96.
10. Jambor, J. L. CANMET, EMR Canada, Report MSL 86-45(IR) 1986.
11. Nickel, E. H.; Ross, J. R.; Thornber, M. R. Economic Geology 1974, 69, 93-107.
12. Thornber, M. R. Chem. Geol. 1975a, 15, 1-14.
13. Thornber, M. R. Chem. Geol. 1975b, 15, 117-144.
14. Thornber, M. R.; Wildman, J. E. Chem. Geol. 1979, 24, 97-110.

15. Burns, R. G.; Fisher, D. S. J. Geophys. Res. 1990, 95, 14415-14421.
16. Fisher, D. S.; Burns, R. G. Geol. Soc. Amer. Ann. Meeting 1990, 22, p 207.
17. Burns, R. G.; Vaughan, D. J.; England, K. E. R. Geol. Soc. Amer. Ann. Meeting 1991, 23, p A146.
18. Goldhaber M. B. Amer. J. Sci. 1983, 283, 193-217.
19. Bannerjee, A. C. Indian J. of Chemistry 1976, 14A, 845-856.
20. Hamilton, I. C.; Woods, R. J. Electroanal. Chem. 1981, 118, 327-343.
21. Ahonen, L.; Hiltunen, P.; Tuovinen, O. H. In Fundamental and Applied Biohydrometallurgy; Lawrence, R.W.; Branion, R.M.R.; Ebner, H.G., Eds.; Elsevier: Amsterdam, 1986; pp 13-22.
22. Pearse, M. J. Mineral Processing and Extractive Metallurgy, 1980; Vol. 89, pp C26-C36.
23. Mehta, M. P.; Murr, L. E. Hydrometall. 1983, 9, 235-256.
24. Arnold, R. G. Can. Mineral 1967, 9, 31-50.
25. Nicholson, R. V.; Gillham, R. W.; Reardon, E. J. Geochim. Cosmochim. Acta 1988, 52, 1077-1085.
26. Lee, B.F.; Stumm, W. J. Am. Water Works Assoc. 1960, 52, 1567-1574.
27. Parks, G. A. In Mineral-Water Interface Geochemistry, Hochella, M.F.; White, A.F., Eds., Reviews in Mineralogy, Vol. 23, Mineral. Soc. Amer.: Washington, D.C., 1990, pp. 133-175.
28. McKibben, M. A.; Barnes, H. L. Geochim. Cosmochim. Acta. 1986, 50, 1509-1520.
29. Wiersma, C. L.; Rimstidt, J. D. Geochim. Cosmochim. Acta 1984, 48, 85-92.
30. Smith, E. E.; Shumate, K. Water Poll. Cont. Res. Rept. 1970, 14010 FPS, AAST-40,
31. Casey, W. H.; Sposito, G. Geochim. Cosmochim. Acta 1991, 56, 3825-3830.
32. Nicholson R. V.; Gillham, R. W.; Reardon, E. J. Geochim. Cosmochim. Acta 1990, 54, 395-402.
33. Moses, C. O.; Nordstrom, D. K.; Herman, J. S.; Mills, A. L. Geochim. Cosmochim. Acta 1987, 51, 1561-1571.
34. Blowes D. W.; Jambor J. L. Appl. Geochem. 1990, 5, 327-346.
35. Tokonami, M.; Nishiguchi, K.; Morimoto, N. Amer. Mineral. 1972, 57, 1066-1080.
36. Levinson, L. M.; Treves, D. J. Phys. Chem Solids 1968, 29, 2227-2231.
37. Vaughan, D. J.; Ridout, M. S. Solid State Comm. 1970, 8, 2165-2167.

RECEIVED August 10, 1993

Chapter 3

Effect of Humidity on Pyrite Oxidation

Sandra L. Borek

Pittsburgh Research Center, U.S. Bureau of Mines, Pittsburgh, PA 15236

The amounts of weathering products formed during abiotic chemical pyrite oxidation was dependent on relative humidity and time. Six pyrites were placed under four relative humidities (RH) (34%, 50%, 70%, and 79% RH). These samples were periodically analyzed using Mössbauer spectroscopy to determine the types and amounts of weathering products formed. Hematite was present in Waldo pyrite samples after 30 days in all experimental humidity conditions. Ferrous sulfates (melanterite and rozenite) were detected in the three sedimentary pyrites (Pittsburgh, Kirby-U, and Kirby-R) after 30 days in 79% RH, and after 90 days in 50%, 70%, and 79% RH conditions. Two hydrothermal pyrites (Iron Mountain and Noranda) displayed no significant weathering over time in any relative humidity.

Pyrite (FeS_2) is an iron disulfide often associated with coal and adjacent strata. Pyrite reacts with water and oxygen to form ferrous iron, sulfate, and acidity:

$$FeS_{2(s)} + 7/2\ O_2 + H_2O \rightarrow Fe^{2+} + 2\ SO_4^{2-} + 2\ H^+ \tag{1}$$

Humidity is an important factor in pyrite oxidation as it has been shown that pyrites weather differently depending on the humidity. The formation of oxidation products is also determined by humidity (1-2). These observations suggest that if pyritic waste materials are not in direct contact with water (i.e., high and dry), the exclusion of water is not guaranteed. High humidity conditions can contribute the water needed for pyrite oxidation. Previous studies performed to determine the effect of humidity on pyrite oxidation rates have shown little dependence on humidities below 98% RH (3), or some dependence on relative humidities at or below 85% (4). Each study included only one pyrite sample. A variety of pyrites must be examined to determine the effect of humidity on oxidation. The variation in ability of different pyrites to utilize water vapor for oxidation could not be realized as a factor in a one-pyrite study.

Mössbauer spectroscopy, which utilizes resonant, recoil-less gamma-ray emissions and absorptions, was used to analyze the products of pyrite oxidation. By examining the position, number, shape, and relative absorbance of spectral lines, the identity of iron minerals in a sample can be determined (5-7). Two quantities routinely used to identify the iron minerals in a sample are isomer shift (IS) and quadrupole splitting (QS).

Isomer shift is caused by an electrostatic interaction between the charge distributions of the nucleus and electrons of an atom. This interaction results in the shifting of ground- and excited-state nuclear levels. IS indicates the oxidation state of the iron atom as well as the characteristics of its substituent groups. Quadrupole splitting occurs because an uneven distribution of electronic charge around the iron nucleus creates an electric field gradient that interacts with the nuclear quadrupole moment of the ^{57}Fe excited state. The excited energy level splits, losing its degeneracy. QS indicates the bonding structure around atoms, isomerization and structure defects.

If the material has magnetic order, a third parameter called magnetic splitting can also be utilized. Magnetic compounds are characterized by a six-peak spectrum. These quantities are used to identify the iron minerals while the area under the peaks can be used to calculate relative abundance of each mineral.

Experimental Methods

The six different pyrite samples used in this experiment are from Iron Mountain, California; Waite-Amulet, Noranda, Quebec; Pittsburgh seam coal, Pennsylvania; Waldo Mine, New Mexico; and two from Kirby, Pennsylvania. Iron Mountain, Noranda, and Waldo are hydrothermal pyrites; Pittsburgh and the two Kirby types are sedimentary pyrites. Each sample was crushed to size fractions between 63μm and 75μm with mortar and pestle. Samples were washed with 1:2 solution of HCl and water to remove any weathered products already present. This was followed by several deionized water rinses. An initial Mössbauer spectrum was recorded for each pyrite sample after the washings. Additional splits from each sample were examined for variations in shape characteristics under 100x power according to standard petrographic procedures (8).

Four different relative humidity environments were set up using saturated salt solutions. Four solutions were selected to maintain a range of relative humidities, from 34% to 79% (Table I).

Temperatures remained relatively constant over the 250 day experiment. Seven-100 mg samples of each pyrite were placed in each of the four

Table I. Average Relative Humidities (RH) and
Temperatures for Saturated Salt Solutions

Salt	n^a	Average RH(%)	Average Temp.(°C)
$CaCl_2 \cdot 6H_2O$	51	34±3	21.7±2.2
$NaBr \cdot 2H_2O$	51	50±3	22.2±2.6
$ZnSO_4 \cdot 7H_2O$	51	70±3	21.7±1.8
$CuSO_4 \cdot 5H_2O$	51	79±2	21.7±1.8

^{a}n = number of readings

polycarbonate desiccators with a salt solution. A combination analog hygrometer/digital thermometer permitted continuous monitoring of the relative humidity and temperature within the chambers. The desiccators were opened only long enough to take samples and remained closed at all other times. Pyrite oxidation did not significantly deplete the oxygen content in each desiccator.

After approximately 30, 60, 90, 120, 150, 200, and 250 days, one sample of each pyrite was removed from each desiccator and analyzed by Mössbauer spectroscopy. The velocity oscillated between -10 and +10 mm/sec. An iron foil standard was run approximately every 30 days to ensure an accurate reference channel for zero velocity. Each sample was run until a clear spectrum was obtained. The resulting spectra were analyzed by a curve-fitting computer program (9) which computed peak areas and positions. Resultant QS and IS values were calculated and used to identify the iron minerals in each sample (Table II). The physical appearance of each sample was also noted for water content and color.

Each pyrite type was tested for the presence of iron-oxidizing bacteria by placing samples of the weathered pyrite in a test tube containing a biomedia solution developed by Cobley and Haddock (10). If the bacteria were present, the ferrous iron would be oxidized to ferric, and an orange color would result. Samples sat for three weeks. No evidence of iron-oxidizing bacteria was detected in any sample at any humidity. The absence of bacteria may be explained by the use of acid to remove existing salts from the pyrite samples. In this experiment, weathering products were formed by abiotic pyrite oxidation.

Results and Discussion

The weathering products identified in this study using the calculated Mössbauer spectroscopy parameters (QS, IS) included two ferrous sulfates, melanterite ($FeSO_4 \cdot 7H_2O$) and rozenite ($FeSO_4 \cdot 4H_2O$), and hematite (α-Fe_2O_3) (7,11). In a few instances, the QS value for the ferrous sulfate salts were lower than the referenced value. A small amount of szomolnokite ($FeSO_4 \cdot H_2O$), or other intermediates, may form upon drying of the sample during the measurement.

Table II. QS and IS Values for Samples in Experimental Relative Humidities

Sample Name	RH%	QS mm/sec		IS mm/sec	
Noranda pyrite	34	.618	(.007)	.342	(.010)
	50	.612	(.020)	.341	(.009)
	70	.617	(.008)	.339	(.013)
	79	.617	(.011)	.342	(.012)
Iron Mountain pyrite	34	.619	(.010)	.344	(.010)
	50	.618	(.012)	.339	(.010)
	70	.608	(.012)	.344	(.010)
	79	.603	(.016)	.342	(.012)
Kirby-R pyrite	34	.625	(.018)	.343	(.010)
	50	.620	(.011)	.338	(.009)
	70	.627	(.013)	.337	(.012)
	79	.619	(.019)	.339	(.015)
melanterite/ rozenite	34	3.21	(.003)	1.16	(.004)
	50	2.98	(.001)	1.20	(.001)
	70	3.23	(.011)	1.30	(.004)
	79	3.06	(.012)	1.30	(.001)
Kirby-U pyrite	34	.610	(.011)	.341	(.009)
	50	.628	(.008)	.340	(.007)
	70	.634	(.015)	.338	(.009)
	79	.624	(.005)	.346	(.009)
melanterite/ rozenite	34	3.36	(.003)	1.20	(.006)
	50	3.05	(.008)	1.25	(.007)
	70	3.31	(.003)	1.32	(.001)
	79	3.21	(.008)	1.31	(.004)
Waldo pyrite	34	.613	(.015)	.342	(.015)
	50	.621	(.008)	.340	(.010)
	70	.611	(.010)	.339	(.012)
	79	.619	(.010)	.346	(.012)
hematite	34	-.110	(.009)	.395	(.025)
	50	-.106	(.014)	.375	(.040)
	70	-.085	(.009)	.367	(.009)
	79	-.113	(.013)	.391	(.025)
Pittsburgh pyrite	34	.627	(.011)	.342	(.016)
	50	.632	(.027)	.336	(.012)
	70	.638	(.012)	.341	(.009)
	79	.609	(.018)	.351	(.014)
melanterite/ rozenite	34	3.02	(.001)	1.22	(.001)
	50	2.98	(.004)	1.24	(.003)
	70	3.28	(.005)	1.24	(.005)
	79	3.24	(.001)	1.33	(.001)

The presence of other iron salts could affect the overall calculations of QS values, but there may not be sufficient quantity for a distinction.

The formation of ferrous sulfate products is one possible end product according to the reaction series for pyrite oxidation explained by Nordstrom (12):

$$FeS_{2(s)} + 7/2\ O_2 + H_2O \rightarrow Fe^{2+} + 2\ SO_4^{2-} + 2\ H^+ \quad (2)$$

$$Fe^{2+} + 1/4\ O_2 + H^+ \rightarrow Fe^{3+} + 1/2\ H_2O \quad (3)$$

$$Fe^{3+} + 3\ H_2O \rightarrow Fe(OH)_{3(s)} + 3H^+ \quad (4)$$

$$FeS_{2(s)} + 14\ Fe^{3+} + 8\ H_2O \rightarrow 15\ Fe^{2+} + 2\ SO_4^{2-} + 16\ H^+ \quad (5)$$

Other species are also possible products of pyrite oxidation. Hematite may have formed by the following dehydration reactions that would follow reaction (4) in the series:

$$Fe(OH)_3 ==> FeOOH + H_2O \quad (6)$$

$$2\ FeOOH ==> Fe_2O_3 + H_2O \quad (7)$$

There were three weathering trends among the six pyrite samples included in this experiment. Noranda and Iron Mountain pyrites displayed no weathering products throughout the experiment. Both Kirby pyrites weathered only at mid to high relative humidities. The Waldo and Pittsburgh pyrites were very reactive and oxidized in every experimental relative humidity.

The Noranda and Iron Mountain pyrites showed no weathered products during the 250 days of exposure in any relative humidity. Mössbauer spectra of Noranda pyrite are shown in Figure 1 both before and after 250 days of exposure at 79% RH. The two distinct peaks are due to pyrite. No peaks indicating oxidation products developed in the Noranda samples throughout the course of the experiment. The Iron Mountain spectra (Figure 2) also show this trend as only two pyrite peaks are present before and after exposure to all experimental relative humidities.

The two pyrite samples from Kirby (designated U and R) both produced iron-sulfate salts (melanterite and rozenite) as weathering products. Kirby-U pyrite displayed detectable weathering products only at the two highest relative humidities, 70% and 79%. As shown in Figure 3, only the third spectrum, after 251 days in 70% RH, contains peaks other than those attributed to pyrite. No weathering products are evident in Kirby-U samples exposed to relative humidities below 70%. The Kirby-R pyrite shows a very small iron sulfate peak after 252 days in 34% RH and 257 days in 50% RH (Figure 4). The peaks due to iron sulfate after 251 days at 70% RH are significantly larger indicating a greater amount of weathering products has accumulated under this condition. Kirby samples produced significantly more iron-sulfate salts at higher relative humidities than in the lower relative humidities.

The samples that reacted most readily under experimental conditions were the Waldo and Pittsburgh pyrites. Although the amount and type of product formed differed, both pyrites tended to weather in all RH conditions. Figure

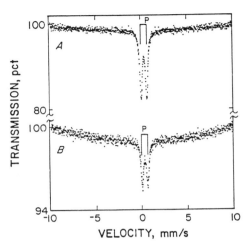

Figure 1. Noranda pyrite (A) before weathering and (B) after 249 days in 79% RH.

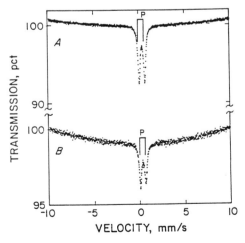

Figure 2. Iron Mountain pyrite (A) before weathering and (B) after 250 days in 79% RH.

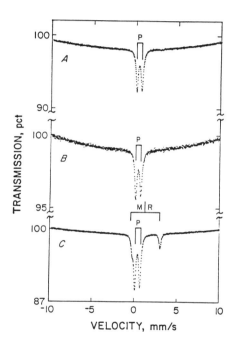

Figure 3. Kirby-U pyrite (A) after 257 days in 34% RH, (B) 256 days in 50% RH and (C) 251 days in 70% RH.

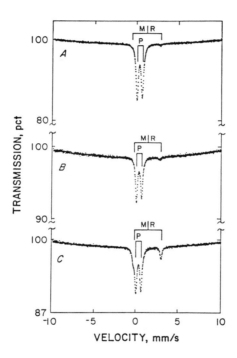

Figure 4. Kirby-R pyrite (A) after 252 days in 34% RH, (B) 257 days in 50% RH and (C) 251 days in 70% RH.

5 shows spectra of Waldo pyrite before and after weathering. Hematite (α-Fe_2O_3) was produced after 30 days in all relative humidities. The Waldo samples weathered to the greatest extent in 34% RH. Figure 6 shows the increase in iron sulfate content of oxidized Pittsburgh pyrite samples. After approximately 250 days, the amount of weathered product increased with increasing humidity. These two pyrites showed oxidation product peaks in all of the experimental humidity conditions.

The amount of water associated with the pyrite samples was also noted over time. The Waldo pyrite remained dry throughout the experiment in the lower three humidities (34%, 50%, and 70%) and contained water drops in the highest humidity, 79% RH. Noranda and Iron Mountain samples remained dry in the two lower humidities and contained water in relative humidities of 70% and 79%. Both samples from Kirby and the Pittsburgh pyrite displayed similar amounts of water present with the samples; the samples were dry in the two lower humidities (34% and 50%) and began displaying water drops in the 70% RH environment. The samples became very wet in 79% RH and had to be air dried before they could be placed in the Mössbauer spectrometer. The Pittsburgh samples in 79% RH environment contained water that was yellow in color whereas Kirby sample water was colorless. Upon drying, the wet samples displayed a small amount of whitish-yellow salts. There was not a sufficient amount of salt available to collect and analyze separately from the sample. It was most likely an iron-sulfate salt.

Areas under the absorption peaks were calculated by the curve-fitting program. These peak areas have been used to represent the relative abundance of the weathering products present in pyrite samples. The percent area of weathering product is plotted over time for each pyrite type in each humidity (Figures 7-10). These figures illustrate the relative reaction rates for the samples in each humidity. Generally, the three sedimentary pyrites weathered more rapidly than the hydrothermal pyrites.

The shape classes (13) for the samples varied from very angular through subrounded: Noranda- angular to subangular; Iron Mountain- subangular to subrounded; Waldo- subrounded; Kirby-R- very angular to angular; Kirby-U- subangular to subrounded; and Pittsburgh- subangular to subrounded. Examination of the sample shape classes relative to the rates of weathering did not show apparent relationships. This may be a result of the sample processing rather than a geochemical phenomenon.

Conclusions

Relative humidity has been shown to be a significant factor in the formation of oxidation products for the sedimentary pyrites in this study. Mössbauer spectra indicated that sedimentary pyrites formed only melanterite and rozenite as oxidation products in all of the humidities tested.

Two hydrothermal pyrites (Iron Mountain and Noranda) produced no detectable weathering products at any humidity tested. However, both pyrites are believed to be responsible for contaminating water at their field sites. This may indicate that more time and/or higher humidities are required if oxidation is to occur to a significant extent. Also, the absence of iron-oxidizing bacteria

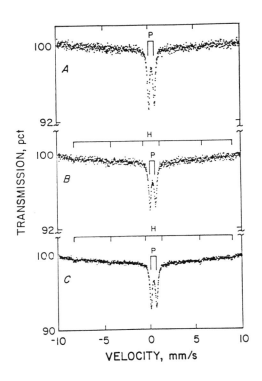

Figure 5. Waldo pyrite (A) before weathering, (B) after 60 days in 34% RH and (C) 250 days in 79% RH.

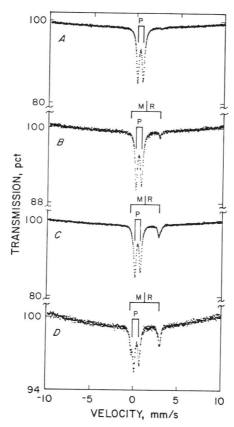

Figure 6. Pittsburgh pyrite (A) before weathering, (B) after 251 days in 34% RH, (C) 252 days in 50% RH and (D) 251 days in 70% RH.

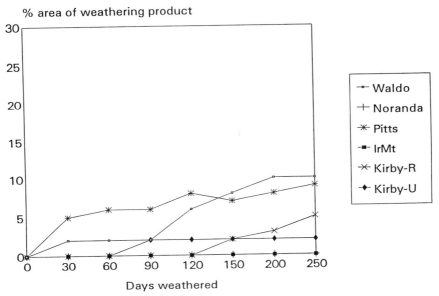

Figure 7. Weathering trends for experimental pyrites over time in 34% RH.

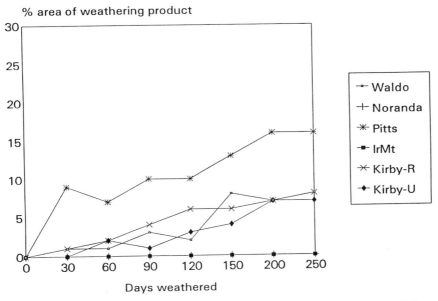

Figure 8. Weathering trends for experimental pyrites over time in 50% RH.

3. BOREK *Effect of Humidity on Pyrite Oxidation*

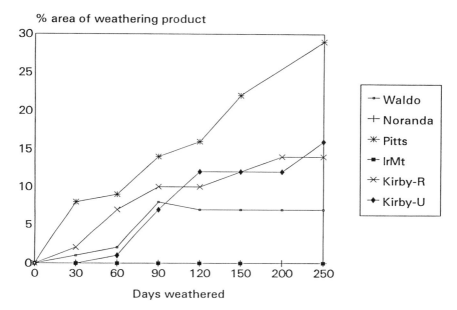

Figure 9. Weathering trends for experimental pyrites over time in 79% RH.

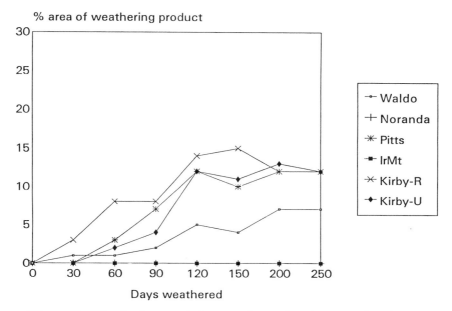

Figure 10. Weathering trends for experimental pyrites over time in 79% RH.

could account for the slow weathering of these pyrites under these experimental conditions, as the oxidation of pyrite in the field can be bacterially catalyzed. The third hydrothermal pyrite, Waldo, produced hematite in all humidities but showed no evidence of iron(III) hydroxide or oxyhydroxide as intermediates. This could be explained by the instability of the these forms and their rapid transformation to hematite.

As previous studies of this type included only one pyrite type, this study may present a more comprehensive view of the effect of humidity on pyrite oxidation. While the amount of product was found to be dependent on the humidity and time, the type of product produced seems dependent on the mode of pyrite formation. The physical properties of pyrite, such as crystal structure, may need to be introduced as contributing factors in oxidation and should be examined like other factors (e.g. oxygen partial pressures, temperature, etc).

Results of this study have demonstrated the utility of Mössbauer spectroscopy as a tool for monitoring pyrite oxidation and identifying oxidation products (5-7). This work has also indicated that humidity control may be an effective means of limiting pyrite oxidation. More work is necessary to determine the effect of humidity on the biotic oxidation of pyrites.

Literature Cited

1. Huggins, F.E.; Huffman, G.P.; Lin, M.C. Int. J. Coal Geol. **1983**, 3, 157-182.
2. Ehlers, E.G.; Stilers, D.V. Am. Mineral. **1965**, 50, 1457-1461.
3. Hammack, R.W. Proceedings: Billings Symposium on Surface Mining and Reclamation in the Great Plains and American Society for Surface Mining and Reclamation; Billings, MT, 1987.
4. Morth, A.H.; Smith, E.E. Fossil Fuel Environmental Pollution Symposium, ACS Division, Fuel Chemistry Preprints, **1966**, 10, 83-92.
5. Huffman, G.P.; Huggins, F.E. Fuel **1978**, 57, 592-604.
6. Narenda, K.J.; Rao, K.R.P.M. Fuel **1979**, 58, 688-689.
7. Pankhurst, Q.A.; McCann, V.H.; and Newman, N.A. Fuel **1986**, 65, 880-883.
8. Folk, R.L. Petrology of Sedimentary Rocks; Hemphill Publishing: Austin, TX, **1974**; 182pp.
9. Patel, A. MOSSFIT Version 2.3, Ranger Scientific, Inc., Burleson, Texas.
10. Cobley, J.G.; Haddock, B.A. FEBS Letters **1975**, 1, 29.
11. Vertes, A.; Zsoldos, B. Acta Chim. Acad. Sci. Hung. **1970**, 65, 261-271.
12. Nordstrom, D.K. In Acid Sulfate Weathering; Kittrick, J.A., et. al., Eds.; Soil Science Soc. of America: Madison, WI, **1982**; 37-56.
13. Powers, M.C. Journal of Sedimentary Petrology. **1953**, 23, 117-119.

RECEIVED October 1, 1993

Chapter 4

Kinetics of Hydrothermal Enrichment of Chalcopyrite

Joon H. Jang[1] and Milton E. Wadsworth

Department of Metallurgical Engineering, University of Utah, Salt Lake City, UT 84112

Chalcopyrite undergoes enrichment in an acid medium under conditions of controlled oxygen injection. When oxygen is fed on a demand basis, the E_h may be maintained at controlled low values resulting in copper enrichment. During the enrichment process, iron is rejected to solution as ferrous ion and a portion of the sulfide sulfur is oxidized to sulfate ion. Copper remains in the solid phase in the form of sulfides, covellite (CuS) and dominantly digenite ($Cu_{1.8}S$). A kinetic model is presented which considers a dynamic balance between anodic enrichment and cathodic oxygen discharge reactions. The particle voltage is controlled by the rate of oxygen injection. Copper remains in the solid phase as long as the E_h is maintained below predicted values. The required rate of oxygen injection for enrichment is related to particle size, the solid to liquid ratio, pH and temperature. Extrapolated to lower temperatures, these reactions are pertinent to the management and environmental control of on-going or discontinued massive dump leaching operations.

The concept of secondary enrichment of porphyry copper deposits was introduced by Emmons (1) in 1900. Later Bateman (2,3) refined the existing concepts and suggested that the copper leached from surface deposits by weathering and groundwater would be reprecipitated below the water table to form various secondary copper sulfide minerals. Most of the early geological studies have dealt with the basic mechanisms and chemistry associated with the changes occurring in ore deposits.

McGauley et al. (4) borrowed the idea of copper enrichment from geologists and patented it for metallurgical purposes. They treated chalcopyrite with cupric sulfate solution in the temperature range of 160 to 230°C. The associated chemical reaction proposed was

[1]Current address: Pohang Iron and Steel Company, Pohang, Kyung Puk, South Korea

$$CuFeS_2 + CuSO_4 \rightarrow 2CuS + FeSO_4 \qquad (1)$$

McKay et al. (5) observed that the above reaction took place at 140°C in the absence of air. For the temperature range of 180 to 200°C, Johnson and Coltrinari (6) proposed the following reaction forming digenite

$$3CuFeS_2 + 6CuSO_4 + 4H_2O \rightarrow 5Cu_{1.8}S + 3FeSO_4 + 4H_2SO_4 \qquad (2)$$

Sohn (7) observed this reaction in the temperature range of 55 to 90°C for finely ground chalcopyrite. He suggested the reaction took place in two steps, first forming covellite by equation (1) and secondly forming digenite by the reaction

$$6CuS + 3CuSO_4 + 4H_2O \rightarrow 5Cu_{1.8}S + 4H_2SO_4 \qquad (3)$$

Peterson (8) studied the kinetics of the enrichment reaction in the temperature range of 125 to 200°C and proposed an electrochemical mechanism for the enrichment process according to reaction (2).

Enrichment to prepare "super concentrates" was proposed by researchers at Anaconda (9,10). In a pilot plant run, it was observed that it was possible to enrich chalcopyrite concentrates by oxygen injection with an overall reaction

$$1.8CuFeS_2 + 4.8O_2 + 0.8H_2O \rightarrow Cu_{1.8}S + 1.8FeSO_4 + 0.8H_2SO_4 \qquad (4)$$

Bartlett (11) subsequently proposed a process for the production of "super-concentrates", based on the kinetics of Peterson (8) for the reaction shown in equation 2.

In this study, the hydrothermal enrichment of chalcopyrite to copper-rich sulfides by direct oxygen injection has been investigated. The main objective was to investigate the fundamental kinetics and mechanisms of *in situ* chalcopyrite enrichment.

Experimental

Equipment. Experiments were carried out in a two-liter Autoclave Engineers autoclave made of 316-stainless steel with a jacket-type heater and a MagneDrive unit for agitation. All stainless steel parts in contact with the solution were replaced by parts made of titanium, and a baffled, cylindrical glass liner was placed inside the body of the autoclave. Agitation speed was monitored with a built-in tachometer. In the conversion tests, oxygen gas was fed into the system at a constant rate. Oxygen flow-rate was controlled with a Brooks Model 5850C mass flow controller in conjunction with a Brooks Model 5876 control unit. The pressure generated during the reaction was monitored with a pressure gauge. A sintered glass frit sampler, connected to a titanium tube, was used to withdraw filtered liquid samples. A schematic diagram of the experimental setup is shown in Figure 1.

Material. The chalcopyrite mineral used in this study was a Kennecott (Bingham) flotation concentrate. By chemical analysis the concentrate was 31.4% Cu, 25.0% Fe and 30.5% S. The concentrate was sized by screening for the preparation of monosize particles: -100/+115 mesh, -170/+200 mesh and -270/+325 mesh. Each size fraction was examined with an optical microscope and by X-ray diffraction for the identification and characterization of the mineral phases contained.

A microscopic point-counting method was employed to determine the mineralogical composition of each size fraction. A description of the method is given by Hausen (12). The results of computer processed point-count data and the mineralogical compositions for three size fractions used in this study are presented in Table I. The method used for data reduction by computer was essentially the same as that described by Odekirk et. al. (13). Details of the computer analysis are presented elsewhere (14). Bornite and chalcocite were predominantly locked in chalcopyrite whereas pyrite existed largely as free particles. Also, as expected, the degree of liberation decreased as the particle size increased.

Table I. Mineralogical Analysis of Kennecott Concentrate

Mineral	Weight percent		
	Screen size, mesh		
	-270/+325	-170/+200	-100/+115
Chalcopyrite	75.26	73.27	69.24
Bornite	9.37	8.55	6.40
Pyrite	7.44	7.51	7.75
Chalcocite	4.63	3.70	3.66
Molybdenite	1.93	2.35	1.64
Gangue	1.37	4.62	11.31

Procedure. Tests were carried out on slurries containing 5 percent of the chalcopyrite concentrate by weight. The solids were pulped with 690 ml of distilled and deionized water and acidified with about 10 ml of sulfuric acid to maintain extracted iron in the soluble form during the course of the reaction. The pulp was agitated and purged with prepurified nitrogen gas for one hour to remove the oxygen remaining inside the system. The autoclave was then heated to the desired temperature. Oxygen was fed into the system at a predetermined rate. During the reaction, the total pressure was measured and samples of about 20 ml were taken. The fraction of chalcopyrite enriched at any time was taken as the fraction of iron released into solution.

Preliminary Conditions. A portion of the iron was acid soluble and was completely dissolved during the heat-up period. For the kinetic study, the fractional conversion of chalcopyrite by oxygen alone was calculated by subtracting the amount of iron released by acid addition from the total amount of iron in solution.

To examine the effect of agitation on the rate of conversion, experiments were performed at three different agitation speeds: 450, 600 and 750 rpm. The rate was independent of the agitation speed. Based on this observation, an agitation speed of 500 rpm was used for subsequent experiments.

Results and Discussion

Chalcopyrite undergoes copper enrichment with the rejection of iron and sulfur at redox potentials slightly above the chalcopyrite stability region. At increasingly more positive potentials, induced by slow oxygen injection, enrichment was observed to occur by sequential anodic reactions in two steps:

Step 1

$$CuFeS_2 + 4H_2O \rightarrow CuS + Fe^{2+} + HSO_4^- + 7H^+ + 8e^- \tag{5}$$

Step 2

$$1.8\,CuS + 3.2\,H_2O \rightarrow Cu_{1.8}S + 0.8\,HSO_4^- + 5.6\,H^+ + 4.8\,e^- \tag{6}$$

$$1.8\,CuFeS_2 + 10.4\,H_2O \rightarrow Cu_{1.8}S + 1.8\,Fe^{2+} + 2.6\,HSO_4^- + 18.2\,H^+ + 19.2\,e^- \tag{7}$$

Coupled with oxygen reduction, equations 5 and 6 become respectively,

$$CuFeS_2 + 2O_2 \rightarrow CuS + FeSO_4 \tag{8}$$

$$1.8\,CuS + 0.8\,H_2O + 1.2\,O_2 \rightarrow Cu_{1.8}S + 0.8\,H_2SO_4 \tag{9}$$

According to equation 8, the covellite (CuS) enrichment stage is expected to be pH independent, while CuS enrichment to digenite (equation 9) is acid producing. Similarly, equation 7, coupled with oxygen reduction is acid producing.

Effect of Oxygen Feed Rate. Figure 2 illustrates the fraction reacted as a function of time for oxygen feed rates of 10, 21 and 30 scc (standard cubic centimeters)/min at 200°C for the -270/+325 mesh size fraction. The enrichment rate increased with a fractional order dependence on oxygen concentration. Figure 3 illustrates the variation of oxygen partial pressure with time. The pressure increased to an initial plateau,

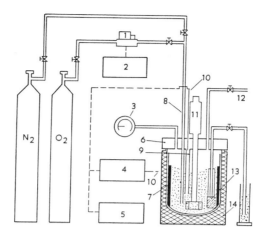

1. Gas Flow Controller
2. Read Out and Control Unit
3. Pressure Gauge
4. Temperature Controller
5. Strip Chart Recorder
6. Autoclave
7. Baffled Glass Liner
8. Gas Inlet
9. Thermowell
10. Thermocouple
11. Magnedrive
12. Vent Line
13. Sintered Glass Sampler
14. Furnace

Figure 1. Schematic diagram of the experimental setup.

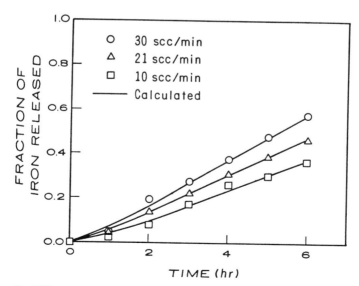

Figure 2. Effect of oxygen feed rate on the fraction of iron released: 200°C; -270/+325 mesh; 0.27 M H_2SO_4.

diminished, and then increased to a second plateau. The two regions correspond essentially to consecutive enrichment processes, forming first covellite (CuS) and then digenite ($Cu_{1.8}S$). Optical examination of reaction products indicated a broad overlap of the two regions. During the latter stage of enrichment, oxygen pressure remained constant, and the rate of oxygen consumption was equal to the rate of oxygen injection. The fraction of copper dissolved was essentially zero for the first 4 hours for 10 scc/min, 2 hours for 21 scc/min and 1 hour for 30 scc/min. Following this initial period the fraction of copper dissolved increased to 0.009, 0.109 and 0.310, respectively, for 10, 21 and 30 scc/min, following closely the increase in oxygen pressure as shown in Figure 3. It should be noted that enrichment occurred essentially without copper dissolution at an oxygen feed rate of 10 scc/min for the 5 percent slurry. The hydrogen ion concentration remained unchanged up to about 2 hours of reaction time, corresponding to the formation (nucleation and growth) of covellite according to equation 8. As the reaction continued beyond two hours, the pH decreased, as expected for digenite formation.

Acid concentration had little influence on the rate of conversion over the range used in this study, 0.27 M to 0.53 M. To examine the effect of back reactions by ferrous ion, experiments were performed at two different initial ferrous ion concentrations. The experimental results without the addition of ferrous ion were compared with those with initial ferrous concentrations of 0.01 M and 0.05 M. The back reaction by ferrous ion had a negligible effect on the rate of conversion.

Effect of Particle Size and Temperature. To investigate the effect of particle size on the conversion rate, experiments were conducted with the three particle sizes prepared for this study. Figure 4 illustrates the results for the three size fractions and an oxygen flow rate of 10 scc/min. The results show an increasing conversion rate with decreasing particle size.

To determine the effect of temperature on the conversion rate, experiments were performed over the temperature range of 172 to 200°C with an oxygen flow rate of 10 scc/min. Rates of conversion for 172, 181, 190 and 200°C are illustrated in Figure 5.

Photomicrograph and X-Ray Examination. Solid samples were taken at 1-hour intervals during the course of the enrichment reaction and examined by x-ray diffraction and with an optical microscope using reflected light with polished specimens. The experimental results were incorporated with the X-ray diffraction and microscopic analysis data to interpret the reaction mechanism. The reaction products formed under constant low oxygen injection rates were identical to those formed under rapid initial oxygen injection rates. The micrographs of the partially enriched chalcopyrite grains clearly indicated that the enrichment reaction takes place in two steps. In the first step chalcopyrite reacted to form columnar, porous covellite. The chalcopyrite grains became totally surrounded with the covellite product layer as the reaction proceeded. The covellite occupied the same volume as the original grain. As the reaction continued, the particle voltage increased into the region where the formation of digenite was favored, step 2. Digenite nucleated at the chalcopyrite-covellite interface and formed a continuous intermediate phase surrounding the

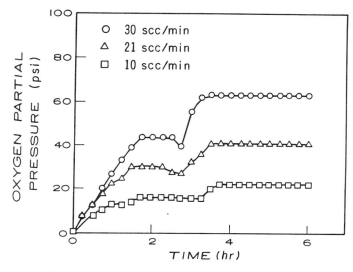

Figure 3. Effect of oxygen feed rate on the oxygen partial pressure: 200°; -270/+325 mesh; 0.27 M H_2SO_4.

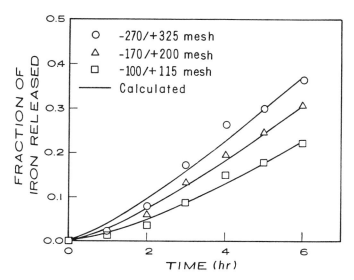

Figure 4. Effect of particle size on the fraction of iron released: 200°C; 10 scc/min O_2; 0.27 M H_2SO_4.

chalcopyrite, with an inner chalcopyrite-digenite interface and an outer digenite-covellite interface. The inner boundary moved toward the center of the particle, consuming chalcopyrite. The digenite-covellite outer boundary moved outwardly, consuming the covellite layer. Also nucleation and growth of digenite occurred at sites throughout the covellite. The digenite filled the volume of the reacted chalcopyrite as a pseudomorph leaving about 60 percent void space. In both cases, the reactions proceed topochemically, with a well defined shrinking core. These porous products provide easy paths for the diffusion of ferrous, sulfate and hydrogen ions away from reaction interfaces. This would indicate that the kinetics may be controlled by the surface reactions. Figure 6 illustrates the layers and proposed boundary reactions.

Rate Equation for the Enrichment Process

A simplified rate expression, based on the electrochemical reactions, as shown in Figure 6, has been developed to simulate the conversion of chalcopyrite to copper sulfides. Because of the highly porous products, permitting the rapid diffusion of ions out through the pore space, it is proposed the rate-controlling reactions are surface-charge-transfer limited. Also, the character of the observed experimental data suggests surface rate control. Chalcopyrite reacts anodically (equations 5 and 6) and is balanced by the cathodic half-cell oxygen discharge reaction (equation 7) occurring at the particle outer surface. These reactions result in a particle mixed potential where the anodic and cathodic total currents are balanced. Butler-Volmer equations for the anodic and cathodic reactions may be simplified, assuming that anodic and cathodic back reactions are negligibly slow and that single electron transfer is favored for both anodic and cathodic charge transfer processes, to give:

anode: $\quad I_a = AZ_a F k_a \exp(\beta_a F E_a / RT)$ (10)

cathode: $\quad I_c = - A_o Z_c F k_c [O_2] \exp(-\beta_c F E_c / RT)$ (11)

where A and A_o are the anodic and cathodic surface areas and F is the Faraday constant and Z_a and Z_c are the total anodic and cathodic charge transferred. Assuming spherical geometry, $A = 4\pi r^2$, where r (cm) is the radius of the chalcopyrite shrinking core at any time t and $A_o = 4\pi r_o^2$, where r_o is the initial radius of the particle. The anodic and cathodic currents (I_a, I_c) are expressed in coulombs hr^{-1} and the voltage-independent rate constants are k_a (mole cm^{-2} hr^{-1}) and k_c (cm hr^{-1}). The transfer coefficients, β_a and β_c, usually have a value of approximately one-half and, in this analysis, are assumed to be equal to 0.5. At the mixed potential, E_M, $E_M = E_a = E_c$ and the anodic and cathodic currents are balanced such that, $I_a = -I_c$. Substitution into equations 10 and 11 yields an expression for the mixed potential,

$$\exp(FE_M/2RT) = (A_o Z_c k_c / A Z_a k_a)^{1/2} [O_2]^{1/2} \qquad (12)$$

The rate of reaction of chalcopyrite is related to the anodic current:

$$dn_{CuFeS_2}/dt = -I_a/Z_a F = -4\pi r r_o ((Z_c/Z_a) k_a k_c)^{1/2} [O_2]^{1/2} \qquad (13)$$

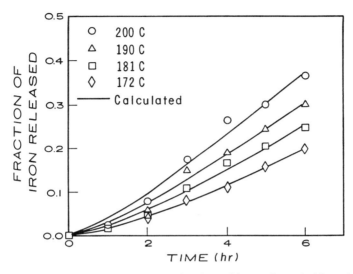

Figure 5. Effect of temperature on the fraction of iron released: 10 scc/min O_2; -270/+325 mesh; 0.27 M H_2SO_4.

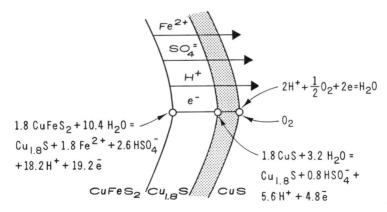

Figure 6. Schematic diagram of proposed mechanism for chalcopyrite enrichment.

Also, from the continuity requirement,

$$dn_{CuFeS2}/dt = (4\pi r^2/v)(dr/dt) \tag{14}$$

where v is the molar volume of chalcopyrite. The radius r is related to the fraction reacted, α, by the equation

$$r = r_o (1-\alpha)^{1/3} \tag{15}$$

Differentiating equation 15 for $d\alpha/dt$, and combining with the above relations yields

$$d\alpha/dt = 3v/r_o ((Z_c/Z_a) k_a k_c)^{1/2} (1-\alpha)^{1/3} [O_2]^{1/2} \tag{16}$$

At any time t, the number of moles of oxygen remaining unreacted, n_{O2}, may be expressed as

$$n_{O2} = n_{O2,g} + n_{O2,l} = n_f t - \sigma n_o \alpha \tag{17}$$

where
- n_f : oxygen feed rate (mole hr^{-1}),
- $n_{O2,g}$: number of moles of oxygen in the gas phase,
- $n_{O2,l}$: number of moles of oxygen in the liquid phase,
- n_o : original moles of chalcopyrite, contained in the mineral feed,
- σ : stoichiometry factor, number of moles of, oxygen reacted per mole of chalcopyrite, and
- t : time (hr).

Introducing the ideal gas law and Henry's law ($[O_2] = k_h P_{O2}$), equation 17 may be expressed in terms of $[O_2]$,

$$[O_2] = (n_f t - \sigma n_o \alpha)/(V_l + V_g/k_h RT) \tag{18}$$

where
- k_h : solubility constant of oxygen (mole cm^{-3} psi^{-1}),
- P_{O2} : partial pressure of oxygen (psi),
- V_g, V_l : volume of gas and liquid (cm^3), respectively.
- R : 1.205 x 10^3 psi cm^3 mole^{-1} K^{-1}.

Equation 16 becomes

$$d\alpha/dt = K_1 (1-\alpha)^{1/3} (n_f t/\phi - K_2 \alpha)^{1/2} \tag{19}$$

and,

$$K_1 ((cm^3 mole^{-1})^{1/2} hr^{-1}) = 3v/r_o ((Z_c/Z_a) k_a k_c)^{1/2} \tag{20}$$

$$K_2 \text{ (mole cm}^{-1}) = \sigma n_o/\phi \tag{21}$$

Also,
$$\phi = V_l + V_g/k_h RT \tag{22}$$

$$K_1 = (K_o / r_o) \exp\{-(E_a + E_c)/2RT\} \tag{23}$$

where,
$$K_o \{(\text{cm}^3/\text{mole})^{1/2}(\text{cm/hr})\} = 3v \{(Z_c/Z_a) f_a f_c\}^{1/2} \tag{24}$$

and f_a and f_c are the Arrhenius frequency factors for k_a and k_c. The value of k_h for the solubility of oxygen in water at various temperatures was calculated using the equation of Zoss et al. (15) for T in °C,

$$k_h = 1.068 \times 10^{-7} - 1.168 \times 10^{-9} T + 6.109 \times 10^{-12} T^2 \tag{25}$$

Equation 19 was solved numerically by a fourth-order Runge-Kutta method and the constants were determined by nonlinear regression analysis. The values of K_1 and K_2 are summarized in Table II.

Table II. Values of K_1 and K_2 for Various Experimental Conditions

Variables	K_1 $(\text{cm}^3 \text{ mole}^{-1})^{1/2} \text{ hr}^{-1}$	$K_2 \times 10^5$ mole cm^{-1}
200°C, 10 scc/min		
-270/+325	52.6	2.98
-170/+200	34.2	3.00
-100/+115	21.1	3.00
200°C, -270/+325		
10 scc/min	52.6	2.98
21 "	57.3	5.32
31 "	66.0	6.40
10 scc/min, -270/+325		
200°C	52.6	2.98
190	40.3	3.00
181	31.4	2.90
172	22.6	2.75

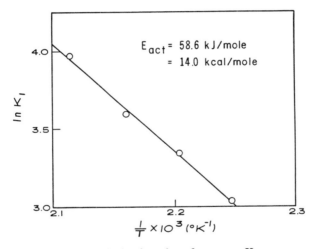

Figure 7. Arrhenius plot of constant K_1.

The values of K_2 increased as the oxygen feed rate increased. This implies that the oxygen introduced to the system is consumed, not only by enrichment reactions, but also by additional anodic reactions such as the dissolution of copper and the oxidation of iron. As expected, K_1 decreased linearly with r_o^{-1} and increased with temperature. Figure 7 is an Arrhenius plot of K_1 against reciprocal temperature. It is apparent from equation 23 that the experimental activation energy is $(E_a + E_c)/2$ or the average (E_{avg}) of the activation energies of the anodic and cathodic charge transfer processes. The average activation energy was found to be 58.6 kJ/mole (14.0 kcal/mole). The calculated value of K_o was 7.59×10^5.

In calculating rates of reaction, an average value of 2.99×10^{-5} was used for K_2 for all conditions except the higher oxygen feed rates (21 and 31 scc/min) where the values listed in Table II were used. Calculated α-t curves for different oxygen feed rates, particle sizes, and temperatures are shown as the solid curves in Figures 2, 4 and 5, respectively. These figures show good agreement between the experimental results and the calculated values.

Conclusions

Based upon experimental rate data under a variety of conditions and upon optical microscopy it is concluded that:
1. Enrichment can occur *in situ* by iron and sulfur rejection with containment of copper in the original solid phase, forming first covellite followed by digenite.
2. The kinetics are moderately rapid in the temperature range of 170 - 200°C. Initial acid concentrations up to 0.53 M H_2SO_4 and ferrous ion up to 0.05 M did not influence the rate.
3. Final products are essentially independent of the rate of oxygen injection.
4. An electrochemical model explained the observed results closely and accounted for temperature, oxygen feed rate, and particle size. The experimental activation energy was 58.6 kJ/mole (14.0 kcal/mole).
5. The results illustrate that even at low temperatures, after sufficient time, chalcopyrite will undergo enrichment to more readily leachable copper sulfides. This can occur entirely by oxygen ingress and does not require exposure of the chalcopyrite to copper bearing solutions. Therefore these reactions add to the many reactions expected in the environmental management of operating and discontinued massive copper sulfide leach dumps and mine sites.

Acknowledgements

This research was supported by the Utah Minerals Resources Research Institute (MRRI) at the University of Utah. The supporting funds for the Institute were provided by the Bureau of Mines, U.S. Department of Interior.

Literature Cited

1. Emmons, W.H. Trans. AIME 1900, 30, 177-217.
2. Bateman, A.M. Economic Mineral Deposits; John Wiley and Sons: NY, 1955, pp. 274-289.

3. Bateman, A.M. The Formation of Mineral Deposits; John Wiley and Sons: NY, 1955, pp. 238-245.
4. McGauley, A.M.; Schaufelberger, F.A.; Roberts, E.S., U.S. Patent 3,755,172, 1956.
5. McKay, D.R.; Swinkels, G.M.; K.R. Szarmes, G.M., German Patent 2,207381, 1972.
6. Johnson, P.K.; Coltrinari, E.L., U.S. Patent 3,957,602, 1976.
7. Sohn, H.J. Ph.D. Thesis, Department of Metallurgical Engineering, University of Utah, 1980.
8. Peterson, R.D. Ph.D. Thesis, Department of Metallurgical Engineering, University of Utah, 1984.
9. Bartlett, R.W.; Wilson, D.B.; Savage, B.J.; Wesely, R.J. In Hydrometallurgical Reactor Design and Kinetics; Bautista, R.G.; Wesely, R.J.; Warren, G.W., Eds.; The Minerals, Metals and Materials Society: Warrendale, PA., 1986; pp. 227-246.
10. Bartlett, R.W. In EPD Congress 1992; Hager, J.P., Ed.; The Minerals, Metals and Materials Society: Warrendale, PA, 1992; pp. 651-661.
11. Bartlett, R.W. Met. Trans. 1992, $\underline{23b}$, 241-248.
12. Hausen, D.M. In Process Mineralogy; Hausen, D.M.; Park, W.C., Eds.; The Metallurgical Society: Warrendale, PA, 1981; pp. 127-142.
13. Odekirk, J.R.; Naruk, S.; Taylor, P. In Process Mineralogy; Hausen, D.M.; Park, W.C., Eds.; The Minerals, Metals and Materials Society: Warrendale, PA, 1981; pp. 253-265.
14. Jang, J.H. Ph.D. Thesis, Department of Metallurgical Engineering, University of Utah, 1992.
15. Zoss, L.M.; Suciu, S.N.; Sibbitt, W.L. Trans. ASME 1954, $\underline{76}$, 69-71.

RECEIVED March 10, 1993

MICROBIAL PROCESSES AFFECTING SULFIDE OXIDATION

Chapter 5

Oxidation of Inorganic Sulfur Compounds by Thiobacilli

Isamu Suzuki, C. W. Chan, and T. L. Takeuchi

Department of Microbiology, University of Manitoba, Winnipeg, Manitoba R3T 2N2, Canada

>Thiobacilli oxidize inorganic sulfur compounds to sulfuric acid and obtain energy for growth from the oxidation. Various species of thiobacilli have different oxidative capabilities, but most can oxidize sulfide, sulfur and thiosulfate. The mechanism of inorganic sulfur oxidation by thiobacilli has been studied for a number of years. It is becoming increasing clear that sulfur and sulfite oxidation are the key reactions in the oxidation of sulfide, elemental sulfur, thiosulfate and polythionates.

Thiobacilli are capable of oxidizing reduced inorganic sulfur compounds such as sulfide, sulfur and thiosulfate to sulfate and of using the energy of oxidation for growth on inorganic nutrients. The physiology and the role of these bacteria in biohydrometallurgy have been extensively reviewed ([1]).

Sulfur Oxidation Scheme

The mechanism of oxidation of these sulfur compounds was formulated in 1974 ([2]) essentially as shown in Figure 1. The scheme still satisfies most experimental results, particularly the accumulated knowledge of enzymes involved. Reaction 1 is the oxidation of sulfide to sulfur which will be discussed later.

$$S^{2-} \rightarrow S + 2e^- \qquad (1)$$

The original proposal was based on the following series of evidence. The S^0-grown *Thiobacillus thiooxidans* ([3]) and *Thiobacillus ferrooxidans* ([4]) and thiosulfate-grown *Thiobacillus thioparus* ([5]) (accumulating sulfur as intermediate) and *Thiobacillus novellus* ([6]) had the sulfur-oxidizing enzyme (reaction 2 in the presence of reduced glutathione, GSH) ([5]).

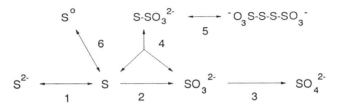

Figure 1. Oxidation of inorganic sulfur compounds. Reactions 1 to 6 are discussed in detail in the text. In the original assignment (2), the enzymes responsible for the reactions were identified as follows: reaction 1, sulfide oxidase; 2, sulfur-oxidizing enzyme; 3, sulfite oxidase or APS reductase; 4, rhodanese (thiosulfate-cleaving enzyme, sulfur transferase); and 5, thiosulfate-oxidizing enzyme.

$$S^0 + O_2 + H_2O \rightarrow H_2SO_3 \tag{2}$$

Sulfite oxidase (reaction 3a) was present in *T. thioparus, T. novellus*, and *T. ferrooxidans* (7-9) and APS(adenosine phosphosulfate) reductase (reaction 3b) in *Thiobacillus denitrificans* and *T. thioparus* (10,11).

$$SO_3^{2-} + H_2O \rightarrow SO_4^{2-} + 2e^- + 2H^+ \tag{3a}$$

$$SO_3^{2-} + AMP \rightarrow APS + 2e^- \tag{3b}$$

Rhodanese (reaction 4) found in *T. denitrificans* and *T. novellus* (12,13) was considered as sulfur transferase.

$$SSO_3^{2-} \leftrightarrow S + SO_3^{2-} \tag{4}$$

In addition the two sulfur atoms of thiosulfate showed a differential behavior in the oxidation (14). Thiosulfate-oxidizing enzyme (reaction 5) was present in *Thiobacillus neapolitanus*, *T. thioparus* and *T. ferrooxidans* (15-18).

$$2SSO_3^{2-} \leftrightarrow {}^-O_3SSSSO_3^- + 2e^- \tag{5}$$

The membrane-bound sulfide oxidase (reaction 1) was studied in *Thiobacillus concretivorus, T. thiooxidans* and *T. thioparus* (19,20). Sulfite readily combined with sulfur chemically to form thiosulfate (reaction 4). Reaction 6 was considered for the interconversion of the single sulfur atom (S) to the elemental sulfur octet (S_8 or S^0).

$$S \leftrightarrow S^0 \tag{6}$$

New Enzymes

New enzymes have been added to the list since 1974. *Thiobacillus ferrooxidans* grown on Fe^{2+} oxidize sulfur or sulfide by reactions 2 and 3, but by transferring electrons first to Fe^{3+} instead of O_2 (21-23) (sulfur:Fe^{3+} oxidoreductase, reaction 7 and sulfite:Fe^{3+} oxidoreductase, reaction 8),

$$S^0 + 4Fe^{3+} + 3H_2O \rightarrow H_2SO_3 + 4Fe^{2+} + 4H^+ \tag{7}$$

$$H_2SO_3 + 2Fe^{3+} + H_2O \rightarrow H_2SO_4 + 2Fe^{2+} + 2H^+ \tag{8}$$

followed by the oxidation of Fe^{2+} by an iron oxidizing system (reaction 9).

$$6Fe^{2+} + 3/2O_2 + 6H^+ \rightarrow 6Fe^{3+} + 3H_2O \tag{9}$$

The overall stoichiometry is the same as in other thiobacilli (reaction 10).

$$S^0 + 3/2 O_2 + H_2O \rightarrow H_2SO_4 \qquad (10)$$

The enzyme oxidizing S^0 requires GSH similar to the enzyme in *T. thiooxidans* and *T. thioparus*, but in substrate quantities and is considered to use sulfide as substrate (24), although GSH is still required and evidence presented does not eliminate glutathione polysulfide (GSS_nH) or glutathione persulfide (GSSH) as a possible substrate. Recently, a sulfide-binding protein containing iron has been isolated from *T. ferrooxidans* membrane (25). The protein acquires a characteristic absorption spectrum upon reaction with sulfide (green color with peaks at 435, 533 and 637 nm) and is considered the sulfide donor for the sulfur (sulfide):Fe^{3+} oxidoreductase. The sulfur (sulfide):Fe^{3+} oxidoreductase has been found in all the *T. ferrooxidans* strains so far tested (26,27) as well as in *Leptospirillum ferrooxidans* strains (26) which are incapable of growth on sulfur.

Versatility of *T. ferrooxidans*

T. ferrooxidans is an extremely versatile organism and in addition to Fe^{2+} and reduced inorganic sulfur compounds the organism can grow on H_2 (28) and HCOOH (29) as oxidizable substrate. The formate-grown cells oxidize both formate and sulfur aerobically and anaerobically with Fe^{3+} (reactions 11 and 12)

$$HCOOH + 2Fe^{3+} \rightarrow CO_2 + 2Fe^{2+} + 2H^+ \qquad (11)$$

$$S + 6Fe^{3+} + 4H_2O \rightarrow SO_4^{2-} + 6Fe^{2+} + 8H^+ \qquad (12)$$

and the anaerobic ferric iron respiration can supply energy for glycine uptake (30). Finally it has been demonstrated recently that *T. ferrooxidans* can successfully grow anaerobically on elemental sulfur using Fe^{3+} as the electron acceptor (31). Since *T. ferrooxidans* cells can leach pyrite with Fe^{3+} (32) the anaerobic growth potential implies that the so-called indirect leaching of sulfide ores by Fe^{3+} may be accelerated by the organism.

The mechanism shown in Figure 1 was questioned when *Thiobacillus versutus*, grown in a chemostat under thiosulfate-limited conditions, produced a thiosulfate-oxidizing enzyme complex which carried out the complete oxidation of thiosulfate to sulfate without any accumulation of intermediates (reaction 13)

$$S_2O_3^{2-} + 2O_2 + H_2O \rightarrow 2SO_4^{2-} + 2H^+ \qquad (13)$$

and none of the protein components separated from the complex showed any enzyme activity shown in the scheme (33-37). This is probably a special example of fully integrated thiosulfate-oxidizing system devoted for maximal derivation of energy forced upon the organism by the growth condition. When *T. versutus* is grown in batch cultures on thiosulfate, on the other hand, cells can oxidize elemental sulfur and sulfite as well as thiosulfate and in fact *T. versutus* can even grow on sulfur if the O_2 pressure is reduced to 2.5 kPa (38). Energy considerations extensively used in evaluating possible mechanisms of oxidation of inorganic sulfur compounds (33) should not be accepted without caution, since the bacterial growth yield does not necessarily correspond to the ATP yield (39) and chemostats lead to the selection of certain predictable types of cells.

Polythionates

Polythionate (^-O_3S-$S_n SO_3^-$) oxidation reactions were extensively studied earlier (14), but with the isolation of long S chain polythionates from tetrathionate-grown *T. ferrooxidans* (40,41) they have again been considered as possible intermediates in the metabolism of inorganic sulfur compounds (41). This situation is reminiscent of controversy over the reduction of sulfite to sulfide by sulfate reducing bacteria with respect to the intermediate formation of trithionate (^-O_3S-S-SO_3^-) (42). In the case of sulfite reduction, both theories had strong enzymatic evidence (42), whereas in the oxidation of sulfur compounds the original thiosulfate-oxidizing enzyme (reaction 5) is the only enzyme clearly established. Evidence for the existence of postulated polythionate hydrolyzing enzymes (reactions 14 and 15)

$$S_3O_6^{2-} + H_2O \rightarrow S_2O_3^{2-} + SO_4^{2-} + 2H^+ \tag{14}$$

$$S_4O_6^{2-} + H_2O \rightarrow HS_2SO_3^- + HSO_4^- \tag{15}$$

is not unequivocal (41, 43-45). The formation of sulfur during the oxidation of sulfide, thiosulfate, trithionate and tetrathionate by *T. ferrooxidans* (41,45) and the inhibition of further oxidation of sulfur by N-ethylmaleimide (NEM) support the mechanism proposed in Figure 1. The O_2 consumption data published on tetrathionate-grown *T. ferrooxidans* (45) in the presence of NEM agree with the stoichiometry expected of the following equations:

$$Na_2S + \tfrac{1}{2}O_2 + H_2O \rightarrow S^0 + 2NaOH \tag{16}$$

$$Na_2S_2O_3 + \tfrac{1}{2}O_2 \rightarrow S^0 + Na_2SO_4 \tag{17}$$

$$Na_2S_4O_6 + \tfrac{1}{2}O_2 + H_2O \rightarrow 2S^0 + Na_2SO_4 + H_2SO_4. \tag{18}$$

Further Support for the Scheme

These results accumulated over many years support the central pathway shown in Figure 1 for the oxidation of inorganic sulfur compounds. The two key intermediates, sulfur and sulfite, are connected by sulfur-oxidizing enzyme system, reaction 2. The aerobic sulfur-oxidizing enzyme has been found also in *Sulfolobus brierleyi* (46) and *Desulfurolobus ambivalens* (47), thermophilic and extremely thermophilic archaebacteria. It is interesting that the enzyme from both organisms had a molecular weight of nearly a half million consisting of one type of subunits with a molecular weight of 35-40,000. The purified enzyme oxidized sulfur to sulfite with O_2 in the absence of GSH at 65°C (46) and 85°C (47). In the latter (47) hydrogen sulfide was also produced particularly in the presence of Zn^{2+} and the enzyme activity was inhibited by thiol-binding reagents. It is possible that elemental sulfur is more reactive at the elevated temperatures for potential hydrolysis (48).

The role of thiosulfate-oxidizing enzyme (reaction 5) may in fact be more than the oxidation of thiosulfate to tetrathionate. The enzyme isolated from a marine heterotroph catalyzes the formation of thiosulfate from tetrathionate, i.e. tetrathionate reductase activity (49). The thiosulfate-oxidizing enzyme of *Thiobacillus tepidarius* (44) is also reversible. Therefore the reaction is now justifiably written as reversible in Figure 1.

Reaction 1 is also shown as reversible (although not necessarily by the same enzyme system) because the anaerobic formation of H_2S from sulfur by *T. ferrooxidans* cells can be as fast as 10% of the rate of aerobic formation of sulfate (50) and the possibility of sulfide as the substrate for the sulfur (sulfide): Fe^{3+} oxidoreductase exists in *T. ferrooxidans* (24,25). In Figure 1 "S" should be considered as "reactive sulfur" which is formed from sulfide (reaction 1), thiosulfate (reaction 4) or elemental sulfur (reaction 6) and is converted to elemental sulfur (S^0 or S_8) when accumulated. It could be equivalent to sulfane sulfur of polysulfanes ($HS-S_n-SH$) or polythionic acids ($HO_3S-S_n-SO_3H$) which are known to be susceptible to nucleophilic attack by cyanide (CN^-), sulfite (SO_3^{2-}) or thiols (RS^-) (14,48). The oxidation pathway is then governed simply by the rates of the sulfur-oxidizing system (reaction 2) and sulfite-oxidizing system (reaction 3). The oxidation of sulfide to sulfur (reaction 1) by thiobacilli is very fast and sulfane sulfur accumulates (6,19,20) when reaction 2 is not fast enough. If reaction 3 is slower than reaction 2, sulfite accumulates and sulfur plus sulfite yield thiosulfate as the oxidation product. Once thiosulfate is formed, reaction 5 can initiate the polythionate pathway. The reversibility of reaction 5 and the copurification of thiosulfate-oxidizing enzyme and trithionate hydrolase activities (44) and the formation of sulfur during tetrathionate oxidation (41,45) suggest the possibility of polythionate oxidation by reactions 5, 4, 2 and 3.

Reaction 3, as the major oxidation step, is catalyzed by sulfite oxidase, APS reductase or sulfite:Fe^{3+} oxidoreductase. The first two enzymes were thought to couple for the reduction of c-type cytochromes (2) and cytochrome oxidase. There is now a clear evidence that sulfite oxidase can enter the electron transfer chain at cytochrome b. Antimycin A or 2-heptyl-4-hydroxyquinoline-N-oxide (HQNO) inhibits sulfite oxidation (43, 51-54) and a b-type cytochrome solubilized from *T. thiooxidans* membrane is reduced by sulfite in the presence of the membrane (54). Inhibition by HQNO of not only sulfite, but also trithionate and tetrathionate oxidations in *T. tepidarius* (43) again supports the idea that polythionate oxidation involves the oxidation of both sulfane (-S-) and sulfonic acid ($-SO_3H$) portions by the enzymes involved in reactions 2 and 3, respectively.

Acknowledgements

We would like to thank the Natural Sciences and Engineering Research Council of Canada and Imperial Oil for research grants.

Literature Cited

1. Rossi, G. Biohydrometallurgy; McGraw Hill: New York, NY; 1990; 609 pp.
2. Suzuki, I. Ann. Rev. Microbiol. 1974, 28, 85-101.
3. Suzuki, I. Biochim. Biophys. Acta 1965, 104, 359-371.
4. Silver, M.; Lundgren, D.G. Can. J. Biochem. 1968, 46, 457-461.
5. Suzuki, I.; Silver, M. Biochim. Biophys. Acta 1966, 122, 22-23.
6. Charles, A.M.; Suzuki, I. Biochim. Biophys. Acta 1966, 128, 510-521.
7. Lyric, R.M.; Suzuki, I. Can. J. Biochem. 1970, 48, 334-343.
8. Charles, A.M.; Suzuki, I. Biochim. Biophys. Acta 1966, 128, 522-534.
9. Vestal, J.R.; Lundgren, D.G. Can. J. Biochem. 1971, 49, 1125-1130.
10. Bowen, T.J.; Happold, F.D.; Taylor, B.F. Biochim. Biophys. Acta 1966, 118, 566-576.
11. Lyric, R.M.; Suzuki, I. Can. J. Biochem. 1970, 48, 334-343.
12. Bowen, T.J.; Butler, P.J.; Happold, F.C. Biochem. J. 1965, 97, 651-657.
13. Tabita, R.; Silver, M.; Lundgren, D.G. Can. J. Biochem. 1969, 47, 1141-1145.
14. Roy, A.B.; Trudinger, P.A. The Biochemistry of Inorganic Compounds of Sulphur; Cambridge Univ. Press: London; 1970; 400 pp.
15. Trudinger, P.A. Biochem. J. 1961, 78, 673-680.
16. Trudinger, P.A. Biochem. J. 1961, 78, 680-686.
17. Lyric, R.M.; Suzuki, I. Can. J. Biochem. 1970, 48, 355-363.
18. Silver, M.; Lundgren, D.G. Can. J. Biochem. 1968, 46, 1215-1220.
19. Moriarty, D.J.W.; Nicholas, D.J.D. Biochim. Biophys. Acta 1969, 184, 114-132.
20. Moriarty, D.J.W.; Nicholas, D.J.D. Biochim. Biophys. Acta 1970, 197, 143-151.
21. Sugio, T.; Mizunashi, W.; Inagaki, K.; Tano, T. J. Bacteriol. 1987, 169, 4916-4922.
22. Sugio, T.; Katagiri, T.; Moriyama, M.; Zhen, Y.L.; Inagaki, K.; Tano, T. Appl. Environ. Microbiol. 1988, 54, 153-157.
23. Sugio, T.; Hirose, T.; Zhen, Y.L.; Tano, T. J. Bacteriol. 1992, 174, 4189-4192.
24. Sugio, T.; Katagiri, T.; Inagaki, K.; Tano, T. Biochim. Biophys. Acta 1989, 973, 250-256.
25. Sugio, T., Suzuki, H.; Oto, A.; Inagaki, K.; Tanaka, H.; Tano, T. Agr. Biol. Chem. 1991, 55, 2091-2097.
26. Sugio, T.; White, K.; Shute, E.; Choate, D.; Blake II, R.C. Appl. Environ. Microbiol. 1992, 58, 431-433.
27. Suzuki, I.; Takeuchi, T.L.; Yuthasastrakosol, T.D.; Oh, J.K. Appl. Environ. Microbiol. 1990, 56, 1620-1626.
28. Drobner, E.; Huber, H.; Stetter, K.O. Appl. Environ. Microbiol. 1990, 56, 2922-2923.
29. Pronk, J.T.; Meijer, W.M.; Hazeu, W.; van Dijken, J.P.; Bos, P.; Kuenen, J.G. Appl. Environ. Microbiol. 1991, 57, 2057-2062.
30. Pronk, J.T.; Liem, K.; Bos, P.; Kuenen, J.G. Appl. Environ. Microbiol. 1991, 57, 2063-2068.
31. Pronk, J.T.; de Bruyn, J.C.; Bos, P.; Kuenen, J.G. Appl. Environ. Microbiol. 1992, 58, 2227-2230.

32. Lizama, H.M., Suzuki, I. Appl. Environ. Microbiol. 1989, 55, 2918-2923.
33. Kelly, D.P. Phil. Trans. R. Soc. Lond. 1982, B298, 499-528.
34. Lu, W.-P.; Kelly, D.P. J. Gen. Microbiol. 1983, 129, 3549-3564.
35. Lu, W.-P.; Kelly, D.P. Biochim. Biophys. Acta 1984, 765, 106-107.
36. Lu, W.-P.; Swoboda, B.E.P., Kelly, D.P. Biochim. Biophys. Acta 1985, 828 116-122.
37. Kelly, D.P. In Autotrophic Bacteria; Bowien, B.; Schlegel, H.G., Ed.; Science Tech Publ.: Madison, WI, 1989; pp. 193-217.
38. Beffa, T.; Berczy, M.; Aragno, M. FEMS Microbiol. Lett. 1991, 84, 285-290.
39. Marr, A.G. Microbiol. Rev. 1991, 55, 316-333.
40. Steudel, R. In Autotrophic Bacteria; Bowien, B.; Schlegel, H.G., Ed.; Science Tech Publ.: Madison, WI, 1989; pp. 289-303.
41. Pronk, J.T.; Meulenberg, R.; Hazeu, W.; Bos, P.; Kuenen, J.G. FEMS Microbiol. Rev. 1990, 75, 293-306.
42. Fauque, G.; LeGall, J.; Barton, L.L. In Variations in Autotrophic Life; Shively, J.M.; Barton, L.L., Ed.; Academic Press: New York, 1991; pp. 271-337.
43. Lu, W.-P.; Kelly, D.P. J. Gen. Microbiol. 1988, 134, 865-876.
44. Lu, W.-P.; Kelly, D.P. J. Gen. Microbiol. 1988, 134, 877-885.
45. Hazeu, W.; Batenburg-van der Vegte, W.H.; Bos, P.; van der Pas, R.K.; Kuenen, J.G. Arch. Microbiol. 1988, 150, 574-579.
46. Emmel, T.; Sand, W.; König, W.A.; Bock, E. J. Gen. Microbiol. 1986, 132, 3415-3420.
47. Kletzin, A. J. Bacteriol. 1989, 171, 1638-1643.
48. Lyons, D.; Nickless, G. In Inorganic Sulfur Chemistry; Nickless, G., Ed.; Elsevier Publ. Co.: New York, 1968; pp. 509-533.
49. Whited, G.M.; Tuttle, J.H. J. Bacteriol. 1983, 156, 600-610.
50. Bacon, M.; Ingledew, W.J. FEMS Microbiol. Lett. 1989, 58, 189-194.
51. Adams, C.A.; Warns, G.M.; Nicholas, D.J.D. Biochim. Biophys. Acta 1971, 235, 398-406.
52. Kodama, A.; Kodama, T.; Mori, T. Plant Cell Physiol. 1970, 11, 701-711.
53. Takakuwa, S. Plant Cell Physiol. 1976, 17, 103-110.
54. Tano, T.; Ito, T.; Takesue, H.; Sugio, T.; Imai, K. J. Ferment. Technol. 1982, 60, 181-187.

RECEIVED October 1, 1993

Chapter 6

Microbial Oxidation of Sulfides by *Thiobacillus denitrificans* for Treatment of Sour Water and Sour Gases

K. L. Sublette[1], Michael J. McInerney[2], Anne D. Montgomery[2], and Vishveshk Bhupathiraju[2]

[1]Center for Environmental Research and Technology, University of Tulsa, 600 South College Avenue, Tulsa, OK 74104
[2]Department of Botany and Microbiology, University of Oklahoma, 770 van Vleet Oval, Norman, OK 73019

It has been demonstrated that the bacterium *Thiobacillus denitrificans* may be cultured aerobically and anoxically in batch and continuous cultures on hydrogen sulfide (H_2S) gas under sulfide-limiting conditions. Under these conditions sulfide concentrations in the culture medium were less than $1\mu M$ resulting in very low concentrations of H_2S in the reactor-outlet gas. Heterotrophic contamination was shown to have negligible effect on reactor performance with respect to hydrogen-sulfide oxidation. In fact, growth of *T. denitrificans* in the presence of floc-forming heterotrophs produced a hydrogen-sulfide-active floc with excellent settling characteristics. Flocculated *T. denitrificans* has been used to remove H_2S from gases and to remove sulfides from sour water at concentrations up to 25 mM. Reactors containing flocculated *T. denitrificans* have been operated for up to nine months continuously.

A sulfide-tolerant strain (strain F) of *T. denitrificans* has also been shown to prevent the net production of H_2S by sulfate-reducing bacteria in both liquid cultures and sandstone cores. As hydrogen sulfide was produced by sulfate-reducing bacteria, it was immediately oxidized to sulfate by *T. denitrificans*. Strain F has also been used to remove dissolved sulfides from formation water pumped from an underground gas storage facility.

The high reaction rates and mild reaction conditions characteristic of microbial processes offer potential for improvement of processes which have historically been purely chemical or physical in nature. One such process, dominated by physicochemical methodology, has been the removal and disposal of hydrogen sulfide (H_2S) from natural gas, biogas, syngas, and various waste gas streams such as the gas stream produced when air or other gases are used to strip H_2S from sulfide-laden (sour) water. However, a microbial process can replace an H_2S-removal system (amine unit), an H_2S disposal unit (Claus or Stretford process), a tail-gas clean-up or an entire conventional gas processing train. Also a microbial process can be used directly to remove dissolved inorganic sulfides from sour water.

We have developed on the bench scale a process for the desulfurization of gases by H_2S oxidation, based on the contacting of a sour gas with a culture of the chemoautotrophic bacterium, *Thiobacillus denitrificans*. The same basic process has also been used to treat sour water and prevent the net formation of H_2S by sulfate-reducing bacteria (SRB). The results of much of this work have been described in detail elsewhere ([1]-[10]). We present here a short review including some of our more recent results.

Thiobacillus denitrificans is an obligate autotroph and facultative anaerobe which can utilize reduced sulfur compounds as energy sources and oxidize them to sulfate. Under anoxic conditions, nitrate is used as a terminal electron acceptor and is reduced to elemental nitrogen (N_2).

Growth of *T. denitrificans* on $H_2S(g)$

In the laboratory we initially grow *T. denitrificans* in 1.5-2.0 L cultures (pH 7.0, 30°C) on thiosulfate as an energy source in the medium described by Table I to a cell density of 10^8-10^9 cells/mL. Following depletion or removal of thiosulfate, H_2S is introduced. The H_2S feed gas generally consisted of 1% H_2S, 5% CO_2 and the balance N_2. It is important to note that this concentration of H_2S is not a technical limitation, merely a safety precaution.

Table I. Growth Medium for *T. denitrificans*

Component	per L
Na_2HPO_4	1.2 g
KH_2PO_4	1.8 g
$MgSO_4 \cdot 7H_2O$	0.4 g
NH_4Cl	0.5 g
$CaCl_2$	0.03 g
$MnSO_4$	0.02 g
$FeCl_3$	0.02 g
$NaHCO_3$	1.0 g
KNO_3 (anoxic)	5.0 g
$Na_2S_2O_3$	10.0 g
Trace metal solution ([1])	15 mL
Mineral water	50 mL

When H_2S was introduced to batch anoxic or aerobic cultures of *T. denitrificans*, the H_2S was immediately metabolized. At an initial loading of 4-5 mmoles $hr^{-1}g^{-1}$ biomass, and with sufficient agitation, H_2S was not detected in the outlet gas (less than 0.05 M). The residence time of a bubble of feed gas (average diameter 0.25 cm) was 1-2 s. Less that 1 μM of total sulfide (H_2S, HS^-, S^{2-}) was observed in the reactor medium. No elemental sulfur was detected; however, sulfate accumulated in the reactor medium as H_2S was removed from the feed gas. Oxidation of H_2S to sulfate was accompanied by growth, as indicated by an increase in optical density and biomass protein concentration and a decrease in the NH_4^+ concentration. Consumption of OH^- equivalents indicated that the reaction was acid-producing. Nitrate was consumed under anoxic conditions. A sample material balance is given in Table II.

These reactors also have been operated continuously on an H_2S-containing feed, at dilution rates of 0.029 hr^{-1} to 0.053 hr^{-1}, for up to five months. Therefore, the biology of the reactor system is considered very stable.

Table II. Sample Material Balances: Aerobic and Anoxic Oxidation of H_2S in Batch Reactors by *T. denitrificans*

	Aerobic	Anoxic
H_2S oxidized	86.0 mmoles	18.3 mmoles
SO_4^{2-} produced	81.8 mmoles	18.8 mmoles
Biomass produced	453 mg	246 mg
NO_3^- consumed	---	27.0 mmoles
NH_4^+ consumed	8.4 mmoles	2.2 mmoles
OH^- consumed	151.3 meq	31.8 meq

Upset and Recovery

Hydrogen sulfide is toxic to most, if not all, forms of life including *T. denitrificans*, even though the organism can use H_2S as an energy source. Therefore, H_2S is said to be an inhibitory substrate for the organism. In the experiments described above, the cultures were operated on a sulfide-limiting basis. In other words, the H_2S feed rate was always less than the maximum rate at which the biomass was capable of oxidizing the H_2S. If this maximum capacity of the biomass of H_2S oxidation is exceeded, inhibitory levels of sulfide will accumulate.

To examine the behavior of a *T. denitrificans* reactor in an upset condition, the H_2S feed rate to aerobic and anoxic batch and continuous-flow reactors, like those described above, was increased in a stepwise manner until H_2S breakthrough was observed. At the point at which breakthrough occurred, N_2O was detected in the outlet gas from anoxic reactors in concentrations approximately equal to that of the H_2S in the feed gas. Analysis of the reactor medium from both aerobic and anoxic reactors also indicated an accumulation of sulfide and elemental sulfur. Sulfur balances for reactors operated under upset conditions showed that all of the H_2S removed from the feed gas could be accounted for in terms of sulfate, elemental sulfur, and sulfide in the medium. It was observed that the upset condition was reversible if the cultures were not exposed to the accumulated sulfide for more than 2-3 hours. Reduction in H_2S feed rate following an upset condition reduced H_2S and N_2O concentrations in the outlet gas to pre-upset levels. In addition, elemental sulfur, which accumulated during upset, was oxidized rapidly to sulfate.

It is important to know at which H_2S loading the specific activity of the *T. denitrificans* biomass will be exceeded, resulting in upset. The maximum loading of the biomass under anoxic conditions was observed to be in the range of 5.4-7.6 mmoles $hr^{-1}g^{-1}$ biomass. Under aerobic conditions, the maximum loading was observed to be much higher, 15.1-20.9 mmoles H_2S $hr^{-1}g^{-1}$ biomass.

Effect of Septic Operation on H_2S Oxidation by *T. denitrificans* Reactor

The medium used in the experiments described above will not support the growth of heterotrophs because there is no organic-carbon source. However, early on in this study it was observed that if aseptic conditions were not maintained, heterotrophic contamination developed in the *T. denitrificans* culture. Evidently *T. denitrificans* releases organic material into the medium in the normal course of growth, or through lysis of non-viable cells, which supports the growth of heterotrophs. To investigate the effect of heterotrophic contamination on the performance of a *T. denitrificans* continuous stirred-tank reactor (CSTR), one anoxic reactor which became contaminated was allowed to operate for an extended time (30 days). The reactor was originally contaminated by two unidentified heterotrophic bacteria with distinctly different colony morphologies when grown on nutrient agar. After 145 hours of operation, the reactor was injected with suspensions of four different heterotrophic bacteria (*Pseudomonas* species) known to be nutritionally versatile. The total heterotroph concentration increased to about 10^8 cells/mL and leveled off. Apparently growth of the contaminants became limited by the availability of suitable carbon sources. The viable count of *T. denitrificans* at steady state was 5.0 x 10^9 cells/mL. The steady-state composition of the culture medium, and the outlet-gas condition, were indistinguishable from that of a pure culture of *T. denitrificans* operated under the same culture conditions. Therefore, the proposed microbial process for H_2S oxidation need not be operated aseptically. These observations led to the efforts to immobilize *T. denitrificans* by co-culture with floc-forming bacteria.

Co-culture of *T. denitrificans* With Floc-forming Heterotrophs

Many microorganisms exist co-immobilized in nature. These associations are often of benefit to all members of the population. Many species of bacteria produce extracellular biopolymers which adsorb and entrap other non-flocculating microbial cells, forming protected environments for the latter, and establishing beneficial cross-feeding. Such immobilized mixed populations are exploited in activated sludge systems, trickling filters, anaerobic digesters, and similar systems for the treatment of waste water.

T. denitrificans has been immobilized by co-culture with floc-forming heterotrophs obtained from activated sludge taken from the aerobic reactor of a refinery waste-water treatment system. *T. denitrificans* cells grown aerobically on thiosulfate and washed sludge were suspended together in fresh thiosulfate medium (Table I) without nitrate. The culture was maintained in a fed-batch mode at pH 7.0 and 30°C with a gas feed of 5% CO_2 in air. This medium was thiosulfate-limiting with respect to the growth of *T. denitrificans*. When thiosulfate was depleted, the agitation and aeration were terminated and the flocculated biomass was allowed to settle under gravity. The supernatant liquid was then removed and discarded. In this way the culture was enriched for *T. denitrificans* cells which had become physically associated with the floc. The volume then was made up with fresh medium, and aeration and agitation restarted. This fed-batch cycle was repeated 5-6 times. Immobilized *T. denitrificans* was used to oxidize H_2S in a CSTR with cell recycle at molar feed rates of up to 6.3 mmoles/hr (2.0 L culture volume) and total biomass concentrations of up to 13 g/L. During five months of continuous operation, the biomass exhibited excellent settling properties; this test demonstrated the long-term stability of the relationship between *T. denitrificans* and the floc-forming heterotrophs (at a biosolids concentration of 3 g/L, 70% compression of the biomass was observed in 10 minutes). No external addition of organic carbon was required at any time.

It seems that the growth of the autotroph *T. denitrificans* was balanced with the growth of the floc-forming heterotrophs through a commensal relationship in which the growth of the heterotrophs was limited by organic carbon derived from *T. denitrificans*. The result was an immobilization matrix which grew with the *T. denitrificans*. This development reduces the proposed microbial process for H_2S oxidation to the level of technical simplicity of an activated sludge system.

Sulfide-tolerant Strains of *T. denitrificans*

We have shown that H_2S is an inhibitory substrate for *T. denitrificans* ([5]). Growth of the wild-type strain (ATCC 23642) on thiosulfate was shown to be inhibited by sulfide (as Na_2S) concentrations as low as 100-200 μM. Complete inhibition was observed at initial sulfide concentrations of 1 mM. Clearly any process for the removal of H_2S from a sour gas or sour water would be more resistant to upset if a sulfide-tolerant strain of *T. denitrificans* were utilized.

Sulfide-tolerant strains of *T. denitrificans* have been isolated by enrichment from cultures of the wild-type. These tolerant strains were obtained by repeated exposure of *T. denitrificans* cultures to increasing concentrations of sulfide. At each step, only tolerant strains survived and grew. Eventually strains were obtained which exhibited growth comparable to controls at sulfide concentrations of up to 2500 μM. These concentrations are lethal to the wild-type.

Microbial Treatment of Sour Water

Inorganic sulfide (H_2S, HS^-, S^{2-}) is often found to contaminate water co-produced with petroleum. Commonly this water is "treated" by air stripping H_2S. However, this practice frequently simply converts a water pollution problem into an air pollution problem. These sour waters may be treated directly by *T. denitrificans*.

Water containing up to 25 mM soluble sulfide has been successfully treated in an aerobic up-flow bubble column (3.5 L) containing 4.0 g/L of a sulfide-tolerant strain *T. denitrificans* immobilized by co-culture with floc-forming heterotrophs. The sulfide-laden water was supplemented with only mineral nutrients. The sulfide-active floc was stable for nine months of continuous operation, with no external organic carbon required to support the growth of the heterotrophs. The floc exhibited excellent settling properties throughout the experiment.

Retention times in the reactor varied from 1.2-1.8 hours. However, the molar sulfide feed rate was more important in determining the capacity of the reactor for sulfide oxidation than either the hydraulic retention time or the influent sulfide concentration. At a biosolids concentration of about 4 g/L the column could be operated at a molar sulfide feed rate of 12.7-15.4 mmoles/hr without upset.

Microbial Control of H_2S Production By Sulfate-reducing Bacteria

As noted earlier, dissolved sulfides often contaminate water co-produced with petroleum. The source of the sulfide generally is the reduction of sulfate by SRB. These bacteria are strict anaerobes which utilize a number of organic compounds, such as lactate, acetate and ethanol, as a source of carbon and energy. These compounds are end-products of the metabolism of fermentative heterotrophs, and are readily available in a consortium of bacteria in an anaerobic environment. SRB, therefore, are ubiquitous to virtually any anaerobic environment conducive to microbial growth (11).

Sulfide production by SRB is directly or indirectly responsible for major damage each year due to corrosion. Sulfide production may be diminished by inhibiting the growth of SRB. For example, in the secondary production of petroleum, water used in flooding operations is treated with a biocide (typically glutaraldehyde) to control SRB growth in the injection well, reservoir, and piping. Because SRB are strict anaerobes, aeration of flooding water can also serve to inhibit sulfide production. These measures are of limited effectiveness, however, because SRB generally are found attached to a solid surface, entrapped with other bacteria in polysaccharide gels produced by "slime-forming" bacteria. Within these gels the SRB find themselves in a somewhat protected environment, in which biocides and oxygen effectively do not penetrate (11,12).

New technology is needed in the control of H_2S production by SRB to address the limitations inherent in the conventional methods described above. For example, the biogenic production of H_2S by SRB may be subject to biological control. A sulfide-tolerant strain of *Thiobacillus denitrificans* (strain F) has been successfully grown in co-culture with the sulfate-reducing bacterium *Desulfovibrio desulfuricans*, both in liquid culture and through sandstone cores, without the accumulation of sulfide (Table III). Microbial sulfide production in an enrichment from an oil-field brine also was controlled by the presence of this sulfide-tolerant strain F. The effectiveness of strain F is due to its ability to grow and utilize sulfide at levels which are inhibitory to the wild-type strain of *T. denitrificans*. There are many sulfide-oxidizing bacteria, but these bacteria are usually inhibited when H_2S concentrations reach a nuisance level. Strain F not only removed sulfide in cultures of *D. desulfuricans* with lactate as the energy source, but it also did so in the presence of a mixture of SRB which use lactate, and products of lactate metabolism, acetate, and H_2 for sulfide production.

Strain F of *T. denitrificans* readily grew through sandstone cores in pure cultures. Its penetration time was roughly 0.4 cm/day, which is much faster than that observed for *D. desulfuricans* and the organisms present in the oil-field brine through the core together. This observation is important because it suggests that strain F uses sulfide as it is being produced, thus preventing a buildup of sulfide.

The slow penetration times of *D. desulfuricans* observed in the cores suggests the *Desulfovibrio* species do not readily grow through sandstone. This observation seems to be a general property of SRB because the various kinds of sulfate-reducers present in the oil-field-brine enrichment also slowly penetrated the cores. If this is the case in a natural environment, sulfide accumulation may occur only near the well-bore. Thus, water-flooding activities may not result in the introduction of SRB deep into the formation. It is interesting to note that plugging of injection wells by biofilm techniques, such as the use of strain F to remove sulfide, may be feasible because only the area near the well-bore needs to be treated.

Table III. Sulfide Production by *D. desulfuricans* Grown With and Without the Wild-type or Sulfide-tolerant Strains of *T. denitrificans*[a]

Culture[b]	*T. denitrificans* Inoculum Size (mL)	Sulfide Concentration After	
		14 days (mg/L)	19 days (mg/L)
DD alone	0	47.4	42.2
DD + wt	0.1	28.0	28.8
	0.2	51.5	43.5
	0.3	22.5	19.5
DD + F	0.1	4.0	<0.1
	0.2	<0.1	<0.1
	0.3	<0.1	<0.1

[a] Cultures were grown anaerobically in *T. denitrificans* thiosulfate medium (Table I) with deletion of thiosulfate and addition of 0.1% (m/v) Na_2SO_4, 0.05% (v/v) sodium lactate syrup (60%) and 0.1% (v/v) Balch vitamin solution (13).
[b] Abbreviations are (DD) *D. desulfuricans*, (wt) *T. denitrificans* wild-type strain, and (F) sulfide-resistant strain of *T. denitrificans*.

The ability of *T. denitrificans* to grow readily through such formations suggests that it could be very useful in the control of sulfide accumulation. Because *T. denitrificans* strain F is a chemoautotrophic bacterium, no additional organic nutrients are needed to support its growth. This lack of organic nutrients will limit the growth of any indigenous organisms. Under anoxic conditions, *T. denitrificans* uses nitrate as an electron acceptor. The addition of nitrate has been shown to inhibit sulfide production in many environments (14). Because it is a facultative anaerobe, *T. denitrificans* could be used not only in strictly anoxic environments, but it may be effective at utilizing H_2S which diffuses into an aerobic zone. As noted above, *T. denitrificans* has been shown to utilize sulfide under aerobic conditions. Anaerobic corrosion by SRB occurs at neutral pH, the optimal pH for growth of strain F.

Limiting factors in the growth of *T. denitrificans*, and its application in the control of the biogenic production of H_2S, include temperature, salt concentration and the concentrations of soluble organics. *T. denitrificans* grows optimally in a thiosulfate medium at 30°C but the growth rate quickly drops when the temperature is higher or lower. The growth rates at 27° and 33°C are approximately 50% and 75%, respectively, of those found at 30°C (1). Temperatures in excess of 40°C are totally inhibitory to growth. However, no significant effect on viability is seen at temperatures as high as 45°C for exposures of up to 5 hours (1). Many thermophilic SRB have been identified and *T. denitrificans* would not be effective in environments where such organisms would be found. Because oil-reservoir depth is related directly to its temperature, the use of mesophilic organisms such as *T. denitrificans* would be restricted to shallow wells. Of the Oklahoma reservoirs, 36% are 45°C or below (15), whereas in Texas, New Mexico, and Wyoming, 23% of the reservoirs have temperatures below 45°C.

The average salt concentration in brines from Oklahoma reservoirs is approximately 9%, which is much too high for the growth of *T. denitrificans*. States such as California, Colorado, and Wyoming have average brine salinities below 3% and, therefore, could be treatable with strain F (16).

Even though strain F of *T. denitrificans* is inhibited by such factors as temperature, salt, and certain organic molecules, it still can be effective in prevention of sulfide accumulation. Growth of *T. denitrificans* is inhibited by approximately 50% when grown in media containing 0.05% lactate, but it is still effective in utilizing the sulfide produced by SRB. This observation indicates that strain F does not have to be growing under optimal conditions to utilize sulfide effectively.

Field Test of a Microbial Process to Control the Production of H_2S by SRB

The ability of *T. denitrificans* strain F to control H_2S production in an experimental system using cores and formation water from an underground gas storage facility also has been investigated. It is important to note that the objective was not to control the concentration of SRB. Strain F does not inhibit the growth of SRB; it simply removes sulfide, the unwanted product of sulfate reduction. The test was therefore considered successful if the sulfide concentration in the effluent of the core treated with strain F was lower than that found before strain F treatment.

The core system contained three cylindrical cores of St. Peter sandstone in series, each with dimensions of 2.5 cm diameter and 7.6 cm length. Each core was mounted in polyvinyl chloride (PVC) tubing. The porosity of St. Peter sandstone is 30%. The core system was injected with formation water, formation water plus 40 mM nitrate, *T. denitrificans* growth medium (Table I minus thiosulfate), strain F cells (10^5 viable cells/mL) plus growth medium, formation water plus nitrate again, and finally formation water plus 10 mM nitrate and nutrients for strain F including (in g/L) KH_2PO_4 (1.8), $MgSO_4 \cdot 7H_2O$ (0.4), NH_4Cl (0.5), $CaCl_2$ (0.03), $NaHCO_3$ (1.0). Prior to injection of 100% nutrient-amended formation water, various mixtures of growth medium and formation water were injected (with increasing percentage of formation water) along with strain F cells.

The experiments were conducted on site at the Northern National Gas Co. gas storage facility in Redfield, Iowa. Formation water was collected daily from the Davis-6 well and contained (in mg/L): iron (0.6), sulfide (9), chloride (420), sulfate (450), phosphate (1.8), hardness (980), alkalinity (660), and total dissolved solids (718). The pH was 7.2, and the average injection rate in the core system was 75 mL/hr. From the porosity of the core and the volume of connecting tubing, the liquid volume of the core system was estimated to be 240 mL giving a hydraulic retention time of 3.2 hours.

Results of these experiments are summarized in Table IV. The addition of nitrate alone to the formation water injected into the core systems resulted in lower effluent-sulfide levels. Concomitant with the decrease in sulfide was the decrease in nitrate concentrations in the core effluent, suggesting the presence of indigenous microbial populations capable of oxidizing sulfide using nitrate as the electron acceptor. The addition of nitrate did not affect the numbers of SRB and acid-producing bacteria. Strain F-like organisms were not detected in the core effluents. The sulfide levels were decreased by about 40%. The sulfide levels in the influent and the effluent before treatments began were similar. This observation suggests that little or no sulfide production occurred within the core system. (No organic nutrients were added to the formation water to support the growth of SRB).

The injection of nutrients of *T. denitrificans* did not stimulate sulfide production in the core systems. Although the numbers of SRB were not affected, the influent and the effluent sulfide levels were low when only medium was injected into the core systems. This again suggests that little or no sulfide production actually occurred within the core system. Significant numbers of strain F cells were detected in the first two cores of the core system after the first treatment with the strain. The number of strain F cells increased with the subsequent treatment with cells followed by medium injection. Thus, cells of strain F were maintained in the core system when growth medium was used.

When the influent was again 100% formation water with nitrate, the levels of strain F decreased, but complete washout of strain F was not observed. Throughout this period, the concentration of sulfide in the effluent of the core system was consistently lower than the influent concentration. There was also a concomitant reduction in nitrate levels in the core system suggesting that these two processes were linked.

The treatment of the core system with strain F, and the subsequent injection of formation water with a lesser nitrate concentration and nutrient amendments, resulted in the re-establishment of strain F in the core system. A decrease in sulfide concentration in the effluent, compared to the influent concentration in the test core, was also observed. The levels of sulfide in the effluent of the core system compared to the influent concentration were decreased by 84 to 99%. There was a substantial decrease in the levels of nitrate and a substantial increase in the levels of sulfate in the effluent compared to the influent of the core system. This observation suggests that, in the core system, strain F was oxidizing the sulfide present in the formation water to sulfate, using nitrate as the electron acceptor. However, the amount of sulfate detected in the effluent of the test-core system was much higher than that expected if strain F completely oxidized only the sulfide present in the formation water. This observation suggests that strain F may have metabolized sulfur compounds that had accumulated within the core system. These sulfur compounds may have been iron sulfides or other sulfide precipitates which accumulated in the core sections during previous experiments. Strain F has been observed to utilize iron-sulfide precipitates, produced by SRB in media containing Fe^{3+} as an energy source (10).

These data support the conclusion that strain F was metabolically active and effective in controlling the level of sulfide in the core system.

Table IV. Summary of the Effects of Strain F Inoculation and Nitrate Addition on Sulfide Production in Core System

Treatment	Effluent Sulfide (μM)	Cell Concentrations (cells/mL)		
		SRB	APB	Strain F
None	160	10^5	10^5	0
Nitrate	110	10^5	10^7	0
Nitrate, strain F and nutrients	3-16	10^7	10^7	10^7

SRB = sulfate-reducing bacteria; APB = acid-producing bacteria

Acknowledgments

The authors wish to express their appreciation to Northern Natural Gas Co. and to Mr. Steven D. Thomas for access to the test site and use of laboratory facilities. We also wish to thank Mr. Rick Gomez of Northern Natural Gas Co. for his assistance in the day-to-day operation of the core system. This work was funded by the Gas Research Institute (Chicago, IL) and ABB Environmental Services (Portland, ME).

Literature Cited

1. Sublette, K.L.; Sylvester N.D. Biotech. Bioeng. 1987, 29, 245-257.
2. Sublette, K.L.; Sylvester N.D. Biotech. Bioeng. 1987, 29, 753-758
3. Sublette, K.L.; Sylvester N.D. Biotech. Bioeng. 1987, 29, 759-761.
4. Sublette, K.L. Biotech. Bioeng. 1987, 29, 690-695.
5. Sublette, K.L.; Woolsey, M.E. Biotech. Bioeng. 1989, 34, 565-569.
6. Oncharit, C.; Dauben, P.; Sublette, K.L. Biotech. Bioeng. 1989, 33, 1077-1080.
7. Ongcharit, Chawan; Sublette, K.L.; Shah, Y.T. Biotech. Bioeng. 1991, 37, 497-504.
8. Lee, C.M.; Sublette, K.L. Water Research 1993 (in press).
9. McInerney, M.J.; Bhupathiraju, V.K.; Sublette, K.L. J. Ind. Micro. 1992, 11, 53-58.
10. Montgomery, A.D.; McInerney, M.J.; Sublette, K.L. Biotech. Bioeng. 1990, 35, 533-539.
11. Postgate, J.R. The Sulfate-Reducing Bacteria, 2nd ed.; Cambridge University Press: Cambridge, MA, 1984.
12. The Role of Bacteria in the Corrosion of Oil Field Equipment; National Association of Corrosion Engineers: Houston, TX, 1976.
13. Balch, W.E.; Box, G.E.; Magnum, L.J.; Woese, C.R.; Wolfe, R.S. Microbiol. Rev. 1979, 43, 260-296.
14. Jenneman, G.E.; McInerney, M.J.; Knapp, R.M. Appl. Environ. Microbiol. 1986, 51, 1205-1211.
15. Clark, J.B.; Munnecke, D.M.; Jenneman, G.E. Dev. Ind. Microbiol. 1981, 22, 695-701.
16. Beardsley, C.W.; Krotinger, N.J.; Rigdon, J.H. Sewage Ind. Wastes 1956, 28, 220-227.

RECEIVED April 5, 1993

Chapter 7

Solid-Phase Alteration and Iron Transformation in Column Bioleaching of a Complex Sulfide Ore

Lasse Ahonen[1] and Olli H. Tuovinen[2]

[1]Geological Survey of Finland, SF–02150 Espoo, Finland
[2]Department of Microbiology, Ohio State University, 484 West 12th Avenue, Columbus, OH 43210–1292

The objective of the work was to characterize solid-phase changes and Fe(III) precipitation during biological leaching of a sulfide ore which contained chalcopyrite, pentlandite, pyrite, pyrrhotite, and sphalerite. The leaching experiments were carried out using bench-scale column reactors which were inoculated with acidophilic Fe- and S-oxidizing thiobacilli. Experimental factors included inoculation, pH, temperature, flood and trickle leaching, aeration, particle size, and mineralogical composition. Secondary solid phases, *viz.* covellite, jarosites, and elemental S, were detected in biologically active columns. Dissolved ferric iron data were pooled from all experiments and compared with solubility curves calculated for jarosites and ferric hydroxides. The data suggested that ferric-iron solubility was controlled by jarosites.

In chemical and biological leaching systems, dissolved ferric iron is an important redox component and is reduced to Fe^{2+} by reaction with sulfide minerals (1,2). The re-oxidation of ferrous iron is very slow under abiotic leaching conditions, unless strong oxidizing chemicals are used. Iron-oxidizing thiobacilli (*Thiobacillus ferrooxidans*), on the other hand, oxidize Fe^{2+}-species at relatively fast rates in the range of pH 1-4 (3,4). *Leptospirillum ferrooxidans*, an iron-oxidizing chemolithotroph, is commonly found in consortia with *T. ferrooxidans* and other acidophilic thiobacilli (5-7). By ferrous iron oxidation, bacteria help maintain high redox-potential values favorable to oxidative dissolution of sulfide minerals (8,9). Thiobacilli also can oxidize directly iron sulfides, resulting in elevated ferric iron concentrations in leach solutions (6,10). The oxidative-dissolution reactions of sulfide minerals may produce or consume acid. The bacterial oxidation of Fe-disulfide (*e.g.*, pyrite) and hydrolysis of ferric iron are acid-producing reactions:

$$4FeS_2 + 15O_2 + 2H_2O \rightarrow 4Fe^{3+} + 8SO_4^{2-} + 4H^+ \tag{1}$$

$$Fe^{3+} + 3H_2O \rightarrow Fe(OH)_3 + 3H^+ \qquad (2)$$

Acid is consumed in the bacterial oxidation of ferrous iron:

$$4Fe^{2+} + O_2 + 4H^+ \rightarrow 4Fe^{3+} + 2H_2O \qquad (3)$$

The oxidation of Fe-monosulfide is also an acid-consuming reaction if the hydrolysis of ferric iron is excluded from the net reaction:

$$4FeS + 9O_2 + 4H^+ \rightarrow 4Fe^{3+} + 4SO_4^{2-} + 2H_2O \qquad (4)$$

Laboratory experiments have demonstrated that the oxidation of non-stoichiometric natural pyrrhotite is characterized by the formation of elemental sulfur (11):

$$4Fe_{(1-x)}S + (3-3x)O_2 + (12-12x)H^+ \rightarrow (4-4x)Fe^{3+} + 4S^0 + (6-6x)H_2O \qquad (5)$$

Contact of leach solutions with carbonate minerals as well as with some silicate phases causes acid consumption. The mineralogical composition of the ore and factors which influence the pH of the leach solution have major influences on ferric iron solubility. At pH range 1.5-3, precipitation of ferric iron in sulfate environments occurs largely in the form of jarosites. Because the Fe^{3+}/Fe^{2+} couple constitutes the major redox system in leach solutions, a decrease in the relative concentration of dissolved ferric iron also causes a decline in the redox potential and thereby in the oxidative dissolution of sulfide minerals (2,8,9).

The dissolution of sulfide phases may lead to the precipitation of secondary minerals (*e.g.*, jarosites, elemental S). The formation of such solid-phase reaction zones may adversely influence the contact of sulfide-mineral surfaces with leach solution and bacteria, replenishment of O_2 at the reaction site, and fluxes of dissolution products (metals, sulfur species). The shrinking-core model describes this situation as a parabolic process, resulting in gradually declining leaching rates because of the lack of proper surface contact or because of an increasing distance between the solution front and the reactive surface (2,12,13). The formation of solid-phase reaction zones has not been characterized well in biological leaching systems, although there is circumstantial evidence for their formation.

The scope of the work was to characterize changes in solid phases during the bacterial leaching of a complex sulfide ore and to examine transformations influencing the solubility of ferric iron in leach solutions. The data presented in this paper are a part of a program which was carried out to evaluate the feasibility of biological leaching techniques to recover metals from a complex sulfide ore. Most leaching experiments were relatively long-term, usually lasting in the range 400-700 days, and represented mostly conditions which were relevant to heap or *in-situ* leaching systems. Data previously published as a part of this research program were concerned with the effect of temperature on the bacterial leaching in column reactors (14) and the effect of Ag^+-addition in column bioleaching experiments in an attempt to stimulate chalcopyrite oxidation (15). Additional results on temperature effects, using this ore material in separate shake-flask and column bioleaching experiments, have also been published previously (9,16).

Materials and Methods

The ore material used in this study was obtained from a Cu-Co-Zn mine in Outokumpu, eastern Finland. The mineralogy of this deposit has been previously described (17,18). The main sulfide minerals were chalcopyrite ($CuFeS_2$), sphalerite (ZnS), Co-pentlandite ($(Ni,Co,Fe)_9S_8$), pyrite (FeS_2), and pyrrhotite ($Fe_{1-x}S$). Several different mixtures of the sulfide ore and gangue minerals were tested in the study. The gangue material was mostly carbonate- and graphite-containing quartzite. In addition, a mixture of skarn (containing diopside and tremolite) and serpentinite, all typical of the deposit, were used. Both pyrite-rich and pyrrhotite-rich samples were mixed with gangue material.

The bacteria used in this work were originally enriched from mine-water samples and used as a mixed culture. The culture was capable of oxidizing ferrous iron and sulfur compounds as the sole source of energy under acidic conditions; therefore, the organisms were designated as acidophilic thiobacilli. Cultures were grown in a mineral salts solution (3.0 mM $(NH_4)_2SO_4$, 2.3 mM K_2HPO_4, and 1.6 mM $MgSO_4 \cdot 7H_2O$ adjusted to initial pH 2.5 with H_2SO_4) which was supplemented with the sulfide-ore material as the sole substrate for energy.

The experiments were carried out with glass-column reactors. The columns contained either 600-800 g or 12 kg of ore sample. The smaller columns (1:1 solid/liquid ratio) were 50 cm high with an internal diameter of 9 cm. In one column leaching experiment with a high grade ore sample, 400 g of ore material was used and the solid/liquid ratio was 1:2. The larger columns (4:1 solid/liquid ratio) were 100 cm high with an internal diameter of 11 cm.

Two different small column types were used. In the first one, a fritted glass base plate supporting the ore was placed at the bottom of the percolator and the ore material was kept submerged in the leach solution (flood leaching). Oxygen supply was ensured by recirculating the leach solution with side-arm airlift. In the second type, the fritted glass base plate was lifted above the level of the leach solution. The solution passing through the ore sample was recirculated either with side-arm airlift or with a peristaltic pump (trickle leaching). After initial experiments, the column design was modified by inserting an open port below the fritted glass base to equalize pressure changes occurring during recirculation of the leach solution. The larger column type (12 kg ore) approximated trickle-leaching conditions: the ore sample was placed on a perforated base plate at the bottom of the column and the leach solution was recirculated with a peristaltic pump.

The initial acid consumption was satisfied with sulfuric acid. Where necessary, sodium hydroxide was used to neutralize excessive acid production. Leach-solution samples removed for chemical analyses were replaced with equivalent volumes of sterile mineral-salts solution. Evaporation losses were compensated for by adding sterile, distilled water.

The oxidative dissolution of metals from sulfide minerals in the leaching experiments was routinely monitored by atomic absorption spectroscopy (AAS). Sulfate formed in the oxidation was not determined because of excessive amounts of sulfuric acid added for pH control.

Partial elemental composition of the ore samples was determined by AAS and by X-ray fluorescence spectrometry (XRF). Solid residues were either dissolved in a HCl-HNO$_3$-mixture and analyzed by AAS or they were analyzed directly by XRF. Total S in solid residues was determined by combustion and IR-detection (LECO IR 32H), elemental S by iodometric titration, and sulfate indirectly by AAS after precipitation of BaSO$_4$. Mineralogical alterations were examined by X-ray diffraction (XRD) and ore microscopy in solid samples collected from column-leaching systems. The concentration of dissolved ferric iron in actual leach solutions was calculated from the measured redox potential and total soluble iron data, on the basis of the Nernst equation. Only FeSO$_4^+$, the predominant species in the pH range of the experiments, was accounted for in the calculation of the relative proportion of ferric iron. Figure 1 shows the theoretical distribution of the major aqueous species of ferric iron in acidic leach solutions. The speciation was calculated without charge balance using the PHREEQE program (19). The calculations were based on the equilibrium constants presented by Nordstrom et al. (20) and on the Debye-Hückel formula for the activity coefficient corrections. The calculations were carried out using equimolar concentrations of ferric iron and sulfate (0.1 M). The actual concentrations have a negligible influence on the distribution of the species. Consequently, changes in iron concentration due to precipitation were not considered.

Results and Discussion

Evolution of H$_2$S was sometimes evident in the columns during pre-leaching stages, suggesting non-oxidative leaching of pyrrhotite:

$$Fe_{(1-x)}S + 2H^+ \rightarrow (1-3x)Fe^{2+} + 2xFe^{3+} + H_2S \tag{6}$$

Microscopic examination of leach residues revealed the presence of covellite (CuS) on pyrrhotite surfaces (Figure 2). Covellite formation was favored at pH >2.5; i.e., under conditions where ferric iron precipitated. At pH <2.5, covellite formation was not detected. These observations suggested that in iron-deficient solution, H$_2$S released from pyrrhotite precipitated CuS:

$$Cu^{2+} + H_2S \rightarrow CuS + 2H^+ \tag{7}$$

Following inoculation, the first changes in the column experiments usually were an increase in acid consumption, redox potential, and dissolution of nickel. These changes were interpreted to result from the partial oxidation of pyrrhotite (reaction 5).

Pyrrhotite particles in leach residues were highly corroded and usually surrounded by a dark zone of elemental sulfur (Figures 2 and 3). Elemental analysis also revealed an elevated concentration of S in these samples. Bacterial and chemical (acid) leaching resulted in comparable morphological changes in the texture of pyrrhotite grains. Similar morphological changes were previously detected in shake-flask leaching experiments with pure pyrrhotite (11), with elemental sulfur as a major component in the leach residues. Sphalerite present in the pyrrhotite-containing specimens also disintegrated (Figure 2). Pentlandite was more resistant under these leaching conditions (Figure 3).

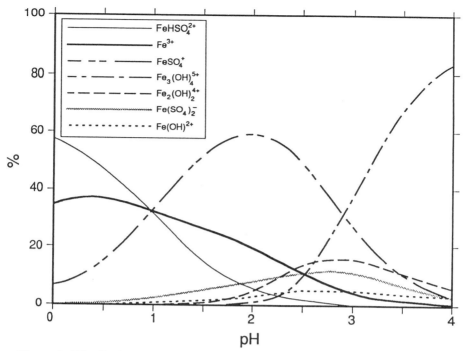

Figure 1. Distribution diagram of aqueous ferric iron species at pH 1.0-4.0, assuming equal concentrations of ferric iron and sulfate.

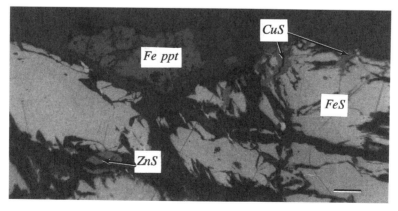

Figure 2. Polished section of a leach residue from column bioleaching after 120 days contact time, showing disintegrated pyrrhotite texture (*FeS*) surrounded by a dark layer of S. Secondary copper sulfide (*CuS*), identified as covellite because of blue color, has precipitated on pyrrhotite surface and in fissures. A part of the pyrrhotite surface is covered by a loosely associated Fe(III) precipitate (*Fe ppt*). A granule of sphalerite (*ZnS*) displays a corroded appearance. Bar, 50 μm.

The bacterial leaching of pyrite appeared to propagate along microfissures and grain boundaries between pyrite and chalcopyrite (Figure 4). New crystalline solid phases were not found on pyrite surfaces. In uninoculated controls, pyrite particles appeared relatively intact.

High concentrations of iron (>10 g/liter) were dissolved during these time courses. In experiments conducted at pH <2.5, Fe(III) precipitates retrieved from partially leached ore particles were yellow. The precipitates were well crystallized, with the lattice parameters $a_0 = 7.324 \pm 0.005$ Å and $c_0 = 16.695 \pm 0.01$ Å. The XRD data indicated that Na-jarosite ($NaFe_3(SO_4)_2(OH)_6$) was the main component. The unit cell dimensions are in agreement with the corresponding parameters of natural Na-jarosite samples (21). The probable source of Na^+ for Na-jarosite was the NaOH used to neutralize excessive acid production. No attempt was made to monitor Na^+ dissolution from the mineral matrix. Some variation in jarosite composition was evident from the broadening of the major XRD reflections, suggesting Na^+-substitution by K^+, NH_4^+, and H_3O^+. The mineral-salts solution was the source of K^+ and NH_4^+; additional K^+ may have dissolved from other mineral constituents (*e.g.*, micas).

At higher pH values the precipitates were brown, but amorphous in XRD analysis. It was concluded that the brown precipitates were cryptocrystalline Fe(III)-oxyhydroxides or Fe(III)-oxyhydroxysulfates. Iron transformations at pH 3-4 may yield schwertmannite, a poorly crystallized Fe(III)-mineral previously termed "mine drainage mineral" (22). This mineral (unit cell formula $Fe_8O_8(OH)_6SO_4$) has been described from natural and biogenic sources involving sulfate environments at pH 3-4 (22,23) and has recently been recognized as a new mineral phase (24). Schwertmannite was not detected in XRD-analysis, but it is relatively poorly crystallized and its XRD reflections may be masked by better-crystallized jarosites and Fe(III)-oxyhydroxides as well as ferrihydrite (24). At higher pH values, rapid hydrolysis of iron would lead to the formation of ferrihydrite which is slowly converted to goethite (α-FeOOH) via dissolution and re-precipitation. Ferrihydrite (unit cell formula $Fe_5HO_8 \cdot 4H_2O$), a poorly crystallized Fe(III)-oxide, is normally found in pH >5 systems (23) and was, therefore, unlikely to exist under the experimental conditions used in the present study.

Solubility product data are not available for schwertmannite. The published log K_{sp} values for ferric hydroxide-type precipitates range between -36.6 and -43.7. The lower log K_{sp} values are more characteristic for X-ray amorphous phases and the higher values are approaching those of crystalline Fe(III)-oxyhydroxides. For example, the log K_{sp} value for goethite is -43 (25). Thermodynamically, goethite formation is the preferred product in these leaching systems. However, goethite was not detected by XRD analysis, suggesting that jarosite formation was kinetically favored.

Figure 5 shows dissolved Fe concentrations as a function of pH values of leach solutions pooled from the experimental results. Dissolved sulfate concentrations were not systematically analyzed because additional sulfate was introduced in the form of sulfuric acid which was used in pH adjustments. The total amount of sulfate added for the pH adjustment averaged approximately 0.5 M in concentration. The complete oxidative dissolution of sulfides in the ore material would produce an additional 0.5 M concentration of sulfate.

Figure 5 also shows the solubility curves of ferric hydroxide in the range of log K_{sp} -37 to -43, based on a sulfate concentration of 0.1 M. Most experimental points are

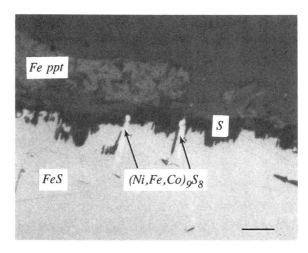

Figure 3. Polished section of corroded pyrrhotite (*FeS*) surface, surrounded by a dark rim of sulfur (*S*) and loosely associated Fe(III) precipitate (*Fe ppt*). Pegs of pentlandite (*(Ni,Fe,Co)$_9$S$_8$*) protrude from pyrrhotite, suggesting that pentlandite solubilization was slower than that of pyrrhotite. The sample was taken from a column bioleaching experiment (contact time 122 days). Bar, 50 μm. Adapted from ref. 10.

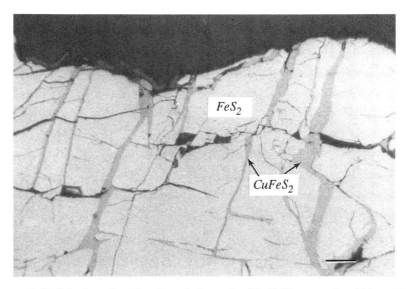

Figure 4. Polished section showing chalcopyrite (*CuFeS$_2$*) network within pyrite (*FeS$_2$*) matrix. Pyrite is more disintegrated than chalcopyrite, and corrosion seems to propagate along phase boundaries. The sample was taken from a column bioleaching experiment after 90 days contact time. Bar, 50 μm.

within the Fe(III) concentration and pH range defined by K_{sp} values of 10^{-39}-10^{-40} (Figure 5), indicating that ferric hydroxide precipitation may control iron concentrations at higher pH-values.

The solubility curves for Na-, K-, H_3O-, and NH_4-jarosites are shown in Figure 6. The jarosite/solution equilibria can be presented with the following equations (equilibrium constants derived from refs. 26-28):

$$NaFe_3(SO_4)_2(OH)_6 + 6H^+ \rightleftharpoons Na^+ + 3Fe^{3+} + 2SO_4^{2-} + 6H_2O \tag{8}$$

$$\log K = -5.3$$

$$KFe_3(SO_4)_2(OH)_6 + 6H^+ \rightleftharpoons K^+ + 3Fe^{3+} + 2SO_4^{2-} + 6H_2O \tag{9}$$

$$\log K = -9.2$$

$$H_3OFe_3(SO_4)_2(OH)_6 + 5H^+ \rightleftharpoons 3Fe^{3+} + 2SO_4^{2-} + 7H_2O \tag{10}$$

$$\log K = -5.4$$

$$NH_4Fe_3(SO_4)_2(OH)_6 + 6H^+ \rightleftharpoons NH_4^+ + 3Fe^{3+} + 2SO_4^{2-} + 6H_2O \tag{11}$$

$$\log K = -6.3$$

The equilibrium for Na-jarosite/ferric hydroxide can be defined as

$$NaFe_3(SO_4)_2(OH)_6 + 3H_2O \rightleftharpoons Na^+ + 3Fe(OH)_3 + 2SO_4^{2-} + 3H^+ \tag{12}$$

$$\log K = -14.3$$

The equilibrium constant for this reaction was inferred from the solubility products of Na-jarosite ($\log K_{sp} = -5.3$) and ferric hydroxide,

$$Fe(OH)_3 \rightleftharpoons Fe^{3+} + 3OH^- \tag{13}$$

$$\log K_{sp} = -39$$

and the dissociation constant of water ($\log K_w = -14.0$). The equilibrium condition in dilute solutions (activity of $H_2O = 1$) is expressed in logarithmic form as

$$\log[Na^+] + 2\log[SO_4^{2-}] - 3pH = -14.3 \tag{14}$$

Assuming activities $[Na^+] = 0.002$ and $[SO_4^{2-}] = 0.01$, it can be calculated that ferric hydroxide and Na-jarosite are in equilibrium at pH 2.5. This pH value shifts to higher pH values with increased concentrations of Na^+ and SO_4^{2-}. The equilibrium between goethite ($\log K_{sp} = -43$) and Na-jarosite predicts goethite to be the stable phase at all positive pH values, in contrast to the XRD data of samples from leach columns.

Conclusions

The initial mechanism in pyrrhotite leaching was concluded to involve non-oxidative dissolution, producing sulfide ion and H_2S with subsequent oxidation to elemental S or precipitation as a secondary Cu-sulfide. Pentlandite was more recalcitrant than pyrrhotite. The leaching of pyrite was oxidative and strongly catalyzed by bacteria.

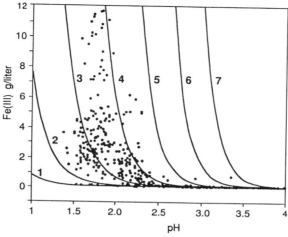

Figure 5. Scatter diagram of dissolved ferric iron concentration as a function of pH in column leaching experiments. Superimposed on the Fe(III) plot are the solubility curves of ferric hydroxide calculated with the following log K_{sp} values: *curve 1*, -43; *curve 2*, -42; *curve 3*, -41; *curve 4*, -40; *curve 5*, -39; *curve 6*, -38; *curve 7*, -37. Sulfate concentration was fixed to 0.1 M concentration for these calculations.

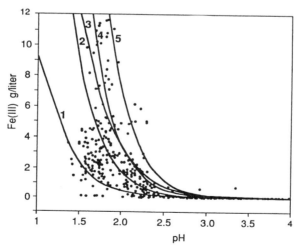

Figure 6. Scatter diagram of dissolved ferric iron concentration as a function of pH in column leaching experiments, with superimposed solubility curves of jarosites. The calculations were based on 0.1 M SO_4^{2-} and either 1 mM K^+ (K-bearing jarosite, *curve 1*), 0.1 M Na^+ (Na-jarosite, *curve 2*), 1 mM H_3O^+ (H_3O-jarosite, *curve 3*), 1 mM NH_4^+ (NH_4-jarosite, *curve 4*), or 1 mM Na^+ (Na-jarosite, *curve 5*).

The concentration of dissolved iron was controlled by secondary precipitates. At pH >2.5, dissolved iron concentrations were low and X-ray amorphous or cryptocrystalline, brown Fe(III)-oxyhydroxide or Fe(III)-oxyhydroxysulfate precipitates were formed. At pH <2.5, dissolved ferric iron concentrations were high and yellow jarosite precipitates were formed. Solubility calculations indicated that jarosite precipitation probably controlled dissolved ferric iron concentrations at low pH.

Acknowledgements

The experimental work was funded by the Ministry of Trade and Industry (Finland). We thank Ms. P. Hiltunen for assistance with the column leaching experiments which were carried out in the Department of Microbiology, University Helsinki. Analytical services by Outokumpu Oy are gratefully acknowledged. Additional financial support was received from Outokumpu Research Oy and the Nordisk Industrifond (O.H.T.). We are also grateful to Dr. J.M. Bigham (Department of Agronomy, The Ohio State University) for helpful comments and for providing unpublished information on schwertmannite.

Literature Cited

1. Ralph, B.J. In Comprehensive Biotechnology; Robinson, C.W.; Howell, J.A., Eds.; Pergamon Press: Oxford, 1985, Vol. 4; pp. 201-234.
2. Rossi, G. Biohydrometallurgy. McGraw-Hill: Hamburg, 1990.
3. Grishin, S.I.; Tuovinen, O.H. Appl. Environ. Microbiol. 1988, 54, 3092-3100.
4. Smith, J.R.; Luthy, R.G.; Middleton, A.C. J. Water Pollut. Contr. Fed. 1988, 60, 518-530.
5. Helle, U.; Onken, U. Appl. Microbiol. Biotechnol. 1988, 28, 553-558.
6. Norris, P.R. In Metal-Microbe Interactions; Poole, R.K.; Gadd, G.M., Eds.; IRL Press: Oxford, 1989; pp. 99-117.
7. Sand, W.; Rohde, K.; Sobotke, B.; Zenneck, C. Appl. Environ. Microbiol. 1992, 58, 85-92.
8. Pesic, B.; Oliver, D.J.; Wichlacz, P. Biotechnol. Bioeng. 1989, 33, 428-439.
9. Ahonen, L.; Tuovinen, O.H. Appl. Environ. Microbiol. 1992, 58, 600-606.
10. Tuovinen, O.H. In Microbial Mineral Recovery; Ehrlich, H.L.; Brierley, C.L., Eds.; McGraw-Hill: New York, 1990; pp. 55-77.
11. Ahonen, L.; Hiltunen, P.; Tuovinen O.H. In Fundamental and Applied Biohydrometallurgy; Lawrence, R.W.; Branion, R.M.R.; Ebner, H.G., Eds.; Elsevier: Amsterdam, 1986; pp. 13-22.
12. Sohn, H.Y. In Rate Processes in Extractive Metallurgy; Sohn, H.Y.; Wadsworth, M.E., Eds.; Plenum Press: New York, 1979; pp. 1-42.
13. Wadsworth, M.E. In Rate Processes in Extractive Metallurgy; Sohn, H.Y.; Wadsworth, M.E., Eds.; Plenum Press: New York, 1979; pp. 133-241.
14. Ahonen, L.; Tuovinen, O.H. Biotechnol. Lett. 1989, 11, 331-336.
15. Ahonen, L.; Tuovinen, O.H. Miner. Engin. 1990, 3, 437-445.
16. Ahonen, L.; Tuovinen, O.H. Appl. Environ. Microbiol. 1991, 57, 138-145.
17. Peltola, E. Econ. Geol. 1978, 73, 461-477.

18. Koistinen, T.J. Trans. R. Soc. Edinburgh, Earth Sci. 1981, 72, 115-158.
19. Parkhurst, D.L.; Thorstenson, D.C.; Plummer, N. U.S. Geological Survey Water-Resources Investigations 80-96, 1980.
20. Nordstrom, D.K.; Plummer, L.N.; Langmuir, D.; Busenberg, E.; May, H.M.; Jones, B.F.; Parkhurst, D.L. In Chemical Modeling of Aqueous Systems II; Melchior, D.C.; Bassett, R.L., Eds.; ACS Symp. Ser. 416: American Chemical Society: Washington, D.C., 1990; pp. 398-413.
21. Alpers, C.N.; Nordstrom, D.K.; Ball, J.W. Sci. Géol., Bull. 1989, 42, 281-298.
22. Bigham, J.M.; Schwertmann, U.; Carlson, L. In Biomineralization Processes of Iron and Manganese; Skinner, H.C.W.; Fitzpatrick, R.W., Eds.; Catena Verlag: Cremlingen, Germany, 1992; pp. 219-232.
23. Bigham, J.M.; Schwertmann, U.; Carlson, L.; Murad, E. Geochim. Cosmochim. Acta 1990, 54, 2743-2758.
24. Murad, E.; Schwertmann, U.; Bigham, J.M.; Carlson, L. 1993; this volume.
25. Macalady, D.L.; Langmuir, D.; Grundl, T.; Elzerman, A. In Chemical Modeling of Aqueous Systems II; Melchior, D.C.; Bassett, R.L., Eds.; ACS Symp. Ser. 416: American Chemical Society: Washington, D.C., 1990; pp. 350-367.
26. Chapman, B.M.; Jones, D.R.; Jung, R.F. Geochim. Cosmochim. Acta 1983, 47, 1957-1973.
27. Zotov, A.V.; Mironova, G.D.; Rusinov, V.L. Geochem. Int. 1973, 10, 577-582.
28. Kashkay, C.M.; Borovskaya, Y.B.; Babazade, M.A. Geochem. Int. 1975, 12, 115-121.

RECEIVED October 11, 1993

Chapter 8

Alteration of Mica and Feldspar Associated with the Microbiological Oxidation of Pyrrhotite and Pyrite

Tariq M. Bhatti[1,4], Jerry M. Bigham[2], Antti Vuorinen[3], and Olli H. Tuovinen[1]

[1]Department of Microbiology, Ohio State University, 484 West 12th Avenue, Columbus, OH 43210–1292
[2]Department of Agronomy, Ohio State University, 2021 Coffey Road, Columbus, OH 43210
[3]Department of Geology, University of Helsinki, P.O. Box 115, SF–00171 Helsinki, Finland

> This work examines mineralogical changes during the bacterial leaching of a black-schist ore material. A mixed culture of acidophilic iron- and sulfur-oxidizing thiobacilli was used in shake flasks containing acid media and finely ground black schist. The main sulfide phases were pyrrhotite and pyrite, with lesser amounts of sphalerite, pentlandite, and chalcopyrite. The solubilization of Cu, Ni, and Zn was enhanced in inoculated suspensions. The major Si-containing phases were quartz, mica (phlogopite), and feldspars (primarily anorthite, and microcline). Pyrrhotite was oxidized faster than pyrite and was associated with elemental S accumulation. The accumulation was transient in inoculated systems because of eventual bacterial oxidation of elemental S. Formation of K-bearing jarosite, resulting from the bacterial oxidation of Fe-sulfides, was coupled with mica alteration to vermiculite. Chemical dissolution also occurred parallel with structural alteration of the mica phase. The results suggested that jarosite was a sink for K released during mica weathering. Jarosite and vermiculite were not detected under comparable abiotic leaching conditions. Gypsum was produced after 100 days, presumably because of the release of Ca from anorthite.

Acidophilic thiobacilli are capable of oxidizing inorganic compounds of Fe and S as the sole source of electrons and energy, with ferric iron and sulfate as the ultimate oxidation products (1-4). These organisms include *Thiobacillus thiooxidans*, a bacterium that oxidizes S compounds at low pH values, *T. ferrooxidans*, an Fe- and S-oxidizing chemoautotroph, and *T. cuprinus*, an S-oxidizing species also capable of

[4]Current address: National Institute of Biotechnology and Genetic Engineering, P.O. Box 577, Faisalabad, Pakistan

growing with organic substrates. Acidophilic thiobacilli are indigenous in acidic mine environments. With their combined action, they can oxidize many different sulfide minerals including those of Fe, Cu, Co, As, Ni, and Zn (5-7). *Leptospirillum ferrooxidans*, an Fe^{2+}- and FeS_2-oxidizing bacterium, is also an important constituent in acidophilic consortia (8). All these bacteria may be used as biological catalysts in mineral-leaching processes for recovery of metals from sulfide minerals (9,10). Commercial applications exist for dump and heap leaching of low-grade Cu ores and for the pretreatment of auriferous pyrite and arsenopyrite concentrates (11-13). Other potential applications based on bacterial oxidation of sulfide minerals include coal desulfurization via biological oxidation of pyritic sulfur (14,15) and solubilization of U in acidic ferric-sulfate leach solutions produced by biological oxidation of iron sulfides (16,17).

Pyrite and pyrrhotite are common mineral constituents in ore materials and provide a source of energy for bacteria in biological leaching systems. Ferric iron thus produced participates in the leaching process, being cyclically reduced by reaction with sulfide minerals and re-oxidized by bacteria (12,18). At pH values used in most bioleaching processes (pH 1.5-2.5), some of the ferric iron precipitates, forming secondary Fe(III) minerals such as jarosite ($KFe_3(SO_4)_2(OH)_6$) or other basic ferric sulfates (19-21). The presence of dissolved Fe(II) in acid leach solutions greatly accelerates biological sulfide leaching processes. Dissolved iron has little effect in abiotic leaching under comparable conditions because, in the absence of a biological catalyst, ferrous iron oxidation is extremely slow (22).

Quartz and other silicate minerals (*e.g.*, micas) are also common constituents of sulfide ore materials used in bioleaching processes. Quartz is relatively inert, whereas some mica minerals may be partially altered or dissolved during leaching. Mica minerals are di- and trioctahedral layer silicates that usually contain K in the interlayer. Feldspars are anhydrous tectosilicates that include both plagioclase (Ca, Na) and alkali (Na, K) mineral groups. Numerous studies have shown that the plagioclase feldspars have faster dissolution kinetics than K-feldspar (23,24).

In the present study, the biological leaching of a black-schist ore sample was evaluated for oxidative dissolution of sulfide minerals. Several solid-phase products of both sulfide and silicate degradation were detected during the oxidative leaching. The objectives of this study were (i) to investigate the dissolution of sulfide minerals during the biological leaching of a black-schist ore and (ii) to identify the solid phases resulting from the structural alteration and dissolution of associated silicate phases (micas, feldspars) in the ore sample.

Materials and Methods

The ore sample used in this study originated from a black-schist deposit in Finland. The sample was ground and a -59 µm particle size fraction was collected for analysis. This fraction had the following elemental composition: 43.10% SiO_2, 10.25% Al_2O_3, 23.56% Fe_2O_3, 3.04% MgO, 2.13% CaO, 0.273% Na_2O, 4.871% K_2O, 0.386% MnO, 1.65% TiO_2, 9.90% S, 0.123% Cu, 0.269% Ni, 0.015% Co, 0.365% Zn. The sample contained pyrrhotite ($Fe_{1-x}S$), pyrite (FeS_2), sphalerite (ZnS), pentlandite (($Ni,Fe,Co)_9S_8$), and chalcopyrite ($CuFeS_2$). Phlogopite, anorthite, microcline, and quartz were the main silicate phases. The ore sample also contained graphite. Figure 1 shows an X-ray diffraction (XRD) scan of the untreated ore sample.

A mixed culture of acidophilic, iron- and sulfur-oxidizing thiobacilli, designated as SB/P-II, was used throughout this work. The bacterial consortium was originally produced by combining several mine-water enrichment cultures and was then maintained with the test-ore material for several years before the present study was initiated. In separate experiments it was established that SB/P-II contained active Fe^{2+}-, elemental S-, and FeS_2-oxidizing bacteria capable of growing at pH $\approx$ 1-4. The use of a mixed culture in this study was in keeping with previous findings that mixed cultures invariably yield higher leaching rates and, in this respect, are often more efficient than pure cultures (5,8,25).

The bacteria were grown with finely ground black-schist ore material in a mineral-salts solution (0.4 g/liter each of K_2HPO_4, $(NH_4)_2SO_4$, and $MgSO_4 \cdot 7H_2O$; initial pH 1.5) at 22°C. For inoculation, cells were harvested by centrifugation, washed with 0.005 M sulfuric acid, and resuspended in the mineral-salts solution. The leaching experiments were carried out in 250 ml shake flasks (150 rev/min) which contained 5 g of ore sample in 100 ml of mineral-salts solution. Abiotic (sterile) controls were included in the experiments. The redox potential values were measured with a Pt electrode against an $Ag^0/AgCl$ reference electrode (4 M KCl).

Samples of leach solutions and suspended solids were taken at various intervals for analysis. Evaporation losses were compensated with double-distilled water. Suspended solids were recovered by centrifugation, air dried, and then gently ground using an agate mortar. XRD analysis of top-fill powder mounts were conducted using $CuK\alpha$-radiation and a vertical, wide-range goniometer (Philips PW 1316/90) equipped with a diffracted beam monochromator and a Θ compensating slit. All specimens were scanned from 3 to 70°2Θ in increments of 0.05°2Θ with a 4 second step time.

Chemical analyses were accomplished by inductively-coupled plasma emission spectroscopy (ICP). A model Jobin Yvon 70+ ICP was used with plasma-torch, cooling, and carrier-gas flow rates of 12.3, 0.2 and 0.4 l/minute, respectively. Si, Al, and Fe were analyzed simultaneously by using a polychromator; all other elements were analyzed sequentially with a monochromator. Aliquots of leach solution (1 ml) in polyethylene tubes were mixed with 1 ml of matrix solution (0.1225 M ammonium oxalate, 0.07 M oxalic acid, 0.3 M hydroxylamine·HCl, 0.6 M acetic acid) to prevent Fe precipitation and to normalize the salt concentration, followed by 10 ml of double-distilled water.

For major elemental analysis of the ore material, a sample (100 mg) was mixed in a Pt-crucible with 0.7 g of $LiBO_2$. Anhydrous Li_2SO_4 (0.3 g) was also added to accelerate the subsequent HCl-dissolution step. The sample was covered and fused for 5 minutes on a bunsen burner. After cooling, the sample was dissolved in 15 ml of 6 M HCl at 90°C on a magnetic stirrer. The cooled sample was made up to 200 ml in double-distilled water containing 0.25 ml of H_2O_2. Standard reference rock samples were used for calibration.

For S, Co, Cu, Ni and Zn analysis, a sample of the ore (500 mg) was mixed in a Pt-dish with 3500 mg of Na_2O_2, covered and sintered for 45 minutes at 480°C. After cooling, 50 ml of double-distilled water were added, followed by 17 ml of 6 M HCl. The dissolved sample was made up to 100 ml in double-distilled water.

8. BHATTI ET AL. *Microbiological Oxidation of Pyrrhotite and Pyrite*

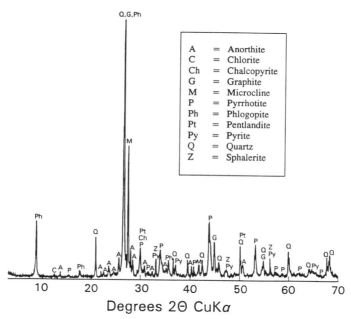

Figure 1. X-ray diffractogram of untreated black-schist ore. Mineral identification is indicated with letter designations.

Results and Discussion

Ni from pentlandite and Zn from sphalerite were completely dissolved in inoculated cultures within 15 days (Figure 2). In contrast, >90% of the Ni and >50% of the Zn remained in the leach residue following abiotic leaching (Table I). Only half of the total Cu was dissolved in the presence of bacteria (Figure 2, Table I), reflecting the recalcitrance of chalcopyrite as compared with pentlandite and sphalerite. The recalcitrant nature of chalcopyrite to biological leaching and its inferior Cu leaching rates are commonly recognized problems in biological leaching processes (26,27). Cobalt was associated mainly with pentlandite, pyrrhotite, and pyrite phases and most of the Co (78%) was dissolved in 25 days. The relative order of metal release from the sulfide matrix was Zn ≥ Ni > Co > Cu. In the black-schist ore sample, chalcopyrite, pentlandite, and sphalerite had only weak XRD peaks (Figure 1) and XRD scans could not, therefore, be used to follow their dissolution.

Table I. Dissolution of Metals from the Black-Schist Ore Sample after 25 Days of Incubation

Experiment	% Recovery			
	Co	Cu	Ni	Zn
Sterile control	25	12	9.1	46
Inoculated with SB/P-II	78	50	100	100

Redox potential values in the inoculated systems indicated active leaching throughout the time course studied. The redox potential increased to >600 mV toward the end of the study (Figure 3), reflecting active Fe oxidation. The redox potential of the parallel abiotic control remained at around 300 mV for 3 weeks, suggesting a lack of Fe oxidation. The subsequent increase in the redox potential to 380 mV on day 25 (Figure 3) was evidence for bacterial activity which was deemed to result from cross-contamination during frequent sampling and was confirmed (i) by the presence of bacterial cells upon microscopic examination; and (ii) by continuing increase in the redox potential. The pH of the abiotic control climbed to >3, whereas the inoculated system remained at around pH 2 (Figure 3), suggesting that the net reactions of acid production and acid consumption more or less balanced each other.

Pyrrhotite and pyrite were the primary sulfide minerals occurring in the black-schist ore sample. A comparison of XRD traces from the original ore and a subsample equilibrated with abiotic control solution over a 10-day time course (Figure 4A and 4B) showed a weakening of the diagnostic diffraction peaks of pyrrhotite. This observation indicated that pyrrhotite was susceptible to dissolution under abiotic conditions. Over an equivalent time period, pyrrhotite decomposition was accelerated in an inoculated system, and it was depleted from the leach residue (Figure 5A). Pyrite was more resistant to oxidation and persisted in the abiotic control (Figure 4B) and 10-day inoculated systems (Figure 5A). However, pyrite was dissolved after 30 days of incubation in the inoculated medium (Figure 5B).

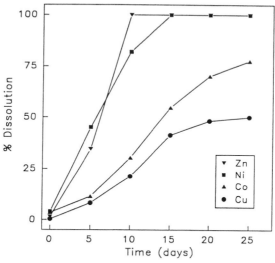

Figure 2. Dissolution of metals from black-schist ore (5% pulp density) in mineral salts medium inoculated with the mixed culture.

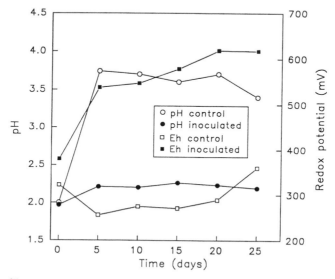

Figure 3. Changes in pH and redox potential values during oxidative dissolution of black-schist ore (5% pulp density) in the control and inoculated systems.

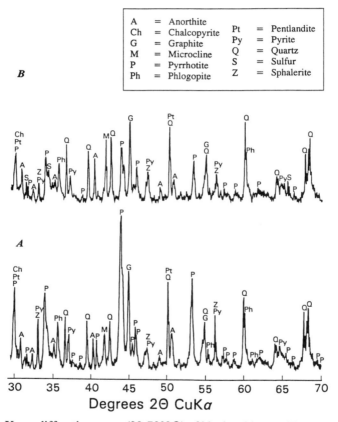

Figure 4. X-ray diffraction scans (30-70°2Θ) of black-schist ore (5% pulp density) initially (A) and after 10 days contact in sterile leach solution (B).

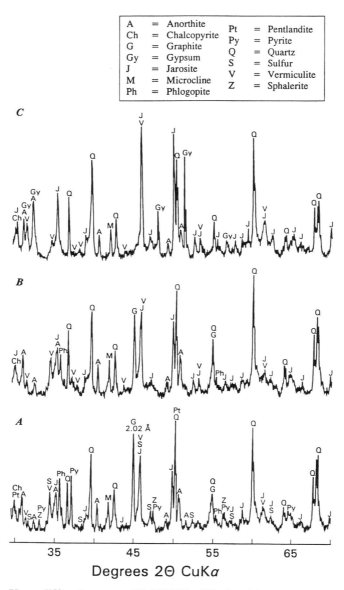

Figure 5. X-ray diffraction scans (30-70°2Θ) of black-schist ore (5% pulp density) after 10 days (*A*), 30 days (*B*), and 100 days (*C*) contact in leach solution inoculated with the mixed culture.

Elemental S was not a constituent of the original black-schist ore (Figure 6A), but its formation has been previously reported for bacterial leaching of pyrrhotite-rich samples (28). In the present study, pyrrhotite decomposition yielded elemental S in both chemical controls and inoculated media (see peak at 3.85 Å in Figures 6B and 7A, respectively); however, the accumulation was transient in inoculated systems due to eventual bacterial oxidation (Figures 7B and C). Accumulations of elemental S were observed during the first 10 days of leaching. During this time, pyrite degradation was negligible, suggesting that its oxidation was not coupled to elemental S formation.

The primary solid-phase product from the biological oxidation of pyrrhotite and pyrite was jarosite, which was readily identified by the sharp 3.11-3.08 Å doublet in XRD scans of the leach residues (Figures 7A-C). Jarosite formed rapidly in the biological systems but was not detected in sterile controls (Figure 6B). In these experiments, because of excess sulfate and K^+, jarosite precipitation was contingent on reaching saturation conditions for ferric iron. Iron dissolved from the black schist remained in the ferrous form in the absence of bacteria, as was evident from differences between the redox potentials of the two systems (Figure 3). Total Fe data indicated faster dissolution in the inoculated system (Figure 8A), but jarosite precipitation was not clearly reflected in the latter part of the Fe time course because of the concurrent dissolution of iron sulfides. The loss of K from the leach solutions (Figure 8B) coincided with K-bearing jarosite precipitation.

Si and Al were progressively dissolved from the black-schist ore during contact with bacterial leach solutions (Figure 9A). Except for the initial release of Si and Al, little additional dissolution took place in the respective abiotic controls. Biological leaching induced major structural alterations in the mica component of the ore. Over a 100-day time course, the mica was transformed to vermiculite (Figure 7C). The observed structural alteration was caused by the expulsion of K^+ from the interlayer region of the mica, coupled with replacement of interlayer-K by hydrated, exchangeable cations. The net result was the formation of an expandable vermiculite with a fundamental repeat distance of 14 Å as compared to 10 Å in the original mica.

The structural alteration of the mica presumably required a net decrease in the layer charge, caused by the oxidation of octahedral Fe or proton attack of framework oxygen-metal bonds. In addition, a sink for K was required to facilitate diffusion from the interlayer regions of the mineral. In this case, the precipitation of K-bearing jarosite provided the required sink as evidenced by the appearance of diagnostic jarosite lines in XRD scans from the leach residues.

Ivarson et al. (29) and Ross et al. (19) reported laboratory experiments where *T. ferrooxidans* was used to oxidize ferrous-sulfate solutions in order to produce jarosites which incorporated monovalent cations from various hydrous micas. K release from the interlayer positions of glauconite resulted in the formation of nontronite (18 Å) as an alteration product. By contrast, illite was more resistant to structural alteration, and the removal of interlayer K appeared to be relatively non-preferential.

In the present study, there was an increase in the concentration of Ca in the leach solutions over time (Figure 9B), coupled with minor decreases in the intensities of diffraction peaks from anorthite (see peaks at 4.03, 3.96, 3.78, 3.48, and 3.30 Å in Figures 7A-C). The dissolution of Ca (and also of Si, Al, Fe) was enhanced in the inoculated flasks, whereas dissolved Mg levels remained more or less constant in both

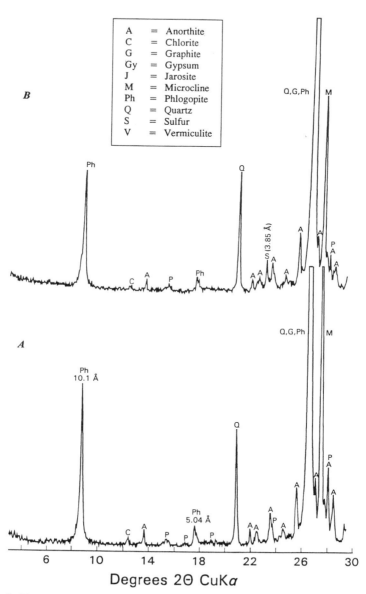

Figure 6. X-ray diffraction scans (3-30°2Θ) of black-schist ore (5% pulp density) initially (A) and after 10 days (B) contact in sterile leach solution.

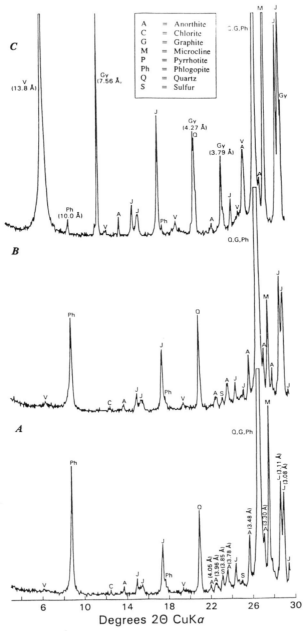

Figure 7. X-ray diffraction scans (3-30°2Θ) of black-schist ore (5% pulp density) after 10 days (*A*), 30 days (*B*), and 100 days (*C*) contact in leach solution inoculated with the mixed culture.

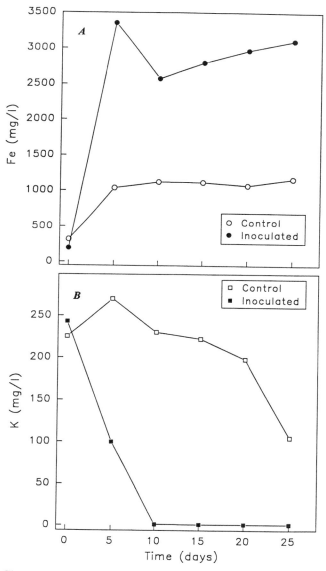

Figure 8. Changes in Fe (*A*) and K (*B*) concentrations during oxidative dissolution of black-schist ore (5% pulp density) in the control and inoculated systems.

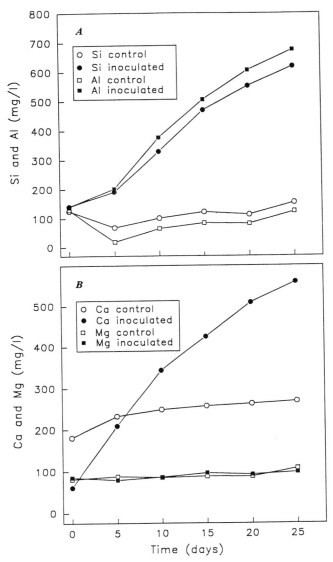

Figure 9. Changes in Si and Al (*A*) and Ca and Mg (*B*) concentrations during oxidative dissolution of black-schist ore (5% pulp density) in the control and inoculated systems.

systems (Figure 9B). The combined XRD and abiotic results suggested that anorthite was unstable in the acidic leach solutions. Dissolved Na concentrations remained <5 mg/l throughout all experiments, suggesting negligible amounts of Na-feldspar in the ore. With prolonged incubation of the inoculated systems, gypsum (see peaks at 7.56, 4.27, and 3.79 Å) precipitated in the leach residues (Figure 7C). Apart from possible Ca impurities in other minerals, anorthite was the sole Ca-bearing mineral in the black-schist sample.

The black-schist ore sample used in this study also contained graphite (see peak at 2.02 Å) as a major component (Figures 5A and B). After 100 days of contact with the inoculated system, the graphite peaks were virtually absent in the leach residue (Figure 5C), suggesting either graphite dissolution or release as a low-density colloid which did not remain in the solids after centrifugation and washing. Graphite dissolution in bioleaching systems has not been previously reported and no attempt was made in this study to investigate the possible mechanism, kinetics, or products of the dissolution reaction.

Concluding Remarks

Elemental S formation was detected with pyrrhotite dissolution in the presence and absence of bacteria. In inoculated flasks, elemental S was depleted upon prolonged incubation. Pyrite was oxidized more slowly than pyrrhotite, and did not produce elemental S in the presence or absence of bacteria.

The weathering of mica in the black-schist ore sample was dependent on the parallel formation of jarosite. Bacteria had a key role in the mica alteration because bacterial oxidation of pyrrhotite and pyrite produced the Fe^{3+} required for jarosite precipitation. In sterile controls, pyrrhotite dissolution took place but the Fe^{2+}-oxidation rates were negligible. Therefore, jarosite was not formed in detectable amounts and mica remained stable.

Mica alteration proceeded via at least two pathways. In the presence of Fe^{3+} at pH ≥2, K was removed from the interlayer regions of the mica to form K-bearing jarosite, and subsequent replacement of K by hydrated cations (*e.g.*, Mg^{2+}) yielded expansible vermiculite. Parallel with this structural alteration, mica dissolution proceeded by proton attack. Ca-feldspar peak intensities were also weakened during prolonged contact time with leach solutions, whereas microcline (K-feldspar) appeared to be much more stable. Gypsum formation coincided with anorthite dissolution in the 100-day samples. Structural alterations of feldspar were not detected, indicating that chemical dissolution was the main mechanism of feldspar weathering.

Acknowledgments

T.M.B. was supported with funds from the U.S. Agency for International Development and the Government of Pakistan. Partial support for ICP analysis was received from Outokumpu Research Oy and the Nordisk Industrifond (O.H.T.). Salary and research support were provided to J.M.B. by state and federal funds appropriated to the Ohio Agricultural Research and Development Center, The Ohio State University. Manuscript No. 305-92.

Literature Cited

1. Harrison, A.P. Annu. Rev. Microbiol. 1984, 38, 265-292.
2. Kelly, D.P. In Autotrophic Bacteria; Schlegel, H.G.; Bowien, B., Eds.; Springer-Verlag: Berlin, 1989; pp. 193-217.
3. Ingledew, W.J. In Microbiology of Extreme Environments; Edwards, C., Ed.; Open University Press: Milton Keynes, U.K., 1990; pp. 33-54.
4. Norris, P.R.; Ingledew, W.J. In Molecular Biology and Biotechnology of Extremophiles; Herbert, R.A.; Sharp, R.J., Eds.; Chapman and Hall: New York, 1992; pp. 115-142.
5. Norris, P.R.; Kelly, D.P. In Microbial Interactions and Communities; Bull A.T.; Slater, J.H., Eds.; Academic Press: New York, 1981; pp. 443-474.
6. Norris, P.R. In Microbial Mineral Recovery; Ehrlich, H.L.; Brierley, C.L., Eds.; McGraw-Hill: New York, 1990; pp. 3-27.
7. Tuovinen, O.H.; Kelley, B.C.; Groudev, S.N. In Mixed Cultures in Biotechnology; Zeikus, J.G; Johnson, E.A., Eds.; McGraw Hill: New York, 1991; pp. 373-427.
8. Sand, W.; Rohde, K.; Sobotke, B.; Zenneck, C. Appl. Environ. Microbiol. 1992, 58, 85-92.
9. Hutchins, S.R.; Davidson, M.S.; Brierley, J.A.; Brierley, C.L. Annu. Rev. Microbiol. 1986, 40, 311-336.
10. Ewart, D.K.; Hughes, M.N. Advan. Inorg. Chem. 1991, 36, 103-135.
11. Pooley, F.D. In Environmental Biotechnology; Forster, C.F.; Wase, D.A.J., Eds.; Ellis Horwood: Chichester, U.K., 1987; pp. 114-134.
12. Rossi, G. Biohydrometallurgy; John Wiley & Sons: Hamburg, 1990.
13. Lindström, E.B.; Gunneriusson, E.; Tuovinen, O.H. Crit. Rev. Biotechnol. 1992, 12, 133-155.
14. Bos, P.; Kuenen, J.G. In Microbial Mineral Recovery; Ehrlich, H.L.; Brierley, C.L., Eds.; McGraw-Hill: New York, 1990; pp. 343-377.
15. Bos, P.; Boogerd, F.C.; Kuenen, J.G. In Environmental Microbiology; Mitchell, R., Ed.; Wiley-Liss: New York, 1992; pp. 375-402.
16. Tuovinen, O.H. Biotechnol. Bioeng. Symp. 1986, 16, 65-72.
17. Bhatti, T.M.; Malik, K.A.; Khalid, A.M. In Biotechnology for Energy; Malik, K.A.; Naqvi, S.H.M.; Aleem, M.I.H., Eds.; Nuclear Institute for Agriculture and Biology and National Institute for Biotechnology and Genetic Engineering: Faisalabad, Pakistan, 1991; pp. 329-340.
18. Tuovinen, O.H. In Microbial Mineral Recovery; Ehrlich, H.L.; Brierley, C.L., Eds.; McGraw-Hill: New York, 1990; pp. 55-77.
19. Ross, G.J.; Ivarson, K.C.; Miles, N.M. In Acid Sulfate Weathering; Kittrick, J.A.; Fanning, D.S.; Hossner, L.R., Eds.; Soil Science Society of America: Madison, WI, 1982; pp. 77-94.
20. Lazaroff, N.; Sigal, W.; Wasserman, A. Appl. Environ. Microbiol. 1982, 43, 924-938.
21. Grishin, S.I.; Bigham, J.M.; Tuovinen, O.H. Appl. Environ. Microbiol. 1988, 54, 3101-3106.
22. Singer, P.C.; Stumm, W. Science 1970, 167, 1121-1123.

23. Chou, L.; Wollast, R. Amer. J. Sci. 1985, 285, 963-993.
24. Blum, A.; Lasaga, A. Nature 1988, 331, 431-433.
25. Lizama, H.M.; Suzuki, I. Hydrometallurgy 1989, 22, 301-310.
26. Rossi, G.; Torma, A.E. In Recent Progress in Biohydrometallurgy; Rossi, G.; Torma, A.E., Eds.; Associazione Mineraria Sarda: Iglesias, Italy, 1983; pp. 185-200.
27. Ahonen, L.; Tuovinen, O.H. Hydrometallurgy 1990, 24, 219-236.
28. Ahonen, L.; Hiltunen, P.; Tuovinen, O.H. In Fundamental and Applied Biohydrometallurgy; Lawrence, R.W.; Branion, R.M.R.; Ebner, H.G., Eds.; Elsevier: Amsterdam, 1986; pp. 13-22.
29. Ivarson, K.C.; Ross, G.J.; Miles, N.M. Soil Sci. Soc. Amer. J. 1978, 42, 518-524.

RECEIVED October 11, 1993

Numerical Modeling of Sulfide Oxidation in Tailings, Waste Rock, and In Situ Deposits

Chapter 9

Rates of Mechanisms That Govern Pollutant Generation from Pyritic Wastes

A. I. M. Ritchie

Environmental Science Program, Australian Nuclear Science and Technology Organization, Private Mailbag 1, Menai 2234, Australia

The environmental impact of pollutants generated by the oxidation of pyrite in mine wastes involves a number of processes which have very different characteristic timescales. In the first instance water quality estimates require the convolution of pollutant generation in and water transport through the wastes followed by convolution with water transport through the aquifer underlying the wastes. As a significant fraction of the wastes may be unsaturated, transit times may be two orders of magnitude greater than those in the saturated aquifer. Such a difference requires care in interpreting the impact of rehabilitation measures. The pollution generation rate within the wastes also depends on a number of interacting processes with greatly differing timescales. For example oxidation rates of pyrite under optimized conditions are typically three orders of magnitude greater than 'high' oxidation rates measured in wastes. In this paper these various rates and the influence that they have on the overall environmental impact is discussed. Data on some important mechanisms is sparse and some indication is given as to how this situation may be rectified.

The fact that the proceedings of the First International Conference on Control of Environmental Problems at Metal Mines (Roros, Norway, June 1988) were contained in one volume while the proceedings of the Second International Conference on the Abatement of Acidic Drainage (Montreal, Canada, September 1991) filled four volumes is indicative of the international focus on the environmental problem posed by oxidation of pyrite in mine wastes. The pollution appearing in surface water near a deposit of pyritic wastes results from a complex of many mechanisms. At the heart of this complex is the oxidation of pyrite which, even in the absence of bacterial catalysis, is a complex of mechanisms itself.

The rate of oxidation of pyrite in a dump is governed by the transport of reactants to oxidation sites in the dump as well as the intrinsic oxidation rate. The oxidation products, which are usually acid and metal sulfates, are pollutants which are transported to the base of the dump and thence to surface water through an aquifer or aquifer system below the deposit. The characteristic timescale of the transport processes will impact on the concentration of pollutants and the way this concentration changes with time in ground or surface waters. The object of rehabilitation measures is to reduce the concentration of pollutants to an acceptable level over the lifetime of oxidation in the mine wastes.

In this paper, I look at various mechanisms and their timescales to note which of them is important in determining pollutant generation rates within the pyritic material and the rate of change of pollutant levels in ground and surface water near the deposits of pyritic waste. I will focus on pyritic waste dumps but many processes are transferable to some tailings dams. I will illustrate the processes with models which, although simple, contain the essential features of important mechanisms. These simple models assist in focusing on what measures to adopt to reduce the environmental impact of oxidation in pyritic wastes and on the parameters which need to be measured to be better able to predict the effectiveness of rehabilitation measures.

The Intrinsic Oxidation Rate

Figure 1 shows a set of equations which describes oxidation in a heap of pyritic material (1). The equations, as they stand, take account of the shape of the heap and the transport of heat, oxygen and water through the heap but not of the changing chemical or microbiological conditions within the heap. The "source" term on the right hand side of the first equation is just a function of the temperature and of the concentration of oxygen and pyritic material. If we added more equations we could take account of the changing chemistry. The source term, which is in fact the oxygen consumption rate, would then become dependent on the water flow rate and the concentration of whatever chemical species were believed to be important. It is convenient to describe this term as the intrinsic oxidation rate of the system.

In any attempt to model the pollution generation rate in a heap of pyritic material it is necessary to have a model for the intrinsic oxidation rate. Both the chemistry and microbiology of pyritic oxidation have been the focus of much experimentation over many years (see reviews by Lowson (2) and Brierley (3)). We would expect such work to be a starting point for our model of intrinsic oxidation rate.

Intercomparison Of Oxidation Rates

Table I shows oxidation rates for pyrite measured by a number of workers with what can be considered as a chemical, microbiological and physical slant. I have deliberately given the rates as quoted to underscore the difficulty in comparing rates measured in one experiment with that in another with nominally similar conditions.

The table is not intended to be exhaustive. The three types of experiment, physics, microbiology and chemistry, have been chosen either because they cover a range of conditions of some importance or they provide sufficient data that allow intercomparisons with the results from one or more of the other types of experiment.

Oxygen Concentration

$$\varepsilon_a \frac{\partial c}{\partial t} + \varepsilon_a \underline{v}^a \cdot \nabla c - \nabla \cdot (D_a \nabla c) = -S_1(c,\overline{c},T)$$

Solid Reactant Concentration

$$\frac{d\overline{c}}{dt} = -S_2(c,\overline{c},T)$$

Temperature

$$\Sigma_\alpha \rho_\alpha c_\alpha \frac{\partial T}{\partial t} + \Sigma_\alpha \rho_\alpha c_\alpha \underline{v}^\alpha \cdot \nabla T - \nabla \cdot (D \nabla T) = S_3(c,\overline{c},T)$$

Macroscopic Pore Velocities

$$\underline{v}^\alpha = -\frac{K k_{r\alpha}(\varepsilon_\alpha)}{\varepsilon_\alpha \mu_\alpha} \nabla(p^\alpha + \rho^\alpha g z)$$

Mass Balance For Air and Water Phases

$$\frac{\partial \rho_\alpha}{\partial t} + \nabla \cdot (\rho_\alpha \underline{v}^\alpha) = 0$$

where $\alpha = a,w$ (air, water; respectively) and ρ_α is related to the intrinsic density ρ^α by $\rho_\alpha = \varepsilon_\alpha \rho^\alpha$.

The volume fractions ε_α, $\alpha = \{a,w,s\}$ must satisfy $\Sigma_\alpha \varepsilon_\alpha = 1$.

Figure 1. Equations describing heap oxidation.

Table I. Comparison of Oxidation Consumption Rates Derived from Different Types of Experiments

Type of experiment	Conditions of experiment	Quoted rate	Normalized rate	Ref.
Microbiological mechanisms	Comparison of results from eight labs 1 g pyrite in 50 mL pH 1.3-3.4 Temperature 28°C	(mg Fe L^{-1}h^{-1}) range 7.8-17.8 average 12.4 abiotic rate average 0.36	(kg·m^{-3}s^{-1}) at a dump pyrite density of 56.3 kg·m^{-3}) 1.9 x 10^{-5} from biotic average 5.6 x 10^{-7} from abiotic average	(4)
Chemical mechanisms	Oxidation in DO saturated solutions and ferric solutions; the data quoted are for DO solutions pH 2.2-9.1 Temperature 22-25°C results normalized to 1 g pyrite in 300 mL	(μM SO$_4$ min^{-1}) range 0.021-0.085 average 0.057	3.2 x 10^{-6} from average rate	(5)
Measurements in waste rock dumps	Inferred from temperature profiles measured in a dump pH 2.0-4.0 Temperature 35-56°C	(kg·m^{-3}s^{-1}) range 0.3-8.8 x 10^{-8}	1.0 x 10^{-8} for model dump	(10)

It should be stressed that many of the chemically and microbiologically based experiments have as their objective the clarification of the mechanisms of pyrite oxidation rather than the provision of data to construct an intrinsic oxidation rate model. It is noteworthy that the abiotic rate in microbiologically based experiments (4) is lower than the chemical rates (5). It is of considerable interest that the lowest laboratory-based oxidation rates are about two orders of magnitude greater than those measured in field experiments.

Modelling Using Simple Intrinsic Rate Models

As discussed above, a heap model as exemplified by the equations in Figure 1 does not describe the effect of changing chemical or microbiological conditions within the heap. In particular, this means that it cannot predict adequately the "lag" phase where the microbial population builds up and acid reacts with alkaline minerals.

Simple Constant Rate Model (SCRM). It is illustrative to consider an intrinsic oxidation rate model where the oxygen consumption rate is independent of pore-gas oxygen concentration and of the pyrite concentration except where these approach zero, where it is assumed that the oxygen consumption rate tends to zero in some way.

Although this model appears very simplistic there is evidence (D. Gibson, written commun., 1992) that it applies to some pyritic material. Let us further assume that this intrinsic oxidation rate applies in a heap with the physical properties given in Table II. Again for simplicity we will assume that the heap was built sufficiently quickly that little or no pore-space oxygen was consumed during the construction phase and further that the moisture content of the heap as built is the equilibrium one for the infiltration rate given.

The first point to note is, that for the stated oxidation rate and assuming the stoichiometry of the first equation in Table III, the initial pore-space oxygen, pyrite and pore-space water in the dump will be used up in about 3 months, 166 years and 2000 years respectively.

It is clear that oxygen needs to be supplied to the system for pyrite oxidation to continue. It is easy to show that oxygen dissolved in water infiltrating the heap is about three orders of magnitude too small to sustain oxidation at the required rate. It has also been shown (1) that unless the air permeability, K, is larger than about 10^{-9} m^2, diffusion dominates over convection as the oxygen transport mechanism. It also follows that in most of the heap one-dimensional transport will be a good description of oxygen transport and the oxygen concentration in the pore space will decrease as indicated in Figure 2a. For the parameters chosen the concentration will fall to zero close to the base of the heap. For higher heaps, it follows that, once the initial pore-space oxygen is used up, the region below 15 m will not contribute to pollution generation until all of the pyrite in the top 15 m is oxidized.

It also follows from our simple model that the concentration of oxidation products (pollutants) in the water infiltrating the heap increases linearly until it reaches the base (see Figure 3a). Again if the heap were higher the concentration of oxidation products in the pore water below 15 m would remain constant. It should be

Table II. Physical Properties of Model Waste Rock Dump

Symbol	Definition	Value	Units
S	The intrinsic oxidation rate for SCRM	1×10^{-8}	kg oxygen $\cdot$ m^{-3}s^{-1}
L	Dump height	15	m
A	Dump area	20	ha
ρ_{rs}	Sulfur density as pyrite	30 (2%)	kg $\cdot$ m^{-3}
ρ	Bulk density of dump material	1500	kg $\cdot$ m^{-3}
q	Infiltration rate	0.5	m $\cdot$ y^{-1}
D_a	Oxygen diffusion through dump pore space	4.1×10^{-6}	m^2 $\cdot$ s^{-1}
c_o	Oxygen concentration in air	0.265	kg $\cdot$ m^{-3}
ε	Mass of oxygen consumed per unit mass of sulfur oxidized	1.75	
ε_w	Volume fraction of water phase	0.1	

Table III. Sulfide Oxidation Reactions

$$FeS_2 + \frac{7}{2}O_2 + H_2O \rightarrow FeSO_4 + H_2SO_4 \qquad \Delta H = 1440 \text{ kJ} \cdot \text{mol}^{-1}$$

$$2FeSO_4 + H_2SO_4 + \frac{1}{2}O_2 \rightarrow Fe_2(SO_4)_3 + H_2O \qquad \Delta H = 102 \text{ kJ} \cdot \text{mol}^{-1}$$

$$FeS_2 + Fe_2(SO_4)_3 + 2H_2O + 3O_2 \rightarrow 3FeSO_4 + 2H_2SO_4$$

$$MS + Fe_2(SO_4)_3 + \frac{3}{2}O_2 + H_2O \rightarrow MSO_4 + 2FeSO_4 + H_2SO_4$$

where MS stands for any metal sulfide.

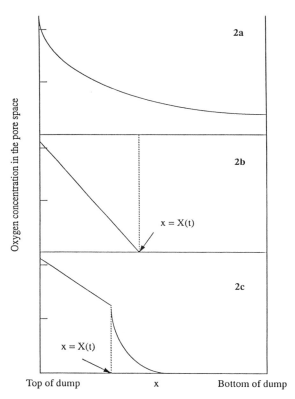

Figure 2. Oxidation concentration profiles for different models of the intrinsic oxidation rate: a) simple constant rate model (SCRM) at low rate; b) simple homogeneous model (SHM); c) shrinking core model (SCM).

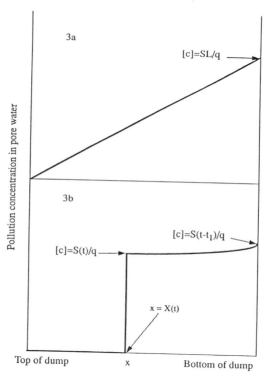

Figure 3. Pollutant concentration in the pore water for different models of the intrinsic oxidation rate: a) simple constant rate model (SCRM); b) simple homogeneous model (SHM) where t_1 is the water transit time from $X(t)$ to the bottom of the dump.

noted that, with this oxidation rate model, the pollutant load from the base of the dump is independent of the infiltration rate. In particular, decreasing the infiltration rate does not decrease the pollution load emanating from the base of the dump. I shall return to this point below.

For the parameters chosen the acid concentration at the base of the heap is about 0.12 molar with a consequent pH of about unity. If the infiltration rate were decreased by about a factor ten the pH would decrease to about 0.2 if the oxidation rate remained unchanged. On the basis of data quoted in the literature, the microorganisms which catalyze the oxidation of pyrite cannot tolerate such a low pH. If the bacteria are important in maintaining this very low oxidation rate we would expect it to decrease at some point above the base, with a consequent decrease in total load.

We could well expect intrinsic oxidation rates to be two to three orders of magnitude higher than the rates assumed for the SCRM on the basis of quoted microbiologically catalyzed and chemical oxidation rates. This is such a large factor that it is as well to consider an infinitely high intrinsic oxidation rate. We can encompass such high rates in another simple model.

Simple Homogeneous Model (SHM). In this model it is assumed that the oxidizable material is uniformly distributed through the heap and the oxidation rate in the heap is limited by the rate at which oxygen can be supplied to an oxidation front which starts at the surface and moves into the heap. In mathematical terms it is a classical moving boundary problem (6) and its properties have been discussed elsewhere (7).

The oxygen concentration in the dump takes the form shown in Figure 2b; the position of the oxidation front is given by,

$$X(t) = \sqrt{\frac{(2D_a c_o t)}{\varepsilon \rho_{rs}}}$$

and the oxygen consumption rate by,

$$S = \delta(x - X(t)) \sqrt{\frac{(D_a c_o \varepsilon \rho_{rs})}{(2t)}}$$

The pollution concentrations then take the form shown in Figure 3b.

Table IV presents some concentrations and loads predicted by applying these simple models to a waste rock dump with the properties given in Table II. The SHM predicts conditions early in the dump's history which are too acid to be consistent with the survival of microorganisms according to the acid tolerances quoted in the literature. Such a model also predicts temperatures in the dump (8) which thermophiles can tolerate but not Thiobacillus spp. It is also clear, however, that oxidation rates high enough to pose a significant environmental problem can be encountered without invoking the catalytic properties of microorganisms.

The Shrinking Core Model (SCM). The infinitely high pollution rates predicted by the SHM at early times are clearly not realistic. The shrinking core model allows such infinities to be circumvented and has the added attraction that the intrinsic oxidation rate predicted by this model decreases as the pore-gas oxygen concentration and the concentration of the oxidizable material decreases (9). Intuitively this is a property we seek in a realistic intrinsic oxidation rate. The SCM gives rise to two moving fronts in the dump; one, above which all the oxidizable material is oxidized; and one, below which there is no oxygen and no oxidation. The resulting oxygen concentrations have the form shown in Figure 2c. The pollution load and the pollutant concentrations predicted by this model at the base of the dump are substantially the same as those predicted by the SHM.

Timescales

These simple models indicate that pollution generation in a typical waste rock dump lasts on a timescale of tens to hundreds of years. It is of interest to examine the equations which describe oxidation in a waste dump (see Figure 1) to see if such timescales are typical of the system. Table V lists the timescales which arise in these equations and evaluates them for typical parameter values.

The longest timescale is that associated with the diffusive transport of oxygen in the dump while the shortest is that for oxidation of a single particle when the shrinking core model is used to describe the intrinsic oxidation rate. Not surprisingly, it has been shown (1) that in a waste dump where diffusional transport dominates, a reduction in the time to oxidize a particle has little effect on the oxidation rate of the dump as a whole.

It has also been shown (1) that, unless the air permeability of the dump is large, convection is not a significant gas transport mechanism even though the timescale associated with convection is so much shorter than that associated with diffusion. One reason is that, initially the only gas transport mechanism is diffusion. The temperature gradients which might potentially drive convection are established on the diffusive timescale and are established at the toe of the dump. If gas were supplied all over the base of the dump then oxidation would proceed at a timescale closer to that associated with convective transport.

Doubling times for the microorganisms which catalyze pyritic oxidation are typically less than a day. This timescale is short compared to those above and it would take less than three weeks for the population to increase by six orders of magnitude and be at a level where the bacterial population was not rate-limiting.

Pollutant Transit Times

Waste rock dumps are generally unsaturated except possibly for the bottom meter or so and unless the climate is very wet or very dry a reasonable value for the infiltration rate is about 0.5 $m \cdot y^{-1}$. With a water-filled porosity of 0.1 the transit time of water from the top to the base of our 15 m high model dump would be about three years unless preferred paths transmit a significant quantity of water. Water flow rates will be very much faster, say 500 $m \cdot y^{-1}$ in the saturated zone at or below the base of

Table IV. Some Indicative Pollutant Concentrations Predicted by SCRM and SHM Assuming No Further Interaction of Oxidation Products with Dump Material

Pollutant	Units	SCRM (At base of dump)	SHM (At 20 years)
Sulfate	$g \cdot m^{-3}$	16,200	23,000
pH		1.08	0.97
Copper (assuming sulfate-to-copper ratio of 50:1)	$g \cdot m^{-3}$	320	460

Table V. Characteristic Timescales

Process	Expression	Timescale
Oxidation of pyrite in a particle	$t_p = \dfrac{\varepsilon \rho_{rs} a^2}{3\gamma D_2 \varepsilon_s c_o}$	0.18 years for 2 mm diameter particle (time to completely oxidize a spherical particle is $t_p/2$)
Convection of gas through a heap with competing chemical reaction	$t_c = \dfrac{\varepsilon \mu_a \rho_{rs} L}{K \rho_\infty^a g \beta T_o c_o}$	1.5 years
Diffusion of gas through a heap with competing chemical reaction	$t_d = \dfrac{\varepsilon \rho_{rs} L^2}{c_o D_a}$	345 years

the dump into which pollution discharges but the distances, at about 500 m, over which this water must travel to traverse the heap will, however, be much greater.

If as discussed above, pollution generation is greatest near the surface of a dump then a toe drain at the base of a dump will not receive pollution for about three years after completion of the dump even if there is no lag time in the initiation of oxidation. The rate of change of pollution levels in the drain at these early times will also reflect a convolution of different travel times from the base of the dump to such a drain.

Implications for Rehabilitation

It follows from the simple models above that measures adopted to reduce the rate of water infiltration into pyritic material may not reduce an existing environmental impact permanently. The reduced infiltration will certainly lead to lower pH within the dump but any consequent inhibition of bacterially catalyzed pyrite oxidation is still likely to leave a chemical rate which is much higher than can be supported by oxygen transport rates if gas transport is dominated by diffusion. The overall oxidation rate will then be determined by oxygen transport rates into the dump. Reduced infiltration at the top surface of the dump will certainly lead to decrease flow from the base of the dump into some underlying aquifer. When steady state has been established after imposition of the reduced infiltration rate, pollutant concentrations will, however, be proportionately increased and the load exiting the base of the dump be the same as before implementation of the rehabilitation measures. The environmental impact will be unchanged.

In practice there will be a drop in the load from the dump immediately after infiltration is reduced as water exiting the dump will contain the "old" pollutant levels. This reduction is a transient. If we assume a tenfold reduction in water infiltration rates then the load exiting the dump will initially be reduced by the same factor but will steadily increase to the steady state value. The timescale for this increase will be of the order of the transit time for water through the waste dump. Using the parameters for our model dump this transit time will increase from 3 years to 30 years with a factor of ten decrease in infiltration rate. The implication is that in 30 years time the environmental impact of the dump will return to its pre-rehabilitation level. A permanent decrease will be obtained if the measures adopted lead to a reduction in the overall oxidation rate in the waste dump.

This argument will hold unless chemical interactions within the wastes place an upper limit on pollutant concentrations in the water-filled pore space. In this case a reduction in the rate at which water infiltrates will lead to a reduction in load; a reduction in the overall oxidation rate in the dump may not lead to a reduction in the load.

Bacterially catalyzed pyrite oxidation rates are very sensitive to pH in the range near neutral to near zero while chemical rates where oxygen rather than ferric ion is the oxidant are much less sensitive over the range 8 to close to zero. Both rates are much higher than the very low rates which can still lead to significant pollution generation from a waste rock dump. It therefore seems likely that pyritic oxidation rates in dumps will be insensitive to changes in pH over a wide range. The implication is that blending of acid-producing material with acid-consuming material may well reduce acidity and some heavy metal production but possibly at the expense of exchanging an AMD problem for a salinity problem.

It is our experience that, although pyritic waste dumps appear to be very heterogeneous, measurements of oxygen and temperature profiles indicate that they behave as if they were homogeneous on the scale of many meters. This implies that an intrinsic oxidation rate can be ascribed to a large volume of dump; it is a measurable quantity and a useful quantity in modelling dump behavior. Again it is our experience that an intrinsic oxidation rate can be measured in a suitably designed column. I believe that the relationship between oxidation rate and the pollution generation rate can also be determined in such experiments.

Conclusions

Intrinsic oxidation rates measured in pyritic waste dumps are much smaller than chemical or bacterially catalyzed pyrite oxidation rates measured in the laboratory. A simple model shows that at the low oxygen consumption rate of 1×10^{-8} $kg \cdot m^{-3} s^{-1}$ the pollution load from a waste dump of typical size and composition will be environmentally significant. If, as is usually the case in such dumps, gas transport into the dump is dominated by diffusion then for intrinsic rates more than ten times this low rate the overall oxidation rate will be largely independent of the intrinsic rate.

This simple model also indicates that, unless the concentration of pollutants in the water-filled pore space is limited by interactions between the oxidation products and gangue minerals, the load exiting the base of the dump under (pseudo) steady-state conditions will be independent of the rate water infiltrates the dump. This means that rehabilitation measures aimed purely at reducing water infiltration rates will reduce the pollution load in the short term but the load will return to the value dictated by the overall oxidation rate in the long term.

Depending on the detailed profile of oxidation within the dump, the timescale for pollution from a new dump to first appear or alternatively the timescale for pollution loads to reach a maximum is dictated by the transit time of water infiltrating the dumps. In these simple models the timescale ranges from the order of a few years for infiltration rates typically due to net precipitation, to tens of years where the infiltration rate has been reduced by rehabilitation measures. These are also the timescales appropriate to the establishment of new steady or pseudo-steady state conditions after a change to the infiltration conditions.

It follows that field observations are required to determine whether or not water transit times in waste dumps are dominated by preferred paths or by diffusive water transport processes.

It also follows that to improve predictive modelling in pyritic waste dumps we require measurements of the intrinsic oxidation rate. This is particularly so if it is low. If it is high then overall oxidation rates will be dictated by oxygen transport rates rather than intrinsic oxidation rates. It further follows that if the correct rehabilitation strategy is to be applied to a particular waste dump then data on the interrelation between oxidation rate and pore water chemistry is required.

Finally more attention needs to be given to the provision of good field data on the effect of specific rehabilitation measures. Both concentration and load information are required and due allowance made for the quantity and timescale of water infiltrating the dump and passing through to the point of collection near the dump base.

Nomenclature

- a: particle radius (m)
- c: oxygen concentration in the pore-space (kg·m^{-3})
- c_o: oxygen concentration in air (kg·m^{-3})
- $\bar{c}$: density of the reactant in the solid phase (kg·m^{-3})
- c_α: specific heat of the alpha phase (J·kg^{-1}K^{-1}) (α=s, solid phase; α=a, air phase; α=w, water phase)
- D_h: coefficient of heat diffusion (J·m^{-1}K^{-1}s^{-1})
- D_a: diffusion coefficient of oxygen in the air in the heap (m^2·s^{-1})
- D_2: diffusion coefficient of oxygen in the particle (m^2·s^{-1})
- g: acceleration due to gravity (m^2·s^{-2})
- K: air permeability of the porous material (m^2)
- L: height of the dump (m)
- p^a: pressure in the air phase (kg·m^{-1}s^{-2})
- T: temperature relative to T_o
- T_o: characteristic temperature, usually annual ambient average (K)
- t_p: characteristic time to oxidize a particle (s)
- t_c: characteristic time for gas convection in a dump where the gas is involved in a chemical reaction (s)
- t_d: characteristic time for gas diffusion in a dump where the gas is involved in a chemical reaction (s)
- v^α: macroscopic velocity of the α phase (m/s)
- β: coefficient of thermal expansion of a gas (°C^{-1})
- γ: a proportionality constant encompassing both Henry's Law and the gas law
- ε: mass of oxygen used per mass of solid reactant in the oxidation reaction
- ε_α: the volume fraction of the α phase
- ρ_α: density of the α phase (kg·m^{-3})
- ρ^α: intrinsic density of the α phase (kg·m^{-3})
- ρ_o^a: density of air (kg·m^{-3})
- ρ_{rs}: initial density of reactant in solid phase (kg·m^{-3})
- μ_a: viscosity of air phase (kg·m^{-1}s^{-1})

Acknowledgments

Much of the above discussion and conclusions spring from access to good field data. I would like to pay tribute to the high level of field experiments which have been conducted by Dr. John Bennett, Dr. Yunhu Tan, Mr. Allan Boyd, Mr. Warren Hart and Mr. Viphakone Sisoutham and the understanding which has followed from discussions of these field results. At the same time I thank Dr. David Gibson and Dr. Garry Pantelis for the insight on mechanisms resulting from discussions with them on how to model the complex of processes involved in the pollution generation from pyritic wastes.

Literature Cited

1. Pantelis, G.; Ritchie, A.I.M. Appl. Math. Model. 1991, 15, 136-143.
2. Lowson, R.T. Chem. Rev. 1982, 82, 461-497.
3. Brierley, C.L. CRC Crit. Rev. Microbiol. 1978, 6, 207-262.
4. Olson, G.J. Appl. and Env. Micro. 1991, 57, 642-644.
5. Moses, C.O.; Nordstrom, D.K.; Herman, J.S.; Mills, A.A. Geochim. Cosmochim. Acta 1987, 52, 1561-1571.
6. Crank, J.C. The Mathematics of Diffusion, 2nd Edition, 1956, Clarenden Press: Oxford, England.
7. Davis, G.B.; Ritchie, A.I.M. Appl. Math. Model. 1986, 10, 314-322.
8. Ritchie, A.I.M. 1977, AAEC/E429.
9. Davis, G.B.; Ritchie, A.I.M. Appl. Math. Model. 1986, 10, 323-330.
10. Harries, J.R.; Ritchie, A.I.M. Water, Air, Soil Pol. 1981, 15, 405-423.

RECEIVED October 20, 1993

Chapter 10

Attempts To Model the Industrial-Scale Leaching of Copper-Bearing Mine Waste

L. M. Cathles

Department of Geological Sciences, Cornell University, Ithaca, NY 14853

The problem of how to match models and reality is fundamental to the safe disposal of hazardous waste and other chemical problems in the natural environment. Research on industrial processes carried out in natural or near-natural settings can provide particularly valuable insights for two reasons: (1) the scale of industrial processes, being intermediate between laboratory and nature, is small enough to allow careful monitoring and post-mortem investigation, and (2) the commercial value of these enterprises, in many cases, already has encouraged and funded extensive model development and testing. Examples include copper leaching in low-grade industrial sulfide-waste dumps, *in situ* leaching of copper from unmined formations, and tertiary oil recovery by steam flooding. Here we describe some studies carried out on copper-heap and *in situ* leaching in the 1970's. The modeling and testing clearly revealed the fundamentals of the heap-leaching process, suggested some new approaches for improving the leach efficiency of the dumps, and provided a basis for optimizing the design of new dumps including completely new kinds of dumps. The work also led to some novel field tests. The accurate, *a priori* prediction of the behavior of a new dump remains elusive, however, despite excellent models and extensive testing.

For the last 50 years or so it has been common practice to leach copper from material that must be mined to access ore but which contains too little copper itself to be processed by crushing, froth flotation, and smelting. This leaching is accomplished by deliberately applying water to the top surface of heaps of waste material. The water exits the dump acidified and copper-laden. The copper is typically removed by passing the solutions through vats containing scrap iron. The copper replaces the iron and the iron-rich solutions are recirculated through the dumps. The resulting copper "precipitate" is removed and sold. In some cases the waste heaps can be very large, filling entire canyons near the mine. Where more land is available, the heaps are generally about 75 ft thick and 300 ft across and are shaped like long fingers, and are called finger dumps.

The dumps are significant economically. In the 1970's Kennecott Copper Corporation obtained about 20% of its copper production (or $100 million of copper per year) from leaching waste dumps. By the 1970's some of Kennecott's dumps had been leached for over 30 years. The initial recovery of copper from the dumps had been spectacular, causing them to be named "Bluewater I", "Bluewater II", etc., after the bright-blue copper-sulfate color of the water. With time, however, the copper concentration in the leach solutions steadily declined. Significantly, this decline was proportional to the inverse of the square root of the time.

These dumps have taught us much about the modeling of natural (or near-natural) processes. Some of what was learned and some pertinent references are given below.

Leaching Copper-Bearing Mine Waste

Quantitative investigation of the leaching process (1) quickly suggested two important controls: The first derives from the fact that sulfide material must be oxidized to be solubilized. The oxidant is supplied by the oxygen in air convecting into the waste dumps. The stoichiometry is such that 20 to 40 times more air than water must circulate through the dump and have its oxygen removed to solubilize the copper observed in the waters exiting from typical dumps. The air convection is driven in large part by the heat generated by the exothermic sulfide-oxidation reactions. The dumps heat to between 55 and 65°C. Convection is assisted (and started) by the fact that oxygen-deficient air is lighter than air with normal oxygen content, and air saturated with water vapor is less dense than drier air.

That air convection is an important factor in dump leaching was confirmed in retrospect by a well-instrumented test dump (1). The Midas Test Dump, operated at Kenecott's Bingham Canyon Mine, had nests of 4 concentric pipes installed and instrumented with thermocouples during construction so that air samples could be drawn from 4 different depths in the 40-ft-thick dump. The dump heated to ~55°C six months after the start of leaching. Oxygen content was near normal at the base of the dump and 50 to 100% depleted at the top, indicating that air was convecting into the high permeability base of the dump and then upward. The base of a dump is more permeable because large boulders roll to the base during construction and are subsequently overlain by finer material as the dump is built outward. Oxygen concentration decreased as the air moved upward. The pattern of oxygen concentration was exactly the opposite of that expected if oxygen was diffusing into the dump; in which case the oxygen content should have been greatest at the top and least at the base.

Air convection is also evident in larger dumps where air visibly (e.g. smoke from a match) moves into some holes drilled into the dump, and out of others. The outward flow of warm, moist air from such holes is sufficient to easily steam a stamp off an envelope. Where a grid of holes has been drilled, a coherent pattern of air convection is observed.

The second factor controlling the rate of leaching that was evident early in our studies was the control of leach rate by the progressive development of "leached rims" on waste fragments in the dumps (1,2). The existence of such leached rims, or rims from which all copper sulfides have been leached, was suggested by the inverse-square-root-of-time decline of copper concentration in dump effluent, and also by kinetic calculations. No one had previously observed such rims. In fact they are not apparent even in sectioned waste fragments. However if the copper-sulfide distribution is imaged by exposing the cut fragment to nitric acid fumes and pressing it on chemically prepared blotter paper, or if the sulfide distribution is point-counted under a microscope, the presence of leached rims is immediately evident. In fact, investigation also showed another interesting phenomenon. Leached copper diffuses into the waste fragment as well as out through the leached rim. This inward diffusion leads to the secondary enrichment of chalcopyrite ($CuFeS_2$) in the interior of the fragments to covellite (CuS) and chalcocite (CuS_2). In dumps permeable enough to support good air convection, the rate of leaching is controlled by the rate of diffusion of oxidant (Fe^{3+}) through the growing leached rim.

These controls provided a basis for the construction of a model (1) of the dump-leaching process that included both kinetic control by diffusion through growing leached rims, and the supply of oxidant by air convection. This model was tested and calibrated against all available Kennecott test-dump data, especially that from the Midas Test Dump. The heating and leaching of the Midas Test Dump could be simulated by the model very well as shown in Figure 1. The modeling of larger dumps, however, suggested that the presence of another control had been missed: the bacterial catalysis of the oxidation of Fe^{2+} to Fe^{3+}. In the absence of this control the model indicated dumps should heat to temperatures close to boiling and not stop at 55 to 65°C as is observed. *Thiobacillus ferrooxidans* bacteria are known to go dormant at about 55°C, with some high-temperature strains remaining active to 65°C. The model was refined to include bacterial catalysis and the model dumps then operated in the temperature range observed in real dumps (2). The model was also extended to two spatial dimensions (3).

The final test of the model was provided by column tests at the New Mexico Institute of Mining and Technology (4,5). The columns used were surplus liquid oxygen (rocket propellant) storage tanks. Each was double walled, 40 ft high, 10 ft diameter, stainless steel, with 1 ft of perlite insulation between the walls. One of these columns was instrumented to our design, and filled with 160 metric tonnes of mine waste form Kennecott's Chino Mines Division at Santa Rita, New Mexico. A schedule of solution flushes and detailed predictions of column temperature and copper recovery were made from the parameters determined from the modeling analysis of four test dumps at Santa Rita. The most unequivocal of the model predictions was that the 160 tonnes of waste would heat up to 55 to 65°C in about 6 months, and then cycle in temperature according to the ambient temperature and solution flush schedule.

The column was leached for two full years, first under contract from Kennecott, and later with funding from the National Science Foundation - Research According to National Needs (NSF-RANN) program. The operation of the column confirmed major aspects of the model predictions, in particular the heating up to 55-65°C. In fact the

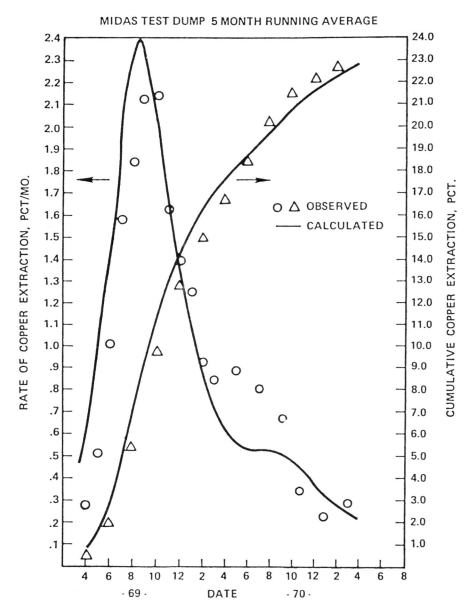

Figure 1. Model predictions of the cumulative extraction and rate of extraction of copper from the 40 ft high Midas Test Dump agree well with the observed rates. Reproduced with permission from ref. 1, copyright 1975 Metallurgical Transactions.

gradual development of the higher temperature strain of bacteria could be seen in the temperature response of the column. Details of the amount and rate of copper leaching were missed by the model, although with recalibration, the model was able to describe all aspects of the leaching very well. In fact the modeling was accurate enough to require inclusion of a finite thermal conductivity of the insulated column walls to simulate the test results accurately.

The column operation provided other insights. At the end of the experiment rodamine-B dye was added to the leach solution, and a careful record kept of the red-stained percentage of the waste as it was unloaded and assayed. Fifty-seven percent of the copper originally in the waste was removed in the two years of leaching whereas only 50% of the waste was stained red. The clear implication of this observation is that the flow pattern shifted with time during leaching. I suspect that this is a general phenomenon. It is required, for example, to explain the very uniform oxygen isotopic alteration of the ocean crust. Shifting of flow patterns so as, on average, to provide uniform rock access, has important implications for the long-term rates of migration of contaminant fronts.

Discussion and Conclusions

The experience of modeling waste dumps is unusual in the intensity and thoroughness with which models were tested. This was allowed by the tractable scale of test dumps, the availability (at the right time) of large columns and the willingness of New Mexico Institute personnel to work on the project, and by the economic importance of the process to industry. Some conclusions of broader interest can be extracted from this work.

First it was possible to develop good models with good predictive capability. The models provided and continue to provide a very useful framework for designing optimum dumps and identifying ways to improve the operation of old dumps (such as air injection) and in assessing the magnitude of expected benefits. Model-guided observation is very useful. For example models directed attention to leached rims and instigated a deliberate search for them. Models led to clarification of the role of bacterial catalysis.

Second the experience indicates the importance of combining modeling and observation and constantly testing and refining the models developed in terms of the objectives of the project. Kennecott formed a task group consisting of engineers, line managers, research scientists, and dump operators for this purpose. This task group met frequently over a four-year period. The interaction was at times sobering. I remember in particular on one occasion explaining to a dump operator that directing water specifically to hot spots on the dump, which was their deliberate practice, was the wrong thing to do. The hot areas should be encouraged because heat promotes faster leaching. Further modeling, in fact the discovery of the importance of bacterial catalysis, proved me quite wrong and the operator quite right. Hot areas cease to leach if the temperature gets too hot. Models with bacterial catalysis included showing that the application of water to overly hot areas will dramatically speed leaching, just as the operator suggested. Sometimes the exchange went the other way. For example as the work progressed it became clear that great amounts of effort had, in retrospect, been wasted trying to fertilize dump bacteria and dissolve iron sulfate and unclog the dumps by applying ammonia bisulfite to the leach solutions. The bacteria have sufficient nutrients in the waste water. Permeability reduction by iron sulfate precipitation is a problem mainly very near the dump surface and is best remedied by plowing or adding acid to the leach solutions. If dumps are oxygen starved, air injection may make economic sense, a possibility that could not have been appreciated in the absence of modeling.

Model development provided a basis for deliberately designing optimum dumps and dump operation. It also suggested the laboratory measurements and procedures that could best lay the groundwork for predicting how new kinds of dumps might operate. These procedures were later applied to investigating whether pyrite could be economically removed from coal by heap leaching (6), or whether copper-waste dumps could be effectively operated in the Andes at over 15,000 ft elevation where temperatures drop below freezing every day of the year. The analysis results (7) are shown Figure 2.

The models of the copper-waste leaching process were not able to predict in detail how a waste of given type would leach, as was evident from the failure of the New Mexico column to match predicted leaching details. Whether better predictions could have been made had the procedures used in predicting coal depyritization by heap leaching (6) been applied to the Santa Rita waste is at present unclear. The New Mexico column waste was not analyzed in this fashion, and the coal leaching predictions were never subjected to the ultimate test of heap construction and operation. My own experience suggests that although models are useful, and often the *only* way to understand and predict a natural process, model predictions of natural processes should not be expected to be capable of fully accurate *a priori* predictions. If, however, an experiment or process can be monitored for a time, good models can be calibrated, and much better predictions made.

Industrial processes of other kinds may also be relevant as natural analogues of hazardous-waste disposal. I have limited myself here to the discussion of just one example. Other examples might be briefly mentioned because of their chemical relevance and the obscurity of the literature in which they are published. One is a tracer experiment performed to assess the accessibility to diffusional leaching of the matrix blocks that lie between the fractures in a porphyry copper deposit. The hope was to leach the matrix with a lixiviant introduced through the fractures. The difference in arrival time between a diffusing (NaCl) tracer and a non-diffusing tracer (in this case 0.5 micron silica beads) provided a direct measure of diffusional accessibility. A field test appeared to confirm this theory (8). Matrix diffusion is important because it is one of the principal factors determining how fast a contaminant front will migrate. Another example of chemical interest is the application of heap leaching models to acid mine drainage. As shown in Figure 3, bacterially catalyzed Fe^{3+} oxidation of pyrite, facilitated by air convection through a coal waste pile before the pile is buried, can proceed at rates similar to those observed in copper waste dumps (~1% per month). This rapid oxidation can produce enough acid (stored as jarosite salts) in one year to acidify 20 inches per year of infiltrating rainfall to pH 2 for 108 years after the waste is

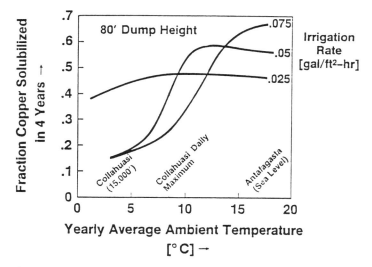

Figure 2. Model calculations show that the copper waste dumps should operate effectively in cold climates provided the average rate at which flush waters are applied to their top surface (the irrigation rate) is reduced to about a third of the optimum rate in warmer climates. Reproduced with permission from ref. 7, Chevron Oil Field Research Company Technical Report.

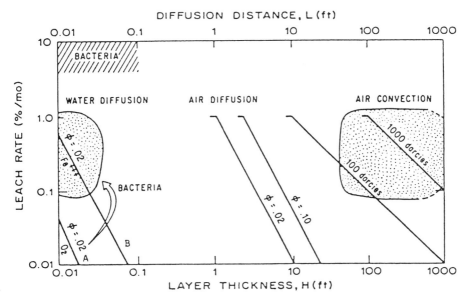

Figure 3. Leaching of copper waste at commercial rates of ~1% per month from >70 ft high dumps is made possible by air convection and bacterial catalysis (two stippled regions). Air convection is necessary to supply oxidant to the interior of the dump at the required rates. Bacterial catalysis is necessary to oxidize Fe^{2+} to Fe^{3+} by O_2. The high concentrations of Fe^{3+} in the dump waters speeds the diffusion of oxidant into the rock fragments in the dump by a factor of ~17 (open arrow, left side of diagram). Reproduced with permission from ref. 9, copyright 1982 The Pennsylvania State University.

buried (9). This is an unconventional view of the origin of acid mine drainage and suggests remedial procedures very different from those in current use.

Acknowledgments

I would like to thank the Kennecott Copper Corporation, The National Science Foundation (RANN Program), The Bureau of Mines, and The Chevron Oil Field Research Company for supporting the work summarized above. Many contributed, but none more than Dennis Bloss, who, knowing he was dying of an incurable disease dedicated the last two years of his life to running the New Mexico leach column experiment, often spending the night on site. This paper is dedicated to Dennis.

Literature Cited

1. Cathles, L.M.; Apps, J.A. Metallurgical Trans. 1975, 6B, 617-624.
2. Cathles, L.M. Math. Geol. 1979, 11, 175-191.
3. Cathles, L.M.; Schlitt, W.J. In Leaching and Recovering Copper from As-mined Minerals; Schlitt, W.J., Ed.; Society of Mining Engineers: 1980; pp. 9-27.
4. Cathles, L.M.; Murr, L.E. In Leaching and Recovering Copper from As-mined Minerals; Schlitt, W.J., Ed.; Society of Mining Engineers: 1980; pp. 29-48.
5. Murr, L.E. Minerals Sci. Engng. 1980, 12, 121-189.
6. Cathles, L.M.; Breen, K.J. Removal of pyrite from coal by heap leaching. Final Report. Bureau of Mines, U. S. Department of Interior 269 pp.
7. Cathles, L.M. Dump Leaching at Collahuasi, Chile Technical Report TM86000366, Chevron Oil Field Research Co., 1986, 52 pp.
8. Cathles, L.M.; Spedden, H.R.; Malouf, E.E. In Proceedings of the Symposium on Solution Mining; Aplan, F.F.; McKinney, W.A.; Pernichele, A.D., Eds.; Am. Inst. Mining, Metall., and Petrol. Eng. Inc.: New York, NY, 1973; pp. 129-147.
9. Cathles, L. M. In Earth and Mineral Sciences; College of Earth and Mineral Sciences, Pennsylvania State University: University Park, PA, 1982; Vol. 51, pp. 37-41.

RECEIVED August 10, 1993

Chapter 11

A Computer Program To Assess Acid Generation in Pyritic Tailings

Jeno M. Scharer[1], Ronald V. Nicholson[2], Bruce Halbert[3], and William J. Snodgrass[4]

Departments of [1]Chemical Engineering and [2]Earth Sciences, University of Waterloo, Waterloo, Ontario N2L 3G1, Canada
[3]SENES Consultants Ltd., 52 West Beaver Creek Road, Richmond Hill, Ontario L4B 1L9, Canada
[4]Department of Civil Engineering, McMaster University, Hamilton, Ontario L8S 4M1, Canada

A computer program known as Reactive Tailings Assessment Program (RATAP) was developed to assist in predicting acid generation and major hydrogeochemical events brought about by the chemical and microbiological oxidation of sulfide minerals. The objective of the program is the application of fundamental kinetic and physical knowledge to field conditions for simulating the rate of acid production with time, simulating porewater quality in space and time.

The kinetics of the abiotic and biological oxidations are key components of the program. The primary variables include the composition of the sulfide minerals, the specific surface area, the partial pressure of oxygen, the temperature, and the pH. In addition, biological reaction rates depend on the sorption equilibrium of the bacteria on the sulfide surface, the moisture content, and the availability of carbon dioxide (carbon source) and other macronutrients.

The tailings soil profile is subdivided into layers comprising the unsaturated zone, the capillary fringe, and the saturated zone. Acid production occurs primarily in the unsaturated zone. A major control on the acid generation rate is the diffusive flux of oxygen in the pore space of the unsaturated zone. The diffusion coefficient, in turn, is related to the moisture content in a given zone.

Computations are carried out at monthly time steps. The program can be run in a deterministic or probabilistic manner. For probabilistic assessment, an additional module known as RANSIM (random simulation), is utilized for the allocation of distributed parameter values. The program has been employed successfully to simulate field data at several tailings sites.

The control of acid mine drainage (AMD) from pyritic tailings is widely recognized as one of the most serious environmental issues facing many base metal, gold and uranium mine operations. While the collection and treatment of acidic drainage is common practice at operating mine sites, it is generally accepted that continued treatment after mine closure is neither desirable nor practical. Besides maintaining an effective treatment system indefinitely, the disposal of the large amount of chemical sludge produced after neutralization is a major operational problem.

Recognizing the magnitude of the problem, Canada Centre for Mineral and Energy Technology (CANMET) initiated several studies on the factors and processes which control mineral oxidation and on developing a predictive modelling tool to simulate acid generation in mine tailings. One of these early studies was a review of the mechanisms and kinetics of sulfide mineral oxidation ([1]). This study also documented the significance of chemolithotrophic bacteria in enhancing the oxidation rates. The first version of the RATAP model was completed by Beak Consultants Ltd. and SENES Consultants Ltd. in 1986. The objectives of the model development were: (1) the application of kinetic data to field conditions; and (2) evaluation of a coupled model to simulate AMD. Since its original conception, the RATAP model has undergone several modifications and has been calibrated extensively on several tailings sites in northern Ontario and in northeastern Quebec. The current version (version 3, released in March, 1992) may be run in a deterministic and probabilistic manner. It also allows the evaluation of a soil cover or underwater disposal options.

Modelling Concepts

In Table I, the reactive sulfide minerals in tailings and their more common oxidation products are shown. The RATAP model includes oxidation of five minerals : pyrite (FeS_2), pyrrhotite ($Fe_{1-x}S$), chalcopyrite ($CuFeS_2$), sphalerite (ZnS), and arsenopyrite ($FeAsS$). Based upon the review of the mineral composition of several tailings sites ([1]), calcite ($CaCO_3$), aluminum hydroxide ($Al(OH)_3$), and ferric hydroxide ($Fe(OH)_3$) are regarded as the principal buffering minerals. Depending on the reaction rate, solubility relationships, and pH, the oxidation may result in secondary mineral precipitation ([2]). The important secondary minerals considered in the program are gypsum ($CaSO_4 \cdot 2H_2O$), ferric hydroxide ($Fe(OH)_3$), jarosite ($KFe_3(SO_4)_2(OH)_6$), siderite ($FeCO_3$), aluminum hydroxide ($Al(OH)_3$), covellite (CuS), malachite ($Cu_2CO_3(OH)_2$), antlerite ($Cu_3SO_4(OH)_4$), smithsonite ($ZnCO_3$), and scorodite/ferric hydroxide coprecipitate ($FeAsO_4 \cdot 2H_2O/Fe(OH)_3$). Thermodynamic data for these solid phases have been compiled from published data ([3-6]).

The physical concept employed in the RATAP model is summarized in Figure 1. The tailings soil profile is subdivided chemically and physically into 22 distinct layers comprising the hydraulically unsaturated zone, the capillary fringe and the saturated zone. Water movement from atmospheric precipitation is downward through the unsaturated zone and the capillary fringe, while a portion of the flow may move horizontally out of the layer in the saturated zone to emerge as seepage passing through or below the perimeters of dams. The remaining portion moves downward into the subsurface aquifer. Exceptionally high mineral oxidation

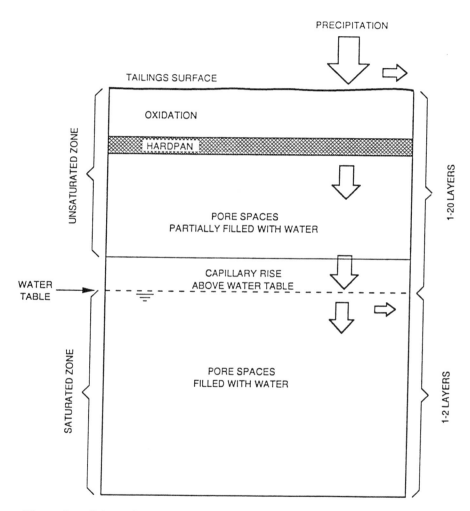

Figure 1. Schematics of the physical model for tailings (adapted from ref. 1)

Table I. Reactive Sulfide Minerals in Tailings and Waste Rock (adapted from references 23 and 40)

SULFIDE MINERALS		OXIDATION PRODUCTS	
FeS_2	pyrite	$CaSO_4 \cdot 2H_2O$	gypsum
FeS	troilite	$Fe(OH)_3$	ferric hydroxide
Fe_7S_8	pyrrhotite	$KFe_3(SO_4)_2(OH)_6$	jarosite
		$Fe^{II}Fe^{III}_4(SO_4)_6(OH)_2 \cdot 20H_2O$	copiapite
$CuFeS_2$	chalcopyrite	$Cu_2CO_3(OH)_2$	malachite
CuS	covellite	CuO	tenorite
Cu_2S	chalcocite	$Cu_3(OH)_4SO_4$	antlerite
$FeAsS$	arsenopyrite	$FeAsO_4 \cdot 2H_2O / Fe(OH)_3$	scorodite/ferric hydroxide (coprecipitate)
ZnS	sphalerite	$ZnCO_3$	smithsonite
$(Zn,Fe)S$	marmatite	$Zn_5(CO_3)_2(OH)_6$	hydrozincite
		$(Zn, Cu)_5(CO_3)_2(OH)_6$	aurichalcite

rates coupled with secondary mineral ($CaSO_4 \cdot 2H_2O$, $Fe(OH)_3$, $Al(OH)_3$, $FeCO_3$, etc.) precipitation can result in hardpan formation in any of the layers. Since the specific volume of the secondary minerals is usually greater than the original sulfides, secondary mineralization effectively reduces the pore space which restricts the downward movement of water at the affected layer (7).

The main modules and subroutines in RATAP are summarized in Figure 2. The geochemical and biochemical formulations employed in constructing RATAP include: (1) dynamic expressions, (2) equilibrium expressions, and (3) empirical expressions. Dynamic expressions are used to model time-dependent processes. These include sulfide oxidation kinetics, mass transport (oxygen, and solute transport), and energy (enthalpy) transport. The dynamic processes are evaluated for monthly time steps in every layer. Equilibrium expressions are used to model solids dissolution, aqueous speciation, neutralization/buffering, ion exchange and sorption processes. It is assumed that these processes, although dynamic in nature, are sufficiently fast so that the time scale may be neglected. These processes are also evaluated monthly in each layer. Empirical expressions are used to model periodic data such as baseline temperature (monthly mean temperature at various depths in the absence of mineral oxidation) and atmospheric precipitation. Some

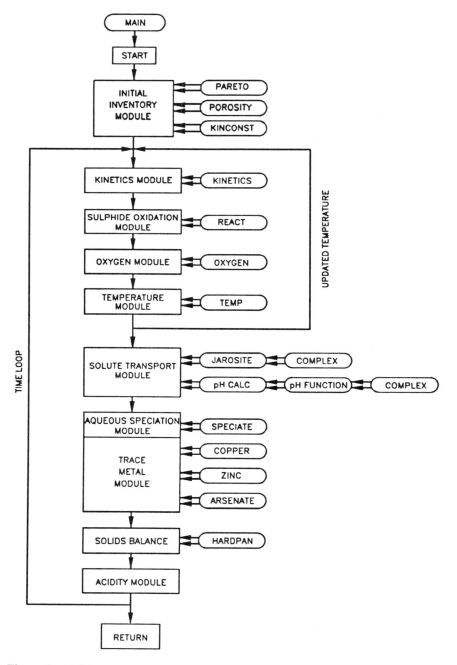

Figure 2. Main component modules and subroutines of the RATAP program (adapted from ref. 1)

parameters of the empirical expressions have no apparent physical significance. These are evaluated for each month and each layer when initiating the program and are not allowed to vary during the program execution.

Initial Inventory Module. In the initial inventory module, the initial values (aqueous and solid phase concentrations) are assigned and the geochemical, kinetic, and physical parameters are calculated. It is well known that the sulfide particle oxidation rate is directly proportional to the surface area ($\underline{8}$, $\underline{9}$). The surface area, in turn, depends on the particle size. Typical particle size distribution in unconsolidated tailings is shown in Figure 3. The cumulative distribution can be conveniently described by a Pareto-type ($\underline{10}$) truncated probability distribution function :

$$F = Y \left(\frac{r}{R}\right)^{\alpha} - (Y - 1) \left(\frac{r}{R}\right)^{\beta} \qquad (1)$$

where:
- F = mass fraction of particles having a radius less than or equal to r
- r = particle radius
- R = maximum particle radius
- Y, α, β = fitted distribution parameters

Analytical data are entered as particle size versus cumulative frequency. A non-linear parameter estimator subroutine ($\underline{11}$) is then employed to obtain the parameter estimates for the distribution. The estimation routine is weighted toward the smaller particle sizes, since the relative contribution of smaller particles to the total surface area is more significant.

The initial inventory module is also used to calculate the oxygen diffusion coefficient through the unsaturated zone and the capillary fringe. Based on the 10 percentile particle size (d_{10}), residual water retention, and the distance to the water table, the water content in each layer is computed by using an empirical expression proposed by Hillel ($\underline{12}$) for silt and sand soils. The effective gaseous oxygen diffusion coefficient (D_e) is then calculated from a model based on experimental data using tailings ($\underline{13}$):

$$D_e = 0.203 \left(\frac{n_a - 0.05}{0.95}\right)^{1.7} + D_w \qquad (2)$$

where:
- n_a = air-filled porosity (m³air/m³)
- D_e = effective diffusion coefficient of oxygen (cm²s⁻¹)
- D_w = diffusion coefficient of oxygen in water (based on gas concentrations)

The Reardon and Moddle equation ($\underline{13}$) was modified so that, for an an air-filled porosity of less than 0.05, the particular layer is assumed to have transport properties identical with a water-saturated layer.

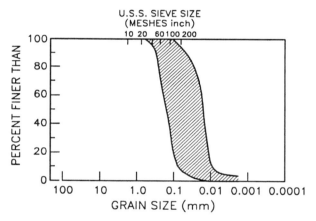

Figure 3. Typical range of grain size distribution in acid generating tailings.

Kinetics Module. The iron sulfides pyrite and pyrrhotite are usually the dominant reactive minerals in tailings. Pyrite and pyrrhotite oxidation reactions can be expressed by the following stoichiometric equations (14).

$$FeS_2 + 3.5O_2 + H_2O \rightarrow Fe^{2+} + 2SO_4^{2-} + 2H^+ \qquad (3)$$

$$Fe_7S_8 + 15.5O_2 + H_2O \rightarrow 7Fe^{2+} + 8SO_4^{2-} + 2H^+ \qquad (4)$$

$$2Fe^{2+} + 0.5O_2 + 2H^+ \rightarrow 2Fe^{3+} + H_2O \qquad (5)$$

$$Fe^{3+} + 3H_2O \rightarrow Fe(OH)_{3(s)} + 3H^+ \qquad (6)$$

$$FeS_2 + 14Fe^{3+} + 8H_2O \rightarrow 15Fe^{2+} + 2SO_4^{2-} + 16H^+ \qquad (7)$$

$$Fe_7S_8 + 62Fe^{3+} + 32H_2O \rightarrow 69Fe^{2+} + 64H^+ + 8SO_4^{2-} \qquad (8)$$

Equations (3) and (4) represent the overall stoichiometry of pyrite and pyrrhotite oxidation to ferrous ion and sulfate. Pyrrhotite is more reactive than pyrite. Pyrrhotite is reported to be highly reactive below at pH values less than 2 but kinetic data lack reproducibility (15). Both may be catalyzed by direct enzymatic oxidation of the sulfide moiety (16-18). Equation (5) is also catalyzed by bacteria, particularly *Thiobacillus ferrooxidans*. This enzymatic oxidation of iron results in indirect biological leaching mechanisms with ferric ion being the principal oxidant in acidic solution. Equations (7) and (8) are abiotic, but they are believed to be the most important acid forming reaction under anoxic conditions. The oxidation of pyrrhotite by ferric ion in Equation (8) is particularly fast and can result in the accumulation of significant amounts of elemental sulfur.

The kinetics of the biological and abiotic oxidation of metal sulfides are key components of the RATAP program. A number of factors have been shown to affect the oxidation rates. The principal controlling factors are the specific surface area, the temperature, and the pH. The chemical oxidation of sulfide minerals is modelled by the following relationship (14):

$$r_c = Ae^{-E_a/RT}[O_2]10^{-xpH} \qquad (9)$$

where:

- r_c = abiotic rate constant (mol·m^{-2}s^{-1})
- $[O_2]$ = dissolved oxygen concentration (mol·m^{-3})
- E_a = Arrhenius activation energy (J·mol^{-1})

T = temperature (K)

The activation energies for the abiotic reactions are reported to be ranging from 42 to 88 kJ·mol⁻¹ (19-22). The value of the pH-dependent parameter, i.e. x in Equation (9), depends on the sulfide mineral. For most minerals, the value of x appears to be near 0.5 (22).

In addition to the controlling factors affecting abiotic oxidation, biological reaction rates are dependent on the sorption of bacteria on the mineral surface, the availability of carbon dioxide as the carbon source, the availability of macronutrients (phosphate, for example), and the presence of inhibitors. Neglecting the secondary effects, the biological reaction rate can be expressed as follows (23):

$$r_b = b \frac{\mu_m \sigma}{Y_{x/s}} e^{-Ea/RT} \frac{[O_2]}{K_o + [O_2]} \frac{1}{1 + 10^{2.5-pH} + 10^{pH-4}} \quad (10)$$

where:

r_b = biological surficial oxidation rate (mol·m⁻²s⁻¹)
b = biological scaling factor
μ_m = specific growth rate (s⁻¹)
σ = specific surface coverage (g·m⁻²)
$Y_{x/s}$ = growth yield (g·mol⁻¹)
K_o = half saturation constant for oxygen (mol·m⁻³)
E_a, T, and $[O_2]$ are defined as in Equation (9)

In equation (10), the biological scaling factor is introduced to fit site-specific data. The specific growth rate of the indigenous bacterial population is evaluated at 30°C and pH = 3.0. The specific growth rate is modulated by taking into account the moisture content (24), the carbon dioxide partial pressure (25, 26), and the availability of nutrients (9).

Typical biological and abiotic oxidation rates and Arrhenius activation energies expected in tailings are summarized in Table II. The reaction rates were obtained with excess oxygen. The specific surface reaction rate constants have been derived from experimental results of Nicholson (21), Mehta and Murr (27), Ahonen et al. (28), Ahonen and Tuovinen (19), and Scharer et al. (23). The nominal surface area has been based on particle size distribution assuming spherical symmetry. The chemical and, to a lesser extent, biological oxidation rates of other sulfides are enhanced by the presence of pyrite (18, 28). This enhancement has been attributed to a galvanic interaction between pyrite and the other minerals. This enhancement was not accounted for explicitly in the RATAP model.

Sulfide Oxidation Module. Sulfides are assumed to occur as distinct, homogeneous particles of relatively pure composition (1). As these particles oxidize, they shrink in size. The oxidizing surface either remains free of secondary mineral precipitation (low-pH conditions) or the deposition is not rate limiting (20). To account for these laboratory and field observations, sulfide oxidation in tailings is based on a concept of "shrinking radius" kinetics. It can be easily shown that the

Table II. Comparison of Specific Reaction Rates of Sulfide Oxidation at 30°C, pH=3.0, 20 kPa P_{O2} (adapted from 18,21,23,27,28,36,37)

COMPONENT	SPECIFIC REACTION RATE ($mol \cdot m^{-2} s^{-1}$)		ACTIVATION ENERGY ($J \cdot mol^{-1}$)
	with bacteria	abiotic	
pyrite	6.64 x 10^{-8}	4.25 x 10^{-9}	52,700
pyrrhotite	9.20 x 10^{-8}	1.10 x 10^{-8}	44,700
chalcopyrite	6.67 x 10^{-9}	1.34 x 10^{-12}	20,000
chalcopyrite with pyrite	1.43 x 10^{-8}	1.88 x 10^{-9}	47,000
sphalerite	7.75 x 10^{-9}	1.9 x 10^{-13}	21,000
sphalerite with pyrite	1.41 x 10^{-8}	6.92 x 10^{-10}	45,000

rate of radial shrinkage of a more-or-less spherical particle is as follows:

$$k_i(t) = \frac{r_{c,i} + r_{b,i}}{\rho_i} \quad (11)$$

where:
$k_i(t)$ = rate of radial particle shrinkage ($m \cdot s^{-1}$)
$r_{c,i}$ = chemical surficial oxidation rate ($mol \cdot m^{-2} s^{-1}$)
$r_{b,i}$ = biological surficial oxidation rate ($mol \cdot m^{-2} s^{-1}$)
ρ_i = molar density of the ith sulfide ($mol \cdot m^{-3}$)

In Equation (11) the subscript i refers to the ith sulfide species. Since the reaction rates are evaluated at monthly intervals, the total shrinkage becomes the sum of the monthly time increments (Δt):

$$X_j = \sum_{t=0}^{t} k_j(t) \Delta t \quad (12)$$

where:
X_j = total shrinkage of particle j to time t (m)

The fraction of unreacted sulfide mineral concentration in any layer can be determined by the combination of Equation (11) with the Pareto size distribution density function, i.e. the differential form of Equation (1). It should be noted that the Pareto distribution refers to an initial (time zero) distribution. It is self evident that no particle with an original radius of X_i or less can exist at any later time, since the particles with lesser radii have shrunk to zero size. Thus, the unreacted sulfide mineral concentration at any time t is given by the following integral expression:

$$M_i(t) = M_{i,o} \int_X^R \left(\frac{r-X_i}{r}\right)^3 \left[Y\alpha\left(\frac{r}{R}\right)^{\alpha-1} - (Y-1)\beta\left(\frac{r}{R}\right)^{\beta-1}\right]\frac{dr}{R} \quad (13)$$

where:
$M_i(t)$ = unreacted sulfide mineral concentration (mol·m^{-3})
$M_{i,o}$ = initial sulfide mineral concentration (mol·m^{-3})

The major advantage of employing a Pareto-type size distribution is that the resulting integral, i.e. Equation (13), is analytic. The integral is a polynomial expression of (X_i/R). The coefficients and the exponents of the polynomial series can be readily determined from the Pareto parameters.

Oxygen Transport Module. In tailings and waste rock, the transport of oxygen through the pore space is regarded as the ultimate limit of the sulfate-generation flux (21, 30-32). The primary modes of oxygen transport in tailings are molecular diffusion through the pore space in the unsaturated zone and, to a much lesser extent, advective transport of dissolved oxygen in the percolating porewater:

$$\frac{\partial C}{\partial t} + D_e\frac{\partial^2 C}{\partial Z^2} + vK_H\frac{\partial C}{\partial Z} = \sum_i \gamma_i\frac{\partial M_i}{\partial t} \quad (14)$$

where:
C = concentration of oxygen in the air-filled pore space (mol·m^{-3})
D_e = effective diffusion coefficient of oxygen (m^{2}s^{-1})
Z = depth into tailings (m)
t = time (s)
v = water infiltration rate (m·s^{-1})
K_H = modified Henry's law constant (mol·m^{-3} oxygen in liquid per mol·m^{-3} oxygen in gas phase)
M_i = concentration of the ith metal sulfide (mol·m^{-3})
γ_i = stoichiometric coefficient relating oxygen uptake to sulfide mineral oxidation

The differential term on the right hand side of Equation (14) is the time differential of Equation (13). It can be shown (23), that the oxygen concentration approaches a steady-state condition within a few days for typical tailings conditions. Consequently, a "monthly steady state" approximation can be derived by neglecting the first term in Equation (14). Moreover, the oxygen dependence of the sulfide oxidation rate (see Equations (9) and (10)) can be resolved as the sum of a zero-order and first-order oxygen-dependent terms, provided that the sulfide oxidation rates approach zero as the oxygen concentration in the pore space becomes zero. A finite difference form of Equation (14) yields a tridiagonal matrix which is easily solved by decomposition followed by forward and backward substitution (11).

Enthalpy (Temperature) Transport Module. The temperature has a profound effect on both the abiotic and the biological oxidation rates. To calculate the temperature in the tailings, a simple enthalpy balance is employed. Because it is difficult to construct a global enthalpy model, the monthly temperature at various depths in the tailings is calculated as a temperature rise resulting from enthalpies of sulfide oxidation reactions:

$$\rho_B C_P \frac{\partial \Delta T}{\partial t} + k \frac{\partial^2 \Delta T}{\partial Z^2} + F_W C_W \frac{\partial \Delta T}{\partial Z} = Q_{RX} - \Delta H_v E_W \qquad (15)$$

where:
- ΔT = temperature rise in the tailings (K)
- ρ_B = tailings bulk density (kg·m^{-3})
- k = thermal conductivity (J·m^{-1}K^{-1}s^{-1})
- F_w = vertical water flux (mol·m^{-2}s^{-1})
- C_w = molar heat capacity of water (J·mol^{-1}K^{-1})
- H_v = enthalpy of evaporation (J·mol^{-1})
- E_w = evaporative water loss (mol·m^{-3}s^{-1})
- Z = depth into tailings (m)
- Q_{RX} = sulfide reaction enthalpy generation (J·m^{-3}s^{-1})
- C_p = specific heat capacity of tailings (J·kg^{-1})

As in case of oxygen transport, the temperature is calculated by assuming monthly steady-state conditions. In addition, the sulfide reaction enthalpy term is linearized by Taylor's expansion. The solution of the equation is iterative as shown in Figure 2. The baseline temperature is taken as the first estimate for the monthly steady-state temperature. Using this baseline, the temperature rise is calculated by employing a finite-element method. The predicted tailings temperatures usually do not rise more than 5 K above the baseline which is consistent with field observations.

Solute Transport Module. Calculations in the solute transport model are based on material balances for each component. Transport calculations include both kinetically controlled reactions (sulfate formation by sulfide oxidation, for example) and equilibrium-controlled reactions (sulfate formation due to gypsum dissolution, for example). Aqueous speciation and the determination of porewater pH are also performed in this module. The calculation procedure is similar to the methodology employed by Parkhurst *et al.* (33). Rather than the Newton-Raphson method, the method of false position (11) is employed as a root-finding algorithm for estimating the pH. The latter method gives generally faster convergence.

Model Simulations

Model calibration and validation studies have been carried out at several tailings areas. In this paper, model simulations are reported for the Nordic tailings management area near Elliott Lake, Ontario. The Nordic tailings cover approximately 100 hectares and contain 12 million metric tons of pyritic tailings. The deposition of tailings in this basin was halted in 1968. The reactive mineral content of the tailings at the cessation of mill operations is estimated to be 6% pyrite (10% in coarse tailings), 5% gypsum, 0.5% calcite, 0.4% ferric hydroxide, 0.2% aluminum hydroxide, and 13.6% sericite. Smyth (34) and Dubrovsky *et al.* (35) collected extensive field data on the site. The data selected for comparison are from two sampling sites (designated as T3 and T5) approximately 0.75 km apart. Location T5 comprises coarse tailings (sands) adjacent to the old tailings discharge pipe. The depth to the water table is 4.0 to 4.5 m. At location T3, the tailings consist of fine material (slimes) and the water table is 6.0 to 6.5 m below the surface. To simulate field conditions, the Pareto parameters for tailings samples at both sites were obtained.

Conditions for 1968 were used as initial conditions and the simulation was run for 13 years. In Figures 4-11, the model output after 13 years of simulation is compared with field data collected in 1981. There was no parameter adjustment between year zero (1968) and year 13 (1981). The measured and predicted gaseous oxygen concentrations with depth are shown in Figure 4 (location T3) and Figure 5 (location T5). The oxygen profile is predicted to fall rapidly to zero at 1.3 m depth at location T3. This is consistent with the higher moisture content, hence lower effective diffusion rate coefficient, of fine tailings. The measured residual oxygen content of 2% persisting to a depth of 6 m represents the detection limit for the method. In contrast, the diffusion of oxygen in coarse tailings is higher and the oxygen concentration profile extends to a greater depth (see Figure 5).

Predicted and observed dissolved calcium concentrations are shown in Figure 6 (location T3) and Figure 7 (location T5). The good agreement between measured and predicted values reflect the presence of gypsum as the controlling solid phase. Figure 8 and Figure 9 show the observed and predicted sulfate concentrations at location T3 and location T5, respectively. As in case of calcium, the simulated results agree fairly well with the observations.

The pH simulations for the porewaters at T3 and T5 are shown in Figure 10 and Figure 11. The predicted pH values are within 1 pH unit of observations, which is reasonable in view of the ionic imbalance noted in the field data. The stepwise profile of the simulations reflect the presence of buffering minerals. Both the measured and simulated pH values are lower in coarse tailings (Figure 11) indicating greater depletion of solid buffers under higher acid flux. This is consistent with the higher gaseous porosity, hence higher rate of oxygen transport in coarse tailings. In fine tailings (Figure 10) the surface pH is slightly higher than at a depth of 1 m. This is due to the initially lower pyrite content and the depletion of pyrite from the upper 0.3 m.

The residual pyrite content after 13 years of oxidation is shown in Figure 12. The oxidizing front (pyrite depletion) at T3 is sharper, which is consistent with the

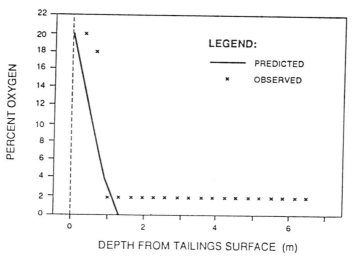

Figure 4. Observed and simulated oxygen concentrations in fine tailings, location T3.

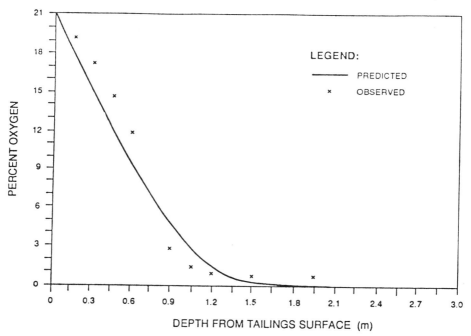

Figure 5. Observed and simulated oxygen concentrations in coarse tailings, location T5.

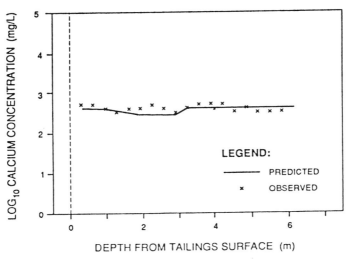

Figure 6. Observed and simulated calcium concentrations in fine tailings, location T3.

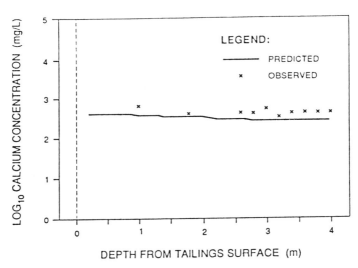

Figure 7. Observed and simulated calcium concentrations in coarse tailings, location T5.

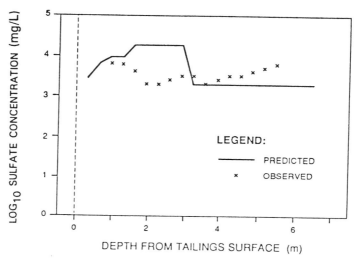

Figure 8. Observed and simulated sulfate concentrations in fine tailings, location T3.

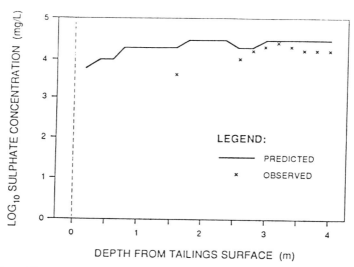

Figure 9. Observed and simulated sulfate concentrations in coarse tailings, location T5.

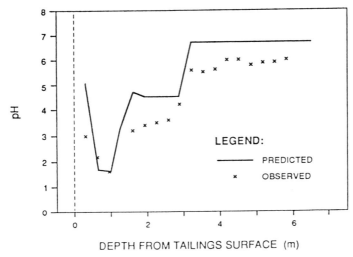

Figure 10. Observed and simulated pH in fine tailings, location T3.

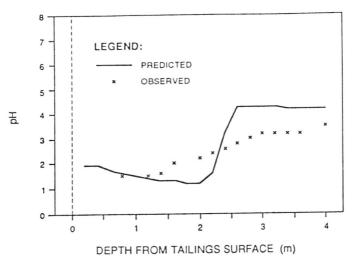

Figure 11. Observed and simulated pH in coarse tailings, location T5.

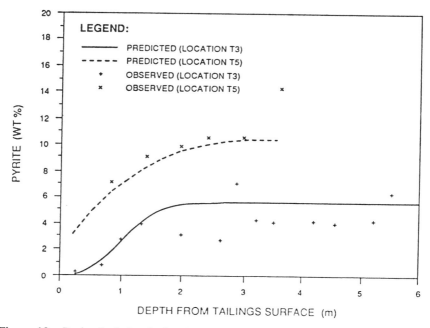

Figure 12. Pyrite depletion in fine (location T3) and coarse (location T5) tailings.

higher moisture content and higher oxygen gradient in the fine-grained tailings. The oxidizing zone after 13 years is approximately 1.5 m deep at T3 and 2.5 m deep at T5. The greater depth of the oxidizing zone in coarse tailings reflects higher oxygen transport.

Random Simulation Module

As with any predictive assessment, probabilistic analysis involves mathematical expressions (models) of some physical system, in the present case, sulfide tailings. Many variables are not known with certainty. The uncertainty arises from natural heterogeneity, normal experimental or measurement errors, or simply the lack of sufficient database. Consequently, it is impossible to describe the input variables by a single value. Rather, the state of knowledge (or ignorance) about any input variable can be described by a subjective probability distribution. The module allows the selection of a number of known statistical distributions ranging from simple triangular distribution to gaussian, log-normal, and beta frequency functions.

In the probabilistic assessment mode, several input parameters are specified as distributions. The distributed parameters are sampled by statistical sampling procedure (either Monte Carlo or latin hypercube sampling techniques) and the values drawn are entered into the component models. The process, which is referred as a trial, is repeated as often as desired. At least 100 trials are required to obtain reasonable outputs for up to 100-year simulations. Figure 13 is an example of probabilistic assessment of pH in tailings porewater.

The output variable (sulfate concentration with depth at a given time, for example) from such an analysis is treated as a sample of a collection of output variables. For these, the statistical mean, the variance, the probability density function, or the cumulative frequency function may be estimated as desired. The information may be interpreted as describing the output in terms of subjective probability. As an example, it may be used to evaluate the probability that the pH drops below a threshold limit in the porewater at a given time and depth.

Conclusions

The goal of the RATAP program is the application of fundamental kinetic and transport data from the laboratory to tailings in the field. In tailings, the predominant reactive minerals are pyrite and pyrrhotite. The mechanism and the kinetics of pyrite oxidation are fairly well known. Although most studies have examined the biological oxidation phenomena from a bioleach processing perspective, the more significant fundamental studies of microbial pyrite oxidation kinetics (28, 29, 36, 37) are relevant to the assessment of biological action on pyrite in tailings. In contrast to the voluminous literature on the oxidation of pyrite, kinetic studies on pyrrhotite, either as a sole reactant or in mixture, are scarce. Although the importance of galvanic interactions between pyrrhotite and other sulfide minerals has been recognized (29, 38, 39), only a single recent work (19) has focussed on the comparative kinetics of pyrrhotite oxidation in sufficient detail to be of any value in estimating biological activity in pyrrhotite containing tailings. Pyrrhotite oxidation kinetics have been a subject of recent laboratory studies by Nicholson and Scharer in this volume.

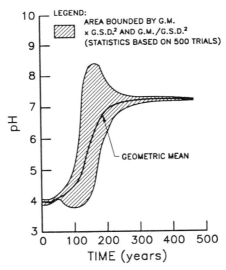

Figure 13. Probabilistic assessment of porewater pH.

RATAP has proven to be a useful analytical tool in assessing acid generation in pyritic tailings. The model has been successfully applied to establish the extent and duration of AMD, at several tailings sites. Comparisons of simulated porewater quality and solids concentrations with field data in the Elliot Lake area have shown excellent agreement. The application of the model at other sites may require scaling of the kinetic parameters. The scale factors can be easily established from laboratory tailings oxidation studies.

We acknowledge the invaluable editorial suggestions and comments by Dr. C. Alpers, of the USGS.

Literature Cited

1. Senes Consultants Ltd., Estimation of the Limits of Acid Generation by Bacterially Assisted Oxidation in Uranium Mill Tailings; CANMET DSS File # 15SQ 23241-5-1712; 1984.
2. Nordstrom, D. K., Aqueous Pyrite Oxidation and the Consequent Formation of Secondary Iron Minerals in Acid Sulphate Weathering, Kittrick, J.A., Fenning, D.S., and Hossner, L.R., Eds. Soils Sci. Soc. Am.: 1982, 37-56.
3. Sillen, L. G., Martell, A. E. Stability Constants of Metal Ion Complexes. Special Publication No. 17. The Chemical Society, London, U.K.: 1964.
4. Smith, R. M.; Martell, A. E. Critical Stability Constants. Inorganic Complexes. Plenum Press, New York and London 1976, Vol. 4, 257 pp.
5. Boorman, R. S.; Watson, D. M. CIM Bulletin 1976, 69, 86-96.
6. Motekaitis, R. J.; and Martell, A. E. The Determination and Use of Stability Constants. VCH Publishers, New York, NY: 1988, 216 pp.
7. Blowes, D. W.; Reardon, E. J.; Jambor, J. L.; Cherry, J. A. Geochim. Cosmochim. Acta 1991, 55, 965-978.
8. Steger, H. F.; Desjardins, L. E. Chem. Geol. 1978, 23, 225-233.
9. Hoffmann, M. R.; Faust, B. C.; Panda, F. A.; Koo H. A.; Tsuchiya, H. M. Appl. Environ. Microbiol. 1981, 42, 259-271.
10. Kendall, M.; Stuart, A. The Advanced Theory of Statistics, Griffin and Co., London, U.K.: 1977, Vol. 2, pp 176-178.
11. Press, W. H.; Flannery, B. P.; Teukolsky, S. A.; VeHerling, W. T. Numerical Recipes: The Art of Scientific Computing, Cambridge University Press, Cambridge, U.K.: 1986, 818 pp.
12. Hillel, D. Fundamentals of Soil Physics, Academic Press, New York, NY: 1980, 282 pp.
13. Reardon, E. J.; Moddle, P. M. Uranium 1985, 2, 111-131.
14. Lowson, R. T. Chemical Reviews 1982, 82, No. 5, 461-497.
15. Pearse, M. J. Mineral Processing and Extractive Metallurgy 1980, 89, C26-C36.
16. Lundgren, D. D.; Silver, M. Ann. Rev. Microbiol. 1980, 34, 262-283.
17. Torma, A. E. Biotechnology Advances 1983, 1, 73-80
18. Torma, A. E.; Banhegyi, I. G. Trends in Biotechnol. 1984, 2, 13-15.
19. Ahonen, L.; Tuovinen, O. H. Appl. Environ. Microbiol. 1991, 57, 138-175.
20. Morth, A. H.; Smith, E. E. Kinetics of the Sulfide-to-Sulphate Reaction. Proc. 151st National Meeting, Am. Chem. Soc., Pittsburgh, PA: 1966.

21. Nicholson, R. V. Pyrite Oxidation in Carbonate-Buffered Systems: Experimental Kinetics and Control by Oxygen Diffusion in Porous Medium. Ph. D. Thesis, University of Waterloo, Waterloo, Ontario, Canada, 1984.
22. Senes Consultants Ltd. and Beak Consultants Ltd. Adaptation of the Reactive Acid Tailings Assessment Program to Base Metal Tailings. CANMET DSS #15SW-2344-7-9208, 1988.
23. Scharer, J. M.; Garga, V.; Smith R.; Halbert, B. E. Use of Steady State Models for Assessing Acid Generation in Pyrite Mine Tailings. Proceedings of the 2nd International Conference on the Abatement of Acidic Drainage, Montréal, CANMET: 1991, Vol. 2, 211-230.
24. Brock, T. D. Appl. Microbiol. 1975, 29, 495-501.
25. Torma, A. E.; Walden, C. C.; Duncan D. W.; Branion, R.W.R. Biotechnol. Bioeng. 1972, 15, 777-786.
26. Harries, J.R.; Ritchie, A.I.M. The Microenvironment Within Waste Rock Dumps Undergoing Pyritic Oxidation. Proc. Int. Symp. on Biohydrometallurgy, Calgary, 1983.
27. Mehta, M.; Murr, L. E. Biotechnol. Bioeng. 1982, 24, 919-940.
28. Ahonen, L.; Hiltunen, P; Tuovinen, O.H. The Role of Pyrrhotite and Pyrite in the Bacterial Leaching of Chalcopyrite Ores. Fundamental and Applied Biohydrometallurgy. Lawrence, R.W.; Branion, R.M.R.; Ebner H.G. Eds., Elsevier, Amsterdam, 1986, 13-22.
29. Tributsch, H.; Bennett, J. C. J. Chem. Tech. Biotechnol 1981, 31, 627-635
30. Cathles, L. M.; Schlitt, W. J. A Model of the Dump Leaching Process that Incorporates Oxygen Balance, Heat Balance and Two Dimensional Air Convection. Proc. of the Symposium on Leaching and Recovering Copper from As-Mined Materials. Schlitt, W.J. Ed., Las Vegas, NV: 1980, 9-15.
31. Davis, G. B.; Ritchie, A.I.M. Applied Math. Modelling 1986, 10, 314-322.
32. Jaynes, D. B. Modelling Acid Mine Drainage from Reclaimed Coal Stripmines. Proc. 2nd International Conference on the Abatement of Acidic Drainage, Vol. 2. Montréal, CANMET: 1991, 191-210.
33. Parkhurst, D. L.; Thorstenson, D. C.; Plummer, L. N. PHREEQE - A Computer Program for Geochemical Calculations. US Geological Survey, Water Resource Investigations, 1980, 80-96.
34. Smyth, D. J. A. Hydrogeological and Geochemical Studies Above the Water Table in an Inactive Tailings Impoundment near Elliot Lake, Ontario. M. Sc. Project Report, University of Waterloo, Waterloo, Ontario, Canada, 1981.
35. Dubrovsky, M.M.; Cherry, J.A.; Reardon, E.J.; Vivyurka, A.J. Can. Geotech. J. 1985, 22, 110-128.
36. Arkesteyn, G.J.M.W. Antonie van Leeuwenhoek 1979, 45, 423-435.
37. Myerson, A. S. Biotechnol. Bioeng. 1981, 23, 1913-1416.
38. Rossi, G.; Trois, P.; Visca, P. In Situ Pilot Semi-Commercial Bioleaching Test at the San Valentino Di Predioi Mine (Northern Italy). In Fundamental and Applied Biohydrometallurgy, Lawrence, R.W.; Brannion, R.M.R.; Ebner, H.G. Eds., Elsevier, Amsterdam, 1985, 175-189.
39. Steffen, Robertson and Kirsten, Ltd. Acid Rock Drainage Technical Guide. Acid Mine Drainage Task Force, Mend Prediction Committee, CANMET, Ottawa, 1985, 285 pp.

RECEIVED September 13, 1993

Chapter 12

Time−Space Continuum Formulation of Supergene Enrichment and Weathering of Sulfide-Bearing Ore Deposits

Peter C. Lichtner[1]

Mineralogisch-petrographisches Institut, Universität Bern, Baltzer-Strasse 1, CH−3012 Bern, Switzerland

A time-space continuum description of solute transport in a porous medium is applied to weathering and enrichment of copper-bearing sulfide ore deposits for three different host rocks consisting of sandstone, granite and limestone. Mineral reactions are described by kinetic rate laws, and homogeneous reactions within the aqueous phase are assumed to be in local equilibrium. The effect of pH buffering by the different host rock minerals is investigated as the protore is oxidized by downward percolating rainwater. The resulting supergene weathering profiles for the sandstone and granite host rock are found to be similar to the case without gangue minerals present. By contrast, the limestone host rock leads to an entirely different supergene profile.

Oxidation of sulfides from operating or abandoned mines is becoming a major environmental problem. Leaching of toxic metals under extremely acidic conditions may result in severe damage to the environment. A thorough understanding of sulfide oxidation in natural systems is therefore essential to better understand and control such processes.

This work focuses on a quantitative description of supergene enrichment of a copper-bearing protore accompanying weathering of different host rocks. Supergene weathering is influenced by a combination of effects including the presence of an unsaturated zone above the water table, as well as downward percolating oxygenated fluid below the water table. The presence of a water table, and its fluctuations with time, overprint and complicate metasomatic effects which are responsible for a continuous, downward movement with time of zones of secondary mineralization. During supergene weathering, primary protore minerals are transformed into secondary copper-bearing minerals which form an enrichment blanket with increased copper grade

[1]Current address: Southwest Research Institute, Center for Nuclear Waste Regulatory Analyses, 6220 Culebra Road, San Antonio, TX 78238−5166

compared to the original protore lying below the water table. The copper grade is observed to increase sharply at the top of the enrichment blanket. Although extensive field observations have been carried out on supergene enrichment occurring in different host rocks (1-12), unfortunately only very few studies have reported detailed mineral modal compositions of reaction zone sequences within the enrichment blanket (2).

Recently Lichtner and Biino (13) investigated supergene enrichment processes in the absence of gangue minerals based on a first-principles, quantitative description of fluid transport and chemical reaction of minerals and aqueous species in a porous medium. Qualitative agreement was obtained between field observations and predictions. These results are extended in this work to include the effect of gangue minerals on enrichment. Comparison is made between a sandstone, granite and limestone host rock. The purpose of this exercise is to show the relative effects of different host rocks for a given set of kinetic rate constants, and not to fit field observations of the individual rock types. Calculations are carried out for pure advective transport using the quasi-stationary state approximation to mass conservation equations representing fluid transport and reactions with minerals (14, 15), based on a continuum representation of a porous medium (16). This approximation leads to a multiple reaction path formulation of transport and reaction, and enables the governing equations to be integrated over geologic time spans for complex, multicomponent systems (14, 15).

Constraints and Controls on Sulfide Oxidation

The general structure of the supergene weathering profile is well-documented by extensive field observations (1-12). The profile consists of a leached zone at the surface, underlain by gossan and oxide zones, a zone enriched in copper lying below the water table, and finally the unaltered protore. The grade and thickness of the enriched zone depend on a number of factors which must be taken into account when modeling the supergene process.

Several factors control the rate of sulfide oxidation. The presence and thickness of a soil zone at the surface of the weathering column determine the amount of oxygen available for oxidation reactions both above and below the water table. The depth of the water table is an important controlling factor during supergene enrichment. Above the water table oxygen can be replenished by transport from the atmosphere, as it is consumed by oxidation reactions. Below the water table, however, oxygen is rapidly consumed. Competition between depletion by sulfide oxidation and diffusion through air-filled pore spaces above the water table control the concentration of dissolved oxygen entering the saturated zone. The gas-phase-oxygen diffusion coefficient is strongly dependent on the water-saturation index and tortuosity of the air-filled pore spaces (17, 18). The rate of diffusion decreases rapidly through the capillary fringe at the water table. In addition the reaction rate for oxidation of pyrite and other sulfides may be proportional to the oxygen fugacity raised to a power as suggested by Nicholson et al. (19).

Although traditionally the presence of a water table and its movement with time has been assumed to be a major factor in the supergene enrichment process, and even ascribed to determining the width of the enrichment blanket itself (5), transport of dissolved oxygen below the water table can also be an important factor when considered over geologic time spans. As demonstrated by Lichtner and Biino (13), the width of the enrichment blanket can be described as a function of the time in response to downward percolating, oxygenated water. Focusing effects of fluid transport caused by regions of high permeability could enhance the local fluid-flow velocity and therefore the amount of dissolved oxygen available to oxidize the protore below the water table.

Another important factor controlling the stability of secondary copper-bearing minerals during enrichment is the pH of the oxidizing fluid. The pH is influenced by the presence of gangue minerals as well as the protore ratio of pyrite (Py) to chalcopyrite (Ccp). For example, a limestone host rock leads to a completely different supergene weathering profile compared to a granitic host rock. The py:ccp ratio affects the pH during oxidation of chalcopyrite above the water table; a greater amount of pyrite present leads to more acid conditions. A smaller Py:Ccp ratio leads to a higher value of the pH during chalcopyrite oxidization. Thus the Py:Ccp ratio affects the pH-dependent stability of mineral alteration products, such as jarosite and various copper-bearing minerals, which form during the enrichment process.

Continuum Representation of Mass Transport and Reaction in a Porous Medium

A continuum representation of solute transport and chemical reaction in a porous medium necessarily involves a significant simplification of the actual physical system. In this formulation the micro-environment is replaced by a spatially averaged macro-system defined in terms of bulk compositions of fluid and rock obtained as averages over a representative elemental volume (REV) which locally characterizes the system (16). Details at a micro-scale involving individual mineral grains are thereby lost. The extent to which this simplification is able to capture the essential features of a natural system can only be decided by comparing the results of numerical calculations with actual field observations.

The formulation of the supergene enrichment process employed here is based on a first-principles description. The theory is formulated in terms of fundamental physicochemical properties including kinetic and thermodynamic constants, Darcy flow velocity, and mineral grain size (surface area). Once initial and boundary conditions are specified for the system, its evolution in time and space is completely determined. In particular the width and grade of the enrichment blanket is completely determined by the governing equations and cannot be externally imposed on the system, as has been done by Ague and Brimhall (20), for example. The formulation employed here, in which no *a priori* assumption is made regarding the size of the enrichment blanket, is referred to mathematically as a moving boundary problem.

Calculations are carried out using the computer code **MPATH** which incorporates pure advective transport coupled to reactions with minerals and aqueous species (15), based on the quasi-stationary state approximation. For pure advective transport the solution to the transport-reaction problem can be represented by a sequence of reaction paths, or stationary states, referred to as the multiple reaction path approach (14, 15). For fluid flow along a one-dimensional streamline each reaction path or stationary state obeys a system of ordinary differential equations, formulated for a complete set of primary species, of the form

$$u\frac{d}{dx}[\theta \Psi_j(x,t)] = -\sum_m v_{jm} I_m(x,t),\qquad(1)$$

where the coordinate x denotes the position along the flow path and the time t enters this equation as a parameter referring to the state of alteration of the host rock. The quantity u denotes the Darcy fluid velocity, θ denotes the water-saturation index defined as the fraction of the total porosity which is filled by water, I_m designates the reaction rate of the mth mineral, and the matrix element v_{jm} refers to the stoichiometric reaction coefficient of the jth primary species in the mth mineral. The quantity Ψ_j refers to the generalized concentration of the jth primary species defined by the expression

$$\Psi_j = C_j + \sum_i v_{ji} C_i,\qquad(2)$$

where C_j and C_i denote the concentrations of the jth and ith species, respectively, and v_{ji} refers to the stoichiometric coefficient of the jth primary species in the ith complex. The generalized concentration may take on positive or negative values (16). Homogeneous equilibrium is assumed within the aqueous phase with the concentration of aqueous complexes related to the concentrations of primary species by the mass action equations

$$C_i = \gamma_i^{-1} K_i \prod_j (\gamma_j C_j)^{-v_{ji}},\qquad(3)$$

where the product is carried out over the primary species, γ_l refers to the activity coefficient of the subscripted species, and K_i denotes the equilibrium constant for the corresponding reaction. An important consequence of this assumption is the implied simultaneous equilibrium of all redox couples within the aqueous phase. This may not be a good assumption for redox reactions involving iron and sulfur if the necessary mediating bacteria are not present.

Alteration of the host rock at each point x is described by the mineral mass-transfer equation

$$\frac{\partial \phi_m}{\partial t} = V_m I_m,\qquad(4)$$

where $\phi_m(x,t)$ denotes the volume fraction of the mth mineral with molar volume V_m. After obtaining the solution of the stationary-state transport equations (1) at time t, equation (4) is integrated explicitly to give

$$\phi_m(x, t+\Delta t) = \phi_m(x,t) + \Delta t V_m I_m(x,t),\qquad(5)$$

determining the alteration of the host rock at time $t+\Delta t$. The time step Δt represents the time interval between successive stationary states. By successively solving equations

(1) and (5), the evolution of the system can be computed to any desired time. Because the time step Δt is not limited by stability considerations, as in transient formulations, it is possible to integrate the transport equations over geologic time spans for arbitrarily complex multicomponent systems.

The stationary-state transport equation at time t is subject to initial and boundary conditions specifying the fluid composition at the inlet and the composition of the host rock, $\phi_m(x, t)$ at each point along the flow path. For the very first reaction path, corresponding to $t=0$, this latter quantity is given by the composition of the unaltered host rock. For subsequent paths it is obtained from the previous reaction path by evaluating the integrated mineral mass-transfer equations (5). The composition of the inlet fluid is taken as rain water, possibly corrected for transport through a soil zone.

The presence of a water table adds an additional complicating factor to a quantitative description of the supergene enrichment process. Above the water table where oxygen can be replenished by oxygen in the atmosphere, either by diffusive transport or barometric pumping, oxidation proceeds more rapidly compared to the water-saturated zone below the water table where the supply of oxygen is limited by that dissolved in downward percolating groundwater. In the following it is assumed that transport of oxygen through air-filled pore spaces is much more rapid than its consumption by oxidation reactions of sulfide minerals. (See (13) for a discussion of the validity of this approximation.) As a consequence the concentration of dissolved oxygen above the water table is considered to be constant buffered by equilibrium with the atmosphere. This is in contrast to leached waste dumps of sulfide minerals where the increased surface area due to crushing may lead to very rapid oxidation rates compared to natural systems, resulting in significant depletion of oxygen and even in elevated temperatures of the dump (see Cathles and Ritchie, this volume). Below the water table, however, no constraint on the oxygen concentration is imposed externally, and oxygen is allowed to become rapidly depleted due to oxidation of sulfide-bearing minerals. The position of the water table, marking the transition between water-unsaturated and -saturated zones, is represented by a sharp boundary. For simplicity, the change in infiltration flux due the reduced water-saturation θ above the water table is not take into account.

The reaction of minerals with an aqueous solution is described by a pseudo-kinetic expression for the reaction rate I_m. The rate is taken to have the form:

$$I_m = -k_m s_m \left[1 - e^{-A_m/RT}\right], \tag{6}$$

based on transition state theory, where k_m denotes the kinetic rate constant, s_m refers to the mineral surface area per unit bulk volume of the porous medium, A_m designates the chemical affinity of the mth mineral, and R and T denote the gas constant and temperature, respectively. The term pseudo-kinetic is employed because, although the rate law given by equation (6) may not describe the actual reaction mechanism, nevertheless it contains local equilibrium as a limiting case and allows deviations from equilibrium to be investigated. The presence of the quantity in brackets containing the chemical affinity is important for ensuring that the reaction rate vanish at equilibrium.

According to this expression, near equilibrium the rate is proportional to the chemical affinity. For the simple form given above, the rate of dissolution for conditions far from equilibrium is equal to a constant and is independent of such compositional parameters as oxygen fugacity f_{O_2} and pH.

The form of the rate expression in equation (6) is assumed to hold for oxidation of chalcopyrite and pyrite above the water table. While certainly an over simplification (15), it leads to the largest possible rate of dissolution for given values of the rate constant and mineral surface area. For mineral precipitation far from equilibrium, the rate grows exponentially with the chemical affinity.

The surface area corresponding to primary minerals is allowed to decrease in proportion to the two-thirds power of the corresponding volume fraction according to the expression

$$S_m = S_m^0 \left(\frac{\phi_m}{\phi_m^0}\right)^{2/3}, \qquad (7)$$

where S_m^0 and ϕ_m^0 denote the initial surface area and volume fraction, respectively. For secondary mineral products the surface area is taken as constant. This result is based on simple geometric arguments and does not take into account the formation of etch pits, for example, which could result in an increase in surface area with mineral dissolution. With this constitutive relation for the change in surface area with reaction, the time τ_m for the mth mineral to dissolve completely is given by the expression (13, 14)

$$\tau_m = \frac{3\phi_m^0}{k_m S_m \overline{V}_m}. \qquad (8)$$

This relation may be used to estimate the time of formation of an enrichment blanket if the kinetic rate constants, surface areas and initial volume fractions are known for the primary protore minerals (13). Conversely, if the time of formation of the enrichment blanket can be estimated from field observations, then the product of the rate constant and surface area can be obtained.

Model Calculations

Model calculations are carried out for three different host rocks consisting of a sandstone, granite and limestone. A list of primary and secondary minerals included in the calculations are presented in Table I along with the corresponding mineral formulas and abbreviations used the figures. Thermodynamic data used in the calculation are taken from the EQ3/6 data base (21), version R10. The modal compositions of the different host rocks together with the effective rate constants κ_m, defined as the product of the kinetic rate constant and the mineral surface area, used in the calculations are given in Table II. Because of the large uncertainty in the surface area of the reacting

Table I. Mineral formulas and abbreviations

mineral	ore formula	abbre.	mineral	gangue formula	abbre.
anilite	$Cu_{1.75}S$	Anl	alunite	$KAl_3(OH)_6(SO_4)_2$	Alu
bornite	Cu_5FeS_4	Bn	annite	$KFe_3AlSi_3O_{10}(OH)_2$	Ann
brochantite	$Cu_4(SO_4)(OH)_6$	Bro	beidellite-Ca	$Ca_{0.165}Al_{2.33}Si_{3.67}O_{10}(OH)_2$	Ca-Beid
chalcocite	Cu_2S	Cc	beidellite-H	$H_{0.33}Al_{2.33}Si_{3.67}O_{10}(OH)_2$	H-Beid
chalcopyrite	$CuFeS_2$	Ccp	beidellite-K	$K_{0.33}Al_{2.33}Si_{3.67}O_{10}(OH)_2$	K-Beid
copper	Cu	Cu	beidellite-Na	$Na_{0.33}Al_{2.33}Si_{3.67}O_{10}(OH)_2$	Na-Beid
covellite	CuS	Cv	calcite	$CaCO_3$	Cal
cuprite	Cu_2O	Cpr	gibbsite	$Al(OH)_3$	Gbs
digenite	$Cu_{1.765}S$	Dig	goethite	$FeOOH$	Goe
djurleite	$Cu_{1.9345}S$	Djr	gypsum	$CaSO_4$	Gyp
malachite	$Cu_2CO_3(OH)_2$	Mal	jarosite	$KFe_3(SO_4)_2(OH)_6$	Jar
tenorite	CuO	Tnr	K-feldspar	$KAlSi_3O_8$	Kfs
pyrite	FeS_2	Py	laumontite	$CaAl_2Si_4O_{12}\cdot H_2O$	Lau
siderite	$FeCO_3$	Sid	muscovite	$KAl_3Si_3O_{10}(OH)_2$	Mus
			kaolinite	$Al_2Si_2O_5(OH)_4$	Kln
			quartz	SiO_2	Qtz
			$SiO_{2(am)}$	SiO_{2am}	SiO_2
			magnetite	Fe_3O_4	Mag
			plagioclase (An_{20})	$Ca_{0.2}Na_{0.8}Al_{1.2}Si_{2.8}O_8$	Plag

minerals, more or less arbitrary effective rate constants were used in the calculations. For calculation involving long time spans this is not as arbitrary as it may seem. Although strictly true only in the absence of scale-dependent features such as the presence of a water table, it can be shown that for sufficiently long time spans the results become independent of the choice of rate constants and scale to the local equilibrium limit (22). However, the calculations presented below involve rather short time spans, and hence these results are probably still sensitive to the choice of rate constants and surface areas. The effective rate constants for secondary minerals are assumed to be equal to the value of 10^{-15} mol cm^{-3} sec^{-1}. The effective rate constant for calcite is taken 10 times larger for the limestone host rock as compared to the sandstone rock. This is reflected in a greater abundance of calcite in limestone compared to sandstone and an assumed smaller grain size. For each rock type the protore contains 4% by volume chalcopyrite and 2% pyrite.

A water table is assumed to be present at a depth of 50 m. A Darcy flow velocity of 1 m y^{-1} is assumed. Homogeneous equilibrium is assumed within the aqueous solution. The composition of the inlet fluid infiltrating at the top of the weathering column is assumed to have a pH of 4.5 with $\log P_{CO_2} = -2$ and $\log P_{O_2} = -0.7$. Dilute concentrations of the species K$^+$, Na$^+$, Ca^{2+}, Al^{3+} and SiO$_2$ are assumed. The concentration of Fe^{2+} is determined by goethite equilibrium, and the assumption of homogeneous equilibrium within the aqueous solution, requiring simultaneous equilibrium of all redox couples. The redox state is specified by the oxygen fugacity.

Variation in pH with Depth. The change in pH with depth can be very different depending on the type of host rock. Shown in Figure 1 is a comparison of the pH profile for sandstone, granite and limestone host rocks, as well as the case without gangue minerals present, plotted as a function of the logarithm of the depth in meters. The profiles correspond to a single reaction path associated with the first packet of fluid to pass through the weathering column. As is apparent from the following discussion, the first path need not be representative of the chemical evolution of the system, even over relatively short time spans. The absence of gangue minerals results in the lowest pH along the flow path because no buffering of the pH takes place. The profiles for the granite and sandstone host rocks are very similar to the case without gangue minerals present. In these three cases the pH first decreases caused by the oxidation of pyrite and chalcopyrite, and then increases further below the water table as the fluid comes to equilibrium with the primary minerals. The sharp decrease in pH at the water table results as secondary iron hydroxide, taken as goethite, ceases to form. This is caused by the reaction path crossing into the aqueous Fe^{2+} window separating the goethite and pyrite stability fields. It is possible to write qualitative overall reactions to describe this situation according to the following reaction with goethite present:

$$\text{FeS}_2 + \frac{15}{4}\text{O}_2 + \frac{5}{2}\text{H}_2\text{O} \rightarrow \text{FeOOH} + 2\text{SO}_4^{2-} + 4\text{H}^+, \tag{9}$$

and the reaction after goethite ceases to precipitate:

Table II. Initial volume fractions ϕ_m^0 and effective kinetic rate constants κ_m (mol cm^{-3}sec^{-1}) for primary gangue minerals corresonding to sandstone, granite, and limestone host rocks used in the calculations for supergene enrichment of a copper-bearing protore

mineral	ϕ_m^0	κ_m	ϕ_m^0	κ_m	ϕ_m^0	κ_m
	sandstone		granite		limestone	
K-feldspar	0.10	5.×10^{-16}	0.20	1.×10^{-15}	—	—
muscovite	0.02	1.×10^{-16}	0.05	1.×10^{-16}	—	—
kaolinite	0.03	4.5×10^{-16}	—	—	—	—
quartz	0.60	1.×10^{-16}	0.20	1.×10^{-16}	0.05	1.×10^{-16}
calcite	0.05	1.×10^{-14}	—	—	0.75	1.×10^{-13}
magnetite	0.02	1.×10^{-15}	0.02	1.×10^{-15}	—	—
plag (An$_{20}$)	—	—	0.35	1.×10^{-15}	—	—
annite	—	—	0.1	1.×10^{-15}	—	—

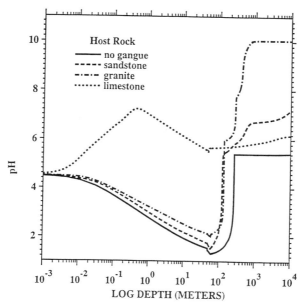

Figure 1. The pH resulting from the oxidation of pyrite- and chalcopyrite-bearing protore plotted as a function of the logarithm of depth for different host rocks corresponding to sandstone, granite, limestone, and in the absence of gangue minerals. A water table is present at a fixed depth of 50 meters.

$$FeS_2 + 14Fe^{3+} + 8H_2O \rightarrow 15Fe^{2+} + SO_4^{2-} + 16H^+. \qquad (10)$$

The latter reaction is assumed not to involve dissolved oxygen, which is completely depleted from solution at this depth in the profile. Clearly the latter reaction results in the larger pH decrease, per mole of pyrite oxidized.

The granite host rock results in the highest final value of the pH of approximately 10, followed by sandstone with a final value for the pH of 7, limestone with a final pH of approximately 6.2, and the case with gangue minerals absent resulting in a final pH value of approximately 5.4. The pH profile for the limestone rock is entirely different. In this case dissolution of calcite results in a rapid increase in pH reaching a maximum of approximately 7, and then the pH begins to decrease to about 5.5 at the water table. The initial increase in pH and maximum value reached above the water table are controlled by the relative reaction rates for calcite, pyrite and chalcopyrite.

Sandstone Host Rock. First the modeling results for a sandstone host rock are considered in detail. The mineral volume fractions are shown in Figure 2 plotted as a function of the logarithm of depth corresponding to an elapsed time of 5,000 years. As oxygenated water percolates down the weather column, chalcopyrite is oxidized releasing copper into solution. Just below the water table, where the oxygen fugacity is no longer buffered by the atmosphere, chalcocite precipitates as the oxygen fugacity plummets to zero caused by further oxidation of chalcopyrite and pyrite producing a redox trap for aqueous copper species. This process results in the formation of an enrichment blanket as the copper mobilized into solution from the oxidation of chalcopyrite reprecipitates in the form of secondary copper sulfides. The top of the enrichment blanket consists of chalcocite and minor amounts of digenite, djurleite and anilite, underlain by covellite and bornite. The chalcocite zone is enriched in copper compared to the protore, whereas in the remainder of the blanket copper is conserved in the solid phases and the copper content (moles of copper per representative elemental volume of bulk rock) is identical to that of the protore.

Secondary pyrite precipitates below the chalcocite zone in the upper portion of the blanket. Siderite begins to precipitate at the base of the enrichment blanket with a maximum in the volume fraction at the top of the unaltered protore. Goethite stops precipitating just below the water table, and continues to precipitate at greater depth but at a greatly reduced rate. Jarosite precipitates above the water table, whereas alunite forms both above and below the water table. Magnetite dissolves both above and below the water table. Kaolinite dissolves at the top of the weathering column and precipitates below the water table. Gibbsite, not shown in Figure 2, forms at the surface of the weathering column. Gypsum precipitates both below and above the water table with a peak in the volume fraction coinciding with the peak in the calcite dissolution rate, marked by the inflection point in the calcite volume fraction.

These results compare favorably with field observations of Alpers and Brimhall (2) of the porphyry copper deposit at La Escondida, Chile. At La Escondida the protore also contains bornite as well as chalcopyrite, and the gangue mineral paragenesis is much more complex (2). These authors observed a similar vertical paragenesis of copper sulfide alteration within the enrichment blanket of Cc+Djr | Dig+Anl | Cv |

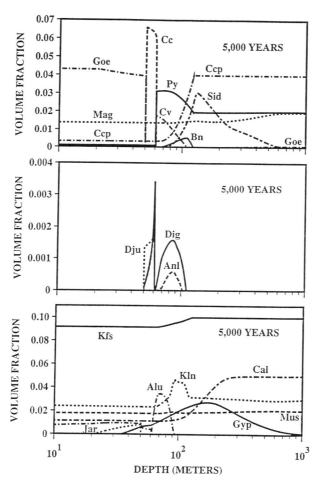

Figure 2. The volume fractions of secondary and primary minerals plotted as a function of the logarithm of depth for supergene enrichment of a sandstone host rock containing chalcopyrite and pyrite with a water table present at a depth of 50 meters.

Cpp+Bn, as obtained in the numerical calculations. In addition they reported formation of jarosite above the water table, and alunite both above and below the water table in agreement with the calculations. Both secondary kaolinite and gypsum were observed consistent with the calculations.

The pH profile is plotted as a function of depth in Figure 3 for different times indicated in the figure. For earlier times the pH first decreases as pyrite is oxidized and then increases as equilibrium is reached with the sandstone host rock. For later times, after pyrite has completely dissolved above the water table, the pH initially increases with increasing depth resulting from silicate hydrolysis reactions. With increasing time a pH hydrolysis front propagates down the weathering column as becomes apparent by comparing the pH profiles for 25,000 and 50,000 years.

The effect of the pyrite rate constant on the pH profile in the supergene weathering of a sandstone host rock is shown in Figure 4 where the pH is plotted as a function of distance for different pyrite rate constants. Calculations are presented for a single reaction path using pyrite kinetic rate constants of -13, -14, -15 and -16 for $\log k_{py}$ with k_{py} in units of mol cm^{-2} s^{-1}. According to equation (8), these values correspond to total dissolution times of τ_{py} = 800, 8,000, 80,000, and 800,000 years, respectively. The rate constants for all other minerals are the same as before. For $\log k_{py}$ smaller than -15 little change in the profile occurs below the water table. Above the water table increasing the pyrite rate constant relative to all other minerals leads to a decrease in the pH.

Granite Host Rock. The situation for a granite host rock is shown in Figures 5a and 5b corresponding to ore and gangue minerals respectively. Mineral volume fractions are plotted as a function of the logarithm of the depth in meters. The profiles for the ore minerals are similar to the case of the sandstone host rock shown in Figure 2. In this case a minor amount of native copper also forms. Secondary weathering products include beidellite (labeled Ca, H, K, and Na) and laumontite. Kaolinite precipitates at the top of the weathering column. Jarosite forms above the water table. However, alunite is present both above and below the water table. Gypsum did not form in this case. These results are also consistent with the field observations of Alpers and Brimhall (2) of the porphyry copper deposit at La Escondida, Chile.

The pH profile for different times is shown in Figure 6. The final pH approaches the value of approximately 10.75, typical for aluminosilicate hydrolysis. This value is similar to that observed in the Stripa granite (23). As in the case of the sandstone host rock the pH first decreases resulting from pyrite oxidation, and then increases after pyrite completely disappears.

Limestone Host Rock. The behavior of the enrichment process for the case of a limestone host rock is completely different from either the sandstone or granite host rocks considered previously. The mineral volume fractions are shown in Figure 7 for an elapsed time of 5000 years. In this case malachite and to a lesser extent tenorite precipitate above the water table. The appearance of malachite and azurite is typical of limestone host rocks (10). Although azurite was included in the calculation it did not form, presumably because of the concentrated copper concentrations ($\approx 0.1 M$) and high

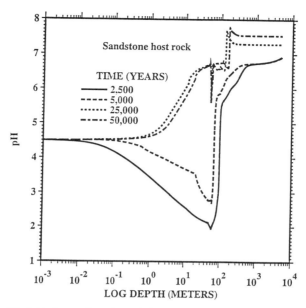

Figure 3. The pH plotted as a function of the logarithm of depth for different times for supergene enrichment of a sandstone host rock for the same conditions as in Figure 2.

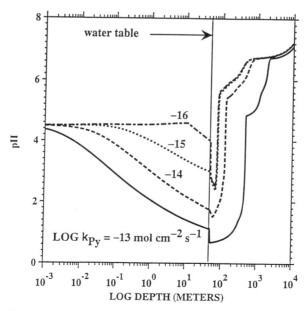

Figure 4. The pH plotted as a function of the logarithm of depth for different pyrite rate constants for supergene enrichment of a sandstone host rock for the same conditions as in Figures 2 and 3.

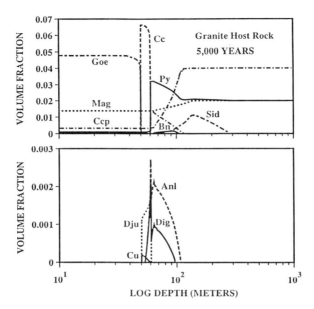

Figure 5a. The volume fractions of ore minerals plotted as a function of the logarithm of depth for supergene enrichment of a granite host rock containing chalcopyrite and pyrite with a water table present at a depth of 50 meters.

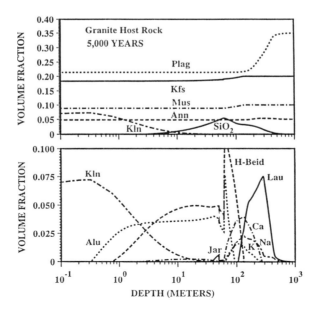

Figure 5b. The volume fractions of gangue minerals plotted as a function of the logarithm of depth for the same conditions as in Figure 5a.

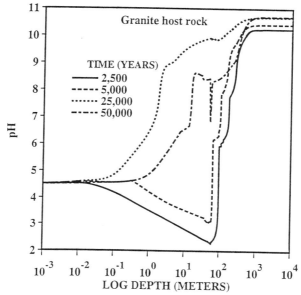

Figure 6. The pH plotted as a function of the logarithm of depth for supergene enrichment of a granite host rock for the same conditions as in Figure 5.

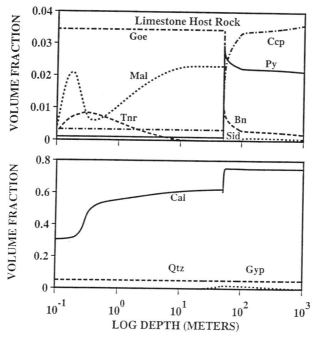

Figure 7. The volume fractions of secondary and primary minerals plotted as a function of the logarithm of depth for supergene enrichment of a limestone host rock containing chalcopyrite and pyrite with a water table present at a depth of 50 meters.

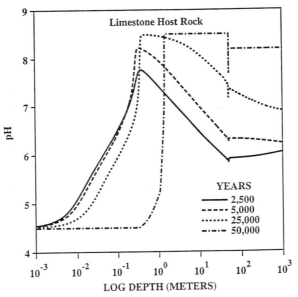

Figure 8. The pH plotted as a function of the logarithm of depth for supergene enrichment of a limestone host rock for the same conditions as in Figure 7.

P_{CO_2} that are required according to solubility considerations. Possibly by decreasing the kinetic rate constant for malachite, allowing the concentration of copper in solution to increase, azurite would form. Bornite forms below the water table as the only secondary copper sulfide. Minor amounts of siderite form below the water table.

The absence of other copper minerals such as chalcocite and covellite, typical for the sandstone and granite host rocks, is related to the steep rise in pH as shown in Figure 8. The pH behavior is dependent on the relative rate constants chosen for minerals pyrite, chalcopyrite and calcite. In this calculation the pyrite:calcite ratio for the effective rate constants is 1:10. By comparison, for the sandstone host rock a ratio of 1:1 was used. Decreasing this ratio could yield precipitation of chalcocite, but its presence would be ephemeral. As chalcopyrite completely dissolves above the water table, with increasing time malachite and tenorite must also eventually completely disappear above the water table. With increasing time a pH front propagates downward through the weathering column, as shown in Figure 8, as calcite dissolves.

Conclusion

Calculations presented for supergene enrichment of three different host rocks appear to be in qualitative agreement with field observations. Both sandstone and granite host rocks result in formation of similar enrichment blanket profiles submerged below the water table. An increase in copper grade occurs at the top of the blanket. These results are, furthermore, similar to the case in which gangue minerals are absent (13). A limestone host rock leads to a completely different profile with the formation of malachite and tenorite above the water table and bornite below the water table. As is apparent from the above calculations, the first reaction path is *not* representative of the chemical evolution of the system, even over relatively short time spans. This is because of the rapid alteration of the host rock and its replacement by secondary alteration products which vary with depth. This affects both the pH and concentrations of solute species, and hence the reaction path followed by the system.

Acknowledgments

The author would like to thank Giuseppe Biino, Victor Balashov and Carl Steefel for helpful discussions, as well as two anonymous reviewers and Charles Alpers for their comments which greatly improved the content of the manuscript.

Literature Cited

1. Alpers, C. N.; Brimhall, G. H. Geol. Soc. America Bull. 1988, 100, 1640-1656.
2. Alpers, C. N.; Brimhall, G. H. Econ. Geol. 1989, 84, 229-254.
3. Anderson, C. A.; Scholz, E. A.; Strobell, J. J. U.S. Geol. Survey Prof. Paper 1955, 278, 103 pp.
4. Anderson, J. A. (1982) In Advances in the geology of porphyry copper deposits, southwestern North America; Titley, S. R., ed.; Univ. Arizona Press: Tucson, AZ; pp. 275-295.

5. Brimhall, G. H.; Alpers, C. N.; and Cunningham, A. B. Econ. Geol. 1985, 80, 1227-1256.
6. Emmons, W. H. U.S. Geol. Survey Prof. Paper 1917, 625, 530 pp.
7. Evans, H. T. Jr. Am. Min. 1981, 66, 807-818.
8. Guilbert, J. M.; Park, C. F. The geology of ore deposits; W.H. Freeman and Co.: San Francisco, CA; 1988; 985 pp.
9. Schwartz, G. M. In Advances in the geology of porphyry copper deposits, southwestern North America; Titley, S. R., ed.; Univ. Arizona Press: Tucson, AZ; pp. 41-50.
10. Schwartz, G. M. Econ. Geol., 1934, 55, 1-61.
11. Simmons, W. W.; Fowells, J. E. In Advances in the geology of porphyry copper deposits, southwestern North America; Titley, S. R., ed.; Univ. Arizona Press: Tucson, AZ; pp. 151-156.
12. Titley, S. R. Econ. Geol. 1978, 73, 765-784.
13. Lichtner, P. C.; Biino G. G. Geochim. Cosmochim. Acta 1992, 56, 3987-4013.
14. Lichtner, P. C. Geochim. Cosmochim. Acta 1988, 52, 143-165.
15. Lichtner, P. C. Wat. Res. Res. 1992, 28, 3135-3155.
16. Lichtner, P. C. Geochim. Cosmochim. Acta 1985, 49, 779-800.
17. Troeh, F. R.; Jabero, J. D.; Kirkham, D. Geoderma 1982, 27, 239-253.
18. Nicholson, R. V.; Gillham, R. W.; Cherry, J. A.; Reardon E. J. Can. Geotech. J. 1988, 26, 1-8.
19. Nicholson, R. V.; Gillham, R. W.; Reardon E. J. Geochim. Cosmochim. Acta 1988, 52, 1077-1085.
20. Ague, J. J.; Brimhall, G. H. Econ. Geol. 1989, 84, 506-528.
21. Wolery, T. J. EQ3NR, a computer program for geochemical aqueous speciation-solubility calculations: theoretical manual, user's guide, and related documentation (version 7.0); Lawrence Livermore Nat. Lab.: Livermore, CA; 1992; UCRL-MA-110662 PT III, 246 pp.
22. Lichtner, P. C. Am. J. Sci. 1993, 293, 257-296.
23. Nordstrom, D. K.; Ball, J. W.; Donahoe, R. J.; Whittemore, D. Geochim. Cosmochim. Acta 1989 53, 1727-1740.

RECEIVED March 26, 1993

Solubility and Sorption Control in Formation of Sulfide-Oxidation Products

Chapter 13

Influence of Siderite on the Pore-Water Chemistry of Inactive Mine-Tailings Impoundments

C. J. Ptacek[1,2] and David W. Blowes[2]

[1]National Water Research Institute, 867 Lakeshore Road, P.O. Box 5050, Burlington, Ontario L7R 4A6, Canada
[2]Waterloo Centre for Groundwater Research, University of Waterloo, Waterloo, Ontario N2L 3G1, Canada

Siderite ($FeCO_3$) precipitation and dissolution can influence strongly the geochemistry of pore water associated with the alteration of sulfide minerals in mine tailings. Laboratory determinations of siderite solubility were made over a pH range of 4.1 - 6.5 and a $FeSO_4$ concentration range of 0.0 - 1.8 m. These conditions are representative of the geochemical zones where siderite has been observed to dominate the chemistry of tailings pore water. The laboratory data were analyzed with the Pitzer ion-interaction model, which is applicable over the large range in $FeSO_4$ concentration studied. The Pitzer model and new constants derived from this solubility study were used to calculate saturation indices for pore-water chemistry data obtained from a tailings impoundment at the Heath Steele mine, New Brunswick, Canada. Sulfide minerals in this impoundment had been oxidizing for approximately 20 years at the time of sampling. The model calculations suggest that the tailings pore water, which contains up to 1 m $FeSO_4$, is at or near saturation with respect to siderite for the majority of the water samples collected. Mineralogical study indicates siderite is present in the tailings solids at approximately the same locations where the pore water approaches saturation with respect to siderite.

The oxidation of sulfide minerals in mine-tailings impoundments releases H^+ and high concentrations of dissolved SO_4, Fe and other metals to the tailings pore water. As these constituents react with dissolved and solid-phase carbonates within the impoundments and the surrounding geologic materials, a series of reactions occurs which controls the concentration and the rate of movement of the sulfide-oxidation products. Among the most important of these reactions is the dissolution of carbonate minerals, which results in an increase in the pore-water pH, and in many cases, the attenuation of dissolved metals. Common carbonate minerals found in the original gangue, and subsequently in the tailings, are calcite ($CaCO_3$), dolomite ($CaMg(CO_3)_2$) and siderite ($FeCO_3$) (Figure 1). These minerals may also be present in geological materials underlying the tailings. Near-equilibrium conditions with respect to siderite have been observed (1-3) at several locations where tailings include siderite within the non-sulfide gangue.

At other locations, the formation of secondary siderite, when Fe(II)-rich water contacts calcite, has been inferred through geochemical calculations (4) or observed (5,6). Information on the solubility of carbonate minerals in waters with compositions typical of those associated with weathered mine wastes is needed to develop more reliable predictions of the future geochemical evolution and potential environmental effects of these wastes.

This paper focuses on the mineral siderite because of *i*) its frequent occurrence as a primary phase in mine tailings, *ii*) its potential importance as a secondary phase, and *iii*) its role in controlling the geochemical evolution of tailings pore water. Saturation indices for siderite are determined using the Pitzer ion-interaction model for pore-water samples collected from a tailings impoundment, and are compared to a mineralogical study delineating the zone of siderite occurrence in the tailings. Constants required to apply the Pitzer model to the field data were evaluated through analysis of data collected in a laboratory study on siderite solubility in $FeSO_4$ solutions.

The Influence of Siderite on Mine-tailings Geochemistry

The oxidation of pyrite (FeS_2), the most abundant sulfide mineral in most tailings impoundments, by atmospheric oxygen produces one mole of Fe^{2+}, two moles of SO_4^{2-} and two moles of H^+ for each mole of pyrite oxidized:

$$FeS_2(s) + \frac{7}{2}O_2 + H_2O \rightarrow Fe^{2+} + 2SO_4^{2-} + 2H^+ \qquad (1)$$

In the presence of O_2, any Fe^{2+} released can oxidize to Fe^{3+}:

$$Fe^{2+} + \frac{1}{4}O_2 + H^+ \rightarrow Fe^{3+} + \frac{1}{2}H_2O \qquad (2)$$

In tailings impoundments, ferric iron can be removed from the pore water by two processes. Under relatively high-pH conditions (pH > 4), accumulation of low-solubility Fe^{3+}-containing hydroxide, oxyhydroxide, or hydroxy-sulfate solids on the surfaces of sulfide grains is favored, resulting in the removal of Fe^{3+} from the tailings pore water. Under low-pH conditions, Fe^{3+} is reduced as the further oxidation of pyrite releases additional SO_4, H^+ and Fe^{2+}:

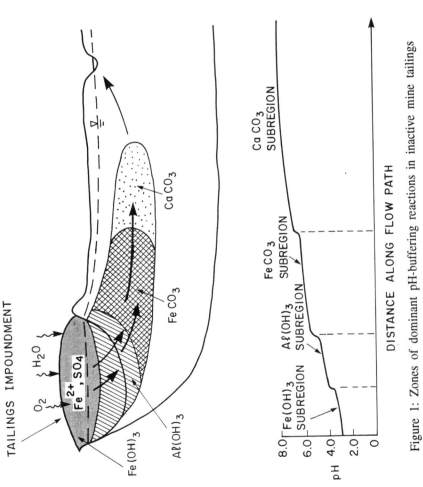

Figure 1: Zones of dominant pH-buffering reactions in inactive mine tailings impoundments.

$$14Fe^{3+} + FeS_2(s) + 8H_2O \rightarrow 15Fe^{2+} + 2SO_4^{2-} + 16H^+ \qquad (3)$$

When moisture contents in the vadose zone and the underlying saturated zone are high, the rate of O_2 diffusion to the surfaces of sulfide grains is low and the rate of Fe^{2+} oxidation (equation 2) is less than the rate of Fe^{3+} reduction (equation 3). A net production of Fe^{2+} results. Because of the high solubility of Fe(II)-bearing minerals in low-pH waters, Fe^{2+} can be displaced downward and laterally into the deeper tailings and the surrounding geologic materials (7).

When pore water containing H^+, Fe^{2+} and SO_4 moves through the deeper tailings and surrounding area, a series of reactions occurs between the pore water and the solids. These reactions include mineral precipitation/dissolution, coprecipitation and adsorption, and complexation reactions. At several inactive mine-waste impoundments, the dissolution of primary siderite has been noted:

$$H^+ + FeCO_3(s) \rightleftharpoons HCO_3^- + Fe^{2+} \qquad (4)$$

which can alter the pore-water pH and Eh (1,2,8). At other sites, it has been postulated that the formation of secondary siderite, by the reaction of dissolved Fe^{2+} with carbonate minerals, principally calcite:

$$Fe^{2+} + CaCO_3(s) \rightleftharpoons FeCO_3(s) + Ca^{2+} \qquad (5)$$

can control pore-water concentrations of Fe^{2+} (4-6,9,10). These conclusions were based on three lines of evidence: *i*) interpretations of pore-water chemistry data indicating that the waters were close to saturation with respect to siderite (1,2,4,8,10), *ii*) direct observation of primary siderite in gangue materials (1,2,8) together with zones of depletion where dissolution of siderite occurred as a result of acid-neutralization reactions, and *iii*) observations of secondary siderite accumulations where $FeSO_4$-rich waters have contacted calcite-rich zones (5,6).

pH-Buffering Mechanisms in Weathered Tailings Impoundments. The H^+ ions generated by sulfide-mineral oxidation are consumed by a number of pH-buffering mechanisms, including the dissolution of hydroxide and aluminosilicate minerals. Morin and Cherry (4) noted a stepwise sequence of these reactions in an aquifer affected by acidic drainage derived from the inactive Nordic Main uranium-mine tailings impoundment, near Elliot Lake, Ontario. The pH-buffering reactions observed follow the sequence: calcite dissolution, siderite dissolution, Al(OH)$_3$ dissolution, and Fe(OH)$_3$ dissolution. Similar pH-buffering mechanisms were noted within the Nordic Main U-tailings impoundment (7), the inactive Waite Amulet Cu-Zn tailings impoundment, Noranda, Quebec (1), and the Copper Cliff Ni-Cu tailings impoundment, Sudbury, Ontario (2).

At all of these sites, the dissolution of calcite contained in the underlying geological materials was favored as low-pH tailings water was displaced downward from the zone of sulfide oxidation (Figure 1). The pH was initially buffered to between 5.4 and 6.9 by calcite dissolution. As calcite dissolved, the precipitation of siderite, amorphous Al(OH)$_3$ and Fe(OH)$_3$ was favored.

After calcite is depleted, the pH decreases to 5.2 - 5.5 and the pore water remains near or slightly undersaturated with respect to siderite (1,4,10); the dissolution of siderite, accompanied by the precipitation of Al(OH)$_3$ and Fe(OH)$_3$, is inferred. Under these moderate pH conditions resulting from siderite dissolution, the precipitation of many dissolved metals as amorphous and crystalline hydroxide phases is favored (1,7,10), which decreases the aqueous concentrations of these elements.

After siderite dissolves, the pH decreases to 4.9 - 5.2, favoring the dissolution of Al(OH)$_3$ and the precipitation of Fe(OH)$_3$. After Al(OH)$_3$ dissolves, the pH falls to < 4.5 and dissolution of Fe(OH)$_3$, or other Fe(III)-bearing solids, is favored. A result of the decrease in pH that accompanies siderite depletion, and later the dissolution of Al(OH)$_3$ and Fe(OH)$_3$, is the dissolution of other metal-bearing hydroxide and carbonate phases and the dissolved concentrations of many metals increase in these zones (2,7).

Efforts to conclusively identify secondary siderite from the aquifer and tailings solids at the Elliot Lake site were unsuccessful (4), possibly because of the small mass of siderite or because of the sampling and sample-handling procedures. At the Waite Amulet and Copper Cliff sites, primary siderite, present as part of the non-sulfide gangue, was isolated; however, secondary siderite, precipitated due to the interaction of Fe(II) with the tailings minerals, was not. At a site where FeSO$_4$-containing waste waters, similar in composition to mine-waste drainage waters, were treated by infiltration through sediments that contained abundant calcite, Ho et al. (5) and Wajon et al. (6) reported the formation of both secondary siderite and ferroan calcite. The formation of secondary siderite was also reported by Thornber and Nickel (9) in the vicinity of a weathered sulfide deposit, as Fe(II) derived from sulfide-oxidation reactions contacted carbonate minerals in the host materials.

Eh-Buffering Mechanisms in Weathered Tailings Impoundments. In most tailings areas Fe, as Fe^{2+} or Fe^{3+}, is the dominant electroactive element. The solid phases that limit the activities of these two Fe species control the Eh of the tailings pore water, and affect the speciation of other electroactive elements such as Mn, Cu, and Cr (1,7). In the high-pH regions of tailings impoundments, the solubilities of Fe(II)- and Fe(III)-bearing phases constrain the pore-water Eh within a narrow range. Specifically, the precipitation and dissolution of siderite, and the accompanying precipitation and dissolution of Fe(III)-bearing phases, including amorphous Fe(OH)$_3$, ferrihydrite, goethite, jarosite or other Fe(III) hydroxy-sulfate solids, control the activities of both Fe^{2+} and Fe^{3+} species. In the low-pH and high-Eh regions of tailings impoundments, there is no readily soluble source of Fe(II), and the pore-water Eh is controlled by the aqueous concentration of Fe(II) and the solubility of the dominant Fe(III)-bearing phases. Specifically, the Fe^{3+} activity seems to be controlled by the precipitation and dissolution of Fe(OH)$_3$, goethite and jarosite. Recent studies have identified a secondary Fe(III)-SO$_4$-OH precipitate, termed "mine-drainage mineral" (11), or schwertmannite (12). This phase may control Fe^{3+} activities at some locations.

Geochemical Modelling of Tailings Water. In the above studies, a primary indication of a siderite solubility control on the geochemistry of tailings water was through calculations of saturation indices. These calculations were conducted using the conventional ion-association model and the Debye-Hückel equation, or one of the extended forms of the Debye-Hückel equation, for activity correction. Application of this model is limited to waters containing relatively low concentrations of dissolved solids. Many waters associated with the oxidation of sulfide minerals can reach high concentrations of dissolved solids due to the intensity of the sulfide-oxidation reactions. An example is the mine-drainage water that originates at Iron Mountain, Shasta County, California (13,14). This water is characterized by concentrations of dissolved solids that exceed 800 g/L. A second example is the water associated with the oxidation of the Heath Steele mine tailings, New Brunswick, which exceeds 200 g/L of dissolved solids (8). At both of these sites, the waters contain predominantly Fe and SO_4, with lesser amounts of H^+ and dissolved metals, including Zn, Cu, Pb and Cd.

Simulation of the chemical behavior of the Iron Mountain or Heath Steele waters, or similar concentrated mine drainage waters, as they contact carbonate minerals requires modifications to the geochemical model to account for the extremely high concentrations of dissolved solids. The Pitzer ion-interaction model (15) is a suitable alternative and can be used to calculate saturation indices of mineral phases present at mine-waste sites characterized by concentrated waters. The Pitzer model was developed extensively to calculate ion activities and mineral solubilities for waters containing the major components of seawater.

In this study the Pitzer model is used to determine saturation indices for siderite in waters containing high concentrations of Fe and SO_4. It is applied first to the results of a laboratory investigation of siderite solubility in $FeSO_4$ solutions, at various alkalinity and pH values, to assess the performance of this model in waters with compositions representative of mine-tailings pore water. It is then applied to the Heath Steele site, where waters with high concentrations of Fe(II) and SO_4 are in contact with carbonate minerals. These simulations are used to assess the role of siderite in controlling the geochemical evolution of tailings pore water that covers a broader range in $FeSO_4$ concentration than has been studied previously.

Methodology

Laboratory Methodology. Determinations of siderite solubility were made in solutions covering a range in $FeSO_4$ concentration, alkalinity and pH. These determinations were made by combining chemically pure, crystalline $FeCO_3$ with purified solutions of $FeSO_4$ in 50 mL glass vials. The mixtures were equilibrated at 25±1°C for more than 12 weeks (16). All sample handling was conducted in an anoxic glovebox to prevent oxidation of Fe(II) during sample preparation and analysis. After the solutions and solids were equilibrated, filtered samples (0.2 μm) were analyzed to determine total Fe(II) concentration, by using the Ferrozine colorimetric technique (17), total inorganic carbonate concentration, by using the method of Stainton (18), pH and Eh, by using electrodes in sealed cells, and alkalinity, by using normalized HCl. Interference studies were conducted for each analytical procedure to assess the influence of the high concentrations of Fe(II) and SO_4 on the analytical determinations.

Field Methodology. Samples of tailings solids, pore water and pore gas were collected *versus* depth at five locations on the old tailings impoundment at the Heath Steele mine (8). Pore water from the vadose zone was sampled using an immiscible displacement technique that isolated the tailings from atmospheric O_2. Pore water from the saturated zone was collected using single-completion drive-point piezometers. Pore gas was collected using stainless-steel drive-point gas samplers. Pore-water pH, Eh, temperature and conductivity were determined at the well site in a sealed flow-through cell that excluded atmospheric O_2. Alkalinity determinations were conducted in sealed vessels using normalized H_2SO_4. Water samples were analyzed by atomic absorption spectroscopy to determine the concentrations of Al, Ag, As, Ca, Cd, Co, Cr, Cu, K, Mg, Mn, Na, Ni, Pb, Sr, and Zn, and by ion chromatography to determine the concentrations of Cl, SO_4, and NO_3. Vadose-zone gas-phase concentrations of O_2 and CO_2 were determined in the field using a Nova model 350LDB gas analyzer. Replicate gas samples were returned to the laboratory for analysis by gas chromatography.

Results

Results of Laboratory Studies. Twenty-one determinations of siderite solubility were made in solutions covering a range in $FeSO_4$ concentration to melanterite ($FeSO_4 \cdot 7H_2O$) saturation (~1.8 m $FeSO_4$) (16). The determinations were made as closed-system dissolution experiments, resulting in different values of pH, concentration of total inorganic carbonate species (C_T) and partial pressure of CO_2 for the different samples examined. The C_T varied from 0.1 - 5.0 mmolal, and the pH varied from 4.1 - 6.5. In general, as the concentration of $FeSO_4$ increased, the solution pH decreased and the C_T, alkalinity and partial pressure of CO_2 increased. These trends indicate that there is greater dissolution of siderite with increasing $FeSO_4$ concentration, because of the greater interaction between Fe^{2+} and SO_4^{2-} at the higher SO_4 concentrations.

The solubility of siderite is dependent on the aqueous activities of Fe^{2+} and CO_3^{2-}, which in turn depend on the total concentrations of dissolved Fe and carbonate species, and the pH and Eh of the water. Solubility products for siderite were calculated from analytical determinations of total Fe (Fe_T), C_T and pH. Intermediate Eh values were measured for all samples (50-200 mV), indicating the dissolved iron was predominantly in the ferrous form. Determinations of alkalinity were used to confirm independently calculated quantities of HCO_3.

Description of Geochemical Model. The Pitzer ion-interaction model was used to determine aqueous concentrations and activities of Fe^{2+} and CO_3^{2-} in the equilibrated mixtures. The ion-interaction model, first introduced by Pitzer (19) and recently summarized by Pitzer (15), has been shown to provide reliable estimates of ion activities and mineral solubilities in complex mixtures of electrolytes (15,20,21). The model is based on statistical mechanical theory and includes terms to account for specific interactions between ions in solution. To calculate ion activities using the Pitzer model, interaction coefficients for all important binary and ternary combinations of species present in the water are used. These coefficients are derived from experimental data and vary as a function of temperature. For strongly associating species, temperature-dependent association constants, may be necessary. Experimentally determined solubility products are required to calculate mineral solubilities and saturation indices. All interaction coefficients, association constants and solubility products must be internally consistent. The geochemical computer code PHRQPITZ (22), which incorporates the Pitzer ion-interaction model and the thermodynamic database developed by Harvie et al. (21), was used for the model calculations. Because the database of Harvie et al. was developed primarily to model the geochemistry of the major components found in seawater, to model the laboratory and field data the parameters for Fe and other transition metals were added to the PHRQPITZ database.

The waters analyzed in the laboratory experiments contained predominantly Fe and SO_4, with lesser amounts of dissolved carbonate species; therefore, additional parameters describing interactions among these species were required. In the intermediate Eh range of the samples, Fe is present principally as Fe(II) and S is present principally as SO_4. Therefore, parameters describing interactions among Fe(II) and SO_4 species were included in the database (23), whereas those describing interactions among Fe(III) and sulfide species were omitted. The samples were slightly acidic, therefore the dominant carbonate species are $CO_2(aq)$ and HCO_3^-, the dominant Fe(II) species is Fe^{2+}, and the dominant SO_4 species is SO_4^{2-}. The hydrolysis of Fe^{2+} to form $FeOH^+$ and the protonation of SO_4^{2-} to form HSO_4^- are less significant. The interaction between Fe and SO_4^{2-} is represented through interaction parameters with the Pitzer model, and not included as the $FeSO_4^0$ ion-pair. The parameters representing mixing between $FeSO_4$ and H_2SO_4, as reported by Reardon and Beckie (23), were added to the PHRQPITZ database to complete the calculations.

Parameters representing the interactions for Fe(II)-OH, Fe(II)-HCO_3, Fe(II)-CO_3 and Fe(II)-H_2CO_3 are not currently available. Harvie et al. (21) derived parameters for the closely related Ca-OH, Ca-HCO_3, Ca-CO_3, and Ca-H_2CO_3, and Mg-OH, Mg-HCO_3, Mg-CO_3, and Mg-H_2CO_3 interactions. Because of the strong interaction for Ca-CO_3, they represented the interactions between Ca-CO_3 using the neutral ion complex $CaCO_3^0$. The interactions between Ca-HCO_3, and Ca-OH are weaker and were represented by interaction coefficients. In deriving parameters for Mg-OH, Mg-HCO_3, and Mg-CO_3, Harvie et al. used interaction coefficients to represent Mg-HCO_3 interaction, and the ion complexes $MgCO_3^0$ and $MgOH^+$ to represent the stronger Mg-CO_3 and Mg-OH interactions. The interactions between Fe(II) and OH, CO_3, and HCO_3 are expected to be similar to those observed for Mg. For this reason, we used conventional association constants to represent Fe(II) interactions with OH, CO_3, and HCO_3 (24), similar to the approach taken by Harvie et al. for Mg. Further interaction of the species $FeOH^+$, and $FeCO_3^0$ with other aqueous components was assumed to be negligible. Ternary interaction of Fe(II) and HCO_3 with SO_4, the dominant anion in the waters, was represented using an interaction coefficient, set to be equivalent to the value reported for Mg-SO_4-HCO_3. The same value was reported for the neutral-ion interaction between $CO_2(aq)$ and Mg, and $CO_2(aq)$ and Ca (21). This value also was used to describe the interaction between $CO_2(aq)$ and Fe(II). This compilation of interaction parameters and association constants (Tables I and II) was added to the computer code PHRQPITZ (22).

Table I: Association Constants and Solubility Products Added to Database of Harvie et al. (21)

Reaction	log K	ΔH (kcal/mol)	Source
$HSO_4^- = SO_4^{2-} + H^+$	-1.979		(23)
$Fe^{2+} + H_2O = FeOH^+ + H^+$	-9.5	13.2	(24)
$Fe^{2+} + HCO_3^+ = FeHCO_3^+$	2.0	-	(24)
$Fe^{2+} + CO_3^{2-} = FeCO_3^0$	4.38	-	(24)

Mineral	Reaction	log K	ΔH (kcal/mol)	Source
Melanterite	$FeSO_4 \cdot 7H_2O = Fe^{2+} + SO_4^{2-} + 7H_2O$	-2.209	-	(23)[a]
Siderite	$FeCO_{3(s)} = Fe^{2+} + CO_3^{2-}$	11.03	-2.48	(24)[b]

a   Temperature relation for K constants as a function of T in K, modified from Reardon and Beckie (23) to be compatible with temperature function used in PHRQPITZ:
log K = 110.851 + 0.016T - 5270.88/T - 40.632 log T + 47257.629/T²

b   log K from this study, enthalpy from Nordstrom et al. (24)

Calculated Solubility Products. From the analyses of Fe_T, C_T and pH, solubility products for siderite were calculated as a function of $FeSO_4$ concentration. The thermodynamic solubility product (K_{sp}) for the dissolution of siderite:

$$FeCO_3(s) \rightleftharpoons Fe^{2+} + CO_3^{2-} \qquad (6)$$

is defined as:

$$K_{sp} = a_{Fe^{2+}} \cdot a_{CO_3^{2-}} = m_{Fe^{2+}} \cdot m_{CO_3^{2-}} \gamma_{Fe^{2+}} \gamma_{CO_3^{2-}} \qquad (7)$$

where a represents the activity, m the molality and γ the activity coefficient for the respective species. If equilibrium conditions are achieved and an accurate model formulation is used, thermodynamic solubility products are expected to be constant, regardless of the concentrations of the dissolved species. The calculated values of K_{sp} showed a fairly large scatter about the mean value (mean $pK_{sp}(\pm 95\%$ confidence interval (CI))= 11.03($\pm$0.26)), but were independent of the concentrations of $FeSO_4$ and C_T, ionic strength, and pH (Figure 2). This constancy in K_{sp} suggests that the ion-interaction model accurately describes the solubility of siderite in $FeSO_4$ solutions covering a range in pH and concentration of dissolved carbonate species.

Calculated values of stoichiometric solubility products (K_{sp}^*), defined as the product of the activity-corrected single-ion concentrations of the siderite dissolution products Fe^{2+} and CO_3^{2-}, or:

$$K_{sp}^* = m_{Fe^{2+}} \cdot m_{CO_3^{2-}} \qquad (8)$$

are also provided in Figure 2. Stoichiometric solubility products are related to the thermodynamic solubility product through:

$$K_{sp}^* = \frac{K_{sp}}{\gamma_{Fe^{2+}} \gamma_{CO_3^{2-}}} \qquad (9)$$

and provide an indication of the factors that most strongly influence the activity coefficients and thermodynamic solubility products. The pK_{sp}^* values decreased with an initial increase in concentration of $FeSO_4$, and then leveled off as melanterite saturation was approached. This behavior suggests that the activity coefficients show the greatest dependence on $FeSO_4$ concentration at low concentrations, with less dependence on $FeSO_4$ at high concentrations.

In the calculations of K_{sp} and K_{sp}^*, several approximations were made that relate to the association constants and interaction parameters for Fe(II) and OH, HCO_3, CO_3, and H_2CO_3. Because the pH of the samples was sufficiently low that hydrolysis of Fe^{2+} and complexation or strong interaction of Fe^{2+} with HCO_3^- or CO_3^{2-} were negligible, the calculated values of the single-ion molality of Fe^{2+} were nearly identical to the analytical concentrations of Fe(II). Any error in the model representation for Fe(II) interaction with OH, HCO_3 and CO_3 should not result in large differences in calculated K_{sp} values. To assess whether error in the estimated

Table II: Ion-Interaction Parameters Added to Database of Harvie et al. (21)

Single-Electrolyte Parameters:

i	j	$\beta_{ij}^{(0)}$	$\beta_{ij}^{(1)}$	$\beta_{ij(2)}$	C_{ij}^{ϕ}	Source
H^+	SO_4^{2-}	0.0298	-	-	0.0438	(23)
Fe^{2+}	SO_4^{2-}	0.2568	3.063	-42.0	0.0209	(23)
Zn^{2+}	SO_4^{2-}	0.1949	2.883	-32.81	0.0290	(15)
H^+	HSO_4^-	0.2065	0.5556	-	-	(23)
Fe^{2+}	HSO_4^-	0.4273	3.48	-	-	(23)

Common-Ion Mixed-Electrolyte Parameters:

i	j	k	θ_{ij}	ψ_{ijk}	Source
HCO_3	SO_4	Fe(II)	0.01	-0.161	a

Neutral-Electrolyte Parameters:

i	j	λ_{ij}	Source
Fe(II)	H_2CO_3	0.018	a

a   estimated from values reported for Mg by Harvie et al. (21)

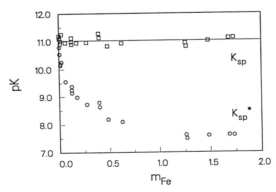

Figure 2: Calculated thermodynamic (K_{sp}) (squares) and stoichiometric (K_{sp}^*) (circles) solubility products for siderite dissolution in solutions containing $FeSO_4$ to melanterite saturation at 25°C.

Fe(II)-SO_4-HCO_3 and Fe(II)-H_2CO_3 interaction parameters substantially influenced the calculated solubility products, values of these parameters were adjusted within the range typically observed for similar cations. These adjustments resulted in insignificant differences in calculated K_{sp} values suggesting that, for the pH range and conditions of the siderite solubility experiments, the constants summarized in Table II, together with the database of Harvie et al., provide an adequate representation of the activities of Fe^{2+} and CO_3^{2-}.

Summary of Laboratory Results. Calculated values of thermodynamic solubility products were nearly constant as a function of $FeSO_4$ concentration, yielding a mean pK_{sp} value of 11.03. This mean pK value for siderite is consistent with those obtained for determinations made in other electrolyte solutions, including carbonic acid (25), $NaClO_4$ (26), and NaCl (27) and Na_2SO_4 (27) solutions, but is slightly higher than those obtained for siderite solubility by other investigators (28,29). The relative constancy in the calculated thermodynamic solubility products suggest that the selected model formulation and parameters provided reasonable estimates of single-ion concentrations and activities of the Fe^{2+} and CO_3^{2-} species. The next step was to extend these model parameters to calculate Fe^{2+} and CO_3^{2-} activities, and thus siderite saturation indices, for the Heath Steele field site where pore-water compositions are also dominated by Fe(II) and SO_4.

Results of Field Studies. The geochemistry of the Heath Steele tailings can be divided into three zones previously identified by Boorman and Watson (30): the Oxidation Zone, the Hard Pan Layer and the Reduction Zone (8). Sulfide oxidation reactions consume gas-phase O_2 within the upper 15 cm of the tailings. These reactions have depleted the solid-phase sulfide content and the solid-phase masses of Fe, Pb, Zn, Cu and As near the tailings surface (31), generating high concentrations of dissolved Fe and SO_4 (Figure 3) and high concentrations of dissolved metals (up to 4 g/l Zn, 15 mg/l Pb and 200 mg/l Cu) (8,31) in the shallow tailings. The pore-water pH in the near surface tailings ranges from 0.8 - 1.5. Mineralogical study indicates that the principal cementing minerals contained in the Hard Pan Layer are the sulfate-bearing minerals gypsum and melanterite. The pH increases sharply to > 5.0 near the depth of the Hard Pan Layer, 35-42 cm below the tailings surface. Measurements of the solid-phase carbonate content show an abrupt increase at the depth of the Hard Pan Layer. Mineralogical study indicates that primary siderite and dolomite are abundant in the deeper tailings but rare in the Oxidation Zone above the Hard Pan Layer (32). Calcite was not detected in the samples of the tailings solids (32). The sharp increase in pH observed near the Hard Pan Layer (Figure 3) corresponds with the observed increase in siderite content, suggesting that dissolution of siderite buffers the pore-water pH near and below the Hard Pan Layer. Dissolution of siderite and other carbonate minerals is indicated by the sharp increase in gas-phase CO_2 concentrations observed immediately below the depth of the Hard Pan Layer.

Additions to Geochemical Model for Field Application. At the Heath Steele site, the tailings water contains high concentrations of Fe(II) and SO_4, with lesser amounts of Ca, Mg, Zn, Al, Fe(III) and other elements. The Heath Steele waters are in the intermediate Eh range with Fe present principally as Fe(II) and S

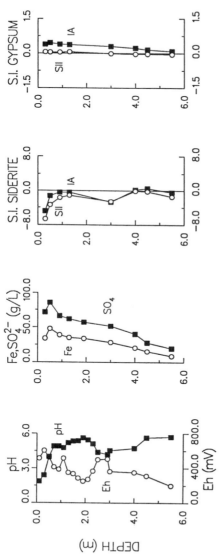

Figure 3: Pore-water pH and Eh, concentrations of Fe and SO_4, and saturation indices (SI), calculated using the specific ion-interaction (SII) and the ion-association (IA) models for gypsum and siderite for piezometer nest OW8 in the Heath Steele old mine-tailings impoundment.

present principally as SO_4, and in the acidic pH range, with dissolved carbonate present principally as $CO_{2(aq)}$ and HCO_3^-. Hydrolysis of Fe^{2+} to form $FeOH^+$ is insignificant as is protonation of SO_4^{2-} to form HSO_4^-, with the exception of the most acidic samples near the tailings surface. The most important parameters required for calculating saturation indices for siderite in the Heath Steele waters are those representing interactions between Fe(II) and SO_4, OH, and the carbonate species, because Fe(II) and SO_4 are the dominant ions in this water and the carbonate species and OH are needed to determine saturation indices for siderite. Therefore, approximations similar to those used to select interaction coefficients for the laboratory data were used for the field data. The pore water at Heath Steele generally ranges between 10 and 15 °C. Terms to account for temperature dependence in the association and solubility constants, and the interaction coefficients, were also included in the PHRQPITZ database (Tables I-III).

In addition to Fe(II), SO_4 and dissolved carbonate species, the next most abundant elements at Heath Steele are Ca, Mg and Zn. Parameters for the interaction of the alkali earth elements Ca and Mg with SO_4, HSO_4, CO_3, HCO_3 and H_2CO_3 as a function of temperature, are included in the database of Harvie et al. Binary interaction parameters for $ZnSO_4$, reported by Pitzer (15), were added to this database. Parameters for other metals present in minor amounts at the Heath Steele site have been reported for sulfate and bisulfate binary mixtures (e.g. 15,33,34), but are not generally available for more complex metal mixtures. The next most important interaction coefficients for the Heath Steele pore water are the ternary coefficients for mixtures involving SO_4, Fe(II) and other cation-forming elements, such as the Fe(II)-Ca-SO_4, Fe(II)-Mg-SO_4 and Fe(II)-Zn-SO_4 mixtures. Although these coefficients are not available, estimates of Fe^{2+} activities required in the calculation of siderite saturation indices are less dependent on the concentrations of Ca, Mg and Zn because of their low concentrations relative to Fe(II). Larger differences are expected in the estimated activities of the elements present at minor concentrations because of the dominance of Fe(II) and SO_4. As new interaction parameters become available, some revisions to the calculations are expected, particularly those relating to the ternary interaction among major and minor components present in the Heath Steele tailings water.

Calculated Saturation Indices for Siderite at the Heath Steele Site. Calculations of saturation indices for siderite *versus* depth (Figure 3), using the ion-interaction model PHRQPITZ together with the additional constants tabulated in Tables I and II, indicate that a near-saturation condition with respect to siderite is attained throughout most of the lower part of the pore-water profile. For the upper part of the tailings impoundment, the calculations indicate undersaturation with respect to siderite, coincident with the strong depletion in the carbonate content of the tailings. At approximately 40 cm depth, where an abrupt increase in siderite content was observed (8), near-saturation conditions with respect to siderite are attained. Similar trends in saturation index *versus* depth were observed for other locations on the Heath Steele tailings impoundment. These results are in close agreement with calculations of saturation indices made using the ion-association model by Blowes et al. (8) using MINTEQA2 (35). The close agreement between the depths where near-saturation with respect to siderite is attained, and the depths where

Table III: Temperature Derivatives of Pitzer Coefficients Used in this Study[a]

Coefficient	c_1	c_2	c_3	c_4	c_5
$\beta^{(0)}$ Fe-SO$_4$	0.257	-0.188×10^4	-0.490×10^1	-0.225×10^{-1}	0.303×10^{-4}
$\beta^{(1)}$ Fe-SO$_4$	3.063	-0.374×10^4	0.922×10^1	-0.304×10^{-1}	-0.587×10^{-4}
$\beta^{(0)}$ Fe-HSO$_4$	0.433	0.731×10^6	0.703×10^4	-0.226×10^2	0.121×10^{-1}
$\beta^{(0)}$ H-SO$_4$	0.0217	0.106×10^5	0.998×10^2	-0.315	0.165×10^{-3}
C^{ϕ} H-SO$_4$	0.0411	-0.285×10^4	-0.258×10^2	0.795×10^{-1}	-0.408×10^{-4}
$\beta^{(0)}$ H-HSO$_4$	0.211	0.770×10^4	0.720×10^2	-0.226	0.118×10^{-3}
$\beta^{(1)}$ H-HSO$_4$	0.532	-0.861×10^5	-0.849×10^3	0.277×10^1	-0.149×10^{-2}

a   Constant terms, c_i, for temperature dependence of Pitzer coefficients, P_{ij}, modified from Reardon and Beckie (23) to be in form compatible with PHRQPITZ:

$$P_{ij} = c_1 + c_2(1/T - 1/T_R) + c_3 \ln(T/T_R) + c_4(T-T_R) + c_5(T^2 - T_R^2)$$

where T is in K, and T_R = 298.15 and T is between 273 and 338.

significant accumulations of siderite were observed, suggests that siderite is an important phase controlling the pore-water chemistry at the Heath Steele site.

Also included in Figure 3 are saturation indices for gypsum ($CaSO_4 \cdot 2H_2O$) *versus* depth, calculated using MINTEQA2 and PHRQPITZ for the same pore-water samples. In contrast to the results for siderite, different results were obtained for gypsum using the different models. The saturation indices for gypsum obtained using PHRQPITZ were close to zero throughout the profile, indicating near-equilibrium conditions. The presence of gypsum throughout the profile was confirmed (8). The near-equilibrium conditions predicted by the Pitzer model are in agreement with the mineralogic observations. In contrast, the ion-association model indicated slightly supersaturated conditions. Because of the ease of formation of gypsum, and because near-equilibrium conditions with respect to gypsum are frequently reported for natural waters in contact with gypsum, the differences in the predicted saturation indices for gypsum are attributed partly to differences in the activity of SO_4^{2-} obtained with the two models. PHRQPITZ calculations indicate the water present near the tailings surface and below the Hard Pan Layer is undersaturated with respect to melanterite, and the water at the depth of the Hard Pan Layer approaches saturation with respect to melanterite. Agreement between the depths where saturation with respect to gypsum, melanterite and siderite is attained and the depth of occurrence of these minerals suggests that the model accurately describes the pore-water geochemistry.

The model calculations indicate near-equilibrium conditions with respect to siderite from the depth of the Hard Pan Layer to the base of the tailings impoundment. The tailings pore water is consistently undersaturated with respect to dolomite and calcite. The presence of siderite and dolomite was confirmed by the mineralogical study (8). Calcite was not observed. Thus, the dissolution of siderite and dolomite apparently are the dominant pH-buffering mechanisms below the Hard Pan Layer. Calculations conducted using data collected at other piezometer nests in the impoundment show similar trends, suggesting that precipitation and dissolution of siderite may affect pore-water pH values over most of the tailings area.

Iron is the most abundant electroactive element in the tailings pore water. It is likely that equilibrium between the Fe^{2+}-Fe^{3+} pair controls the pore-water Eh. Previous geochemical calculations, using the field-measured Eh and MINTEQA2 and a modified version of PHRQPITZ (8), indicated that the water below the Hard Pan Layer approaches equilibrium with respect to amorphous $Fe(OH)_3$ and is slightly supersaturated with respect to goethite. Mineralogical examination indicated the presence of goethite but not ferrihydrite in the tailings solids. These observations suggest that one of the phases, amorphous $Fe(OH)_3$, goethite or ferrihydrite, probably controls the pore-water Fe^{3+} activity. The solubilities of siderite and one of these three ferric-bearing phases probably control the pore-water Eh, and thereby affect the speciation of other electroactive elements contained in the tailings pore water.

Conclusions

The solubility of siderite was measured in solutions over a range in $FeSO_4$ concentration (0.0 - 1.8 m) and a range in pH (4.1 - 6.5). A geochemical model, including Pitzer parameters and association constants to account for the interactions among Fe(II), SO_4, OH, and carbonate species and a constant solubility product for siderite, was used to describe siderite equilibrium in these solutions. The calculated solubility product for siderite is 11.03±0.26 (mean ± 95% CI; standard pK units). This pK value is similar to the solubility product for siderite measured in other electrolyte solutions. The model was used to describe mineral-water equilibrium at the Heath Steele mine-tailings impoundment. The model calculations indicate near-equilibrium conditions with respect to gypsum, melanterite, and siderite at the approximate depth of occurrence of these minerals. The model accurately describes the pore-water geochemistry at the Heath Steele site and can be used to describe mineral-water equilibria at other mine-waste sites where $FeSO_4$-rich water is present. In agreement with previous studies conducted at sites of low ionic-strength tailings water, siderite dissolution is an important mechanism affecting the pH of the concentrated pore water at Heath Steele. Because siderite is a moderately soluble Fe(II)-bearing mineral, it is also likely that siderite precipitation and dissolution also affect the Eh of the tailings pore water.

Acknowledgments

We thank E.J. Reardon, J.L Jambor, J.A. Cherry and H. Veldhuizen for their support during various stages of this study. This manuscript benefited from reviews by C. Cravotta, C.N. Alpers, and D.K. Nordstrom. Funding was provided by Noranda Minerals Inc., the Natural Sciences and Engineering Research Council of Canada, and the CANMET Branch of the Department of Natural Resources Canada.

Literature Cited

1. Blowes, D.W.; Jambor, J.L. Applied Geochem. 1990, 5, 327-346.
2. Coggans, C.J.; Blowes, D.W.; Robertson, W.D. Proc.: Second Internat. Conf. on Abatement of Acidic Drainage; MEND (Mine Environment Neutral Drainage); Montréal, QC, 1991, Vol. 4, pp. 1-23.
3. Germain, M.D.; Tassé, N.; Bergeron, M. In this volume.
4. Morin, K.A.; Cherry, J.A. Chem. Geol. 1986, 56, 117-134.
5. Ho, G.E.; Murphy, P.J.; Platell, N.; Wajon, J.E. J. Environ. Eng. 1984, 110, 828-846.
6. Wajon, J.E.; Ho, G.E.; Murphy, P.J. Water Res. 1985, 19, 831-837.
7. Dubrovsky, N.M.; Morin, K.A.; Cherry, J.A.; Smyth, D.J.A. Can. J. Water Poll. Control Res. 1984, 19, 55-89.
8. Blowes, D.W.; Reardon, E.J.; Jambor, J.L.; Cherry, J.A. Geochim. Cosmochim. Acta 1991, 55, 965-978.
9. Thornber, M.R.; Nickel, E.H. Chem. Geol. 1976, 17, 45-72.
10. Morin, K.A.; Cherry, J.A.; Dave, N.K.; Lim, T.P.; Vivyurka, A. J. Contam. Hydrol. 1988, 2, 271-303.
11. Bigham, J.M.; Schwertmann, U.; Carlson, L.; Murad, E. Geochim. Cosmochim. Acta 1990, 54, 2743-2758.
12. Murad, E.; Schwertmann, U.; Bigham, J.M.; Carlson, L. In this volume.
13. Alpers, C.N.; Nordstrom, D.K. Proc.: Second Internat. Conf. on Abatement of Acidic Drainage; MEND (Mine Environment Neutral Drainage); Montréal, QC, 1991, Vol. 2, pp. 321-342.

14. Alpers, C.N.; Nordstrom, D.K.; Thompson, J.M. In this volume.
15. Pitzer, K.S. In Activity Coefficients in Electrolyte Solutions, Pitzer, K.S. Ed., CRC Press, Inc., 1991.
16. Ptacek, C.J. Ph.D. Thesis, Univ. Waterloo, Waterloo, Ontario, Canada, 1992.
17. Gibbs, M.M. Water Res. 1979, 13, 295-297.
18. Stainton, M.P. J. Fisheries Res. Board Canada 1973, 30, 1441-1445.
19. Pitzer, K.S. J. Phys. Chem. 1973, 77, 268-277.
20. Harvie, C.E.; Weare, J.H. Geochim. Cosmochim. Acta 1980, 44, 981-997.
21. Harvie, C.E.; Møller, N.; Weare, J.H. Geochim. Cosmochim. Acta 1984, 48, 723-751.
22. Plummer, L.N.; Parkhurst, D.L.; Fleming, G.W.; Dunkle, S.A. Water Resources Investigation Report 88-4153, U.S. Geological Survey, 1988.
23. Reardon, E.J.; Beckie, R.D. Geochim. Cosmochim. Acta 1986, 51, 2355-2368.
24. Nordstrom, D.K.; Plummer, L.N.; Langmuir, D.; Busenberg, E.; May, H.M.; Jones, B.F.; Parkhurst, D.L. In Chemical Modeling of Aqueous Systems II, Melchior, C.; Bassett, R.L., Eds., ACS Symposium Series 416, American Chemical Society: Washington, DC, 1990, pp. 390-413.
25. Smith, H.J. J. Am. Chem. Soc. 1918, 40, 879-883.
26. Bruno, J.; Wersin, P.; Stumm, W. Geochim. Cosmochim. Acta 1992, 56, 1149-1155.
27. Ptacek, C.J.; Reardon, E.J. In Proc. 7th Intl. Symp. Water-Rock Interaction - WRI-7, Park City, Utah, 1992, pp. 181-184.
28. Singer, P.C.; Stumm., W. J. Am. Waste Water Assoc. 1970, 62, 198-202.
29. Bardy, J.; Pere, C. Trib. CEBEDEAU 1976, 29, 75-81.
30. Boorman, R.S.; Watson, D.M. Can. Inst. Mining Metallurg. Bull. 1976, 69, 86-96.
31. Appleyard, E.C.; Blowes, D.W. In this volume.
32. Jambor, J.L.; Blowes, D.W. CANMET Div. Report MSL-89-137(IR), CANMET, Energy, Mines and Resources, Canada, 1989.
33. Reardon, E.J. J. Phys. Chem. 1988, 92, 6426-6431.
34. Reardon, E.J. J. Phys. Chem. 1989, 93, 4620-4636.
35. Allison, J.D.; Brown, D.S.; Novo-Gradac, K.J. U.S./E.P.A./600/3-91/021, U.S. Environ. Protection Agency, Athens, GA, 1991.

RECEIVED September 1, 1993

Chapter 14

Mineralogical Characteristics of Poorly Crystallized Precipitates Formed by Oxidation of Fe^{2+} in Acid Sulfate Waters

E. Murad[1,4], U. Schwertmann[1], Jerry M. Bigham[2], and L. Carlson[3]

[1]Lehrstuhl für Bodenkunde, Technische Universität München, D−85350 Freising-Weihenstephan, Germany
[2]Department of Agronomy, Ohio State University, Columbus, OH 43210−1086
[3]Department of Geology, University of Helsinki, SF−00171 Helsinki, Finland

Identification of the minerals produced by the oxidation of sulfides is commonly complicated by the poor crystallinity of the products. A combination of selective chemical extraction procedures and instrumental techniques such as X-ray diffraction and Mössbauer spectroscopy can enable an unequivocal identification of even the most poorly crystalline minerals to be made, and in favorable cases can also provide information on particle dimensions from magnetic blocking temperatures or hyperfine fields at 4.2 K.

The products of sulfide oxidation are controlled by the elements released and the environmental conditions, which favor specific mineral assemblages by controlling the pathways of mineral formation and the kinetics of competing reactions. Well crystallized minerals formed as result of sulfide oxidation can usually be readily identified using standard mineralogical techniques such as X-ray diffraction (XRD) or electron microscopy. Normally, mineral formation takes place at very low pH values resulting from the formation of large amounts of H_2SO_4. If, however, freshwater enters the system, higher pH values and lower SO_4 concentrations ensue. For Fe^{3+}, this leads to a pronounced increase in the rate and degree of hydrolysis, causing rapid precipitation of Fe^{3+} oxides and oxyhydroxides of extremely small particle size ($\leq$ 10 nm). Such particles are difficult to characterize, especially if they are constituents of assemblages of complex mineralogy. Typical examples for such minerals are the poorly crystallized iron oxide ferrihydrite, which has often erroneously been referred to under various names such as "amorphous iron oxide" or "ferric hydroxide", and an iron oxyhydroxysulfate recently identified as a constituent of numerous precipitates of acid sulfate-rich waters all over the world (1). This phase is hereafter referred to as "mine drainage mineral", abbreviated MDM.

Identification of these poorly crystalline products of sulfate oxidation may be possible using XRD carried out by slow step-scanning, where necessary combined with

[4]Current address: Bayerisches Geologisches Landesamt, Concordiastrasse 28, D−96049 Bamberg, Germany

selective dissolution procedures and diffraction line profile analysis. ^{57}Fe Mössbauer spectroscopy has also proven useful for the characterization of iron-bearing minerals of small particle size, especially if these are only minor constituents of a sample. In this paper we show how data acquired by these techniques can be used to characterize selected minerals from acid sulfide oxidation environments, and thus lead to an improved delineation of the processes involved.

Materials and Methods

We collected ochreous precipitates resulting from the bacterial oxidation of $Fe^{2+}_{(aq)}$ in sulfate-rich waters draining abandoned mine works and acid sulfate soils from a variety of localities around the world. This collection of samples was complemented by material synthesized bacterially using Thiobacillus ferrooxidans and abiotically in the presence of sulfate, selenate and chromate in the laboratory. Data on complex samples and samples that have been selected on the basis of purity (i.e. as monomineralic as possible) is given in this paper.

XRD was performed either on a Philips PW1050/70 instrument using CoK_α radiation and a graphite monochromator or on a Huber System 600 Guinier diffractometer using $CoK_{\alpha 1}$ radiation. Random powder samples were counted in steps of 0.02° 2θ for 10 to 50 s per step. Where necessary, silicon was added as an internal standard and peaks were computer-fitted with a combination of Gaussian and Lorentzian lines (Voigt profile) to obtain precise d values and half widths.

Fourier-transform infrared absorption data were obtained from 5 mg of powdered sample mixed with 195 mg KBr. Diffuse reflectance (DRIFT) spectra were collected as the average of 100 sample scans at 1 cm^{-1} resolution using a Mattson-Polaris FTIR spectrometer equipped with a Hanick "Praying Mantis" cell.

^{57}Fe Mössbauer spectra were taken at various temperatures between 295 and 4.2 K using a ^{57}Co/Rh source. The samples were placed in plastic holders to give Fe concentrations between 6 and 12 mg/cm^2. Selected samples were also studied under applied magnetic fields up to 9 T. Spectra were collected in different velocity ranges between ± 2 and ± 13.6 mm/s until sufficiently good statistics had been attained. The spectra were fitted with Lorentzian lines or distributions of Lorentzians. Isomer shifts are given relative to the centroid of the room-temperature spectrum of metallic Fe.

Individual Minerals

Ferrihydrite. The nominal formula of ferrihydrite is given as $Fe_5HO_8 \cdot 4H_2O$ (2). Although the composition may vary somewhat as a function of particle size, recent work (3) has shown that the variation in composition is only in part related to the adsorption of water on particle surfaces. Ferrihydrite is reddish-brown in color (Munsell hue 5 YR - 7.5 YR), and usually consists of minute, aggregated spheres 3 to 6 nm in size. It has a high specific surface area between 200 and 500 m^2/g, and is — in contrast to goethite and hematite — readily soluble in ammonium oxalate at pH 3 (4). Because of the small particle size, XRD diagrams of ferrihydrite consist of only six to two broad, weak bands. The characterization of ferrihydrite by XRD is complicated by the fact that the broad bands go hand in hand with very low count rates. A typical ferrihydrite from an acid drainage environment has an XRD diagram (shown in Figure 1) that consists of four such bands located at 0.25, 0.22, 0.17 and 0.15 nm. When counted for 20 s per step, this ferrihydrite appeared to be to be monomineralic. A prolonged counting time of 50 s, however, revealed additional peaks at 0.42 and 0.34 nm (Figure 1), showing up the presence of minor amounts of goethite and MDM. These admixtures imply a different mineral assemblage, and thus could indicate the crossing of a stability field boundary; they will naturally also contribute to other physical (e.g. magnetic) properties of this sample.

Because of small particle size, natural ferrihydrite is superparamagnetic at room temperature and may remain superparamagnetic down to temperatures as low as 23 K (5). In the superparamagnetic state, Mössbauer spectra of ferrihydrite consist of a broad Fe^{3+} doublet of non-Lorentzian shape. The high line widths indicate a variability of atomic environments, and the spectra consequently have to be fitted with distributions of quadrupole-split doublets rather than with single Lorentzian doublets. The maxima of the distributions, which are higher than those of almost all other (super)paramagnetic iron oxides, shift from 0.62 to 0.80 mm/s with decreasing particle size, and the distribution half-widths simultaneously increase from about 0.70 to 0.86 mm/s. Since the quadrupole splitting of high-spin Fe^{3+} in octahedral coordination is directly proportional to the site distortion, this shows that the site distortion and the variation of nuclear environments increase with decreasing particle size.

For ferrihydrites of different particle size, magnetic blocking temperatures (50 % magnetic order) between 28 and 115 K have been described (5), corresponding to average particle sizes of 3 and 5 nm, respectively. Superparamagnetic particles possess uncompensated surface magnetic moments, and will therefore order magnetically above the magnetic blocking temperature under an applied magnetic field (Figure 2). The magnetic hyperfine fields observed at 4.2 K vary between about 50 T for ferrihydrites with 6 XRD bands and 47 T for such with 2 XRD bands. Because of an increasing randomization of the angle between the magnetic hyperfine field and the electric field gradient, the quadrupole interaction of magnetically ordered ferrihydrites decreases with decreasing particle size, and approaches zero for 2-XRD-band ferrihydrites.

MDM. The major component of all precipitates formed between pH 2.5 and 4.0 is a poorly crystallized, yellowish-brown (Munsell 9-10YR) iron oxyhydroxysulfate with an ideal chemical formula of $Fe_8O_8(OH)_6SO_4$ (1). This component has a fibrous morphology, a high specific surface area between 175 and 225 m^2/g, and is readily soluble in ammonium oxalate at pH 3. The XRD diagram of MDM consists of 8 broad bands for d > 0.15 nm (Figure 1), and indicates a tunnel structure akin to that of akaganéite, β-FeO(OH)$_{1-x}$Cl$_x$ (nominally β-FeOOH). The sulfate is believed to be located in the mentioned tunnels, taking over the role of chloride in akaganéite, but sharing (two) oxygen atoms with neighboring iron atoms. The MDM scan shown in Figure 1 has an additional peak at 0.31, which probably originates from a minor admixture of jarosite.

Mössbauer spectra of MDM taken at room temperature consist, in contrast to those of ferrihydrite, of a broad asymmetric doublet, the low-velocity line of which has a higher dip and is narrower than the high-velocity line (6). To account for the

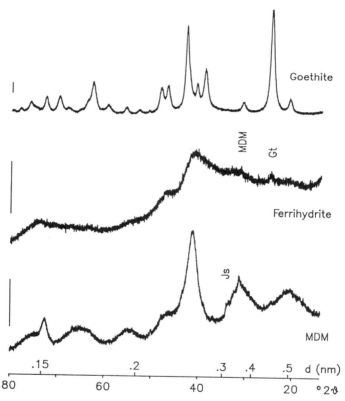

Figure 1. X-ray diffraction diagrams of goethite, ferrihydrite and MDM from acid-drainage environments. Marked foreign mineral admixtures are jarosite (Js), goethite (Gt) and MDM; scale bars correspond to 400 counts per second.

asymmetry, the spectra were fitted with two alternative quadrupole-splitting distribution models that imply either a linear correlation between the quadrupole splitting and the isomer shift, or the existence of two discrete quadrupole-splitting distributions with different isomer shifts. Both fitting models yield average quadrupole splittings of maximum probability of about 0.65 mm/s. The distribution half-widths of MDM (0.46 to 0.52 mm/s for one-distribution fits) were considerably narrower than those of even the best crystallized ferrihydrite (~ 0.70 mm/s), indicating a smaller variation of nuclear environments in MDM than in any ferrihydrite.

Mössbauer spectra show that both natural and synthetic MDM have an average magnetic ordering temperature of 75 K, although the onset of magnetic order is smeared out over about 10 K. The ordering temperature of MDM thus is lower than the Néel temperature of any iron oxide. The application of an external magnetic field of 6 T between 75 and 90 K, however, did not induce magnetic order (Figure 3). This shows the absence of superparamagnetic relaxation, and indicates that the ordering temperature of 75 K is a genuine Néel temperature, and thus that a distribution of Néel temperatures causes the magnetic ordering temperature to be smeared out. The spectra of magnetically ordered MDM are broad and asymmetric, the first (low-velocity) line having a higher dip and being narrower than the sixth line. Therefore fits of hyperfine field distributions that imply a variation of quadrupole splitting as a function of the hyperfine field, or two discrete hyperfine field distributions with different isomer shifts had to be used. Independently of the fit model, the hyperfine fields of maximum probability averaged 45.4 T at 4.2 K. This is lower by about 1.5 T than those of even the most poorly crystallized ferrihydrite.

The asymmetry of the Mössbauer spectra of both paramagnetic and magnetically ordered MDM and the range of magnetic ordering temperatures indicate a diversity of iron sites. The most obvious explanation is that the sulfate incorporated in the crystal structure is responsible for this diversity. Different locations of iron atoms relative to sulfate groups in the structure would cause the development of inequivalent iron sites. The sulfate probably also inhibits magnetic exchange interactions between neighboring iron atoms and thus may be responsible for the low magnetic ordering temperature and hyperfine fields.

MDM can also be synthesized in the presence of selenate and chromate, whereas arsenate and phosphate inhibit the formation of MDM. This is probably due to the fact that arsenate and phosphate have a higher affinity to iron than sulfate and selenate, which prevents the formation of a phase with iron in an ordered, octahedrally coordinated structure (to be published elsewhere). Room-temperature Mössbauer spectra of MDM synthesized in the presence of chromate no longer show the characteristic asymmetry of sulfate MDM. An exchange of arsenate for sulfate can be effected by treating MDM synthesized with sulfate with 0.04 M Na_2HAsO_4 for 24 hours. The effects of this treatment are particularly evident in infrared spectra, in which the SO_4^{2-} bands have almost completely vanished (Figure 4). This is accompanied by a marked attenuation of the (200,111) XRD peak at 0.48 nm, indicating that the anion exchange has also affected the structure.

Goethite. In contrast to ferrihydrite, which is a common mineral in environments where Fe^{2+} is rapidly oxidized, and MDM, which is specific to acid mine drainage

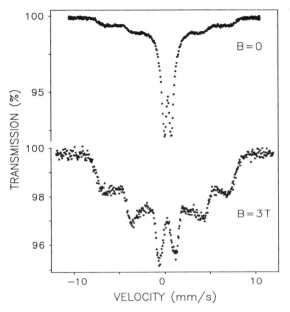

Figure 2. Mössbauer spectra of a ferrihydrite from an acid drainage environment taken at 78 K with and without an externally applied magnetic field.

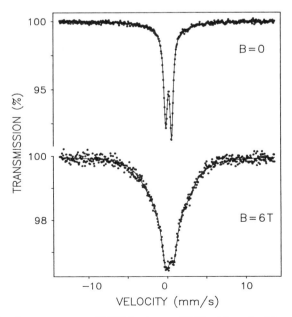

Figure 3. Mössbauer spectra of MDM taken at 75 K with and without an externally applied magnetic field.

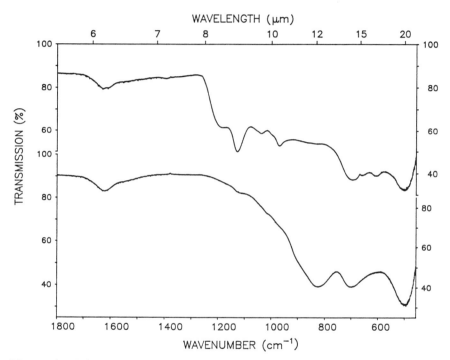

Figure 4. Infrared spectra of synthetic MDM before (top) and after (bottom) treatment with 0.04 M Na_2HAsO_4 for 24 hours at pH 6.0.

environments rich in sulfate, goethite, α-FeOOH, is a very common mineral in a variety of different terrestrial and aquatic environments, for example in soils or sediments. XRD lines of a goethite from acid mine drainage, shown in Figure 1, have widths at half height of about 1° 2θ for CoK_α, corresponding to a particle size of about 15 nm. In contrast to other goethites, which usually have only a very low solubility in ammonium oxalate at pH 3, goethite-rich samples from mine drainage environments have higher ratios of oxalate-soluble to total iron of about 0.3. This oxalate solubility may result in part from the admixture of ferrihydrite and/or MDM, minor amounts of which are difficult to detect by XRD, but which show up clearly in Mössbauer spectra taken at 4.2 K (Figure 5).

Although well crystallized, pure goethite has a Néel temperature of 400 K, room-temperature Mössbauer spectra of goethites from acid drainage environments consist of a doublet. This is probably a result of small particle size, leading to superparamagnetic relaxation. At 4.2 K all studied goethites from acid drainage environments were magnetically ordered, with magnetic hyperfine fields that were, however, noticeably lower than that of pure, bulk goethite (50.6 T). In the absence of other effects that could have an influence on the magnetic properties (for example aluminum-for-iron substitution and/or excess water content), this may be attributed to small particle size. A hyperfine field of 49.88 T, for example, was derived from the spectrum shown in Figure 5. This would correspond to a mean crystallite diameter of 11 nm in the [111]-direction (7), and compares favorably with the crystal dimensions observed by XRD.

Complex Natural and Synthetic Assemblages

Ochreous precipitates of complex mineralogy, collected from streams affected by acid mine drainage in the U.S.A., Australia, and Bosnia (Yugoslavia), were studied by XRD and by Mössbauer spectroscopy at temperatures between 295 and 4.2 K before and after treatment with acid ammonium oxalate.

Figure 5 shows Mössbauer spectra taken at 4.2 K of a precipitate from a stream in southeastern Ohio. Published work indicated this precipitate to consist of goethite and an oxalate-soluble ferrihydrite-like material (8). The spectrum of the untreated sample consists of two sextets, which can be assigned to goethite and MDM in a proportion of about 1:2 (assuming equal recoil-free fractions), and a central doublet with a relative area of about 3%. After a short oxalate treatment of 15 minutes duration (spectrum not shown), the MDM sextet had disappeared. The spectrum now comprised only a goethite sextet and the paramagnetic doublet with a relative area of 4 %. After an oxalate treatment of 2 hours the spectrum still consisted mainly of a goethite sextet, but the relative area of the paramagnetic component had increased to 12 %. This increase in the doublet intensity indicates that the extended oxalate treatment has resulted in a partial removal of goethite (8). The magnetic hyperfine fields of goethite in the untreated sample (49.96 T) and that remaining after the 15-minute (49.99 T) and the 2-hour oxalate treatments (49.69 T) were nevertheless identical within experimental error. Thus the oxalate treatment did not preferentially dissolve goethite of smaller particle size, which should have lower magnetic hyperfine fields. The quadrupole interaction (-0.24 ± 0.01 mm/s) was the same throughout for the component identified

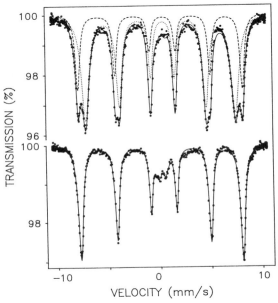

Figure 5. Mössbauer spectra taken at 4.2 K of material from an acid mine drainage containing MDM and goethite before (top) and after (bottom) a 2-hour treatment with ammonium oxalate at pH 3. Subspectra indicated in the untreated sample are goethite (broken line) and MDM (dotted line).

as goethite. This shows that ferrihydrite, which may have a similar hyperfine field but a different quadrupole interaction (between -0.02 and -0.10 mm/s), was absent from all samples. The fact that the relative area of the paramagnetic resonance did not increase significantly following the first oxalate treatment, but showed a strong increase following the second treatment, indicates that this is due to at least two different components. The first of these is as readily soluble in oxalate as MDM, whereas the second is not (or less) soluble in oxalate, and could be due to Fe^{3+} in clay minerals.

XRD showed the complex precipitates from the other mentioned mine drainage environments, New South Wales (Australia) and Bosnia (Yugoslavia), to be made up mainly of MDM, with subordinate admixtures of jarosite in the former and of jarosite and poorly-crystalline goethite in the latter sample. Mössbauer spectra taken at 4.2 K confirmed the presence of MDM in both samples, but because jarosite and poorly-crystalline goethite have similar parameters, a distinction of these in the untreated Yugoslavian sample was not possible. A Mössbauer spectrum of the Australian sample taken at 4.2 K after oxalate treatment showed that MDM had been quantitatively removed, and that only jarosite was left over.

To complement our work on natural precipitates of acid mine drainage and to simulate the conditions under which MDM and goethite coexist in nature, we undertook a long-term laboratory experiment. Synthetic MDM, prepared as described in (1), was dialyzed to remove excess sulfate from the solution and stored in distilled water for up to 1000 days. Aliquots of the suspension were collected at intervals and studied by chemical analysis and XRD. Preliminary results showed that traces of goethite could be detected after 72 days, and that conversion of MDM to goethite was complete after about 200 days. During this process the pH and Fe concentration decreased, whereas the sulfate concentration increased, supporting the reaction

$$Fe_8O_8(OH)_6SO_4 + 2\,H_2O \rightarrow 8\,FeOOH + H_2SO_4.$$

This reaction indicates that MDM is metastable with respect to goethite; it will transform to goethite within a period of several months even in the presence of 2 m mol/l SO_4. Further experiments to determine the conditions under which the conversion of MDM to goethite is inhibited, with the purpose of arriving at a (conditional) solubility product of MDM, are presently under way.

Conclusions

Poorly crystallized minerals formed in acid mine drainage environments and the pathways of their transformations can be characterized by combinations of modern mineralogical techniques such as X-ray diffraction using step-scanned and computer-fitted data, Fourier-transform infrared spectroscopy, and Mössbauer spectroscopy. In this paper we give examples for the characterization of individual minerals by these methods. Further examples are given to show how a combination of these techniques with selective dissolution procedures, for example acid oxalate extraction, also enables the characterization of these minerals to be made in natural and synthetic assemblages of complex mineralogy from acid sulfate oxidation environments.

Addendum. After submission of this article, the phase designated herein as MDM was approved by the Commission on New Minerals and Mineral Names of the International Mineralogical Association as a new mineral with the name "schwertmannite".

Acknowledgments

We are indebted to L.H. Bowen (North Carolina State University, Raleigh, N.C.) for taking some of the low-temperature Mössbauer spectra of MDM and to J.D. Cashion (Monash University, Melbourne) and F.E. Wagner (Technische Universität München, Garching) for providing access to liquid-helium Mössbauer facilities. This study has been supported by the Deutsche Forschungsgemeinschaft and the Alexander von Humboldt Stiftung.

Literature Cited

1. Bigham, J.M.; Schwertmann, U.; Carlson, L.; Murad, E. Geochim. Cosmochim. Acta 1990, 54, 2743-2758.
2. Towe, K.M.; Bradley, W.F. J. Colloid Interf. Sci. 1967, 24, 384-392.
3. Stanjek, H.; Weidler, P.G. Clay Minerals 1992, 27, 397-412
4. Schwertmann, U.; Fischer, W.R. Geoderma 1973, 10, 237-247.
5. Murad, E.; Bowen, L.H.; Long, G.J.; Quin, T.G. Clay Minerals 1988, 23, 161-173.
6. Murad, E.; Bigham, J.M.; Bowen, L.H.; Schwertmann, U. Hyperfine Interact. 1990, 58, 2373-2376.
7. Murad, E.; Schwertmann, U. Clay Minerals 1983, 18, 301-312.
8. Brady, K.S.; Bigham, J.M.; Jaynes, W.F.; Logan, T.J. Clays Clay Minerals 1986, 34, 266-274.

RECEIVED March 16, 1993

Chapter 15

Atomic and Electronic Structure of PbS {100} Surfaces and Chemisorption—Oxidation Reactions

Carrick M. Eggleston[1] and Michael F. Hochella, Jr.[2]

[1]Institute for Water Resources and Water Pollution Control, Swiss Federal Institutes of Technology, CH–8600 Dubendorf, Switzerland
[2]Department of Geological Sciences, Virginia Polytechnic Institute and State University, Blacksburg, VA 24061

> In order to examine oxidation mechanisms at the molecular level, scanning tunneling microscope (STM) images of fresh and oxidized PbS (galena) surfaces are presented, including real-time sequences, and interpreted using electronic structure information from spectroscopic and computational studies. Oxidation is nonuniform; oxidized areas nucleate and spread across the surface. Initial oxidation appears to be autocatalytic. Borders between oxidized and unoxidized areas tend to prefer [110] and equivalent orientations, suggesting that unoxidized S sites along [110] and equivalently oriented borders are less reactive than unoxidized S sites along [100] and equivalently oriented borders. The STM results are valuable in that they focus attention on questions concerning the relative reactivity of specific surface structures.

Natural semiconducting materials (e.g., some sulfides and transition-metal oxides) play three important geochemical roles: they act as electrical connectors between redox couples (i.e. natural electrodes), they act as electron sources or sinks, and they are often important in photochemical processes. Although natural semiconductors are relatively rare, they can play a disproportionate role in local and regional chemistry. A good example is sulfide mineral oxidation (SMO), as this book attests. Here, we focus on microscopic mechanisms of SMO rather than on a particular geological setting for SMO.

As is usually the case when solid minerals interact with their fluid environment, surfaces are the focus of SMO reactions. The surface properties of sulfide minerals, and their alteration by oxidation, have been studied extensively. These studies have clarified the electronic structure of many sulfides and have confirmed the existence of various possible and postulated SMO reaction intermediates and products.

Scanning tunneling microscopy (STM) is known as an atomic-resolution imaging technique, and it has been suggested that a combination of spectroscopic and STM data could advance our understanding of SMO surface reaction mechanisms (*1*). The purpose of this paper is to attempt such a synthesis for PbS (galena), a mineral whose electronic structure is relatively well understood. We examine changes in surface atomic and electronic structure resulting from initial oxidation of PbS {100} surfaces.

Our goal is to understand better oxidation mechanisms, particularly the manner in which surface structure controls oxidation progress. This work extends our previous study of PbS oxidation in relation to reduction of aqueous gold complexes (2).

STM is an electronic structure probe. The extremely local electronic interaction between tip and sample in STM, however, adds the spatial resolution lacked by other electron spectroscopic techniques such as photoelectron spectroscopy (PES). On the other hand, interpretation of STM images depends on knowledge of surface electronic structure, much of which comes from calculations and PES data. The relationship between STM and other electronic structure data is thus close, and the arrangement of this paper reflects this relationship. After a brief methods section, we discuss the electronic structure of "clean" PbS surfaces by incorporating our STM results into a review of electronic structure data from the literature. We then discuss the effects of oxidation as reflected in PES spectra (mostly from the literature) and then in our STM images. The discussion then considers the central question of surface structural controls on reaction progress.

Methods

Information on the electronic structure of PbS comes mostly from PES and electronic structure (molecular orbital and band structure) calculations. For the reader unfamiliar with some of these methods, X-ray PES (XPS) is reviewed in Hochella (3), and UV PES (UPS) in Plummer and Eberhardt (4). Reviews of STM are found in Eggleston and Hochella (5) and Avouris (6), and computational methods are reviewed in Simons (7) and Tossell and Vaughan (8).

We collected UPS spectra for a PbS sample that had been exposed to air for 1 minute before evacuation to 10^{-10} torr. The most common light energies used in UPS are the first and second ionization resonances of helium, He I (21.2 eV) and He II (40.8 eV); we used He II for reasons given below. Many of the oxygen-exposure studies refer to the "Langmuir" surface exposure unit. One second of surface exposure to adsorbing gas at 10^{-6} torr is defined as 1 Langmuir (L); i.e., $1 L = 10^{-6}$ torr sec. One L of gas exposure forms one monolayer at room temperature given a sticking coefficient of unity.

The STM data were obtained using Digital Instruments Nanoscope II and III STMs. We used tungsten tips electrochemically etched in 1 M KOH. STM imaging was done both in air and under oil so as to suppress reaction with air (2, 5, 9). STM, although simple in theory, is highly dependent on the temporally variable shape and electronic properties of the tip and its interaction with sample variability (Fig. 7 shows an example of a tip changing its properties during a scan). Consequently, the probability of obtaining atomically resolved images is relatively small, especially under ambient (rather than vacuum) conditions. This contributes to a problem of STM that is inherent to high-resolution microscpy: although very specific features may be resolved, it is difficult to be sure that they are statistically representative of the surface as a whole.

Electronic structure of clean PbS {100} surfaces

Figure 1 summarizes PES spectra and band-structure calculations from the literature for PbS {100} surfaces, and includes our STM spectroscopic results. Figure 1 (left) shows spectra from (curve A) XPS (10), (curve B) He I UPS (11), and (curves C and D) He I and He II UPS (12). The bottom curve is a density of states (DOS) profile from band-structure calculations (13). The spectra agree well with molecular orbital (MO) calculations (14, 15). The spectra and calculations all show a peak (1') just below the Fermi level (E_F) as a shoulder on peak 1. Peaks 1-1' have been attributed to three S 3p bands (10, 16). Peak 2 is attributed to Pb 6s - S 3p bonding states (14, 15).

We have plotted all the 1-1' peaks with similar size in order to compare peak shapes, but the intensity of peak 1-1' decreases relative to peak 2 in going from He I to He II light. The photoionization cross-sections of electron states depends on the light energy; Figure 2 shows that the cross-section for filled 3p states decreases by more than an order of magnitude between 20 and 40 eV (17), in agreement with observation (compare Fig. 1, left, curves B and C with curve D), supporting the attribution of peaks 1-1' to S 3p bands.

In solids, in contrast to small molecules, electron states form energy-disperse bands. In UPS, electron emission from different parts of a band depends on the light energy and the emission angle (16,18). The small changes within peak 1-1' with light energy in Figure 1, left, are related to band structure (12). Emission-angle dependence of PbS bands has been studied using angle-resolved UPS; the results are in good agreement with calculated band structure (16), and corroborate the assignment of the valence band to S 3p states. The correspondence of calculated valence band (VB) structures with experiment also suggests that the calculations for the conduction band (mixed Pb and S character) are also fairly accurate (13-16).

Figure 1 (right) shows tunneling spectra for PbS taken by varying the STM bias voltage (V) and recording responding variations in tunneling current (I). In the limit of low V, the function (dI/dV)/(I/V) gives the convolved DOS of tip and sample (for a metal-semiconductor junction this primarily reflects the sample DOS) (6). We collected 12 spectra, but plot only four (from a range of tip-sample distances) for clarity. The spectra were quite noisy, but all 12 show a) a large peak at roughly -400 mV, b) indicate an n-type semiconductor (E_F is near the top of the band gap), and c) show conduction band (CB) states above E_F. The band-gap at the surface appears to be about 0.20 - 0.25 eV, in good agreement with a bulk value of 0.29 eV (16) given that the surface value should be less the bulk value (15).

Two of the spectra in Figure 1 (right) show the "clean" bandgap (open symbols), and two spectra (filled symbols) have some DOS within the bandgap. The origin of the mid-gap DOS is unclear, but their intermittent nature suggests that they are impurity states associated with adsorbed contaminants. These spectra differ from our previous spectra taken using a different technique (Eggleston, C. M.; Hochella, M. F. Jr. Am. Mineral., in press.), probably because the voltage modulation used previously may produce "stray" currents that smooth out peaks. The technique used here requires no voltage modulation.

In keeping with the electronic structures discussed above, the peak in the DOS around -400 mV is probably due to S 3p-like states at the top of the VB, and the DOS above E_F is probably due to Pb 6s - S 3p antibonding states of the bottom of the CB. These predictions may be compared with STM images. Figure 3 shows STM images of freshly cleaved PbS {100}. PbS has the NaCl structure. The negative-bias image (A) shows one kind of dominant site; given a VB of mostly S 3p character, these should correspond to S sites. The positive-bias image (B) shows two types of site, one brighter than the other; given a CB of mixed Pb and S character, one of these sites should be due to S and one to Pb (although there is no guarantee that the bright spots correspond exactly to atomic positions). We have shown (5) that the bright sites imaged at both biases coincide, strongly suggesting that the bright sites in both cases are due to S.

Reaction of fresh surfaces with oxygen and air

He I UPS. Figure 4, left, curve A, is a He I UPS spectrum (11) for fresh PbS, replotted for comparison; curve B is a He I UPS spectrum for PbS after exposure to microwave-excited atomic oxygen for 15 minutes at 10^{-5} torr (12). Compared to curve A, curve B has less intensity in the 1 eV region, but other changes are less clear

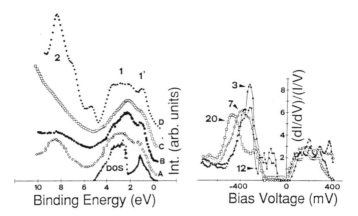

Figure 1. Left: XPS (A) and UPS (B-D) spectra for the valence band of clean PbS {100} surfaces, together with a density of states (DOS) calculation; from (10-13). Right: Tunneling spectra taken with the STM. The original data were digitally differentiated and single-pass 5-point smoothed. Each spectrum is labelled with the current at which the STM feedback was running during data collection; the larger the current, the closer the tip to the sample. Data from the literature in this figure and in Figures 2 and 4 were digitized and replotted; the data points are from digitization and do not necessarily reflect the original data density.

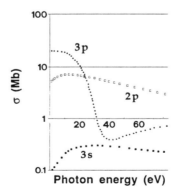

Figure 2. Photoionization cross-sections (s) for 3s, 2p, and 3p electrons in the 0 - 80 eV excitation energy range. Units are in Megabarns and eV. Data from (17).

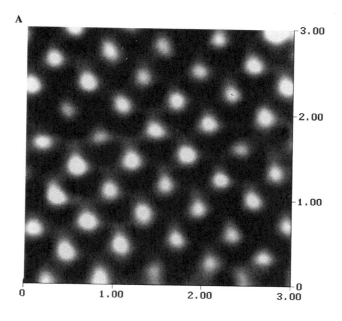

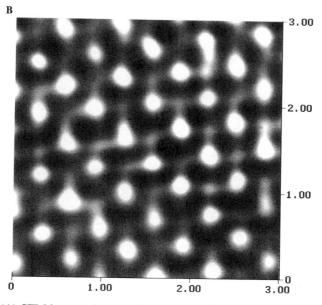

Figure 3. (A) STM image of PbS {100} taken at -580 mV and 1 nA. (B) STM image of PbS {100} taken at 200 mV and 1 nA. The bright sites in (A) and the brighter of the two types of sites in (B) correspond to S sites. Scales are in nanometers.

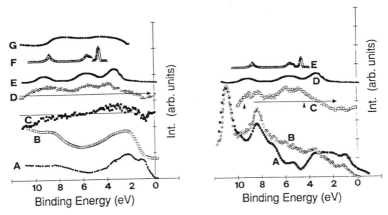

Figure 4. Left: He I UPS spectra (A) for clean (11) and oxygen-exposed (12) PbS surfaces (B), together with difference spectra (11) for oxygen-exposed surfaces (C, D). Curves E-G are He I UPS spectra for SO_4^{2-} in $LiSO_4$, for gas-phase SO_2 (23), and for S_2^{2-} from FeS_2 (27) respectively. Right: He II UPS spectra for (A) clean (12) and air exposed (B) PbS, with difference spectrum (C). Curves D and E are the same as curves E and F at left.

and are more easily illustrated with difference spectra (obtained by subtracting the fresh surface spectrum from that of an oxygen-exposed surface).

Figure 4, left, shows difference spectra from (11) for a PbS surface after 10^9 L (C) and 10^{12} L (D) O_2 exposure, equivalent to about 6 seconds and 2 hours of exposure to (dry) air, respectively. Points in curves C-D above (below) the line correspond to increase (loss) of intensity. In curve C, the 0 to 1.5 eV region has lost, and the 2-5 eV region has gained, intensity. Although the authors (11) claim that O_2 is physisorbed, curve C is not consistent with molecular orbitals of O_2 (18); also, XPS studies (19, 20) show evidence of Pb-O bonding (i.e., chemisorption) for 10^9 L air exposures. The extra intensity is probably due to O 2p states in dissociatively chemisorbed oxygen (18). Pb-O bonding is similar to Pb-S bonding; PbO is also a semiconductor, with a 2.7 eV bandgap (21). Initial O_2 exposure appears to give the surface a PbO-like component; the extra intensity occurs at an energy consistent with a material of roughly 2 eV bandgap. Adsorption probably occurs first at S vacancies (22) and then by slower sorption to other Pb sites. The second difference spectrum (curve D, 10^{12} L O_2) shows three peaks originally attributed to the molecular orbitals of physisorbed O_2 (11); however, the spacings of the peaks are in better agreement (23) with SO_4^{2-} (curve E) or SO_2 (curve F).

Polysulfides have been observed in many XPS studies of PbS oxidation (12, 19-20, 24-26), particularly in aqueous media. The broad band of extra intensity (curve D) between 2 and 8 eV could be attributed to S 3p states in S-S bonding; compare curve D to curve G for the S 3p bands of S_2 in pyrite (27). Thus, we cannot unequivocally assign the VB structure in the He I UPS to one molecular species; most likely, a mixture of species is present.

The important result is that the data show a loss of intensity from the top of the VB, suggesting oxidative stabilization of S 3p states into lower-lying states that are consistent with several known oxidation products. However, core-level XPS data from the surfaces giving curves C and D in Figure 4, left (11) in only one case gave evidence for S oxidation (the one exception was consistent with SO_2). Other core-level XPS studies have found no evidence of oxidized S on the surface for short (1-2 hour) air exposure times (12, 20). The apparent loss of S 3p intensity in the spectra we have discussed could be (and probably partly is) simply an attenuation of the S 3p signal by an overlayer of oxygen-containing species that lack state density in this region. Thus the spectra do not unequivocally prove that oxidation has taken place. This discrepancy is discussed further after the presentation of the He II results because it is important for the interpretation of STM results.

He II UPS. The photoionization cross-sections of 2p and 3p electrons are similar at 21.2 eV (He I), but higher for 2p than for 3p electrons at 40.8 eV (He II; Fig. 2). He II UPS is thus more sensitive to oxygen states at the surface than He I UPS. For this reason, we used He II to study the VB effects of air exposure.

Figure 4, right, curve A, is the He II UPS spectrum for fresh PbS (12). Our He II UPS spectrum (curve B) is for PbS exposed to air for 1 minute; the exposure to O_2 and H_2O is, assuming 150 torr O_2 and 20 torr H_2O, 9×10^9 L O_2 and 1.2×10^9 L H_2O. Curves A and B are plotted with the ~11 eV peaks (Pb 5d) at the same intensity. This peak is not strongly involved in bonding and changes little with oxygen exposure (12). In curve B, peaks 1-1' and peak 2 are still evident, but the relative intensities have changed. The difference spectrum (curve C) is similar to the He I difference spectra, showing loss of intensity at the top of the VB and a broad region of extra intensity deeper in the VB. The extra intensity consists of two broad peaks. OH^- gives similar broad peaks on semiconductor surfaces (23), and is consistent with XPS of PbS after air exposure (20), which shows clear evidence for bonding of surface Pb to OH^-. Sulfate (curve D) is unlikely, but an SO_2-like species (curve E) cannot be ruled out. Curves D and E in Figure 4, right, are from He I UPS data, and we are comparing them to He II data in this plot; the peaks should be at the same energies, but their relative intensities should not be the same; indeed, the lowest binding energy peak of the sulfate spectrum (curve D) is due mostly to O 2p states and should therefore dominate He II UPS spectra, but no such peak is observed.

The loss of intensity at the top of the VB in both the He I and He II spectra after oxygen or air exposure, as well as the changes in the top 10 eV of the VB, suggest that a small amount of S oxidation occurs for short exposure times. This is inconsistent with the conclusions of XPS studies in which the S 2p and S 2s peaks indicate no oxidized sulfur for up to several hours of exposure (12, 20). Let us examine this conclusion in more detail, however. The top layer of atoms on PbS is 3 Å thick. In XPS, using 1254 eV (Mg Kα) X-rays, this layer contributes only 18 % of the total S peak intensity (assuming an electron attenuation length of 15 Å); the rest of the intensity comes from deeper layers. If 10% or less of the S in the surface layer were oxidized, this would affect only 2 % or less of the total S signal, which is below the XPS detection limit (the X-ray cross sections for S 2p and 2s electrons are relatively small). Until more than 20% of the top layer is oxidized, XPS will not detect the oxidation products. For UPS, on the other hand, almost all of the signal comes from the top 5 Å. If 10% of the top layer is oxidized, this will amount to 10% (or more if we are seeing O 2p states and cross-section effects are accounted for) of the signal. We should be able to detect in He I UPS spectra species that cannot be detected with XPS (although definitive species identification is difficult). The discrepancy between UPS and XPS results is thus probably partly a detection limit difference and partly an attenuation effect (as discussed above). It seems reasonable to conclude that some surface S sites are indeed oxidized and decrease the surface S 3p DOS even for short exposure times. This is important, because our STM results apply to relatively short exposure times.

"Static" STM. In STM at negative sample bias, electrons tunnel from the occupied states of the sample VB to the tip. Oxidized sites should therefore appear dark (vacant) because of their local lack of S 3p state density. Figure 5 shows negative-bias STM images of PbS in air (compare to fresh surface in Fig. 3A). Figure 5A shows a PbS surface after air exposure of 20 minutes, equivalent to 1.8×10^{11} L O_2 and 2.4×10^{10} L H_2O; the image shows several individual dark sites, as well as a small "patch" of multiple dark sites (upper right corner). With increasing air exposure time, STM imaging became more difficult, perhaps because of mobile ionic oxidation products interfering with the tunneling current. Figure 5B shows, after three hours of air exposure, an island of unoxidized area (center) in an oxidized area. A structured array of atoms can be seen in the unoxidized area, whereas the oxidized area shows little regular structure (a few individual unoxidized sites remain). Figure 5C shows a negative-bias, 400 x 400 nanometer topographical STM image of a PbS surface after 3 days of air exposure. The sample was aligned with [100] and [010] axes parallel to the edges of the image. It is not immediately clear which areas are oxidized and which are not (the original surface was not perfectly flat), but it is likely that the darkest patches are oxidized areas. The image exhibits many lineations in the [110] and equivalent directions, particularly along the edges of the darkest areas. In contrast, steps on fresh surfaces follow [100] and equivalent directions for electrical neutrality reasons (28, 29). Therefore, the occurrence of [110]-parallel "steps" must be assoicated with the oxidation reaction. Returning to Figure 5B, for example, note that the edges of the unoxidized patch are commonly (though clearly not always, depending on how one classifies "kink" sites) parallel to [110] and equivalent directions.

Figure 6 shows positive-bias STM images of a PbS surface after 30 minutes of air exposure (roughly 2.7×10^{11} L O_2 and 3.6×10^{10} L H_2O). The images resemble the

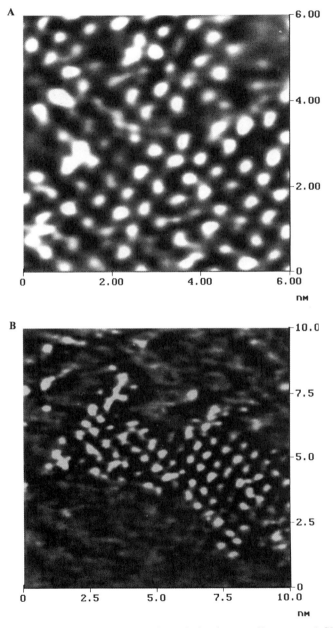

Figure 5. (A) Constant-height (contrast is variation in tunneling current) STM image taken at -405 mV, 2.2 nA after air exposure of 20 minutes, showing several "vacant" sites. These sites correspond to local loss of VB S 3p DOS (oxidation). (B) STM image taken under the same conditions as for (A), except after 3 hours of air exposure. *Continued on next page.*

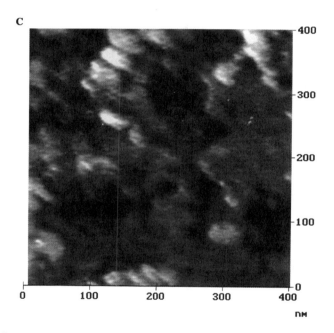

Figure 5. (C) Constant current (contrast is height variation; total relief is 12 Å), 400 x 400 nanometer STM image taken at -480 mV, 1.8 nA after 3 days of air exposure. All scales are in nanometers. Distortion of the square unit cell (see Fig. 10) is due to drift effects.

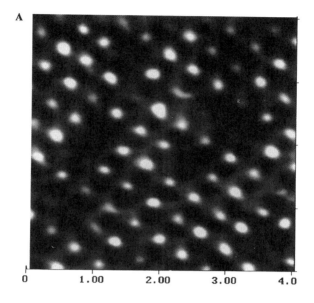

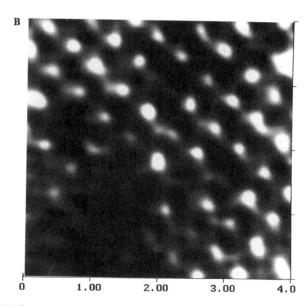

Figure 6. (A) Constant-height STM image of PbS {100} taken at 200 mV and 1.0 nA. Air exposure time was 30 minutes. Several "vacancies" (probable oxidized sites) are apparent. (B) STM image taken under the same conditions as (A), but on a different area. Scales are in nanometers. Slight differences in apparent unit cell size and shape are due to drift effects.

fresh surface in Figure 3B, but also exhibit several vacancies, similar to the negative-bias images (Fig. 5). Real-time image sequences (2, 30) show that such sites tend to coalesce into patches that grow with time. It appears that not only the VB S 3p states are lost upon oxidation, but also CB state density. This makes sense because we would expect that loss of S 3p electrons by oxidation would also result in the breakdown of Pb 6s-S 3p bonding, and antibonding, states; this would result in overall widening of the bandgap, in agreement with the fact that the end product of oxidation, $PbSO_4$, is an insulator. Although individual sites and pairs of sites are oxidized in Figure 6A, it was most common to see small patches of oxidized sites (2), an example of which is seen in the lower left corner of Figure 6B.

Figure 7A shows a surface after exposure to water for 1 minute; again, we see a number of single, double, and triple oxidized sites. Figure 7A also shows how the tip, and its resolving ability, can change abruptly during a single scan. Figure 7B shows a surface after exposure to water for 30 minutes. A narrow unoxidized patch runs from upper left to lower right. The edges of the oxidized area are parallel to both the [110] and [100] directions, though the [110] directions are favored, similar to Figure 5B.

The air and water exposure times we have given should be considered approximate, or at least minima, because it is possible that some of the features in Figs. 5-7 developed during imaging, even for the positive-bias cases in which oxidation reaction was suppressed by covering the surface in non-reactive, non-conductive oil (5); as we shall see below, the oil merely slows the reaction by providing a diffusion barrier between air and the surface.

Real-time STM. In addition to "static" imaging of incipiently oxidized PbS surfaces, we observed PbS oxidation in real time. Some of our real-time results at the atomic scale have been presented elsewhere (2,30). Here, we show microtopographical results.

Figure 8 shows a sequence (A-D) of images of a 200 x 200 nm area spanning three minutes. Image acquisition time was 45 seconds (i.e., the images are not "snap shots"; the tip scans back and forth across the imaged area). Scanning started 3 minutes after cleavage and ran for about 20 seconds before the first complete 45 second scan was begun. A triangular area with a central island is marked in each image. Clearly, the dark patches grow with time.

As one might suspect, the STM influences reaction progress. Figure 8E shows the edge of the area previously scanned by STM in air; outside the previously scanned area, evidence of oxidation was minor. The influence of the STM is electrochemical; the bias voltage is applied so as to promote anodic oxidation of the surface. We suspect that the development of borders between oxidized and unoxidized areas that are parallel to the fast (horizontal) scan direction of the tip (compare Fig. 8A to Fig. 8D) are also due to the STM. Unfortunately, this is also parallel to the [100] direction of the crystal, so it is difficult to separate the influence of the STM from possible crystallographic influences.

In order to test the possibility that the STM physically (rather than chemically) eroded the surface, we repeated the experiment of Figure 8 but covered the freshly cleaved sample in oil to limit air access (air still diffuses through the oil). The resulting sequence is shown in Figure 9, and spans 12 minutes. Dark patches grow over time, but much more slowly and covering a limited portion of the surface relative to Figure 8. If physical erosion of the surface by the tip caused the dark patches, one would expect little difference between air and oil, and would expect that pre-existing topographic features (such as the steps running across the images in Figure 9 from upper left to lower right; the tip scanned from right to left) would be more easily eroded than flat terraces. This is not the case.

Figures 8 and 9 show crystallographic controls on oxidation progress as noted previously (2,30); borders between oxidized and unoxidized areas tend to favor [110]

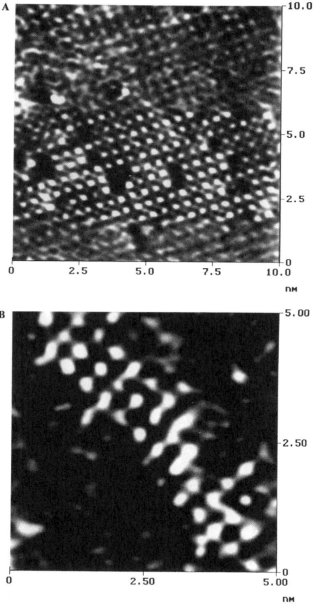

Figure 7. (A) Constant-height STM image of PbS {100} taken at 200 mV and 1.0 nA after 1 minute of exposure to liquid H_2O. The number and arrangement of "vacant" (oxidized) sites is very similar to that in Figure 6A. Note the change in the resolving ability of the tip; this is a very common phenomenon in STM imaging. (B) Constant-height STM image taken under the same conditions as at top, but after 30 minutes water exposure. A small unoxidized area runs diagonally from upper left to lower right between oxidized areas. Scales are in nanometers.

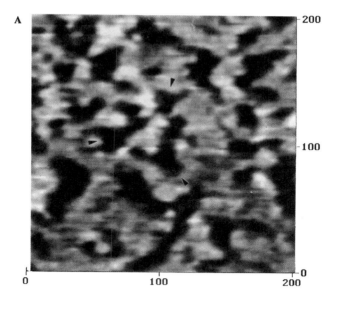

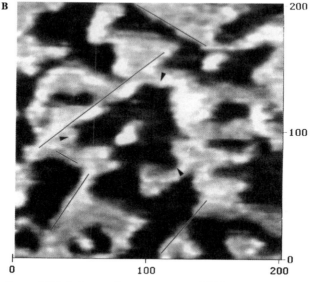

Figure 8. (A through D) A three-minute sequence of 200 x 200 nm STM images of a fresh PbS surface in air (+200 mV, 1 nA). Vertical relief is 20 Å. The [100] and [010] directions are nearly orthogonal to the image axes; [110] runs diagonally from lower left to upper right. The expanding dark patches are oxidized patches whose rate of formation is accelerated by the STM in air (see text). In B we have marked some reaction fronts roughly parallel to [110] and equivalent directions. Arrows mark an expanding triangular region in A through D.

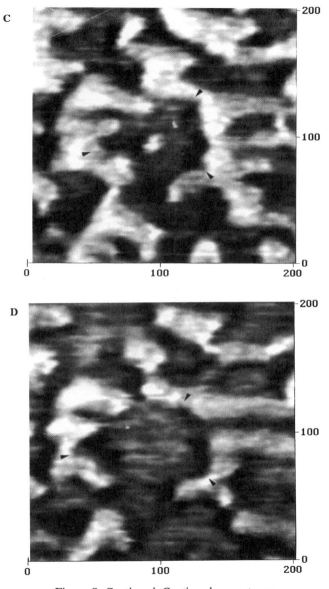

Figure 8. *Continued. Continued on next page.*

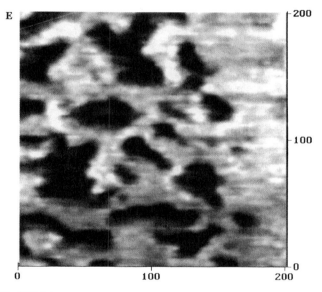

Figure 8E. STM image to the right of the area imaged in B. Outside the primary image area, evidence of oxidation can only be seen at the atomic scale (see Figs. 5–7).

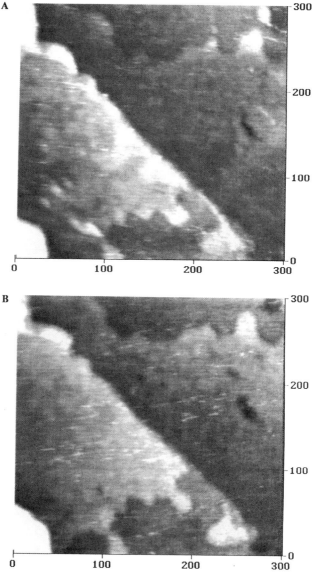

Figure 9. A 12-minute sequence (A through D) of four 300 x 300 nm STM images of a fresh PbS surface in oil (+200 mV, 1 nA). Vertical relief is 20 Å. The large step running diagonally across the image is parallel to [010]. Expansion of several pits can be seen during the sequence; in particular, note the enlargement of the pit on the right side of the images. Also, the step in lower center starts with edges near-parallel to [100] and [010] directions and ends with many edges parallel to [110] and equivalent directions; however, these are parallel and perpendicular to the scan directions (see text). *Continued on next page.*

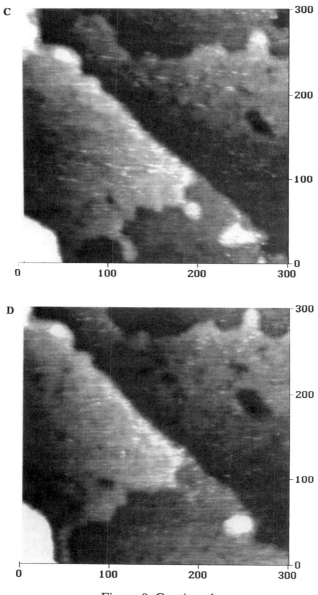

Figure 9. Continued.

and equivalent orientations, although borders parallel to [100] were also found. Examples of this may also be found in the atomic scale images (Figs. 5-7).

Discussion

Kinetic comparisons. The reaction rate implied by the "static" STM results, or observed in the real-time results, can be compared to bulk measurements of PbS oxidation rates. The "static" STM results are maxima for reasons discussed above, but if, as we observe, roughly 10% of the surface sites are oxidized in 30 minutes in air, that corresponds to a rate of 5.2×10^{-10} mol m^{-2} s^{-1}. Steady-state aqueous oxidation rates at room temperature in oxygen saturated water (24) are 5×10^{-9} mol m^{-2} s^{-1} for stoichiometric PbS. Initial dissolution rates (the first 6 hours) (31), including oxidation, are about 9×10^{-9} mol m^{-2}s^{-1}. The rate implied by "static" STM is an order of magnitude or more slower than those measured in aqueous solution; this is reasonable, because PbS oxidizes more quickly in water than in air. XPS results (20) show that long-term (3 months) oxidation in air occurs at a rate of roughly 10^{-12} mol m^{-2} s^{-1}, significantly slower than our STM result for initial air oxidation. These results are all within expectations, and show that for a slowly reacting material such as PbS, ex-situ "stop-action" STM imaging can be useful in following reaction progress.

Our real-time STM results may be compared with other measures of PbS oxidation rates. If we assume that the growing dark areas in Figure 8 represent complete oxidation of a PbS surface layer, then the oxidation rate within the image area is about 5×10^{-8} mol m^{-2}s^{-1}. This is comparable with voltammetry results (32), but an order of magnitude faster than aqueous oxidation rates and several orders of magnitude faster than air oxidation rates. This supports the conclusion that the reaction we observe is electrochemically accelerated by the STM. The comparisons we have made are approximate because they compare rates measured in different conditions and using different techniques; they are only intended to show that the STM results are not wildly divergent from expectation.

[110]-parallel reaction borders and mechanism constraints. The STM results are consistent with the idea that individual S sites oxidize very slowly at first on perfect surfaces but faster at defects such as dislocation outcrops (as imaged with STM; 29) and Pb or S vacancies (24). Such disperse oxidized sites are seen in our STM results (Figs. 5A, 6A) for short exposure times. For longer exposure times, we see larger oxidized "patches" (Figs. 8, 9), suggesting that initial oxidation can be likened to surface nucleation and growth of oxidized areas. Aqueous PbS oxidation shows an initial induction period, a brief acceleratory period, and then steady-state reaction (24); this correlates well with, and has been modelled as, a nucleation and growth process (24).

The STM results show directly that oxidation is non-uniform. This allows us to consider questions concerning the mechanism(s) by which oxidized areas spread across a surface. Nucleation and growth of oxidized areas suggests that unoxidized sites next to an oxidized area are more likely to oxidize than sites within unoxidized areas (i.e., that oxidation is autocatalytic); otherwise one would expect oxidation of randomly dispersed individual sites across the surface.

The STM results, here and in our previous work (2, 30), show that borders between oxidized and unoxidized areas are often parallel to [110] and equivalent directions, and less often parallel to [100] and equivalent directions. Although the STM results are subject to the statistical limitations of small data sets, and are subject in many cases to influences of the STM tip itself, they at least bring us to the point of asking specific questions related to the relative reactivity of [110] and [100]-parallel boundaries. For example, what determines the relative reactivity of S sites along these two kinds of borders?

Electron transfer (ET) rates, such as the oxidation of S sites via ET from S 3p states to O 2p states, are directly related to the overlap of the wavefunctions of the occupied and unoccupied states involved in the ET. Generally, this overlap is an exponential function of distance because radial atomic wavefunctions decay exponentially. This suggests that nearest-neighbor interactions (e.g., between an unoxidized S site and nearest neighbor oxidants) will be of most importance. Figure 10 shows a 0.597 x 0.597 nm PbS cleavage unit cell and a 0.422 x 0.422 nm primitive unit cell, and shows that S sites on [110]-parallel borders are "exposed" to only one nearest-neighbor oxidized site, whereas S-sites on [100]-parallel borders are "exposed" to two nearest-neighbor oxidized sites (and are structurally identical to kink sites on [110]-parallel borders). If, as we have argued above, oxidized sites are the important influence on the reactivity of neighboring unoxidized sites, this relationship provides a qualitative explanation for the STM observation of more [110]-parallel borders than [100]-parallel borders. That is, if [100] borders are more reactive than [110] borders, they will retreat across the surface more quickly, residually enriching the surface in the slower-reacting [110] borders. Also, oxidation of sulfide must be accomplished in a series of electron transfer steps, requiring a series of reaction intermediates. Among the experimentally observed intermediates are various molecules with S-O bonding (see above) and polysulfides. Figure 10 shows that the [110] and equivalent directions are those in which S atoms are closest to each other. These directions may thus have mechanistic significance in the formation of S chains (polysulfides). Water may play a role in mobilizing Pb^{2+} ions, allowing singly oxidized S atoms to bond to each other. This may favor polysulfide formation in aqueous systems, as often observed experimentally (19, 20, 26).

C. O. Moses and E. S. Ilton (Lehigh University, pers. comm., 1993) have pointed out that in considering the relative abundance of borders between unoxidized and oxidized areas oriented in the [100] and [110] directions, one must consider not only the relative reactivity of sites along the different borders, but also the length of the borders (i.e., the relative amounts of each kind of reactant). For example, the length of a [110]-parallel border may be limited by the probability that one of the S sites along it will oxidize to produce two kink sites travelling in opposite directions. In the STM images, the number of sites along [110] and [100]-parallel borders in the images presented here and in our previous publications (2, 30), restricting ourselves to atomic-resolution images containing oxidized patches big enough that the distinction between [100] and [110] oriented borders is physically meaningful, give roughly three times as many sites along [110] borders as along [100] borders. Most of the sites considered to be along [100] borders are kink sites along [110] borders; rarely are [100]-parallel borders more than one or two sites long. If a [110] border grows to a length of 7 sites before undergoing oxidation of the middle site to give 6 sites and 2 [100] (kink) sites, a 3-to-1 ratio of sites on [110] and [100] borders would result, in rough agreement with the STM data. Such a configuration would involve oxidation of 6 [100] sites for every [110] site oxidized, suggesting that sites along [100] borders are roughly 6 times more likely to oxidize than sites along [110] boundaries.

Although the foregoing discussion is approximate and based on rather limited data, it illustrates the use of STM to compare detailed structural models with real-space observations; STM observations allow us to concentrate on specific surface sites that appear to be important in reaction progress. Specific structures can then be modeled in detail, for example using molecular orbital calculations, so as to understand which aspects of the local electronic structure probably control the reactivity. Models, such as BCF theory for crystal growth modified to suit PbS oxidation, or Monte-Carlo simulations, may then be used to test the calculations by comparing simulated reaction progress at the atomic scale to observed progress in STM studies. Specific, rate-controlling elementary mechanisms may thus be identified and studied. Such information could lead to 1) more precise explanations of a variety of other experimental results, and 2) the construction of models built upon physically realistic mechanisms that are able to predict oxidation rates or other properties more reliably than is currently the case.

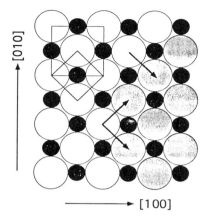

Figure 10. Model of the PbS {100} surface. Open and shaded large circles represent S sites, filled small circles represent Pb sites. In upper left, the standard and primitive surface unit cells are drawn. In lower right, shaded large circles represent oxidized S sites; an unoxidized S site on a [100] or [010]-parallel border between unoxidized and oxidized areas has two oxidized nearest neighbors (arrows), whereas an unoxidized S site on a [110]-parallel border has only one oxidized nearest neighbor.

Summary

Our re-examination of available electronic structure data, in conjunction with STM results, shows that the PbS valence band has S 3p character and that loss of intensity from the top of the valence band in UPS spectra of oxygen-exposed PbS surfaces, despite claims to the contrary based on XPS results, is partly an effect of incipient oxidation of the surface. This information is used to interpret STM results for fresh and reacted PbS surfaces. Oxidation depopulates the 3p bands, broadening the surface bandgap; this manifests itself as dark areas in STM images, which can therefore be used to follow 2-dimensional oxidation progress across the surface. STM shows the crystallographic constraints on oxidation progress; the rates are in relatively good agreement with bulk measurements of oxidation rates, although the STM electrochemically accelerates oxidation in air when it is polarized as a cathode. Surface oxidation of PbS is non-uniform and appears to proceed initially as autocatalytic nucleation and growth of oxidized areas. Reaction borders between unoxidized and oxidized areas spread across the surface as oxidized areas grow, and are preferentially oriented in [110] and equivalent directions, although [100]-oriented borders are also seen.

Acknowledgments

We thank the Department of Geology, the Department of Applied Earth Sciences, and the Center for Materials Research at Stanford University for support during portions of this research. We thank Werner Stumm and Barbara Sulzberger of EAWAG for their support of CME for the bulk of this research, as well as the National Science Foundation (grant EAR-9105000 to MFH). Thoughtful reviews by Drs. Thomas Sharp, Carl Moses, and Eugene Ilton were most helpful in improving the original manuscript.

Literature Cited

1. Bancroft, G.M.; Hyland, M. M. In Mineral-Water Interface Geochemistry; Hochella, M.F. Jr.; White A.F., Eds.; Mineralogical Society of America Reviews in Mineralogy, 1990, 23, pp 511-558.
2. Eggleston, C. M.; Hochella, M. F. Jr. Science 1991, 254, 983-986.
3. Hochella, M. F. Jr. In Spectroscopic Methods in Mineralogy; Hawthorne F.C., Ed.; Mineralogical Society of America Reviews in Mineralogy, 1988, 18, pp 573-637.
4. Plummer, E. W.; Eberhardt, W. In Advances in Chemical Physics, Prigogine I. and Rice J., Eds.; Wiley: New York, 1982, 49; pp 533-656.
5. Eggleston, C. M.; Hochella, M. F. Jr. Geochim. et Cosmochim. Acta 1990, 54, 1511-1517.
6. Avouris, P. J. Phys. Chem. 1990, 94, 2246-2256.
7. Simons, J. J. Phys. Chem. 1991, 95, 1017-1029.
8. Tossell, J. A.; Vaughan, D. J. Theoretical Geochemistry: Applications of Quantum Mechanics in the Earth and Mineral Sciences, Oxford University Press, New York, 1992, 514 pp.
9. Eggleston, C. M.; Hochella, M. F. Jr. Am. Mineral. 1992, 77, 221-224.
10. McFeely, F. R.; Kowalczyk, L.; Ley, L.; Pollak. R. A.; Shirley, D. A. Phys. Rev. B. 1973, 7, 5228-5236.
11. Hagstrom, A. L.; Fahlman, A. Appl. Surf. Sci. 1978, 1, 455-470.
12. Evans, S.; Raftery, E. J. Chem. Soc.; Faraday Trans 1 1982, 78, 3545-3560.
13. Tung, Y. W.; Cohen, M. L. Phys. Rev. 1969, 180, 823.
14. Hemstreet, L. A. Phys. Rev. B. 1975, 11, 2260-2270.
15. Tossell, J. A.; Vaughan, D. J. Canadian Mineral. 1987, 25, 381-392.
16. Grandke, T.; Ley, L.; Cardona, M. Phys. Rev. B 1978, 18, no. 8, 3847-3871.
17. Samson, J. A. R.; In Electron Spectroscopy: Theory, Techniques and Applications; Brundle, C. R., Baker, A. D., Eds.; Academic Press, New York, 1981, 4, 361-396.
18. Zangwill, A. Physics at Surfaces, Cambridge University Press, Cambridge, 1988, 454 pp.
19. Brion, D. Appl. Surf. Sci. 1980, 5, 133-152.
20. Buckley, A. N.; Woods, R. Appl. Surf. Sci. 1984, 17, 401-414.
21. Stumm, W. Chemistry of the Solid-Water Interface; Wiley-Interscience, John Wiley & Sons, New York, 1992, 428 pp.
22. Grandke, T.; Cardona, M. Surf. Sci. 1980, 92, 385-392.
23. Kurtz, R. L.; Henrich, V. E. Phys. Rev. B 1987, 36(6), 3413-3421.
24. Eadington, P.l Prosser, A. P. Trans. Inst. Mining Metall. 1969, C82, 74-82.
25. Manocha, A. S.; Park, R. L. Appl. Surf. Sci. 1977, 1, 129-141.
26. Hyland, M. M.; Bancroft, G. M. Geochim. Cosmochim. Acta 1989, 53, 367-372.
27. Li, E. K.; Johnson, K. H.; Eastman, D. E.; Freeouf, J. L. Phys. Rev. Lett. 1974, 32(9), 470-472.
28. Cotterill, G. F.; Bartlett, R.; Hughes, A. E.; Sexton, B. A. Surf. Sci. Lett. 1990, 232, L211-L214.
29. Zheng, N. J.; Wilson, I. H.; Knipping, U.; Burt, D. M.; Krinsley, D. H.; Tsong, I. S. T. Phys. Rev. B 1988, 38, 12780-12782.
30. Casey, W. H.; Eggleston, C. M.; Johnsson, P. A.; Westrich, H. R.; Hochella, M. F. Jr. Mat. Res. Soc. Bull. 1992, 17, no. 5, 23-29.
31. Hsieh, Y. H.; Huang, C. P. J. Coll. Interf. Sci. 1989, 131, no. 2, 537-549.
32. Gardner, J. R.; Woods, R. J. Electroanal. Chem. 1979, 100, 447-459.

RECEIVED August 27, 1993

Transport of Sulfide-Oxidation Products in Surface Waters

Chapter 16

Effects of Instream pH Modification on Transport of Sulfide-Oxidation Products

Briant A. Kimball[1], Robert E. Broshears[2], Diane M. McKnight[2], and Kenneth E. Bencala[3]

[1] U.S. Geological Survey, 1745 West 1700 South, Room 1016, Salt Lake City, UT, 84104
[2] U.S. Geological Survey, Box 25046, MS 415, Denver, CO 80225
[3] U.S. Geological Survey, 345 Middlefield Road, MS 496, Menlo Park, CA 94025

We studied the dynamic response in aqueous concentrations of metals and sulfate to an experimental increase of pH in a mountain stream affected by acidic mine drainage. Downstream from mine spoils in St. Kevin Gulch, Colorado, U.S.A., ambient pH was 3.5; filtered concentrations of Al, Cu, Fe, and SO_4^{2-} were 3.1 , 0.18 , 1.1 , and 128 mg/L, respectively. Injection of Na_2CO_3 caused pH to increase to 4.2 and eventually to 5.9 at a site 24 meters downstream from the injection. Filtered Al decreased to 0.07 mg/L; Cu, to 0.12 mg/L; Fe, to 0.41 mg/L; and SO_4^{2-} to 122 mg/L. Particulate metal concentrations increased as filtered concentrations decreased, indicating processes of partitioning. Measured concentrations compared favorably with concentrations calculated by a geochemical equilibrium model that simulated precipitation of amorphous Fe and Al hydroxysulfates and sorption of Cu to the precipitating Fe phase. Differences between measured values and concentrations simulated with a conservative solute transport model indicated the substantial buffering capacity of streambed sediments. Particulate concentrations of Al, Cu, and Fe decreased downstream because of sedimentation to the bed. After the injection, concentrations of filtered Al and Cu exceeded background levels. At a site 24 meters downstream, 100 percent of the particulate Al that had settled from the water column was returned after the injection; 53 percent of the Cu was returned, but only 19 percent of the Fe was returned. These observations are attributed to dissociation of the settled polynuclear complexes of Al and to desorption of Cu from the settled Fe phases.

Diverse interactive physical, chemical, and biological processes affect the mobility and ultimate fate of sulfide-oxidation products in surface water. Advection, solute dispersion, and inflow mixing contribute to patterns of instream concentrations downstream from contaminant sources such as acidic mine drainage. In addition,

This chapter not subject to U.S. copyright
Published 1994 American Chemical Society

patterns in solute concentrations are affected by biological cycles and by chemical processes including complexation and speciation of metals leading to precipitation-dissolution, sorption-desorption, and oxidation-reduction reactions. While field studies are necessary to characterize physical transport in streams, the determination of chemical equilibria and rates of chemical and biological reactions often is relegated to the controlled conditions of the laboratory. A complete understanding of natural conditions, however, requires that interactive processes be studied in the field, where the complexities of stream history, variable boundary conditions, and distributions of residence times in biogeochemical microenvironments combine in ways not duplicated in the laboratory.

Field studies of the transport and fate of sulfide-oxidation products in surface water have been conducted by several investigators. Dilution by inflows, precipitation of amorphous Fe and Al hydroxides, and sorption of Cu to Fe precipitates were indicated in two mountain streams affected by acidic mine drainage (1,2). Linked watershed, geochemical, and solute transport codes were used to simulate the measured behavior of Cu mobilized from mine wastes (3). The importance of uptake by periphyton and sorption to sediments on the mobility of Cu in a stream was demonstrated in field experiments (4). Radio-labeled Fe in limnocorrals helped to determine the partitioning of Fe among dissolved, non-settling colloidal, and settling particulate phases (5). Instream measurement of Fe^{2+} has shown the importance of diel fluctuations due to photoreduction (6). Photosynthetically-induced changes in pH affected instream As concentration (7). Chapman (8) used a geochemical speciation code coupled with a physical transport model to simulate the responses of Zn, Al, Cu, and Fe to an experimental increase of pH in a stream. Elevation of instream pH from 3 to 10 was accompanied by formation of chemical precipitates and base-neutralizing surface reactions. Experimental decrease of instream pH was used to detail Fe transformations and transport in an acidic, mountain stream (9).

This paper provides a preliminary analysis of an experiment conducted in St. Kevin Gulch, an acidic, mountain stream in Colorado, U.S.A. (Figure 1). St. Kevin Gulch is a pool-and-riffle stream receiving acidic, metal-rich water of pH 2.6 from mine drainage. To study responses to pH changes, we injected Na_2CO_3 to raise pH from a background value of 3.5 to a maximum value of 5.9. Our experiment differed from the experiment of Chapman (8) in that: (1) the pH range induced by the experiment was of more environmental relevance because it resembles the pH changes the stream encounters during downstream and seasonal changes; (2) both filtered and unfiltered samples were obtained to evaluate variations of filtered and particulate concentrations; and (3) the injection period was sustained for 6 hours at two pH levels rather than a slug exceeding pH 10. Data from our experiment will be used to evaluate a computer code that is being developed to couple kinetic effects of hydrologic transport and chemical equilibrium. The purposes of this paper are to describe the chemical variations during the instream experiment, to propose likely processes causing those variations, and to test the hypothesis that chemical equilibrium calculations and physical transport can account for these changes.

Remediation of streams and lakes affected by metals requires an understanding of the geochemical reactions that determine responses to varying environmental conditions. Among the metals sensitive to changes in pH, Fe tends to

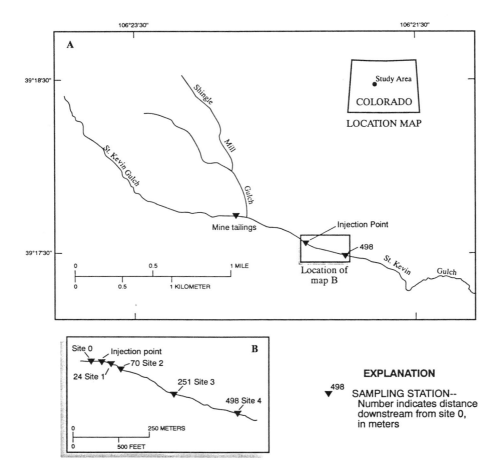

Figure 1.--Location of St. Kevin Gulch, Colorado, showing locations of sampling sites.

form hydrous Fe oxide precipitates with extensive reactive surface area that can influence the mobility of other metals in streams. Thus, Fe can function as a controlling variable as it responds to changes in pH. Mobility of Al and Cu are sensitive to pH and the cycling of Fe (10-13). They are also of biogeochemical importance in acidic waters because of their effects on aquatic organisms (14, 15).

METHODS

Modification experiment. Instream pH of St. Kevin Gulch was increased from 3.5 to 5.9 by pumping a concentrated solution of Na_2CO_3, with NaCl included as a conservative tracer, into the stream. As the pulse of increased pH was transported downstream, the response of major ions and trace metals was documented by collecting samples at four sites located 24, 70, 251, and 498 m from the injection site (Figure 1). These sites are numbered 1 through 4, respectively, in this paper. A control or background site (site 0) was sampled upstream from the injection point. The changes in pH (Figure 2) defined three steps in the experiment. For site 1, 24 m downstream, step 1 began at 0900 hours with the initial injection of base and continued for 3 hours. Step 2 began at 1200, when pH was brought above 5.3 and eventually to 5.9 at site 1. Step 3 began at 1500 when the injection of base ended. Appearance of the three steps occurred successively later downstream because of the time of transport.

The experimental modification added Na, which was geochemically conservative in the stream throughout the measured range of pH in the experiment. Thus, aqueous, filtered Na served as a tracer in defining physical transport through the study reach. Injection of Na from Na_2CO_3 and NaCl sufficiently elevated instream concentrations of Na above background values so that a distinct concentration pulse delineated the injection period (Figure 3). The sequential arrival of the Na pulse at each sampling site permitted definition of subreach travel times, while the asymmetry of the arriving shoulders and departing tails of the pulse allowed calibration of parameters of transient storage in immobile zones adjacent to the main stream channel (16, 17). Instream flow ranged from 12.3 L/s at site 1 to 15.9 L/s at sites 3 and 4. Travel time through the reach was approximately 100 minutes. Parameters of transient storage were comparable to those calibrated in studies of similar mountain streams (18). Instream concentrations of Al, Cu, Fe, and SO_4^{2-} experienced the same regime of physical transport documented for Na. By accounting for these physical processes in a rigorous manner, we are better able to address the chemical and biological processes that affect the reactive metals as they move downstream.

Sampling and analytical methods. In describing the behavior of Al, Cu, and Fe, we distinguish between two operationally defined phases. First, a 0.1-μm-filtered phase was obtained by using pressure filtration through a nitro-cellulose membrane. Second, a particulate phase was obtained by subtracting the 0.1-μm-filtered concentrations from the concentrations in an unfiltered, acidified sample. The particulate concentration represents an acid-soluble concentration, principally consisting of aggregated colloids of hydrous Fe or other metal oxides greater than 0.1

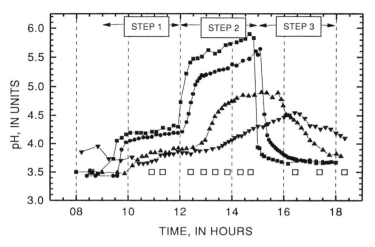

Figure 2.--Experimental response of pH to the injection of base into St. Kevin Gulch. At site 1 (solid squares), step 1 was from 0900 to 1200 hours, step 2 was from 1200 to 1500 hours, and step 3 was after 1500 hours when the injection of base ended. Arrival of these modifications at downstream sites is delayed by travel time to the sites (site 0, open squares; site 2, solid circles; site 3, upward triangles; site 4, downward triangles).

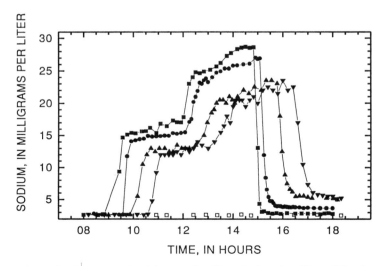

Figure 3.--Concentration history of sodium during the pH modification experiment. Instream sodium concentration was from injection of sodium carbonate and sodium chloride. Site 0, open squares; site 1, solid squares; site 2, solid circles; site 3, upward triangles; site 4, downward triangles.

μm (19). Although this membrane pore size may not remove all the colloidal Fe (20), it provides an operationally defined, dissolved-metal concentration that was necessary for rapid sampling during the experiment. Metal concentrations were determined by inductively coupled argon plasma atomic emission spectroscopy on samples acidified to less than pH 2.0 with ultrapure nitric acid in the field (21). Determination of SO_4^{2-} was by ion chromatography. Detection limits, accuracy, and precision for chemical determinations are presented elsewhere (22). Relative standard deviation for Al was 5.4 percent, for Cu was 5.3 percent, for Fe was 3.7 percent, and for SO_4^{2-} was 1.1 percent. To account for the Fe^{2+}/Fe^{3+} redox couple in thermodynamic calculations, Fe^{2+} was determined colorimetrically using a bipyridine method (23).

RESULTS AND DISCUSSION

In interpreting the results of the experiment, we asked two questions. First, can the initial effects of the pH modification as measured at site 1 be explained in terms of an equilibrium model? Second, are concentration histories at sites 2-4 simply the result of physical transport of the concentrations at site 1, or are other reactions involved? Results will be presented to address these questions.

Initial effects of pH modification. Prior to the injection, background pH values were 3.5, with an increase to 3.8 at site 4. Pre-injection concentrations of filtered Al were nearly equal at each of the sites, with a concentration of about 3.1 mg/L. Filtered Cu concentrations before the injection were approximately 0.18 mg/L; no trend of concentration with downstream distance was evident. Pre-injection concentrations of filtered Fe declined gradually from 1.1 mg/L at site 1 to 0.75 mg/L at site 4; however, background particulate Fe concentration was essentially the same at each site, at about 0.06 mg/L. Thus, particulate Fe likely was forming in the stream prior to the injection, and there were approximately equal rates of aggregation of Fe precipitates and physical settling of Fe particulates along the reach.

With the injection of base at 0900 hours, pH increased to 4.2 at site 1 (Figure 2). An increased injection rate begun at 1200 hours eventually resulted in a pH of 5.9 at this site. Temporal changes that occurred for filtered and particulate Al, Cu, and Fe as pH increased at site 1 were substantial (Figure 4). The primary changes occurred during step 2 from 1200 to 1500 hours, when filtered concentrations decreased and particulate concentrations increased. Total concentrations (filtered plus particulate) declined only slightly at site 1, consistent with the formation of a precipitate in the water column. A slight decrease in SO_4^{2-} during the same time interval indicated that the precipitate might also contain SO_4^{2-}. At the end of the injection, filtered concentrations of Al and Cu briefly increased above background levels. This rebound or spike became more pronounced at each site downstream.

Geochemical reactions. The substantial loss of filtered Al, Cu, and Fe upstream from site 1 can be evaluated in terms of chemical equilibrium by plotting metal activities versus pH (Figure 5). Each plot includes a line that represents equilibrium of the aqueous phase with respect to a solid phase. Phases were amorphous $Al(OH)_{3(s)}$ for Al^{3+} activity, $Cu(OH)_{2(s)}$ for Cu^{2+} activity, and ferrihydrite, $Fe(OH)_{3(s)}$,

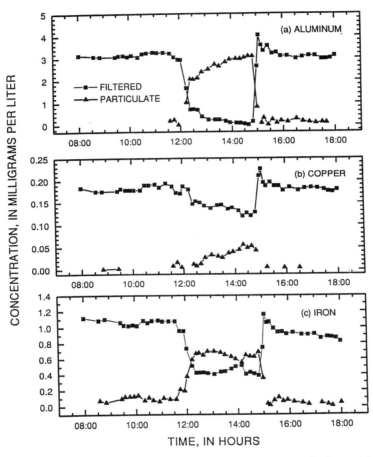

Figure 4.--Temporal variations of filtered and particulate (a) aluminum, (b) copper, and (c) iron at site 1. Filtered concentrations (solid squares) are from 0.1-μm pressure filtration. Particulate concentrations (upward triangles) are from unfiltered concentrations less the 0.1 μm filtered concentration.

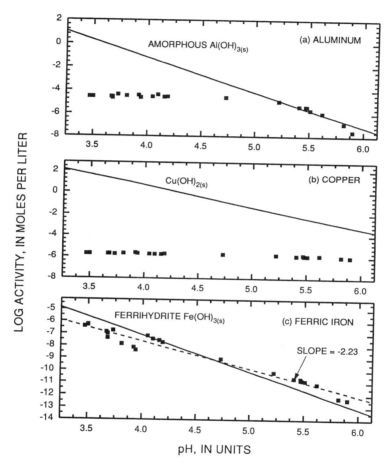

Figure 5.--Variation of calculated activities of (a) Al^{3+}, (b) Cu^{2+}, and (c) Fe^{3+} with pH. Activities were calculated with the WATEQ4F chemical equilibrium model.

for Fe^{3+} activity; thermodynamic data from the chemical equilibrium model WATEQ4F were used (24). Higher pH caused Al^{3+} activity to approach equilibrium; the slope of the points changed above pH 5.0. Values of Cu^{2+} activity never approached equilibrium. Activity of Fe^{3+} trended along the equilibrium line over the entire range of pH. The slope of a regression line for Fe^{3+} activity is -2.23, substantially different from the -3.0 slope that would result from precipitation of a pure hydrous Fe oxide phase. Values of an Fe:S ratio for Fe precipitates in acidic mine drainage have been measured from 8 to 5 (25). The Fe:S ratio of a precipitating Fe phase from the slope of the regression line is about 2.6, indicating substantial sulfur in the Fe phase. Perhaps the initial precipitation of the phase is high in SO_4 because the initial Fe:S ratio in the water column is 0.015.

To answer our first experimental question, we suggest plausible reactions to account for the chemical changes. Overall, the chemical changes from background conditions to the conditions during step 2 (from 1200 to 1500 hours) may be described by the following chemical reactions:

$$Na_2CO_3 + H_2O \Leftrightarrow 2Na^+ + HCO_3^- + OH^- \tag{1}$$

$$NaCl \Leftrightarrow Na^+ + Cl^- \tag{2}$$

$$SO_4^{2-} + (OH)_2:Oxide \Leftrightarrow SO_4:Oxide + 2OH^- \tag{3}$$

$$Al^{3+} + 1.96H_2O + 0.52SO_4^{2-} \Leftrightarrow Al(OH)_{1.96}(SO_4)_{.52} + 1.96H^+ \tag{4}$$

$$Cu^{2+} + H_2:Oxide \Leftrightarrow Cu:Oxide + 2H^+ \tag{5}$$

$$Fe^{3+} + 2.23H_2O + 0.385SO_4^{2-} \Leftrightarrow Fe(OH)_{2.23}(SO_4)_{.385} + 2.23H^+ \tag{6}$$

Changes in concentration, caused by these reactions, are quantified according to the matrix equation (26):

$$\Delta m_{T,i} = m_{T,i(final)} - m_{T,i(initial)} = \sum_{p=1}^{P} \alpha_p b_{p,i} \tag{7}$$

where $\Delta m_{T,i}$ is the net change in total solution concentration of the ith component, in mM/L;

$m_{T,i(final)}$ is the final solution composition of the ith component, in mM/L;

$m_{T,i(initial)}$ is the initial solution composition of the ith component, in mM/L;

α_p is the mass-transfer coefficient for dissolving (positive) or precipitating (negative) of the pth solid phase, in mM/L; and

$b_{p,i}$ is the stoichiometric coefficient for the ith component in the pth phase, dimensionless.

The 6 components are aqueous Na, Cl, SO_4^{2-}, Al, Cu, and Fe; the 6 phases are in equations 1-6.

Table I presents a summary of the mass-transfer calculation, listing compositions for the initial and final solutions and values of α_i for phases in Equations 1-6. The initial water had the chemical composition at site 1 at 0835 hours (Table I). The final water had the average chemical composition at site 1 for samples from 1420 to 1450 hours. Reactions 1 and 2 represent the amounts of Na_2CO_3 and NaCl that were added to bring about the pH modification during step 2. Reaction 3 represents SO_4^{2-} sorption to the metal-oxide phases. Reactions 4-6 represent the formation of Al and Fe hydroxysulfates and sorption of Cu.

Table I. Summary of mass-transfer calculations for initial chemical reactions upstream from site 1

Component	Reaction	$m_{T.i(initial)}$	$m_{T.i(final)}$	$\Delta m_{T.i}$	α_p
Na	1	0.113	1.24	1.13	0.30
Cl	2	.085	.609	.526	0.526
SO_4^{2-}	3	1.33	1.27	-0.06	0.000
Al	4	.115	.003	-0.112	-0.112
Cu	5	.003	.002	-0.001	-0.001
Fe	6	.019	.007	-0.012	-0.012

Although reaction 3 might account for a decrease in SO_4^{2-}, anion sorption is less likely as pH increases. Thus, the set of equations was solved to achieve $\alpha = 0$ for reaction 3. The stoichiometry for the Fe hydroxysulfate in reaction 6 was derived from the slope of Fe^{3+} activity versus pH (Figure 5c). The amount of sulfur included in the Al hydroxysulfate was adjusted to completely account for the remaining loss of SO_4^{2-} from the water column. The adjusted Al:S ratio is 1.92, which is intermediate between ratios of 1 and 4 that occur in the minerals jurbanite and basaluminite, respectively (27). These results suggest that about 10 times more of the SO_4^{2-} was removed with the Al phase than with the Fe phase. The amounts of Al and Fe phases precipitated were substantial in comparison to the amount of Cu sorbed.

Variations of filtered Al concentration were not likely due to sorption for two reasons. First, total Al generally was constant as the pH increased, indicating that the reaction did not involve sorption to the bed, but occurred in the water column. Second, the moles of particulate Fe in the water column were not sufficient to provide surface sites to sorb 3 mg/L of Al (Figure 4). Assuming 0.2 moles of sorption sites per mole of particulate Fe (28), the molar concentration of filtered Al removed exceeded the molar concentration of sorption sites on particulate Fe in the water column. Time of travel to site 1 from the injection was about 6 minutes, and the Al removal occurred during this time (Figure 4a). This time compares favorably with observations that the initial precipitation of amorphous $Al(OH)_{3(s)}$ may occur on the order of seconds (29). Initial Al precipitation may result from the formation of a polynuclear Al complex, such as $Al_{13}O_4(OH)_{24}(OH_2)_{12}^{7+}$, that contains some SO_4^{2-} replacing OH^- (30-32). Such a polynuclear complex has been identified in water from St. Kevin Gulch (Vivet, D., U.S. Geological Survey, written communication, 1992). The complex likely grew to a size that prevented it from passing through a 0.1-μm filter. The dissociation of the polynuclear complex after the injection caused an increase in filtered Al and also occurred in less than 6 minutes.

Variation of filtered Cu (Figure 5b) did not appear to relate to a control by $Cu(OH)_{2(s)}$. However, unlike Al, there was a sufficient number of sorption sites on the particulate Fe in the water column to accommodate the mass of Cu that was lost, assuming no competition for surface sites (Table I). Thus, reaction 5 is written as a sorption of Cu onto hydrous Fe oxide, displacing H^+.

Geochemical simulation. Given the plausible reactions, the next step was to simulate the geochemical reactions by using an equilibrium model. We used the geochemical speciation code MINTEQA2 (33), modified with the thermodynamic data complied for WATEQ4F (24) and the generalized two-layer adsorption model (28). Water-sediment partitioning studies conducted on sediments from St. Kevin Gulch have shown the successful application of this model to this site (34). For these simulations, increases in Na concentration were attributed to Reactions 1 and 2. This increase was equivalent to injecting 0.301 mM/L of CO_3^{2-}. Geochemical simulations assumed equilibrium conditions between water and the solid phases in Reactions 4 and 6. Equilibrium constants for the hydroxysulfates are not available, so calculated instream ion activity products of $10^{0.316}$ and $10^{-2.34}$ were used for the Fe and Al hydroxysulfates, respectively. These solubilities may only apply to conditions at this site. Simulated values of pH agreed well with measured values (Table II). Simulated concentrations of Al also were comparable. Including SO_4^{2-} in the Al phase accounted for the decrease of SO_4^{2-} but the slope of Al^{3+} activity versus pH does not agree with the inclusion of SO_4^{2-} or any other anionic species in the phase (Figure 5a).

Concentration of filtered Cu was overestimated. The additional Cu may be sorbed to the bed sediment. Also, the difference could reflect continued diffusion of Cu^{2+} into aggregated Fe hydroxysulfate, a process that has been studied for sorption of As to ferrihydrite (35). If some part of the Cu were removed during the formation of the Al complex, even as a trace amount, it could account for the Cu removed from the water column.

The measured decline in Fe concentration at site 1 was simulated well, but required the introduction of an oxidation reaction for conversion of Fe^{2+} to Fe^{3+} during precipitation. Photoreduction caused the majority of Fe to be present as Fe^{2+}, but as pH increases above 4.0, the abiotic oxidation of Fe^{2+} is accelerated (36).

Thus, the reaction of Fe involves two steps: (1) abiotic Fe^{2+} oxidation and (2) Fe hydroxysulfate precipitation. These steps are combined in the reaction:

$$Fe^{2+} + 0.25O_2 + 1.73H_2O + 0.385SO_4^{2-} \Leftrightarrow Fe(OH)_{2.23}(SO_4)_{.385} + 1.23H^+ \quad (8)$$

Table II. Comparison of measured and simulated concentrations during part of step 2 (1420 to 1450 hours) at site 1. Simulation is with the chemical equilibrium model MINTEQA2 (33), assuming equilibrium with Fe and Al hydroxysulfates and Cu sorption onto the iron phase

Constituent, in mg/L	Measured	Simulated
pH, in units	5.8	6.0
Al	.07	.09
Cu	.12	.16
Fe (total filtered)	.41	.42
SO_4^{2-}	122	122

Loss of Fe at pH values above 4.5 was also attributed to oxidation of Fe^{2+} and ferrihydrite precipitation during a pH modification experiment in the Snake River (9). Overall, considering the comparison of measured and simulated values in Table II, the effects of the pH modification are explained in large part by an equilibrium model that assumes the precipitation of Al and Fe hydroxysulfates.

Effects of transport. To answer our second question, solute transport simulations were used to evaluate profiles of pH and concentrations of filtered Al, Cu, and Fe at the three downstream sites. Simulations were conducted with a one-dimensional solute transport code developed for small mountain streams (16, 37). Measured pH values and filtered concentrations at site 1 provided upstream boundary conditions, and parameters describing physical transport were calibrated from the conservative tracer data. No additional chemical reactions were simulated. The simulations thus tested the hypothesis that downstream profiles represented only the physical transport of chemical conditions at site 1.

pH. Measured and simulated values of pH are shown in Figure 6. While the general shapes of the profiles agree, measured pH at sites 3 and 4 was consistently lower than simulated pH. Under the low pH of background conditions, hydrous Fe oxides on the bed are highly protonated. When overlying water of higher pH passes over this reservoir of protons, acid-buffering reactions would be expected to occur and pH in the water column would decrease. This conclusion is similar to that of Chapman (8) with respect to the role of the streambed. The rate of decrease in pH might be kinetically limited by restraints on vertical mixing required for equilibration between water column and bed chemistry. During step 3 (after 1500 hours), pH at downstream sites remained above values simulated under the

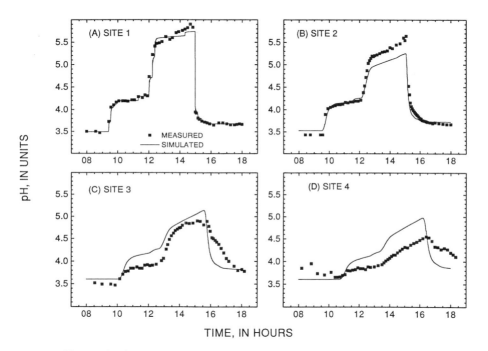

Figure 6.--Comparison of measured (squares) and simulated (solid line) values of pH at downstream sites. The simulation assumes conservative transport of conditions at site 1.

assumption of conservative transport. During this step, pH in the water column may remain high while exchange sites on the streambed are being restored to their original state of protonation.

Aluminum. Hydrologic transport accounted for most of the variation in filtered Al concentration from site 1 to 2 (Figures 7a and 7b). Departure from simulated transport was more substantial for sites 3 and 4, where filtered Al was higher than simulated Al (Figures 7c and 7d). At site 4, pH did not exceed values that may cause a change in filtered Al from conservative to reactive transport (13). The relative increase of filtered Al at downstream sites may have resulted from dissociation of the polynuclear Al complexes at lower pH.

The Al mass in the post-injection spike increased with each successive downstream site, mostly due to increased filtered Al (Figures 7a-7d). The spike is consistent with the dissociation of an Al complex that had settled to the bed during the period of elevated pH. Much of the Al mass that settled from the water column was returned after the injection (Table III). The percent returned after the injection was a substantial part of the amount lost at all 4 sites.

Copper. Partitioning of Cu^{2+} to the particulate phase occurred at each site during the latter stages of the pH perturbation (Figures 7e-7h). Partitioning was less pronounced at each successive site, likely reflecting the lower pH values downstream. Filtered Cu was lowest at site 4, perhaps reflecting upstream sorption to the streambed. Post-injection spikes of filtered Cu occurred at all sites. The mass contained in these spikes was substantial, but less than the mass removed from the upstream water column during the period of elevated pH (Table III).

Several factors may have contributed to the incomplete return of filtered Cu to the water column at downstream sites following the injection. Sampling continued at site 1 for a period equivalent to several travel times through the first subreach. Instream chemistry was restored to pre-injection conditions at site 1 during this interval. Sampling at downstream sites continued for shorter periods relative to travel times to these sites, and pre-injection conditions had not been restored to these sites when sampling ended. Moreover, the Cu removed in the first subreach may have been more available for desorption because filtered Cu removed upstream from site 1 likely was sorbed to the surfaces of particulate Fe large enough to settle quickly to the bed. Filtered Cu sorbed to smaller, nonsettling particles could be incorporated into the interior of larger Fe particles as the latter formed by aggregation. When these particles eventually settled at downstream locations, the Cu contained therein would be less available for rapid release when pH conditions favorable for desorption were restored (34).

Iron. Filtered Fe continued to decrease in the downstream direction throughout the experiment (Figures 7i-7l). The pattern of Fe is distinct from patterns of filtered Al and Cu. First, there was more removal during step 1, and second there was less return after step 2; there was no spike. These observations are consistent with increased formation of particulate Fe immediately downstream from the injection point and subsequent settling of this particulate Fe during the travel interval to sites 3

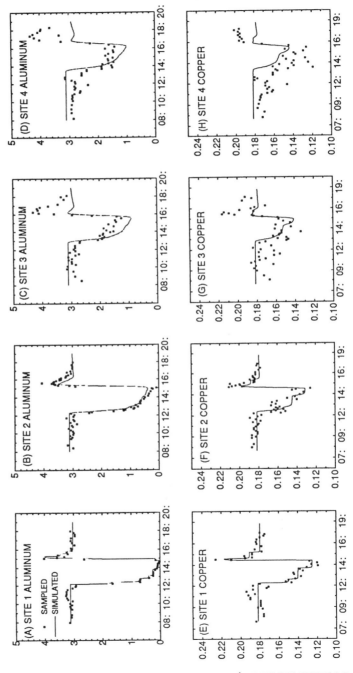

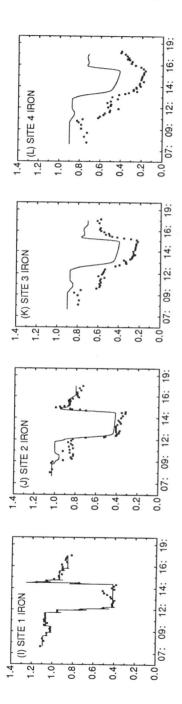

Figure 7.--Comparison of measured (solid squares) and simulated (solid line) concentrations of filtered aluminum (A-D), copper (E-H), and iron (I-L) with time. Conservative transport of conditions at site 1 is assumed. Measured concentrations below the simulated line indicate mass lost; measured concentrations above the simulated line indicate mass gained.

Table III. Comparison of the total (filtered plus particulate) metal mass lost from the water column during the pH modification with the total metal mass gained during the post-injection spike

[All quantities of loss and gain in grams; NG, no gain of metal concentration; --, not calculated]

Site	Al			Cu			Fe		
	Lost	Gained	Percent returned	Lost	Gained	Percent returned	Lost	Gained	Percent returned
1	27.0	28.0	>100	1.5	0.8	53	32.2	6.0	19
2	67.7	47.2	70	3.0	1.0	33	36.6	NG	--
3	159.8	>97.4	>61.	4.1	>1.7	>41	119.9	NG	--
4	215.0	>80.3	>37	5.9	>1.1	>19	142.6	NG	--

and 4. Compared with Al and Cu, much less Fe was returned to the water column after the injection.

CONCLUSIONS

Filtered Al, Cu, Fe, and SO_4^{2-} showed dynamic variation in response to a stepped modification of instream pH that was within the range of possible seasonal pH variations for many mountain streams affected by acid mine drainage. Particulate Al, Cu, and Fe initially remained in the water column, but settled out of the water column downstream. Variations in filtered Al likely resulted from complexation of a polynuclear hydroxysulfate species, aggregation, settling, and then dissociation. Filtered Cu likely changed in response to sorption and desorption on the Fe hydroxysulfate surfaces. Changes in filtered Fe involved abiotic oxidation of Fe^{2+} and subsequent precipitation of Fe hydroxysulfate. These variations occurred during the period of diel variation of filtered Fe due to the photoreduction of Fe^{3+} and oxidation of Fe^{2+}.

These chemical changes are partly explained by a geochemical equilibrium model, but kinetic aspects involving oxidation and photoreduction reactions with streambed sediment during transport are also important. This preliminary assessment of the instream processes gives an indication of which chemical processes must be included in a reactive transport model.

ACKNOWLEDGMENTS

We thank Donald Campbell, Jerry Duncan, John Gray, Ben Kelly, Wendy Maura, Bryan Nordland, William Van Liew, Katie Walton-Day, Greg Wetherbee, and Gary Zellweger for their assistance in the sample collection. Chris Fuller and David Naftz provided helpful suggestions for improving the technical content of this paper. This work was supported by the Toxic Substances Hydrology Program of the U.S. Geological Survey.

LITERATURE CITED

1. Amacher, M.C.; Brown, R.W.; Sidle, R.C.; Kotuby-Amacher, J. EOS 1991, 72, 177.
2. McKnight, D.M.; Bencala, K.E. Water Res. Res. 1990, 26, 3087-3100.
3. Brown, K.P.; Hosseinipour, E.Z. J. Environ. Sci. Health 1991, A26(2), 157-203.
4. Kuwabara, J.S.; Leland, H.V.; Bencala, K.E. J. Environ. Eng. 1984, 110, 646-655.
5. Vezina, A.F.; Cornett, R.J. Geochem. Cosmochim. Acta 1990, 54, 2635-2644.
6. McKnight, D.M.; Kimball, B.A.; Bencala, K.E. Science 1988, 240, 637-640.
7. Fuller, C.C.; Davis, J.A. Nature 1989, 340, 52-54.
8. Chapman, B.M. Water Res. Res. 1982, 8, 155-167.
9. McKnight, D.M.; Bencala, K.E. Geochim. Cosmochim. Acta 1989, 53, 2225-2234.

10. Baes, C.F.; Mesmer, R.E. The Hydrolysis of Cations; John Wiley & Sons: New York, 1976.
11. Hem, J.D. Study and Interpretation of the Chemical Characteristics of Natural Waters, 3rd Ed.; U.S. Geological Survey Water-Supply Paper 2254; U.S. Geological Survey: Washington, D.C., 1985.
12. Leckie, J.O.; James, R.O. In Aqueous-Environmental Chemistry of Metals; Rubin, A.J., Ed.; Ann Arbor Science Publishers, Inc: Ann Arbor, MI, 1976, 1-76.
13. Nordstrom, D.K.; Ball, J.W. Science 1986, 232, 54-56.
14. Gough, L.P.; Shacklette, H.T.; Case, A.A. Element Concentrations Toxic to Plants, Animals, and Man; U.S. Geological Survey Bulletin 1466; U.S. Geological Survey: Washington, D.C., 1979.
15. Forstner, U.; Wittmann, G.T.W. Metal Pollution in the Aquatic Environment; Springer-Verlag: New York, 1979.
16. Bencala, K.E.; Walters, R.A. Water Res. Res. 1983, 19, 718-724.
17. Stream Solute Workshop. J. North Am. Benth. Soc. 1990, 9, 95-119.
18. Broshears, R.E.; Bencala, K.E.; Kimball, B.A.; McKnight, D.M. Tracer-dilution experiments and solute-transport simulations for a mountain stream, Saint Kevin Gulch, Colorado; U.S. Geological Survey Water Resources Investigations Report 92-4081; U.S. Geological Survey: Denver, CO, 1993.
19. Ranville, J.R.; Smith, K.S.; Macalady, D.L.; Rees, T.F. In Geological Survey Toxic Substances Hydrology Program--Surface-Water Contamination; Proceedings of the technical meeting, Phoenix, Arizona, September 26-30, 1988, Mallard, G.E., and Ragone, S.E., Eds., U.S. Geological Survey Water-Resources Investigation Report 88-4220; U.S. Geological Survey: Washington, D.C., 1989, 111-118.
20. Kimball, B.A.; McKnight, D.M.; Wetherbee, G.A.; Harnish, R.A. Chem. Geol. 1992, 96, 227-239.
21. Garbarino, J.R.; Taylor, H.E. App. Spectroscopy 1980, 34, 584.
22. Wetherbee, G.A.; Kimball, B.A.; Maura, W.S. Selected Hydrologic Data for the Upper Arkansas River Basin, Colorado, 1986-89; U.S. Geological Survey Open-File Report 91-528; U.S. Geological Survey: Denver, CO, 1991.
23. Brown, E.; Skougstad, M.W.; Fishman, M.J. Methods for collection and analysis of water samples for dissolved minerals and gases. Techniques for Water Resources Investigations of the United States Geological Survey, Bk. 5, Chap. A1; U.S. Geological Survey: Washington, D.C., 1970, 101-105.
24. Ball, J.W.; Nordstrom, D.K. User's munual for WATEQ4F, with revised thermodynamic data base and test cases for calculating speciation of major, trace, and redox elements in natural waters; U.S. Geological Survey Open-File Report 91-183; U.S. Geological Survey: Washington, D.C., 1991.
25. Bigham, J.M.; Schwertmann, U.; Carlson, L.; Murad, E. Geochim. Cosmochim. Acta 1990, 54, 2743-2758.
26. Parkhurst, D.L.; Plummer, L.N.; Thorstenson, D.C. BALANCE -- A Computer Program for Calculating Mass Transfer for Geochemical Reactions in Ground Water. U.S. Geological Survey Water-Resources Investigations 82-14; U.S. Geological Survey; Washington, D.C., 1982.
27. Nordstrom, D.K. Geochim. Cosmochim. Acta 1982, 46, 681-692.

28. Dzombak, D.A.; Morel, F.M.M. Surface Complexation Modeling: Hydrous Ferric Oxide; John Wiley & Sons: New York, 1990.
29. Lydersen, E.; Salbu, B.; Poleo, A.B.S.; Muniz, I.P. Water Res. Res. 1991, 27, 351-357.
30. Hem, J.D.; Roberson, C.E., In Chemical Modeling of Aqueous Systems II; Melchior, D.C.; Bassett, R.L., Eds.; American Chemical Society Symposium Series 416, American Chemical Society: Washington, D.C., 1990.
31. May, H. In Water-Rock Interaction: Volume 1, Low Temperature Environments; Kharaka, Y.F., Maest, A.S., Eds.; A.A. Balkema: Rotterdam, 1992.
32. Furrer, G.; Trusch, B.; Muller, C. Geochim. Cosmochim. Acta 1992, 56, 3831-3838.
33. Allison, J.D.; Brown, D.W.; Novo-Gradac, K.J., MINTEQA2 / PRODEFA2, a geochemical assessment model for environmental systems: Version 3.0 User's Manual.; U.S. Environmental Protection Agency, Athens, GA, 1991.
34. Smith, K.S.; Ranville, J.F.; Macalady, D.L. In U.S. Geological Survey Toxic Substances Hydrology Program--Proceedings of the technical meeting, Monterey, California, March 11-15, 1991, Mallard, G.E., and Aronson, D.A., Eds., U.S. Geological Survey Water-Resources Investigation Report 91-4034; U.S. Geological Survey: Washington, D.C., 1991, 380-386.
35. Fuller, C.C.; Davis, J.A.; Waychunas, G.A. Geochim. Cosmochim. Acta 1993, 57, in press.
36. Singer, P.C.; Stumm, W. Science 1970, 167, 3921.
37. Runkel, R.L.; Broshears, R.E. One-Dimensional Transport with Inflow and Storage (OTIS): A Solute Transport Model for Small Streams; Center for Advanced Decision Support for Water and Environmental Systems: Boulder, CO, 1992.

RECEIVED April 12, 1993

Chapter 17

Transport and Natural Attenuation of Cu, Zn, As, and Fe in the Acid Mine Drainage of Leviathan and Bryant Creeks

Jenny G. Webster[1], D. Kirk Nordstrom[2], and Kathleen S. Smith[3]

[1]Institute of Environmental Health and Forensic Science, P.O. Box 92021, Auckland, New Zealand
[2]U.S. Geological Survey, 3215 Marine Street, Boulder, CO 80303
[3]U.S. Geological Survey, MS 973, Denver Federal Center, Denver, CO 80225

> The Leviathan and Bryant Creek (LBC) drainage system, on the border of California and Nevada, flows through overburden and waste from a former open-pit sulfur mine. The drainage contains acid mine waters with high concentrations of several trace elements, including Cu, Zn, and As, derived from oxidative weathering of sulfides in the wastes and altered bedrock. In June and October, 1982, the mainstream and tributary flows of the LBC drainage were measured and the waters sampled and analyzed for major and trace elements. Empirical mass flow and metal attenuation rates were determined, and chemical models were used to examine mechanisms of trace element removal during downstream transport. In June the flow in the mainstream was 2-5 times greater than in October, and with higher contributions from the acid mine effluent. Seasonal variations in the attenuation rates of Cu, Zn, and As were directly related to this increase in acid mine-effluent production, and to the consequent increase in the acidity of the mainstream drainage. Although As concentrations immediately below the mine site were high in June, As was readily removed from solution by adsorption onto an assumed iron(III) oxyhydroxysulfate precipitate, whereas Cu was incompletely adsorbed and Zn remained unaffected by adsorption. In October, the smaller discharge of acidic LBC drainage waters were more readily diluted (and neutralized) by other regional tributaries. Arsenic concentrations remained low, and both Cu and Zn were removed from solution by adsorption onto iron(III) oxyhydroxysulfate in the lower regions of the LBC drainage system.

Leviathan and Mountaineer Creeks are major tributaries of Bryant Creek, near Markleeville in Alpine County, California. Bryant Creek crosses the border of

California to join with the East Fork Carson River in Nevada (Figure 1). Although Mountaineer Creek flows through an area disturbed by roads, revegetation projects and waste piles, it contains uncontaminated waters whereas Leviathan Creek is significantly contaminated by seepage from the waste dumps of the Leviathan copper-sulfur mine and from an unstable landslide to the north of the Leviathan mine open pit. Prior to remediation activities in 1984, contaminated drainage from the open pit also reached Leviathan Creek. Consequently, the waters of Leviathan Creek had become acidic and contained anomalously high concentrations of sulfate and trace elements, including As, Co, Cr, Cu, Fe, Mn, Ni, and Zn. Trout kills and cattle deaths, observed by local ranchers and caretakers of the local fish hatchery, have been attributed to the contaminated stream waters.

In June and October of 1982 hydrologic and water quality data were collected for the Leviathan and Bryant Creek (LBC) drainage basin as part of a pollution abatement project undertaken by the California Regional Water Quality Control Board. Trace metal concentrations from the study by Ball and Nordstrom (1) have been used, together with auxiliary data from a previous study (2), to examine attenuation processes for Cu, As, and Zn from the site of drainage contamination to the junction of Bryant Creek with the East Fork Carson River (ca. 20 km). These three metals were chosen because their geochemical behavior was expected to be quite different, ranging from strongly reactive for arsenic to strongly non-reactive for zinc. Mass flows (mol/s) and mass balances have been calculated for each metal, and the chemical speciation and adsorption models, WATEQ4F and MINTEQA2, respectively, have been used to interpret the geochemical behavior of these metals and their attenuation in the LBC drainage.

Many previous studies of trace-metal attenuation have been based on trace-metal concentrations in a river system (e.g., 3), rather than on trace-metal mass flow, hence certain terms used in this study need to be clearly defined. "Attenuation" is defined here as a decrease in concentration, density or some other measurable quantity over a specified distance. "Chemical attenuation" relates to the mass loss of a dissolved constituent from a flowing body of water by hydrobiogeochemical processes such as oxidation, reduction, hydrolysis, gas exchange, mineral precipitation, adsorption, storage in hydrologic "dead zones", and biological uptake. Dilution at tributary confluences does not affect the mass flow or attenuation of a metal. The "chemical attenuation rate," referred to later in the text, is the mass loss of a dissolved constituent over a unit distance per unit time (μmol/km/s has been used here).

Methods

Sampling and Analytical Techniques. Revised chemical analyses are given by Ball and Nordstrom (1) for the samples taken in June and October 1982, from the sites shown on Figure 1. Temperature, pH, Eh, and specific conductance were determined on site. Water samples were filtered on site through 0.1 μm, 142 mm-diameter membranes in a pre-cleaned acrylic plastic filter assembly. Samples for Fe determinations were acidified with 4 mL of ultrapure HCl per 500 mL sample, and samples for other trace and major metals were acidified with 2 mL of ultrapure

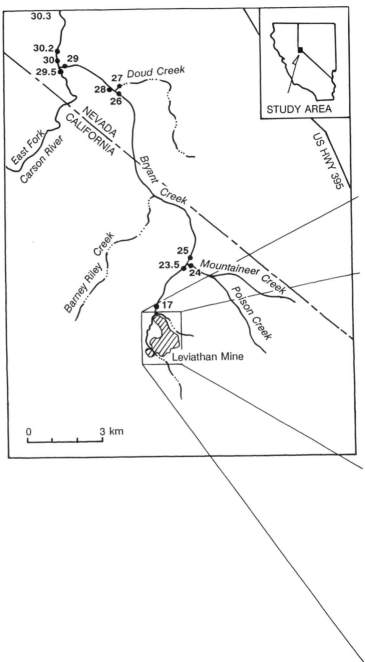

Figure 1. Locality map of features and sampling sites of the Leviathan/Bryant Creek drainage system.

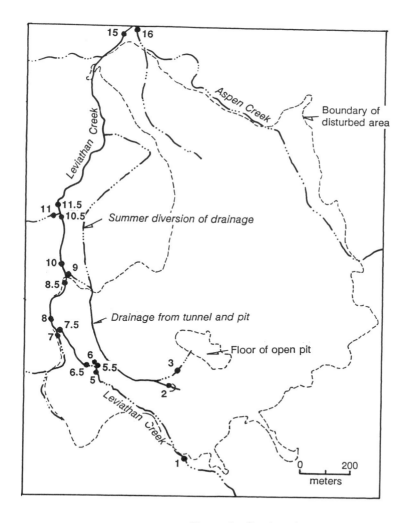

Figure 1. Continued.

HNO$_3$ per 250 mL sample. Metal analyses were done by simultaneous multielement atomic emission spectrometry (both by direct-current and inductively-coupled plasma techniques), flame and graphite furnace atomic absorption techniques and by ferrozine colorimetry for ferrous ion determinations. Sulfate determinations were made using ion chromatography. Sampling and analytical techniques are described in more detail by Ball and Nordstrom (1, and written communication). The analytical and physical data used in this study are shown in Table I. As these analyses had previously been subjected to a critical review process a few alterations to the revised data set were necessary (see Appendix).

Flow Rates. Stream discharges were determined by conventional procedures using a pygmy flow meter over a measured cross sectional area of the stream. The discharges of main and tributary drainages must be known and must be consistent to do useful mass-balance calculations for each component across and between confluences. During an initial attempt to balance SO$_4$ across confluences, shortcomings in the accuracy of some of the measured discharge data became evident. The most conservative ions occurring at concentrations high enough to be easily detected after dilution are SO$_4$, Li, Ca, and Mg so these ions were used to establish consistent discharge values. If the mass balance of each ion was significantly improved by a reasonable change in the discharge rate and if this change was compatible with adjacent discharge rates in the drainage, then the new value was adopted for inclusion in Table I (revised flow rates allow the mass balance of conservative ions to within 10% across each confluence). Only in a few cases did this result in major revision of the original discharge value (see Appendix).

June discharges ranged from 25 to 240 L/s for the main drainage compared to 3 to 110 L/s for the October discharges. Discharges and pH values are plotted in Figure 2a for June and in Figure 2b for October. In June, summer diversion of the open-pit drainage caused an increase in discharge between sites 11.5 and 15. An unsampled tributary and seepage from the toe of the landslide also contributed to the mainstream flow near the confluence of Aspen and Leviathan Creeks. The theoretical discharge for these seeps and surface flows and for Barney Riley Creek, which meets Bryant Creek between the confluences with Mountaineer and Doud Creek, can be calculated but their effect on trace-metal concentrations in the LBC drainage can not be quantified because there were no samples taken in the vicinity of these confluences.

Computational Methods. The computer codes WATEQ4F (4) and MINTEQA2 (5) were used with the analytical data to evaluate the main mechanisms of trace-metal removal. WATEQ4F calculates the distribution of aqueous species in solution based on thermodynamic data and the assumption of equilibrium. WATEQ4F also calculates a 'free' metal-ion concentration and saturation indices log(IAP/K$_{sp}$) for mineral phases. A summary of the computational methods and data used in WATEQ4F is given by Ball and Nordstrom (4).

Adsorption modelling was performed with MINTEQA2 (5), using the Generalized Double-Layer Sorption Model. Details of this adsorption model, which

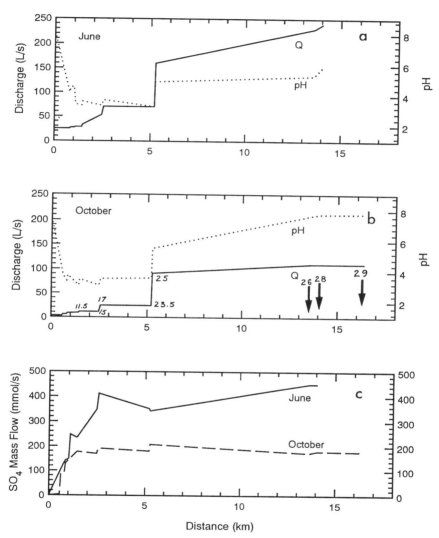

Figure 2. a. Variation in discharge and pH with distance downstream for the sampling of June 14-17, 1982. b. Variation in discharge and pH with distance downstream for the sampling of October 4-6, 1982. Numbers inside diagram box refer to site locations found in Figure 1. c. Mass flow for dissolved sulfate with distance downstream for both June 1982 and October 1982 sampling.

Table I. Chemical analyses (mg/L) and physical parameters (discharge, Q, in L/s; temperature in °C; Eh in volts) for the Leviathan and Bryant Creeks

Site No.	Q	Temp.	pH	Eh	SO_4	Al	As	Cu	Fe^{II}	Fe^{TOT}	Mn	Zn
June, 1982:												
1	25	12.5	8.10	.380	5.3	.018	.004	.0015	.0076	.0088	.0187	.0001
7	2	18	6.85	.238	276	.038	.002	.0005	6.03	6.38	2.13	.0182
7.5	25	18	4.18	.488	517	21.7	.008	.18	90.0	91.1	3.87	.137
8	27	18	4.50	.471	504	16.0	.005	.165	77.8	80.0	3.85	.155
8.5	27	18	4.58	.461	530	15.0	.007	.153	79.4	81.3	4.31	.146
9	1.5	18	2.10	.601	5730	355	27	5.43	1070	1210	15.4	1.29
10	28.5	18	3.40	.591	833	39.9	1.5	.447	142	150	5.05	.205
10.5	28.5	18.5	3.32	.622	790	37.5	1.2	.348	123	141	5.25	.192
11	4.5	13.5	8.00	.384	57.2	.195	.005	.002	.0081	.0091	.0247	.0013
11.5	33	18	3.58	.612	686	32.2	1.0	.261	103	117	4.31	.147
15	53	14.5	3.31	.644	631	28.8	.45	.260	66.6	83.3	3.84	.136
16	17	13	7.98	.304	283	.109	.003	.012	.0177	.0207	1.26	<.0001
17	70	14	3.68	.628	564	20.5	.35	.231	47.9	55.7	3.06	.109
23.5	70	19.5	3.25	.689	483	19.8	.018	.231	9.01	18.4	3.04	.100
24	90	12.5	8.85	.379	1.89	.01	.002	.001	.0086	.0099	.0052	.0001
25	160	16	4.90	.692	206	5.06	.01	.093	4.44	4.72	1.26	.0429
26	230	21	5.30	.398	188	.73	.003	.0555	5.12	5.52	1.04	.0327
27	8.5	17.5	8.41	.340	38.9	.007	.003	.002	.0236	.0242	.0115	.0002
28	240	20	5.88	.338	180	.202	.003	.0385	4.84	5.17	.961	.0335
29	71	23	5.50	.392	189	.048	.002	.0445	4.29	4.59	.957	.0331

Table I. (continued): Chemical analyses (mg/L) and physical parameters (discharge, Q, in L/s; temperature in °C; Eh in volts) for the Leviathan and Bryant Creeks

Site No.	Q	Temp.	pH	Eh	SO_4	Al	As	Cu	Fe^{II}	Fe^{TOT}	Mn	Zn
Oct., 1982:												
1	3.1	6	7.50	.279	7.21	.029	.0007	≤.0005	.0005	.0041	.0085	.0046
5	3	8.2	5.08	.437	143	2.13	.001	.026	11.1	11.4	.775	.0346
5.5/6	2.7	7.9	3.13	.546	2850	106	.48	.398	616	625	22.6	.694
6.5	5.7	7.8	3.78	.547	1450	51.0	.023	.192	278	280	11.1	.332
7	2.5	14.3	6.00	.263	1200	.45	.006	<.0005	37.6	38.1	11.4	.051
7.5	5.7	11.5	3.43	.574	1570	55.6	.032	.209	299	308	12.0	.383
8	8.2	12	3.78	.555	1520	46.9	.019	.196	264	266	12.6	.32
10.5	9	9.1	3.10	.612	1870	58.1	.032	.224	254	277	15.2	.385
11	2	4	7.15	.463	130	.016	.001	.0005	.0053	.0127	.0109	.0003
11.5	11	6	3.52	.607	1550	47.8	.032	.213	215	233	12.7	.332
15	11	7	3.11	.658	1480	45.0	.019	.202	112	174	11.8	.307
16	14	11.9	7.62	.235	245	.088	.005	.0015	.0045	.0123	.95	.0015
17	24	7.8	3.55	---	764	18.8	.012	.101	45.6	56.2	5.53	.125
23.5	24	8.3	3.60	.662	723	19.9	.011	.108	23.6	35.5	5.48	.129
24	67	7	8.20	.344	1.26	.01	.003	.002	.0059	.011	.0052	.0001
25	91	7	5.53	.370	219	.62	.004	.0175	6.94	7	1.57	.0388
26	110	12.7	7.69	.404	152	.036	.002	.002	.0018	.0091	1.04	.0019
27	5.7	12.2	8.20	.439	44.7	.009	.003	.0015	.0155	.0389	.0075	.0006
28	110	13.5	7.78	.394	158	.107	.003	.0015	.0013	.0033	.928	.0017
29	110	14	7.80	.423	156	.14	.003	.0015	.0019	.0066	.923	.0032

is a surface complexation model, and the accompanying set of internally consistent surface complexation constants for adsorption onto hydrous iron oxide are given by Dzombak and Morel (6). The properties for hydrous Fe oxide used in this study are the same as those used in (6) to determine the surface complexation constants. For modelling purposes, it is assumed that the suspended Fe particulate is the sole adsorbent and is present as hydrous Fe oxide. Calculations for binding site concentrations and total amorphous Fe concentrations are based on these assumptions. MINTEQA2 was used to calculate trace-metal adsorption onto Fe oxide over selected intervals of the LBC drainage, using upstream water chemistry and downstream pH.

The suspended material (> 0.10 µm) was not measured or analyzed in the water samples on which this study is based. An estimate of the amount of freshly precipitated, suspended Fe particulates was made for each interval of the LBC drainage, using the difference in mass flow of dissolved Fe. The decrease in the mass of soluble iron is assumed equal to the mass of suspended particulate iron. This estimate provides a concentration and a mass flow of particulate iron. It is then assumed that the particulate iron has the same properties (surface area, i.e. 600 m^2/g, and adsorption site density) as those of freshly precipitated hydrous oxide described by Dzombak and Morel (6). Two site densities are used, 0.005 mol/mol Fe for the high-affinity site densities and 0.20 mol/mol Fe for the low-affinity site densities. A further assumption is that the iron from the mass balance results is used as $Fe_2O_3 \cdot H_2O$ with a formula weight of 89 g/mol Fe (6). A comparison of the Fe content of the suspended load, for the few sites which were previously sampled simultaneously (2), suggests that this method provides a reasonable approximation to the concentrations of suspended Fe particulate in the river. Variation of the total concentration of Fe-rich particulates or the equilibrium pH (using upstream rather than downstream pH values) in the MINTEQA2 model does not affect greatly the qualitative outcome of these results.

Results: Trace-Metal Attenuation

Stream Description. Iron-rich, brown-to-orange precipitates were particularly evident on the stream bed in the tributary entering the LBC drainage in three locations: at site 7, near the confluence of Mountaineer Creek, and near the confluence with the East Fork Carson River. During periods of high rainfall the LBC mainstream becomes turbid and appears to carry a high concentration of suspended Fe(III) precipitate. Often the stream-bed precipitates are mixed with white precipitates of Al hydroxysulfates, particularly below the Mountaineer Creek confluence. Analyses of the mixed Al- and Fe-bearing precipitates from the LBC stream bed downstream from Mountaineer Creek indicated that the precipitate contained 1-24 wt% Fe. Analyses of bed sediments in this region of the drainage (at sites 17, 25, and 26, collected in May 1983; 2) are consistent, identifying an Al-rich sediment with < 1 wt% Fe.

Sites of trace-metal removal from solution. Partitioning of metals between the dissolved and solid phases can be identified by mass-flow calculations for Cu, As,

Zn, and Fe at each sampling site on the main drainage. The mainstream mass flow of a metal will not decrease as a result of dilution of the mainstream at a confluence, but will decrease if the metal is removed from solution by precipitation or adsorption, or increase if the mainstream mixes with a metal-rich tributary. The sulfate mass flow only increases from acid mine water inflows and then remains constant (Figure 2c). The mass flow of each metal (in mmol/s and μmol/s) is shown as a function of distance from site 1 in Figure 3a-d. An average error of 10% in the discharge rates and 5% in the analyses has been assumed, giving a total error of 15% in the metal mass flow shown.

Iron and Arsenic. In June 1982, the As and Fe content of the LBC drainage increased dramatically between site 5 and site 10 (Figure 3a-b), where the mainstream is contaminated mainly by drainage from the open pit (site 9). The mainstream also appeared to have been contaminated by Fe over this interval by the tributary at site 7 and by seepage from waste dumps near sites 5 and 6.5. On the other hand, the As present at site 10 was derived solely from the surficial tunnel drainage entering LBC drainage at site 9 and was attenuated to background levels similar to those upstream from the Leviathan mine site, above site 23.5 just above the Mountaineer Creek confluence (Figure 3b).

Iron was removed from solution between site 10 and 10.5, but replaced between site 11.5 and 15. This recharge may have been caused by input from the summer diversion of the open-pit drainage as shown in Figure 1 and/or seepage into the stream from the landslide which occurs in the northwest area of the disturbed ground surrounding the mine. A similar input can be seen for SO_4, Cu, and Zn in this section of the drainage. As the groundwaters in the toe of the landslide have generally low concentrations of Cu, As, and Zn (2), it seems likely that the open-pit drainage diversion caused the observed increase in Fe mass flow between sites 11.5 and 15. More than 80% of the dissolved Fe present at site 15, just upstream from the Aspen Creek confluence, was removed before site 25, just below the Mountaineer Creek confluence. However, there also appears to have been an increase in Fe mass flow below site 25 (Figure 3a), similar to that noted for SO_4 mass flow. This increase may have been due to either the photoreduction of Fe(III) in the flat, exposed section of the drainage (7), or to the oxidation of pyrite aggregates carried through the faster sections of the drainage. Pyrite oxidation is more likely because it explains the coincident increase in both Fe and SO_4 mass flows observed in this region of the LBC drainage.

In October, 1982, the mass flow of Fe was less than that of June, but showed a similar trend within the drainage. Leviathan Creek was contaminated primarily by seepage from waste dumps (e.g., at sites 5.5 and 6) and not by surficial flow from the open pit in October. Approximately 75% of the dissolved Fe present at site 11.5 was removed upstream from the Mountaineer Creek confluence, and the remaining 25% was removed between the confluences with Mountaineer and Doud Creeks. The As mass flow was low (Figure 3b; note the different vertical scales for June and October As mass flow) and although variations are apparent, there is over 100% variability on repeat analyses and between different analytical techniques for As concentrations < 0.01 mg/L (1). In October

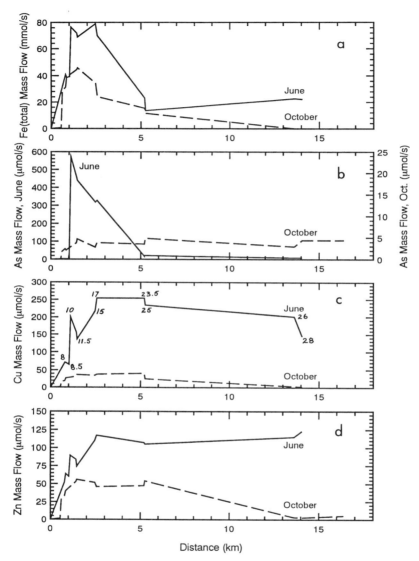

Figure 3. a. Mass flow for total dissolved iron with distance downstream, June and October 1982. The discharge and mass flow at site 29 for June have been omitted because summer diversion of this water for irrigation greatly decreases the discharges. b. Mass flow for dissolved arsenic with distance downstream where the mass flow for June is shown on left-hand scale and the mass flow for October is shown on the right-hand scale. c. Mass flow for dissolved copper with distance downstream, June and October 1982. Numbers inside diagram box refer to site locations found in Figure 1. d. Mass flow for dissolved zinc with distance downstream, June and October 1982.

mainstream As concentrations ranged from 0.0007 to 0.032 mg/L, so there is considerable analytical error which is further propagated by the mass flow calculations. In both June and October, As and Fe were primarily removed between confluence points, even at low pH.

Copper and Zinc. In June the mass flows of Cu and Zn were similar to that of Fe above the Aspen Creek confluence (Figure 3c and 3d, above site 17). For example, a small amount of the dissolved Cu and Zn appears to have been removed from solution between sites 10 and 10.5, immediately after the input of open-pit effluent at site 9. This effect is more significant for Cu (22% removed) than for Zn (6.3%), as the latter is within the calculation error. Like Fe, Cu and Zn mass flows increased between site 11.5 and 15, where seepage from the landslide and the diverted drainage from the open pit found their way into the LBC mainstream. However, Cu and Zn were not removed from solution between the Aspen Creek and Mountaineer Creek confluences (sites 17 to 25). Although approximately 42% of the dissolved Cu present at site 15 was removed further downstream, between Mountaineer and Doud Creek (sites 25 to 28), Cu and Zn mass flows remained high in the lower drainage region. At site 29, Cu and Zn concentrations were respectively 30 and 330 times those of the uncontaminated LBC drainage at site 1.

In October the Cu and Zn mass flows were significantly less than those of June, and had decreased to near background levels by the Doud Creek confluence (site 26). Of the Cu and Zn introduced by contamination of the LBC mainstream at the mine site, 92% of the dissolved Cu was removed between sites 23.5 and 26, and 94% of the dissolved Zn between sites 25 and 26. In both June and October, therefore, Cu and Zn failed to show significant attenuation in waters of pH < 5. At pH > 5 Cu and Zn were partially removed from solution, mainly between confluence points.

Chemical Attenuation Rates. It is evident that Fe, As, Cu, and Zn were each removed from solution over different intervals of the LBC drainage, and at different rates. When the drainage is arbitrarily divided into two intervals, from the mine site (site 10) to Mountaineer Creek confluence (site 23.5) and from the Mountaineer Creek confluence to site 28, attenuation rates can be calculated for Fe, Cu, As, and Zn for these intervals (see Table II).

Between the mine site and Mountaineer Creek, the attenuation rate of Fe was higher in June than in October, reflecting the greater quantities of Fe reaching the LBC drainage from the mine site in June. In June, Fe was removed from solution over this interval at more than 100 times the rate of attenuation downstream from the Mountaineer Creek confluence. Arsenic had the highest attenuation rate of the three metals (As, Cu, and Zn) between the mine site and Mountaineer Creek confluence. There was a net increase in the mass flow of Cu over this interval in both June and October, and a net increase in the mass flow of Zn in June. Downstream from the Mountaineer Creek confluence, both Cu and Zn were removed from solution at a similar, low rate of attenuation in October, and Cu was removed at a higher rate in June.

Table II. Chemical attenuation rates (μmol/km/s) for trace metals in the LBC drainage system

Site Interval	Length (km)	As (J)	Cu (J)	Cu (O)	Zn (J)	Zn (O)	Fe (J)	Fe (O)
10.5-23.5	3.79	116	+	+	+	1.5	12900	7750
23.5-28	8.85	0.8	12.3	4.4	+	5.0	95	1720

'+' indicates a net increase in the mass flux of a metal between these sites; 'J' and 'O' represent June and October.

Chemical Modelling. The analytical data in Table I has been used with the geochemical speciation models WATEQ4F (4) and MINTEQA2 (5) to indicate whether mineral-phase precipitation or adsorption processes were limiting trace-metal concentrations in the LBC drainage in 1982.

Iron. Ferrihydrite ($Fe(OH)_3$) saturation indices computed by WATEQ4F show saturation or slight supersaturation for all samples. Supersaturation with respect to jarosite is also indicated for waters of the upper drainage (pH < 6). The solubility-limiting mineral phase for Fe may be deduced from plots of the free Fe^{3+} activity against pH (8). For waters of pH > 7, such as the uncontaminated LBC drainage and LBC drainage near and after the East Fork Carson River confluence, log $a_{Fe^{3+}}$ plotted against pH has a slope of -3.0, suggesting equilibrium with ferrihydrite. In the waters of pH < 6, however, a slope of -2.3 is observed. This slope is inconsistent with that expected for waters at equilibrium with ferrihydrite, goethite, or jarosite.

It has been proposed (9) that a poorly crystalline oxyhydroxysulfate of Fe (unit cell formula of $Fe_{16}O_{16}(OH)_{12}(SO_4)_2$ or $Fe_{16}O_{16}(OH)_{10}(SO_4)_3$ is a primary component of the Fe-rich precipitates formed in acid-sulfate mine waters of pH 2.5 to 4. However, if Fe^{III} is in equilibrium with a phase of either stoichiometry, slopes of -2.8 or -2.6 respectively may be expected. A slope of -2.3 appears to be consistent with a more sulfate-rich iron oxyhydroxide: $Fe_{16}O_{16}(OH)_5(SO_4)_{5.5}$ which may also be written as the iron oxyhydroxsulfate: $Fe(OH)_{2.3}(SO_4)_{0.35}$. Similar problems have been experienced characterizing orange amorphous precipitates from an acid mine drainage in the West Shasta mining district in California (10), which have been described as either a mixture of jarosite [$KFe_3(SO_4)_2(OH)_6$] and ferric hydroxide or ferric hydroxide with adsorbed sulfate.

Arsenic. There appears to be a close correlation between the removal of As and Fe from solution in the LBC drainage, although any stoichiometric relationship between the quantities of Fe and As removed from solution is obscured by the abundance of Fe. All saturation indices for scorodite ($FeAsO_4 \cdot 2H_2O$) show undersaturation. An alternative mechanism for As attenuation is the adsorption and (or) coprecipitation of As with the suspended or stream-bed iron phases. The adsorptive capacity of a ferrihydrite or ferric oxyhydroxysulfate phase is high for both As(III) and As(V), even at a pH as low as 4 ([11](#)). Arsenic in the oxidized, acid mine effluent will be predominantly present as As(V), which is adsorbed more strongly at lower pH where it exists as an anionic complex. Adsorption onto the surface of suspended particulate Fe can account for the observed removal of As, for both June and October, based on MINTEQA2 results shown in Table III. In June, the amount of As actually removed over most intervals in the upper drainage is less than that predicted by MINTEQA2. This may be due to the sulfate-rich nature of the adsorbent or perhaps to the lack of time in the fast-flowing stream. Also, the As/Fe weight ratio of the stream-bed sediment is 0.25 to 0.15 at sites 17, 25, and 26 ([2](#)), and is significantly higher than the similar ratio for As/Fe removed from solution which is 0.04 between sites 10 and 17. Therefore, As may also be adsorbed by existing stream-bed precipitate.

Table III: Metal adsorbed over specific intervals in the LBC drainage, expressed as a percentage of the upstream metal mass flow

	Site Interval	pH	Fe (mg/L)[#]	As (m)[§]	As (c)[§]	Cu (m)[§]	Cu (c)[§]	Zn (m)[§]	Zn (c)[§]
June:	10-10.5	3.4-3.3	9	31	20	0	22	0	6
	11.5-15	3.6-3.3	9	33	28	0	-	0	-
	17-23.5	3.7-3.3	37	70	95	0	-	0	8
	23.5-25	3.3-4.9	7	59	-[*]	4	8	0	2
	25-26	4.9-5.3	7	84	57[*]	21	14	0	-
Oct:	11.5-15	3.5-3.1	59	70	41[*]	0	5	0	8
	23.5-25	3.6-5.5	8	77	-[*]	33	39	0	-
	25-26	5.5-7.7	7	99	40[*]	97	86	77	94

[#] 'Fe' is the amount of Fe precipitated, calculated as the difference in mass flow (mg/sec) over the interval, divided by upstream discharge (L/sec).

[§] Adsorption predicted using the MINTEQA2 model is shown as (m) and the calculated percentage decrease in the mass flow of the metal as (c). '-' indicates that metal load has increased or remained unchanged over the interval.

[*] The uncertainties in the analyses for these samples cause a high degree of error in these values.

Copper and Zinc. All of the LBC drainage waters were undersaturated with respect to the Cu and Zn sulfate and hydroxysulfate salts such as antlerite ($Cu_3(SO_4)(OH)_4$), brochantite ($Cu_4(SO_4)(OH)_6$) and goslarite ($ZnSO_4 \cdot 7H_2O$), which potentially could precipitate from the sulfate-rich mine waters. Saturation with respect to tenorite (CuO), malachite ($Cu_2(OH)_2CO_3$), zincite (ZnO) and $Zn(OH)_2$ was approached in the waters of pH > 7. Tenorite and zincite are phases unlikely to form in this environment, however phases such as malachite, hydrozincite ($Zn_5(CO_3)_2(OH)_6$) and hemimorphite ($Zn_4Si_2O_7(OH)_2 \cdot H_2O$) may form; unfortunately the thermodynamic and kinetic data are not available to evaluate adequately their role in metal attenuation. For the waters of the LBC drainage with pH < 6 neither WATEQ4F saturation indices nor free-ion activities can be used to identify a likely Cu- or Zn-bearing precipitate.

Although the attenuation trends for Cu and Zn in the LBC drainage are not as clearly defined as for Fe and As, the removal of dissolved Cu and Zn in the lower drainage (23.5 to 26) in October is consistent with adsorption onto hydrous Fe oxide (Table III). Likewise, the removal of dissolved Cu over the same interval in June is very likely to have been due to adsorption processes. Because Cu is adsorbed more readily onto hydrous Fe-oxides than Zn at pH < 6 (6,12), Cu was removed preferentially between sites 23.5 to 25 in October and sites 25 and 26 in June 1982, while Zn remained in solution. A further increase in pH, as occurred between sites 25 and 26 in October, would have been required for Zn adsorption.

Summary and Discussion

A geochemical model is proposed to explain trace-metal attenuation in a mountainous stream receiving acid mine drainage, based on the precipitation of a ferric oxyhydroxysulfate and consequent trace-metal attenuation by adsorption onto this particulate phase. The model can account for variations in trace-metal concentration, both with distance from the mine site and with seasonal changes in drainage flow and acidity. Adsorption onto suspended ferric oxyhydroxysulfate appears to be the major mechanism regulating the concentrations of As, Cu, and Zn downstream from the mine site in the LBC drainage system. In June 1982, when the concentrations of As, Cu, and Zn in the mainstream were high, close to the mine site, all the dissolved As was adsorbed onto Fe particulates within 5 km of the mine site. Although a fraction of the dissolved Cu similarly was adsorbed further downstream, between Mountaineer and Doud Creeks, the Cu and Zn concentrations remained high in the lower drainage and may have contaminated the East Fork Carson River at this time of year. In October 1982, As concentrations in the LBC drainage were not increased significantly by seepage from the mine site, and dissolved Cu and Zn were removed effectively by adsorption between the Mountaineer and Doud Creek confluences.

Seasonal differences in metal attenuation were caused by increased discharge from the mine site during the spring and early summer months. This increased As concentrations in the LBC, because the surficial drainage from the open pit of the Leviathan Mine had high As concentrations. The elevated mine-site discharge also increased the acidity of the lower drainage (below Mountaineer Creek). The LBC drainage between Mountaineer Creek and Doud Creek (pH = 4.5 to 5.3) was considerably more acid in June than in October (pH = 5.5 to 7.7). Because Zn is not adsorbed as readily onto the surface of Fe-oxide as copper at pH < 6 (6,12), Zn remained in solution in the lower drainage in June, but was adsorbed in October. Similar controls have been reported for Cu and Zn concentrations in other rivers and estuarine waters receiving acid mine drainage. Cu and Zn concentrations in the Canon River and Restrongnet Creek in England, for example, also were interpreted to be controlled by binding to amorphous Fe oxyhydroxides (13); a process which is pH-dependent and consistent with laboratory studies.

There are other factors which may affect trace metal attenuation in the LBC drainage which have not been considered here. For example, the role of other adsorbents, including organic material such as the algal mats which often occur with precipitated Fe particulates, Al hydroxide and Mn oxides, has not been fully investigated. Neither has the effect of Cu and Zn ferrite formation (14) on Cu and Zn attenuation been assessed. Future research should investigate these factors.

Appendix: Modifications of the original data

The following alterations were made to the original analytical data (1). Inconsistencies which may have been caused by sample contamination or analytical errors have been identified from anomalous metal/metal or metal/anion ratios. For the October samples, water composition for sites 5.5 and 6 have been averaged for the mass balanced calculations. A value of 0.398 mg/L (that of site 6) has been adopted for Cu as the Cu concentration for site 5 appears to be inconsistent with other trace and major ion concentrations in the sample. Also, a Cu concentration of 0.0015 mg/L has been adopted for site 28. In June, Zn concentrations of 0.0001 mg/L and 0.100 mg/L have been adopted for sites 24 and 23.5 respectively, and an Al concentration of 0.01 and a Mn concentration of 0.0052 mg/L have been adopted for site 24.

Several minor alterations have been made to the discharge data reported (1), based on the mass balance of conservative ions (as explained in the text). The following major alterations have also been made; June discharge at site 10.5 was altered from 41 L/s to 28.5 L/s, and at site 15 was altered from 40 to 53 L/s. Note that the discharge at site 29 is low owing to summer diversion of the mainstream for agricultural use. October discharge at site 23.5 was altered from 40 to 24 L/s and at site 24 from 57 to 67 L/s. An error in the discharge at site 23.5 was substantiated by K.E. Bencala (pers. comm., 1992), who calculated a discharge of 29 L/s, based on unpublished revisions of a tracer injection experiment.

Acknowledgments

The authors wish to thank Jim Ball and Reini Leinz of the USGS for their analytical assistance, and Briant Kimball, Ken Bencala and Jim Ball for reviewing this manuscript. The project has been funded by the USGS Water Resources Division and by the Institute of Environmental Health and Forensic Science (formerly DSIR Chemistry) of New Zealand.

Literature Cited

1. Ball J.W.; Nordstrom D.K. U.S.G.S. Water-Res. Invest. Rept. 89-4138 1989, 46 pp.
2. Hammermeister D.P.; Walmsley S.J. U.S.G.S. Open File Rept. 85-160 1985, 120 pp.
3. Chapman B.M.; Jones D.R.; Jung R.F. Geochim. Cosmochim. Acta. 1983, 7, 1957-1973.
4. Ball J.W.; Nordstrom D.K. U.S.G.S. Open File Rept. 91-183 1991, 188 pp.
5. Allison J.D.; Brown D.S.; Novo-Gradac K.J. U.S. Environ. Protect. Agency, 1991, EPA/600/3-91/021, 106 pp.
6. Dzombak D.A.; Morel F.M.M. Surface Complexation Modeling: Hydrous Ferric Oxide. Wiley Interscience, New York, 1990, 393 pp.
7. McKnight D.M.; Bencala K.E. Geochim. Cosmchim. Acta 1989, 53, 2225-2234.
8. Nordstrom D.K. In Proceedings of the U.S.G.S. Toxic Substances Hydrology Program; Mallard G.E. and Aronson D.A., eds. U.S.G.S. Water Res. Invest. Report 91-4034, pp 534-538.
9. Bigham J.M.; Schwertmann U.; Carlson L.; Murad E. Geochim. Cosmochim. Acta 1990, 54, 2743-2758.
10. Filipek L.H.; Nordstrom D.K.; Ficklin W.H. Environ. Sci. Tech. 1987, 21, 388-396.
11. Pierce M.L.; Moore C.B. Water Res. 1982, 16, 1247-1253.
12. Benjamin M.M.; Leckie J.O. J. Coll. Interfac. Sci. 1981, 79, 209-221.
13. Johnson C.A. Geochim. Cosmochim. Acta 1986, 50, 2433-2438.
14. Hem J.D. Geochim. Cosmchim. Acta 1977, 41, 527-538.

RECEIVED September 23, 1993

Chapter 18

Acid Mine Drainage in Wales and Influence of Ochre Precipitation on Water Chemistry

Ron Fuge, Fiona M. Pearce, Nicholas J. G. Pearce, and William T. Perkins

Geochemistry and Hydrology Research Group, Institute of Earth Studies, University of Wales, Aberystwyth, Dyfed SY23 3DB, United Kingdom

> A long history of metalliferous mining has left Wales with a legacy of abandoned mines and attendant spoil tips which are continuing sources of environmental contamination. Several of these old mines in the ore fields of mid and north Wales are sources of highly polluted acid drainage resulting from the oxidative weathering of pyrite and marcasite (FeS_2). The acid mine drainage which contains elevated levels of many elements, including Al, Fe, Cu, Zn, As, Cd and the SO_4 anion, is characterized by red-brown precipitates of ochre on stream beds and in lakes intercepting the drainage. Precipitation of the ochre has a marked influence on the chemistry of these acid waters with many elements concentrated in the ochre. The precipitation of Pb and Ba is controlled by the SO_4^{2-} content of the waters whereas elements such as As, Mo, Sn and Tl appear to be coprecipitated or adsorbed by the hydrated iron oxide. The degree of precipitation of Al, Mn, Fe, Cu, Zn and Cd is controlled by pH; lower acidity and higher pH favor precipitation.

Wales has a long history of metalliferous mining, ores having been extracted since at least Roman times with the industry at its peak during the eighteenth and nineteenth centuries. The extensive ore fields of mid and north Wales were mainly Pb, Zn and Cu producers with Ag also being extracted from the PbS ore. Barium, Mn and Fe were produced in some areas whereas Au mines were important in the Dolgellau gold belt ([1]) and are a continuing source of interest.

Although some metal mining interest has persisted to the present day, most mines had ceased operation by the 1920s and many abandoned mines and spoil tips scar the old ore fields. During the peak period of mining, techniques of ore extraction were very inefficient and there was little environmental control of the operations. Many of the old mines occurred in fault-controlled river valleys and as a result many river systems were badly polluted and were devoid of fish and insect life ([2-4]). In addition, flood-plain soils were contaminated and are found to contain high levels of heavy metals (5,6).

Cessation of much of the metalliferous-mining operations in the early part of this century has seen a marked improvement of river water quality but old mine workings and spoil heaps are continuing sources of pollution (7). Several of the mines intersected lodes containing pyrite and marcasite (FeS_2) and, as a result of the oxidative weathering of these minerals, some of the abandoned mines and associated spoil heaps have acidic drainage waters. These acid waters are characterized by hydrated iron oxide precipitation resulting in red-brown drainage channels and tip areas.

In this paper we consider the chemistry of some of these acid waters and their attendant ochres. In addition, some consideration is given to the role of ochre precipitation as a control on the chemistry of the pollutant drainage.

The Study Areas

The specific study sites are marked on Figure 1. The Cwmrheidol and Y Fan mines are typical of the mid Wales ore field, working ENE-trending lodes within Lower Paleozoic sediments. Both mines were worked primarily for lead and zinc, however, the Y Fan mine also produced some barite. While the Cwmrheidol deposit contains abundant pyrite and marcasite, the Y Fan deposit contains only subordinate amounts (8). Both mines, with extensive underground workings drained by adits driven in from valley sides, were abandoned early this century and have been continuing sources of significant pollution (9,10). Drainage channels from both mines are coated with rust-colored ochre, this being particularly so at Cwmrheidol where the stream draining the mine is known locally as "the red stream."

Gwynfynedd mine which was originally worked as a lead mine became a gold producer in the mid 19th century. Situated at the northern end of the Dolgellau Gold Belt (1), it produced 40,000 ounces of gold between 1888 and 1916 (11). Closed in 1938, it re-opened during the 1980s and was abandoned again in 1989. The mineralization within this mine occurs in veins cutting Lower Paleozoic sediments and intrusive rocks and consists of the sulfides pyrite, arsenopyrite, chalcopyrite, galena and sphalerite with gold occurring rarely in the veins. Mining at Gwynfynedd was from a series of shafts, open cuts and adits the lowest of which also serves as a drainage conduit. The mine drainage is characterized by ochre precipitation.

Level Goch in Snowdonia, north Wales, was one of several copper mines in that area. Little is known of its history but it is thought to have been mined more than 200 years ago (12). The geology of the area is comprised of Lower Paleozoic sedimentary and volcanic rocks. The old adit is drained by an ochreous stream.

Parys Mountain is known to have been mined by the Romans but the peak of extraction was during the late 18th and the 19th centuries when it was a major copper producer (13); mining ceased in 1911 (14). Although chiefly worked for copper, the deposit is polymetallic and there is current interest in the area for Pb, Zn and Cu mining (14). The mineralization includes chalcopyrite, sphalerite, galena and arsenopyrite but according to Pointon and Ixer (15) it is dominated by pyrite; these workers also list Bi sulphosalts and barite as occurring. The deposit is intimately

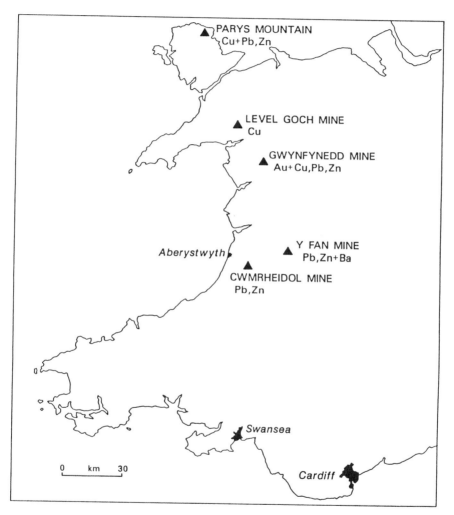

Figure 1. Wales with sampling sites marked.

associated with mid-Ordovician rhyolitic volcanism and is overlain by Silurian shales (16); it has been described as a Kuroko-type massive sulfide deposit (17).

Much of the drainage of the surface workings at Parys Mountain flows into the old pits and other hollows creating a series of lakes with brown-colored waters. In addition, the area had several ochre settling ponds during its working life and these also intercept surface and subsurface drainage (Figure 2).

Methodology

Water samples were collected in acid-washed polypropylene bottles and refrigerated until analyzed. Analyses were performed within 48 h of collection. Samples were filtered through a 0.45 μm filter but were not acidified. Initial results from both acidified and non-acidified samples were found to be identical when analyses were performed within 48 h of collection so it was deemed unnecessary to acidify all other samples.

Mine drainage was sampled several times over periods of from several months to more than a year during 1990 to 1992. Results quoted are mean values for all of these replicates.

Lake waters were sampled twice during the summer of 1992. Samples were collected from the surface, bottom and mid-level depths; the samples were analyzed independently.

Ochre samples were collected by gently removing surface material (generally to 1 cm depth). The samples were dried at a temperature of 40°C and disaggregated in an agate pestle and mortar. Samples were prepared for analysis with a variety of acids (40 ml) involving extraction of 500 mg of duplicate samples with:
1. Concentrated HCl
2. Concentrated HNO_3
3. A 4:1 mixture of concentrated HNO_3 and $HClO_4$

Sample extractions 1 and 2 gave results which were identical within analytical error for all elements other than Fe, HCl giving the more consistent results. Method 3 gave lower results for several metals including Pb and As. Therefore, for this work analyses were performed on duplicate samples leached with HCl.

Analyses were performed using a VG Instruments inductively coupled plasma - mass spectrometer (ICP-MS). The analytical technique used was based on a single calibration method using indium as the internal standard (18). In addition two 10 g samples of ochre powder were mixed with 1 ml of 1000 ng g^{-1} indium solution, dried, reground and bricketted before being analyzed by laser ablation ICP-MS, using artificial standards made from analytical grade chemicals (see 19,20). Results obtained by this method gave values which agree closely with HCl solution analysis.

Results and Discussion

Some representative analyses of acid mine drainage waters and ochres are presented in Tables I, II, and III. Results in Table I are for mine drainage waters and associated ochres from mid and north Wales and Table II for drainage samples from the mine and tip areas of Parys Mountain. Table II presents data for waters and ochreous precipitates

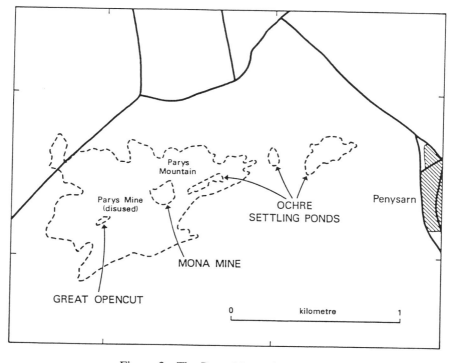

Figure 2. The Parys Mountain minesite.

Table I. Chemistry of mine water and ochre

	Cwmrheidol mine		Gwynfynedd mine		Y Fan mine	
	Water (mg/l)	Ochre (mg/kg)	Water (mg/l)	Ochre (mg/kg)	Water (mg/l)	Ochre (mg/kg)
pH (units)	2.9		3.5		5.0	
SO_4	850		280		285	
Li	0.014	8.7	0.028	4.1	0.22	4.0
Mg	25.6	2140	8.03	450	10.3	712
Al	14.2	14000	0.77	3840	1.05	6890
Ca	32.2	117	27.1	700	77.3	519
Mn	2.50	688	5.26	267	0.92	184
Fe	36.2	26.7%	21.0	40.9%	0.68	20.9%
Co	0.14	10	0.034	52	0.043	8
Ni	0.76	22	0.057	120	0.073	14
Cu	0.063	44	0.12	527	0.13	1520
Zn	53.3	764	3.79	608	64.5	24600
As	0.014	13	0.13	102	0.16	44
Cd	0.071	2.4	0.009	3.0	0.42	160
Sb	0.001	14	0.009	6.0	0.001	110
Ba	0.004	52	0.005	3.3	0.116	28
Pb	0.017	127	0.033	891	1.28	76900
Bi	0.002	0.3	0.003	0.3	0.005	110

Table II. Chemistry of typical acid lake water and ochre from Parys Mountain

	Lake, Great Opencut		Lake, Mona mine		Ochre settling ponds	
	Water (mg/l)	Ochre (mg/kg)	Water (mg/l)	Ochre (mg/kg)	Water (mg/l)	Ochre (mg/kg)
pH (units)	2.3		2.4		2.5	
SO_4	5200		4100		3800	
Li	0.97	4.0	1.2	2.7	2.1	2.3
Mg	59.6	183	68.2	88	155	187
Al	108	794	91.2	432	174	514
Ca	21.6	143	10.9	3460	45.0	74
Mn	4.80	13	17.5	20	23.0	100
Fe	1300	14.4%	783	13.3%	934	35.9%
Co	0.38	15	0.37	6.0	0.63	16
Ni	0.16	3.0	0.38	30	0.30	67
Cu	69.6	1790	57.3	1910	74.4	636
Zn	54.0	247	47.6	246	82.2	314
As	0.77	667	0.16	1510	0.20	623
Mo	0.005	56	nd	29	nd	14
Ag	nd	17	nd	29	nd	3.0
Cd	0.118	0.9	0.098	1.0	0.17	1.3
Sn	nd	44	nd	24	nd	31
Sb	0.002	53	0.004	59	0.001	48
Ba	nd	22.0	nd	3.0	nd	7.0
Hg	nd	6.0	nd	10	nd	2.6
Tl	0.006	17	0.001	26	0.005	6.0
Pb	0.53	24300	0.056	80300	0.11	6450
Bi	0.001	275	nd	172	nd	16
U	0.009	0.4	0.006	0.5	0.010	0.2

nd = not detected (detection limits = .001 mg/l for water; .001 mg/kg for ochres)

Table III. Chemistry of some waters flowing into Parys Mountain lakes

	1 (mg/l)	2 (mg/l)	3 (mg/l)	4 (mg/l)
pH	2.1	2.7	2.2	2.0
SO_4	7200	480	4800	7400
Li	0.33	0.038	0.05	0.28
Mg	41	9.7	6.9	31
Al	48.3	19.7	27.8	60.1
Ca	51.6	16.4	0.7	5.0
Mn	1.6	0.70	1.6	7.1
Fe	1070	13.5	649	1290
Co	0.40	0.043	0.34	0.54
Ni	0.08	0.025	0.32	0.14
Cu	60.7	48.0	71.2	108
Zn	52.7	26.7	66.8	45.6
As	2.6	0.004	0.60	2.9
Mo	0.035	nd	0.034	0.029
Ag	nd	0.001	nd	nd
Cd	0.129	0.058	0.120	0.080
Sn	nd	nd	nd	nd
Sb	0.013	nd	0.010	0.008
Ba	0.003	0.011	0.002	0.003
Hg	nd	nd	nd	nd
Tl	0.024	0.011	0.002	0.012
Pb	0.508	1.64	0.466	0.325
Bi	0.010	0.001	0.004	0.007
U	0.016	0.002	0.017	0.217

nd = not detected (detection limits = 0.001 mg/l)

from the lakes and pools which intersect the surface and subsurface drainage of Parys Mountain.

Water chemistry. There are obvious differences in the pH values of the different waters. The highest values recorded were for the Y Fan mine where only minor FeS_2 is associated with the mineralization, whereas the lowest pH values are recorded for Parys Mountain where there is ubiquitous pyrite and arsenopyrite. Generally, the pH values are reflected in the SO_4^{2-} contents of the waters; Parys Mountain samples contain up to 7400 mg/l and those of Y Fan contain only 285 mg/l.

As would be expected the Al content of the waters reflects the acidity, with the content ranging from 1.05 and 0.77 mg/l in the Y Fan and Gwynfynedd waters (pH 5.0 and 3.5 respectively) to 60.1 mg/l in a sample draining Parys Mountain (pH 2.0). Iron contents of the waters also reflect the acidity ranging from 0.68 mg/l at Y Fan to 1970 mg/l in a sample from Parys Mountain. Magnesium while being more variable is generally present in higher concentrations in the more acidic waters.

The Pb content of the waters would seem to be roughly reflecting the levels of the SO_4^{2-} anion. At Y Fan where that anion is relatively low the highest level of Pb is recorded (5.07 mg/l), similarly at Parys Mountain the highest Pb level (1.64 mg/l) is found in a sample with only 480 mg/l of sulfate. The low lead contents of high SO_4 waters suggests solubility control by $PbSO_4$ and the precipitation of $PbSO_4$ due to the common ion effect. A similar pattern emerges for Ba but the highest value is recorded for Y Fan where the mineral barite occurs.

From sample K_{sp} calculations it is apparent that the Y Fan water together with some of the Parys Mountain waters contain more Pb and Ba than would be expected from the SO_4^{2-} content. While samples were filtered through a 0.45 μm filter it is possible that solid $PbSO_4$ and $BaSO_4$ could exist in a very fine form that passes through this filter.

For several metals analyzed in the present study the content of the drainage waters reflects the source. Thus at Cwmrheidol, where there is little copper mineralization, low levels of that element occur in the drainage waters. At Parys Mountain, a copper mine, very large copper concentrations are recorded in the waters. Similarly, the variable levels of Ca are likely to reflect the amounts of calcite gangue.

The relatively high levels of U and Mo recorded for surface runoff at Parys Mountain probably reflect the presence of these elements within the deposit of Silurian black shales. Such shales occurring in other parts of north Wales are enriched in these elements ([21]; R. Fuge, unpublished data).

There are some marked differences in the chemistry of the lake waters and surface runoff at Parys Mountain. Aluminum is considerably higher in the lake waters than in the runoff; the same is true for Mg, Mn and possibly Ca. However, Pb, Bi, Tl, Ba, Sb, Mo and As are generally lower in the lake and pool waters. The relative enrichment of Al, Mg, Mn and Ca in the lake waters is possibly due to evaporation where these elements are held in solution due to the relatively high acidity. It is perhaps pertinent to note that the alkali metal Li is also much enriched in the lake waters (range 0.97 to 2.1 mg/l) relative to surface runoff (range 0.038 to 0.33 mg/l); in addition this element is almost totally absent from the ochreous precipitates (maximum value 4 mg/kg). The group of elements which are lower in concentration in the lake water are possibly lost from the standing waters by mineral precipitation (see following sections).

The smaller lakes of Parys Mountain are generally about 0.5 to 1 m in depth. To assess if any stratification occurs samples were collected from the top, bottom and mid depth of each lake. Only Pb exhibited any difference in concentration. It was slightly elevated in bottom waters of 3 of the lakes by about 10 to 20%. Similar elevated values for lead were found in deeper waters from the Great Opencut pit.

During the course of this work, samples of mine drainage were analyzed on a monthly basis for more than one year. From this work it is apparent that there are large variations in composition through time with, generally, highest values recorded for the winter months. For example, in Gwynfynedd outfall Zn ranges from 0.4 to more than 14 mg/l (20).

Ochre chemistry. The Fe content of the ochreous sediments is somewhat variable, ranging from more than 40% to 13.3%. Dry ochre from disused ochre settling pits was found to contain between 38 and 42% Fe. Of the ore metals, Cu is highly concentrated in the Parys Mountain ochres whereas Zn and Cd tend to be somewhat richer in the ochres from the Pb-Zn mines.

Several elements are strongly enriched in the ochres, including As, Mo, Ag, Sn, Sb, Ba, Hg, Tl, Bi and Pb. Whereas highest values of most of these elements were found in the Parys Mountain samples the highest level of Sb was found at Y Fan. The highest Mo content found was for the Level Goch mine where values of up to 240 mg/kg were recorded along with 42 mg/kg of W, which was undetected in any other samples. Aluminum is enriched in the ochres; the highest concentration detected was 1.4% in the Cwmrheidol samples. The Parys Mountain samples are lower in aluminum than are those from the adit drainage waters. Magnesium and Mn are also considerably more enriched in the adit ochres than in the ochreous lake sediments of Parys Mountain.

Relationship of water and ochre chemistry. A comparison of water and ochre chemistry shows that, in general, As, Mo, Ag, Sn, Sb, Ba, Hg, Tl, Pb and Bi are strongly concentrated into the solid phase (see Figure 3). In the case of Ba and Pb, the very high levels of the sulfate anion in the acid waters together with the low solubilities of the respective sulfates would suggest that these are the major controlling factors resulting in precipitation. Certainly in the waters, the highest levels of Ba and Pb are generally found where SO_4 contents are lowest (see earlier section).

In the case of the elements other than Pb and Ba that are enriched in the solid phase, it is likely that coprecipitation and adsorption by hydrated oxides of iron are responsible for their concentration in the ochreous precipitates. The adsorption and coprecipitation of metals by hydrated Fe oxides has long been known to be of importance in explaining soil and water geochemistry (22). Several workers have suggested that the adsorption of heavy metals and As by precipitating hydrated iron oxides is a major controlling influence on the mine drainage (23-25). The strong enrichment of As and Mo in ochres has been demonstrated by several workers (24,26,27). However, the high levels of Ag in the ochres would seem to contradict the findings of other workers (28).

Although the present data for elements such as As, Mo and Bi suggest that scavenging by hydrated Fe oxides is unaffected by pH variations in waters, several of the elements appear to show differing degrees of scavenging in differing pH regimes.

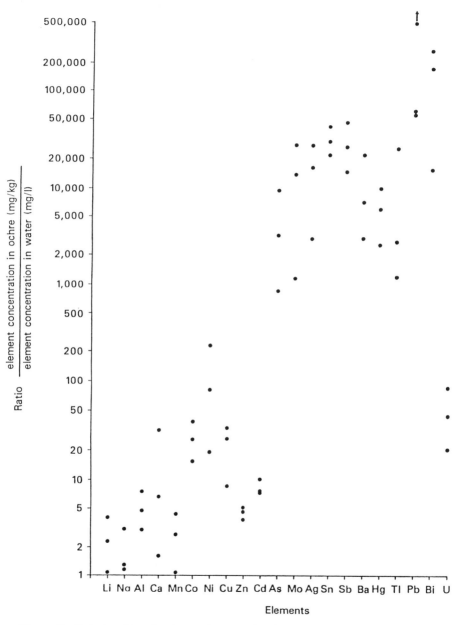

Figure 3. Relationship of ochre and water chemistry in Parys Mountain lakes.

Table IV lists concentration factors for ochres relative to the corresponding water, together with the pH of the waters. The pH ranges from 5 to 2; from Table IV it is apparent that with decreasing pH the concentration factors for Fe, Cu, Zn and Cd decrease markedly. Thus the greatest relative concentrations of all 4 of these elements is in the Y Fan mine ochres while the lowest is found in the Parys Mountain ochres. For Mg and Mn there is not a correlation with decreasing pH but both are strongly enriched in ochres from the less acidic adit drainage. The results strongly suggest that the pH of the metal-rich ochre-precipitating solutions has a dramatic effect on the degree of scavenging of some metals by hydrated iron oxide.

The increased adsorption of Zn and Cu by Fe oxyhydroxide in higher pH waters has been previously demonstrated for the acid mine drainage impacted River Carnon, Cornwall, England (29). It is possible that the degree of adsorption of Zn, Cu and Cd is controlled by the nature and structure of the ochreous precipitate, as this has been found to vary with pH (30,31).

The relative behavior of Zn and Cd is of interest as they are generally considered to be geochemically similar and highly mobile (25). However, it has been suggested that ochreous mine waters have relatively high Zn/Cd ratios compared to those which are non-ochreous (32). It is apparent from Table IV that the relative concentration of Cd to Zn in the ochres compared to the waters is 1 for Y Fan but ranges from 0.4 to 0.6 for all of the other samples considered. Thus it seems likely that in near-neutral waters Cd and Zn are adsorbed to a similar degree into the ochres whereas in waters of pH 3.5 and below Cd is preferentially adsorbed.

For Co and Ni there appears to be little effect of pH on scavenging; the degree of enrichment in ochre varies from locality to locality.

The relative enrichment of U in the Parys Mountain ochres suggests this element is being adsorbed by the hydrated Fe oxides. However, the degree of adsorption is relatively small which is possibly due to the low pH of the Parys Mountain waters making U generally soluble (33).

The precipitation of Al from acid mine drainage due to increasing pH has been demonstrated by several workers (24,34). Although Nordstrom (34) has suggested that Al tends to precipitate as amorphous or poorly crystalline $Al(OH)_3$ at a pH of 4.6 and above, the data in the present study would suggest that appreciable precipitation occurs at lower pH (see Table IV). The relatively high levels of Al in the ochres from the low-pH waters suggest precipitation in some other phase. In this context it is of interest that Tardy and Nahon (35) showed that Al could substitute for Fe in hydrous Fe oxides in soils.

Conclusions

1. The composition of mine drainage waters in Wales is strongly influenced by pH with highest concentrations of Mg, Al and Fe occurring in the most acid waters. The SO_4 anion is enriched in the most acid waters and, from simple calculation of the K_{sp} for $BaSO_4$ and $PbSO_4$, it is probable that this anion exerts a major controlling influence on the solubility of Ba and Pb which consequently become enriched in the precipitated ochres.

Table IV. Concentration factors for selected elements in ochre

Locality[1]	1	2	3	4	5	6
pH	5.0	3.5	2.9	2.5	2.4	2.3
	----------Concentration factor[2]----------					
Mg	69	56	87	1.2	1.3	3.1
Al	6660	4990	985	3.0	4.7	7.4
Mn	200	51	275	4.3	1.1	2.7
Fe	307000	19500	7380	384	169	111
Cu	11500	4390	700	8.5	33	26
Zn	382	160	14	3.8	5.2	4.6
Cd	381	333	34	7.6	10	7.6

[1] Localities: 1 = Y Fan mine; 2 = Gwynfynedd mine; 3 = Cwmrheidol mine; 4, 5, 6 = Parys Mountain.

[2] Concentration factor = concentration of element in ochre (mg/kg) / concentration in water (mg/l).

2. The ochres are also strongly enriched in As, Mo, Ag, Sn, Sb, Hg, Tl and Bi, probably by coprecipitation and/or adsorption by the hydrated Fe oxides.

3. The degree of precipitation of Fe, Cu, Zn and Cd is controlled by pH; ochres precipitated from the most acid waters show less relative enrichment in Cu, Zn and Cd than those from higher pH waters.

4. Aluminum precipitates in response to high pH levels and is relatively enriched in ochres from moderate- to high-pH waters.

Literature Cited

1. Andrew, A.R. Geol. Mag. 1910, 7, 159-171, 201-211, 261-271.
2. Erichsen Jones, J.R. Ann. Applied Biol. 1940, 27, 368-378.
3. Erichsen Jones, J.R. J. Animal Ecology 1940, 9, 188-201.
4. Newton, L. Ann. Applied Biol. 1944, 31, 1-11.
5. Davies, B.E.; Lewin, J. Environ. Pollut. 1974, 9, 49-57.
6. Fuge, R.; Paveley, C.F.; Holdham, M.T. Environ. Geochem. Health 1989, 11, 127-135.
7. Abdullah, M.I.; Royle, L.G. Nature 1972, 238, 329-330.

8. Jones, O.T. The mining district of north Cardiganshire and west Montgomeryshire; Geological Survey Memoir, Special Report on the Mineral Resources of Great Britain; H.M.S.O.:London, 1922; No. 20.
9. Johnson, M.; Roberts, D.; Firth, N. Sci. Total Environ. 1978, 10, 61-68.
10. Fuge, R.; Laidlaw, I.M.S.; Perkins, W.T.; Rogers, K.P. Environ. Geochem. Health 1991, 13, 70-75.
11. Hall, G.W. The Gold Mines of Merioneth, Griffin: Kington; 1986, 2nd. edit.
12. Bick, D. The Old Copper Mines of Snowdonia. Alan Sutton Publishing: Gloucester, 1982.
13. Manning, W. The Future of Non-Ferrous Mining in Great Britain and Ireland; IMM: London; 1959, pp. 313-328.
14. Westhead, S.J. Geol. Today 1991, July-August, 130-133.
15. Pointon, C.R.; Ixer, R.A. Trans. Inst. Mining Metall. 1980, 89, 143-155.
16. Ixer, R.A.; Pointon, C.R. Conf. Proceedings European Copper Deposits; Belgrade, 1980.
17. Ixer, R.A.; Gaskarth, J.W. Abstract; Mineral Deposit Studies Group; Geol. Soc. Lond.: Leicester, 1975.
18. Pearce, F.M. Environ. Geochem. Health 1991, 13, 50-55.
19. Perkins, W.T.; Fuge, R.; Pearce, N.J.G. J. Anal. At. Spectrom. 1991, 6, 445-449.
20. Pearce, N.J.G.; Perkins, W.T.; Fuge, R. J. Anal. At. Spectrom. 1992, 7, 595-598.
21. Cave, R. Geol. Jour. 1965, 82, 865-903.
22. Levinson, A.A. Introduction to Exploration Geochemistry; Applied Publishing: Wilmette, IL, 1980.
23. Chapman, B.M.; Jones, D.R.; Jung, R.F. Geochim. Cosmochim. Acta 1983, 47, 1957-1973.
24. Rampe, J.J.; Runnells, D.D. Appl. Geochim. 1989, 4, 445-454.
25. Davis, A.; Olsen, R.L.; Walker, D.R. Appl. Geochem. 1991, 6, 333-348.
26. Hem, J.D. U.S. Geol. Surv. Water-Supply Paper 2254, 1985.
27. Horsnail, R.F.; Elliot, I.L. Geochem. Exploration, CIM Spec. 1971, 11, 166-175.
28. Nowlan, G.A. J. Geochem. Explor. 1976, 6, 193-210.
29. Johnson, C.A. Geochem. Cosmochim. Acta 1986, 50, 2433-2438.
30. Bigham, J.M.; Schwertmann, U.; Carlson, L.; Murad, E. Geochim. Cosmochim. Acta 1990, 54, 2743-2758.
31. Bigham, J.M.; Schwertmann, U.; Carlson, L. Catena, 1992, Supplement 21, 219-232.
32. Fuge, R.; Pearce, F.M.; Pearce, N.J.G.; Perkins, W.T. Appl. Geochem. 1993, Supplement 2, 29-35.
33. Langmuir, D. Geochim. Cosmochim. Acta 1978, 42, 547-569.
34. Nordstrom, D.K.; Ball, J.W. Science 1986, 232, 54-56.
35. Tardy, Y.; Nahon, D. Amer. J. Sci., 1985, 285, 865-903.

RECEIVED October 1, 1993

Transport and Storage of Sulfides and Oxidation Products in Sediments

Chapter 19

Stratigraphy and Chemistry of Sulfidic Flood-Plain Sediments in the Upper Clark Fork Valley, Montana

David A. Nimick[1] and Johnnie N. Moore[2]

[1]Water Resources Division, U.S. Geological Survey, Helena, MT 59626
[2]Department of Geology, University of Montana, Missoula, MT 59812

> Sulfide-rich tailings deposited by historic floods have contaminated large areas of the upper Clark Fork flood plain in western Montana. About 704,000 m^3 of mine wastes are spread over 274 ha of the flood plain along a 10-km reach near the river's headwaters. The tailings deposits primarily are fine-grained overbank deposits and point-bar deposits containing reworked mixtures of tailings and other sediment. Analyses of total and weak-acid extracts of the sediments show that As, Cu, Fe, Mn, Pb, and Zn are released into solution by oxidation of the sulfides contained in the tailings. These constituents either move upward to the land surface and precipitate as hydrated metal sulfates or move downward to be concentrated in weak-acid-extractable phases in reduced tailings or pre-mining flood-plain deposits. Erodible flood-plain tailings are a major source of trace elements to the river.

Large quantities of sulfidic, flood-deposited tailings rich in As, Cd, Cu, Fe, Mn, Pb, and Zn are spread over the upper Clark Fork flood plain in western Montana (1-4). The Clark Fork is the head of the Pend Oreille River in the Columbia River basin. The tailings originated primarily from 1864 to about 1915 from the uncontrolled disposal of wastes produced from mining and smelting sulfide ores into Clark Fork tributaries draining Butte and Anaconda (5). Copper was the principal metal mined from ores containing chalcocite (Cu_2S), bornite (Cu_5FeS_4), enargite (Cu_3AsS_4), and chalcopyrite ($CuFeS_2$). Other associated sulfide minerals were pyrite (FeS_2), sphalerite (ZnS), galena (PbS), greenockite (CdS), arsenopyrite ($FeAsS$), and acanthite (Ag_2S) (6-8). Floods carried tailings to the Clark Fork and caused widespread deposition (8) of tailings on the Clark Fork flood plain downstream to Milltown (2,9). The Clark Fork flood plain between Warm Springs and Milltown has been designated as a Superfund site by the U.S. Environmental Protection Agency.

Several studies have examined trace-element concentrations in streambank deposits along the river channel (1,3,4). Those studies found considerable variability in trace-element concentrations and only little evidence of downstream decreases in trace-element concentrations in flood-plain tailings. In the one previous study of the chemistry and stratigraphy of flood-plain tailings (10), concentrations of trace elements in pore water, ground water, and sediments in a small area near Racetrack (Figure 1) were used to identify the movement of trace elements within and from tailings. Although ground water is a potential receptor of trace elements released from flood-plain tailings, only limited contamination attributable to flood-plain tailings has been documented in the coarse-grained alluvium that underlies flood-plain tailings (10; Nimick, D. A., unpublished data). Contaminant plumes have been identified downgradient from the thick tailings deposits found in impoundments near Warm Springs (5) and Milltown (11,12).

As the flood-plain sediments are eroded into the river with time, potentially toxic trace elements are released to the aquatic environment. Biological effects attributed to the elevated concentrations of dissolved and particulate-bound metals include reduced diversity of benthic invertebrates, limited trout populations, and fishkills (13-17). Although little is known currently about the distribution and thickness of the flood-plain tailings and the concentrations and partitioning of arsenic and heavy metals in the tailings, an understanding of these factors is critical for assessing the extent of contamination and for planning effective remedial activities. This paper describes the thickness and distribution as well as the chemical data for a 10-km reach of the Clark Fork (Figure 1) that characterize the flood-plain contamination. The data add to a limited but improving understanding of the sources and quantity of trace elements in large contaminated river systems.

Sampling and Analytical Methods

Tailings thickness was mapped from about 680 measurements in soil pits and streambanks. Flood-plain deposits were sampled at 20 of these sites. Acid extracts of sediments were made by combining 0.6 g of sample with 20 ml 5% HCl. The mixture was shaken for 2 hours and filtered through a 0.45-µm filter. Total extracts used a $HF-HNO_3-HCl$ microwave digestion (18,19). All extracts were analyzed for As, Cd, Cu, Fe, Mn, Ni, Pb, and Zn by inductively coupled argon plasma emission spectrometry. All analyses are reported in µg/g, dry weight basis. Sampling, analytical, and quality-control methods are described more fully elsewhere (20).

Although partial extractions frequently are used to determine partitioning of trace elements in sediments, their use is problematic (21-23). The weak-acid extract for this study was not used to determine partitioning, but rather was designed to determine which trace elements might be released to the river and biota. In particular, it was used to provide a crude approximation of concentrations of bioavailable trace elements (24). The cold dilute HCl used in this extract can dissolve Fe and Mn oxide coatings and associated trace elements, dissolve carbonates, replace trace-element ions adsorbed on organic and inorganic materials, and dissolve amorphous or diagenetic sulfides in reduced sediment (25,26).

Stratigraphy

The Clark Fork has been subjected to increased loads of metal-contaminated sediment since mining began in Butte in 1864 and the construction of smelters in Anaconda in 1884. Prior to the early 1900's, no sediment-control structures existed on Silver Bow or Warm Springs Creeks, which are the Clark Fork tributaries into which mine wastes were dumped. Wastes were transported to the Clark Fork by at least four major floods in the 1890's and the largest flood on record in 1908 (27). Stratigraphic evidence indicates that these few floods deposited considerable quantities of essentially pure tailings as they built a "mining terrace" (28) over the pre-existing flood plain and low terraces in the Clark Fork valley. Although the flood plain aggraded during this period, analysis of the relative position of present (1988) and pre-mining flood-plain deposits indicates that the present channel is near its pre-

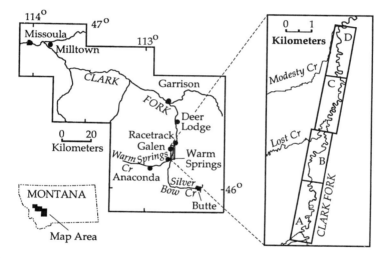

Figure 1. Location of study area (patterned). Letters indicate reaches of the Clark Fork shown in Figure 2.

mining altitude. Bank height of the mining terrace is 1-2 m above the river. Even though migrating channel meanders and channel avulsions have eroded parts of the mining terrace, original tailings deposits still overlie pre-mining flood-plain deposits in large areas of the valley.

Sediments on the valley floor are separated into three categories: overbank tailings, reworked tailings, and pre-mining flood-plain deposits (Table I). Most of the tailings are in widespread overbank deposits, whereas reworked tailings have a more limited distribution near present and former locations of the Clark Fork.

Table I. Description of Flood-Plain Deposits along the Clark Fork

Unit Name	Description
Overbank Tailings	
Upper tailings	Very fine sandy silt with interbedded planar and ripple cross laminae; tan with orange and gray staining; as much as 35 cm thick.
Lower tailings	Lenticular deposits of fine to medium silty sand; discontinuous planar and ripple cross laminae; orangish light brown to medium brown with mottled reddish-orange and gray iron staining; as much as 60 cm thick.
Reduced tailings	Similar to lower tailings except gray to black; less than 40 cm thick.
Reworked Tailings	
Reworked tailings	Fluvial deposits containing reworked mixtures of tailings and cleaner sediments found as silty fine sand in overbank deposits, silty sand and sand in channel-accretion deposits, and sand and gravel in point bars and some levees; generally light brown with little iron staining; support growth of grass and willows; overbank deposits commonly rootbound.
Pre-Mining Flood-Plain Deposits	
Buried "A" horizon	Dark-brown to black organic-rich soil horizon marking top of pre-mining flood plain; as much as 40 cm thick.
Buried subsoil	Dark-brown thin-bedded silt and sand overlying sand and gravel.

Overbank tailings have been divided into three units. The upper two units lie above the saturated zone and are oxidized, whereas the lower unit is reduced. The top unit, called upper tailings, is essentially pure tailings and contains little, if any, uncontaminated sediment. Iron hydroxide coatings were observed on nearly all grains when upper tailings were viewed with an energy-dispersive X-ray scanning electron microscope (SEM-EDX). Goethite crystals compose a large part of these coatings. Large organic fragments (leaves and sticks) are present on some foreset beds. Smelter slag or glass composes about 15% of the tailings. If exposed at the surface, the upper tailings form barren areas devoid of vegetation. A crust of efflorescent salts forms on these barren areas during warm dry periods (28). Throughout the study area, the upper and lower tailings are separated by a thin (<10 cm), massive silty clay bed that may represent deposition from one flood that carried an extensive quantity of very finely ground tailings.

The middle unit in the overbank tailings is called lower tailings. Generally, this unit is similar to the upper tailings but is more variable in color and grain size. Organic fragments generally are present. In some profiles, the lower tailings contain thin (<10 cm) beds of well-sorted sand. These sand layers appear more permeable than the tailings above and below and have a pervasive green color.

Reduced tailings underlie the lower tailings. This unit is similar to the lower tailings except for its gray color and the presence of finely disseminated organic matter.

Reworked tailings encompass all river sediments deposited since probably about 1911, when the first settling pond was built on Silver Bow Creek (5). Although tailings continued to be carried downstream, the construction of sediment controls decreased the influx of pure tailings to the Clark Fork and allowed uncontaminated sediments to dilute the tailings. Reworked tailings are found in overbank, crevasse-splay, and natural-levee deposits on top of the mining terrace, in point bars, and in channel-accretion deposits along the margins of channels that were once wider but have narrowed naturally.

The original or pre-mining flood-plain deposits buried by tailings consist of fine-grained overbank deposits on top of stream gravels. Pre-mining overbank deposits are easily distinguished from overbank tailings by the dark-brown color and more sandy texture of the pre-mining deposits. Pre-mining sediments are divided into a buried "A" horizon, if present, and buried subsoil.

Distribution of Tailings

Four map units (Figure 2) were used areally to delineate thin tailings (0-10 cm), moderate tailings (10-30 cm or 30-50 cm), and thick tailings (>50 cm). Measured thicknesses include primarily overbank tailings but may also include a thin surficial layer of reworked overbank tailings. Thick reworked channel tailings, such as those present in point bars, are mapped separately. The actual thickness of contaminated sediments may be greater than just the tailings thickness, because trace elements have moved from the tailings into underlying pre-mining sediments in many places. The width of tailings-contaminated land is generally 180-490 m but ranges from 90 to 900 m. About 704,000 m^3 of tailings are estimated to be spread over 274 ha along the 10-km study reach (Table II). The thickest tailings (30-50 cm and >50 cm) are mostly in a narrow band near the river but some are near the course of the late-1800's channel; these tailings account for two-thirds of the total volume. Thick tailings are extensive at the upstream end of the study area and where the flood plain is narrow (Figure 2). Tailings 10-30 cm thick are extensive where the flood plain is wide. Thin tailings are found away from the river at the margins of the flood plain, mostly west of the river. East of the river, several alluvial fans of intermittent tributaries limited the eastward extent of Clark Fork flooding. Some agricultural lands east of the river may have been contaminated by floods or irrigation water diverted from the river, but any tailings deposited have been reworked and masked by plowing.

The volumes of flood-plain tailings measured during this study indicate that previous estimates of the volume may have been conservative. Johnson and Schmidt (15) estimated that 1,000,000 m^3 of tailings are present along 24 km of the Clark Fork valley between Warm Springs and Deer Lodge. On the basis of this study and field reconnaissance of the area between the downstream end of the study area and Deer Lodge, the volume of flood-plain tailings between Warm Springs and Deer

Table II. Extent and Volume of Flood-Deposited Tailings

Map Unit (cm)	Average Thickness (cm)	Area Covered (ha)	Volume (m^3)
0-10	5	94.6	47,300
10-30	20	90.6	181,200
30-50	40	49.4	197,600
>50	70	39.7	277,900
Total		274.3	704,000

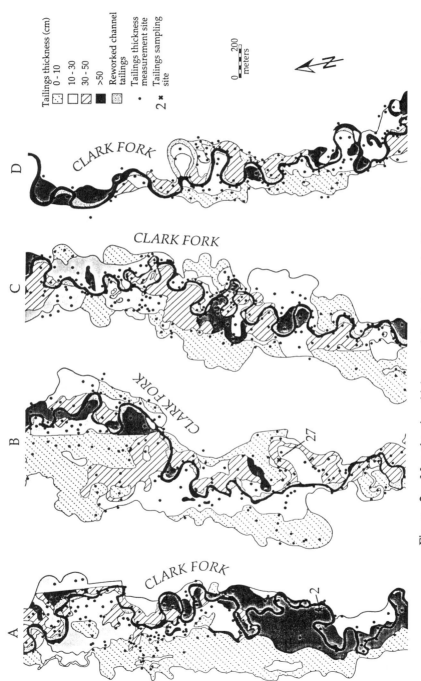

Figure 2. Map showing thickness of flood-deposited tailings on the Clark Fork flood plain. The locations of the four reaches (labeled A, B, C, and D) are shown in Figure 1.

Lodge probably is about 1,500,000 m^3. The estimate by Moore and Luoma (2) of 2,000,000 m^3 of flood-plain tailings between Warm Springs and Milltown also could be small.

Sediment Chemistry

Flood-plain tailings contain high concentrations of arsenic and base metals. Ranges of concentrations for each stratigraphic unit are shown in boxplots (Figure 3). The bed-sediment chemical-concentration data (29) in Figure 3 are for fine-grained samples from a 25-km reach of the river that includes the study area. Median concentrations of total Cu, Mn, and Zn are greater than 1,000 µg/g in most of the tailings deposits and more than 2,000 µg/g in many of the units. Total Cu concentrations are very high--more than 10,000 µg/g in more than half of the reduced tailings samples and as much as 49,300 µg/g in the green sand of the lower tailings. Median concentrations of total As and Pb in tailings deposits are slightly less and generally range from 100 to 2,000 µg/g. Although most total Cd concentrations are less than 10 µg/g in samples of all sediment types, generalization is difficult because Cd concentrations in many of the samples are less than the detection limit. The ranges of trace-element concentrations reported by Brooks and Moore (10) for flood-plain tailings at a site near Racetrack are similar to those determined during this study.

Ni concentrations are consistently low in all sediment types. Total Ni concentrations generally range from 10 to 30 µg/g uniformly in all sediment types. Brook (29) found similar total Ni concentrations in bed sediments of both the Clark Fork and Clark Fork tributaries and concluded that sediments in the Clark Fork system are not enriched in Ni. This conclusion is consistent with the lack of Ni in ores mined at Butte (6).

Fe concentrations vary little between contaminated and uncontaminated sediments or within the various types of tailings. Most total concentrations range from about 20,000 to 50,000 µg/g in tailings and buried "A" horizons and 9,000 to 28,000 µg/g in buried subsoil. None of these values is much higher than the average concentration of 13,700 µg/g reported for flood-plain sediments from tributary valleys (4). Therefore, flood-plain sediments are not nearly as enriched with Fe as with other metals. The enrichment that does exist probably reflects the original abundance of iron sulfides in the tailings. The lack of anomalous Fe concentrations and the variability through all the sediment types are consistent with the results of other Clark Fork studies (4,10).

Total concentrations of As, Cu, Fe, Mn, Pb, and Zn in tailings deposits and in underlying pre-mining sediments differ considerably but the concentration patterns in vertical profiles generally are consistent. Profiles of trace-element concentrations at two representative sites are shown in Figure 4. The highest concentrations generally are found either at the ground surface or at depth near the bottom of the tailings deposits. Concentrations generally are less in the upper tailings and pre-mining sediments. The high concentrations of Cu, Mn, and Zn at the surface are caused by hydrated metal sulfates that precipitate during warm weather (29). These concentrations, which were measured by Nimick (20) in surficial (0-2 cm) samples, represent minimum values because they are water-soluble rather than total concentrations.

Trace-element concentrations tend to be lowest in the tailings just below the surface, whether the tailings are part of the upper tailings (eg., lower part of Figure 4) or lower tailings (upper part of Figure 4). Concentrations generally increase with depth below the near-surface tailings, and the highest subsurface concentrations are found within the lower tailings, reduced tailings, or buried "A" horizon. Locally,

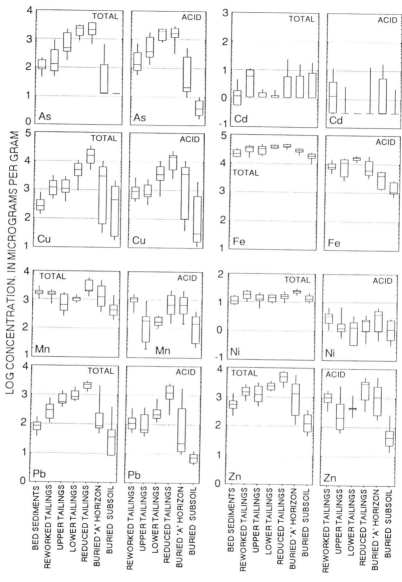

Figure 3. Boxplots showing 10th (lower), 25th, 50th (median), 75th, and 90th (upper) percentiles of total and acid-extractable trace-element concentrations in major stratigraphic units along the Clark Fork. Data for bed sediments are from Brook (29); data for tailings and flood-plain deposits are from Nimick (20). Concentrations are expressed as log units.

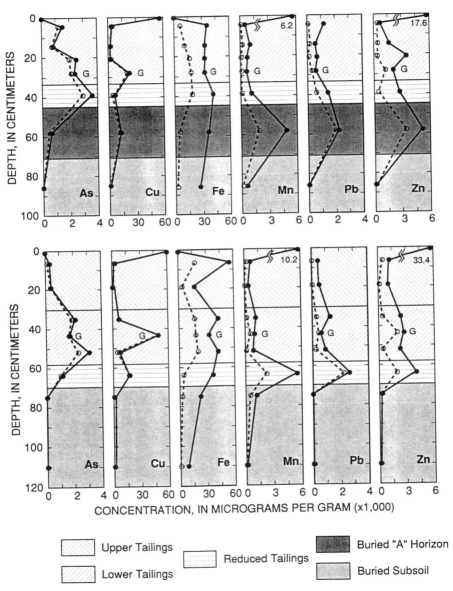

Figure 4. Profiles of total (●) and acid-extractable (○) trace-element concentrations in pits having 45 cm of tailings (top) and 70 cm of tailings (bottom). Not all stratigraphic units are present at each site. Data for surficial material (0-2 cm) are from Nimick (20) and are water-soluble concentrations. "G" indicates green sand layer in lower tailings. Locations of site 27 (top) and site 2 (bottom) are shown in Figure 2.

subsurface Cu concentrations are highest in the thin green sand layer of the lower tailings.

The degree of contamination in flood-plain sediments from mine wastes can be estimated by comparing trace-element concentrations in Clark Fork sediments to concentrations reported (4) for flood-plain sediments from Clark Fork tributaries where flood-deposited tailings are largely absent. Cu enrichment is greatest; typical total Cu concentrations in the three tailings units are 30 to 960 times greater than those in tributary sediments. The highest subsurface Cu concentration was found in a green sand layer within the lower tailings and is 1,825 times greater than the average tributary concentration. As, Pb, and Zn also are highly enriched, with some mainstem concentrations being more than 100 times the tributary values. Concentrations of Fe and Mn, which exhibit relatively little enrichment, are typically as much as 3 and 11 times, respectively, concentrations in tributary sediments.

Trace-element concentrations in the pre-mining flood-plain sediments along the mainstem vary widely, commonly by more than two orders of magnitude. Concentrations in the buried "A" horizon can be nearly as high as the highest concentrations in any of the tailings units or as low as concentrations in tributaries (Figure 3). Concentrations in buried subsoil generally are substantially less than in the buried "A" horizon; however, some enrichment with respect to tributary values is observed.

Vertical profiles of acid-extractable concentrations follow the same trends exhibited by total concentrations, but the relative magnitude of the acid-extractable concentrations differs for each constituent (Figures 3 and 4). Acid-extractable concentrations of As and Cu generally are at least 75% of the total concentration. Acid-extractable concentrations of Pb and, to a certain extent, Zn are a large percentage of total concentrations in the reduced tailings and buried "A" horizon. These large percentages indicate that As, Cu, Pb, and Zn are no longer in primary sulfide minerals. Acid-extractable concentrations of Fe and Mn are a smaller percentage, generally 15-50%, of the respective total concentrations.

The patterns of total and acid-extractable concentrations indicate that trace elements released by sulfide oxidation in the upper and lower tailings have been mobilized within the tailings and pre-mining flood-plain sediments. The high concentrations of metals at the ground surface have been explained by precipitation of hydrated metal sulfates from soil moisture that is drawn to the surface and evaporated during warm weather (28). Compared to lower and reduced tailings, upper tailings are depleted of trace elements owing to sulfide oxidation and vertical migration of arsenic and metals. Total concentrations in reduced tailings generally are higher than in either of the two oxidized tailings units, because oxidation of sulfides is not likely and because several reactions are likely to immobilize trace elements that move downward into this unit. Metal carbonates are likely to be precipitated where neutral pH conditions prevail. Dissolved carbonate probably is derived primarily from the dissolution of carbonate minerals enhanced by acid production from sulfide oxidation. Carbonate-rich water might also move laterally into tailings areas in more permeable layers such as the green sand. Copper-carbonate coatings were identified by means of SEM-EDX analysis of the green sand. Formation of organo-metallic complexes could also immobilize Cu, Mn, Pb, or Zn in organic-rich zones such as the reduced tailings or buried "A" horizon (10). SEM-EDX analysis demonstrated that Cu, Mn, and Zn are concentrated in small organic fragments such as rootlets. Lastly, trace metals released from oxidized tailings can be fixed in secondary sulfides in reduced tailings by reaction with the H_2S produced from sulfate reduction or, in the case of copper, by secondary or supergene enrichment.

Differences in trace-element concentrations in original overbank tailings, bed sediments, and reworked tailings provide some insight into the fate of trace elements as they move through the fluvial system (Figure 3). With the exception of Mn, median total trace-element concentrations in bed sediments are less than average concentrations in any of the overbank tailings units, indicating that flood-plain tailings lose trace elements or are diluted as they are eroded into the river and deposited on the streambed. The increase in Mn concentrations in bed sediments may be due to interactions between bed sediment and ground water. Slightly reduced ground water rich in Mn discharges to the river in the study area (Nimick, D.A., unpublished data). Mn oxide coatings may develop as the Mn oxidizes. The relatively high acid-extractable concentration in reworked tailings also indicates the presence of Mn oxides. Figure 3 also shows that median concentrations of As, Cu, Pb, and Zn (but not Mn) are higher in deposits of reworked tailings than in bed sediments. This enrichment of As, Cu, Pb, and Zn in reworked tailings might be occurring by upward movement of trace elements from underlying weathered tailings.

Sources of Trace Elements in the Clark Fork

The Clark Fork receives trace elements from two principal sources. The first source is runoff produced during intense summer thunderstorms on unvegetated areas of streamside tailings. Precipitation dissolves hydrated metal sulfates and produces acid and high concentrations of dissolved metals that periodically have caused acute toxicity to fish (9,29). The second source is contaminated flood-plain sediments. Although the two major headwaters streams, Silver Bow and Warm Springs Creeks, drain the Butte/Anaconda mining and smelting district, the large areas of wastes in these drainages generally are located away from streams and probably are much less important as sources of contaminants to the Clark Fork than erosion of flood-plain tailings along the mainstem. Erosion of these sediments increases concentrations of dissolved and particulate trace elements in the river and increases the potential for chronic toxicity to aquatic organisms. For example, Lambing (30) estimated that the total recoverable concentration of Cu in the Clark Fork near Galen and at Deer Lodge exceeds the U.S. Environmental Protection Agency (31) aquatic-life criterion for chronic toxicity about 70-80% of the time. Flood-plain tailings in this reach presumably are the primary source of the metals contamination.

Several lines of reasoning support the assertion that flood-plain tailings are a significant source of contamination. Phillips (16) noted that the total recoverable trace-element concentrations in the Clark Fork were highest at the onset of runoff in response to initial bank erosion caused by rising floodwaters. Also, trace-element concentrations were high for longer periods and trace-elements loads were highest at downstream sites, indicating that the downstream effect caused by the increasing length of source banks is cumulative. Phillips concluded that the increases in downstream trace-element loading resulted from erosion of contaminated bank material.

Suspended-sediment and total-recoverable concentrations of trace elements determined over 6 years at three Clark Fork sites (30) also indicate that flood-plain tailings may be a significant source of trace elements to the river. Total-recoverable concentrations were determined using a hot dilute acid (HCl) extraction. Lambing (30) estimated that 83% of the suspended-sediment load and 50-75% of the total-recoverable load of As, Cu, Fe, Pb, and Zn measured at Deer Lodge come from sources in the reach between Galen and Deer Lodge. The tributaries in this reach are unlikely to have any significant trace-element load; therefore, the increase in load between Galen and Deer Lodge most likely is due to bed and bank sediment. Bank sediment must be an important contributor, because the median estimated trace-element concentrations in the suspended sediment at Deer Lodge (30) are higher than median total concentrations in bed sediments (Figure 3) measured (29) between Warm Springs and Deer Lodge. The estimated concentrations in suspended sediment are similar to the total concentrations in flood-plain tailings measured during this study. This similarity indicates that flood-plain tailings are an important source of trace elements to the river.

The last line of reasoning linking flood-plain tailings to elevated trace-element concentrations in the Clark Fork is the large percentage of banks that are actively eroding. A bank survey conducted by Nimick (20) showed that typically 16-21% of the river banks in a 8-km reach of the valley near Galen are actively eroding. Meander-migration rates as high as 1 m/yr were also measured. Practically all eroding banks in the study area contain moderate to thick tailings (30-120 cm). Cattle grazing in riparian areas probably is responsible, at least in part, for these erosive conditions. Land use and, therefore, probably hydrologic conditions are similar at least as far downstream as Garrison.

Conclusions

Tailings deposited during decades of releases of mining and smelter wastes have resulted in flood-plain aggradation and arsenic and base-metal contamination of large areas of the upper Clark Fork valley. The original tailings are present primarily in fine-grained overbank deposits that contain trace elements that are being weathered, mobilized, and reconcentrated at the ground surface or in reduced tailings or pre-mining flood-plain sediments. Mixtures of cleaner sediments and tailings have been deposited by overbank flows on top of tailings and in channel-accretion deposits and point bars. The river continues to rework the tailings, which causes trace elements to be released to the river as sediment is cycled back and forth between the channel and the flood plain.

Major conclusions of this study are:

1. Within the study area, flood-deposited tailings as thick as 120 cm overlie the pre-mining Clark Fork flood plain. About 704,000 m^3 of tailings are spread over 274 ha along the 10-km study reach. In most areas, the tailings are less than 30 cm thick; however, the thickest tailings are generally near the river, where the probability of being eroded into the river is greatest.

2. Flood-plain sediments, including tailings, reworked tailings, and buried soils, are contaminated with As, Cu, Fe, Mn, Pb, and Zn and are enriched as much as 1,825 times over concentrations found in relatively uncontaminated tributaries. Except for metal salts at the ground surface, the highest trace-element concentrations are no longer in near-surface tailings. Oxidation of sulfides releases trace elements that migrate downward where they are trapped as diagenetic sulfides and organic complexes in reduced, organic-rich sediments.

3. Acid-extractable concentrations of trace elements in flood-plain tailings are high, particularly for As, Cu, Pb, and Zn. Therefore, flood-plain tailings could be a major cause of potential chronic toxicity to aquatic organisms in the river.

4. Flood-plain tailings probably are the most important source of arsenic and base metals to the Clark Fork.

Acknowledgments

Donna Pridmore, Karianne Schumacher, and Redmond Wyatt provided laboratory assistance. Reviews by Ellen V. Axtmann and John H. Lambing are appreciated. Funding was provided by the Montana Water Resources Center and the University of Montana Geology Department.

References

1. Axtmann, E. V.; Luoma, S. N. Applied Geochem. 1991, 6, 75-88.
2. Moore, J. N.; Luoma, S. N. Environ. Sci. Tech. 1990, 24, 1278-1285.
3. Moore, J. N. Source of Metal Contamination in Milltown Reservoir, Montana: An Interpretation Based on the Clark Fork River Bank Sediment; Unpublished report to the U.S. Environmental Protection Agency: Helena, MT, 1985, 108 pp.
4. Moore, J. N.; Brook, E. J.; Johns, C. Environ. Geol. Water Sci. 1989, 14, 107-115.
5. Montana Department of Health and Environmental Sciences Feasibility Study for the Warm Springs Ponds Operable Unit: Helena, MT, 1989, 951 pp.

6. Meyer, C.; Shea, E.P.; Goddard, C.G. In <u>Ore Deposits of the United States, 1937-1967: The Granton Sales Volume</u>; Ridge, J., Ed., American Institute of Mining, Metallurgical, and Petroleum Engineers Inc.: New York, NY, 1968; 1375-1415.
7. Miller, R. N. In <u>Guidebook for the Butte Field Meeting of the Society of Economic Geologists</u>; Miller, R. N., Ed.; Anaconda Company: Butte, MT, 1973; Chapter F.
8. Weed, H. W. <u>Geology and Ore Deposits of Butte District, Montana</u>; Prof. Paper 74; U.S. Geological Survey: Washington, DC, 1912; 257 pp.
9. Moore, J. N.; Luoma, S. N. In <u>Proceedings of the 1990 Clark Fork River Symposium</u>; Watson, V., Ed.; University of Montana: Missoula, MT, 1990; 163-188.
10. Brooks, R.; Moore, J. N. <u>GeoJournal</u> 1989, <u>19</u>, 27-36.
11. Moore, J. N.; Ficklin, W. H.; Johns, C. <u>Environ. Sci. Tech</u>. 1988, <u>22</u>, 432-437.
12. Woessner, W. W.; Moore, J. N.; Johns, C.; Popoff, M. A.; Sartor, L. C.; Sullivan, M. L. <u>Final Report--Arsenic Source and Water Supply Remedial Action Study, Milltown, Montana</u>; University of Montana: Missoula, MT, 1984; 448 p.
13. Averrett, R. C. <u>Macroinvertebrates of the Clark Fork River, Montana</u>; Montana Department of Health, Water Pollution Control Report 61-1, 1961, 27 pp.
14. Axtmann, E. V.; Cain, D. J.; Luoma, S. N. In <u>Proceedings of the 1990 Clark Fork River Symposium</u>; Watson, V., Ed.; University of Montana: Missoula, MT, 1990; 1-18.
15. Johnson, H. E.; Schmidt, C. L. <u>Clark Fork Basin Project: Draft Status Report and Action Plan</u>; Clark Fork Basin Project, Office of the Governor: Helena, MT, 1988.
16. Phillips, G. R. <u>Proceedings-Clark Fork Symposium</u>; Carlson, C. E.; Bahls, L. L., Eds.; Montana Academy of Sciences, Montana College of Mineral Science and Technology: Butte, MT, 1985; 57-73.
17. Phillips, G. R. and Spoon, R. In <u>Proceedings of the 1990 Clark Fork River Symposium</u>; Watson, V., Ed.; University of Montana: Missoula, MT, 1990; 103-118.
18. Nadkarni, R. A. <u>Anal. Chem</u>. 1984, <u>56</u>, 2233-2237.
19. Brook, E. J.; Moore, J. N. <u>The Science of the Total Environment</u> 1988, <u>76</u>, 247-266.
20. Nimick, D. A. <u>Stratigraphy and Chemistry of Metal-Contaminated Floodplain Sediments, Upper Clark Fork River Valley, Montana</u>; M.S. thesis; University of Montana: Missoula, MT, 1990; 118 pp.
21. Belzile, N. P.; Lecomte, P.; Tessier, A. <u>Environ. Sci. Tech</u>. 1989, <u>23</u>, 1015-1020.
22. Gruebel, K. A.; Davis, J. A.; Leckie, J. O. <u>Soil Sci. Soc. Am. J</u>. 1988, <u>52</u>, 390-397.
23. Kheboian, C.; Bauer, C. F. <u>Anal. Chem</u>. 1987, <u>59</u>, 1417-1423.
24. Luoma, S. N. <u>Hydrobiologia</u> 1989, <u>176/177</u>, 379-396.
25. Chao, T. T.; Zhou, L. <u>Soil Sci. Soc. Am. J</u>. 1983, <u>47</u>, 225-232.
26. Rapin, F.; Tessier, A.; Campbell, P. G.; Carignan, R. <u>Environ. Sci. Tech</u>. 1986, <u>20</u>, 836-840.
27. CH2M Hill, Inc. <u>Silver Bow Creek Flood Modeling Study</u>: Boise, ID, 1989.
28. Nimick, D. A.; Moore, J. N. <u>Applied Geochem</u>. 1991, <u>6</u>, 635-646.
29. Brook, E. J. <u>Particle-size and Chemical Control of Metals in Clark Fork River Bed Sediment</u>; M.S. thesis; University of Montana: Missoula, MT, 1988; 126 p.
30. Lambing, J. H. <u>Water-Quality and Transport Characteristics of Suspended Sediment and Trace Elements in Streamflow of the Upper Clark Fork Basin from Galen to Missoula, Montana, 1985-90</u>; Water-Resources Invest. Rep. 91-4139; U.S. Geological Survey: Helena, MT, 1991, 73 pp.
31. U.S. Environmental Protection Agency <u>Quality Criteria for Water 1986</u>; Office of Water Regulations and Standards, EPA 440/5-86-001: Washington, DC, 1986, 433 pp.

RECEIVED March 26, 1993

Chapter 20

Release of Toxic Metals via Oxidation of Authigenic Pyrite in Resuspended Sediments

John W. Morse

Department of Oceanography, Texas A&M University, College Station, TX 77843

During early diagenesis in anoxic sediments many reactive trace metals are coprecipitated with authigenic pyrite. Metals of potential environmental concern, such as As and Hg, can have in excess of 80% of their reactive fraction incorporated into authigenic pyrite within the top 10 cm of many sediments. Resuspension of sediment into overlying oxic waters can result in the oxidation of pyrite leading to release of the coprecipitated metals. Experimental studies indicate that from about 20% to 50% of the pyrite can oxidize in a day. The loss of trace metals from the pyrite fraction relative to pyrite-Fe is highly variable during the oxidation reaction, but generally is close to or greater than that of pyrite-Fe.

Factors controlling the availability of toxic metals to biota in aquatic environments are of major environmental interest ([1]). Sediments are usually the ultimate sink for these metals. Consequently, chemical reactions that control the burial or remobilization of toxic metals occurring near the sediment-water interface, during early diagenesis, are especially important in determining their fate.

Fine-grained terrigenous sediments are commonly found in shallow-water environments, such as lakes, rivers, estuaries and near oceanic coasts. Such aquatic environments are also those that are likely to receive major contaminant toxic metal inputs and be otherwise perturbed by human activities. A common chemical characteristic of these sediments is that there is a sharp transition from oxic to anoxic conditions within centimeters of the sediment-water interface. This transition is dominantly driven by bacterially-mediated oxidation of organic matter within the sediments ([2]).

After the oxygen in pore waters has been exhausted, bacteria utilize other oxidants such as nitrate, iron and manganese oxides, and dissolved sulfate to metabolize organic matter. This sequence of reactions has several important consequences. Sediments become increasingly reducing and pH drops with increasing depth within the sediment. The dominant initial pools for reactive trace metals are most commonly organic matter and metal oxides. These sediment components are significantly decomposed during diagenesis, releasing the associated trace metals. As sulfate reduction becomes a dominant oxidative process, significant amounts of hydrogen sulfide are produced which reacts with reduced iron, via a complex pathway, to produce a variety of metastable iron-sulfide minerals and thermodynamically stable pyrite (3).

Although it has long been recognized that many different metals can coprecipitate with pyrite (4), until recently it has not been possible to quantify this process for authigenic pyrite in sediments (5). We have recently (6) conducted an extensive investigation of the composition of authigenic pyrite from a wide range of aquatic environments. Our general observations indicate that the coprecipitation of Hg, As, Mo, and Cr with pyrite is a major process in anoxic sediments. The reactive fraction of most class B and heavy metals appears to generally undergo only minor pyritization. The chemical pathways leading to the coprecipitation of metals with pyrite are currently not known, but may involve initial coprecipitation with metastable precursor iron-sulfide phases such as mackinawite (FeS).

Experimental studies of the oxidation kinetics of sedimentary pyrite in seawater indicate that the pyrite can be divided into two major fractions (7). The first fraction oxidizes rapidly, in a few days or less, and probably consists of submicron individual pyrite grains. The second fraction oxidizes slowly and can persist in oxic seawater for months to years. It probably consists mostly of pyrite framboids that are typically 10 to 30 µm in diameter. The first fraction has been observed to comprise about 15% to 50% of sedimentary pyrite in near-interfacial anoxic sediments from shallow-water environments (7).

The observation that very fine-grained, rapidly oxidizable pyrite is often a significant fraction of the total authigenic pyrite, in near-interfacial anoxic sediments, raises the possibility that if such sediments are resuspended in the water column oxidation of pyrite may lead to a release of coprecipitated metals. Resuspension of near-interfacial sediments is a common occurrence in many shallow-water environments. This may be caused by natural processes, such as storms, or by human activities, such as bottom trawling and dredging.

This paper reports the results of an experimental investigation of the loss of trace metals from sedimentary pyrite when anoxic sediments are exposed to oxic waters.

Study Area

Sediments from Galveston Bay, Texas, were chosen for study because of concerns about pollution of this estuary and related studies of toxicant metals in this bay (8). It has a surface area of 1600 km^2, and is one of the largest embayments on the U.S. coastline. The Bay water is shallow, averaging only about 2 m in depth and is largely cut off from the Gulf of Mexico by the Bolivar Peninsula and Galveston Island. A mean residence time for waters in the Bay of about 40 days has been estimated from its average salinity of 15 and an average river flow of 12 km^3 y^{-1} (9).

Thirty to fifty percent of the U.S. chemical production and oil refineries are situated around Galveston Bay. The Bay receives more than half of the total permitted waste-water discharges for the state of Texas. The combination of large population, high industrialization, shallow depth and restricted water exchange gives Galveston Bay the potential for serious trace-metal contamination problems.

Samples were collected from a soft-bottom shallow-water area where extensive bottom trawling for shrimp resuspends sediments, and from ship channels and a boat basin where maintenance dredging and ship traffic also frequently result in resuspension of anoxic sediments. Additional samples were collected from a dredge-spoil bank on Pelican Island immediately after its formation.

Methods and Procedures

Sample Collection. Samples were collected from the RV Roman Empire (except for the dredge-spoil bank) either by hand coring with a precleaned plastic core tube or with an epoxy-coated grab sampler in deeper waters. Only the top ~10 cm of sediment was used. This sampling depth was used because it represents "near-surface" sediments. The sediment was homogenized upon collection in sealed bags from which air was excluded to prevent oxidation.

Metal Extraction and Analysis. Metals were extracted from the sediments using leaching techniques. Briefly, the sequential extraction procedure involves digestion of the sediment sample with 1M HCl (reactive fraction), 10M HF (silicate fraction) and concentrated HNO_3 (pyrite fraction). A more complete explanation of the sequential-extraction procedures and the development of the separation method is given elsewhere (5).

Trace metals (As, Cr, Cu, Fe, Hg, Mo, Ni, Pb and Zn) were determined by flame atomic absorption (FAA) using a Perkin Elmer model 2380 spectrophotometer. Metals below the detection limit of this instrument were analyzed by direct injection into a Hitachi model 170-70 graphite furnace atomic absorption (GFAA) spectrophotometer with Zeeman background correction. The analytical precision (± relative standard deviation) was normally between 5 and 10% for FAA analyses and between 10 and 15% for GFAA analyses. Salt matrix effects for As determination by GFAA were partially overcome by using a Ni-Pd-ascorbic acid matrix modifier. Determination of Pb by GFAA in samples containing high Fe/Pb ratios (>250) was carried out following the procedure developed by Shao and Winefordner (10).

Mercury was determined with a Laboratory Data Control UV monitor equipped with a 30-cm path-length cell, using the cold-vapor technique. For samples suspected of having high dissolved organic matter (DOM) concentrations, Hg was measured after destruction of the DOM with bromine monochloride.

All reagents were ACS reagent grade or better. Milli-Q water was used for the preparation of all aqueous solutions. Acidic working standard solutions were always freshly prepared. Materials were carefully cleaned using established acid leaching procedures. The use of the sequential extraction procedures precluded comparisons with standard reference materials for which only total metal concentrations are generally available.

Oxidation Kinetics Experiments. Samples that were chosen for the oxidation kinetics experiments were handled exclusively in a glove bag under an Ar atmosphere using deaerated water and fluid reagents to avoid oxidation prior to the experiment. Analyses were made on a subsample to establish the initial portion of reactive metal in the pyrite fraction. A second subsample was then stirred in a covered beaker containing natural seawater with a salinity of 15 (average for Galveston Bay) at room temperature (~23°C) for 1 day. A paddle stirrer was used to avoid grinding of the sample and air was continuously bubbled through the solution.

Results and Discussion

Pyritization of Metals. Berner (11) introduced the concept of degree of pyritization (DOP) to describe the extent of reactive-Fe transformation to pyrite. DOP is defined as:

$$DOP = \frac{Pyrite\text{-}Fe}{Pyrite\text{-}Fe + Reactive\text{-}Fe} \quad (1)$$

where pyrite-Fe is assumed equal to 0.5 x (total inorganic reduced sulfur - acid volatile sulfides), based on the 1:2 stoichiometry of Fe:S in pyrite. Subsequently, Huerta-Diaz and Morse (5) expanded this concept to metals (Me) other than Fe defining the degree of trace metal pyritization (DTMP) similarly to DOP for iron. By comparing DTMP with DOP it is possible to relate the pyritization of a given trace metal to that of Fe, which is the dominant metal that is pyritized.

$$DTMP = \frac{Pyrite\text{-}Me}{Pyrite\text{-}Me + Reactive\text{-}Me} \qquad (2)$$

The relationships between the DTMP of trace metals and DOP in near interfacial Galveston Bay sediments are shown in Figure 1. The plots are arranged in approximate order of increasing tendency of metals to become pyritized. All metals have a significant degree (>10%) of pyritization in at least some samples. However, Ni, Zn and Pb are only slightly pyritized in most samples. They have, therefore, not been considered in the experiments to study the release of trace metals during pyrite oxidation. It is important to note that three of the metals that undergo extensive pyritization, Cu, As, and Hg, are ones that have often been of major environmental concern because of their toxicity in aqueous ecosystems (1).

Oxidation of Pyrite and Release of Metals.

Results of the experiments on the oxidative release of metals from pyrite are summarized in Table I. For the four metals studied (As, Cu, Hg and Mo), 88% of the results indicated at least a 25% oxidation of the pyritized metal. However, the extent of metal oxidation was highly variable for both a given metal between different samples and for different metals within a single sample.

The relationship between the initial degree of pyritization of a metal and its percent oxidation is shown in Figure 2. There is a broad scatter in this relationship. However, if the relationship is examined for individual metals it is observed that they can be divided into two groups. Fe and As oxidation are not correlated ($r^2 = 0.00$) with their initial extent of pyritization. However, Cu, Hg and Mo exhibit a moderately good correlation ($r^2 = 0.58, 0.55$ and 0.52, respectively) between extent of oxidation and initial extent of pyritization. The correlations are positive for Cu and Mo, but negative for Hg.

The extent of loss of As, Cu, Hg and Mo from the pyrite fraction is compared with the extent of pyrite oxidation, as represented by loss of Fe from the pyrite fraction, in Figure 3. Again there is a wide scatter in the results. Generally (in 80% of the samples), the oxidation of metals from the pyrite fraction is greater than the extent of pyrite oxidation. About a third of the samples exhibit particularly high (>70%) extents of metal oxidation for As and Cu.

One possible reason for the large scatter in the data presented in Figure 3 is that "apples and oranges" are perhaps being mixed, in that near surficial (top 10 cm) sediments and deeper sediments from the dredge spoil site are both included. It is reasonable to expect the deeper, and therefore older, sediments from the dredge spoil site may contain a smaller fraction of highly reactive, very fine-grained pyrite and the behavior of trace elements might also differ. This hypothesis is supported by the data in Table II. Fe and Cu are about twice as reactive, and Mo is about three times as reactive in the surficial sediments as in the dredged sediments.

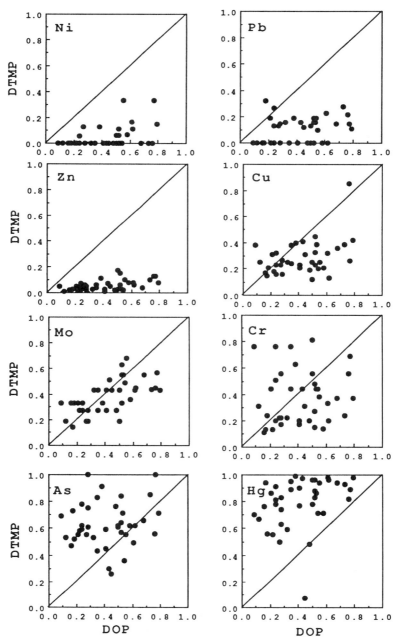

Figure 1. The relationship between DTMP and DOP for trace metals in Galveston Bay sediments. The line on the plots represents an ideal relationship.

Table I. Results of oxidation experiments. For the type of site SC = ship channel, MB = shallow muddy bottom, BB = boat basin and DS= dredge spoil. Ini and Fin are the initial and final degrees of metal pyritization. %Δ is the percentage change in metal pyritization

Site	Type	Fe Ini	Fe Fin	Fe %Δ	As Ini	As Fin	As %Δ	Cu Ini	Cu Fin	Cu %Δ	Hg Ini	Hg Fin	Hg %Δ	Mo Ini	Mo Fin	Mo %Δ
1	SC	.27	.22	19	.73	.37	49	.91	.46	50	.99	.55	44	.63	.26	58
2	SC	.25	.14	43	-	-	-	.80	.32	60	.89	.73	18	.66	.43	36
3	SC	.22	.15	35	.83	.51	38	.81	.09	89	-	-	-	.70	.27	62
4	SC	.21	.10	53	.75	.43	42	.63	.19	71	.91	.62	32	.63	.53	16
5	MB	.13	.08	40	.73	.33	55	.67	.13	80	.57	.30	48	.68	.21	69
6	BB	.21	.17	19	.83	.58	30	.74	.27	64	.76	.45	40	.72	.45	37
7	DS	.28	.23	17	.89	.18	80	.07	.05	25	.85	.50	41	.53	.38	29
8	DS	.13	.11	18	.63	.07	87	.09	.05	40	.85	.48	44	.52	.48	8
9	DS	.16	.13	18	.89	.07	93	.10	.07	33	.59	.14	76	.46	.42	9

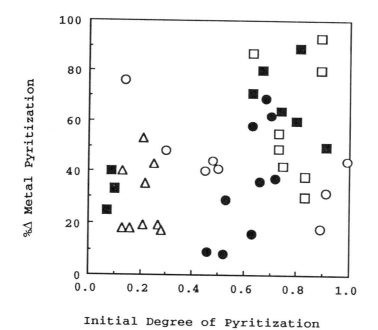

Figure 2. The relationship between the change in extent of metal oxidation and the initial degree of metal pyritization. Fe = open triangles; As = open squares; Cu = closed squares; Hg = open circles; Mo = closed circles.

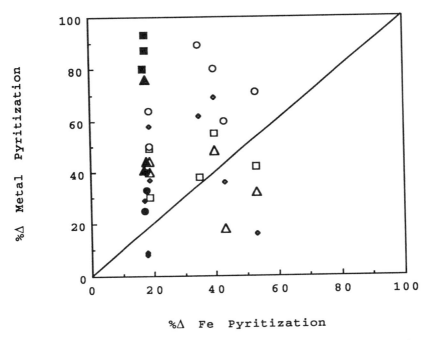

Figure 3. The relationship between percentage loss of metals from the pyrite fraction versus Fe loss. As = squares; Cu = circles; Hg = triangles; Mo = diamonds. Samples from the dredge spoil site are solid symbols and other sites are open symbols. The solid line represents a 1-to-1 relationship.

Table II. Average percent of metal lost from pyrite during one day of oxidation ± numbers are standard deviations

Sediment	Fe	As	Cu	Hg	Mo
Surficial	35±14	44±13	69±14	36±12	46±20
Dredged	18±1	87±7	33±8	54±19	15±12
Ratio	1.9	0.5	2.1	0.7	3.1

Surprisingly, As and Hg are not as reactive in surficial sediments as in the dredged sediments. These two metals have the highest ratios of release relative to Fe in the dredged samples (As/Fe = 4.8; Hg/Fe = 3; Cu/Fe = 1.8; Mo/Fe = 0.83). A possible explanation for this observation is that these metals are more strongly associated with the surface of pyrite in the dredged samples than the other metals and are, therefore, preferentially released during early oxidation of the pyrite.

Conclusions

The results of this study demonstrate that incorporation of trace metals in authigenic pyrite is an important early diagenetic process. Traditional techniques for assessing concentrations of reactive trace metals do not generally include the pyrite fraction. However, a substantial portion of pyrite-associated metals is reactive if sediments are exposed to oxidizing waters. This may occur if sediments are resuspended by storms or human activities such as bottom trawling and dredging.

Acknowledgments

This publication was partially supported by Grant#NA16RGO457-01 from the National Oceanic and Atmospheric Administration through the National Sea Grant Program. The views expressed herein are those of the author and do not necessarily reflect the views of NOAA or any of its sub-agencies. Additional funding was provided by Texas A&M University and the Chemical Oceanography Program of the National Science Foundation.

Literature Cited

1. Mance, G. Pollution Threat of Heavy Metals in Aquatic Environments; Elsevier Applied Science: New York NY, 1990, 372 pp.
2. Berner, R.A. Early Diagenesis: A Theoretical Approach; Princeton University Press: Princeton, NJ, 1980, 241 pp.
3. Morse, J.W.; Millero, F.J.; Cornwell, J.; Rickard, D. Earth-Sci. Rev. 1987, 24, 1-42.
4. Vaughan, D.J.; Craig, J.R. Mineral Chemistry of Metal Sulfides; Cambridge Univ. Press: Cambridge, UK, 1978, 492 pp.
5. Huerta-Diaz, M.A.; Morse, J.W. Marine Chem. 1990, 29, 119-144.
6. Huerta-Diaz, M.A.; Morse, J.W. Geochim. Cosmochim. Acta 1992, 56, 2681-2702.
7. Morse, J.W. Geochim Cosmochim. Acta 1991, 55, 3665-3668.
8. Morse, J.W.; Presley, B.J.; Taylor, R.J.; Benoit, G.; Santschi, P. Mar. Envir. Res. 1993, 36, 1-37.
9. Armstrong N.E. In Estuarine Comparisons; Kennedy, E.S., Ed.; Academic Press: New York, NY, 1982: pp. 103-120.
10. Shao, E.Y.; Winefordner, J.D. Microchem. J. 1989, 39, 229-234.
11. Berner, R.A. Amer. J. Sci. 1970, 268, 1-23.

RECEIVED October 11, 1993

Chapter 21

Mobilization and Scavenging of Heavy Metals Following Resuspension of Anoxic Sediments from the Elbe River

W. Calmano, U. Förstner, and J. Hong[1]

Technical University of Hamburg-Harburg, Eissendorfer Strasse 40, D–2100 Hamburg 90, Germany

Resuspension of sediments from the Elbe River can significantly decrease pH values due to high acid-producing potential and low neutralizing capacity. Metals can be released into the dissolved phase, but may subsequently be readsorbed or precipitated in part to solid phases. To evaluate the potential for metal release and scavenging on a regional and long-term scale, experiments were undertaken at stable neutral-pH values. Rates of total metal release and metal scavenging were extrapolated from time series of net-release values measured over 630 hours. The order of total release from the sediment was Cd (5%) > Zn (1.5%) > Cu (1%) > Pb (0.7%). The scavenging of released metals was in the reverse order: Pb (86%) > Cu (53%) > Zn (34%) > Cd (30%). Dominant processes are adsorption on organic substances, adsorption/coprecipitation by fresh Fe-Mn oxides and precipitation of metal phosphates originating from the decomposition of organic matter. Based on our experimental results and relevant published data, a four-stage interaction model is developed for metals in anoxic sediments. This model describes the behavior of heavy metals in the system subsequent to oxidation.

Many investigations have shown that the sediments of the Elbe River have been severely contaminated by heavy metals and other pollutants during the last decades (1-4). The sediments of the Hamburg harbor, situated at the downstream end of the Elbe River, are dredged to meet the needs of shipping. Dredging produces 2 million m³ of material per year consisting of about 0.8 million m³ of sand and approximately 1.2 million m³ of contaminated harbor mud (5). During dredging,

[1]Current address: Department of Chemical Engineering, Tulane University, New Orleans, LA 70118

the anoxic bottom sediments are suspended. These sediments are exposed to the overlying water, which is rich in dissolved oxygen. A pollution problem arises when heavy metals bound or adsorbed to solids are released into the aqueous phase or transformed into more bioavailable forms during the suspension of the anoxic sediments. Similar problems occur if the polluted sediments are deposited onto spoil areas and exposed to air. Such effects may also occur at the sediment-water interface under different redox conditions during early diagenesis.

Oxidation of the sediments from the Hamburg-Harburg section of the harbor can decrease the pH due to their high acid-producing potential and low acid-neutralizing capacity (6,7). During suspension, anoxic sediments are oxidized by oxygen in water. A lack in understanding of the opposed and interacting effects of certain of these processes originates from two views of the effects of changing redox potential. One view is that the presence of sulfide under reduced conditions will precipitate toxic metals, resulting in very low metal concentrations; the oxidation of sulfides to sulfates under oxidizing conditions will release these metals into the overlying water (8-12). The opposite view is that Fe and Mn compounds will tend to sorb or coprecipitate heavy metals under oxidized conditions. These hydrous oxides are more soluble under reduced conditions due to the reduction of Fe and Mn hydrous oxides (13-17). The effect of sulfides, oxides and hydroxides of Fe and Mn in controlling the solubility of trace metals in the sediment/water systems (interstitial waters) is greatly modified by the presence of organics (18-22).

A study of the release of Cd and Pb from dredge spoil suspended in estuarine water was carried out by Prause et al. (23). A considerable release of Cd from polluted sediments of the Bremen harbor was observed in long-term experiments. Hirst et al. (24) did not observe similar results for Cu, Zn, Fe, and Mn during experimental resuspension and reoxidation of polluted anoxic sediments collected from the Mersey estuary, northwest England. Hirst et al. concluded that, in the estuarine environment, the resuspension of metal-rich (polluted) anoxic sediments is exceedingly unlikely to result in increased metal concentrations in surface waters. It was stressed that such sediments may in fact act as additional scavengers of trace elements.

Khalid et al. (25) and Wallmann (8) found an increase in dissolved Cu, Cd and Pb concentrations with time during the oxidation of an anoxic estuarine sediment in seawater and fresh water. In their experiments, pH was not controlled and became progressively lower. Although metal mobilization was investigated intensively in controlled pH-Eh systems (26-28), these studies only focused on the transformation (including dissolved phase) of metals in sediment/water systems without consideration of interacting processes.

For polluted sediments, some important questions remain:
- Can heavy metals in the sediments be released into solution during oxidation under stable neutral-pH conditions?
- If so, which metals and what concentrations can enter the aqueous phase?
- Which release process is dominant during suspension/oxidation?

This study was designed to answer these questions. Based on the results, a model is developed to explain the behavior and interacting processes controlling the concentrations of heavy metals in sediment/water systems during suspension of anoxic sediments.

Materials and Methods

Sediment Sampling. A sediment core was collected from the Mühlenberger Loch, which is situated at the downstream Elbe near the Hamburg harbor (Figure 1). The sedimentation rate here is about 8 cm·y^{-1} (29). The core was sealed without air space. Care was taken to avoid exposure to atmospheric O_2 during transportation and storage prior to laboratory studies. In an O_2-free glove box, under a nitrogen atmosphere, the sediment core was cut and divided into sections. The sediment at the depth 7-10 cm was used for the suspension experiment. Some of the sediment properties are presented in Table I.

Table I. Characteristics of the Sediment

Parameter	Value		Parameter	Value	
Organic carbon	3.83	mass %	Extractable Fe (1N HCl)	222	μmol·g^{-1}
Loss on ignition	10.2	mass %	Extractable Fe^{2+} (1N HCl)	184	μmol·g^{-1}
Protein	24.17	mg·g^{-1}	Extractable Fe^{3+} (1N HCl)	38	μmol·g^{-1}
CaCO$_3$	7	mass %	Total Cu	87	mg·kg^{-1}
Water content	65.4	mass %	Total Zn	538	mg·kg^{-1}
Acid volatile sulfide	3.8	μmol·g^{-1}	Total Pb	64.6	mg·kg^{-1}
Pyrite	42.8	μmol·g^{-1}	Total Cd	2.59	mg·kg^{-1}

Laboratory Simulation Experiment. Wet sediment material was transferred to a 180 ml reaction cell. 135 ml of artificial river water (2 mM CaCl$_2$ and 5 mM NaCl) was added to the reaction cell. Together with the sediment material the total suspension volume in the reaction cell was 140 ml. The solid:liquid ratio was 1:64. The suspension was stirred with a magnetic stirrer. Air was continuously bubbled through the system. pH was automatically adjusted with 0.1 N NaOH or 0.1 N HCl by a control system (30). During the experiment, the pH of the system was maintained at 7.5. The experiments were carried out under exclusion of light. The temperature during the experiments was not controlled, but was recorded. The variation was between 21-23°C.

Analysis. The suspension was sampled at regular intervals. The suspension samples were centrifuged and filtered through a 0.2 μm membrane. The same volume of artificial river water used to wash the centrifugation tube was added into the reaction cell in order to maintain a constant suspension volume. Part of the filtrate was acidified with concentrated nitric acid to pH 1.5 for determination of concentrations of heavy metals by atomic absorption spectroscopy (AAS). Another part of the filtrate was used for nitrate and sulfate determinations by ion chromatography (IC). Sulfide and pyrite were analyzed by polarography. The details can be found elsewhere (31). After digestion of the sediment with concentrated nitric acid at 80°C for 6 hours, the total content of each metal was determined.

Results

Data Treatment. Because the solution was sampled at regular time intervals and artificial river water was added into the suspension to keep the volume constant, part of the dissolved metals were removed from the system and the original suspension was diluted. Release of heavy metals from the sediment phase and scavenging from the aqueous phase are determined by the following equation:

$$\alpha = C_i^m - C_{i-1}^m \left[\frac{V_0 - V_i}{V_0} \right] \quad (1)$$

where C_i^m and C_{i-1}^m are measured dissolved metal concentrations at ith time and (i-1)th time, respectively; V_0 and V_i are total volume and sampled volume, respectively; α is a discriminating factor.

If

$$f = \frac{V_0 - V_i}{V_0} \quad (2)$$

we get

$$\alpha = C_i^m - fC_{i-1}^m \quad (3)$$

where f is the residual coefficient of the solution. If $\alpha < 0$, metals are scavenged from the solution; if $\alpha > 0$ metals are released into the solution. If $\alpha = 0$, the system is stable with constant metal concentrations in solution.

For the whole process the release of metals from the sediment can be calculated as follows:

$$C^{re} = \sum \left[C_i^m - fC_{i-1}^m \right] \quad (4)$$

where C^{re} is the total released concentration.

The scavenging concentration can be calculated by equation 5:

$$C^{sc} = \sum \left[C_{i-1}^{net} - C_i^{net} \right] \quad (5)$$

where C^{sc} is the scavenging concentration, C_i^{net} and C_{i-1}^{net} are net-release concentrations at ith time and (i-1)th time, respectively. The net-release concentration is calculated by equation 6:

$$C_i^{net} = C_{i-1}^{net} \left[\frac{C_i^m}{C_{i-1}^m \left[\frac{V_0 - V_i}{V_0} \right]} \right] \quad (6)$$

where C_i and C_{i-1} are the concentrations at ith time and $(i-1)$th time. Equation 6 assumes that the metals removed by sampling would participate in interacting processes and would be scavenged proportionally, if they had not been removed from the operational procedure.

If

$$\beta = \frac{C_i^m}{C_{i-1}^m \left[\frac{V_0 - V_i}{V_0} \right]} \quad (7)$$

we get

$$C_i^{net} = \beta C_{i-1}^{net} \quad (8)$$

Combining equations 5 and 8 we can calculate the total scavenged concentration:

$$C^{sc} = \sum \left[(1 - \beta) C_{i-1}^{net} \right] \quad (9)$$

The total concentration changes of net release are:

$$C^{net} = C^{re} - C^{sc} \quad (10)$$

That part of released metal, which was sampled at regular time intervals, is not considered to be scavenged by sediment phase. So:

$$C^{net} = C_i^m + \frac{\sum C_{i-1}^m V_{i-1}}{V_0} \quad (11)$$

If there is only release but no scavenging in the suspension, equation 11 is suitable. If there is both release and scavenging from the aqueous phase, the first method is better to describe the interacting processes. In our experiments we found both release and scavenging. Therefore, we used equations 4, 9, and 10 to calculate release, scavenging and net release during the suspension.

Cadmium Release. Continuous release of Cd in the suspension was observed. As seen in Figure 2, there was a rapid release at the beginning of the experiment, then the release slowed. After about 200 hours the rate of the net release (C^{net}) increased again. The highest release rate occurred after about 300 hours and later.

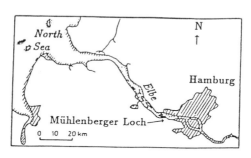

Figure 1. Sampling site.

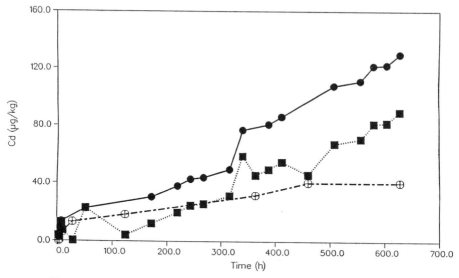

Figure 2. Cadmium release. (•) - total release; (⊕) - scavenging (adsorption/coprecipitation); (■) - net release.

During the entire experiment (over 628 hours) Cd release reached 0.129 mg per kilogram sediment, or 4.99% of the total Cd content in the sediment. A portion of the Cd released was scavenged. Finally only 0.090 mg Cd·kg^{-1} sediment remained in solution, which was about 3.5% of the total Cd content in the sediment (Table II). At the end of the experiment, the concentration of Cd in solution was 1.40 μg·l^{-1}.

Table II. Release Percentage of Heavy Metals During the Suspension

	Total Content (mg·kg^{-1})	Total Release (mg·kg^{-1})	(%)	Net Release (mg·kg^{-1})	(%)	Final Conc. in Solution (μg·l^{-1})
Cd	2.585	0.129	4.99	0.090	3.5	1.40
Zn	538	8.28	1.54	5.48	1.02	85.6
Cu	87	0.824	0.95	0.388	0.45	6.06
Pb	65	0.441	0.68	0.064	0.10	0.99

This study confirms that Cd in the sediment of the Elbe River can be mobilized during oxidation in early diagenesis, suspension and dredging activities under stable pH conditions. Unfortunately there exist no field data from the Elbe River. More experimental work is needed to explain the unusual release occurring after 300 h. This release is probably due to organic-matter decomposition.

Lead Release. Pb release from the sediment was significantly different from that of Cd and other metals. The dissolved Pb concentration increased at the beginning of the experiment and later decreased continuously (Figure 3). Readsorption during the suspension process is a possible explanation for the concentration decrease. After 10 hours, the highest release concentration reached 0.36 mg·kg^{-1}. After about 100 hours, an approximate steady state between release and readsorption processes was attained, and the total concentration of Pb released from the sediment remained relatively constant at a value between 0.15-0.2 mg·kg^{-1}. In the whole suspension process, the mass of Pb released from the sediment was 0.441 mg·kg^{-1} (Table II), i.e. 0.68% of the total Pb content of the sediment, however, 85.5% of the Pb released was scavenged by the solids. Net release was only 0.1% of the total Pb content in the sediment.

Copper Release. Cu release is shown in Figure 4. There was a significant increase at the beginning of the experiment. Continuous increase of the dissolved Cu concentration was found until about the 350th hour. Although there was still a slow release, the scavenging rate of Cu was faster than the release rate of Cu at this stage. This process resulted in a decrease of dissolved Cu concentration. As can be seen from Figure 4, the concentration remained relatively constant. At the end of the experiment, 0.824 mg·kg^{-1} of Cu was released into the solution, which

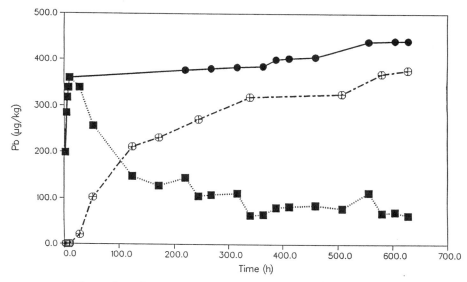

Figure 3. Lead release. (•) - total release; (⊕) - scavenging (adsorption/coprecipitation); (■) - net release.

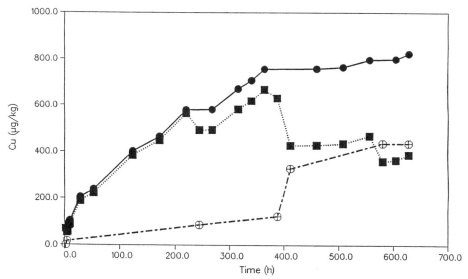

Figure 4. Copper release. (•) - total release; (⊕) - scavenging (adsorption/coprecipitation); (■) - net release.

corresponded to about 0.95% of the total Cu content in the sediment. The net release was only 0.388 mg·kg^{-1}, i.e. 0.45% of the total copper content. At the end of the experiment the Cu concentration in the solution was 6.06 μg·l^{-1}.

Zinc Release. The Zn release was similar to that of Cu at the start of the experiment but Zn was not strongly scavenged at the later stage. Figure 5 shows a continuous increase. The released Zn concentration was relatively high, as a result of the high Zn content in the sediment (538 mg·kg^{-1}). During the experiment, 8.28 mg·kg^{-1} of Zn was released, or 1.54% of the total Zn content. The net-release concentration and net-release percentage were 5.48 mg·kg^{-1} and 1.02%, respectively. A relatively high Zn concentration of 85.6 μg·l^{-1} was found in the solution at the end of the experiment.

Readsorption/Coprecipitation of Metals. The scavenging of different heavy metals during the suspension experiment is listed in Table III. Scavenging is mainly attributed to readsorption and coprecipitation. Approximately one-third of the released Cd (30%) and Zn (33.8%) was scavenged from the solution, mainly by readsorption onto the solid surface. Readsorbed/coprecipitated Cu exceeded half of the released Cu (52.9%). Up to 85.5% of released Pb was scavenged during the suspension. Very little Pb remained in the solution at the end of the experiment. The order of scavenging percentage for metals is: Pb > Cu > Zn > Cd. This order is the inverse of the mobilization percentage for metals (Table II): Cd > Zn > Cu > Pb. However, both orders are different from the total amount of release (mg·kg^{-1}) for the metals: Zn > Cu > Pb > Cd, which is the same order as the original total contents of these metals (Table II).

Table III. Release and Readsorption/Coprecipitation of Heavy Metals During Suspension

	Total Release (mg·kg^{-1})	Readsorption/coprecipitation (mg·kg^{-1})	(% of total)	Net Release (mg·kg^{-1})	(% of total)
Cd	0.129	0.039	30	0.090	70
Zn	8.28	2.80	33.8	5.48	66.2
Cu	0.824	0.436	52.9	0.388	47.1
Pb	0.441	0.377	85.5	0.064	14.5

Discussion

Results from Other Studies. Cd release has also been observed in other investigations of polluted harbor sediments. Prause et al. studied the release of Cd from dredged material of the Bremen Europahafen in the harbor environment (23). During a long-term experiment extensive Cd remobilization was recorded. Polluted

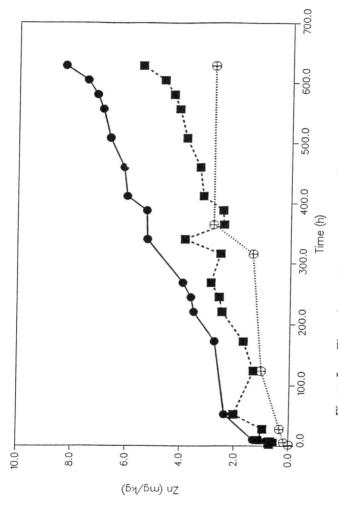

Figure 5. Zinc release. (•) - total release; (⊕) - scavenging (adsorption/coprecipitation); (■) - net release.

fresh-water sediment disposed into seawater at pH 7.9 caused release of 1-2 μg Cd per gram solids (about 25% of total content) in a 50-day experiment.

In a study of periodic oxidation-reduction at pH 6.0, Cd increase in solution during oxidation has also been observed (7). In some investigations, without pH control, oxidation often results in pH decrease and causes Cd release (7,8,25,30,32). Calmano et al. found that in a 3-week suspension experiment using fresh-water sediments treated with seawater, 15.9% of Cd was released (33). In an oxidation experiment without pH control, Cd release can reach 35.6% (pH 3.5) (30).

Gambrell et al. studied Pb mobility in Mobile Bay sediment suspensions (26) spiked with carrier-free ^{210}Pb to determine the dissolved Pb concentration in systems with varying pH and redox potentials. No more than 0.15% of labelled Pb could be found in the solution at a pH range from 5.0 to 8.0. Furthermore, Prause et al. found no release of Pb from the sediments in their long-term and short-term experiment (23). But in an oxidation experiment without pH control (30), 1.6% of the total Pb in the sediment was mobilized when the pH in the system decreased below 3.5. Similar results were found by Khalid et al. (25).

Förstner et al. reported that in a 3-week suspension experiment, 1.3% of the total Cu was released from dredged mud of the Hamburg harbor when treated with seawater, and 0.48% of the total Cu was in solution at the end of the experiment (34). The net-release percentage is very similar to the result of the present study (0.47%). In a 30-day experiment without pH control (30), the Cu mobilization reached 6.8% of the total Cu content at the end of the experiment (pH 3.5).

A different result was observed by Hirst and Aston (24). During resuspension and reoxidation of polluted anoxic sediments, they did not find any release of Cu. After recalculating their results, however, we found that their results and the current study are not contradictory. In their experiment they used a very dilute suspension, where the sediment:water ratio in suspension was 1:2000. If we assume that there was the same net-release proportion as in our experiment (0.45%), the net release of Cu in their experiment would be only 0.2 μg·l^{-1}. This concentration is insignificant compared with the total-release concentration in our experiment. On the other hand, their experiment was performed only for 72 hours, while the net Cu release of 0.45% in our experiment was determined after 628 hours. In another suspension experiment (34) the redox potential after 100 hours was still below -50 mV. Little or no metal release is expected under these conditions.

Calmano et al. observed that 9.1% of the total Zn in the dredged mud was released during the suspension experiment when river sediments were treated with seawater (35). In a 30-day oxidation experiment without pH control (30), 24.6% of the total Zn was found in the solution at pH 3.5.

Again, Hirst et al. found no release of Zn from anoxic sediment during the suspension in their experiments (24). From the data listed in Tables 3 and 4 of their paper, no Zn was released during suspension at 5°C in all four group experiments. At 20°C, they observed an increased tendency for mobilization during the suspension. These results can be explained by two possibilities. One possibility is that limited microbial activity at relatively low temperatures (5°C).

Prause et al. found strong influence of microbial activity on metal release (23). The second possible explanation is the low sediment:water ratio discussed above.

Sources of the Released Metals. Sulfide oxidation provides the primary source of metal release. The solubility of amorphous sulfide minerals is very low, but they are very unstable in aqueous environments when exposed to oxidants such as dissolved oxygen (DO) or Fe(III). Oxidation will result in remobilization or redistribution of metals. From the data of our experiment Cu release as well as Zn release were correlated with sulfate concentration during the suspension experiment (Figures 6,7), especially at the first stage. Calculations show that there is a marked correlation between sulfate concentration ($[SO_4^{2-}]$) and total metal concentration ([T-Me]) in the solution. The correlation coefficients with $[SO_4^{2-}]$ are 0.98, 0.98, 0.93, and 0.92 for Cu, Zn, Cd, and Pb, respectively. It is not surprising that Cu is more closely correlated to SO_4^{2-}. As can be seen from Table IV, the solubility constant (K_{sp}) of CuS is much lower than that of other metals. In other words, Cu will be the first sulfide precipitated if any sulfide ion (S^{2-}) is present. In a sequential extraction procedure, Cu in the sulfidic/organic fraction can often reach a very high percentage. For example, a sediment sample from the Harburg section of the Hamburg harbor contains more than 80% of total potentially mobile Cu (7).

Table IV. K_{sp} and Metal Sulfide Fraction in Two Hamburg Harbor Sediments

Metal	No. 1*	No. 2**	Sulfide	K_{sp}
Cu	87%	86.9%	Cu_2S	2.5×10^{-48}
			CuS	6.3×10^{-36}
Pb	79%	63%	PbS	8.0×10^{-28}
Cd	66%	58%	CdS	8.0×10^{-27}
Zn	38%	39%	α-ZnS	1.6×10^{-24}
			β-ZnS	2.5×10^{-22}

*No. 1: from Hong et al. (30). In the percentage calculation, the total content does not include the content in the residual fraction treated with concentrated HNO_3.

**No. 2: from Kersten and Förstner (36).

Earlier studies, which are similar to this investigation, have assumed certain mechanisms that result in metal release. Although sulfide oxidation is often discussed, limited evidence has been provided. The increased sulfate concentration with increasing heavy-metal concentration (especially for Cu and Zn) under stable-pH conditions observed in our experiments is probably the first evidence for assessing heavy-metal behavior in anoxic sediment during suspension and oxidation. In addition to metal-sulfide oxidation, pyrite (FeS_2) oxidation is also an important source of metals because pyrite can strongly bind heavy metals during

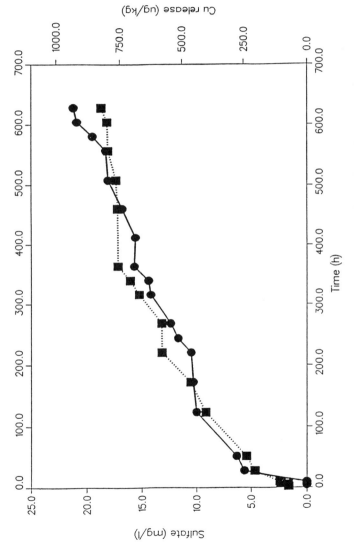

Figure 6. Sulfate concentrations and copper release into the suspension. (•) - sulfate; (■) - copper.

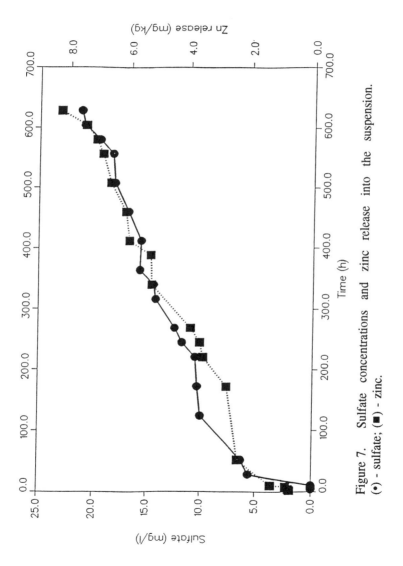

Figure 7. Sulfate concentrations and zinc release into the suspension. (●) - sulfate; (■) - zinc.

formation. A recent study (37) has found that the binding capacity for divalent heavy-metal cations by pyrite increased in the order $Co^{2+} < Cd^{2+} < Mn^{2+} < Ni^{2+} < Zn^{2+}$. When pyrite is oxidized, heavy metals bound by pyrite can be released into the solution. Pyrite oxidation during the suspension experiment is shown in Figure 8.

Displacement by inorganic cations and complexation by Cl⁻ is known to increase the solubility of trace elements, in particular of Cd, if the concentration of Cl⁻ ions is high enough to compete with hydroxide formation (38), and complexation with ligands on the organic-solid surface. In our artificial estuarine water, Cl⁻ was the only complexing anion. Hem (39) suggested the presence of a mixed complex, $Cd(OH)Cl^{o}$, when the Cl⁻ concentration exceeds 0.1 M, with a solubility greater than for $Cd(OH)_2^{o}$. Khalid et al. pointed out that the probable formation of $Cd(OH)Cl^{o}$ is reasonable in Barataria Bay sediment suspensions containing a 0.3 M sodium-chloride, sodium-sulfate solution.

As can be calculated by the use of stability constants, the most dominant Cd species in waters of low salinity and containing low concentrations of dissolved organic ligands are the positively charged Cd^{2+} and the mono-chloro-Cd-complex (see Table V). With increasing salinity, the positively charged species decrease and the neutral or negatively charged species increase. This changing species distribution has a direct influence on the extent of Cd-sorption on negatively-charged organic surfaces. In addition to competition with positive alkali ions, the decreased Cd sorption at higher salinity may be due to this change in the dissolved Cd speciation.

Table V. Calculated Cd-species Distribution (%) at Different Salinities

	(‰) Salinity			
	5	10	15	20
$[Cd^{2+}]$	19.63	11.43	7.65	5.41
$[CdCl^{+}]$	65.46	61.97	55.95	49.71
$[CdCl_2^{o}]$	13.74	22.15	29.40	33.44
$[CdCl_3^{-}]$	0.72	2.37	4.42	6.56
$[CdCl_4^{2-}]$	0.12	0.84	2.39	4.72
$[Cd(OH)Cl^{o}]$	0.21	0.19	0.17	0.15
Positively charged species	85.1	73.4	63.6	55.1
Neutral or negatively charged species	14.9	26.6	36.4	44.9

These effects could not be observed in earlier experiments in the multichamber device, where Cd had been mobilized from polluted dredged

sediments (35). Cd from dredged mud did not behave like ionic Cd. The reason for this behavior could be that:
- the pore water of the sediment suspension contained high concentrations of dissolved organic ligands which complex Cd;
- the organically-complexed Cd had a high affinity for the organic surfaces of the biological material. Increasing salinity did not prevent sorption on these surfaces.

A number of investigations have focused on studies of organic-bound metals in sediment/water systems. A general assumption is that organic carbon is oxidized continuously by a sequence of energy-yielding reactions with decreasing energy production per mole of organic matter oxidized. The calculated ratios of the metal/C flux support the argument for metal mobilization during organic-carbon oxidation. For example, Wallmann (8) found that a considerable percentage of the total Cu content exists as a labile organic complex during suspension of anoxic sediment.

Interstitial-water studies provide many examples to analyze the relationship between metals and organic ligands in sediment-water systems. The data of Elderfield (40) indicated that approximately 80% of the total Cu was organically bound in the interstitial water of surface sediments. Most of the organically-associated heavy metals are "colloidal" and equivalent in molecular weight to humic substances. Hart and Davies (41) reported that most of the Cd in the interstitial water from a sediment sample of an urban creek at Melbourne, Australia, was associated with colloidal labile species and more strongly bound by organic complexes. Most of the Zn and Pb was associated with the colloidal phase, 20% in labile forms and 10% in more tightly bound (organic) forms. Van den Berg et al. studied the organic complexation capacity of Zn with samples collected from interstitial waters in sediments of the upper Gulf of Thailand and the estuary of the river Mersey in England (42). These authors concluded that between 93 and 98% of the total Zn is complexed by organic material in the interstitial water, as a result of high ligand concentrations. These results indicate that the dissolved Zn concentrations may increase as a result of competition between dissolved ligands and ligands on the particle surface. Ligand production by bacterial or algal decay may, in part, explain increases in dissolved heavy-metal concentrations in solution of sediment-water systems.

Scavenging of Heavy Metals from the Aqueous Phase. In addition to organic and biological material, $Fe(III)$ and $Mn(IV)$ oxide minerals are often considered the most important factors affecting the distribution of heavy metals in sediment-water systems. Freshly precipitated oxyhydroxides are more effective in scavenging trace metals than aged crystalline materials because of greater reactive surface area.

As can be seen from Table VI, at least half of the iron in the sediment exists as ferrous minerals. The oxidation kinetics of some species, such as vivianite, siderite and $Fe(II)$ in clay minerals, are relatively unknown. Oxidation rates of sulfidic-bound $Fe(II)$ have been determined. The change in pyrite content of the sediment during oxidation of the suspension was determined and will be reported in a later paper. Here we give preliminary results in Figure 9.

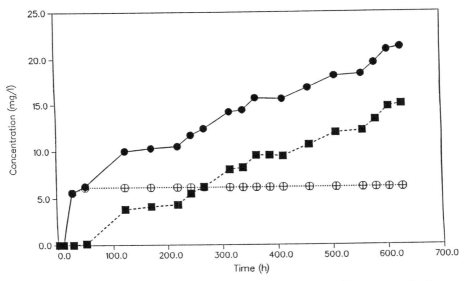

Figure 8. Changes in concentrations of sulfate from different contributing sources. (•) - total sulfate concentration; (⊕) - sulfate from oxidation of acid-volatile sulfide; (■) - sulfate from pyrite oxidation.

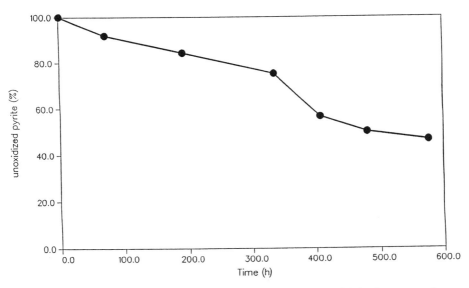

Figure 9. Change in percentage of unoxidized pyrite (FeS_2) in the suspension.

Table VI. Fe Species in Sediments

Iron Species	% of Total Fe
Vivianite ($Fe_3(PO_4)_2$)	20
Siderite ($FeCO_3$)	8
Fe(II) in Ca-Mg carbonate	4
Fe(II) in humic matter	1
Fe(II) in clay minerals	18
Fe(III) and sulfide-bound Fe(II)	49

These results suggest that the suspension system provided ample Fe(II). This Fe(II) was oxidized to Fe(III)-hydroxides. Freshly formed Fe(III)-hydroxides can adsorb and coprecipitate metals released to the solution. Considerable readsorption of Cd, Zn and Cu on Fe-hydroxide, Mn-oxide and algal cell walls has been observed in our experiments with a multi-chamber device, where competition effects of these compounds were studied (35).

Fe(III) and Mn(IV) oxyhydroxides and organic matter, which all are considered heavy-metal adsorbents, tend to form "coatings" and biofilms on mineral surfaces, which show a high affinity to some heavy metals. Studies of adsorption of heavy metals on oxides indicate that Cd, Cu, Pb, Zn, and some other metals are adsorbed more strongly by Mn(IV) oxides than by Fe(III) oxides (43-45). The order of binding stability and affinity of goethite (α-FeOOH) and amorphous Fe(III) oxide for heavy metals is listed as follows (46-50):

$$Pb \sim Cu > Zn > Cd.$$

The order of adsorption is similar for birnessite (δ-MnO_2) and Mn(IV) oxyhydroxide (46,47,51):

$$Pb > Cu > Cd > Zn.$$

The affinity of marine oxic sediments with high Mn(IV) contents has been given by Balistrieri and Murray (49):

$$Pb > Cu > Zn > Cd.$$

The investigations of both Fe-Mn oxide-rich sediments and Fe-Mn oxides as absorbents suggest that the order of affinity of these compounds is the same as the scavenging affinity observed in our experiment (Table III).

The released metals can also be readsorbed by solid organic matter in sediments. Jonasson (52) established a probable order of binding strength for a number of metal ions onto humic or fulvic acids: $Cu^{2+} > Pb^{2+} > Zn^{2+}$. This result has been demonstrated by an investigation on "model" sediment components (34). For example, 1.3% of the total Cu content of the sludge was released during the suspension, but only one third remained in the solution. Two thirds of the released Cu was readsorbed to the various model substrates. Quartz had 3 mg·kg^{-1} of copper, bentonite clay contained about 15 mg·kg^{-1}, iron hydroxide retained about 80 mg·kg^{-1}, and manganese oxide retained 100 mg·kg^{-1}, whereas the cell walls, a minor component of the "model" sediment, accumulated nearly 300 mg·kg^{-1} of the Cu. A similar result for Cd has also been reported (35).

Release and Scavenging Processes: A Four-Stage Interaction Model. Based on this experiment and other supporting materials the behavior of heavy metals during suspension and oxidation of anoxic polluted sediment can be described as four-stage interactions illustrated in Figure 10. These stages are: (I) release stage, (II) transition stage, (III) scavenging stage, and (IV) steady-state stage.

Release Stage. This stage includes amorphous-sulfide oxidation, some pyrite oxidation and some organic-detritus decomposition (Figure 11). At the beginning of the experiment, amorphous-sulfide oxidation is a dominant chemical process. The metals, which are strongly bound with acid-volatile sulfides, will not be released at this stage. This hypothesis is reasonable if we compare the K_{sp} of CuS and Cu$_2$S as well as other metal sulfides (Table IV). We have found that the oxidation of cadmium sulfide was not significant during stage I compared with the entire release process.

The original sediment samples consisted of very complex compounds with different porosities. During suspension, the degree and rate of sulfide oxidation depend on sediment composition and, to a certain degree, on the stirring speed.

The second contribution is characterized by mineralization of ammonium and pyrite oxidation. In this respect, Prause et al. (23) stressed the effects of microbial activity resulting in decomposition of organic matter. In our experiment Cu showed a relatively high dissolved concentration, perhaps due to organic complexing. An increase in sulfate concentration combined with pyrite decrease, provides an indication of pyrite oxidation in this stage (Figures 6-9). Sulfate is formed from oxidation of both acid-volatile sulfide and pyrite. Assuming acid-volatile sulfide is first oxidized, sulfate formed from pyrite oxidation can be observed in Figure 8. The rate of sulfate accumulation would not necessarily reflect the stoichiometric relationship of pyrite oxidation, however, because of the occurrence of intermediate sulfoxy species such as thiosulfate and polythionates. The relative analysis, based on the sulfate results, is still useful. The release of metals bound with FeS$_2$, such as Cu, Zn and Cd, is dominant at this stage.

Transition Stage of Release/Scavenging. At this stage, acid-volatile sulfur compounds have been oxidized; pyrite oxidation and organic-detritus decomposition continues. At the same time, metal readsorption becomes stronger. Phosphate, derived from the decomposed organic detritus, accumulates and then

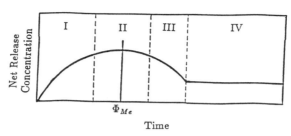

Figure 10. Schematic diagram showing the change in the net metal-release concentration in the four-stage process. I - release stage; II - transition stage; III - scavenging stage; IV - steady-state stage.

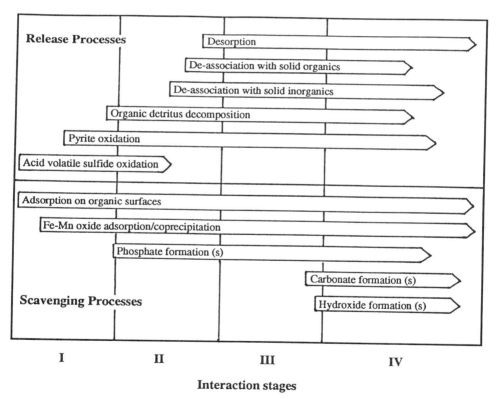

Figure 11. Sequence of chemical and biochemical reactions involved in metal release and scavenging in suspension/oxidation processes. "Acid volatile sulfide oxidation" includes solid monosulfides.

precipitates with some metals. Release and adsorption/precipitation occur together. At this stage neither the release nor the scavenging process is dominant (Figure 10).

Scavenging Stage. At this stage, all release processes decline including pyrite oxidation, organic-detritus decomposition, and metal release from solid organic and inorganic compounds. The formation of Fe-Mn oxide continues and scavenging is a dominant process. This process leads to a decrease in the concentration of metals in the aqueous phase.

Steady-state Stage. Release and scavenging processes have reached a new equilibrium. At this stage, the release rate is equal to the scavenging rate. The metal concentrations in solution will remain constant, however, metal-solid "speciation" and relevant distribution change greatly.

Release and scavenging of Pb, Cu, Zn, and Cd in anoxic sediments during oxidation/suspension can be generalized and are illustrated in Figure 12. For different metals, different stages appear at different times.

Characteristic Value of Net Metal Release, Φ_{Me}, at "the Worst Situation". The time at which the highest value (peak value) of net metal release is found can be designated as the characteristic value of net metal release and is represented by Φ_{Me} in hours. The Φ_{Me} is situated at the peak position of stage II (Figure 10). There are different Φ_{Me} values for different metals, depending on their physicochemical properties and the conditions of suspension. When the oxidation time is $< \Phi_{Me}$, the dominating process is release of metals. When the oxidation time is $> \Phi_{Me}$, the dominating process is scavenging or relative equilibrium. In our experiment, Φ_{Pb} was < 10 h, $\Phi_{Cu} \sim 350$ h, both Φ_{Zn} and Φ_{Cd} were > 600 h. Φ_{Me} reflects how quickly metals in the system may reach "the worst situation" or the highest concentration in the solution. From Figure 12 we can see that Φ_{Pb} is very small. After a relatively short time of release, rapid scavenging of Pb occurred. Φ_{Zn} and Φ_{Cd} were much higher than Φ_{Pb}.

Conclusions

In this study, the behavior of heavy metals in anoxic sediment during oxidation/suspension has been investigated. Cd, Pb, Cu, and Zn can be released into the aqueous phase during sediment resuspension. The sources of metal release are metal-sulfide oxidation, organic decomposition, and desorption. Different metals showed different release rates. About 5% of the total Cd was mobilized. For Pb, the percentage mobilized was only about 0.7%, and for Cu and Zn, 1 and 1.5%, respectively. The order of net percentage release of total metals from the sediments is Cd > Zn > Cu > Pb. For the total mass released, however, the order is Zn > Cu > Pb > Cd; the same as the order of total concentration in the sediment. At the end of the experiment, the order of net release, and the magnitude of released metals, were similar to that observed in the natural waters of the Elbe River at Hamburg. The released metals are scavenged to a different extent. The order of the scavenging percentage is Pb (86%) > Cu (53%) >

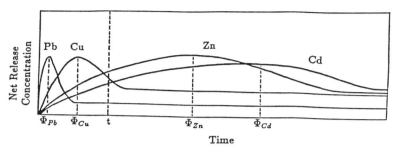

Figure 12. Schematic diagram showing generalized change in relative concentrations for metals during suspension/oxidation. "t" refers to experiment time in this study (628 h).

Zn (34%) > Cd (30%). Dominant processes are sorption by freshly precipitated Fe-Mn oxides and organic matter, and precipitation of some metal phosphates, which result from the decomposition of organic matter.

Based on our experimental results, a four-stage interaction model for metals in anoxic sediment during oxidation has been developed to describe heavy-metal behavior. An index, Φ_{Me}, has been developed to assess the timing of "the worst situation", the maximum net metal release under certain conditions. The results suggest that Pb and Cu can cause only a short-term problem, while Cd and Zn result in a long-term impact on the system if the assessment is based on the metal concentrations in the aqueous phase, which is the most important factor for bioavailability.

The illustration of this study seems to explain some controversial and potentially confusing results of earlier studies. During the first stage, metals can be released to the aqueous phase due to sulfide oxidation and other mobilizing processes; during subsequent stages the metals released will then be readsorbed or precipitate again to different extents for different metals. These metals will reach maximum concentrations at different times, represented by Φ_{Me} values.

Literature Cited

1. Mart, L.; Nürnberg, H.W.; Rützel, H. J. Sci. Total Environ. 1985, 44, 35-49.
2. Förstner, U.; Schoer, J.; Knauth, H.-D. J. Sci. Total Environ. 1990, 97/98, 347-368.
3. Lichtfuss, R.; Brümmer, G. Naturwiss. 1977, 64, 122-125.
4. Knauth, H.-D.; Schroeder, F.; Irmer, U. In Proc. 2nd Int. Conf. Environmental Contamination, Amsterdam; Ernst, W.H.O., Ed.; CEP Consultants: Edinburgh, 1986; pp. 186-188.
5. Tent, L. Hydrobiologia 1987, 149, 189-199.
6. Förstner, U.; Calmano, W.; Hong, J.; Kersten, M. In Proc. 5th Int. Symp.

on River Sedimentation; Larsen, P.; Eisenhauer, N., Eds.; Karlsruhe, 1992; Vol. IV, pp. 1229-1234.
7. Calmano, W.; Hong, J.; Förstner, U. Vom Wasser 1992, 78, 245-257.
8. Wallmann, K. Ph.D. Thesis, Technical University of Hamburg-Harburg, 1990.
9. Engler, R.P.; Patrick, W.H. Jr. J. Soil Sci. 1975, 119, 217-221.
10. Chen, K.Y.; Gupta, S.K.; Sycip, A.Z.; Lu, J.C.S.; Knezevic, M. The effects of dispersion, setting, and resedimentation on migration of chemical constituents during open water disposal of dredged materials, Contract Report U.S. Army Engineer Waterways Experiment Station, Vicksburg, MI, Contract No. DACW39-74-C-0077, 1976, 221 pp.
11. Gobeil, C.; Silverberg, N., Sundby, B.; Cossa, D. Geochim. Cosmochim. Acta 1987, 51, 589-596.
12. Giblin, A.E.; Luther III, G.W.; Valiela, I. Estuarine Coastal Shelf Sci. 1986, 23, 477.
13. Klinkhammer, G. Earth Planet Sci. Lett. 1980, 49, 81-101.
14. Sawlan, J.J.; Murray, J.W. Earth Planet. Sci. Lett. 1983, 64, 213-230.
15. Jenne, E.A. In Trace Inorganics in Water; Gould, R. F., Ed.; ACS Advances in Chemistry Series; American Chemical Society: Washington, DC, 1968; pp. 337-387.
16. Brooks, B.B.; Presley, J.J.; Kaplan, I.R. Geochim. Cosmochim. Acta 1968, 32, 397-414.
17. Palsma, A.J.; Loch, J.P.G. In Heavy Metals in the Environment; Farmer, J.G., Ed.; CEP Consultants: Edinburgh, 1991; Vol. 2, pp 20-23.
18. Morel, F.; McDuff, R.E.; Morgan J.J. In Trace Metals and Metal Organic Interactons in Natural Water; Singer, P.C., Ed.; Ann Arbor Science Publ.: Ann Arbor, MI, 1973; pp. 157-200.
19. Lindberg, S.E.; Harriss, R.C. Environ. Sci. Technol. 1974, 8, 459-462.
20. Nissenbaum, A.; Swaine, D.J. Geochim. Cosmochim. Acta 1976, 40, 809.
21. Piemontesi, D.; Baccini, P. Environ. Technol. Lett., 1986, 7, 577-592.
22. Huynh-Ngoc, L.; Whitehead, N.C.; and Boussemart, M. Mar. Chem. 1989, 26, 119-132.
23. Prause, B.; Rehm, E.; Schulz-Baldes, M. Environ. Technol. Lett. 1985, 6, 261-266.
24. Hirst, J.M.; Aston, S.R. Estuarine Coastal Shelf Sci. 1983, 16, 549-558.
25. Khalid, R.A.; Patrick, W.H. Jr.; Gambrell, R.P. Estuarine Coastal Mar. Sci. 1978, 6, 21-35.
26. Gambrell, R.P.; Khalid, R.A.; Patrick, W.H. Jr. Environ. Sci. Technol. 1980, 14, 431-436.
27. Hong, J.; Wang, T. Environ. Sci. Lett. 1984, 6, 48-56.
28. Brannon, J.M.; Patrick, W.H. Jr. Environ. Sci. Technol. 1987, 21, 450-459.
29. Petersen, W.; Knauth, H.-D.; Pepelnik, R. Sci. Total Environ., 1990 97/98, 531-547.
30. Hong, J.; Calmano, W.; Wallmann, K.; Petersen, W.; Schroeder, F.; Knauth, D.-H.; Förstner, U. In Heavy Metals in the Environment; Farmer, J.G., Ed.; CEP Consultants: Edinburgh, 1991; pp 330-333.
31. Hennies, K. Diplomarbeit, Bayerische Julius-Maximilians-Universität zu Würzburg, 1991.

32. Calmano, W. Schwermetalle in kontaminierten Feststoffen. Verlag: TÜV Rheinland, Köln, 1989; 237 pp.
33. Calmano, W.; Ahlf, W.; Bening, J.-C. Hydrobiologia 1992, 235/236, 605-610.
34. Förstner, U.; Ahlf, W.; Calmano, W. Mar. Chem. 1989, 28, 145-158.
35. Calmano, W.; Ahlf, W.; Förstner, U. Environ. Geol. Water Sci. 1988, 11, 77-84.
36. Kersten, M.; Förstner, U. Mar. Chem. 1987, 22, 299-312.
37. Kornicker, W.K.; Morse, J.W. Geochim. Cosmochim. Acta 1991, 55, 2159-2171.
38. Hahne, H.C.H.; Kroontje, W. J. Environ. Qual. 1972, 2, 444-450.
39. Hem, J.D. Water Resour. Res. 1972, 8, 661-679.
40. Elderfield, H. Am. J. Sci. 1981, 281, 1184-1196.
41. Hart, B.T.; Davies, S.H.R. Aust. J. Mar. Freshwater Res. 1977, 28, 105.
42. Van den Berg, C.M.G.; Dharmvanij, S. J. Limnol. Oceanogr. 1984, 29, 1025-1036.
43. Jenne, E.A. In Molybdenum in the Environment; Chappell, W.R.; Petersen, K.K., Eds.; Dekker: New York NY, 1977; Vol. 2, pp 425-553.
44. Lion, L.W.; Altmann, R.S.; Leckie, J.O. Environ. Sci. Technol. 1982, 16, 660-666.
45. Davies-Colley, R.J.; Nelson, P.O.: Williamson, H.J. Environ. Sci. Technol. 1990, 18, 491-499.
46. McKenzie, R.M. Aust. J. Soil Res. 1980, 18, 61-73.
47. Gerth, J. Ph.D. thesis, Christian-Albrecht-Universität Kiel, 1985.
48. Tessier, A.; Rapin, F.; Carignan, R. Geochim. Cosmochim. Acta 1985, 49, 183-194.
49. Balistrieri, L.S.; Murray, J.W. Geochim. Cosmochim. Acta 1986, 50, 2235-2243.
50. Leckie, J.O. In The Importance of Chemical "Speciation" in Environmental Processes; Dahlem Conference 1986, Bernhard, M.; Brinckman, F.E.; Sadler, P.J., Eds.; Springer Verlag: Berlin - Heidelberg, 1986; 237 pp.
51. Murray, J.W. Geochim. Cosmochim. Acta 1975, 39, 505-519.
52. Jonasson, I.R. In The Fluvial Transport of Sediment-Associated Nutrients and Contaminants; Shear, H.; Watson, A.E.P., Eds.; IJC/PLUARG: Windsor, Ont., 1977, 255 pp.

RECEIVED October 11, 1993

EFFECTS OF SULFIDE-OXIDATION PROCESSES ON GROUND-WATER GEOCHEMISTRY

Chapter 22

Seasonal Variations of Zn/Cu Ratios in Acid Mine Water from Iron Mountain, California

Charles N. Alpers[1], D. Kirk Nordstrom[2], and J. Michael Thompson[3]

[1]Water Resources Division, U.S. Geological Survey, Federal Building, Room W-2233, 2800 Cottage Way, Sacramento, CA 95825
[2]Water Resources Division, U.S. Geological Survey, 3215 Marine Street, Boulder, CO 80303
[3]Geologic Division, U.S. Geological Survey, Branch of Volcanic and Geothermal Processes, MS-910, 345 Middlefield Road, Menlo Park, CA 94025

> Time-series data on Zn/Cu weight ratios from portal effluent compositions [$(Zn/Cu)_{water}$] at Iron Mountain, California, show seasonal variations that can be related to the precipitation and dissolution of melanterite [$(Fe^{II},Zn,Cu)SO_4 \cdot 7H_2O$]. Mine water and actively forming melanterite were collected from underground mine workings and chemically analyzed. The temperature-dependent solubility of Zn-Cu-bearing melanterite solid solutions was investigated by heating-cooling experiments using the mine water. Rapid kinetics of melanterite dissolution and precipitation facilitated reversed solubility experiments at 25°C. Non-reversed solubility data were obtained in the laboratory at 4° and 35°C and at ambient underground mine conditions (38° and 42°C). Copper is partitioned preferentially to zinc into melanterite solid solutions at all temperatures investigated. During the annual dry season, values of $(Zn/Cu)_{water}$ in the Richmond portal effluent increase to values between 8 to 13, consistent with formation of melanterite during this period. During the annual wet season, the onset of high discharge from the mine portals is characterized by a significant decrease in $(Zn/Cu)_{water}$ to values as low as 2. This phenomenon may be caused by dissolution of melanterite with values of $(Zn/Cu)_{solid}$ ranging from 1.5 to 3.5.

The rapid oxidation rate of sulfide minerals in abandoned mines and associated waste materials results in the formation of metal-rich sulfuric acid solutions that pose a threat to human health and the environment. Cycles of wetting and drying determined by local climate and hydrology can influence soil-water and ground-water composition (1). In particular, the progress of sulfide oxidation reactions and the composition of resulting acid mine drainage responds to wetting-drying cycles by transient storage of metals and acidity in the form of secondary sulfate minerals (2-5). Seasonal variations in the relative concentrations of metals, measured in certain mine waters, may be controlled in general by cycles of precipitation and dissolution of soluble secondary sulfate salts. Melanterite, a Zn-Cu-bearing hydrated ferrous-sulfate salt [(Fe^{II},Zn,Cu)SO$_4$·7H$_2$O], may be a major contributor to this phenomenon, based on its propensity to incorporate Cu in preference to Zn in solid solution and the observation that it is generally the first mineral to form when Fe(II)-rich acid mine water is evaporated.

The primary dissolved components of most acid mine drainage (AMD) are iron and sulfate, resulting from oxidation of the minerals pyrite (FeS$_2$), marcasite (FeS$_2$), and pyrrhotite (Fe$_{1-x}$S). Massive sulfide deposits hosted in volcanic rocks are a principal contributor to AMD problems because of the high concentrations of sulfide minerals and the relative lack of neutralizing agents found in carbonate-hosted massive sulfide deposits (for example, Leadville, Colorado). Other base metals typically associated with iron in massive sulfide deposits include Cd, Cu, Pb, and Zn. These elements occur most commonly in the minerals chalcopyrite (CuFeS$_2$), galena (PbS), and sphalerite [(Zn,Fe,Cd)S)]. Lead concentrations in sulfate-rich water are low because of the limited solubility of anglesite (PbSO$_4$) (6,7). As a result, Cd, Cu, and Zn, are most frequently the primary base metals of environmental concern in AMD from massive sulfide deposits. An improved knowledge of geochemical mechanisms controlling the composition of AMD is required to anticipate how metal concentrations in surface and ground waters would be affected by proposed remediation efforts at abandoned mine sites.

Melanterite, a common efflorescent salt, is often the first mineral deposited from solutions at sites of pyrite oxidation (2,8,9). Melanterite has a crystal structure that can accommodate large proportions of Cu and Zn substituting for Fe^{II} in solid solution. A general formula for the melanterite solid solution can be expressed as (Fe$^{II}_{1-x-y}$Zn$_x$Cu$_y$)SO$_4$·7H$_2$O. Maximum values of Zn and Cu mole fractions recorded for natural samples are x = 0.307 and y = 0.654, respectively (10-11). In this study, actively forming melanterite and associated mine water near to melanterite saturation were collected from underground mine workings. These water samples were used in heating-cooling experiments to show that Zn/Cu partitioning into melanterite can explain cyclic seasonal variations in mine-water chemistry. Although numerous other sulfate salts of Cu, Fe, and Zn form in similar environments, melanterite is a common, moderately soluble mineral in mine workings and mine tailings containing abundant pyrite (12). Therefore, the formation and dissolution of melanterite can have a dominant effect on the chemistry of mine effluent, which responds to seasonal cycles of wetting and drying.

Hydrogeologic Setting and Mining History at Iron Mountain, California

Mine drainage from the pyritic massive sulfide deposits at Iron Mountain, California (Figure 1), is the most acidic and metal-rich reported anywhere in the world (4,14). The AMD leaving the Richmond adit (Figure 2) ranges in pH from 0.02 to 1.5, and contains more than 100,000 mg/L of total dissolved solids (4,13). Effluent from mine workings mixes with metal-rich runoff and with seeps from sulfidic waste-rock piles and flows into Boulder and Slickrock Creeks, which are tributaries to Spring Creek (Figure 1). The metal-rich water of Spring Creek (pH ~3.0) is impounded in Spring Creek Reservoir behind the Spring Creek Debris Dam, built by the U.S. Bureau of Reclamation (USBR) in 1963 to prevent sediment buildup at the Spring Creek Powerplant. Effluent from Spring Creek Reservoir is released into Keswick Reservoir,

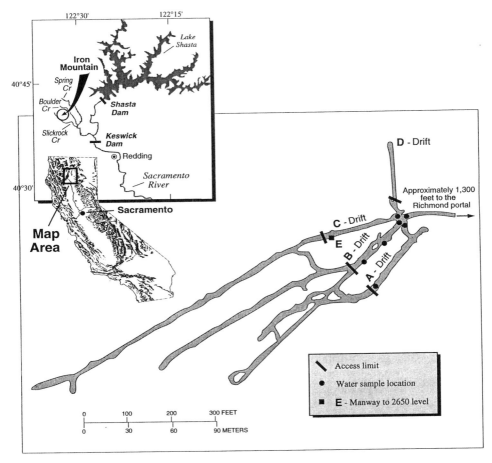

Figure 1. Location of Iron Mountain and plan view of mine workings on the 2600 level of the Richmond Mine. (Adapted from Alpers and others, 13)

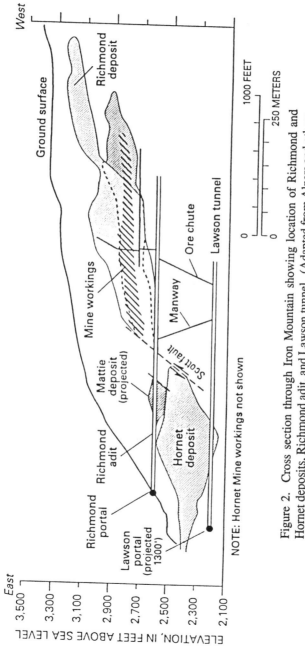

Figure 2. Cross section through Iron Mountain showing location of Richmond and Hornet deposits, Richmond adit, and Lawson tunnel. (Adapted from Alpers and others, 13)

where mixing occurs with relatively dilute water released from the Spring Creek Powerplant and from Shasta Dam. Releases from the Spring Creek Debris Dam, the Spring Creek Powerplant, and Shasta Dam are controlled by the USBR to ensure that dissolved metal concentrations at Keswick Dam meet criteria for Cu, Zn, and Cd designed to protect beneficial uses, including fish and wildlife habitats.

The Iron Mountain area was placed on the National Priority List of the U.S. Environmental Protection Agency (EPA) for cleanup under the Superfund program in September 1983 (15). Renovation of underground workings in the Richmond Mine by the EPA in 1989-91 provided access to stopes, drifts, and raises that had been inaccessible for 35 years. Samples of mine water and efflorescent sulfate minerals were collected and analyzed by the U.S. Geological Survey (USGS) to document the storage of metals and acidity in the underground workings. Drip water was found to give extreme measurements of pH, including the first negative pH values ever documented from such a setting (4,14). Measured values of pH from mine-tunnel drips were found to range from -0.45 to -3.4, using methods described in a later section.

The massive sulfide deposits at Iron Mountain (Figure 1) constitute a part of the West Shasta mining district. Host rocks are hydrothermally altered volcanic rocks of Devonian age, including the keratophyric Balaklala Rhyolite and the spilitized Copley Greenstone (16,17). A cross section through Iron Mountain (Figure 2) shows the presence of two sulfide deposits, the Richmond and Hornet deposits, that originally were a continuous lens of massive sulfide cut later by normal faults (16). The pyrite-rich Brick Flat deposit (not shown on Figure 2) was also originally connected to the Richmond deposit and offset by normal faulting (16). Gossan, the oxidized top part of a massive sulfide deposit consisting of oxidized iron minerals (goethite and hematite) and residual silica, is found on the top of Iron Mountain and also in the area of Boulder Creek near the Lawson portal (Figure 2). The gossan was mined for its silver content during the 1880's. The Hornet deposit was mined for its Cu content by underground methods from 1907 to 1926. The Richmond deposit was discovered about 1915 but was not mined for Cu and Zn on a large scale until the 1940's. The Brick Flat deposit was mined principally for its pyrite content by open-pit methods from 1950 to 1962. As part of a remedial action mandated by the EPA in 1985-86 (18), the Brick Flat pit was equipped with a liner and used as a disposal site for pyritic tailings. Future plans for site remediation include preparation of the Brick Flat pit for disposal of high-density sludge from lime treatment of the AMD now flowing from the Richmond and Lawson portals (19). The total metal load from the combined effluent from the Richmond adit and the Lawson tunnel represents about 75% of the Cu and 90% of the Cd and Zn released to surface waters in the area (20).

The massive sulfide deposits at Iron Mountain consist almost entirely (90 to 99 percent by volume) of pyrite, sphalerite, and chalcopyrite. Small amounts of bornite (Cu_5FeS_4), arsenopyrite (FeAsS), and tennantite-tetrahedrite (($Cu,Ag,Fe,Zn)_{12}(As,Sb)_4S_{13}$) have been reported (16). Trace amounts of gangue minerals, also present in the ore, include quartz, calcite, muscovite, and chlorite (16). The wall rock for the Richmond deposit is Balaklala Rhyolite, a dacitic unit which was affected by sea-floor metasomatism (keratophyrization; 17) and hydrothermal alteration. The wall rock consists mainly of albite, chlorite, epidote, quartz, and muscovite. The dissolution of these minerals contributes Al, Ca, K, Mg, and Si to the mine water in proportions determined by mass-balance calculations described by Alpers and others (13,21).

The AMD at Iron Mountain is among the most acidic ever reported because the massive sulfide deposits have remained in the unsaturated zone since underground mining activity ceased in the 1950's. The Richmond adit and Lawson tunnel at elevations of about 2,600 and 2,200 feet above sea level (Figure 2), provide effective drains for the Richmond and Hornet deposits. A comparison of discharge rates from the Richmond and Lawson portals (mine openings), based on weekly monitoring data from 1983 to 1991 (13), shows a generally positive correlation, but differences between the hydrologic response of the two portals are significant. Minimum flow rates during dry-season conditions are about twice as high at the Lawson portal as at the Richmond portal (Figure 3, Table I), which indicates that the Lawson tunnel receives a greater proportion of steady ground-water inflow than does the Richmond Mine. Discharge rates at both portals increase during the wet season, generally from October through April (Figure 4). The total range in discharge (Table I) is much greater for the Richmond portal (0.5 to 50 l/s) than for the Lawson portal (0.8 to 15 l/s), indicating that flows from the Richmond Mine are affected more by rapidly infiltrating surface waters.

Geochemistry of Acid Mine Drainage

Effluent from the Richmond and Lawson portals has been monitored weekly during the wet seasons and monthly during the dry seasons since 1983 by the California Regional Water Quality Control Board, in cooperation with the EPA. In addition to discharge measurements, water samples were analyzed for pH and dissolved Cd, Cu, and Zn. A summary of the data for discharge, pH, Cu, and Zn for 1983 to 1991 is presented in Table I. Metal concentrations generally are about three times higher in the Richmond portal effluent than in the Lawson portal effluent, and the pH is about 1 unit lower.

Table II presents analytical data for major and trace elements in four water samples from the Richmond Mine. Samples 90WA103, 90WA108, and 90WA109 were collected at extreme low-flow conditions during September 1990, after 5 consecutive years of below-average precipitation. Sample 91WA111 was collected in June 1991. Sample 90WA103 is from the Richmond portal effluent, which represents the sum of all water leaving the Richmond Mine workings on the 2600 level (Figures 1 and 2). Samples 90WA108, 90WA109, and 91WA111 represent drip water collected on the 2600 level either from timbers (108) or from melanterite stalactites (109 and 111).

The dissolved metals in the Richmond portal effluent clearly are derived from oxidation of sulfide minerals in the Richmond Mine; however, it remained unclear whether the dissolved constituents in the Lawson portal effluent were also derived from the Richmond mine and not from the more proximal Hornet deposit (Figure 2). Mass-balance analysis using 11 dissolved constituents in water samples collected from the Richmond and Lawson portals during low-flow conditions of September 1990 showed that at least 94% of the dissolved metals in the Lawson portal effluent were derived from sulfide oxidation in the Hornet deposit (13). Thus, the Richmond and Lawson portal effluents represent two distinct hydrogeochemical reactors in the Richmond and Hornet deposits, respectively.

The gravimetric concentration ratios of Zn to Cu from the two portal effluents

Table I. Characteristics of Portal Effluents, 1983-91 (--, no data)

	Richmond portal effluent		Lawson portal effluent	
	Mean	Range	Mean	Range
Overall				
Discharge (liters/second)	4.4	0.5 - 50	2.5	0.8 - 15
pH (units)	0.8	0.02 - 1.5	1.6	0.6 - 2.8
Zinc (mg/l)	1,600	700 - 2,600	540	280 - 840
Copper (mg/l)	250	120 - 650	90	50 - 150
Zinc/copper (weight ratio)	7.5	2 - 13	6.2	2 - 10
Wet season				
Discharge (liters/second)	--	6 - 12+	--	3 - 6+
Zinc/copper (weight ratio)	--	2 - 6	--	2.5 - 6
Dry season				
Discharge (liters/second)	~ 0.9	< 1.3	~ 1.3	< 1.9
Zinc/copper (weight ratio)	--	8 - 13	--	4 - 10

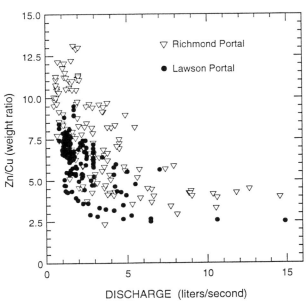

Figure 3. Gravimetric concentration ratios of Zn to Cu as a function of discharge rate. Richmond and Lawson portal effluents, 1983-91. (Adapted from Alpers and others, 13)

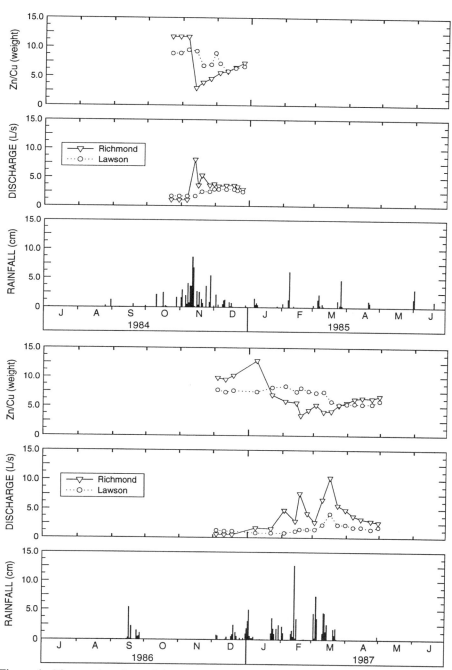

Figure 4. Time-series plots of the gravimetric concentration ratio of Zn to Cu and the discharge of Richmond and Lawson portal effluents, and rainfall at Shasta Dam. A. July 1984 to June 1985. B. July 1986 to June 1987. (Adapted from Alpers and others, 13).

Table II. Selected Analytical Data for Major and Trace Elements in Ground-Water Samples from the Richmond Mine (-- no data; data from Nordstrom and Alpers, written commun., 1990, and CH2M Hill, written commun., 1991)

Sample number	90WA103	90WA108	90WA109	91WA111
Water temperature (°C)	34.8	41.8	32.2	28 ±2
pH (field)	0.48 ±0.02	-0.35 ±0.05	-0.7 ±0.1	--
Element		(mg/l)		
Aluminum	2,210	4,710	6,680	6,470
Antimony	4.0	11	16	15
Arsenic(III)	8.14	27.2	38.0	74.0
Arsenic(total)	56.4	169	154	850
Barium	0.068	0.25	0.10	< 0.10
Beryllium	0.026	0.10	0.10	< 0.1
Boron	1.5	2.5	2.5	--
Cadmium	15.9	43.0	48.3	370
Calcium	183	424	330	443
Chromium	0.12	4.5	0.75	2.6
Cobalt	1.3	2.2	15.5	3.6
Copper	290	578	2,340	9,800
Iron(II)	18,100	50,800	79,700	--
Iron(total)	20,300	55,600	86,200	68,100
Lead	3.6	4.3	3.8	8.3
Magnesium	821	1,380	1,450	2,560
Manganese	17.1	41.8	42.1	119
Molybdenum	0.59	1.0	3.7	2.3
Nickel	0.66	2.8	2.9	6.3
Potassium	261	704	1,170	11.1
Selenium	0.42	2.1	2.1	< 2.8
Silica (as SiO_2)	170	69	34	--
Silver	0.16	0.49	0.65	0.70
Sodium	251	355	939	44.0
Strontium	0.25	0.30	0.49	--
Sulfur (as SO_4)	118,000	420,000	360,000	--
Thallium	0.44	0.15	0.15	1.6
Tin	1.6	6.5	15	--
Titanium	5.9	8.6	125	--
Vanadium	2.9	17	11	28
Zinc	2,010	6,150	7,650	49,300

show a similar range during wet-season conditions (values of 2 to 6); however, during dry seasons the values of $(Zn/Cu)_{water}$ in the Richmond portal effluent (8 to 13) are higher than those from the Lawson portal (4 to 10) (Table I, Figure 3). (The gravimetric Zn/Cu ratio is related to the molar Zn/Cu ratio by a multiplication factor of 0.97, so the two ratios are essentially interchangeable within analytical error.) The different values of $(Zn/Cu)_{water}$ during the dry season could be caused by different $(Zn/Cu)_{solid}$ values in the primary sulfide ore, but more likely are caused by a more complete fractionation of Zn relative to Cu because of a greater proportion of sulfate-salt precipitation in the Richmond Mine workings, where the more concentrated mine water approaches melanterite saturation.

Figure 4 compares the $(Zn/Cu)_{water}$ ratios for the Richmond and Lawson portal effluents with the discharge rates from these portals on time-series plots of the weekly monitoring data for parts of the 1984-1985 and 1986-1987 water years. The $(Zn/Cu)_{water}$ ratio is significantly reduced when the discharge increases at each portal in response to rainfall events, indicating the likelihood that the dissolution of melanterite and other sulfate salts during the wet season has a strong influence on the Zn/Cu ratio of the mine drainage.

Secondary Sulfate Minerals

A complex assemblage of sulfate salts has formed in the underground workings of the Richmond Mine. Ideal formulas for the sulfate minerals identified to date are listed in Table III. Melanterite and its less hydrated equivalents, rozenite and szomolnokite, are the only iron-bearing sulfate salts in Table III that contain exclusively Fe^{II}; all others are made up of Fe^{III} or a mixture of Fe^{II} and Fe^{III}. In the presence of atmospheric oxygen, the bacterium *Thiobacillus ferroxidans* will oxidize $Fe^{2+}_{(aq)}$ to $Fe^{3+}_{(aq)}$ very rapidly by facilitating electron transfers between dissolved O_2 and Fe^{II} (22,23). However, the principal reaction responsible for pyrite oxidation at pH < 5 in the sulfide deposits is accomplished by dissolved Fe^{III}:

$$FeS_2 + 14\ Fe^{3+} + 8\ H_2O \implies 15Fe^{2+} + 2\ SO_4^{2-} + 16\ H^+ . \qquad (1)$$

The rapidly flowing AMD in close proximity to the actively oxidizing massive sulfide in the Richmond mine has a high proportion of ferrous iron ($Fe^{2+}/Fe_{total} = 0.9$) (4), consistent with reaction (1). This rapidly flowing water is associated spatially with melanterite, which forms stalactites and stalagmites as the water drips into the 2600 level of the mine (Figures 1 and 2). The ferric-bearing sulfate minerals are found in areas of the Richmond Mine workings where water flows less rapidly or is stagnant and can be considered representative of hydrologic "dead-ends" where the Fe(II) has time to oxidize to Fe(III) before the minerals are formed (4). Therefore, the Fe^{II}-rich melanterites have the most effect on the composition of AMD leaving the mine workings through the Richmond portal because they form along the major flow paths. Additional evidence pointing to the influence of melanterite on AMD chemistry in the Richmond Mine workings is the observation that melanterite is the first mineral to form when the mine water is allowed to evaporate at 25°C (24; C. Maenz, written commun., 1993).

Table III. Ideal Formulas of Sulfate Minerals Found in the Richmond Mine

Mineral	Formula
Melanterite	$Fe^{II}SO_4 \cdot 7H_2O$
Rozenite	$Fe^{II}SO_4 \cdot 4H_2O$
Szomolnokite	$Fe^{II}SO_4 \cdot H_2O$
Römerite	$Fe^{II}Fe_2^{III}(SO_4)_4 \cdot 14H_2O$
Halotrichite-Bilinite	$Fe^{II}(Al,Fe^{III})_2(SO_4)_4 \cdot 22H_2O$
Copiapite	$Fe^{II}Fe_4^{III}(SO_4)_6(OH)_2 \cdot 20H_2O$
Voltaite	$K_2Fe_5^{II}Fe_4^{III}(SO_4)_{12} \cdot 18H_2O$
Kornelite	$Fe_2^{III}(SO_4)_3 \cdot 7H_2O$
Coquimbite	$Fe_2^{III}(SO_4)_3 \cdot 9H_2O$
Rhomboclase	$(H_3O)Fe^{III}(SO_4)_2 \cdot 3H_2O$
Gypsum	$CaSO_4 \cdot 2H_2O$
Chalcanthite	$CuSO_4 \cdot 5H_2O$

Field and Experimental Studies of Zn-Cu Partitioning Between Melanterite and Water

Methods of Sample Collection and Analysis. The partitioning of Zn and Cu into melanterite [$(Fe^{II}_{1-x-y}Zn_xCu_y)SO_4 \cdot 7H_2O$] was investigated using water and mineral samples collected from the Richmond Mine in September 1990 and June 1991. Sample 90WA109 was a collection of drips from an actively forming melanterite stalactite. Cooling of the unfiltered water sample from the *in situ* temperature of collection (~32°C) to ambient conditions in a ventilated part of the mine (~25°C) resulted in crystallization of several grams of melanterite in a 125 ml flint-glass bottle. Further cooling to 4°C yielded considerable additional melanterite crystals (approximately 30 grams per 125 ml original volume). Sample 90WA108 was collected from drips (~42°C) from timbers; however, melanterite did not form actively in the immediate vicinity at the time of collection in September 1990. (A subsequent visit to this site during October 1992 revealed newly formed melanterite.) Sample 90WA108 did not form melanterite when cooled to 25°C, but it did form a small amount when cooled to 4°C. Sample 91WA111, collected in June 1991, consisted of drips collected from melanterite stalactites in the Richmond adit near the Mattie ore body. The small volume collected (less than 1.0 ml) necessitated field dilution. Filtration of water samples 90WA108 and 90WA109 was attempted in the field using membranes with a pore diameter of 0.2 µm with negative results, probably because melanterite precipitated on the filter membranes and blocked the pores.

Dilution of unfiltered samples was carried out in the field using calibrated autopipettes (500 and 1,000 µL) and a graduated cylinder (100 ml). The uncertainty in absolute concentrations from this dilution procedure, estimated to be as much as 5%, did not effect the Zn/Cu ratios. Separately diluted splits of the water samples were acidified with ultrapure HCl and HNO_3. The HCl-acidified split was used for iron speciation analysis and the HNO_3-acidified split was used for other cations. An unacidified, diluted split was retained for anion analysis and one or more raw (unfiltered, undiluted, unacidified) splits also were collected, which formed secondary minerals on cooling. Samples were chilled in an ice chest and analyzed within 1 week for most constituents. Major and trace elements (Al, Ca, Cd, Cu, K, Mg, Na, Pb, Mn, Si, Sr, and Zn) were determined by inductively coupled plasma - atomic emission spectrometry (ICP-AES) at the CH2M Hill laboratory in Redding, California. Sample analysis conformed to the quality assurance and quality control procedures mandated by the EPA, which include spiked recoveries, duplicates, blanks, and chain-of-custody precautions. Fe(II) and Fe(total) were determined at the USGS laboratory in Menlo Park, California, by UV-visible spectrophotometry using ferrozine as the coloring agent; Fe(III) was calculated by difference. Sulfate was determined by ion chromatography; no other anions (Cl, NO_3, PO_4) were detected.

The following unstable parameters were measured in the field: temperature, specific conductance, pH, and Eh. Measurement of pH values less than 1.0 requires an acceptable definition of pH, standard buffers, acceptable levels of precision, and corrections for interferences, non-ideal behavior, and liquid-junction potentials. The Pitzer method was used as a basis for defining pH (25) and the PHRQPITZ program (26) was used to calculate values of pH as a function of H_2SO_4 molality. Plots for pH as a function of H_2SO_4 molality were constructed for three temperatures that span the measured temperatures in the underground workings of the Richmond Mine: 25°, 35°, and 47°C (4,14). Calibration curves based on 10 standardized sulfuric acid solutions were determined for the same temperature range. The sulfuric acid standard solutions were prepared by dilution from a primary standard of 10.196 M. The molality of the three most dilute standards was determined by titration with a Na_2CO_3 solution using an autotitrator. This method gave excellent agreement among the three measurements for the recalculated molality of the primary standard. The density of all 10 standard solutions was determined and compared with tabular data for sulfuric acid solutions (27) as an additional check. The electromotive force of the sulfuric acid standard solutions measured using two different combination pH electrodes was plotted as a function of the pH taken from the PHRQPITZ calculations. Corrections to the pH standard curves for Fe(II) and liquid junction potential are minimal (C. Ptacek and D. Blowes, written commun., 1993).

Samples of melanterite and other secondary sulfate minerals were collected in glass jars with mineral oil added to prevent dehydration. Mineral identification was confirmed by powder X-ray diffractometry. Mineral oil was removed using hexane prior to dissolution of about 50 mg of each sample in 100 ml of 0.1 M HCl. Chemical composition of the dissolved minerals was determined using the analytical procedures described previously for the water samples.

Methods for Heating-Cooling Experiments. The temperature dependence of Zn-Cu partitioning in the melanterite solid solution was investigated by heating-cooling experiments using mine water collected from the Richmond Mine. The compositions of the water samples used in the experiments are given in Table II. Samples 90WA108-F and 90WA109-F were raw splits of samples 90WA108 and 90WA109, respectively, that were kept in 125 ml flint-glass bottles. Four equivalent raw splits of sample 90WA109 (-X1 through -X4) were kept in 60 ml polypropylene bottles and run in parallel with sample 90WA109-F. Results of the duplicate experiments 90WA109-F and 90WA109-X4 are given in Tables IV and V and plotted in Figure 5(A-D); results from duplicate experiments 90WA109-X1, -X2, and -X3 (not shown) were very similar to those from 90WA109-X4.

The water samples were first placed in a temperature-controlled water bath at 25°C. Several days were allowed for equilibration. Samples of water and co-existing melanterite were collected and analyzed. This procedure was repeated at 35° and 43°C and then the water bath was cooled progressively to 35°, 25°, and 4°C; samples of water and melanterite were collected at each temperature, sample volume permitting. Water samples collected from the heating-cooling experiments were analyzed by atomic absorption spectroscopy for Cu, Zn, and Mg, by the ferrozine procedure for Fe(II) and Fe(total), and by ion chromatography for SO_4. Solid samples were digested in 0.1 M ultrapure HCl and analyzed by ICP-AES for Cu, Fe, Mg, Zn, and other elements.

Results: Temperature Dependence of Zn-Cu Partitioning in Melanterite. Results of the heating-cooling experiments are given together with pertinent data from field samples in Tables IV and V. Figure 6 shows the Zn/Cu weight ratio of the mine water as collected in the field (open squares) and the composition of the coexisting melanterite stalactites from which the water samples were collected (solid squares). In cases where melanterite was actively forming in the mine workings (water samples 90WA109 and 91WA111), the values of $(Zn/Cu)_{solid}$ in melanterite (solid squares) are

Table IV. Composition of Mine Waters Associated with Melanterite from the Richmond Mine Workings

Water sample number	Temp. (°C)	Conditions	Phases Present	Fe(total) (g/l)	Fe(II) (g/l)	Zn (g/l)	Cu (g/l)	Mg (g/l)	SO$_4$ (g/l)	Zn/Cu (mass)	density (g/ml)
90WA108	43	field	W	55.6	50.8	6.15	0.578	1.38	420	10.6	--
90WA108-X1	43	heating	W	52.5	46.4	5.96	0.558	1.31	--	10.7	1.241
90WA108-F	25	heating	G,W	52.7	44.7	6.22	0.482	1.33	261	12.9	1.250
	25	cooling	G,W	52.3	45.3	5.84	0.558	1.32	264	10.5	--
	4	cooling	G,M,W	40.3	33.1	5.60	0.370	1.20	242	15.1	1.231
90WA109	38	field	M,W	86.2	79.7	7.65	2.34	1.45	360	3.27	--
90WA109-F	43	heating	W	91.8	82.4	8.94	2.55	1.65	--	3.51	1.359
	35	heating	M,W	72.8	66.6	8.32	2.07	1.50	318	4.02	1.324
	35	cooling	M,W	85.6	77.4	--	--	--	--	--	--
	25	cooling	M,W	57.6	45.9	7.00	1.38	1.35	310	5.07	1.296
	25	cooling	M,W	57.7	48.9	7.22	1.50	1.39	306	4.81	--
	4	cooling	M,W	33.7	26.9	6.34	.798	1.17	283	7.94	1.234
90WA109-X4	43	heating	W	80.3	69.3	8.30	--	1.55	--	--	1.349
	35	heating	M,W	76.0	65.9	8.00	2.50	1.42	324	3.20	1.336
	25	heating	M,W	59.5	47.4	7.62	1.56	1.40	309	4.88	1.305
	25	cooling	M,W	60.9	50.1	7.42	1.65	1.32	306	4.50	--
	4	cooling	M,W	36.3	27.4	6.24	.808	1.17	281	7.72	1.263
91WA111	28	field	M,W	68.1	--	49.3	9.80	2.56	--	5.03	--

Water samples 90WA108 and 90WA109 collected on September 11 and 12, 1990, respectively; Water sample 91WA111 collected on June 26, 1991; --, no data; G, Gypsum; M, melanterite; W, water; g/l, grams per liter; g/ml, grams per milliliter.

Table V. Compositions of Melanterite from the Richmond Mine and from Heating-Cooling Experiments

Mineral sample number	Co-existing water sample number	Temperature of formation (°C)	Fe	Zn	Cu	Mg	Zn/Cu (moles)	Zn/Cu (mass)
			------ Mole fraction ------					
90WA108-FT4	90WA108-F	4 (lab)	0.959	0.032	0.009	--	3.42	3.52
90RS118-1	90WA109	38 (field)	0.929	0.036	0.019	0.016	1.90	1.95
90RS118-2	90WA109	38 (field)	0.930	0.036	0.018	0.016	1.97	2.03
90RS118-A1	90WA109	38 (field)	0.931	0.034	0.019	0.015	1.78	1.83
90WA109-FT3	90WA109-F	35 (lab)	0.923	0.038	0.021	0.017	1.78	1.83
90WA109-FT2	90WA109-F	25 (lab)	0.917	0.039	0.026	0.018	1.50	1.54
90WA109-FT4	90WA109-F	4 (lab)	0.912	0.041	0.027	0.020	1.48	1.52
90WA109-X4T4	90WA109-X4	4 (lab)	0.917	0.038	0.027	0.018	1.44	1.48
91RS209-A1	91WA111	28 (field)	0.647	0.219	0.106	0.029	2.08	2.14
91RS209-B1	91WA111	28 (field)	0.525	0.294	0.137	0.044	2.15	2.21
91RS209-B2	91WA111	28 (field)	0.544	0.268	0.147	0.042	1.82	1.87

Compositions of coexisting water samples in Table IV; --, no data.

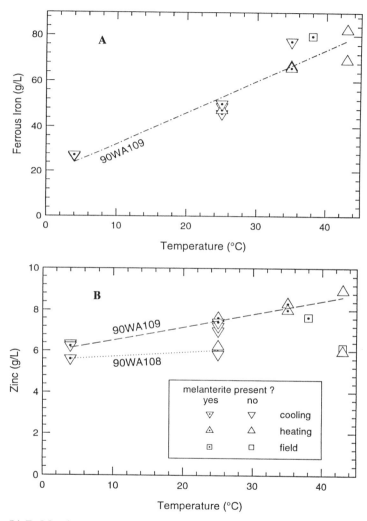

Figure 5A,B. Metal concentrations in mine water samples 90WA108 and 90WA109 as a function of temperature in heating-cooling experiments. Open symbols with dots indicate presence of melanterite. A. Fe(II) concentrations (sample 90WA109 only). B. Zn concentrations. *Continued on next page.*

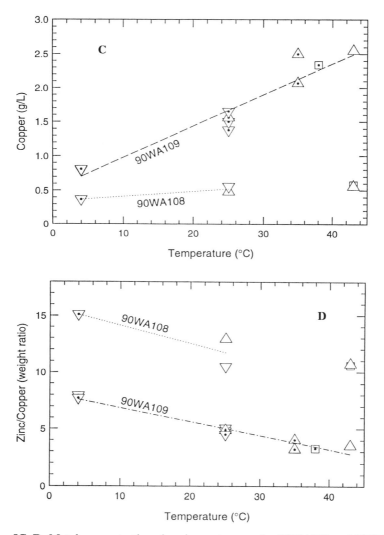

Figure 5C, D. Metal concentrations in mine water samples 90WA108 and 90WA109 as a function of temperature in heating-cooling experiments. Open symbols with dots indicate presence of melanterite. C. Cu concentrations. D. Gravimetric concentration ratios of Zn to Cu.

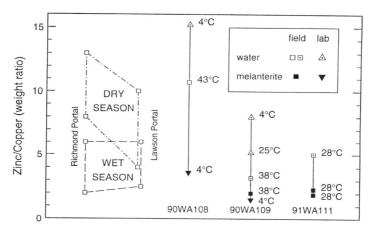

Figure 6. Gravimetric concentration ratios of Zn to Cu in melanterite and coexisting mine water from Richmond Mine. Filled symbols indicate melanterite, open symbols indicate mine water; open symbols with dots indicate water samples with melanterite present. Squares represent samples collected *in situ* at temperature indicated; triangles represent products from heating-cooling experiments using mine water samples at 4° and 25°C. Ranges of values of concentration ratio of Zn to Cu for Richmond and Lawson portal effluents during wet and dry seasons shown for comparison.

lower than values of $(Zn/Cu)_{water}$ for coexisting mine water (open squares with dots). Raw water samples 90WA109-F and 90WA109-X4 were saturated with melanterite solid solution at the time of collection; melanterite formed on cooling to 25° and 4°C. The compositions of the residual water and the melanterite formed on cooling are given in Tables IV and V, and the Zn/Cu ratios are plotted in Figures 5 and 6. As the temperature decreased in experimental runs 90WA109-F and 90WA109-X4, the $(Zn/Cu)_{solid}$ ratio of the precipitates decreased (in 90WA109-F) from an initial values between about 1.8 and 2.0 at 38°C to a value of 1.5 at 4°C (Table V); the values of $(Zn/Cu)_{water}$ increased from an initial value of 3.3 at 38°C (in 90WA109) to a value of 7.9 in the residual water (of 90WA109-F) at 4°C (Table IV). The Zn/Cu ratios of coexisting water and melanterite are shown in Figure 6. At the lower temperatures, more extreme differences in Zn/Cu between water and solid were observed. Because each experiment was a closed system, except for the subsamples collected for analysis (1 ml for each analysis from 125 ml total volume), Cu was depleted to a great degree in the water as the solid melanterite crystallized on cooling (Figure 5C). These results indicate that Zn/Cu partitioning favors Cu into the solid phase at all temperatures, and that this tendency is accentuated at lower temperatures.

The compositional data for Fe(II), Zn, and Cu in the experimental solutions are plotted in Figures 5(A-D) as a function of temperature. Dots in open symbols in Figure 5 indicate the presence of solid melanterite. Vertical triangles indicate that the temperature was reached by heating and inverted triangles indicate cooling. Reversed solubility data (temperature approached from both directions) were obtained only at 25°C. Figure 5A indicates the strong temperature dependence of melanterite solubility, as documented by Reardon and Beckie (28). Figures 5B and 5C show a relatively minor decrease in Zn concentrations and a strong decrease in Cu concentrations, respectively, with decreasing temperature in the residual mine water as melanterite crystallized on cooling. The significant increase in the Zn/Cu ratio in the residual water at the lower temperatures (Figure 5D) indicates that Cu is partitioned into the melanterite solid solution preferentially to Zn.

Rapid kinetics of melanterite dissolution and crystallization facilitated collection of reversed solubility data at 25°C. It is assumed that equilibrium was reached between the aqueous phase and the surface of the melanterite. The homogeneity of the melanterite was not tested because poor polishing properties and its tendency to dehydrate under vacuum made this material not suitable for electron microprobe analysis. Therefore, it is not known whether or not the melanterite crystals came to homogeneous equilibrium with the aqueous phase at each temperature.

Also shown in Figure 6 are the ranges in values of $(Zn/Cu)_{water}$ for effluent mine water from the Richmond and Lawson portals. The lower values of $(Zn/Cu)_{water}$ for mine effluent in the wet season are similar to the values of the melanterite crystals. The higher values of $(Zn/Cu)_{water}$ during the dry season for both portals are similar to the values of the residual waters after melanterite precipitation. These similarities indicate that cyclic precipitation and dissolution of Zn-Cu-bearing melanterite solid solutions during the annual wetting-drying cycles may be at least partly responsible for the observed annual cycles in the Zn/Cu ratios in mine effluents.

Conclusions

Acid mine water issuing from the Richmond and Lawson portals of Iron Mountain show significant and systematic variations in Zn/Cu weight ratios related to seasonal patterns of climate. High discharges during the wet season result in low Zn/Cu ratios (values of 2 to 6 in both portal effluents) whereas Zn/Cu ratios obtained during the dry season are consistently higher (values of 8 to 13 in the Richmond portal effluent and 4 to 10 in the Lawson portal effluent). The underground mine workings contain abundant masses of highly soluble, efflorescent salts. Melanterite, a common efflorescent mineral and usually the first to form during evaporation of Fe(II)-rich mine water, is a good candidate to affect changing patterns of Zn and Cu concentrations in this AMD because of its large compositional range.

An experimental study, designed to investigate the temperature-dependent solubility of melanterite precipitated from actual mine water samples of low pH (-0.7 to 3.0), demonstrated strong preferential partitioning of dissolved Cu into the melanterite solid solution. This partitioning provides a mechanism to explain the correlation of Zn/Cu variations in the portal effluent with climatic fluctuations. The Zn/Cu ratios in the melanterite are relatively low (2 to 4) due to the enrichment in Cu. When melanterite dissolves during high flow conditions, the effluent mine water ratios decrease to similar low values. The dry season conditions promote formation of melanterite, causing a decrease in Cu concentration of the residual mine water and a marked increase in Zn/Cu ratios in the AMD to values above 4 (Lawson portal) and above 8 (Richmond portal).

Additional work is needed to establish the thermodynamic properties of both melanterite solid solutions and the concentrated acid-sulfate brines from which they precipitate. The specific ion interaction approach of Pitzer (25) is the most appropriate method by which to compute the non-ideal behavior of the most concentrated AMD (e.g. 28). When thermodynamic data become available for the interactions of Zn and Cu with the Fe(II)-SO_4-HSO_4-H_2O system, it will be possible to derive mixing parameters for melanterite solid solutions of the general formula $(Fe^{II}_{1-x-y}Zn_xCu_y)SO_4 \cdot 7H_2O$. Until then, the empirical data presented here will be useful in defining the solubility relations in the $ZnSO_4$-$CuSO_4$-$Fe(II)SO_4$-H_2SO_4-H_2O system.

Acknowledgments

The authors thank Rick Sugarek of the U.S. Environmental Protection Agency for providing financial support and access to the underground workings at Iron Mountain. We also thank the following individuals for assistance with underground sampling: Dale Smith (CH2M Hill), Larry Salhaney (formerly of Engineering International), Cathy Maenz (McGill University), and Nick Waber (NAGRA, Switzerland). Our interpretation of hydrology and geochemistry at Iron Mountain benefitted from discussions with Rick Sugarek (EPA); Daryl Greenway, Dick Coon, Jim Mavis, and Pete Lawson (CH2M Hill); and Jo Burchard (USGS). The manuscript was improved by thoughtful reviews by R. Fujii (USGS), J. Hem (USGS), E. Reardon (Univ. of Waterloo), and an anonymous reviewer. A portion of this work was done while CNA was a National Research Council/Resident Research Associate at the U.S. Geological Survey in Menlo Park, CA. Funding was also obtained from the National Science and Engineering Research Council (NSERC) of Canada and from the Fonds pour la Formation de Chercheurs et L'Aide à la Recherche (FCAR) of Québec, while C.N. Alpers was an Assistant Professor at McGill University.

Literature Cited

1. Drever, J.I.; Smith, C.L. Amer. Jour. Sci., 1978, 278, 1448-1454.
2. Nordstrom, D.K.; Dagenhart, T.V., Jr. Geological Society of America, Abstracts with Programs, 1978, 10, 464.
3. Dagenhart, T.V., Jr., M.Sc. Thesis, Univ. Virginia, Charlottesville, VA, 1980
4. Alpers, C.N.; Nordstrom, D.K. In Proceedings, Second International Conference on the Abatement of Acidic Drainage; Ottawa: MEND (Mine Environment Neutral Drainage); 1991, 2, 321-342.
5. Cravotta, C.A., III. In this volume.
6. Dubrovsky, N.L. Ph.D. Thesis, Univ. of Waterloo, Waterloo, Ontario, 1986, 373 p.
7. Blowes, D.W.; Reardon, E.J.; Jambor, J.L.; Cherry, J.A. Geochimica et Cosmochimica Acta, 1991, 55, 965-978.
8. Buurman, P. Geologie en Mijnbouw, 1975, 54, 101-105.
9. Nordstrom, D.K., In Acid Sulfate Weathering; Kittrick, J.A.; Fanning, D.S.; Hossner, L.R., Eds.; Soil Science Soc. of Amer.: Madison, WI; Spec. Pub. 10, 1982, pp. 37-55.
10. Palache, C.; Berman, H.; Frondel, C. Dana's System of Mineralogy; J. Wiley & Sons: New York, 1951.
11. Glynn, P.D., In Chemical Modeling in Aqueous Systems II; Melchior, D.C. and

Bassett, R.L., Eds.; ACS Symposium Series, No. 416; American Chemical Society: Washington, DC, 1990; 74-86.
12. Blowes, D.W.; Jambor, J.L. Applied Geochemistry, 1990, 5, 327-346.
13. Alpers, C.N.; Nordstrom, D.K.; Burchard, J.M. U.S. Geol. Survey Water-Resources Investigations Report 91-4160, 1992, 173 p.
14. Nordstrom, D.K.; Alpers, C.N.; Ball, J.W. Geological Society of America, Abstracts with Programs, 1991, 23, (3), A383.
15. Biggs, F.R. U.S. Bureau of Mines Information Circular 9289, 1991, 15 p.
16. Kinkel, A.R., Jr.; Hall, W.E.; Albers, J.P. U.S. Geol. Survey Professional Paper 285, 1956, 156 p.
17. Reed, M.H. Economic Geology, 1984, 79, 1299-1318.
18. U.S. Environmental Protection Agency. Final remedial investigation report, Iron Mountain mine, near Redding, California, Prepared by CH2M Hill: Redding, CA; EPA WA No. 48.9L17.0, 1985.
19. U.S. Environmental Protection Agency. Public Comment Remedial Investigation Report, Boulder Creek Operable Unit, Iron Mountain Mine, Redding, California; Prepared by CH2M Hill, Redding, CA; EPA WA No. 31-01-9N17, 1992.
20. U.S. Environmental Protection Agency. Public Comment Feasibility Study, Boulder Creek Operable Unit, Iron Mountain Mine, Redding, California; Prepared by CH2M Hill, Redding, CA; EPA WA No. 31-01-9N17, 1992.
21. Alpers, C.N.; Nordstrom, D.K. In Acid Mine Drainage - Designing for Closure; Gadsby, J.W.; Mallick, J.A.; Day, S.J., Eds.; Bi-Tech Publishers Ltd.: Vancouver, BC, 1990, 23-33.
22. Nordstrom, D.K. Selected papers in the Hydrologic Sciences 1985, U.S. Geological Survey Water-Supply Paper 2270, 1985, 113-119.
23. Suzuki, I.M., Takenuchi,____. In this volume
24. Alpers, C.N.; Maenz, C.; Nordstrom, D.K.; Erd, R.C.; Thompson, J.M. Geol. Soc. Amer. Abstracts with Programs 1991, 23, (5), A382.
25. Pitzer, K.S.; Roy, R.N.; Silvester, L.F. Jour. Amer. Chem. Soc., 1977, 99, 4930-4936.
26. Plummer, L.N., Parkhurst, D.L.; Fleming, G.W.; Dunkle, S.A. U.S. Geol. Survey Water-Resources Investigations Report 88-4153, 1988.
27. Lide, D.R., Ed., CRC Handbook of Chemistry and Physics, 73rd edition; CRC Press: Ann Arbor, MI; 1992, p. 15-17.
28. Reardon, E.J.; Beckie, R.D. Geochimica et Cosmochimica Acta, 1987, 51, 2355-2368.

RECEIVED September 23, 1993

Chapter 23

Secondary Iron-Sulfate Minerals as Sources of Sulfate and Acidity

Geochemical Evolution of Acidic Ground Water at a Reclaimed Surface Coal Mine in Pennsylvania

C. A. Cravotta III

U.S. Geological Survey, 840 Market Street, Lemoyne, PA 17043

> Despite negligible concentrations of dissolved oxygen (O_2) in ground water, concentrations of sulfate (SO_4^{2-}) and acidity increase with depth along paths of ground-water flow from the water table in mine spoil through underlying bedrock at a reclaimed surface coal mine in the bituminous field of western Pennsylvania. This trend can result from the oxidation of pyrite (FeS_2) in the unsaturated zone, transport of oxidation products, and additional oxidation in the saturated zone. FeS_2 can be oxidized by O_2 or by ferric ions (Fe^{3+}) in the absence of O_2. In acidic water, Fe^{3+} is more soluble than O_2. Iron-sulfate hydrates, including römerite [$Fe^{II}Fe_2^{III}(SO_4)_4 \cdot 14H_2O$], copiapite [$Fe^{II}Fe_4^{III}(SO_4)_6(OH)_2 \cdot 20H_2O$], and coquimbite [$Fe_2^{III}(SO_4)_3 \cdot 9H_2O$], are associated with FeS_2 in coal from the mine. These soluble salts can form on the surface of oxidizing FeS_2 in unsaturated mine spoil, coal, and overburden, and can dissolve in ground water thereby releasing SO_4^{2-} and Fe^{3+}. Subsequent oxidation of pyritic sulfur by Fe^{3+} and (or) hydrolysis of Fe^{3+} will produce acid under water-saturated conditions, even if O_2 is not present in the system.

Acidic mine drainage (AMD) can result from the accelerated oxidation of iron-disulfide minerals (FeS_2), mainly pyrite, that are exposed to oxygen (O_2) during coal mining (_1-6_). Accordingly, various schemes have been proposed to reduce the contact between O_2 and FeS_2, including submerging pyritic mine spoil in water (_7_, _8_) or adding organic waste as an O_2-consumptive barrier (_9_). Because oxidation is rapid, however, secondary iron-sulfate minerals can form during mining and can supply acidity and solutes even in anoxic systems.

During 1986-89, the U.S. Geological Survey, in cooperation with the Pennsylvania Department of Environmental Resources, conducted an investigation at a reclaimed surface coal mine in western Pennsylvania to evaluate effects on ground-water quality from the addition of municipal sewage sludge to mine topsoil. The primary goal of the sludge application was to promote revegetation; a secondary goal was to create an O_2-consumptive barrier. If O_2 in recharge water could be

This chapter not subject to U.S. copyright
Published 1994 American Chemical Society

depleted by decay of the sludge, FeS_2 oxidation in underlying mine spoil could be abated. Although concentrations of dissolved O_2 (DO) were negligible [less than 0.034 mmol/kg (millimoles per kilogram)] in ground water from sludge-treated and untreated mine spoil, concentrations of sulfate (SO_4^{2-}) and acidity increased with depth below the water table. This trend can result from the oxidation of FeS_2 in the unsaturated zone, transport of oxidation products, and additional oxidation in the saturated zone. This paper reviews geochemical reactions involving FeS_2 and iron-sulfate hydrate minerals that are likely to produce acidic ground water in O_2-limited systems, and presents results of geochemical mass-balance calculations. The purpose is to evaluate whether availability of O_2 is a limiting factor in acidity generation along ground-water flow paths through mine spoil and bedrock. This evaluation incorporates revised sulfur data used for previous modeling at the site (10), which allows for more narrow constraint of the mass-balance models.

Secondary Iron-Sulfate Minerals as Sources of Acidity

The complete oxidation of pyrite in weakly acidic to neutral systems (pH 4-7) (3-6) can be written as

$$FeS_2 + 3.75\ O_2 + 3.5\ H_2O \rightarrow Fe(OH)_3 + 2\ SO_4^{2-} + 4\ H^+. \quad (1)$$

In acidic weathering environments, however, iron-sulfate hydrate minerals can form instead of ferrihydrite [$Fe(OH)_3$] (5, 6, 11-16). For example, römerite can form by the oxidation of pyrite,

$$3\ FeS_2 + 11\ O_2 + 16\ H_2O \rightarrow Fe^{II}Fe_2^{III}(SO_4)_4 \cdot 14H_2O + 2\ SO_4^{2-} + 4\ H^+. \quad (2)$$

For each mole of FeS_2 oxidized, 1/3 of the quantities of dissolved SO_4^{2-} and H^+ produced in reaction 1 are produced in reaction 2; the remaining 2/3 of the "acidity" is stored as unhydrolized, partly oxidized iron (Fe) in the solid phase. In addition to römerite, copiapite [$Fe^{II}Fe_4^{III}(SO_4)_6(OH)_2 \cdot 20H_2O$] and coquimbite [$Fe_2(SO_4)_3 \cdot 9H_2O$] are associated with FeS_2 in high-S coal [S greater than 2 wt % (weight percent)] from the study area. These and other iron-sulfate hydrate minerals can form as oxidation products on the surface of iron-sulfide minerals in stockpiled coal, reclaimed mine spoil, and dewatered, unmined rock. They also can precipitate from oxidizing and evaporating Fe-and-SO_4-rich solutions at AMD discharge zones or at the capillary fringe in the unsaturated zone (6, 14-16).

Because iron-sulfate hydrate minerals are highly soluble (6), they provide an instantaneous source of acidic water upon dissolution and hydrolysis, as shown by the dissolution of römerite

$$Fe^{II}Fe_2^{III}(SO_4)_4 \cdot 14H_2O \rightarrow 2\ Fe(OH)_3 + Fe^{2+} + 4\ SO_4^{2-} + 6\ H^+ + 8\ H_2O. \quad (3)$$

Subsequent oxidation of the ferrous ion (Fe^{2+}) and hydrolysis of the ferric ion (Fe^{3+}) at pH greater than 2 will produce additional acidity according to the overall reaction

$$Fe^{2+} + 0.25\ O_2 + 2.5\ H_2O \rightarrow Fe(OH)_3 + 2\ H^+. \quad (4)$$

Hence, iron-sulfate hydrate minerals are important as both sinks and sources of AMD by storing acid, Fe, and SO_4^{2-} in a solid phase during dry periods and by releasing the solutes when dissolved during wet periods (10, 14, 17).

Storage of Fe^{3+} in sulfate minerals and its subsequent release into ground water also are an important aspect of oxidation below the water table, because coupled Fe^{3+} reduction and FeS_2 oxidation can occur in anoxic systems (2, 18-22) as follows:

$$FeS_2 + 14\ Fe^{3+} + 8\ H_2O \rightarrow 15\ Fe^{2+} + 2\ SO_4^{2-} + 16\ H^+. \quad (5)$$

Thus, as the water table rises or as recharge water percolates through the unsaturated zone of mine spoil or unmined rock, dissolution of iron-sulfate minerals and subsequent ferric oxidation of FeS_2 can result. For example, the following overall reaction of pyrite and römerite generates acid in the absence of O_2:

$$FeS_2 + 7\ Fe^{II}Fe^{III}_2(SO_4)_4 \cdot 14\ H_2O \rightarrow 22\ Fe^{2+} + 30\ SO_4^{2-} + 16\ H^+ + 90\ H_2O. \quad (6)$$

Other iron-sulfate hydrate minerals, including copiapite and coquimbite, can also provide Fe^{3+} for FeS_2 oxidation in anoxic, acidic systems. Ferric oxyhydroxides are not a likely source because they are relatively insoluble and will neutralize acid upon dissolution. For example, the overall reaction combining the oxidation of pyrite and dissolution of ferrihydrite consumes H^+ in the absence of O_2:

$$FeS_2 + 14\ Fe(OH)_3 + 26\ H^+ \rightarrow 15\ Fe^{2+} + 2\ SO_4^{2-} + 34\ H_2O. \quad (7)$$

Equivalent reactions with goethite or hematite also consume H^+.

Mine Location, Geology, and Hydrology

The study area consists of a reclaimed 60-ha (hectare) surface coal mine in the bituminous field of western Pennsylvania that includes a steep-sided, broad hilltop bounded by two small tributary streams that flow northward and westward (Figure 1). Acidic seepage from the mine-spoil banks and the exposed, underlying bedrock flows into these tributaries. Land-surface altitudes at the mine range from greater than 450 m (meters) to less than 360 m above sea level (National Geodetic Vertical Datum of 1929).

Three 0.8-m-thick seams of bituminous coal were mined by use of a dragline and front-end loader to remove the overburden and coal, respectively. Because about half of the area was mined only for the lower Kittanning coal (hilltop) and the other half only for the upper and lower Clarion coals (hillside), a highwall was created that separated upper and lower benches (Figure 2). During mining, the mine was backfilled with overburden and coal-waste material. At the completion of mining in 1981, the mine spoil was regraded into a terraced configuration corresponding to the upper and lower benches. Thickness of the mine spoil averages about 15 m on the upper bench and 21 m on the lower bench. During spring 1986, a mixture of composted wood chips and municipal sewage sludge was spread at a density of about 135 T/ha (tonnes per hectare) over 24 ha of the mine-spoil surface on the lower bench. The sludge-covered area was seeded with grasses and legumes (Figure 1).

After reclamation, nests of monitor wells were installed in drill holes to evaluate the geology, hydrology, and hydrochemistry of the study area (Figure 1). In particular, two well nests were located along a line roughly parallel to the principal direction of ground-water flow (Figures 1 and 2). Nest 14 consists of 4 wells (14, 14A, 14B, 14C) in the untreated zone and nest 15 consists of 3 wells (15, 15A, 15B) in the sludge-covered zone. Each well was constructed of polyvinyl chloride (PVC) pipe in separate drill holes.

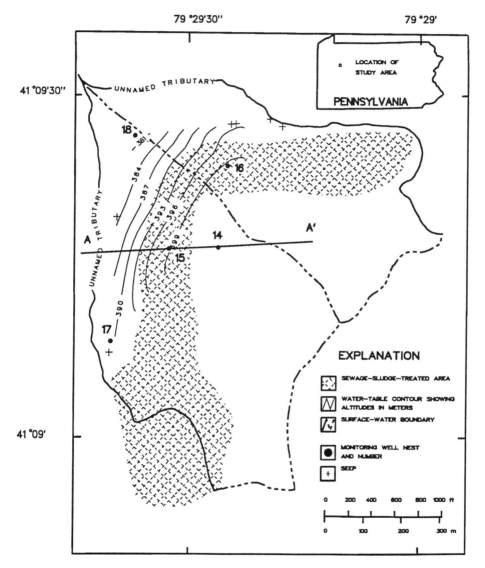

Figure 1. Location of reclaimed surface coal mine, sewage-sludge-covered area, hydrologic monitoring sites, and configuration of water table. Line A-A' is trace of geologic section shown in Figure 2.

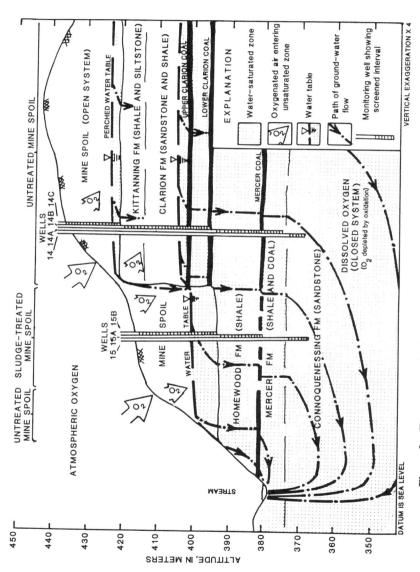

Figure 2. East-west geologic section through the mine showing water tables and simulated flow directions reported by Saad and Cravotta (10, 25). Trace of section shown in Figure 1.

Geology. The study area is underlain by a series of sandstones, shales, and coals (Figure 2) of Pennsylvanian age that comprise (from oldest to youngest) the Connoquenessing, Mercer, and Homewood Formations of the Pottsville Group, and the Clarion and Kittanning Formations of the Allegheny Group (23, 24). Fine-grained, white-to-gray sandstone predominates in the overburden, and gray, interbedded shale and siltstone lie above and below the coals. Limestone is not present at the mine.

Pyrite, marcasite, and iron-sulfate minerals are present primarily in the coals, which contain 3-17 wt % total S and 2-20 wt % Fe, and have molar ratios of 1.2-2.1 inorganic S/Fe (10). The iron-sulfate minerals, including römerite, copiapite, coquimbite, and jarosite, are present as white and yellow efflorescences concentrated mainly in the upper and lower Clarion coals, but also are present in lesser quantities in the Mercer coals. Most FeS_2 oxidation probably occurred during mining when the water table in the highwall was lowered to a level at or below the base of the lower Clarion coal (Figure 2). The local abundance of FeS_2 and iron-sulfate minerals in the Clarion coals indicates that both spoiled and unmined, high-S coals can be sources of acid, Fe^{3+}, and SO_4^{2-}.

Ground-Water Hydrology. Ground water in the mine spoil and adjacent bedrock is recharged within the study area and flows westward and northward to streams that form the mine boundary (Figures 1 and 2). The mine spoil on the upper bench forms a perched aquifer with a saturated thickness of about 2.3 m. The spoil on the lower bench has a saturated thickness of about 5 m and is part of the water-table aquifer that includes the adjacent unmined rock in the highwall (Figures 1 and 2). Figure 2 shows simulated, steady-state ground-water-flow paths in the mine spoil and bedrock.

Ground-water-flow simulations were conducted by Saad and Cravotta (10, 25) using the steady-state option of the finite-difference computer program, MODFLOW (26). The resultant flow paths were computed and plotted using the program, MODPATH (27). The water table was simulated as a constant-flux (recharge) boundary; the eastern and western divides and base of the flow system were simulated as no-flow boundaries. Hydraulic conductivity [K, in cm/s (centimeter per second)] was estimated from rising-head slug tests (28) and was varied as follows: unconsolidated spoil and alluvium, $K = 10^{-2.6} - 10^{-2.0}$; coal and sandstone, $K = 10^{-3.4} - 10^{-3.0}$; and shale and underclay, $K = 10^{-6.0} - 10^{-5.6}$. Values of K were adjusted within these ranges until the differences between simulated and median values of measured heads were within 1 m. Figure 2 shows that simulated flow paths are dominantly horizontal in permeable units (spoil, coal, and sandstone) and downward in confining layers (shale and underclay). However, ground-water flow is upward near the stream-discharge zone at the western boundary of the modeled area.

Hydrochemical Data Collection and Evaluation

From December 1986 through September 1989, hydrochemical data were collected quarterly from the seven monitoring wells in nests 14 and 15 (Figure 2). Measurements of temperature, pH, DO, and platinum-electrode potential (Eh) of ground water were conducted at the wellhead following methods of Wood (29). Ground-water samples for analysis of cations (Fe, Mn, Al, Mg, Ca, Na, K) and silica (SiO_2) were filtered through 0.45-micrometer pore-size filters, transferred to acid-rinsed polyethylene bottles, and preserved with nitric acid. Unfiltered samples for analysis of SO_4^{2-}, acidity, and alkalinity were stored in polyethylene bottles and preserved on ice. Concentrations of cations and SiO_2 were measured by inductively coupled plasma emission or atomic absorption spectrometry, and alkalinity (endpoint pH = 3.9) and acidity (endpoint pH = 8.3) were measured by titration following methods of Skougstad et al. (30). Concentration of SO_4^{2-} was measured by colorimetry, by use of the methylthymol blue method (30), which has an error in precision of about 10% attributed to dilution steps (Lynn Shafer, Pennsylvania Department of Environmental Resources, Bureau of Laboratories, oral commun., 1992). Charge balance errors were as large as 12.4%.

Table I reports the median and range of values of temperature, pH, Eh, and solute concentrations in the water from each well. Medians in Table I were used in the computer program WATEQ4F (31) to compute mineral saturation indices and in NEWBAL (32) to compute mass balance. Medians were used because travel times along ground-water flow paths were not precisely known; hence, the chronological relation of individual samples from different locations was difficult to establish. First, SO_4^{2-} concentrations were adjusted by 2-13% of the median values to achieve a charge balance; this adjustment attempts to correct for potential errors associated with the analytical method for SO_4^{2-} and the use of median values instead of reported data for individual water samples. After adjusting SO_4^{2-} concentrations, WATEQ4F was used to compute concentrations of the solutes, including Fe^{2+} and total CO_2, as mmol/kg. Computations of total CO_2 concentrations required alkalinity data, which were not available for waters with pH less than 3.9; thus, a concentration of C = 6.78 mmol/kg (corresponding to P_{CO_2}, of approximately 0.1 atmosphere) was assumed. Lastly, solute concentrations were used to compute the operational redox state (RS), which is used in NEWBAL to account for electron transfer (32).

The saturation index (SI) indicates the thermodynamic potential for dissolution or precipitation of a solid phase by the ground water (3, 33-35) and provides a constraint for mass-balance models indicated by:

Initial Solution 1 + Initial Solution 2 + Reactant Phases - Product Phases → Final Solution.(8)

The initial and final solutions are the molal compositions of waters from respective upflow and downflow zones along a flow path. More than one initial solution indicates mixing of source waters. Equation 8 can be simplified to include only one initial solution. The reactant and product phases are solid or aqueous phases that can feasibly react to produce or remove elements and balance electrons. If SI is a negative number, the water is undersaturated with the solid phase and can dissolve it as a reactant. If SI is a positive number, the water is supersaturated and cannot dissolve the solid phase, but can potentially precipitate it as a product. Table II shows the computed values of SI for selected solid phases. For some minerals, such as muscovite, even though SI is positive, precipitation under surface temperature and pressures is not likely because of kinetic factors. Nevertheless, SI values for muscovite and other silicate minerals are shown in Table II to indicate potential for their dissolution. Thermodynamic data were not included in WATEQ4F for microcline, römerite, copiapite, or coquimbite, so SI values for these minerals could not be calculated. However, ground water at the mine is consistently undersaturated with respect to adularia ($KAlSi_3O_8$; SI < -1), albite ($NaAlSi_3O_8$; SI < -3), melanterite ($FeSO_4 \cdot 7H_2O$; SI < -2), and ferric sulfate ($Fe_2(SO_4)_3$; SI < -22) included in WATEQ4F.

Table I. Median and range of field measurements, major-element concentrations, and operational redox state in ground-water samples collected during 1986-89 [a]

[units are millimoles per kilogram of water, except as noted; values are median (low/high)]

Field measurements	Well 14	Well 14A	Well 14B	Well 14C	Well 15	Well 15A	Well 15B
Head (m)	401.5 (400.8/402.3)	401.5 (400.8/402.3)	401.5 (400.8/402.3)	423.1 (422.5/423.4)	384.4 (378.9/386.8)	392.9 (392.3/398.7)	399.0 (397.2/399.6)
Temp (°C)	12.0 (7.0/15.0)	12.0 (9.0/14.0)	12.0 (8.0/14.0)	11.0 (8.0/14.5)	11.5 (10.0/14.0)	12.0 (10.0/13.5)	12.0 (10.5/15.5)
NetAlk (g/kg)[b]	-.084 (-.244/-.020)	-.045 (-.233/-.010)	.034 (-.060/.143)	-.319 (-.469/-.299)	-1.71 (-3.60/-1.10)	-3.00 (-7.60/-.30)	-1.50 (-6.50/-.61)
pH (units)	5.80 (5.10/6.20)	5.90 (5.50/6.10)	5.92 (5.60/6.50)	3.50 (3.20/3.80)	3.12 (2.50/3.50)	2.73 (2.40/3.40)	2.75 (2.50/3.40)
Eh (volts)	.36 (.34/.41)	.33 (.28/.35)	.31 (.29/.34)	.64 (.64/.64)	.58 (.55/.60)	.59 (.53/.60)	.60 (.55/.62)
DO	.010 (.007/.023)	.011 (.007/.014)	.008 (.005/.019)	.008 (.008/.008)	.010 (.005/.012)	.008 (.004/.030)	.013 (.005/.034)
Elements[c]							
Ca	8.684 (6.986/10.479)	8.796 (7.735/9.356)	8.646 (5.988/10.479)	1.904 (1.497/2.183)	9.479 (6.986/12.974)	9.355 (7.236/10.978)	8.522 (5.739/11.477)
Mg	11.080 (9.461/15.220)	11.160 (9.872/12.135)	10.810 (7.404/11.929)	3.752 (3.332/4.114)	11.740 (10.284/13.986)	13.560 (10.695/18.100)	12.950 (9.461/15.631)
Na	.303 (.217/.413)	.263 (.226/.335)	.333 (.252/.357)	.117 (.096/.178)	.272 (.139/.370)	.184 (.117/.391)	.197 (.135/.365)
K	.223 (.197/.460)	.199 (.182/.371)	.221 (.002/.307)	.179 (.002/.207)	.177 (.128/.307)	.143 (.084/.358)	.146 (.092/.179)
C[b]	8.370	8.170	9.400	14.400	[6.780]	[6.780]	[6.780]
S (adjusted)[d]	21.630 (15.615/24.984)	21.590 (14.574/22.902)	20.760 (17.697/22.902)	8.849 (7.495/10.410)	37.953 (29.148/57.776)	49.000 (22.902/104.102)	34.930 (18.738/83.281)
Fe	1.821 (1.200/2.865)	1.776 (1.343/2.596)	1.614 (1.289/2.149)	1.762 (1.522/1.970)	9.277 (6.446/14.861)	14.030 (.788/44.763)	5.954 (.895/34.020)
Fe^{2+}[e]	1.763	1.748	1.600	1.370	9.036	13.500	5.674
Mn	.732 (.564/.946)	.735 (.637/.892)	.734 (.528/.892)	.323 (.237/.364)	1.103 (.819/1.274)	1.138 (.746/1.547)	1.123 (.619/1.456)
Si	.259 (.153/.350)	.236 (.152/.333)	.255 (.166/.516)	1.016 (.899/1.115)	.926 (.599/1.165)	1.458 (.400/1.831)	1.046 (.516/1.515)
Al	.015 (.002/.032)	.009 (.002/.041)	.009 (.002/.115)	.549 (.445/.890)	3.800 (1.705/7.413)	6.510 (1.371/20.015)	3.522 (1.186/15.567)
RS[f]	168.424	167.270	166.870	115.256	275.821	351.986	251.134

a. Quarterly data collected by USGS for 3 years during 1986-89, which resulted in a total of 12 values for most constituents to determine the median and range. DO and Eh were measured only during 1988-89, which resulted in a total of 4 data values. Only one water sample could be obtained from well 14C in 1988-89, which resulted in a total of 8 data values for most constituents and 1 value for DO and Eh.
b. NetAlk = alkalinity - acidity, in g/kg as $CaCO_3$; acidic water defined as NetAlk < 0, and alkaline water defined as NetAlk > 0. Alkalinity and acidity were titrated in the laboratory. Total CO_2, in mmol/kg as C, calculated in WATEQ4F (31) by use of alkalinity and pH; value of 6.78 mmol/kg assumed where alkalinity below detection.
c. For element concentrations in milligrams per liter, only two or three significant figures were input to WATEQ4F (31); however, concentrations in millimoles per kilogram are reported to three decimal places to avoid rounding errors in mass-balance calculations.
d. Median sulfate adjusted to attain ionic charge balance within 1%.
e. Ferrous iron computed in WATEQ4F (31) by use of measured total dissolved iron and Eh.
f. Operational redox state (RS) is treated as an element in mass-balance calculations (32, 34), where: RS = 6·S + 4·C + 2·Fe^{2+} + 3·Fe^{3+} + 2·Mn.

Table II. Saturation indices[a] for water samples collected during 1986-89
[values are unitless; --, insufficient data for computation]

Solid-phase name	Chemical formula	Well number						
		14	14A	14B	14C	15	15A	15B
Calcite	$CaCO_3$	-1.7	-1.5	-1.4	-6.5	--	--	--
Gypsum	$CaSO_4 \cdot 2H_2O$	-.2	-.2	-.2	-.9	-.1	-.1	-.1
Rhodochrosite	$MnCO_3$	-.8	-.6	-.6	-5.3	--	--	--
Manganite	$MnOOH$	-6.0	-6.3	-6.6	-8.2	-10.1	-11.1	-10.8
Pyrolusite	MnO_2	-11.0	-11.6	-12.2	-10.5	-13.9	-15.0	-14.5
Siderite	$FeCO_3$	-.4	-.2	-.2	-4.6	--	--	--
Ferrihydrite	$Fe(OH)_3$	2.2	2.0	1.7	.3	-1.3	-2.1	-2.2
Goethite	$FeOOH$	6.2	5.9	5.6	4.2	2.6	1.8	1.7
Melanterite	$FeSO_4 \cdot 7H_2O$	-2.9	-2.9	-2.9	-3.1	-2.1	-1.9	-2.3
Ferric sulfate	$Fe_2(SO_4)_3$	-32.7	-33.8	-34.6	-23.6	-23.2	-22.4	-23.0
Hydronium jarosite	$(H_3O)Fe_3(SO_4)_2(OH)_6$	4.1	2.9	1.9	6.9	4.4	3.6	3.1
Jarosite	$NaFe_3(SO_4)_2(OH)_6$	6.0	4.9	4.0	6.2	3.6	2.2	1.8
Natrojarosite	$KFe_3(SO_4)_2(OH)_6$	9.6	8.5	7.5	10.1	7.1	5.9	5.4
Alunite	$KAl_3(SO_4)_2(OH)_6$	5.9	5.3	5.3	-.8	-1.2	-2.9	-3.4
Alum	$KAl(SO_4)_2 \cdot 12H_2O$	-9.5	-10.0	-10.0	-7.5	-6.4	-6.3	-6.6
Al-hydroxysulfate	$AlOHSO_4$	.1	-.2	-.3	.1	.5	.3	.1
Al-hydroxysulfate	$Al_4(OH)_{10}SO_4$	6.0	5.4	5.4	-6.9	-9.0	-12.0	-12.5
Albite	$NaAlSi_3O_8$	-3.6	-3.8	-3.5	-8.8	-9.8	-10.7	-11.2
Adularia	$KAlSi_3O_8$	-1.0	-1.1	-.9	-5.9	-7.2	-8.1	-8.6
Muscovite	$KAl_3Si_3O_{10}(OH)_2$	8.4	8.0	8.3	-5.2	-8.1	-10.8	-11.5
Illite	$K_{.6}Mg_{.25}Al_{2.3}Si_{3.5}O_{10}(OH)_2$	1.2	.9	1.1	-9.4	-11.7	-13.6	-14.3
Chlorite 14Å	$Mg_5Al_2Si_3O_{10}(OH)_8$	-17.3	-16.7	-16.3	-49.2	-53.1	-57.9	-58.2
Kaolinite	$Al_2Si_2O_5(OH)_4$	3.7	3.4	3.5	-3.8	-5.5	-7.0	-7.4
Allophane (am)	$[Al(OH)_3]_{(1-x)}[SiO_2]_x$	.1	.0	.1	-1.0	-1.0	-.8	-.9
Gibbsite (c)	$Al(OH)_3$	1.1	1.0	1.0	-3.3	-4.1	-5.0	-5.1
Al-hydroxide (am)	$Al(OH)_3$	-.7	-.8	-.8	-5.0	-5.8	-6.8	-6.8
Chalcedony	SiO_2	.1	.1	.1	.7	.7	.8	.7
Silica (am,L)	$SiO_2 \cdot nH_2O$	-.4	-.5	-.4	.2	.2	.3	.2

a. Saturation index, $SI = \log(IAP/K_{sp})$. Computations performed using WATEQ4F (31) with data in Table I, which consist of median chemical concentrations, temperature, pH, and Eh.

The concept of water-unsaturated, oxygenated zones (open system) versus water-saturated, O_2-limited zones (closed system) provides a basis for evaluating the location and extent of oxidation reactions in the subsurface (36). For example, DO in water at 6 °C (degrees Celsius), the minimum temperature of ground water at the site, can attain a saturation concentration of 0.38 mmol/kg (33). As this O_2-saturated water flows downward from the water table, oxidation reactions will consume the O_2. In an open system, DO will be replenished and oxidation reactions will not be limited by O_2 supply. In a closed system, however, DO will be depleted at some point along the flow path (Figure 2). Thus, if pyrite is oxidized in a closed system according to reaction 1, where 1 mol of FeS_2 reacts with 3.75 mol of O_2 to produce 2 mol of SO_4^{2-}, then reaction with 0.38 mmol/kg O_2 can produce only 0.20 mmol/kg SO_4^{2-}. Additional and potentially much greater quantities of SO_4^{2-} may be produced in the absence of DO by oxidation of FeS_2 by Fe^{3+}.

Geochemical Evolution of Acidic Ground Water

This section describes the hydrochemical trends along flow paths through the mine spoil and bedrock and evaluates the roles of O_2, pyrite, and iron-sulfate minerals in producing acidic ground water.

Hydrochemical Trends. Water from wells in nests 14 and 15, from untreated and sludge-treated zones, respectively, contains negligible concentrations of DO (< 0.034 mmol/kg) and has differing net alkalinity, pH, and Eh (Table I). Water from wells in nest 15 is more acidic, has lower pH (< 3.5) and higher Eh (> 0.5 volts), and contains greater overall solute concentrations relative to water from equivalent horizons at nest 14. At nests 15 and 14, water from the mine spoil (wells 15B and 14C) is less mineralized than the water from underlying bedrock (wells 15A and 14B), which contains elevated concentrations of SO_4^{2-} and base cations. Along flow paths, concentrations of S in ground water generally do not increase in stoichiometric proportion with Fe (S/Fe > 2) relative to pyrite (S/Fe = 2). These trends suggest, firstly, that acid produced by the oxidation of FeS_2 in the mine spoil is neutralized by reactions with silicate and carbonate minerals in the underlying bedrock. Secondly, because of neutralization along the flow path, Fe can be removed from solution by precipitation of jarosite and (or) ferrihydrite, since the water in bedrock is supersaturated with one or both of these minerals. Thirdly, intermediate concentrations of major solutes in water from well 15 relative to those farther up the flow path may result by mixing (dilution) of highly mineralized water from overlying material with less mineralized water from the east.

Supplemental data for ground-water head, pH, and sulfate concentrations during the period 1982-91, which were collected by the coal-mining company, are presented as time series in Figure 3. These data indicate the potential for transport of a previously formed plume of acidic, mineralized water from upgradient to downgradient zones and for additional oxidation reactions along the flow path(s). However, they do not indicate a simple trend of decreasing FeS_2 oxidation over time. Through 1989, pH generally was unchanged or decreased and sulfate concentration increased in the mine spoil (wells 14C and 15B) and adjacent bedrock (wells 14B, 15A, and 15), while the saturated thickness of mine spoil was relatively constant. During 1988-89, the most acidic and mineralized water was sampled from zones beneath the sludge-covered area (wells 15B and 15A). The absence of a time lag between peak sulfate concentrations in upgradient (well 15B) and downgradient (well 15A) waters and the larger sulfate concentrations in downgradient waters support the argument for sulfide oxidation along the flow path.

Inverse Modeling. Inverse modeling is used to identify and quantify the geochemical processes that cause ground-water-quality trends. Such modeling assumes that paths of ground-water flow are well defined, the stoichiometries of reactants and products are known, and the composition of the upgradient water when it began moving along the flow path is known (34, 35). For calculations herein, the composition of the upgradient water, as determined by sampling during the study, is assumed to be that of the initial water. By producing a set of models to explain changes among several locations along a flow path, the models begin to account for the sequence of reactions over distance or time. Because mass-balance models typically are not unique--many different minerals can be reacted or produced to

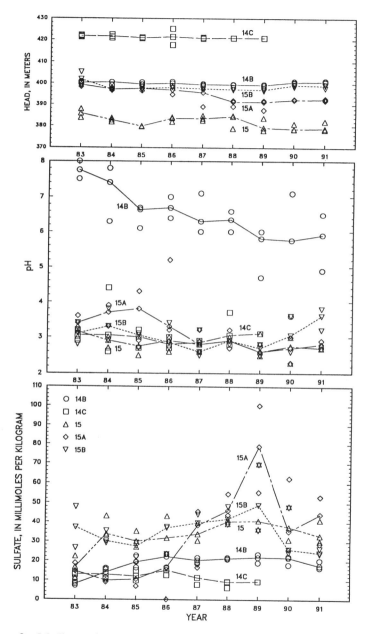

Figure 3. Median and range of values for ground-water head, pH, and sulfate concentrations, summarized by year (e.g. "83" represents June 1982 - May 1983). For a particular sampling location, lines connect annual medians. Data were collected quarterly by the coal-mining company and are comparable with data reported in Table I, which were collected at the same locations but on different dates by USGS.

achieve the same water composition--the results are constrained by thermodynamic equilibrium criteria. Specifically, acceptable mass-balance models were constrained by considering (1) the computed SI values for solid phases (Tables II and III), to allow dissolution or precipitation of appropriately saturated phases and (2) the maximum concentration of DO (= 0.38 mmol/kg), if a closed system. Furthermore, kinetic criteria were used implicitly to constrain models to dissolve or precipitate solid phases that are "active" under the conditions. For example, under conditions of increasing pH, kinetics favor the precipitation of Fe^{3+} as amorphous $Fe(OH)_3$ or ferrihydrite, even though goethite (FeOOH) and hematite (Fe_2O_3) are less soluble (3, p. 234-237).

On the basis of simulated, steady-state flow paths in Figure 2, the following four inverse modeling scenarios were evaluated to assess the feasibility of FeS_2 oxidation in different zones, which can be open or closed with respect to O_2:

I. Open--Flow from untreated spoil (well 14C) to underlying bedrock (well 14B);

II. Open--Waters from untreated spoil (well 14C) and unmined coal (well 14B) mixed beneath sludge-treated spoil (well 15B);

III. Closed--Flow from sludge-treated spoil (well 15B) to underlying bedrock (well 15A);

IV. Closed--Waters from bedrock below untreated (well 14B) and sludge-treated (well 15A) areas mixed in deep bedrock below sludge-treated area (well 15).

Table I shows the concentrations of elements in ground water from the wells, and Table III shows the stoichiometry of potential reactant or product minerals and gases, including O_2 and CO_2. Table IV summarizes the results of the mass-balance calculations, which indicate the integrated, average quantities of reactants and products involved in reactions along the flow path(s). Minerals indicated in Table II were confirmed to be present in the mine spoil and bedrock (10), with the exception of pyrolusite, which likely is present in joints and fractures in the weathered zone. Römerite is the iron-sulfate reactant used in the models, because mixtures of ferrous and ferric sulfate hydrates are common in coal spoil (6, 11, 14), and the composition of römerite is equivalent to a 1:1 mixture of coquimbite and melanterite. In this evaluation, römerite dissolution represents the effects on the ground water from past pyrite oxidation in zones which were previously unsaturated, but which are presently below the water table and along the flow path.

I. Open system: Downward flow from untreated spoil to underlying bedrock. Perched ground water can flow downward from untreated mine spoil (well 14C) to the underlying bedrock aquifer (well 14B). This flow path is not water-saturated, and thus is open to O_2. The water in spoil is acidic; however, the water in underlying bedrock is alkaline, has greater concentrations of S, Mn, Ca, Mg, Na, and K, and has lesser concentrations of Fe, Si, and Al (Table I). The bedrock water is supersaturated with ferrihydrite, jarosite, natrojarosite, gibbsite, chalcedony (silica phase), and kaolinite (Table II), which can precipitate and remove Fe, Na, K, Si, and Al from solution.

Table III. Stoichiometry,[a] reactivity,[b] and operational redox state[c] (RS) of potential reactant and product phases used in mass-balance computations

[units are millimoles per millimole; +, dissolution or ingassing only; -, precipitation or outgassing only]

O_2 Gas	+RS	4.00									
Pyrite	+Fe	1.00	S	2.00	RS	.00					
Römerite	+Fe	3.00	S	4.00	RS	32.00					
Ferrihydrite	Fe	1.00	RS	3.00							
Natrojarosite	-Na	1.00	Fe	3.00	S	2.00	RS	21.00			
Jarosite	-K	1.00	Fe	3.00	S	2.00	RS	21.00			
Pyrolusite	+Mn	1.00	RS	4.00							
Siderite	+Fe	.865	Mn	.015	Mg	.02	Ca	.10	C	1.00	RS 5.76
Calcite	+Ca	1.00	C	1.00	RS	4.00					
Chlorite	+Mg	5.00	Al	2.00	Si	3.00					
Microcline	+K	.85	Na	.15	Al	1.00	Si	3.00			
Kaolinite	Al	2.00	Si	2.00							
Amorph. silica	-Si	1.00									
CO_2 Gas	-C	1.00	RS	4.00							

a. Stoichiometry excludes hydrogen and oxygen (*32*). Siderite stoichiometry reported by Morrison et al. (*37*).
b. Reactivity defined by saturation index (SI, Table II): '+' indicates reactant (SI < 0); '-' indicates product (SI > 0); no sign indicates SI could be positive or negative depending on the particular sample.
c. Redox state for minerals computed considering redox active components (S, C, Fe, Mn, O) as defined by Parkhurst et al. (*32*). See footnote f in Table I.

Four models were developed that satisfy mineral-solubility constraints and can be used to explain the differences in water chemistry between zones sampled by wells 14C and 14B (Table IV). The major distinction among the models is the source of 11.7 mmol/kg SO_4^{2-} as S and of 0.41 mmol/kg Mn added along the flow path. Production of S from pyrite and Mn from impure siderite (*37*) would require 32.6 mmol/kg of O_2 (model 1). However, if römerite and pyrolusite react (model 4), the same quantity of S can be generated with only 0.63 mmol/kg O_2. In all four models, calcite, chlorite, and microcline also are reactants; ferrihydrite, jarosite, kaolinite, amorphous silica, and CO_2 are products.

II. Open system: Waters from untreated spoil and unmined coal (highwall) mixed beneath sludge-treated spoil. Ground water can flow laterally from the intact Clarion coal horizon (well 14B) and flow vertically from the perched aquifer (well 14C) to the sludge-covered mine spoil on the lower bench (well 15B). Because the water table defines this path of ground-water mixing, the system is probably open with respect to O_2. The water of the lower bench contains greater concentrations of S, Fe, Mn, Mg, Si, and Al, intermediate concentrations of Ca and Na, and lesser concentrations of K relative to source waters upgradient along the flow paths (Table I); mixing of the source waters can produce the intermediate Ca and Na concentrations. The water of the lower bench is supersaturated with jarosite, natrojarosite, and amorphous silica (Table II), which can precipitate and remove Fe, Na, K, and Si from solution.

Table IV. Summary of mass-balance results

[All mineral and gas mass-transfers are in millimoles per kilogram of water; negative for precipitation or outgassing, positive for dissolution or ingassing; "•" indicates no reaction occurred. Input data are median values in Table I; mineral stoichiometry and reactivity in Table III]

Well and model no.[a]	Mixing ratio	O_2 gas	Pyrite	Römerite	Ferrihydrite	Natro-jarosite	Jarosite	Pyrolusite	Siderite	Calcite	Chlorite	Microcline	Kaolinite	Amorphous silica	CO_2 gas
I. Well 14B															
1	•	32.63	7.14	•	-27.44	•	-1.18	•	27.40	4.00	1.30	1.44	-2.29	-4.40	-36.40
2	•	26.50	7.14	•	-3.74	•	-1.18	0.41	•	6.74	1.41	1.44	-2.40	-4.51	-11.74
3	•	6.76	•	3.57	-31.01	•	-1.18	•	27.40	4.00	1.30	1.44	-2.29	-4.40	-36.40
4	•	.63	•	3.57	-7.31	•	-1.18	.41	•	6.74	1.41	1.44	-2.40	-4.51	-11.74
II. Well 15B	14B:14C														
1	98:2	31.12	8.62	•	•	-.32	-1.11	.40	•	•	.45	1.22	.69	-5.62	-2.71
2	98:2	16.36	4.50	2.75	•	-.52	-2.28	.40	•	•	.45	2.60	•	-8.37	-2.71
3	94:6	37.22	10.11	•	•	-.50	-2.19	.41	2.62	•	.50	2.49	•	-8.20	-5.53
4	78:22	38.01	10.48	•	•	-.37	-1.68	.48	•	1.39	.74	1.90	•	-7.32	-5.13
III. Well 15A															
1	•	23.19	6.64	0.21	•	-.01	-.00	•	1.00	.73	.12	•	1.38	-2.69	-1.73
2	•	16.70	4.91	1.07	•	-.01	-.00	.02	•	.83	.12	•	1.34	-2.70	-.83
3	•	•	.16	4.52	•	-.34	-1.84	•	1.00	.73	.12	2.16	.30	-7.01	-1.73
4	•	•	.25	4.18	•	-.25	-1.32	.02	•	.83	.12	1.55	.60	-5.80	-.83
IV. Well 15	15A:14B														
1	31:69	22.28	6.00	•	•	-.28	-1.52	.19	3.70	.24	•	1.77	•	-5.01	-5.75
2	34:66	14.03	4.00	•	•	-.03	-.15	.23	•	.59	•	.15	.72	-1.63	-2.33
3	34:66	•	.07	2.71	•	-.25	-1.38	.23	.23	.57	•	1.60	•	-4.54	-2.54
4	34:66	•	.09	2.61	•	-.23	-1.25	.23	•	.59	•	1.46	.07	-4.24	-2.33

a. Model ground-water flow paths shown in Figure 2. Well number indicates "final" water: I. well 14B ← well 14C; II. well 15B ← well 14B + well 14C; III. well 15A ← well 15B; and IV. well 15 ← well 15A + well 14B. Model number arbitrarily assigned to facilitate discussion of results.

Four mixing models were developed to evaluate this scenario (Table IV). Although the mixing ratios are different in each model, all indicate a larger contribution of total flow from the unmined coal in the highwall (78-98%) than from the perched aquifer. All four models involve O_2, pyrite, and pyrolusite as reactants to produce at least 14.2 mmol/kg S and 4.2 mmol/kg Fe. Chlorite and microcline also are reactants, and jarosite, natrojarosite, amorphous silica, and CO_2 are products in all four models. Models 1, 3, and 4 involve pyrite and 31 to 38 mmol/kg O_2 as reactants; and model 2 involves pyrite, römerite, and 16.4 mmol/kg O_2 as reactants. The difference between models 1 and 2 is whether römerite or kaolinite is a reactant; the difference between models 3 and 4 is whether siderite or calcite is a reactant. All four models involve large quantities of O_2.

III. Closed system: Downward flow from sludge-treated spoil to underlying bedrock.

Ground water can flow downward from the water table in sludge-treated mine spoil (well 15B) to the underlying bedrock (well 15A). This flow path is water-saturated and probably closed to O_2. The bedrock water contains greater concentrations of S, Fe, Mn, Al, Ca, Mg, and Si, lesser concentrations of Na and K (Table I), and is supersaturated with jarosite, natrojarosite, and amorphous silica (Table II).

Four models were developed to evaluate this scenario (Table IV). Pyrite and römerite are reactants in all four models. Oxygen is a reactant in only models 1 and 2. The difference between models 1 and 2 and between models 3 and 4 is whether siderite or pyrolusite is a reactant. Additional reactants include calcite, chlorite, microcline, and kaolinite; products include jarosite, natrojarosite, amorphous silica, and CO_2. Only models 3 and 4 meet the closed-system criterion for acceptability (O_2 ≤ 0.38 mmol/kg). Both closed-system models involve pyrite and römerite as reactants under anoxic conditions to produce 13.9 mmol/kg S and 8 mmol/kg Fe along the flow path. Therefore, substantial quantities of SO_4^{2-} can be generated by the dissolution of previously formed ferric-sulfate salts and by the oxidation of pyrite by Fe^{3+}, even under anoxic conditions.

IV. Closed system: Waters from bedrock below untreated and sludge-treated areas mixed in deep bedrock below sludge-treated area.

Ground water can flow downward and laterally from the intact Clarion coal horizon in the highwall (well 14B) and downward from the overlying shale (well 15A) to the deep sandstone bedrock beneath sludge-covered mine spoil (well 15). This path of ground-water mixing is water-saturated and closed to O_2. The resultant water contains intermediate concentrations of all the major ions, except for Ca, which is present in concentrations greater than those in the source waters (Table I). Mixing of the source waters can produce the intermediate concentrations; reactions with calcite and (or) impure siderite can produce the increased Ca concentration. The resultant water is supersaturated with jarosite, natrojarosite, and amorphous silica (Table II).

Four mixing models were developed to evaluate this scenario (Table IV). The mixing ratios indicate a larger contribution of total flow from the weathered, unmined coal in the highwall (66-69%). Pyrite and pyrolusite are reactants in all four models. Oxygen is a reactant in only models 1 and 2, and römerite is a reactant in only models 3 and 4. The difference between models 1 and 2 and between models 3 and 4 is whether siderite or kaolinite is a reactant. Calcite and microcline also are reactants; jarosite, natrojarosite, amorphous silica, and CO_2 are products. Models 3 and 4, which involve pyrite, römerite, and pyrolusite as reactants under anoxic conditions, meet the closed-system criterion ($O_2 \leq 0.38$ mmol/kg). The argument for acid-generating reactions by ferric ions and pyrite under anoxic conditions is supported by these models. An alternative model involves the mixing of water from the deep bedrock zone beneath the unsludged area (well 14) with water from the deep shale (well 15A), which results in similar geochemical reactions to those above (Table IV). However, oxidized minerals are not likely to be abundant, and may not be present, along deep flow paths.

Discussion. Despite negligible concentrations of DO in the ground water, concentrations of SO_4^{2-} and acidity increased with depth along flow paths from the water table in mine spoil through underlying bedrock. One interpretation is that the downgradient water quality only reflects a mineralized plume of older water that formed in the upgradient zone under unsaturated conditions, and the rate of sulfide oxidation in the upgradient zone has decreased over time as the system became water saturated. An alternative explanation, which is evaluated in this paper, is that the downgradient water quality results from hydrogeochemical mass-transfer reactions that occur along the flow path from the upgradient zone(s). In an open system, one cannot reject a mass-balance model that involves reaction with more O_2 than was measured, because it is the integrated quantity of O_2 involved in geochemical reactions that produced the solutes in the water sample. However, in a closed system, the saturation concentration of O_2 (0.38 mmol/kg or less) is the maximum quantity of O_2 available for the integrated reactions, and the measured DO is the absolute limit of O_2 available for continued reactions. Assuming that a closed system is in effect, one can reject mass-balance models that require extreme quantities of O_2 to generate the additional SO_4^{2-}. Mass-balance models indicate that either (a) extremely large quantities of O_2, as much as 100 times the O_2 solubility, can generate observed concentrations of dissolved SO_4^{2-} from FeS_2, or (b) under anoxic conditions, Fe^{3+} from iron-sulfate minerals, such as römerite, can oxidize FeS_2 along closed-system ground-water flow paths. It is likely that in the unsaturated zone and in the vicinity of the water table, which is open to O_2, oxidation of FeS_2 by O_2 accounts for most SO_4^{2-}, Fe^{2+}, and Fe^{3+} in acidic ground water. However, in the saturated zone, where the aqueous solubility of O_2 is limiting, dissolution of iron-sulfate hydrates and oxidation of FeS_2 by Fe^{3+} can explain the increase in concentration of SO_4^{2-} with increasing depth below the water table. These findings are consistent with kinetic studies ([2], [21], [22]), which show rates of pyrite oxidation by Fe^{3+} are faster than oxidation by O_2 in acidic systems, and with oxygen isotopic measurements of SO_4^{2-} by Taylor et al. ([20], [38]), which indicate 25-100% of SO_4^{2-} in AMD can result from oxidation of FeS_2 by Fe^{3+}.

The computed masses of pyrite and römerite involved in the mass-balance models (Table IV) are reasonable. The concentrations of total sulfur in a rock core collected from the location of well 14 range from <0.01 to 17.4 wt%, which is primarily in the form of iron-sulfide and iron-sulfate minerals ([10]); mass-weighted-average concentrations of total sulfur are 0.60, 0.61, and 0.57 wt% in the rock to the base of the screened intervals of wells 14B, 14A, and 14, respectively. The largest masses of pyrite and römerite involved in the mass-balance models are 10.48 mmol (672.0 mg S) and 4.52 mmol (579.6 mg S), respectively, per 1 kg water (Table IV). If the aquifer rock has a porosity of 25%, or a bulk density of 2.0 g/cm³ (grams per cubic centimeter), and a particle density of 2.67 g/cm^3, then 1 kg of water (=approximately 1 liter) can occupy the pores in 10.7 kg of rock. Hence, pyrite as 10.48 mmol/kg water is equivalent to a sulfur concentration of 63 mg/kg rock or 0.0063 wt%, and römerite as 4.52 mmol/kg water is equivalent to a sulfur concentration of 54 mg/kg rock or 0.0054 wt%. These concentrations are only a small fraction of the average total sulfur concentrations of 0.6 wt%.

The precipitation of jarosite and natrojarosite as a potential sink for Fe^{3+}, SO_4^{2-}, Na^+, and K^+ ions is suggested because the ground water from all wells is supersaturated with respect to these minerals. This indicates that S-bearing minerals such as pyrite and römerite react at rates greater than S could precipitate. Subsurface formation of jarosite or natrojarosite has not been confirmed, however, and their supersaturation may imply kinetic barriers to precipitation. Furthermore, Na can substitute for K in jarosite solid solution (39), and the solubility of jarosite solid solutions [$(K,Na,H_3O)Fe_3(SO_4)_2(OH)_6$], jarosite, and natrojarosite are such that relatively large concentrations of S and Fe can remain at equilibrium in water containing low concentrations of K and Na.

In previous models for the same mass-transfer scenarios, Cravotta (10) did not adjust SO_4^{2-} concentrations to correct for charge imbalances, although he noted SO_4^{2-} concentrations probably were underestimated in the elevated-concentration (> 35 mmol/kg) range. Thus models involving reactions with more than 0.38 mmol/kg O_2 in a closed system could not be ruled out because of the propagation of errors in the computed O_2. This problem has been avoided herein by adjusting SO_4^{2-} concentrations. The overall conclusions are unchanged.

Conclusions

During and after mining, pyrite oxidation forms aqueous SO_4^{2-}, Fe^{2+}, Fe^{3+}, and iron-sulfate minerals. Römerite and other iron-sulfate hydrate minerals in mine spoil and unmined coal can be a relatively important source of SO_4^{2-} and Fe^{3+}. The iron-sulfate minerals can dissolve in recharge water or in ground water, and Fe^{3+}, which is substantially more soluble than O_2 in acidic waters, can be transported to pyrite-oxidation sites below the water table where DO is absent. Subsequent coupled reduction of Fe^{3+} and oxidation of pyritic S and (or) hydrolysis of Fe^{3+} will produce acid. Negligible concentrations of DO (< 0.034 mmol/kg), high Eh (> 0.55 volt), low pH (< 3.5), and large concentrations of SO_4^{2-} and Fe^{2+} in solution are consistent with coupled reduction of Fe^{3+} and oxidation of FeS_2. Thus, aqueous SO_4^{2-} can form under anoxic conditions by dissolution of previously formed iron-sulfate minerals and by reactions between Fe^{3+} and FeS_2.

The timing and extent of formation of soluble iron-sulfate hydrate minerals in mine spoil and in unmined rock and the potential for sustained reactions between Fe^{3+} and FeS_2 are questions that have not been entirely addressed herein. Iron-sulfate minerals do not cause AMD--they are intermediate products of FeS_2 oxidation by O_2. The oxidation products can form as a transient during mining, and they can form continuously, as a result of a sustained O_2 transport to pyritic material in unsaturated mine spoil or by evaporation of AMD. Dissolution of iron-sulfate minerals can result from infiltration of recharge water or rising ground water. However, little information is available on the subsurface formation and occurrence of the sulfate salts or on the magnitude of their effect on water quality.

To prevent AMD, some researchers have suggested that pyritic materials should be submerged below a permanent water table where low O_2 solubility will limit the extent of acid-forming oxidation reactions. However, the possibility for dewatering and in situ oxidation of unmined coal during mining and oxidation of mine spoil prior to submergence of pyritic material complicates matters. An alternative surface coal-mining method involves the segregation of pyritic rock during mining and the placement of pyritic spoil high in the backfill, to be above the water table after reclamation. Placement of pyrite above the water table allows the oxidation of pyrite by humid, oxygenated air in the unsaturated zone; however, placement of pyritic rock below the water table enables leaching of oxidation products formed during mining and allows the continued oxidation of pyrite by DO and dissolved Fe^{3+} ions. Other reclamation practices to stop FeS_2 oxidation, such as the addition of O_2-consumptive barriers, alkaline additives, or bactericides to mine spoil, also are not likely to prevent AMD if previously formed oxidation products are present.

Hydrological, mineralogical, and rock-chemical data are critically important for evaluating ground-water-quality trends. At the mine studied, more densely spaced samples of ground-water and rock within the unsaturated and saturated zones would help to locate zones of active oxidation. Missing data on heat and gaseous fluxes, pyrite and ferric-sulfate mineral distribution, and recharge infiltration through unsaturated mine spoil would help to define the rate of oxidation of pyrite above the water table. In general, data could be obtained for identical locations over a period of years including the earliest stages of mining and extending several years after mine closure. These data would provide a basis for mapping the historical sequence of reactions and would enable simulation of the coupling of the physical and chemical systems to explain the evolution of acidic ground water.

Literature Cited

1. Temple, K. L.; Koehler, W. A. Engineering Experiment Station Research Bulletin 25; West Virginia University: Morgantown, WV, 1954, 35 p.
2. Singer, P. C.; Stumm, W. Science, 1970, 167, 1121-1123.
3. Stumm, W.; Morgan, J.J. Aquatic Chemistry--An Introduction Emphasizing Chemical Equilibria in Natural Waters (2nd); John Wiley & Sons, Inc.: New York, NY, 1981.
4. Kleinmann, R. L. P.; Crerar, D. A.; Pacelli, R. R. Mining Eng. 1981, 33, 300-306.
5. Nordstrom, D. K.; Jenne, E. A.; Ball, J. W. In Chemical Modeling in Aqueous Systems; Jenne, E. A., Ed.; ACS Symposium Series 93; American Chemical Society: Washington, DC, 1979, pp 51-79.
6. Nordstrom, D. K. In Acid Sulfate Weathering; Kittrick, J. A.; Fanning D. S.; Hossner L. R., Eds.; Soil Sci. Soc. of America: Madison, WI, 1982, pp 37-63.
7. Watzlaf, G. R. In Proceedings 9th Annual National Meeting American Society for Surface Mining and Reclamation; American Society for Surface Mining and Reclamation: Princeton, WV, 1992, pp 191-205.
8. Rahn, P. H. Environ. Geol. Water Sci. 1992, 19, 47-53.
9. Backes, C. A.; Pulford, I. D.; Duncan, H. J. U.S. Bur. Mines Information Circular 9183; U.S. Gov. Printing Office: Washington, DC, 1988, Vol. 1, pp 91-96.

10. Cravotta, C. A. III. In Proceedings of the 1991 National Meeting of the American Society for Surface Mining and Reclamation; Oaks, W. R.; Bowden, J., Eds.; American Society for Surface Mining and Reclamation: Princeton, WV, 1991, pp 43-68.
11. Dixon, J. B.; Hossner, L. R.; Senkayi, A. L.; Egashira, K. In Acid Sulfate Weathering; Kittrick, J. A.; Fanning D. S.; Hossner L. R., Eds.; Soil Sci. Soc. of Am.: Madison, WI, 1982, pp 169-191.
12. Karathanasis, A. D.; Evangelou, V. P.; Thompson, Y. L. J. Environ. Qual. 1988, 17, 534-543.
13. Karathanasis, A. D.; Evangelou, V. P.; Thompson, Y. L. J. Environ. Qual. 1990, 19, 389-395.
14. Olyphant, G. A.; Bayless, E. R.; Harper, D. J. Contam. Hyd. 1991, 7, 219-236.
15. Alpers, C. N.; Maenz, C.; Nordstrom, D. K.; Erd, R. C.; Thompson, J. M. GSA Abstracts with Programs 1991, 23, 382.
16. Alpers, C. N.; Nordstrom, D. K. In Proceedings 2nd International Conference on the Abatement of Acidic Drainage; MEND (Mine Environment Neutral Drainage): Montreal, Quebec, Canada, 1991, Vol. 2, pp 321-342.
17. Williams, J. H.; Pattison, K. L. GSA Abstracts with Programs 1987, 19, 65.
18. Temple, K. L.; Delchamps, E. W. App. Microbiol. 1953, 1, 255-258.
19. Garrels, R. M.; Thompson, M. E. Am. J. Sci. 1960, 258, 57-67.
20. Taylor, B. E.; Wheeler, M. C.; Nordstrom, D. K. Nature 1984, 308, 538-541.
21. McKibben, M. A.; Barnes, H. L. Geochim. Cosmochim. Acta 1986, 50, 1509-1520.
22. Moses, C. O.; Nordstrom, D. K.; Herman, J. S.; Mills, A. L. Geochim. Cosmochim. Acta 1987, 51, 1561-1571.
23. Williams, E. G. J. Paleontology 1960, 34, 908-922.
24. Berg, T. M.; McInerney, M. K.; Way, J. H.; MacLachlan, D. B. Pennsylvania Geol. Survey General Geology Report 75; 1983.
25. Saad, D. A.; Cravotta, C. A. III. In Proceedings 1991 National Meeting American Society for Surface Mining and Reclamation; Oaks, W. R.; Bowden, J., Eds.; American Society for Surface Mining and Reclamation: Princeton, WV, 1991, p 545.
26. McDonald, M. G.; Harbaugh, A.W. U.S. Geol. Surv. Techniques of Water-Resources Investigations 6-A1; U.S. Gov. Printing Office: Washington, DC, 1988.
27. Pollack, D. W. U.S. Geol. Surv. Open-File Report 89-381; U.S. Gov. Printing Office: Washington, DC, 1989.
28. Bouwer, Herman. Ground Water 1989, 27, 304-309.
29. Wood, W. W. U.S. Geol. Surv. Techniques of Water-Resources Investigations 1-D2; U.S. Gov. Printing Office: Washington, DC, 1976.
30. Skougstad, M. W.; Fishman, M. J.; Friedman, L. C.; Erdmann, D. E.; Duncan, S. S. U.S. Geol. Surv. Techniques of Water-Resources Investigations 5-A1; U.S. Gov. Printing Office: Washington, DC, 1979.
31. Ball, J. W.; Nordstrom, D. K.; Zachman, D. W. U.S. Geol. Surv. Open-File Report 87-50; U.S. Gov. Printing Office: Washington, DC, 1987.
32. Parkhurst, D. L.; Plummer, L. N.; Thorstenson, D. C. U.S. Geol. Surv. Water-Resources Investigations Report 82-14; U.S. Gov. Printing Office: Washington, DC, 1982.

33. Drever, J. I. The Geochemistry of Natural Waters; Prentice-Hall, Inc.: Englewood Cliffs, NJ, 1982.
34. Plummer, L. N.; Parkhurst, D.L.; Thorstenson, D.C. Geochim. Cosmochim. Acta 1983, 47, 665-686.
35. Plummer, L.N. In First Canadian/American Conference on Hydrogeology, Practical Applications of Ground Water Geochemistry; Hitchon, B.; Wallick, E.I., Eds.; National Water Well Association: Worthington, OH, 1984, pp 149-177.
36. Champ, D. R.; Gulens, R. L.; Jackson, R. E. Can. J. Earth Sci. 1979, 16, 12-23.
37. Morrison, J. L.; Atkinson, S. D.; Scheetz, B. E. In Proceedings 1990 Mining and Reclamation Conference and Exhibition; West Virginia University: Morgantown, WV, 1990, Vol. 1, pp 249-256.
38. Taylor, B.E.; Wheeler, M.C. In this volume.
39. Alpers, C. N.; Nordstrom, D. K.; Ball, J. W. Sci. Géol., Bull. 1989, 42, 281-298.

RECEIVED August 23, 1993

Chapter 24

Limit to Self-Neutralization in Acid Mine Tailings

The Case of East Sullivan, Quebec, Canada

M. D. Germain, N. Tassé, and M. Bergeron

Institut National de la Recherche Scientifique–Géoressources, P.O. Box 7500, Sainte-Foy, Quebec G1V 4C7, Canada

>Near-surface pore waters of the East-Sullivan tailings are enriched in H^+, Fe^{2+}, and SO_4^{2-} by sulfide oxidation. As the pore waters percolate downward, calcite dissolution results in an increase in pH and a release of Ca^{2+}, bringing the pore water in the stability field of siderite and gypsum with concomitant precipitation of Fe^{2+}, SO_4^{2-} and heavy metals. Locally, however, insufficient amounts of carbonate minerals are dissolved and high concentrations of metals and SO_4^{2-} reach the saturated zone. Negative environmental effects are observed in the tailings-water discharge area around the impoundment; Fe^{2+} is oxidized by atmospheric O_2 and precipitated as $Fe(OH)_3$ (or a similar phase), generating acidic waters. The situation is expected to worsen, because the jarosite cements, precipitated in the early stages of tailing alteration and limiting O_2 diffusion, are now undergoing dissolution. Furthermore, the calcite reservoir within the tailings decreases with the steady progress of sulfide oxidation.

Oxidation of sulfide-bearing tailings often generates low-pH water, mobilizing heavy metals that can contaminate groundwaters and surface waters. The East-Sullivan tailings impoundment, near Val-d'Or, Québec (Figure 1), makes up one such problematic site. In order to understand how contamination is generated and dispersed, a detailed hydrogeochemical study was undertaken.

The impoundment studied results from the exploitation, from 1949 to 1966, of a massive-sulfide ore body containing about 50% FeS_2 and $Fe_{1-x}S$. The impoundment totals 15 Mt of tailings forming a plateau extending over 136 ha, rising 4 to 5 m above the surrounding terrain. The thickness, determined from 50 drill holes, varies between 2.1 and 13.5 m, with an average of 7.3 m. The tailings overlay an impermeable substratum composed of clay sediments and bedrock. Pore gases, pore waters and solids were analyzed at eight stations. One of these was located in spilled tailings 150 m south of the impoundment (Figure 1).

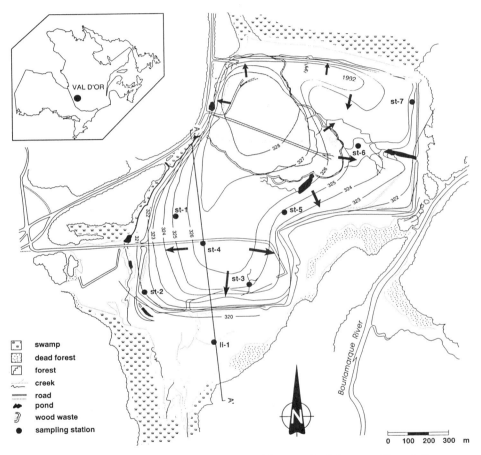

Figure 1. Location of the East-Sullivan tailings and of the sampling stations. Elevation contours for the water table in meters. Arrows indicate groundwater flow direction.

Methods of Investigation

Pore Gases. Interstitial gases were sampled in plastic disposable syringes via a small-diameter stainless-steel tube with a perforated tip. Gas-sampling procedures are identical to those described by Tassé et al. (1). Gas composition was determined using a Varian 3300 gas chromatograph equipped with a thermal conductivity detector. All analyses were performed the same day as the sampling. Detected gases included O_2, CO_2, CH_4, and N_2.

Pore Waters. Pore waters were sampled from tailing cores in the vadose zone and from piezometers in the saturated zone, following the method developed by Blowes and his collaborators (2,3).

Vadose-Zone: Sampling. Tailings were collected in 7.6 cm (outer diameter) aluminium tubes driven 60 to 70 cm at a time into the tailings with a hollow-stem auger. The cores were sectioned into 20 cm lengths for pore-water extraction. A plunger and a base-plate with a filter assembly and sampling port were fitted to each end of the 20 cm cores, and the water was extracted by squeezing under a pressure of up to 17,000 kPa. The water was filtered in line at the sampling port with a 0.45 µm filter and collected in 60 ml plastic syringes. In some cases, an immiscible liquid (Rohm and Hass epoxy plasticizer G-62) was needed to displace the pore water of coarse material.

Saturated Zone: Sampling. Pore waters were collected from drive-point polyvinyl chloride piezometers, 3 cm in diameter, through a 50-µm porous membrane enclosed in a 22-cm-long screened interval. Water samples were collected with a peristaltic pump, filtered twice with 0.45-µm filters connected in series to the pump tubing, and stored in 60 ml plastic syringes. A total of 70 piezometers were used to monitor the water-table surface and determine the groundwater-flow direction.

Measurements. pH, Eh, and temperature were measured as soon as the samples were collected with an Orion potentiometer Model 250A. The response of the pH electrode (Orion Triode 91-57BN) was calibrated against 4.01 and 7.00 pH buffers before each series of analyses. The response of the redox electrode (Orion 9678BN) was checked with a solution of potassium dichromate (Eh of 200 mV). Two aliquots of each water sample were acidified with 0.1 M H_2SO_4 solution for ferrous-iron determinations. Concentrations were measured in duplicate with a Mettler DL-25 automatic titrator equipped with a Mettler DM 140-SC electrode, using potassium dichromate as titrant. All titrations were done in a temporary laboratory the day of collection. Alkalinity was determined in the field using an Orion Total Alkalinity Test Kit. As the test requires 100 ml of sample, this analysis was restricted to samples from the saturated zone. The accuracy and precision of the alkalinity determinations was periodically verified with a standard (0.001 M Na_2CO_3). Water samples for cation analysis were acidified immediately after their collection with 1 ml of 12 M HCl and sent for analysis at the Québec Ministry of Energy and Resources laboratory. Cation concentrations were determined by inductively coupled plasma atomic emission spectrometry. The concentration of SO_4^{2-} was estimated from the former total sulfur analysis. Given the pH and Eh measured, and the geochemical speciation calculations (see below), it can be safely assumed that SO_4^{2-} is the dominant sulfur species in solution.

Tailings. Cores for mineral analysis were collected in aluminium tubes similar to those used for pore-water sampling and frozen immediately. Core samples were thawed at room temperature and examined megascopically for any special features. Eighty samples were selected in intervals of either coarse or fine-grained material, dried at 50°C, and split for geochemical analysis, X-ray diffraction (XRD) (bulk samples), and grain-size analysis by the Québec Ministry of Energy and Resources laboratory. Major elements were determined by X-ray fluorescence, inorganic C as CO_2 by combustion of both total sample and HCl-leached samples, and total S by combustion. Water-soluble sulfate and total sulfate were determined by liquid chromatography on leachates obtained respectively by shaking 2 g sample in 100 ml of water (water-soluble SO_4) and by digesting 2 g sample in 100 ml of 10 % m/v sodium carbonate solution (total SO_4) ([2]). Grain-size distributions were determined by combined sieve and photometric techniques. Jarosite identification by XRD was confirmed by electron microprobe. Its composition was evaluated by energy-dispersive analysis.

Geochemical Calculations. Geochemical-equilibrium conditions were evaluated using PHREEQE (version 1990, including revision 1.8; [4]). Equilibrium constants for jarosite (K, Na, and H_3O end-members) were imported from the WATEQ4F data base ([5]). For K-Na-H_3O solid solution, the ideal Gibbs free energy of formation was calculated from Alpers et al. ([6]). The input parameters used were: temperature, pH, Eh, alkalinity, and concentrations of major elements (Al, Ca, Fe, K, Mn, Mg, Na, and S) and trace metals (Cu, Pb, and Zn). Aluminium was considered, despite two water samples that could possibly be contaminated by the sampling tubes at st-6. For samples from the vadose zone, alkalinity was estimated from the CO_2 content of pore gases, assuming that the gas, liquid and solid phases were in equilibrium and that the total alkalinity was carbonate alkalinity. The mean charge imbalance of all samples was less then 3.5%.

Results

The tailings are composed of interlayered fine and coarse particles. The fine material is a fine to very-fine silt (median diameter (d_{50}) = 3.5 to 17 µm, average of 8.5 µm) in layers 1 to 15 cm thick, with an average of 4.5 cm. The coarse material is composed mostly of very-fine sand and coarse silt (d_{50} = 11 to 153 µm, average of 70 µm) in layers 1 to 17 cm thick, with an average of 7.0 cm. Grain size does not vary dramatically from one sampling site to another (Table 1). The proportion of coarse to fine tailings, however, varies more significantly, with ratios around 0.30 at basin centers (st-3 and st-6), reaching 3.12 at the margin of the impoundment (st-7). Two sites within the impoundment (st-4 and st-5) also show high ratios of coarse to fine tailings, but are located next to older dikes that are now included within the present impoundment.

Table 1. Grain-Size Characteristics of East-Sullivan Tailings

Station	Median Diameter (μm)		Thickness Ratio
	Fine-grained Layers	Coarse-grained Layers	(Coarse-grained/fine-grained)
1	6.9	58.3	0.60
2	10.6	86.6	1.05
3	7.7	85.4	0.31
4	9.0	125.0	2.71
5	8.7	96.1	3.17
6	10.5	60.8	0.29
7	11.2	78.0	3.12

The hydraulic conductivity of the tailings, estimated by falling-head tests, ranges from 1.2×10^{-6} m/s to 9.1×10^{-6} m/s. Low hydraulic conductivities occur in fine-grained tailings, at the center of the impoundment (st-3 and st-6), whereas high conductivities occur in coarse-grained zones, at the impoundment edge (st-2 and st-7) and locally within the impoundment (st-4 and st-5). Hydraulic gradients indicate that groundwaters within the impoundment generally move downward and outward, mainly toward the east and south (Figure 1). In the vicinity of station li-1, groundwater is discharging at a rate of 3.1×10^{-4} m/s/m^2 (Figure 2).

For simplicity, the results from st-1, st-4, st-6, st-7, and li-1 only are presented. These stations are representative of the main hydrogeochemical regimes observed on the site: the poorly drained area at the center of the impoundment (st-6), well drained areas at the impoundment edge (st-7) and within the impoundment (st-4), an intermediate case between the periphery and the center of the impoundment (st-1), and a groundwater discharge area in spilled tailings, outside the impoundment (li-1).

Data discussed here were collected in the autumn of 1991, between October 14 and October 29. Pore-water and pore-gas concentration profiles are generally smooth, with some variability (Figure 3). This variability reflects the fact that water samples from piezometers come mainly from relatively well drained, coarse-grained layers, whereas squeezed pore waters come preferentially from less-permeable, fine-grained layers with slightly different geochemical properties.

Pore Gases. The O_2 concentration in the pore gases declines abruptly to < 5 vol. % below 40 to 60 cm depth (Figure 3). The CO_2 content of pore gas increases from 0.15 vol. % at the tailings surface to concentration as high as 24 vol. % at 2 m depth at st-1 and st-7. At st-4, the maximum CO_2 concentration is 4 vol. %. At st-6, only two depths could be sampled. The high water level at li-1 prevented gas sampling.

370   ENVIRONMENTAL GEOCHEMISTRY OF SULFIDE OXIDATION

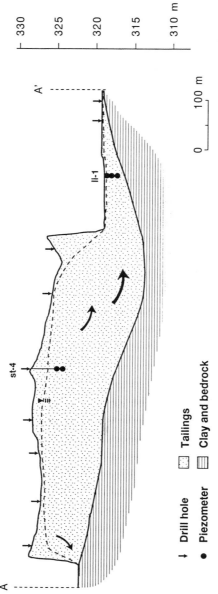

Figure 2. Cross-section A-A' through the East-Sullivan tailings impoundment. Arrows indicate locations of drill holes to the bedrock or clay basement. Dots represent piezometer locations for water sampling.

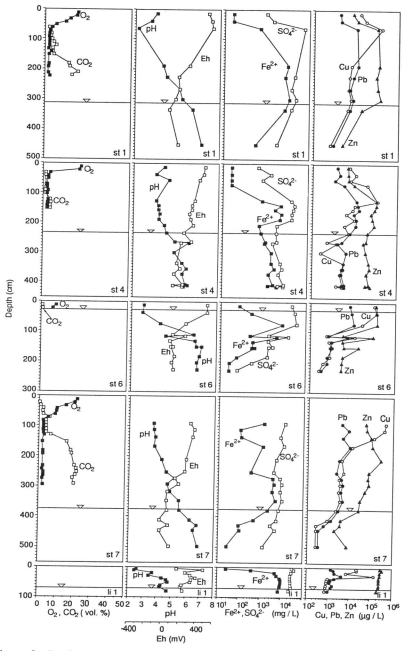

Figure 3. Profiles of O_2 and CO_2 in pore gases and of pH, Eh, Fe^{2+}, SO_4^{2-}, Cu, Pb, and Zn in pore-waters at the five selected stations located in Figure 1. The line with the triangle indicates the average water-table depth during the sampling period.

Pore Waters. Eh values decrease gradually from 500 mV near the surface to values less than 100 mV in the saturated zone (Figure 3). The pH values increase with depth from pH 2 near the tailings surface to near-neutral values at the bottom of the profiles, except at st-4, where the mean pH value in the saturated zone is lower (pH ≈ 5.6). Large variations in the concentrations of Fe^{2+} and SO_4^{2-} were observed over the sampled depths. Profiles for both ions are characterized, at st-1, st-6, and st-7, by an increase within the first meter, followed by a plateau, and then a decrease in the saturated zone, from 7,000 mg/L to < 300 mg/L Fe^{2+} and from 17,000 mg/L to < 3,000 mg/L SO_4^{2-}. At st-4, the plateau of high Fe^{2+} and SO_4^{2-} concentrations continues down into the saturated zone. The trace-metal profiles behave in a similar fashion to those of Fe^{2+} and SO_4^{2-}: an increase, followed by a stabilization down to the vicinity of the water table, and then a decrease (st-1, st-6 and st-7). At st-4, the profiles reach plateaus which are also maintained throughout the entire depth investigated. The profiles at station li-1, outside the impoundment, are comparable to those within the tailings. However, pH values at li-1 do not rise above 4.8, and Fe^{2+} concentrations range between 7,000 and 9,000 mg/L only 40 cm below the surface. Sulfate concentrations are also high compared to other stations, with concentrations in the 20,000 to 37,000 mg/L range.

Tailings Solids. Results for analysis done on the tailings solids are summarized in Figure 4. The primary mineral phases identified by XRD are alumino-silicate minerals. Quartz, chlorite, micas, and feldspars are present in all samples, with amphibole as a trace mineral in about 20% of the samples. The only sulfide mineral present at concentrations high enough to show up on bulk sample diffraction is pyrite. It was detected in about 80% of the samples, essentially missing in the near-surface specimens. Near-surface depletion of sulfide minerals is demonstrated well by the ratio of sulfur as sulfates to total sulfur, showing the strong speciation of sulfur to sulfates in the upper sections. Sulfate minerals (jarosite and gypsum) were observed as minor or trace minerals in more than 35% of the samples, especially near the surface. Limited electron microprobe analyses on jarosite suggest the composition $K_{.21}Na_{.43}(H_3O)_{.36}Fe_3(SO_4)_2(OH)_6$. Total SO_4 in solids decreases downward from the central part of the vadose zone. The proportion of water-soluble SO_4 is lower in the upper part of the section, which might reflect, similar to Blowes *et al.* ([3](#)), the relatively high proportions of less-soluble jarosite over gypsum and other more-soluble sulfates in the near-surface layers. Low concentrations and/or poorly diffracting properties prevent detection and identification of any sulfate beneath the 1 to 1.5 m upper tailings. Calcite, dolomite, or both carbonate minerals were detected as traces in about 20% of the samples, especially near the bottom of the sampled sections. Inorganic C as CO_2 profiles (Figure 4) show more clearly the depth-controlled distribution of these carbonate minerals. Goethite and hematite were detected in only one sample.

Discussion

Sulfide Oxidation. The dominant sulfide mineral in the East-Sullivan tailings is pyrite. Pyrite oxidation occurs through many reaction steps, some of which are

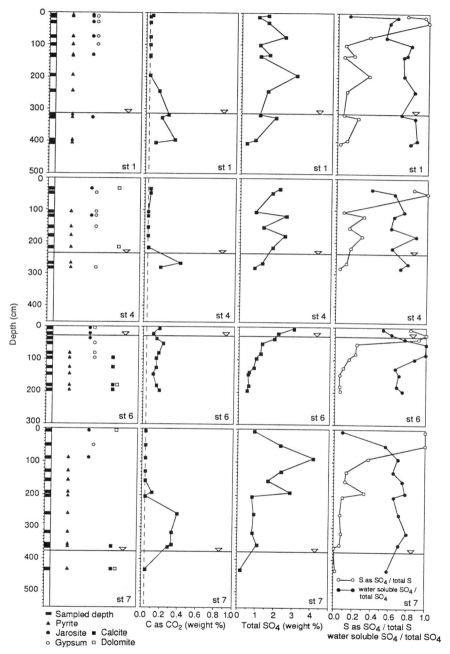

Figure 4. Mineral composition, inorganic C as CO_2, total sulfate content, ratios of sulfur as sulfate to total sulfur and of water-soluble sulfate to total sulfate in the East-Sullivan tailings. The dashed line in the C as CO_2 diagrams is the detection limit.

catalyzed by sulfide-oxidizing bacteria of the genus *Thiobacillus* (2,3,7), and are typically described by the following reactions:

$$FeS_2 + 7/2O_2 + H_2O \rightarrow Fe^{2+} + 2SO_4^{2-} + 2H^+ \qquad (1)$$
$$Fe^{2+} + 1/4O_2 + H^+ \rightarrow Fe^{3+} + 1/2H_2O \qquad (2)$$
$$FeS_2 + 14Fe^{3+} + 8H_2O \rightarrow 15Fe^{2+} + 2SO_4^{2-} + 16H^+ \qquad (3)$$

Most parameters measured in the pore gases, pore water, and tailings solids show the effect of sulfide oxidation (Figures 3 and 4). The observed decrease with depth in pore-gas O_2 concentrations, in the upper 80 cm of tailings, indicates oxygen consumption by pyrite and ferrous-iron oxidation (Equations 1 and 2). Low sulfur as sulfide concentrations in near-surface tailings is the consequence of sulfide oxidation by oxygen and ferric iron. The observed low pH results from acid generation (Equations 1 and 3). The rusty color of the upper 50 centimeters of the tailings results from the hydrolysis and precipitation of ferric iron generated in Equation 2. Equations 1 and 3 produce Fe^{2+} and SO_4^{2-}, resulting in increased concentrations of these constituents.

The increased Fe^{2+} and SO_4^{2-} concentrations extend beyond the lower limit of atmospheric O_2 diffusion (*e.g.* st-1 and st-4, Figure 3). This observation is likely the result of downward pore-water migration. This enrichment can also result from oxidation processes that exclude gaseous O_2, or from dissolution of iron-bearing sulfates. Dissolution of a sulfate mineral, such as jarosite, is possible, because that mineral, which probably formed during the early stages of the tailings alteration, is no longer stable according to geochemical calculations (Figure 5). Jarosite dissolution could increase the concentration of both dissolved Fe^{2+} and SO_4^{2-}, as the ferric iron released by dissolution is an important oxidant involved in pyrite oxidation (Equation 3).

A deep water table and consequently thick vadose zone allow sulfide oxidation to proceed deeper. Thus, large quantities of Fe^{2+} can be released to the underlying oxygen-depleted zones. Conversely, fine-grained material, with a low hydraulic conductivity, supports a high water-table position, and a more extensive capillary fringe, minimizing pyrite oxidation. Station li-1, outside the impoundment, is exceptional in that the very high concentrations of dissolved Fe^{2+} and SO_4^{2-} are observed at shallow depth, despite a nearly saturated porous medium. This observation is not compatible with oxidation throughout a deep vadose zone. Because station li-1 is in a discharge area, the source of Fe^{2+} is likely from the impoundment. The concentrations observed, being higher than those in the saturated zone of the impoundment, require some enrichment, which is likely provided by summer evaporation.

Acid-Neutralization Processes. Acidic water, generated by near-surface pyrite oxidation, percolates through the vadose zone and becomes neutralized gradually by geochemical reactions (Figure 3). These reactions can include dissolution of carbonate minerals, dissolution of alumino-silicate minerals, dissolution of Fe- and Al-hydroxide minerals, and H^+ adsorption on mineral surfaces (8). The most common acid-consuming reaction is probably calcite dissolution, which can be represented by the following reaction (for pH < 5.0):

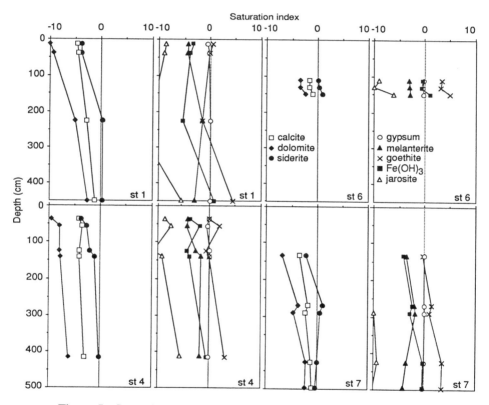

Figure 5. Saturation indices computed by the PHREEQE program (4).

$$CaCO_3(s) + 2H^+ \rightarrow Ca^{2+} + H_2CO_3^° \qquad (4)$$

where $H_2CO_3^°$ can dissociate to H_2O and $CO_2(g)$. This reaction, occurring during downward migration of the pore water, increases the pH, releases Ca^{2+} to pore waters, and increases the CO_2 concentration in pore gases. Dissolution of dolomite has the same effect, but also releases Mg^{2+}. Several effects of the dissolution of carbonate minerals are evident within the East-Sullivan tailings. Changes in the CO_2 concentration in the pore gas, in the inorganic C as CO_2 in the tailings solids, and in the qualitative calcite and dolomite distributions in core material (Figures 3 and 4), show that the carbonate minerals initially present in the vadose zone are being leached by acid-neutralizing reactions. Moreover, geochemical calculations with PHREEQE show that near-surface pore waters are undersaturated with respect to calcite, dolomite and siderite (Figure 5).

Pore-gas CO_2 concentrations at st-4 and st-6 are low. At st-6, carbonate leaching is minor, and little CO_2 is released. Pyrite oxidation and acid generation are limited by the high water table, only a few tens of centimeters below the ground surface, and by a thick capillary fringe. Carbonate minerals are relatively abundant from the base to the top of the cored section. Piezometer nest st-4 is located in an area composed of coarse-grained material, with a water table about 2.6 m below surface. The coarse-grained tailings favor a deeper water table and rapid O_2 diffusion. Consequently, pyrite oxidation and acid generation occur, leading to higher calcite- and dolomite-dissolution rates and to the present depletion of carbonate minerals, as shown by the low inorganic C content (Figure 4). Profiles of inorganic C in the tailings of st-1 and st-7 suggest that the high CO_2 pore-gas contents result from dissolution of carbonate minerals found in the lower half of the vadose zone.

Precipitation of Secondary Minerals. The main environmental problem associated with the East-Sullivan mine tailings is the potential mobility of Cu, Pb, Zn, Ni, Cd and Co in low-pH waters. Understanding the processes which control the mobilities of these metals is required to predict the future release of these metals to the environment. These metals may precipitate as discrete mineral phases, coprecipitate with other metals and/or become adsorbed onto solid phases. The formation of secondary minerals is in turn controlled by the concentrations of the major ions Ca, Fe, K, Mg, Na, Al, Si and SO_4^{2-} in pore waters, and the Eh, pH, alkalinity, and temperature.

The main alteration minerals observed and detected by XRD are goethite, jarosite and gypsum. At several locations, precipitation of these minerals was extensive enough to cement the tailings, forming a hard layer 10 to 20 cm thick just below the oxidation zone. No secondary minerals, however, could be detected by bulk XRD below 1 to 1.5 m, except gypsum at 3 m, at st-4. Nevertheless, sulfate minerals are present, as shown by profiles of sulfates in solids (Figure 4). A variety of other minerals are stable, according to geochemical equilibrium calculations (Figure 5). These minerals probably escaped detection due to their low abundance and/or their lack of crystallinity.

Besides goethite, many crystalline precipitates formed by the oxidation of ferrous iron are probably present but were not detected by XRD (*e.g.* Murad *et al.*,

9). Saturation indices calculated with PHREEQE indicate saturation with respect to goethite from the top to the bottom of the sections, except in the middle of st-1 (Figure 5), and saturation with respect to amorphous $Fe(OH)_3$ at the bottom of the sections. Nevertheless, significant amounts of iron oxyhydroxides are not expected to precipitate below near-surface tailings, given the very low concentrations of ferric iron in pore waters.

Jarosite, containing ferric iron, is observed only in near-surface tailings. As the pore waters are undersaturated with respect to jarosite (Figure 5), its formation probably dates back to the early years of alteration, when large amounts of sulfide minerals were available for oxidation. The decreasing rate of sulfide oxidation likely results in the observed degree of undersaturation. This observation is important because hard layers are thought to be important in limiting downward O_2-diffusion (e.g. 3). Evolution of the chemistry of pore waters at East-Sullivan suggest that a hard layer may become metastable on the long term, and prone to dissolution, if cemented dominantly by jarosite.

Pore waters are saturated or near-saturated with respect to gypsum from the top to the bottom of the sections (Figure 5). Close examination of SO_4^{2-}, inorganic C and total SO_4 profiles at st-1, st-4 and st-7 suggest that the source of calcium is calcite and dolomite. In the upper vadose zone, there is little carbonate mineral remaining as shown by inorganic C profiles (Figure 4). The amounts of Ca^{2+} released to the pore waters are thus minor. Consequently, dissolved SO_4^{2-} concentration reaches a maximum and forms a plateau (Figure 3). However, calcite and dolomite increase in mass down the section, making Ca^{2+} available in larger quantities, except at st-4, where the CO_2 profile shows that the amounts dissolved are low. Gypsum precipitation results in lower SO_4^{2-} concentrations. Comparison of inorganic C and total SO_4 profiles (Figure 4) with matching carbonate-poor/sulfate-rich and carbonate-rich/sulfate-poor segments show the net results of this precipitation process over years.

Another effect of carbonate dissolution is the increase in pH that brings the pore waters to saturation with respect to siderite (Figure 5). Precipitation of siderite decreases Fe^{2+} concentrations in all stations but st-4, where the amounts of dissolved carbonate are low.

Pore waters become undersaturated with respect to melanterite, and the formation of less hydrated forms of ferrous sulfate is unlikely. The strong covariation of Fe^{2+} and SO_4^{2-} in pore-water profiles can thus be misleading if interpreted as evidence of precipitation of $FeSO_4 \cdot nH_2O$, if not cross-checked with geochemical equilibrium models. Actually, precipitation of two distinct mineral phases, gypsum and siderite, controls Fe^{2+} and SO_4^{2-} concentrations.

A variety of mechanisms can explain the decreasing contents with depth of Cu, Pb, and Zn in the pore waters. Copper shows the lowest mobility. The pore waters are undersaturated with respect to copper-sulfate and copper-carbonate minerals. However, Blowes et al. (3) discuss replacement reactions by which iron and zinc of pyrrhotite ($Fe_{1-x}S$) and sphalerite (ZnS) are replaced by copper. This mechanism could also occur at the East-Sullivan tailings but was not confirmed. The pore waters are repeatedly near-saturation or at saturation with respect to anglesite ($PbSO_4$). Pore waters are undersaturated with respect to zinc sulfate or carbonate minerals. Adsorption or coprecipitation are probably the main control on Zn concentrations (2).

No detailed mineralogical work was done on the spilled tailings outside the impoundment. However, ferrous iron which percolates into the saturated zone is discharged in these areas. In contact with atmospheric O_2, Fe^{2+} oxidizes, then hydrolyzes and precipitates as $Fe(OH)_3$ (or other poorly crystalline hydrated ferric oxides, e.g. 9) according to the reactions:

$$Fe^{2+} + 1/4O_2 + H^+ \rightarrow Fe^{3+} + 1/2H_2O \qquad (2)$$
$$Fe^{3+} + 3H_2O \rightarrow Fe(OH)_3 + 3H^+ \qquad (5)$$

The latter reaction generates low-pH conditions. These reactions probably best explain the low pH values observed at station li-1 (Figure 3), where the nearly saturated porous medium and thin vadose zone (approximately 50 cm) do not allow extensive sulfide oxidation.

Calcite and dolomite dissolution was recognized as the prime factor in the precipitation of gypsum and siderite within the impoundment. It can be predicted that the exhaustion of these carbonate minerals by neutralization of acidic waters will result in pore-water profiles similar to those observed at st-4, where the buffering capacity is nearly completely consumed. It follows that the high H^+, Fe^{2+}, and heavy metal concentrations released by sulfide oxidation will be transferred directly to the groundwaters, and added to the acid-generating processes summarized in Equations 2 and 5.

Conclusions

Sulfide oxidation in the East-Sullivan tailings is a near-surface process that releases Fe^{2+} and SO_4^{2-} to the pore waters. Some iron is oxidized to the ferric state and hydrolyzed. Jarosite precipitates were observed, but jarosite does not seem to be stable in the alteration profiles studied, according to geochemical equilibrium calculations. Hard layers cemented by jarosite are thus metastable.

Neutralization of low-pH waters by calcite and dolomite dissolution allows precipitation of gypsum by increasing Ca^{2+} activity and stabilization of siderite by increasing pH, with a net decrease in Fe^{2+} and SO_4^{2-} in the pore waters. In some instances, however, the neutralizing capacity of the tailings has been exceeded over the depth investigated, and is expected to be eventually exceeded in other areas underlain by coarse-grained tailings, especially at the periphery of the impoundment. At some locations, high concentrations of Fe^{2+} (4,000 mg/L) are present in the saturated zone. This Fe^{2+} may be discharged in the surrounding environment. In contact with O_2, Fe^{2+} ions are oxidized, then hydrolyzed, and precipitated as $Fe(OH)_3$ (or similar phase), adding acidity. It is likely, therefore, that the surface waters will be more and more acidic with time.

Acknowledgements

This study was funded by the Québec Ministry of Energy and Resources. We thank David Blowes, University of Waterloo and Pierre Glynn, U.S.G.S. (Reston), for critically reviewing the manuscript, and Robert Tremblay, Québec Ministry of Energy and Resources and Martin Piotte, INRS-Géoressources, for helpful discussions. Assistance in the collection of field data was provided by Ms. K. Oravec and M.-A. Cimon. Ms. K. Oravec performed the microprobe analysis.

Literature Cited

1. Tassé, N.; Germain, M.D.; Bergeron, M. In this volume.
2. Blowes, D.W.; Jambor, J.L. Applied Geochemistry 1990, 5, 327-346.
3. Blowes, D.W.; Reardon, E.J.; Jambor, J.L.; Cherry, J.A. Geochimica et Cosmochimica Acta 1991, 55, 965-978.
4. Parkhurst, D.L.; Thorstenson, D.C.; Plummer, L.N. Water-Resources Investigations Report 1980, 80-96 210 pp.
5. Ball, J.W.; Nordstrom, D.K. In User's Manual for WATEQ4F, with Revised Thermodynamic Data Base and Test Cases for Calculating Speciation of Major, Trace, and Redox Elements in Natural Waters.; U.S. Geol. Survey Open-File Report 91-183, 1991, 189 pp.
6. Alpers, C.N.; Nordstrom, D.K.; Ball, J.W. Sciences Géologiques, Bulletin 1989, 42, 281-298.
7. Nordstrom, D.K. In Acid Sulfate Weathering; Soil Sci. Soc. Am., Special Publication No. 10, 1982, pp. 37-62.
8. Dubrovsky, N.M.; Cherry, J.A.; Reardon, E.J.; Vivyurka, A.J. Canadian Geotech. J. 1984, 22, 110-128.
9. Murad, E.; Schwertmann, U.; Bigham, J.M.; Carlson, L. In this volume.

RECEIVED October 11, 1993

SULFIDE-OXIDATION PROCESSES IN WETLANDS AND THE OCEANS

Chapter 25

Attenuation of Acid Rock Drainage in a Natural Wetland System

Y. T. J. Kwong and D. R. Van Stempvoort[1]

National Hydrology Research Institute, 11 Innovation Boulevard, Saskatoon, Saskatchewan S7N 3H5, Canada

At an abandoned mine site on Mount Washington in central Vancouver Island, British Columbia, natural attenuation of acid rock drainage is occurring in a small pond 50 m downstream from a shallow open pit. During the relatively dry summer months, water at the outlet of the pond system usually exhibits a pH value about one-half unit higher than that at the inlet. Dissolved concentrations of Cu and Zn and sulfate are concomitantly reduced to values lower by factors of 7, 15 and 3, respectively. Analyses of sediment cores from the pond reveal two distinct geochemical settings. Under oxidizing, deltaic conditions at the pond inlet, elemental copper and ferric hydroxide precipitate. Apparently, mixing of surface and ground waters provides the appropriate chemistry for the precipitation reactions to proceed. Near the pond outlet, reducing conditions predominate. Iron monosulfides and framboidal pyrite form *in situ*. Sulfur isotope analyses confirm that sulfate reduction is occurring in the pond sediments. In addition to formation of authigenic minerals, sorption of trace elements onto the iron hydroxides and sulfides is likely an important attenuation mechanism.

Acid rock drainage derived from oxidation of sulfide minerals is one of the most serious and costly environmental problems facing the mining industry today. Whereas recent research has greatly improved our knowledge on various aspects of acid mine drainage, abatement technology is still developing (e.g. [1]). Though many active schemes for treating acid rock drainage are readily available, they are generally costly and require ongoing maintenance. In addition, sludges produced during conventional treatment (e.g. lime addition) often require costly storage and disposal. Passive or "walk-away" solutions are more desirable because of the possibility of eliminating long-term maintenance costs.

The natural capacity of wetlands to attenuate contaminant transport has been recognized for a long time and has been successfully utilized to treat municipal and

[1]Current address: Saskatchewan Research Council, 15 Innovation Boulevard, Saskatoon, Saskatchewan S7N 2X8, Canada

industrial wastes (e.g. 2). However, the efficiency of passive wetland treatment to ameliorate metal-contaminated mine effluents is still largely unknown. For passive treatment to be effective, the long-term stability of natural wetlands, which tend gradually to fill with sediments, will have to be considered. Though recent research has clearly demonstrated the applicability of the concept under controlled conditions, using engineered wetlands (e.g. 3-5), there are only a few studies that provide information on the metal amelioration effect in natural systems (e.g. 6). Furthermore, the inorganic or biogeochemical mechanisms of metal attenuation in engineered or natural wetlands generally have not been described in detail.

The purpose of this paper is two-fold: 1) to document the attenuation of acid mine drainage products in a natural wetland system; and, 2) to investigate possible geochemical reactions involved in the metal attenuation process.

Site Location and Description

The abandoned Mount Washington Mine is located on the north side of Mount Washington (Figure 1) at an elevation of about 1320 m above mean sea level. The mine operated for about two years, starting in December, 1964. The ore consisted mainly of stockworks of chalcopyrite in fracture and breccia zones close to the contact of a Tertiary dioritic laccolith with Cretaceous meta-sediments (7). The abandoned mine site contains two remnant open pits, designated respectively as the North and the South Pit, each with associated waste rock dumps. The North Pit renders acidic drainage enriched in copper which apparently has had a detrimental effect on the salmonoid population downstream, notably in the Tsolum River (8, 9).

Figure 1b shows a map of the North Pit and the drainage leaving the site. The main generators of acidic drainage are the East Dump and the highly fractured bedrock in the shallow open pit which still contains abundant sulfides including chalcopyrite. The drainage flows to Pyrrhotite Creek, one of the headwater streams of the Tsolum River. The site has a steep slope; the vertical distance between the southern boundary of the pit and the toe of the East Dump near Weir #1 is 60 m. Snow melt is the main source of surface runoff, with peak discharge occurring usually in late May. Groundwater flow at the site is fracture-controlled with the hydraulic gradient largely dictated by the local topography.

Based on petrographic evidence, Kwong and Ferguson (10) suggested that acid generation in the North Pit was initiated with the stepwise oxidation of pyrrhotite. The relative reactivity of the prevalent sulfide minerals is in the following order:

marcasite > pyrrhotite > chalcopyrite > pyrite = arsenopyrite

Marcasite and elemental sulfur are intermediate alteration products of pyrrhotite, and are in turn oxidized to sulfate and hydrogen ions in solution and ferric hydroxide precipitates. A detailed discussion of the acid generation reactions operative at the site has been presented by Kwong (11). Chalcopyrite is dissolved by the acid generated during oxidation of the iron sulfides. During spring freshet and heavy fall rainstorms, flushing of concentrated acid solutions adhered to mineral or rock surfaces and dissolution of alteration minerals including soluble sulfate and carbonate salts (10) produce high acidity and metal loadings in drainage leaving the mine site. From June to September, the flux of surface drainage from the open pit decreases to a negligible value due to summer drought. In contrast, seepage from the East Dump directed through Weir #1 (Figure 1b) persists, although the flow rate may drop by more than 50%.

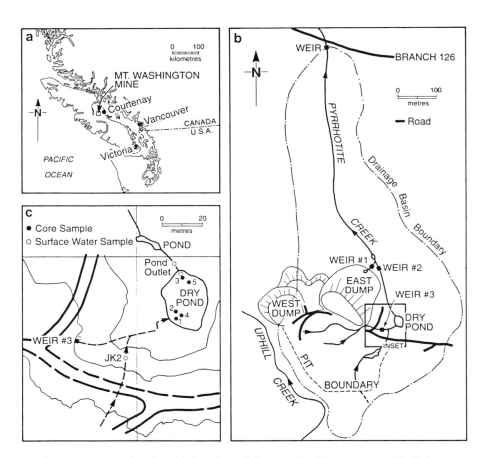

Figure 1. Maps showing (a) location of the Mt. Washington mine, (b) drainage leaving the North Pit (adapted in part from ref. 13) and (c) sample locations in the vicinity of Dry Pond.

The Dry Pond - A Natural Wetland System

General Setting. The Dry Pond, located 50 m downstream from the open pit, is a natural pond with an area of about 700 m² and a maximum depth of approximately 1 m. It drains into the perennial Pyrrhotite Creek system. Surface runoff from the North Pit accounts for the bulk of drainage water volume in the pond during the wet seasons (early spring and late fall), although ground-water seepage prevents it from drying out during summer drought. The pond inlet is characterized by a deltaic environment in which a mixture of yellowish-brown ferric hydroxide sludge and other fine-grained sediments are deposited. Unless stated otherwise, in this paper ferric hydroxide refers to undifferentiated ochreous precipitates of Fe-hydroxide, Fe-oxyhydroxides (ferrihydrite and goethite) and, possibly, minor Fe-oxyhydroxysulfate (12). Near its outlet, the pond has a steep slope and the shore is covered by forest vegetation. Often, an odor of hydrogen sulfide can be detected near the outlet. Between the pond outlet and Weir #2 near the toe of the East Dump (Figure 1b), Pyrrhotite Creek flows through a few short intervals of swampy areas where a soil profile is covered by peat, sphagnum moss or sedges. Elsewhere, the creek bed is rocky with prominent coatings of yellowish-brown ferric hydroxide.

Despite its small size, the Dry Pond system between Weirs # 3 and 2 apparently has a high capacity for attenuation of acid mine drainage products. Through most of the year, the water draining through Weir #2 has a higher pH and lower concentrations of dissolved Cu, Zn, and sulfate relative to Weir #3 (13). To investigate the fate and transport of the acid mine drainage products in the system, a series of surface-water samples and five sediment cores (Figure 1c) were collected in late September (a typical low-flow period), 1990, for various analyses. To shed light on possible seasonal variations in water chemistry, analyses of surface waters collected in early July, 1990 were compared to the September data.

Sampling and Analytical Methods. Surface-water samples were pumped through a 0.45 μm cellulose nitrate membrane filter into a flow cell equipped with a temperature probe, pH and Eh electrodes and a conductivity cell. After the system was thoroughly flushed, all measurements were recorded and two 125 ml (for metal and sulfate analyses) and one 250 ml (for isotope analysis) samples were collected. The samples for metal analysis were preserved with 1 ml of ultra-pure nitric acid.

Three sediment cores were sampled near the inlet and two near the outlet of the pond (Figure 1c) using 6 cm (diameter) PVC tubes. While still under water, the bottom of each core tube was sealed with a rubber stopper. Excess tube was cut off and the core was capped with another rubber stopper and tightly wrapped with duct tape for transportation.

In the laboratory, the cores were sectioned in an argon atmosphere to minimize oxidation (e.g. precipitation of ferric hydroxide). The pore water from each segment was squeezed out using a stainless steel pneumatic pressure cell. Metal concentrations in subsamples of each squeezed core segment were determined by induction coupled plasma analysis. Selected core segments were also examined using a JEOL JSM-840A scanning electron microscope equipped with a KEVEX energy dispersive X-ray analyzer. Mineralogical analyses were conducted using a SCINTAG XDS 2000 X-ray diffractometer with a copper anode. The dissolved Fe(total), Cu and Zn contents of the pore waters and the surface-water samples were determined by atomic absorption spectroscopy and dissolved sulfate by ion chromatography. For sufficiently large water samples, the dissolved sulfate was precipitated as barium sulfate and the sulfur isotope composition determined by mass spectrometry (14).

Observations. The analyses of the September 1990 suite of surface-water samples (Table I) show a pH increase of nearly 0.7 units, and reduction of dissolved concentrations of Cu, Zn and sulfate to values lower by factors of 7, 15 and 3, respectively, in the outflow (Weir #2) relative to the inflow (Weir #3). The total dissolved Fe concentrations also tend to decrease downstream. The anomalously low Fe value for the September Weir #3 sample may have resulted from either precipitation of ferric hydroxide or an analytical error. Along the flow path, dissolved sulfate is also progressively enriched in ^{34}S relative to ^{32}S suggesting that sulfate reduction has occurred.

Table I. Aqueous and Sulfur Isotope (dissolved sulfate) Geochemistry of Surface Water in the Dry Pond System, Mt. Washington Mine

Sample	Sampling Date (1990)	pH	Cu	Zn	Fe(total)	SO_4	$\delta^{34}S$ (SO_4)
					mg/L		‰ CDT
JK2	July 5	6.20	0.33	0.14	<0.03	126	-0.1
Weir 1	Sept. 11	3.61	12.8	0.65	2.17	794	+0.7
Weir 2	Sept. 11	4.49	3.26	0.11	0.31	172	+1.4
Weir 3	July 5	3.50	20.7	0.76	4.78	422	-0.1
	Sept. 11	3.83	20.7	1.67	0.84	581	+0.1
Dry Pond	July 5	3.70	14.9	0.51	1.41	274	+0.2
	Sept. 11	3.98	12.0	0.65	1.67	323	+1.0
Pond Outlet	Sept. 11	4.02	6.32	0.32	0.19	249	+1.1

A comparison of the July and September 1990 Dry Pond samples suggests that the extent of sulfate reduction and metal attenuation is inversely proportional to the inflow volume. September was drier than July and thus the downstream attenuation effect was more prominent in September. In addition, the July data indicate that drainage from south of the Dry Pond (JK2) was near neutral in pH and low in metals. The surface drainage from JK2 has little influence on the overall chemistry of the Dry Pond system because its discharge is generally a small fraction of that from Weir #3. However, the composition of the JK2 drainage can be used as a proxy for uncontaminated ground water in the vicinity of the Dry Pond.

Table II shows the pore-water geochemistry of the Dry Pond sediments. With depth, pH values tend to increase and dissolved concentrations of Cu, Zn, Fe and sulfate tend to decrease . Also, the pore-water sulfate becomes more ^{34}S-enriched with depth. Moreover, just beneath the water-sediment interface, pore-water sulfate near the pond outlet is more enriched in the heavier sulfur isotope than that from near the pond inlet.

The three cores sampled at the pond inlet are dominantly yellowish-brown reflecting the abundance of ferric hydroxide; streaks of black organic matter become more prominent with depth. When examined under a scanning electron microscope equipped with an energy-dispersive X-ray analyzer, elemental copper was observed in close association with ferric hydroxide (Figure 2) in the 10-20 cm section of Core #1 and about 10 cm below the water-sediment interface in Core #2. The presence of native copper was later confirmed by X-ray diffraction analysis. The ferric hydroxide, however, is apparently transparent to X-rays and the energy dispersive spectrum

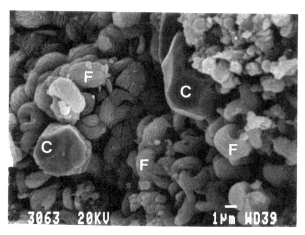

Figure 2. A scanning electron photomicrograph of the ferric hydroxide (F) - elemental copper (C) assemblage observed in shallow sediments near the pond inlet.

suggests the absence of sulfate. Overall, the total copper contents of these two core segments (after squeezing) are 1.2% and 0.93%, respectively.

Table II. Pore-water Geochemistry of the Dry Pond Sediments, Mt. Washington Mine (Date of sampling: September 11, 1990)

Core # (water depth cm)	Core Depth (cm)	pH	Cu	Zn	Fe(total)	SO_4	$\delta^{34}S$ (SO_4)
					mg/L		‰ CDT
1 (0)	0-4	4.87	0.74	0.34	96.6	413	ND*
	4-10	4.80	3.90	0.48	68.4	399	ND*
	10-20	5.05	1.06	0.14	50.1	301	ND*
2 (13)	0-10	3.31	1.40	0.18	40.4	246	+2.1
	10-18	6.30	0.16	0.03	21.4	118	+7.0
	18-28	6.65	0.02	<0.01	4.95	51.8	+10.7
4 (10)	0-32	6.23	0.04	0.09	14.9	132	+1.7
3 (15)	0-8	3.45	1.25	0.17	12.4	312	ND*
5 (20)	0-6	6.08	0.06	0.08	3.10	168	+8.0

*ND = not determined

The two cores sampled near the pond outlet are greyish-black in color, reflecting the abundance of organic debris and reducing conditions. Both cores contain framboidal pyrite in association with iron monosulfide (Figure 3). Whereas the presence of pyrite has been confirmed by X-ray diffraction analysis, the mineralogy of the iron monosulfide has not been determined.

Mechanisms of Metal Attenuation

Based on the hydrologic setting of the Dry Pond, the surface- and pore-water chemistries and the mineral phases observed in the sediments, three processes are apparently responsible for metal attenuation in the Dry Pond system: i) mixing of surface drainage with near-neutral pH ground water that seeps into the pond, ii) precipitation of metals as discrete mineral phases, including copper in the shallow sediment near the pond inlet, and iii) sorption onto surfaces of minerals including ferric hydroxide in the pond inlet area and iron sulfides in the pond outlet area. In addition, as suggested by Machemer and Wildeman (5), sorption of metals onto organic compounds may be an important attenuation mechanism.

Sorption of trace elements onto precipitating iron oxyhydroxides and pyrite is a well-documented phenomenon (e.g. 15, 16). The large surface areas observed in the ferric hydroxide and framboidal pyrite grains (Figure 3) identified from the Dry Pond sediments indicate that abundant active sites are available for sorption of trace metals. However, based on the present data, we could not determine the relative importance of sorption of metals in the Dry Pond system.

Over the long term, precipitation is likely the dominant process for metal removal from acid mine drainage (5). Thus, it is worthwhile to consider possible geochemical reactions occurring in the Dry Pond system that have resulted in the formation of authigenic minerals. Generally speaking, the pond inlet area is an oxidizing

Figure 3. Scanning electron photomicrographs of (a) framboidal pyrite and (b) iron monosulfide (arrows) observed in shallow sediments near the pond outlet.

environment marked by relatively rapid sedimentation under shallow water conditions. Geochemical reactions that can account for the formation of the observed minerals include the following:

$$2Fe^{2+} + Cu^{2+} \Leftrightarrow 2Fe^{3+} + Cu^0_{(s)} \qquad (1)$$

$$Fe^{3+} + 3H_2O \Leftrightarrow Fe(OH)_{3(s)} + 3H^+ \qquad (2)$$

Combining Equations 1 and 2, the overall reaction is:

$$2Fe^{2+} + Cu^{2+} + 6H_2O \Leftrightarrow 2Fe(OH)_{3(s)} + Cu^0_{(s)} + 6H^+ \qquad (3)$$

In contrast, the pond outlet has a relatively reducing environment as evidenced by the dark color of the sediment and hydrogen sulfide odor. Using CH_2O as a proxy for all organic matter, the following geochemical reactions (cf. [17]) can explain the formation of the observed iron sulfides:

$$2H^+ + SO_4^{2-} + 2CH_2O \Rightarrow 2CO_2 + 2H_2O + H_2S \qquad (4)$$

$$4Fe(OH)_{3(s)} + CH_2O + 8H^+ \Rightarrow 4Fe^{2+} + CO_2 + 11H_2O \qquad (5)$$

$$4H_2S + 4Fe^{2+} \Rightarrow 4FeS_{(s)} + 8H^+ \qquad (6)$$

$$CO_2 + 4FeS_{(s)} + 4H_2S \Rightarrow 4FeS_{2(s)} + CH_4 + 2H_2O \qquad (7)$$

In all cores analyzed (Table II), sulfate at depth is ^{34}S-enriched relative to Dry Pond surface water (Table I). These isotope data provide further evidence that sulfate reduction is taking place in the natural wetland system, including locally at depth in the pond inlet area.

The formation of pyrite and iron monosulfide is common in many modern reducing sedimentary environments including constructed wetlands used to treat mine drainage ([5], [18]). In contrast, although native copper is a common supergene alteration mineral (e.g. [19]), its formation in modern sediments has not been reported. Therefore, it is useful to consider whether reactions 1, 2 and 3 could have occurred in the Dry Pond environment. In particular, one should examine the change in Gibbs free energy of the overall reaction (Equation 3) accounting for the coexistence of ferric hydroxide and native copper.

Under standard conditions, based on free energy of formation data tabulated by Wagman et al. ([20]), the reaction as described by Equation 3 should proceed to the left. However, the direction of the reaction is influenced by the relative activities of H^+, Fe^{2+} and Cu^{2+} as evident in the following expression of free energy change for the reaction:

$$\Delta G_r = \Delta G_r^0 + RT\ln\{[H^+]^6/[Fe^{2+}]^2[Cu^{2+}]\} \qquad (8)$$

where ΔG_r and ΔG_r^0 are the change in Gibbs free energy under ambient and standard conditions, respectively; R is the gas constant, T the absolute temperature in Kelvins, and, [ ] denotes aqueous activity.

Using the measured pore-water chemistry corresponding to the depth of copper occurrence in Core #1 (i.e. pH = 5.0; dissolved Cu = 1.0 mg/L, assumed all to be in cupric state; and, dissolved Fe = 50 mg/L, assumed all to be in ferrous state) and assuming aqueous activity coefficients = 1, at 25°C and atmospheric pressure, the Gibbs free energy change of reaction 3 is still positive indicating that the reaction will again proceed to the left. However, if we assume that the pore water of Core #1 has been affected by mixing and/or diffusion during sample transport and storage and that the hydrogen ion activity observed in Core #2 is more representative of the field conditions, an entirely different conclusion is reached. For pH = 5.5, dissolved Cu^{2+} = 1.0 mg/L and dissolved Fe^{2+} = 30 mg/L (all values interpolated from the top of the sediment to the depth of native copper occurrence in Core #2), reaction 3 has a negative Gibbs free energy change, making the forward reaction possible.

Similarly, Equation 1 represents a reaction that is dependent on the relative ion activities, Fe^{3+}, Fe^{2+}, and Cu^{2+}. The mixing of surface and ground water in the pond inlet area may have produced the right conditions for the reaction to proceed to the right. For Eh = 0.2v and T = 25°C, pH = 5.5, dissolved Cu^{2+} = 1.0 mg/L and dissolved Fe^{2+} = 30 mg/L as above, reaction 1 has a negative change in Gibbs free energy favoring the precipitation of native copper.

Thus we conclude that the proposed reactions 1, 2 and 3 are thermodynamically feasible, given the right conditions. In absence of sufficient data, notably *in situ* Eh measurements and Fe(II) and Fe(total) analyses, a detailed speciation analysis is not warranted.

As shown in Tables I and II, Zn, relative to Cu, is only a minor component of the surface- and pore-water samples. The attenuation of Zn in the Dry Pond system could be explained by sorption and/or precipitation of Zn sulfide. However, we do not have direct evidence to determine which process is dominant.

The extent of attenuation of metals and acidity by various mechanisms is flow-dependent. In the Dry Pond system, acid attenuation by sulfate reduction and the accompanied immobilization of dissolved metals are minimal during high surface flow periods because the rate of flow of H^+ and metals through the system is much greater than the rate of consumption by attenuation reactions. Larger wetland systems occur downstream from the mine site. These could perhaps be developed to manage peak flow conditions to produce a stepwise reduction in acidity and dissolved metals along the drainage flow path.

Summary and Conclusion

In this paper, we present evidence for attenuation of acid mine drainage products in a natural wetland system at the abandoned Mount Washington Mine. In particular, we document the precipitation of authigenic minerals including elemental copper and iron sulfides. The precipitation of elemental copper may be related to mixing with neutral ground water. Another important metal attenuation process may be sorption of dissolved metals onto surfaces of minerals and organic phases. Given that the acidity and metal contents of drainage from the Mount Washington mine site are orders of magnitude lower than those prevailing at other sites, such as Equity Silver or Gibraltar (21), and that sulfate reduction is actively occurring in the Dry Pond, passive wetland treatment may be a viable option for the reclamation of the site. The major challenge lies in how peak flows can be managed effectively during the wet seasons, to prolong water retention in the wetland system so that sulfate reduction can proceed. Perhaps a series of wetland systems could be developed downstream from the site to enhance sulfate reduction and metal attenuation. Furthermore, we recommend that future investigations should examine in more detail the seasonal dynamics of metal fluxes from surface water to sediments.

Acknowledgments

We would like to thank M. Galbraith of British Columbia Ministry of Energy, Mines and Petroleum Resources for his co-operation and assistance in the field sampling for this study. C. Casey, D. Schill and K. Supeene of the National Hydrology Research

Institute assisted in the SEM-EDX analyses, core preparation and chemical analysis, respectively. Constructive suggestions by C. Alpers, D. Blowes and two anonymous reviewers have improved the manuscript.

Literature Cited

1. MEND (Mine Environment Neutral Drainage Program, Canada). Proceedings of the Second International Conference on the Abatement of Acidic Drainage, Montreal, September 16-18, 1991. 4 volumes.
2. Reed, S.C.; Middlebrooks, E.J.; and Crites, R.W. Natural Systems for Waste Management and Treatment; McGraw-Hill: New York, NY, 1988; 308 pp.
3. Girts, M.A.; Kleinmann, R.L.P. National Symposium on Surface Mining, Hydrology, Sedimentology and Reclamation; University of Kentucky: Lexington, KY, 1986; pp 165-171.
4. Kalin, M. In Ecological Engineering: An Introduction to Ecotechnology; Mitsch, W.J.; Jorgensen, S.E., Eds.; Wiley & Sons: New York, NY, 1989; pp 443-461.
5. Machemer, S.D.; Wildeman, T.R. J. Contaminant Hydrology 1992, 9, 115-131.
6. Emerick, J.C.; Huskie, W.W.; Cooper, D.J. Proceedings of a Conference on Mine Drainage and Surface Mine Reclamation, Vol.I: Mine Water and Mine Waste; U.S. Bur. Mines Info. Circ. IC-9183, pp 345-351.
7. Carson, D.J.T. Canadian Institute of Mining and Metallurgy, Transactions 1969, 72, pp 116-125.
8. Kangasniemi, B.J.; Erickson, L.J. A preliminary assessment of acid drainage from an abandoned copper mine on Mount Washington; B.C. Ministry of Environment: Victoria, B.C. 1986.
9. Erickson, L.J.; Deniseger, J.H. Impact assessment of acid drainage from an abandoned copper mine on Mt. Washington; B.C. Ministry of Environment & Parks: Nanaimo, B.C. 1987.
10. Kwong, Y.T.J.; Ferguson, K.D. In Acid Mine Drainage - Designing for Closure; Gadsby, J.W.; Malick, J.A.; Day, S.J., Eds.; BiTech Publishers: Vancouver, BC, 1990, pp 217-230.
11. Kwong, Y.T.J. Proceedings of the Second International Conference on the Abatement of Acidic Drainage, Montreal, September 16-18, 1991; MEND: Ottawa, ON, 1991; Tome 1, pp 175-190.
12. Bigham, J.M.; Schwertmann, U.; Carlson, L.; Murad, E. Geochim. et Cosmochim. Acta 1990, 54, 2743-2758.
13. Galbraith, D.M. Proceedings of the Second International Conference on the Abatement of Acidic Drainage, Montreal, September 16-18, 1991; MEND: Ottawa, ON, 1991; Tome 2, pp 145-161.
14. Ueda, A.; Krouse, H.R. Geochem. J. 1986, 20, 209-212.
15. Winland, R.L.; Traina, S.T.; Bigham, J.M. J. Environ. Qual. 1991, 20, 452-460.
16. Kornicker, W.A.; Morse, J.W. 1991. Geochim. et Cosmochim. Acta 1991, 55, 2159-2171.
17. Berner, R. A. Amer. J. Sci. 1970, 268, 1-23.
18. Hedin, R.S.; Hyman, D.M.; Hammack, R.W. Proceedings of a Conference on Mine Drainage and Surface Mine Reclamation, Vol. I.: Mine Water and Mine Waste U.S. Bur. Mines Info. Circ. IC-9183, 1988; pp 382-388.
19. Guilbert, J.M.; Park, C.F. Geology of Ore Deposits; W.H. Freeman: San Francisco, CA, 1985; 985 pp.
20. Wagman D.D.; Evans, W.H.; Parker, V.B.; Schumm, R.H.; Halow, I.; Bailey, S.M.; Churney, K.L.; Buttall, R.L. J. Phys. Chem., Ref. Data 11, suppl 2:392, 1982.
21. Errington, J.C..; Ferguson, K.D. Proceedings of the Acid Mine Drainage Seminar/Workshop, Halifax, Nova Scotia, March 23-26, 1987, Environment Canada: Ottawa, ON, 1987; pp 67-87.

RECEIVED October 8, 1993

Chapter 26

Kinetics of Oxidation of Hydrogen Sulfide in Natural Waters

Jia-Zhong Zhang and Frank J. Millero

Rosenstiel School of Marine and Atmospheric Science, University of Miami, 4600 Rickenbacker Causeway, Miami, FL 33149–1098

Recently we have studied the oxidation of H_2S with O_2 in natural waters as a function of pH (4 to 10), temperature (278.15 to 338.15 K) and salinity (0 to 36). The major products formed from the oxidation of H_2S were SO_3^{2-}, $S_2O_3^{2-}$ and SO_4^{2-}. A kinetic model was developed to predict the distribution of the reactants and products over a wide range of conditions. Dissolved and particulate metals have a significant effect on the rates of oxidation and the product formation. Field measurements made in the Black Sea, Framvaren Fjord, Chesapeake Bay and Cariaco Trench are in reasonable agreement with the values predicted from laboratory studies at the same concentration of Fe^{2+}.

The formation of hydrogen sulfide occurs in a variety of natural waters. The production in the pore water of sediments and stagnant basins (seas, lakes, rivers and fjords) is due to biological processes while the production in hydrothermal systems is an abiotic process. In anoxic environments organic matter can be oxidized by bacterial anaerobic respiration using various oxidants as an electron acceptor. The preference of oxidant is related to the greatest free energy yield per mole of organic carbon oxidized (1). Molecular oxygen is the thermodynamically most favorable electron acceptor which, if available, will be used preferentially in any ecosystem. If the supply of organic matter exceeds that of oxygen, other electron acceptors (in the order of MnO_2, NO_2^-, NO_3^-, Fe_2O_3 and SO_4^{2-}) are used when oxygen has been depleted. When one oxidant is depleted, the oxidant with the next highest energy yield is consumed until every oxidant is removed, or until all the metabolizable organic carbon has been depleted. Dissimilatory sulfate reduction

$$SO_4^{2-} + 2CH_2O \rightarrow HS^- + 2CO_2 + H_2O + OH^- \qquad (1)$$

is most commonly observed in marine environments where water circulation, consequently oxygen availability, is limited, but where sulfate is easily available because of its relatively high concentration in seawater (≈ 0.029 M).

The production of hydrogen sulfide also occurs in hydrothermal systems. In hot vent waters hydrogen sulfide may be leached from crustal basalts or produced by reduction of sulfate from seawater coupled with oxidation of Fe^{2+} from basalt to Fe^{3+}. Part of hydrogen sulfide so produced reacts with metal ions depositing metal sulfide minerals, mainly as pyrite, the remainder stays in the vent solution.

$$3SO_4^{2-} + 6H^+ + 17Fe_2SiO_4 \rightarrow H_2S + FeS_2 + 2H_2O + 11Fe_3O_4 + 17SiO_2 \qquad (2)$$

Once the hydrogen sulfide has been formed in natural waters, the oxidation of hydrogen sulfide is an important pathway for its removal. When water containing H_2S mixes with oxygenated water at the oxic and anoxic interface, the hydrogen sulfide can be oxidized by a number of oxidants. This oxidation is frequently coupled to changes in the redox state of metals (2) and nonmetals (3). Dissolved oxygen, however, is the most important and abundant oxidant. This oxidation involves a complex mechanism that results in the formation of several sulfur species (i.e., SO_3^{2-}, $S_2O_3^{2-}$, S and S_n^{2-}) as well as SO_4^{2-}. Although the formation of the resultant products has been studied by a number of workers (4-7), only a few of these product studies have been made over a wide range of experimental conditions and reaction media.

In recent years we have studied the oxidation of H_2S (8, 9) and H_2SO_3 (10) with O_2 in water and seawater in the laboratory (9, 10) and in the field (11-14). We have attempted to characterize how the rates and distributions of products are affected by trace metals (15). A kinetic model has been developed (16) to predict the rates of oxidation and formation of products. The results of these studies are briefly reviewed in this paper.

Overall Rate of Oxidation of H_2S

The overall rate equation for the oxidation of hydrogen sulfide can be represented by

$$- d[H_2S]/dt = k_{S(-II)} [H_2S]^a [O_2]^b \qquad (3)$$

where $k_{S(-II)}$ is the overall rate constant, a and b are the order of reaction and the brackets represent molar concentrations (9). When the concentration of oxygen is in excess the rate equation can be reduced to

$$- d[H_2S]/dt = k' [H_2S]^a \qquad (4)$$

where k' is related to $k_{S(-II)}$ by

$$k' = k_{S(-II)} [O_2]^b \qquad (5)$$

The order with respect to sulfide was determined by fitting the data to various rate equations with different values of a (9, 16). Plots of ln $[H_2S]$ versus time were found (9, 16) to give straight lines and indicated that a is equal to 1 or the reaction is first order with respect to the concentration of H_2S in agreement with earlier studies (4-7). The order of the reaction with respect to the oxygen concentration was determined from the values of k' at different concentrations of oxygen in seawater (16). These results gave a first-order dependency with respect to the concentration of oxygen. Our measured order of the oxidation with respect to oxygen is in close agreement with the value (0.8) obtained at mM levels of initial sulfide concentrations (7), but higher than the value of 0.56 found by Chen et al. (5). In summary all our studies indicated that the rate equation for the oxidation of H_2S in water and seawater is given by

$$- d[H_2S]/dt = k_{S(-II)} [H_2S][O_2] \qquad (6)$$

The values of $k_{S(-II)}$ for the oxidation of H_2S in NaCl solution and seawater have been measured as a function of pH (1 - 12), temperature (278.15 - 338.15 K), and ionic strength (I, 0 - 6M) (9). At pH 8.0, the rate constant ($k_{S(-II)}$, M^{-1} min^{-1}) in equation (6) is given by (T, K)

$$\log k_{S(-II)} = 10.00 - (3.0 \times 10^3)/T + 0.44\, I^{0.5} \qquad (7)$$

($\sigma = 0.18$ in log $k_{S(-II)}$). At 298.15 K the half time for the oxidation of H_2S with O_2 was found to be 50 ± 16 h in water and 25 ± 9 h in seawater (9). These results are in good agreement with the results of Chen and Morris (5) and O'Brien and Birkner (7).

The effect of pH in water at 328.15 K was found to be represented by (9)

$$k_{S(-II)} = (k_{H2S} + k_{HS}\, K_1/[H^+])/(1 + K_1/[H^+]) \qquad (8)$$

where $k_{H2S} = 1.33 \pm 0.28$ M^{-1} min^{-1} for the oxidation of H_2S and $k_{HS} = 5.73 \pm 0.12$ M^{-1} min^{-1} for the oxidation of HS$^-$

$$H_2S + O_2 \xrightarrow{k_{H2S}} \text{products} \qquad (9)$$
$$HS^- + O_2 \xrightarrow{k_{HS}} \text{products} \qquad (10)$$

The value of K_1 is the thermodynamic constant for the ionization of H_2S (8). The effect of temperature and ionic strength on the rate constants k_{H2S} and k_{HS} have been given by (9)

$$\log k_{H2S} = 7.44 - (2.4 \times 10^3)/T \qquad (11)$$
$$\log k_{HS} = 8.72 + 0.16\, pH - (3.0 \times 10^3)/T + 0.44\, I^{0.5} \qquad (12)$$

These equations are valid from pH = 4 to 8, T = 278.15 to 338.15 K, and I = 0 to 6 M.

Effect of Metals on the Rate of Oxidation

As will be discussed later, field measurements (11-14) made on the oxidation of H_2S in natural waters yielded half times that were much faster than determined in the laboratory on Gulf Stream seawater (9). To determine if this increase was due to trace metals, we have measured the rates of oxidation of H_2S in seawater with added transition metals (15). These studies have shown that at total dissolved concentrations below 300 nM, the rates are only affected by Fe^{2+}, Cu^{2+} and Pb^{2+}. At higher metal concentrations, the rates of oxidation of H_2S increase for all the metals except Zn^{2+}. The order of increase in the rates at higher concentrations for these metals is $Fe^{2+} > Pb^{2+} > Cu^{2+} > Fe^{3+} > Cd^{2+} > Ni^{2+} > Co^{2+} > Mn^{2+}$ (15).

Only Fe^{2+} and Mn^{2+} have levels in anoxic basins high enough to affect the oxidation of H_2S. The relative effect of metals on the oxidation of H_2S with oxygen at 298.15 K and pH 8.1 can be estimated (15) from (Figure 1)

$$\log(k_{S(-II)}/k_{S(-II)}^\circ) = a + b\, \log[M] \qquad (13)$$

where $k_{S(-II)}$ and $k_{S(-II)}^\circ$ are the rate constants, respectively, with and without added metal and

a = 6.55, b = 0.820 for Fe(II) from 10^{-8} to $10^{-5.3}$M
a = 5.18, b = 0.717 for Fe(III) from $10^{-7.2}$ to $10^{-3.3}$M
a = 1.68, b = 0.284 for Mn(II) from $10^{-5.9}$ to $10^{-3.3}$M.

The larger effect of Fe^{2+} is probably related to the formation of dissolved Fe^{3+} from the rapid oxidation of Fe^{2+} with oxygen (17)

$$Fe^{2+} + O_2 \rightarrow Fe^{3+} + O_2^- \qquad (14)$$

The oxidation of Fe^{2+} provides a higher initial concentration of Fe^{3+} than can be added from a stock solution of Fe^{3+} (which may be locally supersaturated). The peroxide generated by the oxidation of Fe^{2+}

$$O_2^- + H^+ \rightleftarrows HO_2 \tag{15}$$
$$HO_2 + HO_2 \rightarrow H_2O_2 + O_2 \tag{16}$$

can also increase the rates since it has a higher rate of oxidation with sulfide than oxygen (18). The Fe^{2+} formed from the reaction of dissolved or particulate Fe^{3+} can regenerate Fe^{2+} to complete the catalytic cycle. The effect of Fe^{3+} on the rates at low concentrations may be related to the reduction of Fe^{3+} to Fe^{2+} with HS^- and resultant oxidation of Fe^{2+} and generation of O_2^-. We presently are investigating this reduction process in the absence of O_2.

The effect of the metals on the rates of oxidation of H_2S below the observable precipitation of metal sulfides (which may be a slow process) can be attributed to the formation of ion pairs (15)

$$M^{2+} + HS^- \rightarrow MHS^+ \tag{17}$$

The overall rate constant is given by

$$k_{S(-II)}[HS^-]_T = k_{HS}[HS^-] + k_{MHS}[MHS^+] \tag{18}$$

where k_{HS} and k_{MHS} are the rate constants for the oxidation of HS^- and MHS^+. From the mass balance of $[HS^-]_T$ and $[M^{2+}]_T$ and the stability constant for the formation of MHS^+

$$\beta_{MHS} = [MHS^+]/[M^{2+}][HS^-] \tag{19}$$

we have

$$k_{S(-II)}[HS^-]_T = k_{HS}[HS^-] + k_{MHS}[M^{2+}]_T/(1 + 1/\beta_{MHS}[HS^-]) \tag{20}$$

If the value of β_{MHS} is large enough (19), then $(1 + 1/\beta_{MHS}[HS^-])$ is close to 1. Thus, a plot of $k_{S(-II)}$ versus the total metal concentration can be used to estimate k_{MHS}. These plots give $k_{MnHS} = 14.7 \pm 1.9$ M^{-1} min^{-1}, $k_{CdHS} = 53.5 \pm 1.6$ M^{-1} min^{-1} and $k_{CuHS} = 201.7 \pm 16.5$ M^{-1} min^{-1} (15). For the metals Co^{2+}, Pb^{2+}, and Ni^{2+} the rates did not increase until visual precipitation of metal sulfides formed in the solutions.

The decrease (15) in the rate of oxidation with added Zn^{2+} above 2 μM may be related to the formation of zinc sulfide ion pairs, $ZnHS^+$ (19). Unlike the ion pairs of the other metals, $ZnHS^+$ may be more stable than HS^-. If we assume that the $ZnHS^+$ species is non-reactive, we obtain

$$k_{S(-II)}/k_{S(-II)}^\circ = \alpha_{HS} = [HS^-]_F/[HS^-]_T = (1 + \beta_{ZnHS}[Zn^{2+}]_F)^{-1} \tag{21}$$

where α_{HS} is the fraction of free HS^-, β_{ZnHS} is the stability constant for the formation of $ZnHS^+$ and $[Zn^{2+}]_F$ is the free zinc concentration. The experimental results gave $\log \beta_{ZnHS} = 8.2 \pm 0.5$, which is higher than the value given by Dyrssen (19) of $\log \beta_{ZnHS} = 6.5$. These difference could be related to the formation of a kinetically stable product that is not in equilibrium.

The presence of Fe and Mn in natural waters not only increases the rate of oxidation of sulfide, but also can have an effect on the oxidation of intermediates such as sulfite. This catalysis can change the distribution of the products formed during the oxidation. The effects of metals on the formation of products during the oxidation of H_2S are discussed in the next section.

Products from the Oxidation of H_2S

The final product from the oxidation of sulfide is sulfate, the sulfur compound having the highest oxidation state and the most stable compound in oxic waters.

Various intermediates, such as sulfite and thiosulfate, also can be formed during the course of the reaction. The products formed from the oxidation of H_2S in seawater have been studied (16) as a function of pH, temperature, salinity, and reactant concentration. To examine the mass balance of sulfur compounds during the oxidation the experiments were done (16) in pure water where SO_4^{2-} formed from the oxidation could be measured by an ion chromatographic technique. The major products formed were found to be SO_4^{2-}, SO_3^{2-} and $S_2O_3^{2-}$ (Figure 2). Elemental sulfur or polysulfides were not found by spectroscopic techniques (5). The total equivalent sulfur of the products and reactants was constant indicating that SO_4^{2-}, SO_3^{2-} and $S_2O_3^{2-}$ are the main products. The distribution of products from the oxidation of H_2S in seawater (Figure 3) is similar to the results in water.

Sulfite has been proposed to be the initial product from the oxidation of sulfide with oxygen in alkaline solutions (4). Unfortunately sulfite was not measured in the earlier studies due to the limitation of methods used. Our measurements in water and seawater at pH = 8.2 support the contention that sulfite is the initial oxidation product (16). The concentration of sulfite increases rapidly at the beginning of the reaction and decreases slowly when the production is slower than the rate of oxidative removal (see Figure 2). The major product is SO_4^{2-} in all of the seawater runs. The concentration of sulfate is higher at higher concentrations of oxygen as one might expect. Because we determined SO_4^{2-} in seawater by difference, some of the assigned values could be elemental sulfur or polysulfides present at levels below the detection limit of the spectroscopic method used (1 μM).

The concentration of thiosulfate increases slowly throughout the reaction after an initial lag period. This observation suggests that thiosulfate is not the initial product of the oxidation and supports the previously made assertion that SO_3^{2-} serves as the initial oxidation product (16). In earlier studies, the formation of sulfite and thiosulfate was treated as a parallel reaction based on 8-hour experiments (7). Under those experimental conditions (equal moles of initial sulfide and oxygen at 298.15 K) the sulfide oxidation was very slow and only 30% of the initial sulfide was oxidized. A longer period of reaction is necessary to show the true pattern of the intermediate products. This requirement is, of course, the reason that we made most of our measurements at 318.15 K. Thiosulfate is a stable product in the absence of bacteria and little oxidation occurs over 80 hours. This observation is in agreement with the finding of earlier studies (4, 7). The effect of pH on the distribution of products has also been examined and the results have been attributed to changes in the rates of the individual reaction steps (16).

The effect of metals (Fe^{2+}, Fe^{3+}, Mn^{2+}, Cu^{2+}, Pb^{2+}) and solids (FeOOH and MnO_2) on the distribution of products has also been studied (16). The intermediates formed (Figure 4) during the oxidation of Cariaco Trench waters (350 nM Fe^{2+}) clearly show that dissolved and particulate metals not only increase the rate of oxidation of H_2S, but also change the distribution of the products.

Oxidation of the Intermediate S(IV)

Because the rate of oxidation of sulfite to sulfate affects the distribution of products formed during the oxidation of sulfide, we have studied its oxidation (10). The overall rate equation for the oxidation of sulfite in seawater can be written as (10)

$$- d[S(IV)]/dt = k_{S(IV)} [S(IV)]^2 [O_2]^{0.5} \qquad (22)$$

where $k_{S(IV)}$ is the overall rate constant; the brackets represent concentrations. All the kinetic measurements in seawater and seasalts (10) showed that the oxidation reaction was second order with respect to sulfite. This relationship is in agreement with previous measurements in NaCl solutions (20) and in seawater (21). The order of the oxidation with respect to oxygen was found to be 0.5($\pm$

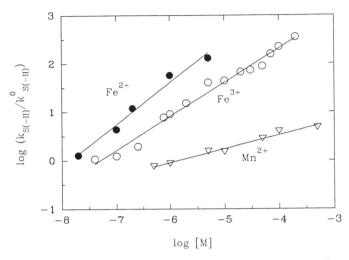

Figure 1. The effect of Fe and Mn on the rate of oxidation of H_2S in seawater at 298.15 K and pH 8.1 (15).

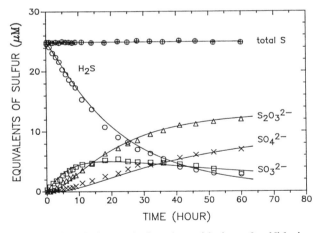

Figure 2. The sulfur balance during the oxidation of sulfide in water at 318.15 K and pH = 8.2. The smooth curves are calculated from the model (16).

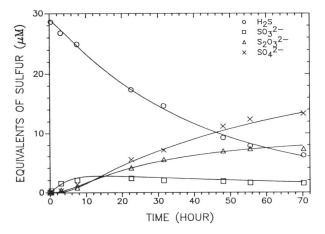

Figure 3. The distribution of products from the oxidation of sulfide in seawater at S = 35.0, pH = 8.2 and T = 298.15 K. The smooth curves are calculated from the model (16).

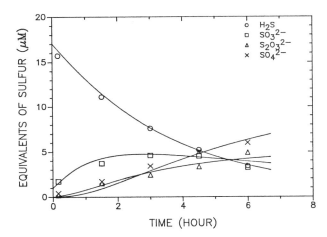

Figure 4. The distribution of products from the oxidation of sulfide in the Cariaco Trench (mixture of surface and deep water) at 298.15 K. The smooth curves are calculated from the model (16).

0.03) (10). The values of log $k_{S(IV)}$ as a function of temperature and ionic strength has been fitted to (10)

$$\log k_{S(IV)} = 19.54 - 5069.47/T + 14.74\ I^{0.5} - 2.93\ I - 2877.0\ I^{0.5}/T \tag{23}$$

where I is the molal ionic strength, T is the temperature (K), k is in $M^{-1.5}\ min^{-1}$. The standard error is 0.05 in log $k_{S(IV)}$. This equation should be valid for most estuarine and sea waters. The effect of ionic strength on the energy of activation is in agreement with the earlier measurements made in NaCl solution (20).

The major ionic components of seawater also have an effect on the rates of oxidation of sulfite (10). The rates measured in 0.57M NaCl solution were found to be higher than the rates in seawater. Measurements made in the major sea salts solution indicate that Ca^{2+}, Mg^{2+} and SO_4^{2-} added to NaCl solution cause the decrease. Measurements made in artificial seawater (Na^+, Mg^{2+}, Ca^{2+}, Cl^- and SO_4^{2-}) were found to be in good agreement with the measurements in real seawater (10).

The effect of pH on the rate of oxidation was found to be significant (10). The rate increased from pH 4 to a maximum at pH 6.5 and decreased at higher pH. The effect of pH on the rates was attributed to the rate-determining step involving the combination of HSO_3^- and SO_3^{2-}. This yields

$$k_{S(IV)} = k''\ \alpha(HSO_3^-)\ \alpha(SO_3^{2-}) \tag{24}$$

where $\alpha(i)$ is the molar fraction of species i versus total S(IV). Values of $k'' = 6.66 \pm 0.06$ and 6.17 ± 0.17 were found for NaCl solution and seawater respectively (10).

Kinetic Model for the Oxidation of H_2S

A kinetic model was formulated (16) based on the concentration-time dependence of the reactants (H_2S and O_2) and products (SO_3^{2-}, $S_2O_3^{2-}$ and SO_4^{2-}) in an abiotic system. The validity of the model was evaluated by comparing the model predictions with the experimental measurements of reactants and products. At low concentrations the overall oxidation of HS^- with O_2 is given by (4)

$$HS^- + 1.5\ O_2 \rightarrow HSO_3^- \tag{25}$$
$$SO_3^{2-} + 0.5\ O_2 \rightarrow SO_4^{2-} \tag{26}$$
$$SO_3^{2-} + HS^- + 0.5\ O_2 \rightarrow S_2O_3^{2-} + OH^- \tag{27}$$
$$S_2O_3^{2-} + 0.5\ O_2 \rightarrow SO_4^{2-} + S \tag{28}$$

With these overall reactions in mind, one can attribute the formation of SO_3^{2-} to the oxidation of HS^- and the formation of SO_4^{2-} to the oxidation of SO_3^{2-}. The formation of SO_4^{2-} from the oxidation of $S_2O_3^{2-}$ can be neglected (4) for solutions devoid of bacteria (13). The formation of $S_2O_3^{2-}$ can be attributed to the overall reaction of SO_3^{2-} and HS^- with O_2. This leads to the following overall rates of oxidation

$$H_2S + O_2 \xrightarrow{k_1} Products\ (SO_3^{2-}) \tag{29}$$
$$H_2SO_3 + O_2 \xrightarrow{k_2} Products\ (SO_4^{2-}) \tag{30}$$
$$H_2S + H_2SO_3 + O_2 \xrightarrow{k_3} Products\ (S_2O_3^{2-}) \tag{31}$$

The overall rate equations for H_2S, SO_3^{2-}, $S_2O_3^{2-}$, SO_4^{2-} are given by

$$d[H_2S]/dt = -k_1[H_2S][O_2] - k_3[H_2S][SO_3^{2-}][O_2] \tag{32}$$
$$d[SO_3^{2-}]/dt = k_1[H_2S][O_2] - k_2[SO_3^{2-}]^2[O_2]^{0.5} - k_3[H_2S][SO_3^{2-}][O_2] \tag{33}$$

$$d[S_2O_3^{2-}]/dt = k_3[H_2S][SO_3^{2-}][O_2] \quad (34)$$
$$d[SO_4^{2-}]/dt = k_2[SO_3^{2-}]^2[O_2]^{0.5} \quad (35)$$

where [i] is the total concentration of i. It should be pointed out that these are overall rate equations and they do not represent the mechanism of the reactions. The order of the rates of oxidation of H_2S (equation 29) and H_2SO_3 (equation 30) are assumed to be equal to the order in equations 6 and 22, respectively. The order of the rate of formation of $S_2O_3^{2-}$ has been taken from the work of Avrahami and Golding (4)

These rate equations have been integrated simultaneously to evaluate the values of k_1, k_2 and k_3 using the experimental time dependence concentrations of all the reactants and products (16). The values of k_1, k_2 and k_3 determined from these studies as a function of temperature, salinity and pH in water and seawater are given elsewhere (16). The experimentally measured concentrations of H_2S, SO_3^{2-}, $S_2O_3^{2-}$ and SO_4^{2-} were found to be in good agreement with the model predictions up to reaction times of 80 hours (Figure 2 and 3). The values of k_2 in seawater needed to fit the data were slightly smaller than the values determined in our previous study (10), especially at higher temperatures. This difference is probably due to the inhibition of the oxidation of sulfite in the presence of sulfide (5). This finding is supported by the previous observations (6) that sulfite in the presence of H_2S is more stable in seawater than predicted by its rate of oxidation. Because the oxygen is in excess, the competition for oxidant is unlikely to cause this difference. The complexation of HS- with trace metals which can catalyze the oxidation of sulfite in seawater may be a more likely cause. Trace metals have higher tendency to complex with HS- than SO_3^{2-} (16).

We found (16) that the formation of thiosulfate could be best represented with a rate equation which is zero-order with respect to oxygen, a finding in agreement with earlier work (7). The effect of changes in the ratio of H_2S to O_2 on the product distribution over the range of our measurements can be attributed to the order in the rate equations which is independent of the rate constants.

The values of k_1, k_2 and k_3 from the experiments in pure water as a function of pH (4 to 10) have been fitted to smooth equations of pH (T = 318.15 K) (16)

$$\ln k_1 = -4.71 + 0.914 \text{ pH} - 0.0289 \text{ pH}^2 \quad (36)$$
$$\ln k_2 = 3.87 + 1.51 \text{ pH} - 0.103 \text{ pH}^2 \quad (37)$$
$$\ln k_3 = -9.09 + 3.01 \text{ pH} - 0.177 \text{ pH}^2 \quad (38)$$

Assuming that the pH dependences of rate constants are independent of ionic strength and temperature (9, 10), these equations can be used to estimate the rate constants for other natural waters. Further measurements are needed to examine the pH dependence of rate constants over a wide range of temperature and ionic strength to test the validity of this assumption.

The values of k_1, k_2 and k_3 as a function of salinity (S) and temperature (T, K) have been fitted to the equations (pH = 8.2)

$$\ln k_1 = 26.90 + 0.0322 \text{ S} - 8123.21/T \quad (39)$$
$$\ln k_2 = 14.91 + 0.0524 \text{ S} - 1764.68/T \quad (40)$$
$$\ln k_3 = 28.92 + 0.0369 \text{ S} - 8032.68/T \quad (41)$$

These equations should be valid for most estuarine and sea waters. This kinetic model can be used to predict the product distribution for the oxidation of sulfide in natural waters with low concentrations of trace metals.

The agreement between the model and the observed distribution of reaction products does not provide conclusive proof that the reaction pathways of the overall model actually describe the series of elementary reactions that occur in an abiotic environment. The detailed mechanisms might involve many elementary reaction steps. At present there are two detailed mechanisms for the oxidation of H_2S with O_2 in aqueous solutions (5, 22), the polar mechanism and the free radical chain mechanism. The polar mechanism for the formation of HSO_3^- is given below (22):

$$HS^- + O_2 \rightarrow HSO_2^- \tag{42}$$
$$HSO_2^- \rightarrow H^+ + SO_2^{2-} \tag{43}$$
$$SO_2^{2-} + O_2 \rightarrow SO_2^- + O_2^- \tag{44}$$
$$SO_2^- + O_2 \rightarrow SO_2 + O_2^- \tag{45}$$
$$SO_2 + H_2O \rightarrow HSO_3^- + H^+ \tag{46}$$

The overall reaction is given by

$$HS^- + 3O_2 + H_2O \rightarrow HSO_3^- + 2HO_2 \tag{47}$$

The slow step in the oxidation of HS^- to HSO_3^- is given by equation 42. The superoxide ion formed can react with itself to form hydrogen peroxide which can also react with HS^- (18, 23).

$$HO_2 + HO_2 \rightarrow H_2O_2 + O_2 \tag{16}$$

The intermediate HSO_2^- or SO_2^{2-} can react with HS^- and form thiosulfate

$$HSO_2^- + HS^- \rightarrow S_2O^{2-} + H_2O \tag{48}$$
$$S_2O^{2-} + O_2 \rightarrow S_2O_3^{2-} \tag{49}$$

The initial reaction can also result in the formation of elemental sulfur

$$HS^- + O_2 \rightarrow S + HO_2^- \tag{50}$$

which can react with sulfite to give thiosulfate

$$S + SO_3^{2-} \rightarrow S_2O_3^{2-} \tag{51}$$

or hydrogen sulfide to give polysulfides

$$nS + HS^- \rightarrow HS_n^- \tag{52}$$

The formation of sulfate comes from the oxidation of sulfite with oxygen

$$2SO_3^{2-} + O_2 \rightarrow 2SO_4^{2-} \tag{53}$$

or hydrogen peroxide

$$HSO_3^- + H_2O_2 \rightarrow HSO_4^- + H_2O \tag{54}$$

which involves the formation of the intermediate $^-O_2SOOH$ (23).

In the free radical mechanism, oxidation is initiated by an outer-sphere electron transfer from HS^- to oxygen to form the $HS\cdot$ radical.

$$HS^- + O_2 \rightarrow HS\cdot + O_2^{-\cdot} \tag{55}$$

In the presence of trace metals in seawater, electron transfer from HS^- to the transition metal ions to form $HS\cdot$ radical is more favorable.

$$HS^- + M^{n+} \rightarrow HS\cdot + M^{(n-1)+} \tag{56}$$

The $HS\cdot$ radical is further oxidized to sulfite by a free radical chain sequence involving oxygen

$$HS\cdot + O_2 \rightarrow HSO_2\cdot \tag{57}$$
$$HSO_2\cdot \rightarrow H^+ + SO_2^{-\cdot} \tag{58}$$
$$SO_2^{-\cdot} + O_2 \rightarrow SO_2 + O_2^{-\cdot} \tag{59}$$
$$SO_2 + H_2O \rightleftarrows HSO_3^- + H^+ \tag{60}$$
$$HSO_3^- \rightleftarrows SO_3^{2-} + H^+ \tag{61}$$

The sulfite formed can react with HS$^-$ to produce thiosulfate

$$HS^- + SO_3^{2-} \rightarrow HS_2O_3^{2-} \quad (62)$$
$$HS_2O_3^{2-} + O_2 \rightarrow HS_2O_3^- + O_2^{-\cdot} \quad (63)$$
$$HS_2O_3^- \rightleftarrows S_2O_3^{2-} + H^+ \quad (64)$$

The sulfite formed can be further oxidized to sulfate through the free radical mechanism ($\underline{10}$)

$$M^{n+} + SO_3^{2-} \rightarrow M^{(n-1)+} + SO_3^{\cdot-} \quad (65)$$
$$SO_3^{\cdot-} + O_2 \rightarrow SO_5^{\cdot-} \quad (66)$$
$$SO_5^{\cdot-} + SO_3^{2-} \rightarrow SO_5^{2-} + SO_3^{\cdot-} \quad (67)$$
$$SO_5^{2-} + SO_3^{2-} \rightarrow 2SO_4^{2-} \quad (68)$$

Depending upon the assumptions made for the nature of the rate-determining propagation and termination steps, different rate equations can be obtained. Although we cannot propose a unique mechanism for the oxidation of H_2S and formation of the resultant products in natural water, it is useful to make some comparison of our findings with the proposed mechanisms given above. The first-order dependence of the oxidation of H_2S with respect to H_2S and O_2 can be attributed to the slow steps in the oxidation being equation 42 in polar mechanism or equation 57 in free radical mechanism. The slow step in the formation of thiosulfate can be attributed to equation 62 which will give the correct orders with respect to sulfide and sulfite. As discussed elsewhere ($\underline{10}$) the termination steps in the oxidation of sulfite such as

$$SO_5^- + SO_3^- \rightarrow S_2O_6^{2-} + O_2 \quad (69)$$
$$2HO_2 \rightarrow H_2O_2 + O_2 \quad (16)$$

would give the appropriate oxygen dependence for the oxidation of sulfite. Further experimentation is needed to prove these speculations. Beside the stable intermediate determined in this study, identification of those extremely reactive, short-lived intermediates, such as free radical or transition state complexes, would provide evidence for the possible mechanism.

As mentioned earlier, trace metals can have a significant effect on the oxidation of H_2S and H_2SO_3. We have determined the values of k_1, k_2 and k_3 as a function of added Fe and Mn. These individual rate constants give a reasonable fit of the experimental measurements (Figure 5). As discussed in the next section these results can be used to explain the distribution of products in natural waters.

Oxidation of H_2S in Anoxic Environments

Black Sea. The Black Sea is the largest anoxic basin in the world. The interface between oxic and anoxic water in the central basin is near 100 m and coincides with the permanent halocline that results from the large input of fresh waters and the precipitation greatly exceeding the evaporation. The rates of oxidation of aerated surface Black Sea waters with added NaHS gave a $t_{1/2} = 22 \pm 2$ h ($\underline{13}$). These results are in good agreement with our earlier measurements ($\underline{9}$) made on Gulf Stream waters ($t_{1/2} = 25 \pm 9$ h). The rates of oxidation of mixtures of surface and deep Black Sea waters were measured at 298.15 K. The values of $t_{1/2}$ vary from 1.2 to 2.9 h (Figure 6). Within the experimental error of the measurements, the half times obtained for unfiltered (2.0 ± 0.4 h) and filtered (2.3 ± 0.5 h) mixtures were the same.

These measurements indicate that the faster rates for the mixtures are caused by dissolved constituents in the deep waters. Since the fraction of deep waters (X_D) used was about 0.3, the oxidation rates of the deep waters are three times

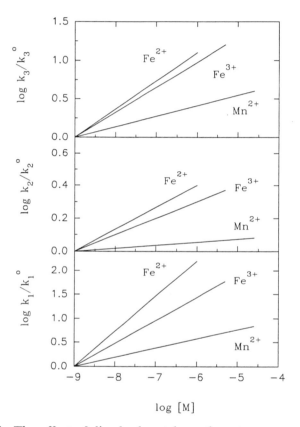

Figure 5. The effect of dissolved metals on the rate constants for the oxidation of sulfide (k_1), sulfite (k_2) and formation of thiosulfate (k_3) in seawater (16).

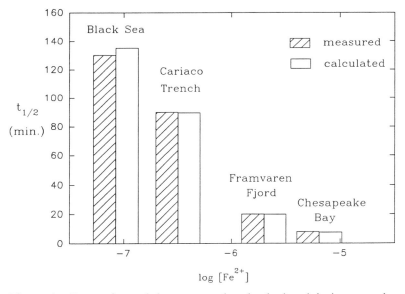

Figure 6. Comparison of the measured and calculated (using equation 13) half times for the oxidation of H_2S by O_2 in deep waters from the Framvaren Fjord (11), the Chesapeake Bay (12), the Black Sea (13), and the Cariaco Trench (14).

faster than surface waters mixed with deep waters. The adjusted half time ($t_{1/2} \cdot X_D$) is 32 ± 7 min for the deep water. The effect of metals on the oxidation of H_2S with oxygen can be estimated from equation 13. This equation can be used to make an estimate of the expected rates for waters of different concentrations of Fe^{2+} and Mn^{2+}. The rates predicted (see Figure 6) for the half times for the oxidation of the deep Black Sea waters, at the same concentration of Fe, were in good agreement (13).

Framvaren Fjord. Of the anoxic basins studied the Framvaren Fjord has the highest concentration of H_2S (6 mM). The fjord has an extremely shallow sill of 2 m. The large vertical salinity gradient is responsible for the pycnocline that separates the surface and deep waters. The waters are permanently anoxic below a depth of 20 m. The concentration of H_2S in the deep waters is 25 times greater than in the Black Sea.

The oxidation of unfiltered surface waters with added H_2S gave a half time $t_{1/2}$ = 900 ± 65 min (11). Measurements of surface water with added H_2S from deep waters (5230 μM) gave a similar half time of $t_{1/2}$ = 938 ± 60 min. The small amount of deep water added to the surface waters in this experiment had little effect on the rates of oxidation. These rates are faster than found earlier when H_2S was added to surface waters of the Gulf Stream ($t_{1/2}$ = 25 h) (9) and the Black Sea ($t_{1/2}$ = 22 h) (13) apparently due to the Fe in the surface waters (24). The measured half times ($t_{1/2}$ = 30 to 42 min) found for the mixtures (50 % of each) of surface and 25 m waters were slightly greater than the half times for the oxidation of the 22 m waters ($t_{1/2}$ = 19 to 20 min). The rates for the aerated deep waters are 200 times faster than the surface waters (Figure 6). This increase in the rates is largely due to the increased levels of Fe(II) in the deep waters. Since the rates of oxidation of the deep waters poisoned and unpoisoned with sodium azide ($\approx$ 1 mM) were the same, these large increases in oxidation rates appear to be abiotic.

A concentration of 98 nM of Fe^{3+} has been reported in the surface waters of the Framvaren (24). This level of Fe^{3+} would yield rates that are 1.4 times faster than waters without Fe. This observation is in reasonable agreement with the measured ratio of the half times of 1.6 (24). The levels of Fe(II) in the deep waters lead to predicted rates 56 to 150 times faster than waters without Fe, which is in reasonable agreement (see Figure 6) with the measured increases of 41 to 75 times relative to Gulf Stream surface water.

The products ($S_2O_3^{2-}$, SO_3^{2-} and SO_4^{2-}) formed during the oxidation in the Framvaren also have been examined and used to determine k_1, k_2 and k_3. The measured results are compared in Figure 7 with the estimated values at the same level of Fe. The agreement is quite good.

Chesapeake Bay. During the summer months the deep waters of the northern Chesapeake Bay become anoxic due to the stratification of the water column. The build up of H_2S to micromolar levels has been attributed to a diffusive flux from the anoxic sediments and to sulfate reduction in the water column. The concentration of H_2S in the deep water can be as high as 12 μM and is quite variable during the day at a given station (12). The interface between the oxic and anoxic waters in our recent studies was found to be at 15-meter depth (12).

The rate of oxidation of surface waters (12) with added H_2S gave a halftime of 25 ± 2 h in good agreement with the results for surface waters of the Gulf Stream ($t_{1/2}$ = 25 h) and the Black Sea ($t_{1/2}$ = 22 h). The oxidation of H_2S in aerated deep waters gave an average halftime $t_{1/2}$ = 8.3 ± 1.2 min (Figure 6). The deep waters have rates that are 186 times faster than the surface waters (12).

The concentrations of Fe^{2+} and Mn^{2+} can reach levels of 6 to 10 μM in the deep waters. The levels of Fe^{2+} (6.5 μM) and Mn^{2+} (10.7 μM) at 25 m would yield rates, respectively, that are 200 and 2 times faster than waters with

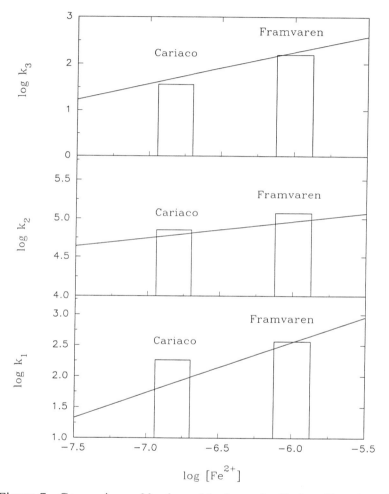

Figure 7. Comparison of k_1, k_2 and k_3 from the Cariaco Trench and Framvaren Fjord with the values predicted at the same levels of Fe^{2+} (16). The lines are the rate constants predicted from laboratory measurements at various concentrations of Fe^{2+} (16).

nanomolar levels of Fe and Mn. This increase in the rate is in reasonable agreement with the measured ratio of 186 (see Figure 6).

The rate of oxidation of deep anoxic waters in absence of O_2 gave halftimes that varied from 16 to 23 min. These faster rates probably were due to the presence of MnO_2 or FeOOH (25, 26). The oxidation could be abiotic or due to bacterial oxidation. Unfortunately we were not able to do experiments under sterile conditions and the filtration through a 0.45 μm filter does not eliminate the possibility of bacterial oxidation. Our results are, however, in agreement with the earlier work (27), suggesting that non-oxygen oxidation occurs in the Chesapeake Bay waters. More controlled experiments are needed to fully elucidate these effects. In the presence of O_2 the abiotic oxidation of H_2S appears to dominate.

Cariaco Trench. The Cariaco Trench is an anoxic basin located on the continental shelf north of Venezuela. The Cariaco Trench has been widely used as a natural laboratory for the study of anaerobic processes since the deep waters of the Cariaco Trench were found to be anoxic in 1954. In an attempt to compare the rates of oxidation in different anoxic environments, we made direct measurements of the oxidation of H_2S with oxygen in the Cariaco Trench in 1990 (14). We also have examined the products ($S_2O_3^{2-}$, SO_3^{2-} and SO_4^{2-}) formed during the oxidation in the Cariaco Trench.

The rate of oxidation for the surface waters gave a $t_{1/2}$ = 17.3 ± 0.1 h. These results are in reasonable agreement with our earlier measurements (9) made on Gulf Stream and Black Sea surface waters ($t_{1/2}$ = 22 to 25 h). The rates of oxidation of deep waters (800 m) and 50% mixtures of deep and surface waters also were measured after air saturation. Values of $t_{1/2}$ = 3.0 ± 0.1 h and 1.5 ± 0.1 hours were found, respectively, for the mixtures and the deep waters (Figure 6). These measurements indicate that the faster rates for the mixtures are caused by dissolved constituents in the deep waters. The rates in the mixtures are 1/2 the values obtained from the deep waters, which agrees with the mixing ratios.

In the Cariaco Trench only Fe^{2+} (28) is at sufficiently high concentrations (320 nM) to affect the rates of oxidation. At the maximum level of Fe^{2+} in the Cariaco Trench, one would expect the rates of oxidation to be 17 times faster due to Fe^{2+}. These estimates are the same order of magnitude as found in our direct measurements. The calculated half times from concentration of Fe^{2+} ($t_{1/2}$ = 17.2, 2.7, and 1.5 h), respectively for the surface, mixed and deep waters, are in good agreement with the measured values ($t_{1/2}$ = 17.3, 3.0, and 1.5 h) (see Figure 6).

Summary

The oxidation of H_2S with O_2 in natural waters has been studied in the laboratory as a function of pH (4 to 10), temperature (278.15 to 338.15 K) and salinity (0 to 36). A kinetic model developed can be used to predict the distribution of the reactants and products over a wide range of conditions in natural waters. The field measurements of the rates of oxidation of H_2S were found to be in good agreement with those estimated from laboratory studies at the same concentration of Fe^{2+}. The concentrations of Fe^{2+} are high enough in most anoxic environments to increase the rate of oxidation of H_2S. A kinetic model has been used to analyze the distribution of products (SO_3^{2-}, $S_2O_3^{2-}$, SO_4^{2-}) formed during the oxidation in the Framvaren Fjord and the Cariaco Trench (11, 14). The rate constants for the production of SO_3^{2-} (k_1), the production of SO_4^{2-} (k_2) and the production of $S_2O_3^{2-}$ (k_3) estimated for these waters are in reasonable agreement with the predicted values at the same concentration of Fe^{2+} (Figure 7).

The values of k_2 estimated for the Framvaren Fjord and Cariaco Trench are slightly higher than the predicted values. This deviation could be due to errors in our estimation of the concentration and form of iron in these waters. Direct measurements of iron and manganese in the anoxic waters should be made in all future kinetic studies to avoid this problem. The estimates based upon only the concentration of iron may also lead to some errors. The concentrations of Mn^{2+} below the oxic-anoxic interface reach levels of 0.5 and 15 μM in the Cariaco Trench and Framvaren Fjord, respectively. This Mn^{2+} comes from reduction of MnO_2 sinking from above the oxic-anoxic interface. Reoxidation of Mn^{2+} to MnO_2 by bacteria occurs when Mn^{2+} diffuses up to the oxic layer. This cycling of manganese between Mn^{2+} and MnO_2 is an important feature of the oxic-anoxic interface and probably affects the distribution of products of H_2S oxidation in the field.

Acknowledgments

The authors gratefully acknowledge the support of the Ocean Sciences Division of the National Science Foundation (OCE89-22580) and the Office of Naval Research (N00014-90-J-1225). We also thank C. N. Alpers, C. O. Moses and an anonymous reviewer for their comments.

Literature Cited

1. Richards, F. A. In Chemical Oceanography; Riley, J. P. and Skirrow, G., Eds.; Academic: New York, 1965, Vol. 1; pp 215-243.
2. Jacobs, L.; Emerson, S. Earth Planet. Sci. Lett. 1982, 60, 237-252.
3. Zhang, J.-Z.; Whitfield, M. Mar. Chem. 1986, 19, 121-137.
4. Avrahami, M.; Golding, R. M. J. Chem. Soc. 1969, (A) 647-651.
5. Chen, K. Y.; Morris, J. C. Environ. Sci. Technol. 1972, 6, 529-537.
6. Cline, J. D.; Richards, F. A. Environ. Sci. Technol. 1969, 3, 838-843.
7. O'Brien, D. J.; Birkner, F. G. Environ. Sci. Technol. 1977, 11, 1114-1120.
8. Millero, F. J. Mar. Chem. 1986, 18, 121-147.
9. Millero, F. J.; Hubinger, S.; Fernandez, M.; Garnett, S. Environ. Sci. Technol. 1987, 21, 439-443.
10. Zhang, J.-Z.; Millero, F. J. Geochim. Cosmochim. Acta 1991, 55, 677-685.
11. Millero, F. J. Limnol. Oceanogr. 1991, 36, 1007-1014.
12. Millero, F. J. Estuar. Coastal Shelf Sci. 1991, 33, 521-527.
13. Millero, F. J. Deep-Sea Res. 1991, 38, S1139-S1150.
14. Zhang, J.-Z.; Millero, F. J. Deep-Sea Res. 1993, in press.
15. Vazquez-G. F.; Zhang, J.-Z.; Millero, F. J. Geophys. Res. Lett. 1989, 16, 1363-1366.
16. Zhang, J.-Z.; Millero, F. J. Geochim. Cosmochim. Acta 1993, in press.
17. Millero, F. J.; Sotolongo, S.; Izaguirre, M. Geochim. Cosmochim. Acta 1987, 51, 793-801.
18. Millero, F. J.; Laferriere, A. L.; Fernandez, M.; Hubinger, S.; Hershey, J. P. Environ. Sci. Technol. 1989, 23, 209-213.
19. Dyrssen, D. Mar. Chem. 1988, 24, 143-153.
20. Clarke, A. G.; Radojevic, M. Atm. Environ. 1983, 17, 617-624.
21. Clarke, A. G.; Radojevic, M. Atm. Environ. 1984, 18, 2761-2767.
22. Hoffmann, M. R.; Lim, B. C. Environ. Sci. Technol. 1979, 13, 1406-1414.
23. Hoffmann, M. R.; Edwards, J. O. J. Phys. Chem. 1975, 79, 2096-2098.
24. Haraldsson, C.; Westerlund, S. Mar. Chem. 1988, 23, 417-424.
25. Aller, R. C.; Rude, P. D. Geochim. Cosmochim. Acta 1988, 52, 751-765.
26. Burdige, D. J.; Nealson, K. H. Geomicrobiol. J. 1986, 4, 361-387.
27. Luther, G. W.; Ferdelman, T.; Tsamakis, E. Estuaries 1988, 11, 281-285.
28. Jacobs, L.; Emerson, S.; Huested, S. S. Deep-Sea Res. 1987, 34, 965-981.

RECEIVED March 16, 1993

Recent Advances in Analytical Methods

Chapter 27

Determination of Hydrogen Sulfide Oxidation Products by Sulfur K-Edge X-ray Absorption Near-Edge Structure Spectroscopy

Appathurai Vairavamurthy, Bernard Manowitz, Weiqing Zhou, and Yongseog Jeon[1]

Department of Applied Science, Brookhaven National Laboratory, Upton, NY 11973

> The application of synchrotron-radiation-based XANES spectroscopy is described for the determination of the products formed from the oxidation of aqueous sulfide. This technique allows simultaneous characterization of all the different forms of sulfur both qualitatively and quantitatively. Thus, it is superior to other commonly used techniques, such as chromatography, which are usually targeted at specific compounds. Since use of the XANES-based technique is relatively new in geochemistry, we present here an overview of the principles of the technique as well as the approach used for quantitative analysis. We studied the oxidation of hydrogen sulfide under conditions of high sulfide-to-oxygen ratios using 0.1 M sulfide solutions, and the catalytic effect of Ni^{2+} was also examined. Results obtained from this study are presented to illustrate the value of the XANES technique for the determination of the products formed from the oxidation of sulfide at high concentrations.

Recently there has been widespread interest in sulfur cycles from a biological, geochemical, biogeochemical, and environmental viewpoint. Sulfur transformations and speciation exhibit a great deal of complexity because sulfur exists in a number of oxidation states between -2 and +6, and forms a large variety of organic and inorganic species. Extreme changes in sulfur speciation are prominent at the redox boundaries (oxic-anoxic interface) in marine systems, including organic-rich sediments and anoxic basins such as the Black Sea and Cariaco Trench (1). In anoxic marine environments, hydrogen sulfide is the major reduced sulfur compound, and is formed as a result of bacterial reduction of sulfate associated with anaerobic oxidation of organic matter. The formation of hydrogen sulfide in hydrothermal systems probably occurs by a thermochemical reduction mechanism, although high-temperature bacterial reduction is also possible (2,3). At the oxic-anoxic interface, when water containing H_2S mixes with oxygenated water, the sulfide is oxidized by oxygen resulting in the formation of

---

[1]Current address: Department of Physics, Jeonju University, Hyojadong 3-ga 1200, Wansan-gu, Jeonju, Seoul 560-759, Korea

several partially oxidized sulfur species, for example polysulfides, elemental sulfur, sulfite, thiosulfate, tetrathionates as well as sulfate (4-6). In natural waters, redox transformations of some metals and nonmetals are also frequently coupled to the oxidation of sulfide. For example, Fe^{3+} is the dominant oxidant of reduced-sulfur species in mine-drainage environments (7). However, in general, dissolved oxygen is the principal oxidant (8). In addition to the chemical mechanism, bacteria also play an important role in sulfide oxidation, and add to the complexity of the sulfur cycle in natural waters (9).

Although hydrogen sulfide oxidation has been the subject of several past studies (10-18; reviews: 19-20), there are still many uncertainties regarding rates, mechanisms of oxidation, and the formation of the oxidation products. Serious drawbacks in these studies have been analytical limitations involved in the measurement of the different sulfur species. The methods widely used for the determination of various sulfur species are based on colorimetry or chromatography. In most sulfide-oxidation studies, Urban's colorimetric method was used for thiosulfate (21), and sulfite was determined by the West and Gaeke colorimetric method (22); sulfate was estimated by difference. Although selective, colorimetric methods used for the determination of most inorganic sulfur species are usually prone to interferences. For example, in Urban's method for thiosulfate, sulfide causes a positive interference and must be removed by stripping with nitrogen at low pH. Recently, thiosulfate and sulfite were measured by a high performance liquid chromatographic (HPLC) method after precolumn derivatization with 2,2'-dithiobis(5-nitropyridine) (DTNP) for reversed-phase separation and UV detection (23). Methods based on ion-exchange separation have also been described for the determination of most oxyanions of sulfur, including sulfite, thiosulfate and polythionates in natural waters (24,25). However, ion chromatographic methods may not be suitable for analysis of ionic media such as seawater. Although a choice is possible from either colorimetric or chromatographic methods for determining sulfite and thiosulfate, suitable methods are still lacking for other sulfur intermediates, for example, polysulfides. In both the colorimetric and chromatographic methods that are commonly used for determining various sulfur intermediates, there is a serious concern about alteration of sample composition in the lengthy sample preparation procedures or during the analysis.

Recently, synchrotron-radiation-based X-ray absorption near-edge structure (XANES) Spectroscopy has emerged as an important tool for determining the speciation of sulfur in a variety of geochemical samples including coal, petroleum, and sedimentary rocks (26-29). In general, the XANES spectrum provides characteristic "fingerprint" information of a sulfur compound because of the sensitivity of XANES to the electronic structure, oxidation state, and the geometry of the neighboring atoms (30). Consequently, the XANES spectroscopy has proved to be a valuable tool for qualitative characterization of sulfur-containing compounds based on spectral comparison with model compounds. The recent popularity of XANES spectroscopy for sulfur-species determination is mainly due to the fact that accurate quantitative information of the different sulfur constituents can be obtained with this technique. An important advantage of XANES is that it allows non-destructive analysis of either solid or liquid samples.

In this paper, we describe the use of XANES spectroscopy as an analytical technique for the determination of the products formed from the oxidation of aqueous sulfide. Because XANES is a relatively new technique in geochemistry, we also discuss the principles of X-ray absorption spectroscopy as well as the approach used to obtain quantitative information with XANES analysis. In this study, sulfide oxidation was examined with excess sulfide and limiting oxygen concentrations, conditions typically present in reducing marine sediments. At high concentrations, sulfide causes interference in the determination of the oxidation products with either chromatography or colorimetry (for example, in the determination of sulfite and thiosulfate by the HPLC method involving DTNP derivatization), and may be a reason for the sparsity of studies under these conditions. We discuss here some significant results from sulfur XANES analysis which illustrate the value of this technique for the determination of the products formed from the oxidation of sulfide at high concentrations.

Principles of X-ray Absorption Spectroscopy

An X-ray absorption spectrum is produced by the transition of a photoelectron from a core level to symmetry-available outer empty states and to the continuum (i.e., beyond the sphere of influence of the atom). The absorption spectrum is rather featureless until one observes an abrupt increase in the absorption coefficient, known as the "absorption edge", which occurs when the X-ray photon has sufficient energy to excite a core-level electron in the atom. The fine structure usually appearing as peaks (the so-called "white lines") at the high-energy side of each main absorption edge reflects the bound-state transitions between the absorber core orbital and outer unoccupied atomic or molecular orbitals (e.g. 1s --> 3p). Continuum transitions contribute to subsequent oscillations in absorption (Figure 1) (31). Conventionally, the structured absorption occurring within ca. 50 eV of an element's absorption edge is referred to as the X-ray Absorption Near-Edge Structure (XANES). Thus, XANES includes both bound-state transitions and "continuum-resonance" transitions. The region of the spectrum beyond this range is the Extended X-ray Absorption Fine-Structure (EXAFS), which consists typically of relatively broad and weak sinusoidal oscillations that diminish in intensity with increasing energy above the edge, but which may persist for as much as 1,000 eV above the edge (Figure 2). It is important to note here that the absorption spectra are usually measured as fluorescence excitation spectra with a fluorescence detector. This measurement is possible because the fluorescent X-ray emission is proportional to the absorption coefficient for dilute samples.

The absorption fine structure due to continuum transitions is influenced mainly by the nature and geometry of the absorber's environment. When the absorbing atom is surrounded by other atoms, as in any condensed phase, the outgoing photoelectron can be backscattered by the surrounding atoms, where the backscattered waves can interfere either constructively or destructively with the outgoing wave. No backscattering will occur in the absence of near neighbors around the absorber (e.g., monoatomic gas). In the XANES region, the final state of the continuum transition is a low-energy photoelectron. As a result of its long mean-free path, the low-energy photoelectron can undergo multiple scattering interactions. This effect makes XANES structure sensitive to the three-dimensional arrangement of atoms around the absorber, and potentially allows the use of XANES to determine geometry. However, despite recent progress on XANES theories, there is still much uncertainty in establishing a one-to-one correlation between "continuum resonances" and a specific geometric structure. Thus, continuum resonances are most often used as a fingerprint to compare a model compound and an unknown sample. In contrast, the spectral oscillations in the high-energy EXAFS region can be modeled well by considering only the single scattering interactions of the excited photoelectron (32). In fact, the EXAFS oscillations provide useful data to determine the local radial distribution around the absorbing atom.

Sulfur Speciation and Edge Shift. In the XANES spectrum, the fine structure corresponding to bound-state transitions (i.e., the absorption maxima in the edge region) is rich in information about the electronic structure, charge density, and the bonding environment of the absorbing atom. This is mainly because the orbitals involved in the bound-state transitions are the regions of interatomic orbital overlap. For sulfur, the edge correlates with the 1s --> 3p transition which occurs as an intense, well-resolved transition corresponding to the "white-line" maximum of the XANES spectrum. Several recent studies clearly show that both the energy and the intensity of the sulfur 1s --> 3p transition are sensitive to the oxidation state of sulfur (28,29,33). This dependence is reflected by the fact that the "white-line" maximum changes by ca.

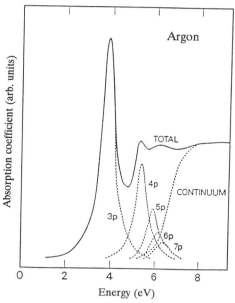

Figure 1. Absorption spectrum for gaseous argon resolved into transitions to allowed bound states and to the continuum. (source: reference 31).

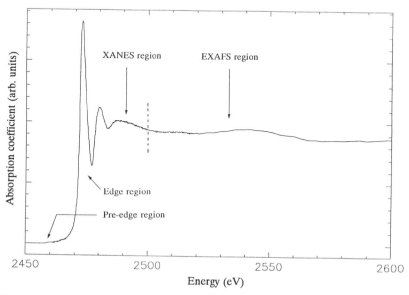

Figure 2. High intensity X-ray absorption fine structure spectrum of elemental sulfur showing phenomenological energy ranges.

10 eV between elemental sulfur and sulfate, with formal oxidation states of 0 and +6 respectively. It is shown in Figure 3 that each of the XANES spectra of the variety of sulfur forms is richly endowed with characteristic features, including the edge energy, which allow ready qualitative recognition among various oxidation states and structures.

XANES spectroscopy provides an experimental technique of inferring the charge density (and hence the oxidation state) of an atom in a molecule (30,34). The evidence in the X-ray spectrum for a change in atomic charge density is the shift in the position of an X-ray absorption edge towards either the positive or the negative side relative to that in the pure element (note: the edge shift is also referred to as the chemical shift). Kunzl (35) was one of the first workers who suggested a linear relationship between edge shift and valence based on a critical study regarding the shift of the K-absorption discontinuities of the oxides of several elements. For sulfur, previous studies (28,33) have shown a nearly linear correlation between edge energy and (formal) oxidation state, especially with inorganic sulfur species (Figure 4). Thus, XANES spectroscopy can be a useful technique for determining the oxidation state of sulfur using Kunzl's law. In fact, in a recent study (36), this correlation was the basis for assigning new oxidation states for the two different sulfur atoms in thiosulfate (-1 and +5 for the terminal and inner sulfur atoms respectively).

Quantitative Analysis

The Approach. The use of XANES for quantitative characterization of the different forms of sulfur present in a sample is a relatively new development. Several quantitative methods were attempted by a number of workers and are briefly discussed below. In one approach, a differential XANES treatment was used to distinguish and quantify different forms of sulfur in petroleum asphaltenes. In this method, the heights of certain features in the third derivative of the XANES spectrum were considered proportional to the relative amounts of the different organic sulfur forms in the sample (27). In a second method, developed and applied by Huffman and co-workers (29,37) for the analysis of sulfur forms in coal, a least-squares fitting of the XANES spectrum with a series of artificial functions representing the spectral features of the different sulfur forms (for example pyrite, sulfide, thiophene, sulfoxide, sulfone, sulfate) is used for quantitative characterization. An arctangent function is typically used to model the step-function feature in the spectrum corresponding to 1s excitations into continuum states. The s --> p transition peaks (in the sample spectra) are fitted by peak functions (which are 50% Lorentzian and 50% Gaussian) specific to different forms of sulfur and converted to weight percentage of sulfur using calibration constants derived from standard-compound mixtures. A third method for quantitative analysis is that developed by Waldo and co-workers at the University of Michigan (28) which is currently used by us for sulfur analysis. Essentially, it involves a non-linear least-squares fitting procedure which uses linear combinations of normalized spectra of model compounds to obtain quantitative information of the different forms of sulfur.

The fact that the XANES spectra are usually measured as fluorescence excitation spectra presents a problem in quantitative analysis for concentrated samples. As pointed out in Section on Principles of X-ray Absorption Spectroscopy, a linear correlation between fluorescence signal and central-atom absorption is only true for dilute samples and for thin-film samples. In the case of concentrated (or thick) samples, the fluorescence signal is considerably distorted (reduced amplitude especially) due to self-absorption or the thickness effect. Self-absorption is especially important for elements of relatively low atomic number, such as sulfur, because X-ray absorptions are typically quite large at the low energy range (2 - 4 keV). Ideally, self-absorption effect can be avoided by diluting the sample; however, this can be difficult or impossible for insoluble samples and will also degrade the signal-to-noise ratio. The numerical method developed by Waldo and Penner-Hahn allows the fluorescence-excitation spectra to be corrected for the self-absorption effect prior to least-squares deconvolution of the spectrum (28,38).

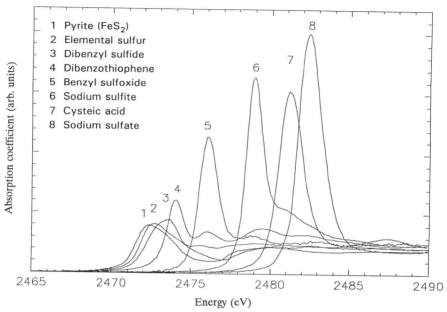

Figure 3. Normalized XANES spectra of various sulfur compounds showing characteristic features including edge energies.

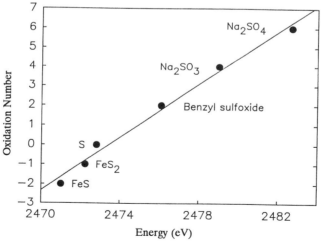

Figure 4. Plot of white-line peak energy (eV) versus sulfur oxidation state for some sulfur compounds fitted with a linear least-squares line.

There are four main steps in the method by Waldo et al. (28): (1) background absorption correction (the correction is determined first by low-order polynomial fitting in the pre-edge region and then extrapolation to the post-edge region followed by subtraction of the background curve from the entire spectrum); (2) normalization of the spectrum in the post-edge region to fit a tabulated X-ray-absorption cross section (McMaster table) (39); (3) correction for attenuation of the fluorescence signal due to the self-absorption effect; and (4) fitting the normalized spectrum with linear combinations of absorption spectra of model compounds using a non-linear least-squares procedure. The use of experimentally obtained spectra of model compounds for fitting has a unique advantage over the other methods when a model compound has several major peaks, for example, thiosulfate. Furthermore, even if the edge energies overlap and do not provide adequate species-specific information, other spectral features (for example, the multiple-scattering resonance peaks at higher energies) can be of use for distinguishing among the different sulfur forms because the entire spectrum is used for fitting. The fitting range is usually selected so that it covers the white-line maxima for all of the model compounds. In fitting, the adjustable parameters used for each model spectrum (equation 1) are an optional energy offset (ΔE) and a scaling factor (c_i) which is constrained to $\sum_i c_i = 1$.

$$\mu_{sample}(E) = \sum_i c_i \, \mu_{model}(E + \Delta E_i) \tag{1}$$

Under ideal operational conditions, no energy offset is required; however, in practice, there is some uncertainty which is usually less than 0.5 eV. The scaling factor directly reflects the quantitative amount of each sulfur form.

In the XANES technique, the basis for quantification, which is reflected in the scaling factor described above, is a linear correlation between step height in the spectrum and the mass of sulfur in the corresponding sample. We verified this relationship between step height and sulfur mass using sodium sulfate solutions in different concentrations. The Na_2SO_4 solutions were each mixed with an exact amount of NaCl for normalization of the sulfur absorption coefficient against that of chlorine which was assigned a value of unity in each spectrum. Thus, the step heights in Figure 5 represent values normalized against chlorine X-ray absorption peak (2833.4 eV). The data shown in Figure 5 reaffirm the linear relationship between step height and the concentration of sulfur. It is important to note that step height is invariant of the self-absorption effect which affects the shape of peaks in the edge region in fluorescence excitation spectra.

An Illustration of Self-Absorption Correction. Figure 6 shows the sulfur K-edge fluorescence excitation spectra of two different Na_2SO_4 solutions of 0.025 M and 0.1 M concentrations prepared in deionized water. The peak height of the more concentrated solution (0.1 M) is lower than that of the less concentrated solution (0.025 M) after pre-edge subtraction and post-edge normalization to the same step height (McMaster X-ray absorption cross-section value) which is consistent with the self-absorption effect. After applying the self-absorption correction (using the numerical expression for fluorescence intensity as a function of sample elemental composition, weight percentage or concentration, and sample thickness), the corrected spectrum of the 0.1 M solution matches well with that of the 0.025 M solution (see the top two curves in Figure 6). This excellent agreement strongly supports the view that self-absorption correction is an essential step in the analysis of X-ray fluorescence spectra distorted by the concentration effect.

Analysis of Standard Mixtures. We examined the validity of the XANES method for quantitative analysis using test mixtures prepared in deionized water with sulfite,

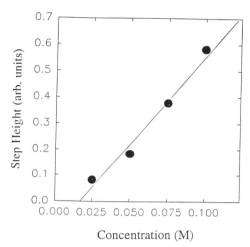

Figure 5. Correlation between X-ray absorption step height and the concentration of sulfur using Na_2SO_4 solutions prepared in deionized water.

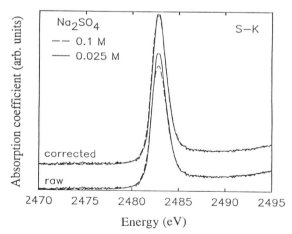

Figure 6. The sulfur K-edge fluorescence excitation spectra of two different Na_2SO_4 solutions (0.025 M and 0.1 M in deionized water) with and without self-absorption correction.

thiosulfate, and sulfate. The approach involved preparing standards of accurately known composition, collecting the spectra (at beam line X-19A) and analyzing the data (computer fitting). The XANES spectrum of a solution containing $Na_2S_2O_3$ (50 mM), Na_2SO_3 (50 mM), and Na_2SO_4 (50 mM) after spectral deconvolution is shown in Figure 7 which indicates excellent agreement of the sum of basis model spectra with the experimental spectrum. Both the scaling and the energy-offset parameters were varied in a non-linear least-squares fit. The results with different test solutions are summarized in Table I. It is clear from these data that XANES analysis combined with the spectral deconvolution method provides a reliable method for quantifying the different forms of sulfur used in this study.

In general, the quantitative analysis using XANES spectroscopy is valid if correct model spectra are used for fitting. Thus, as shown in the preceding paragraph, accurate quantitative results can be obtained with samples of known qualitative composition. With unknown samples, the spectral features are a major guide in selecting the appropriate model spectra for analysis. However, this selection may be a problem for components present in low concentrations, especially at sub-millimolar levels. At these concentrations, spectral noise also may affect the fitting. Currently, with a Lytle fluorescence detector, the lower end of the working range at the X-19A beam line at the NSLS is about one millimolar for the analysis of most oxyanions of sulfur in solutions. This lower limit can be improved to sub-millimolar concentrations by using a more sensitive detector (for example, a 13-element-solid-state detector). In the future, the detection capabilities can be improved even further with the availability of third-gereration synchrotron light sources (for example, the Advanced Photon Source at Argonne National Laboratory) producing high-energy X-ray beams.

Analysis of Sulfide-Oxidation Products

Experimental Methods. Time-series experiments were conducted using a 0.1 M NaHS solution in deionized water and in 0.7 M NaCl solution at room temperature (25 ± 1 °C). The initial pH was adjusted to be in the 8.0 to 8.5 range using a 3 M HCl solution. The effect of Ni^{2+} on the rate of oxidation was examined because of the well-known catalytic properties of this metal (15). Sulfide and its oxidation products were both detected and quantified using XANES spectroscopy performed at the National Synchrotron Light Source (NSLS) X-19A beam line. The 0.1 M sulfide solution used in this study was prepared with nitrogen-sparged Milli-Q water, and was maintained in a nitrogen atmosphere. The oxidation-time-series samples were prepared by exposing ca. 5 mL portions of the stock sulfide solution to different periods of air oxidation in a 50 mL beaker. Our previous observations suggest that in such small volumes of solution, oxygen concentrations rapidly reach equilibrium values with air through diffusion. $NiCl_2$ was added at a concentration of 100 µM. For X-ray analysis, the samples were prepared by packaging ca. 1-2 mL portions in thin Mylar film bags. Sulfur standards were analyzed in the same way. Deionized water was produced with a Milli-Q water purification system (Millipore, Bedford, MA). Anhydrous NaHS was obtained from Johnson Matthey Co. (Ward Hill, MA); other sulfur standards were obtained from Aldrich Chemical (Milwaukee, WI). The Mylar film (2.5 µm) was obtained from Chemplex Industries (Tuckahoe, NY), and the bags were prepared in the laboratory by impulse sealing.

Beam-line Setup. In the X-19A beam line at the NSLS, the X-ray beam emitted from the storage ring is collimated by an adjustable vertical slit (1 mm) and is diffracted by a double-crystal monochromator [Si(111)], which passes a narrow energy band. The overall energy resolution was estimated to be 0.5 eV at 2.5 keV photon energy. The monochromator was detuned by ca. 80% to minimize higher order-harmonics in the X-ray beam. The incident beam intensity (I_o) was monitored by an ion chamber which absorbs a few percent of the beam to ionize a gas such as helium. The

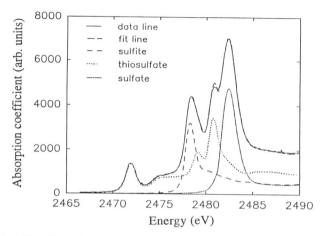

Figure 7. Non-linear least-squares fit of the XANES spectrum of an aqueous solution containing sulfite, thiosulfate and sulfate with spectra of model compounds.

Table I. XANES analysis of different test mixtures containing aqueous solutions of sodium sulfite, sodium thiosulfate and sodium sulfate. The values represent ratios of concentrations in g atom sulfur units. Initial concentrations are given in molar units within parentheses

	Sulfite	Thiosulfate	Sulfate
Test mixture	0.33 (50 mM)	0.67 (50 mM)	-
XANES analysis	0.33	0.67	-
Test mixture	-	0.67 (50 mM)	0.33 (50 mM)
XANES analysis	-	0.71	0.29
Test mixture	0.25 (50 mM)	0.50 (50 mM)	0.25 (50 mM)
XANES analysis	0.22	0.52	0.26

data were collected as fluorescence excitation spectra using a Lytle fluorescence detector (40) placed at 90 degrees to the X-ray beam. A typical beam-line setup is shown in Figure 8. Samples were run in a helium atmosphere to minimize the attenuation of the X-ray beam by air. The spectra were recorded in such a way that the scanning procedure yields sufficient pre-edge and post-edge data for precise background determination needed for analysis. The X-ray energy was calibrated using XANES spectra of elemental sulfur measured between sample runs, assigning 2472.7 eV to the "white-line" maximum of elemental sulfur spectrum. The uncertainty of the energy calibration, determined by comparing the spectra of model compounds obtained at different times, was less than ± 0.15 eV.

The Products of Sulfide Oxidation: Results and Discussion. Figure 9 shows the XANES spectra of 0.1 M sulfide in deionized water after different periods of exposure to air. We obtained excellent fitting of these spectra using sulfide, sulfite, thiosulfate and sulfate as the bases. The XANES spectra of the various sulfur standards used for fitting are shown in Figure 10. Thiosulfate was detected as the predominant product of sulfide oxidation with sulfite and sulfate amounting to < 10 % of the product sulfur. In a 0.7 M NaCl medium, the pattern of product formation was similar to that of pure deionized water medium but the rate of oxidation was accelerated (Table II). In previous studies, which have been conducted mostly under pseudo-first order conditions with a low initial ratio of sulfide to oxygen, sulfate was found as the major product; sulfite and thiosulfate contributed as minor products (4,12). Our results suggest that thiosulfate is the favored product at high sulfide concentrations with a high ratio of sulfide to oxygen. These results are in agreement with a previous study which describes thiosulfate as the major product of sulfide oxidation in strongly reducing marine sediments containing millimolar levels of sulfide (42). Furthermore, these results support the suggestion that the ratio of sulfide to molecular oxygen is an important factor, in addition to other factors (e.g. pH), in controlling the type and extent of formation of different sulfur species (12).

Our fitting results indicate significant formation of polysulfides in the Ni^{2+} series which contrasts with the other series (Figure 11). Thus, it appears that catalysis by transition elements such as nickel is probably important for the formation of polysulfides in marine sediments. A striking feature in the Ni^{2+} series is the presence of a peak at 2476.1 eV in the spectra of the initial oxidation period (Figure 12). After about 100 hours of air oxidation, this peak completely disappeared in both deionized water and 0.7 M NaCl solution media. Although we obtained excellent fitting of the spectra for the period after ca. 100 hours of air exposure with sulfide, polysulfide, thiosulfate, and sulfate (Figure 11), the same standards could not be used to fit the spectra of the initial oxidation period mainly due to the peak at 2476.1 eV. Based on the linear correlation between oxidation state and peak energy discussed previously, the position of this peak corresponds to an oxidation state of +2. However, there is no inorganic sulfur compound with a +2 oxidation state of sulfur, although in organic sulfoxides sulfur is present with a relative charge density of +2 (Figure 4). The unknown peak cannot be due to an organic sulfoxide because no organic compound was added to the system. However, we used an organic sulfoxide (benzyl sulfoxide) as a surrogate to obtain a quantitative estimate of this unknown species and such a fitting is shown in Figure 12. Previous studies postulate that oxidation proceeds probably through the formation of an intermediate HSO_2^- having a +2 oxidation-state sulfur (14), but so far there has been no experimental evidence for its formation. The peak at 2476.1 eV probably indicates the formation of such a species. Further studies are being carried out to determine the structure of this unknown species.

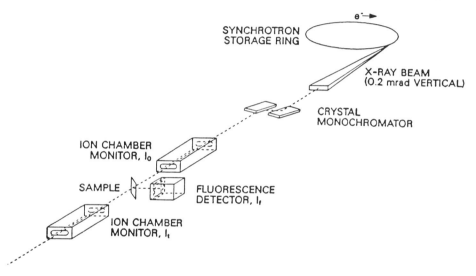

Figure 8. Schematic diagram of a typical X-ray absorption beam-line setup. The storage ring-to-sample distance at the NSLS X-19A is 20 m. (source: reference 41).

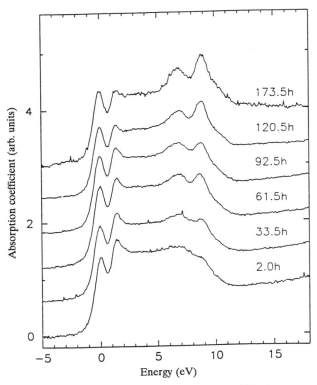

Figure 9. Sulfur K-edge XANES spectra of 0.1 M sulfide in water after different periods of exposure to air. Conditions: $t = 25 \pm 1$ °C; pH = 8; $[O_2]_o = 255$ μM.

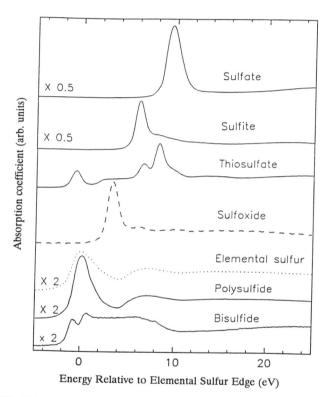

Figure 10. XANES spectra of the various sulfur standards used for fitting. Benzyl sulfoxide was used in the solid form; all other standards were prepared as 10 - 50 mM liquids.

Table II. Percentage of products formed from the oxidation of aqueous sulfide in different media

Medium	Co-substance	[H$_2$S] (mM)	[O$_2$] (mM)	t$_{1/2}$ (hour)	Product ratio after 48 hours				
					SO$_3^{2-}$	S$_2$O$_3^{2-}$	SO$_4^{2-}$	S$_n^{2-}$	S(II) species
water	none	100	0.255	385	9	83	8	n.d.	n.d.
water	100 μM NiCl$_2$	100	0.255	40	n.d.	67	1	20	12
0.7 M NaCl	none	100	0.232	289	10	90	n.d.	n.d.	n.d.
0.7 M NaCl	100 μM NiCl$_2$	100	0.232	35	n.d.	58	2	24	16

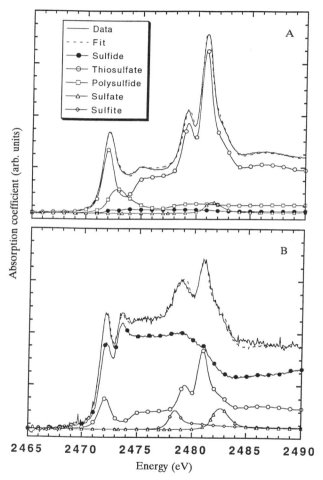

Figure 11. Non-linear least-squares fit of the XANES spectra of 0.1 M aqueous sulfide in deionized water after 172 hours of air oxidation: (A) with added $NiCl_2$ (100 μM) and (B) without $NiCl_2$. Note that polysulfides are formed only in the presence of $NiCl_2$.

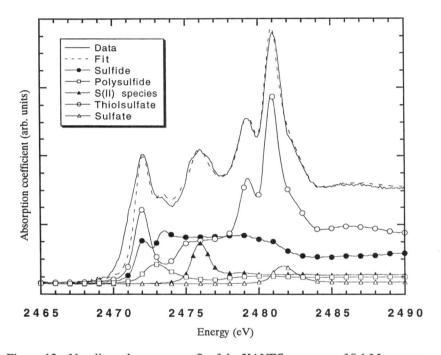

Figure 12. Non-linear least-squares fit of the XANES spectrum of 0.1 M aqueous sulfide in deionized water with added $NiCl_2$ (100 μM) after 63 hours of air oxidation. Benzyl sulfoxide was used as a surrogate to quantify the unknown sulfide oxidation intermediate with peak at 2476.1 eV.

Conclusions

The results discussed above clearly indicate the potential of XANES spectroscopy for the determination of the different sulfur species formed from the oxidation of aqueous sulfide. In this technique, quantitative measurements of the different sulfide-oxidation products, including polysulfides, are not affected by the presence of high concentrations of sulfide (as high as 0.1 M) because of the lack of sulfide interference. Other techniques in current use such as chromatography may not have detected the presence of the +2 oxidation-state species because they are usually targeted at specific compounds. The ability to obtain both qualitative and quantitative information of all the sulfur forms present in a sample provides a unique advantage of using XANES spectroscopy over other techniques, including chromatography. Furthermore, since this technique involves no cumbersome sample preparation, the problems of artifacts formation and the alteration of sample composition usually associated with lengthy sample preparatory procedures are greatly minimized.

Acknowledgments

We thank Charles Alpers and David Blowes for their editorial reviews. Grant Bunker and Martin Schoonen are acknowledged for their valuable comments. This research was performed under the auspices of the U.S. Department of Energy Division of Engineering and Geosciences of the Office of the Basic Energy Sciences under Contract No. DE-AC02-76CH00016 (KC-04).

Literature Cited

1. Millero, F. J. Deep-Sea Research 1991, 38, Suppl. 2, S1139-S1150.
2. Orr, W. L. AAPG Bull 1974, 58, 263-276.
3. Jorgensen, B. B.; Isaksen, M. F.; Jannasch, H. W. Science 1992, 258, 1756-1757.
4. Cline, J. D.; Richards, F. A. Environ. Sci. Technol. 1969, 3, 838-843.
5. Luther, G. W. III; Church, T. M.; Giblin, A. E.; Howarth, R. W. In: Organic Marine Geochemistry; Sohn, M. L., Ed.; ACS Symposium Series 305; American Chemical Society, Washington, DC, 1986; pp 340-357.
6. Vazqez, F; Zhang,J. -Z.; Millero, F. J. Geophys. Res. Lett. 1989, 6, 1363-1366.
7. Moses, C. O.; Herman, J. S. Geochim. Cosmochim. Acta 1991, 55, 471-482.
8. Jacobs, L.; Emerson, S. Earth Planet. Sci. Lett. 1982, 60, 237-252.
9. Jorgensen, B. B.; Fossing, H.; Wirsen, C. O.; Jannasch, H. W. Deep-Sea Research 1991, 38, Suppl. 2, S1083-1103.
10. Avrahami, M.; Golding, R. M. J. Chem. Soc. (A) 1968, 647-651.
11. Chen, K. Y.; Morris, J. C. Environ. Sci. Technol. 1972, 6, 529-537.
12. O'Brien, D. J.; Birkner, F. G. Environ. Sci. Technol. 1977, 11, 1114-1120.
13. Hoffman, M. R. Environ. Sci. Technol. 1977, 11, 61-66.
14. Hoffman, M. R.; Lim, B. C. Environ. Sci. Technol. 1979, 13, 1406-1414.
15. Weres, O.; Tsao, L.; Chhatre, R. M. Corrosion-NACE 1985, 41, 307-316.
16. Millero, F. J.; Hubinger, S.; Fernandez, M.; Garnett, S. Environ. Sci. Technol. 1987, 21, 439-443.
17. Millero, F. J.; Zhang, J.-Z. Geochim. Cosmochim. Acta 1993, 57, in press.
18. Zhang, J.-Z.; Millero, F. J. In: The Environmental Geochemistry of Sulfide Oxidation, Alpers, C.; Blowes, D., Eds.; Amer. Chem. Soc., this volume.
19. Millero, F. J. Mar. Chem. 1986, 18, 121-147.
20. Morse, J. W.; Millero, F. J.; Cornwell, J. C.; Rickard, D. Earth Sci. Rev. 1987, 24, 1-42.
21. Urban, P. J. Z. Anal. Chem. 1961, 179, 415-426.

22. West, P. W.; Gaeke, G. C.   Anal. Chem. 1956, 28, 1816-1819.
23. Vairavamurthy, A.; Mopper, K.   Environ. Sci. Technol. 1990, 24, 333-336.
24 Moses, C. O.; Nordstrom, D. K.; Mills, A.   Talanta 1984, 31, 331-339.
25 Takano, B.; McKibben, M. A.; Barnes, H. L.   Anal. Chem. 1984, 34, 1594-1600.
26. Spiro, C. E.; Wong, J.; Lytle, F; Greegor, R. B.; Maylotte, D.; Lampson, S. Science 1984, 226, 48-50.
27. George, G. N.; Gorbaty, M. L.   J. Am. Chem. Soc. 1989, 111, 3182.
28. Waldo, G. S.; Carlson, R. M. K.; Moldowan, J. M.; Peters, K. E.; Penner-Hahn, J. E.   Geochim. Cosmochim. Acta 1991, 55, 801-814.
29. Huffman, G. P.; Mitra, S.; Huggins, F. E.; Shah, N.; Vaidya, S.; Lu, F.   Energy & Fuels 1991, 5, 574-581.
30. Bart, J. C. J.   In: Advances in Catalysis; Eley, D. D.; Pines, H.; Weisz, P. B., Eds.; Academic Press: Orlando, Florida, 1986, Vol. 34; pp. 203-296.
31. Parratt, L. G.   Rev. Mod. Phys. 1959, 31, 616-645.
32. Bart, J. C. J.; Vlaic, G.   In: Advances in Catalysis; Eley, D. D.; Pines, H.; Weisz, P. B.; Eds.; Academic Press: Orlando, Florida, 1987, Vol. 35; pp 1- 138.
33. Frank, P.; Hedman, B.; Carlson, R. M. K.; Tyson, T. A.; Roe, A. L.; Hodgson, K. O.   Biochemistry 1987, 26, 4975-4976.
34. Wong, J.; Lytle, F. W.; Messmer, R. P.; Maylotte, D. H.   Phys. Rev. 1984, B30, 5596-5607.
35. Kunzl, V.   Collect. Czech. Commun. 1932, 4, 213-224.
36. Vairavamurthy, A; Manowitz, B; Luther, G. W. III; Jeon, Y. J.   Geochim. Cosmochim. Acta 1993, 57, 1619-1623.
37. Huffman, G. P.; Huggins, F. E.; Francis, H. E.; Mitra, S.; Shah, N.   In: Processing and Utilization of High Sulfur Coals III, Markuszewski, R.; Wheelock, T. D., Eds.; Elsevier: Amsterdam, 1990, pp 21-32.
38. Waldo, G. S.   Ph.D. Thesis, University of Michigan, 1991.
39. McMaster, W. H.; Kerr Del Grande, N.; Mallett, J. H.; Hubbell, J. H.   In: Compilation of X-ray Cross Sections, National Technical Information Service: Springfield, VA, 1969.
40. Stern, E. A.; Heald, S. M.   Rev. Sci. Instrum. 1979, 50, 1579-1582.
41. Gordon, B. M.; Jones, K. W.   In: Biological Trace Element Research, American Chemical Society Symposium Series No. 445, Subramanian, K. S.; Iyengar, G. V.; Okamoto, K. A., Eds.; American Chemical Society, Washington, DC, 1991, pp 290-305.
42. Jorgensen, B. B.   Science 1990, 249, 152-154.

RECEIVED September 3, 1993

Chapter 28

Evolved-Gas Analysis

A Method for Determining Pyrite, Marcasite, and Alkaline-Earth Carbonates

Richard W. Hammack

Pittsburgh Research Center, U.S. Bureau of Mines, P.O. Box 18070, Cochrans Mill Road, Pittsburgh, PA 15236

An evolved-gas analysis technique has been developed for the simultaneous determination of pyrite and alkaline-earth carbonates in geologic materials. The technique can also be used to estimate oxidation rates for coal pyrite and therefore may improve the prediction of acid discharges resulting from mining activity. A programmable tube furnace was used to heat crushed-rock samples in a 10% oxygen atmosphere. The evolution of sulfur dioxide and carbon dioxide was monitored with respect to time and temperature using a quadrupole mass spectrometer. Sulfur-dioxide peaks attributable to the oxidation of sedimentary pyrite occurred between 380°C and 440°C; sulfur-dioxide peaks attributable to the oxidation of hydrothermal pyrite were present between 475°C and 520°C. Two CO_2 peaks resulting from the decomposition of calcium carbonate (calcite) occurred at 380-370°C and at 550-650°C. Dolomite ($CaMg(CO_3)_2$) decomposed to yield CO_2 at 780-900°C. Calibration curves were prepared by plotting SO_2- and CO_2-evolved gas peak areas versus the concentration of pyrite and calcium carbonate, respectively. The detection limit for pyrite was found to be about 3% pyrite, with a working range that extended beyond 30% pyrite. The detection limit for calcite was less than 0.5% calcite, with a working range to about 14% calcite. Evolved-gas analysis cannot distinguish between hydrothermal pyrite and sulfur bound to aromatic organic compounds. This limitation may prevent the use of this technique in the analysis of anthracite coal and related strata.

Acidic mine drainage (AMD) is one of the most persistent and serious sources of industrial pollution in the United States. Premining prediction of acidic mine drainage alerts mine operators to potential sources of acid discharge and allows them to plan mining operations and reclamation to minimize water-quality degradation. In geographical areas or coal seams that have historically been sources of AMD, mine operators are required by law (1) to identify possible hydrologic consequences prior to opening a new mine. Concern in the United States regarding the possibility of acidic drainage has led to the requirement that an assessment of acid-discharge potential accompany each new mine-permit application. Overburden analysis to determine acid-discharge potential includes chemical tests to quantify the acidic or alkaline weathering products from each stratum overlying or directly underlying the coalbed to be mined.

There are many proposed overburden analysis techniques ranging from direct chemical determinations to simulated weathering methods (2). The acid/base account, a direct chemical technique, is the most widely used because of its simplicity and low cost. The method is based on measuring the total-sulfur content of each lithologic unit and converting that value to an acid potential based on the stoichiometry of complete pyrite oxidation. The neutralization potential is determined for each lithology by its ability to neutralize strong acid. The two values, acid potential and neutralization potential, are both represented as calcium-carbonate equivalents and the net excess or deficiency of neutralizers is calculated.

The acid/base account was originally developed as a quick method of identifying acid or alkaline weathering material for revegetation planning. Use of the acid/base account to predict drainage quality from heterogeneous mine spoils without considering the various other contributory factors (mine type, extent of reclamation, climate, and hydrology) is a serious overextension of the original intent of the method. Erickson (2) indicated that the acid/base account, as typically applied to overburden analysis, has little ability to predict drainage quality.

The tendency of the acid/base account to overestimate both the acid potential and the neutralization potential may be partly responsible for its poor predictive capability. The acid potential is based on the total-sulfur content even though all sulfur compounds do not contribute acidity. For example, the sulfur in gypsum does not react to form acidity. The overestimation of the neutralization potential results from the fact that, although all carbonate forms are soluble to some degree in strong acid, not all forms are readily available in the mine environment. In addition, the dissolution of manganese and iron carbonates will provide no net neutralization if the mine-waste drainage is ultimately exposed to an aerobic environment, resulting in the oxidation of manganese and iron.

This study was undertaken to develop an analytical method that would improve the predictive capability of the acid/base account. The first step was to improve the selectivity of analyses typically employed by the acid/base account. An ideal analytical procedure would be one that quickly and directly determined the sulfur species contributing acidity, and the carbonate species contributing alkalinity under field conditions. A previous study (3) indicated that the technique conventionally used for total-sulfur determinations (combustion-furnace ignition with infrared SO_2 detection) can be made more selective by operating at lower temperatures. For example, at 500°C, only SO_2 from the combustion of pyritic and organic sulfur is detected because sulfate sulfur is stable at this temperature. In noncarbonaceous samples, where the organic-sulfur content is negligible, the low temperature technique results in the direct determination of pyritic sulfur. Later studies (4,5) indicated that samples with low SO_2 initiation temperatures were more reactive and generated more acid in laboratory weathering tests.

The evolved-gas analysis (EGA) technique used in this study is capable of simultaneously quantifying pyrite and various carbonate minerals in a single run. The EGA method monitors the evolution of SO_2 and CO_2 produced by the following reactions:

$$2FeS_2(s) + 11/2\ O_2(g) \rightarrow Fe_2O_3(s) + 4SO_2(g) \quad (1)$$

$$Me_xCO_3(s) \rightarrow Me_xO(s) + CO_2(g) \quad (2)$$

where Me = monovalent or divalent cation.

The type of gas and temperature of evolution is characteristic of the original mineral or compound. Therefore, minerals can be tentatively identified by the temperatures at which certain gases are evolved. The amount of a particular mineral present in a sample is proportional to the partial pressure of the gas evolved.

EGA may improve the predictive capability of the acid/base account by:

1. determining pyritic sulfur more quickly than conventional ASTM sulfur forms speciation; and

2. providing neutralization potentials based only on carbonates species that are known to generate alkalinity under oxidizing field conditions.

Methods

An instrument designed specifically for EGA was assembled for this study (Figure 1). The design was adapted from an evolved-gas instrument constructed by LaCount and others (6). Major components include an electronic mass-flow controller/gas blender, a programmable tube furnace, a quadrupole mass spectrometer, a programmable analog-to-digital (A/D) converter, and a microcomputer.

Three hundred mg of -60 mesh sample was diluted with 3 g of tungsten oxide to aid uniform heating. The sample was then placed in a 2.54 cm diameter by 50 cm long quartz tube and secured with either glass wool or quartz wool, depending upon the maximum temperature of the run. A 32 mm (1/8 in), Type K thermocouple was inserted into the sample and the whole tube assembly was placed in the furnace. A 10.0% oxygen/90.0% nitrogen gas mixture was introduced into the tube furnace where it flowed through the crushed sample material at a flow rate of 100 mL/min.

The tube furnace used for this study was capable of performing two heating ramps with selectable heating rates, dwell temperatures, and dwell times. The sample was heated at a rate of 6°C/min from about 70°C until 380°C was attained. The heating rate was then decreased to 3°C/min until the run was terminated.

Evolved gases were detected with a quadrupole mass spectrometer. The capillary inlet to the mass spectrometer was placed in the gas stream, immediately downstream from the sample. This placement minimized lag time between gas evolution and detection. The mass spectrometer could simultaneously monitor the ion current at 12 user-selected mass to charge ratios (M/e). For each gas, the ion current was then converted to partial pressure by a calibration factor. Gases typically monitored included: SO_2 (M/e = 64), CO_2 (M/e = 44), COS (M/e = 60), H_2O (M/e = 18), H_2S (M/e = 34), O_2 (M/e = 32), and C_2H_6 (M/e = 30). The partial pressures of all monitored gases were transmitted to a microcomputer at each 1.5°C increase in temperature. Run time, temperature, and gas partial pressures were converted to ASCII files and written to floppy disk. Periodically, these files were transferred to a mainframe computer where graphics, Gaussian peak fitting, and peak integration were performed using library functions.

Results and Discussion

Pyrite Determination. Initial efforts were directed toward resolving SO_2 peaks resulting from the oxidation of pyrite or marcasite from those attributable to the combustion of organic sulfur. Figure 2 is a thermogram of Rasa Coal, from the former Yugoslavia, which contains 8.67% organic sulfur and 0.08% pyritic sulfur. Because the pyritic-sulfur content is two orders of magnitude less than the organic-sulfur content in this sample, it provides an EGA thermogram of organic sulfur in coal without the usual pyrite interferences. The thermogram (Figure 2) of the Rasa Coal exhibits coincident CO_2, H_2O and SO_2 peaks at 325°C and 475°C. Aliphatic compounds, due to their higher hydrogen content, would be expected to evolve more water during combustion than aromatic compounds. In Figure 2, the H_2O peak area relative to the CO_2 peak area is much greater at 325°C than at 475°C. This difference indicates that the peaks at 325°C and 475°C are likely due to the combustion of aliphatic and aromatic compounds, respectively. Less SO_2 evolution accompanies CO_2 evolution at 325°C (aliphatic hydrocarbons) than at 475°C (aromatic hydrocarbons). LaCount and others (7) have shown that the evolution of SO_2 at 320°C is due to the oxidation of sulfur in nonaromatic structures, whereas the SO_2 evolution near 480°C is due to the oxidation of thiophenic and aryl sulfide type structures.

Standards of sedimentary and hydrothermal pyrite were tested to determine the temperature range for sulfur dioxide evolution (Figure 3). Hydrothermal pyrite from the Noranda Mine in Québec, Canada evolved SO_2 between 430°C and 520°C, with the predominant peak at 485°C (Figure 3B). For hydrothermal pyrite, the SO_2 peak-evolution temperature decreased as the pyrite content increased. Because hydrothermal pyrite evolves SO_2 in the same temperature range as organic sulfur, the two species could not be distinguished if present in the same sample. Fortunately, organic sulfur and hydrothermal pyrite are found together only in the most thermally mature types of coal, such as the anthracite.

Coal pyrite standards were obtained as the sink fraction from the heavy media (bromoform) separation of pyritic coal. Combustion furnace analysis indicated that these concentrates contained greater than 90% pyrite. Two SO_2 peaks attributable to

28. HAMMACK   *Evolved-Gas Analysis*

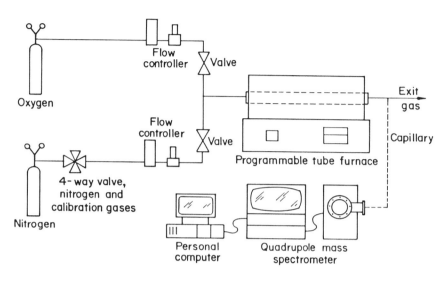

Figure 1. Schematic of the evolved gas analysis system used in this study.

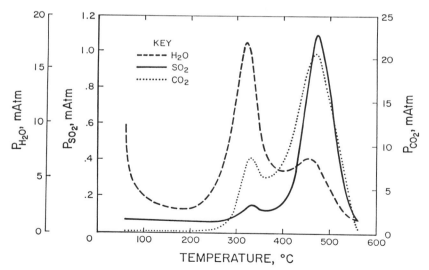

Figure 2. Evolved gas thermogram of the Rasa Coal from Yugoslavia.

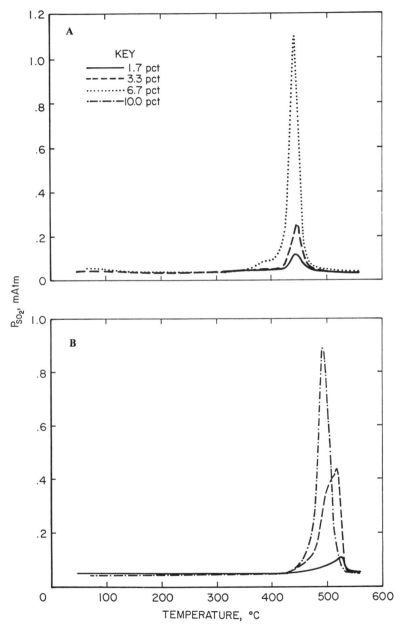

Figure 3. SO$_2$ thermograms for sedimentary pyrite standards (A) and hydrothermal pyrite standards (B).

the oxidation of sedimentary pyrite (Figure 3A) occurred at about 380°C and 430°C. Unlike hydrothermal pyrite, these SO_2 peak evolution temperatures were not dependent on the pyrite content of the sample. One or both peaks may be present but the 430°C peak generally predominated. Exceptions were noted where the 380°C peak was the sole peak (Figure 4A) or the predominant peak (Figure 4B). EGA studies of bituminous coal (8) and pyritiferous shale (9) showed similar results despite different experimental conditions (33% O_2, 300 mL/min flow).

Initially, samples in this study were heated at a constant rate of 6°C/min. At this heating rate, the SO_2 peak due to organic sulfur was not resolved from sedimentary pyrite peaks. By reducing the heating rate to 3°C/min at temperatures above 380°C, the organic-sulfur component at 500°C was partially resolved (Figure 4C). The application of a Gaussian peak-fitting routine permitted the resolution and quantification of two pyritic-sulfur peaks and one organic-sulfur peak (results not shown). Since the completion of this study, R. LaCount (personal commun., 1993) has been able to resolve pyritic sulfur from organic sulfur completely by increasing the oxygen content of the carrier gas to 100%.

The pyrite content can be determined from EGA data in two ways. The first method is to determine the amount of SO_2 evolved that is attributable (based on temperature of SO_2 evolution) to pyrite and using that value, calculate the amount of pyrite present. The advantage of this method is that only an SO_2 standard gas is required for calibration. However, this method does not account for the SO_2 adsorption/desorption effects from the sample matrix, tube walls, or capillary. Adsorption/desorption effects are compensated for in the second calibration technique. Using this technique, a set of pyrite standards was prepared and run in the same manner as actual samples. A calibration curve (Figure 5) was made by plotting the SO_2 peak area attributable to pyrite versus the pyrite content. Originally, a concave upward curve was obtained (Figure 5A) because of SO_2 sorption onto the walls of the silica capillary. After a capillary heater was installed, the plot approached linearity over the range of 3-30% pyrite (Figure 5B). The low-level detection limit of the mass-spectrometer detector used in this study (about 3% pyrite) was not adequate for many geologic samples. However, recent work by R. LaCount (personal commun., 1993) indicated that replacing the mass-spectrometer detector with a fourier-transform infrared (FTIR) detector would lower the detection limit to 0.1% pyrite. With this modification, EGA should be appropriate for the analysis of all pyrite-containing materials.

Oxidation rates measured by X-ray photoelectron spectroscopy (XPS) for sedimentary pyrite, hydrothermal pyrite, and hydrothermal marcasite (10) are plotted versus SO_2 peak temperatures in Figure 6. As expected, the more reactive pyrites evolved SO_2 at lower temperatures than the more stable pyrites. The relationship between pyrite reactivity and SO_2 evolution temperature can be used as a quick method to estimate pyrite reactivity, especially for sedimentary pyrites that exhibit a wide range of reactivities. A similar relationship may exist for marcasite, although only two marcasite-containing samples have been examined to date. The incorporation of a reactivity estimate into the acid/base account would likely improve its ability to predict the water quality of drainage from disturbed strata containing sedimentary pyrite and possibly marcasite. The hydrothermal pyrites we have studied

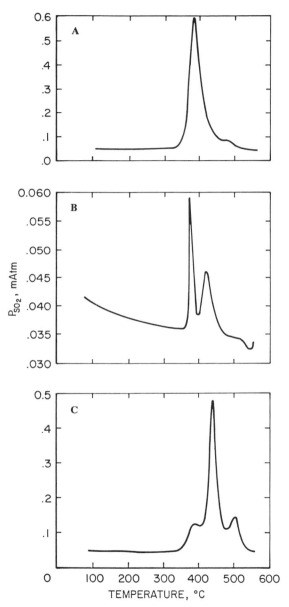

Figure 4. SO$_2$ thermograms exhibiting a single 380°C pyrite peak (A), a predominant 380°C pyrite peak (B), and an organic sulfur peak at 500°C (C).

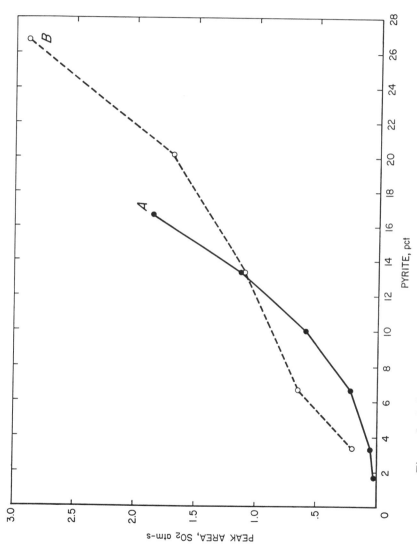

Figure 5. Calibration curves for sedimentary pyrite without (A) and with (B) a capillary heater installed.

weather at similar rates. There is little reason, therefore, to measure the reactivity of each. Pyrite-containing samples must be prepared identically when SO_2 evolution temperatures are used to estimate pyrite reactivity. Figure 7 contains two SO_2 thermograms of the same pyrite. One split was run immediately after washing the sample with 4 N HCl, whereas the second split was run after the sample had been exposed to air for several weeks. The accumulation of oxidation products on the pyrite surfaces shifted SO_2-evolution temperatures to higher values. Because of this temperature shift upon exposure to air, inconsistent sample preparation could result in an inaccurate estimate of pyrite reactivity.

Carbonate Determination. EGA can be used to distinguish alkaline-earth carbonates, that thermally decompose to yield CO_2 primarily above 520°C, from transition-metal carbonates, which evolve CO_2 between 220°C and 520°C (Figure 8). Only the alkaline-earth carbonates (such as calcite ($CaCO_3$) and dolomite ($CaMg(CO_3)_2$)) are important to the acid/base account because transition-metal carbonates (such as siderite ($FeCO_3$) and rhodochrosite ($MnCO_3$)) provide no net neutralization under oxidizing field conditions.

The presence of carbonaceous material and transition-metal carbonates may interfere with the EGA determination of calcite. Some CO_2 is evolved from calcite between 300°C and 500°C, the same temperature range as siderite and rhodochrosite decomposition (Figure 8), and organic combustion (Figure 9). However, most of the CO_2 evolution during calcite decomposition occurs at temperatures above 550°C, where there is only a slight CO_2 contribution from carbonaceous material and transition-metal carbonates. The calcite decomposition peak at 600°C is the only peak that can be used for the quantitative analysis of calcite without significant interference. If only the high-temperature peak is used, then not all of the CO_2 evolution attributable to calcite would be measured. Therefore, accurate quantitation can only be achieved if the CO_2 peak at 600°C is proportional to the calcite content.

The relationship between the peak area for the 600°C CO_2 peak and the calcite content (Figure 10) indicated that EGA may be used for calcite determinations between 2% and 14% calcite. Saturation of the mass-spectrometer detector above 14% calcite may be avoided by diluting the sample or by using an FTIR detector that has a wider working range.

The quantitative determination of dolomite is also important to the acid/base account. Dolomite decomposes to yield CO_2 at 800-900°C, well above the decomposition temperatures for transition-metal carbonates. Dolomite standards were not run in this study, but the quantitation of dolomite should be simple because there are no interferences (R.W. Hammack, unpublished data).

Conclusions

Evolved-gas analysis may provide a relatively simple and quick method for determining the pyrite and alkaline-earth carbonate content of overburden samples. All analyses are performed on a single sample in one run. The method can discriminate species that are critical to the acid/base account (iron disulfides, calcite, and dolomite) from species that should be excluded (organic sulfur and transition-

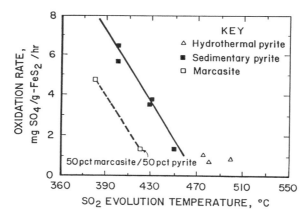

Figure 6. Plot of oxidation rate versus SO_2 evolution temperatures for hydrothermal pyrite, sedimentary pyrite, and hydrothermal marcasite.

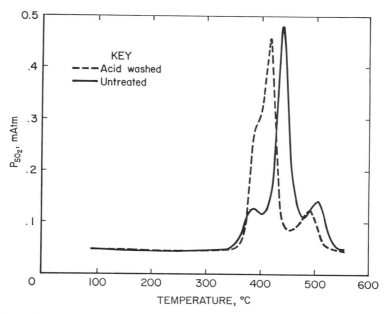

Figure 7. SO_2 thermograms of a sample containing sedimentary pyrite and organic sulfur that show peak shift in response to the removal of oxidation products by acid washing.

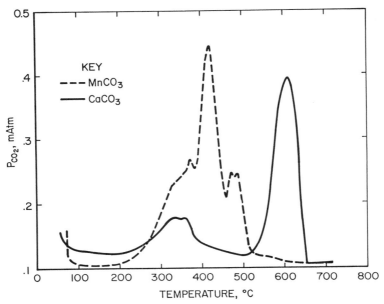

Figure 8. CO_2 thermograms of rhodochrosite ($MnCO_3$) and calcite ($CaCO_3$).

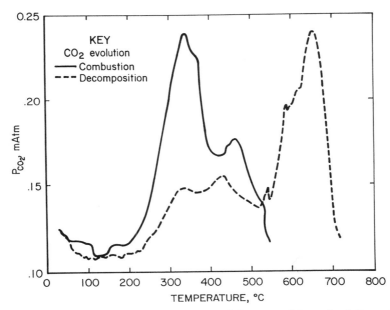

Figure 9. CO_2 thermograms of the thermal decomposition of calcite and the combustion of carbonaceous material in coal refuse.

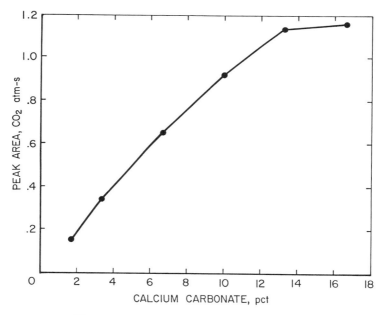

Figure 10. Calibration curve for calcite.

metal carbonates). Use of the EGA technique may improve the predictive capability of the acid/base account.

Although the setup used in this study was not optimized for evolved-gas analysis, it demonstrated the potential application of EGA to acid-drainage prediction. The substitution of a fourier-transform infrared detector for the mass spectrometer detector used in this study would improve the sensitivity and working range of the analysis. Work by others, subsequent to this study, has shown that increasing the oxygen content of the carrier gas will resolve sedimentary and hydrothermal pyrite from sulfur in aromatic compounds. Recently, an EGA instrument has been developed and tested (R. LaCount, personal commun., 1993) that will simultaneously analyze four samples, greatly increasing sample throughput. Future research will be directed toward the EGA determination of sulfide minerals affecting water quality at metal mines.

Literature Cited

1. U.S. Code of Federal Regulations. Title 30--Minerals Resources; Chapter VII--Office of Surface Mining, Department of the Interior; Subchapter G--Surface Coal Mining and Reclamation Operation Permits; July 1, 1984.
2. Erickson, P.M. In Proceedings of the National Mined Land Reclamation Conference; St. Louis, MO, 1986.
3. Erickson, P.M.; Hammack, R.W.; Kleinmann, R.L.P. In Proceedings of a Technology Transfer; U.S. Bureau of Mines IC 9027, 1985; 8 pp.
4. Hammack, R.W. In Proceedings of the 1985 National Symposium on Surface Mining, Hydrology, Sedimentology, and Reclamation; Lexington, KY, 1985; pp. 139-145.
5. Hammack, R.W. In Proceedings of the Seventh Annual West Virginia Surface Mine Drainage Task Force Symposium; Morgantown, WV, 1986.
6. LaCount, R.B.; Anderson, R.R.; Helms, C.A.; Friedman, S.; Romine, W.B. U.S. Department of Energy DOE/PETC/TR-83/5, 1983, 22 pp. plus appendices.
7. LaCount, R.B.; Anderson, R.R.; Friedman, S.; Blaustein, B.D. Fuel 1987, 66, 909-913.
8. Warne, S. Thermochimica Acta 1985, 93, 745-748.
9. Morgan, D.J. J. of Thermal Analysis 1977, 12, 245-263.
10. Hammack, R.W.; Lai, R.W.; Diehl, J.R. Proceedings of the 1988 Mine Drainage and Surface Mine Reclamation Conference; U.S. Bureau of Mines: Pittsburgh, PA, 1988; Vol. 1, pp. 136-146.

RECEIVED April 2, 1993

Stable Isotope Fractionation and Equilibration in Oxidizing Sulfide Systems

Chapter 29

Controls of $\delta^{18}O$ in Sulfate
Review of Experimental Data and Application to Specific Environments

D. R. Van Stempvoort[1,3] and H. R. Krouse[2]

[1]National Hydrology Research Institute, 11 Innovation Boulevard, Saskatoon, Saskatchewan S7N 3H5, Canada
[2]Department of Physics and Astronomy, University of Calgary, 2500 University Drive Northwest, Calgary, Alberta T2N 1N4, Canada

Comparison of experimental and field data indicates that the dominant controls of the $\delta^{18}O$ of environmental $SO_4^{2-}{}_{(aq)}$ are S redox reactions, and $\delta^{18}O$ of ambient H_2O. Several other factors play a role in determining $\delta^{18}O_{SO4}$, including variable S oxidation rates, dissolution of mineral sulfate, and transport and mixing of $SO_4^{2-}{}_{(aq)}$. In most environments, O exchange between $SO_4^{2-}{}_{(aq)}$ and H_2O, and isotope fractionation during sorption, precipitation, and diffusion processes apparently have negligible effects on $\delta^{18}O_{SO4}$.

Atmospheric SO_4^{2-} has many sources, including sea spray, primary ^{18}O-enriched SO_4^{2-} from anthropogenic emissions, and secondary SO_4^{2-} formed in the atmosphere by oxidation-hydrolysis of anthropogenic SO_2 and biogenic sulfides. In many soil, surface-water and groundwater environments, $\delta^{18}O_{SO4}$ is compatible with *in situ* S oxidation. However, some of these environments have ^{18}O-enriched $SO_4^{2-}{}_{(aq)}$ that has been affected by dissolution of evaporite mineral sulfate, or by microbial reduction, which selectively consumes ^{16}O.

Sulfate is a key component of the global biogeochemical S cycle. The concentration and isotopic composition of $SO_4^{2-}{}_{(aq)}$ may fluctuate in response to environmental change, for example: influx of atmospheric pollution ("acid rain"), weathering of sulfide-bearing mine wastes, or enhancement of $SO_4^{2-}{}_{(aq)}$ reduction in a landfill site. In the past two decades, measurements of $\delta^{34}S$ (°/$_{oo}$ CDT) and $\delta^{18}O$ (°/$_{oo}$ SMOW) in sulfate (1) have aided the interpretation of the origin, reactivity and mobility of $SO_4^{2-}{}_{(aq)}$ in various environments. However, the basic controls of environmental $\delta^{18}O_{SO4}$ values are only partially understood (2, 3). The aims of this paper are: 1) to review the pertinent laboratory experiments, 2) to assess the relative importance of the various factors that can affect $\delta^{18}O_{SO4}$ either during formation of $SO_4^{2-}{}_{(aq)}$ or by other processes, 3) to compare data collected in various Earth-surface and near-surface

[3]Current address: Saskatchewan Research Council, 15 Innovation Boulevard, Saskatoon, Saskatchewan S7N 2X8, Canada

environments to the laboratory findings, and 4) to provide working guidelines for the interpretation of $\delta^{18}O_{SO4}$ in environmental waters. Values of $\delta^{34}S_{SO4}$ provide relevant parallel information, but their consideration lies outside the scope of this paper.

Much of the oxygen in $SO_4^{2-}{}_{(aq)}$ in the present environment was obtained during biologically mediated S oxidation. The oxygen isotope composition of $SO_4^{2-}{}_{(aq)}$ may be affected by at least six interacting factors during its formation:

1) the type(s) of oxidation reaction(s) in which the $SO_4^{2-}{}_{(aq)}$ formed from one or more reduced S species, either (a) gas-phase oxidation or (b) aqueous oxidation and hydrolysis;

2) the relative amounts of $SO_4^{2-}{}_{(aq)}$ oxygen incorporated from the various oxygen sources (notably H_2O and $O_{2\,(aq)}$);

3) $\delta^{18}O$ of ambient water, $O_{2\,(aq)}$ and any other source(s) of $SO_4^{2-}{}_{(aq)}$ oxygen;

4) the specific oxidizing agent(s), catalyst(s) or enzyme(s) involved in $SO_4^{2-}{}_{(aq)}$ formation, whether biological or abiotic;

5) ambient physiochemical conditions, notably pH, P_{O2}, and temperature (T); and

6) equilibrium or rate-dependent (kinetic) isotope enrichment associated with incorporation of oxygen from various sources during formation of $SO_4^{2-}{}_{(aq)}$ or precursor S oxyanions.

Figure 1 illustrates the theoretical variation of $\delta^{18}O$ in $SO_4^{2-}{}_{(aq)}$ which is a consequence of the above factors. To some degree, one can control factors 1 through 5 independently in experiments, but the significance of kinetic effects (factor 6) during S oxidation is usually difficult to determine, and not well documented.

In the next section, the above six factors are discussed with reference to pertinent experimental S oxidation studies. The following section examines other non-oxidative factors that may affect $\delta^{18}O_{SO4}$: exchange of oxygen between $SO_4^{2-}{}_{(aq)}$ and H_2O, hydrolysis of organic sulfates, and microbial reduction of sulfate. Finally, environmental $\delta^{18}O_{SO4}$ data are discussed, with reference to the experimental findings.

Aqueous Oxidation of Various S Compounds to Sulfate: Experimental Evidence.

In this section, $\delta^{18}O$ of $SO_4^{2-}{}_{(aq)}$ (S valence +6) formed by oxidation of sulfite (valence +4) is considered first because this is the shortest path with 2 electrons transferred. Subsequently, data for oxidation of thiosulfate (valence -2, +6), elemental S (0) and sulfides (-1, -2) are considered. Oxidation of these lower valence S compounds occurs stepwise, with only one or two electrons transferred at a time. Finally, $\delta^{18}O_{SO4}$ data for oxidation of reduced organic S are considered briefly.

Oxidation of sulfite. Experimental data on oxidation of dissolved SO_2 (SO_3^{2-}, HSO_3^-, H_2SO_3), referred to below as sulfite or $SO_3^{2-}{}_{(aq)}$, provide information on the dependence of the oxidation rate on various catalysts, the rate of oxygen exchange between $SO_3^{2-}{}_{(aq)}$ and water, and the $\delta^{18}O$ of sulfate formed.

Reaction Rates: Abiotic oxidation of $SO_3^{2-}{}_{(aq)}$ by $H_2O_{2\,(aq)}$, $O_{3\,(aq)}$, $O_{2\,(aq)}$ or other electron acceptors occurs at rates dependent on pH and the presence of catalysts (4, 5). Data suggest that in the atmosphere, cloud droplet $SO_3^{2-}{}_{(aq)}$ is predominantly oxidized by $H_2O_{2\,(aq)}$ at low pH (4, 6), but by $O_{3\,(aq)}$ or $O_{2\,(aq)}$ (catalyzed by metals) at pH values greater than 4-5 (5). In aerobic soils, surface waters and groundwaters the concentrations of catalysts are likely higher, and thus abiotic oxidation by $O_{2\,(aq)}$ may be dominant over a wide spectrum of pH conditions. Abiotic oxidation of $SO_3^{2-}{}_{(aq)}$

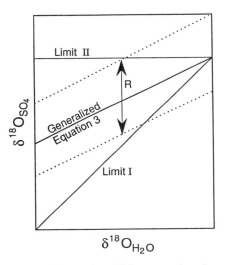

Figure 1. Theoretical diagram illustrating $\delta^{18}O_{SO_4}$ as a function of $\delta^{18}O_{H_2O}$. Equation 3 is given in text; special cases of this equation are Limit I, where all oxygen is derived from H_2O ($m = 1$), and Limit II, where all oxygen is from other sources ($m = 0$). Range R, shown for generalized Equation 3, is due to isotopic fractionation during incorporation of oxygen from H_2O or O_2 and to variations in $\delta^{18}O_{O_2}$. Range R also applies to Limits I and II.

may be dominant in some soils (7), but other evidence points to the importance of microbial oxidation of $SO_3^{2-}{}_{(aq)}$ in soils (8).

Oxygen Isotope Exchange between Sulfite and Water. Experiments at 25°C by Betts and Voss (9) indicate that sulfite-water oxygen exchange is very rapid at pH < 9 ($t_{1/2} = 1.3$ min at pH 8.9), but slower at elevated pH values ($t_{1/2} = 25.3$ h at pH 10.5; no observed exchange after 17 h at pH 12.7). Similarly, Holt et al. (10, 11) observed that, in closed containers with water at 24°C and pH < 7, or with water vapor at 22°C and pH ~ 7, the $\delta^{18}O$ of SO_2 was completely controlled by water oxygen. In other words, either directly or indirectly, water contributed all of the oxygen present in the SO_2. They inferred rapid oxygen exchange between $SO_3^{2-}{}_{(aq)}$ and water. At 22°C the equilibration produces a $\delta^{18}O$ in SO_2 enriched by ~ 24 ‰ relative to $\delta^{18}O$ in H_2O vapor (11).

$\delta^{18}O$ of Sulfate Formed by Abiotic Oxidation of Sulfite. The most complete data on the control of the $\delta^{18}O$ in $SO_4^{2-}{}_{(aq)}$ formed by oxidation of $SO_3^{2-}{}_{(aq)}$ is the work of Holt et al. (10) who carried out abiotic oxidation of $SO_3^{2-}{}_{(aq)}$ under various conditions. By varying $\delta^{18}O_{H2O}$, Holt et al. found that during uncatalyzed oxidation of $SO_3^{2-}{}_{(aq)}$ in water exposed to air, the water contributed (indirectly or directly) approximately 75 % of the $SO_4^{2-}{}_{(aq)}$ oxygen (Equation 1):

$$\delta^{18}O_{SO4} = 0.75(\delta^{18}O_{H2O}) + K \quad (1)$$

where intercept K was dependent on experimental conditions.

Thus $SO_3^{2-}{}_{(aq)}$, and indirectly water via equilibrium exchange with $SO_3^{2-}{}_{(aq)}$, contributed 3 of the 4 oxygens to $SO_4^{2-}{}_{(aq)}$, while one oxygen was apparently derived from $O_{2\ (aq)}$ (Equation 2).

$$SO_3^{2-}{}_{(aq)} + \frac{1}{2}O_{2\ (aq)} \Rightarrow SO_4^{2-}{}_{(aq)} \quad (2)$$

In other tests of air ($O_{2\ (aq)}$) oxidation of $SO_3^{2-}{}_{(aq)}$, with or without Fe^{3+} or Cu^{2+} catalysts, water contributed approximately 55 to 75 % of the $SO_4^{2-}{}_{(aq)}$ oxygen (10, 12). A significant amount of ^{18}O-labelled $O_{2\ (aq)}$ was incorporated into the $SO_4^{2-}{}_{(aq)}$ produced (12). The variation in oxygen contribution by water probably occurs because of several different pathways of $SO_3^{2-}{}_{(aq)}$ oxidation to $SO_4^{2-}{}_{(aq)}$, which may involve disulfate ($S_2O_7^{2-}{}_{(aq)}$) or other intermediates (13). Thus, Equation 2 should only be considered a summary reaction, not necessarily reflecting oxygen by source stoichiometrically. The majority of the experimental $SO_3^{2-}{}_{(aq)}$ oxidation data of Holt et al. (10) fall into Area A of Figure 2. Boundaries for Area A are based on regressions of experimental $SO_3^{2-}{}_{(aq)}$ oxidation data of Holt et al. (10; some data plot beyond the range of Figure 2). The upper boundary, $\delta^{18}O_{SO4} = 0.74(\delta^{18}O_{H2O}) + 15.4$, is their curve A, where O_2 was added in stoichiometric quantity, per Equation 2; the lower boundary, $\delta^{18}O_{SO4} = 0.71(\delta^{18}O_{H2O}) + 8.2$, is their curve I, with air in excess.

Area A in Figure 2 has ~ 7 ‰ range in $\delta^{18}O_{SO4}$ for a given $\delta^{18}O_{H2O}$, far in excess of measurement precision ($\sigma \approx 0.1$ and 0.5 ‰ respectively for H_2O and $SO_4^{2-}{}_{(aq)}$). This $\delta^{18}O_{SO4}$ range is due mainly to variations in K, as defined in Equation 1, during $SO_3^{2-}{}_{(aq)}$ oxidation. For each experiment, the value of K apparently is dependent on equilibrium fractionation between water and $SO_3^{2-}{}_{(aq)}$, $\delta^{18}O_{O2}$, and kinetic oxygen isotope fractionation during reaction of $O_{2\ (aq)}$ and $SO_3^{2-}{}_{(aq)}$. The more general form of Equation 1 is (14):

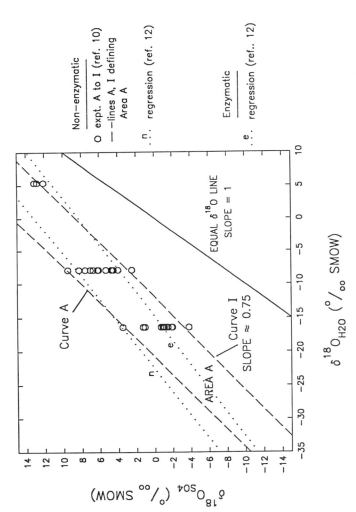

Figure 2. Experimental data for $\delta^{18}O_{SO_4}$ as a function of $\delta^{18}O_{H_2O}$ during abiotic or enzyme-catalyzed oxidation of $SO_3^{2-}{}_{(aq)}$. Regression lines from Krouse et al. (12) are based on experiments using waters with $\delta^{18}O$ values ranging from -11 to +200 ‰ whereas the field shown represents the approximate maximum range of $\delta^{18}O$ in natural waters. The upper boundary of Area A is curve A of Holt et al.(10) where O_2 was added in stoichiometric quantity based on Equation 2; the lower boundary is curve I (10) with air in excess.

$$\delta^{18}O_{SO4} = m(\delta^{18}O_{H2O} + \varepsilon_{H2O}) + (1-m)(\delta^{18}O_{O2} + \varepsilon_{O2}) \tag{3}$$

where m = fraction of oxygen contributed by H_2O, ε = kinetic or equilibrium isotope enrichment factor for each oxygen source.

The ε values may be pH, P_{O2} and/or temperature dependent. This general Equation 3 is illustrated in Figure 1. For a given value of m in the linear Equation 3 (i.e., aqueous oxidation with H_2O contribution fixed), a range in K values, R, can be substituted for a single K value, yielding a zone rather than a line as a solution of Equation 3 (Figure 1). The vertical range of Area A in Figure 2 is an example of R.

During hydrogen peroxide oxidation of $SO_3^{2-}{}_{(aq)}$, Holt and Kumar (15) found that approximately 60% of $SO_4^{2-}{}_{(aq)}$ oxygen was contributed by water, and 40% by $H_2O_{2\ (aq)}$. They postulated a transitory reaction intermediate ($H_2O_2.SO_3^{2-}{}_{(aq)}$), in which $SO_3^{2-}{}_{(aq)}$ contributed 3 of the oxygens, which were in exchange equilibrium with $\delta^{18}O_{H2O}$. Thus, an even more general equation than Equation 3 could be constructed to account for oxygen sources other than water and O_2. However, the $\delta^{18}O$ of H_2O_2 in Earth's atmosphere is unknown.

Water contributed (directly or indirectly) all of the $SO_4^{2-}{}_{(aq)}$ oxygen during alkaline oxidation of $SO_3^{2-}{}_{(aq)}$ by bromine (10). The 8 ‰ ($\delta^{18}O$) enrichment in the $SO_4^{2-}{}_{(aq)}$ oxygen relative to water may reflect the fact that $SO_3^{2-}{}_{(aq)}$, in equilibrium with water, was the direct source of 3 of the oxygens. This enrichment is likely temperature dependent; Caron et al. (16) hypothesized that, similar to bicarbonate-water-oxygen exchange, the enrichment decreases by 0.21 ‰ °C^{-1}.

Holt and Kumar (17) oxidized SO_2 in humidified air containing O_2 at high temperature (475 to 500°C), using platinum as a catalyst. The $SO_4^{2-}{}_{(aq)}$ that formed had $\delta^{18}O$ around +40 to +45 ‰, enriched in ^{18}O relative to all reactants, with little dependence on initial $\delta^{18}O$ of water vapor. This observation apparently is due to rapid oxygen exchange between water vapor and O_2 at these elevated temperatures, and kinetic ^{18}O enrichment of the $SO_4^{2-}{}_{(aq)}$ produced. The $\delta^{18}O$ in $SO_4^{2-}{}_{(aq)}$ versus that of initial water vapor (not shown) plots well above Area A of Figure 2. Holt and Kumar suggested that this ^{18}O-enriched $SO_4^{2-}{}_{(aq)}$ may be similar isotopically to "primary" SO_4^{2-} that forms from anthropogenic SO_x emissions (industrial stacks, automobile exhaust, etc.), and contributes a major portion of sulfate in the atmosphere.

$\delta^{18}O$ of Sulfate Formed by Microbial Oxidation of Sulfite.

Chemolithotrophic S-oxidizing bacteria, including the thiobacilli, cannot utilize energy from oxidation of S atoms that react directly with $O_{2\ (aq)}$ molecules (18). These bacteria employ enzymatic electron-transport chains that shuttle electrons from S atoms in sulfite to O_2 molecules, extracting energy in the process. Kelly (18) has postulated that, in some cases, the $SO_4^{2-}{}_{(aq)}$ that forms in such a enzyme-catalyzed reaction may derive 3/4 of its oxygen from $SO_3^{2-}{}_{(aq)}$, and 1/4 from ambient water (H_2O).

As discussed above, studies (9, 10) indicate that $\delta^{18}O$ of $SO_3^{2-}{}_{(aq)}$ may be completely controlled by oxygen exchange with water. Thus, if Kelly's (18) model is correct, during chemolithotrophic bacterial oxidation of $SO_3^{2-}{}_{(aq)}$, water will contribute 100 % of the oxygen to $SO_4^{2-}{}_{(aq)}$ (see sections on oxidation of elemental S and sulfides below):

$$\delta^{18}O_{SO4} = \delta^{18}O_{H2O} + \varepsilon_{H2O} \tag{4}$$

Equation 4 (Limit I on Figure 1) is the special case of Equation 3 where all oxygen in sulfate is derived from H_2O (m = 1). However, the experimental evidence of Peck and Stulberg (19) indicates that during enzymatic $SO_3^{2-}{}_{(aq)}$ oxidation by the chemolithotroph *Thiobacillus thioparus*, 20 % of $SO_4^{2-}{}_{(aq)}$ oxygen was derived from ^{18}O-labelled phosphate. These authors inferred an oxidation pathway in which APS is an intermediate (20) (Equations 5 & 6).

$$Ad.PO_3{}^{18}O + SO_3{}^{2-} \Leftrightarrow Ad.PO_3{}^{18}OSO_3 + 2e^- \qquad (5)$$

$$Ad.PO_3{}^{18}OSO_3 + PO_4{}^{3-} \Leftrightarrow ADP + SO_3{}^{18}O^{2-} + e^- \qquad (6)$$

where ADP = adenosine diphosphate, $Ad.PO_3{}^{18}O$ = labelled AMP (adenosine monophosphate), $Ad.PO_3{}^{18}OSO_3$ = labelled APS (adenosine 5'-phosphosulfate).

The stoichiometry of these reactions suggests that one in four (25 %) of $SO_4^{2-}{}_{(aq)}$ oxygens is derived from APS, but the less than 25 % labelled $SO_4^{2-}{}_{(aq)}$ oxygen observed by Peck and Stulberg (19) is possibly due to enzyme-catalyzed phosphate-water or phosphyl-water oxygen exchange (21), prior to the reaction of $SO_3^{2-}{}_{(aq)}$ and AMP. It is unclear whether phosphate or phosphyl-water oxygen exchange contributes water oxygen to $SO_4^{2-}{}_{(aq)}$ indirectly during enzyme-catalyzed oxidation of $SO_3^{2-}{}_{(aq)}$.

Krouse et al. (12) found that during oxidation of $SO_3^{2-}{}_{(aq)}$ catalyzed by a sulfite oxidase enzyme taken from chicken liver, water contributed 56% of the $SO_4^{2-}{}_{(aq)}$ oxygen (his Series W data - see Figure 2), and that a considerable amount of ^{18}O-labelled $O_2{}_{(aq)}$ was incorporated in the $SO_4^{2-}{}_{(aq)}$ produced. The role of abiotic oxidation in this and other experiments noted above, parallel to enzyme-catalyzed oxidation, is unknown. One can safely assume that abiotic oxidation of $SO_3^{2-}{}_{(aq)}$ is a ubiquitous process, but may be minor in some of these enzyme experiments.

In contrast, various thiobacilli (*T. novellus*, *T. thiooxidans*, *T. denitrificans*, *T. thioparus*) employ AMP-independent electron "shuttle" enzymes to oxidize $SO_3^{2-}{}_{(aq)}$, using either $O_2{}_{(aq)}$ or nitrate $_{(aq)}$ as electron acceptors (22, p. 467). The oxygen in the $SO_4^{2-}{}_{(aq)}$ produced by these reactions may be 100 % water-derived, due to hydrolysis, as predicted by the Kelly (18) model.

There is no detailed information available on the sulfite oxidation mechanisms used by heterotrophic bacteria and fungi (23), which may be the dominant S oxidizers in soils (24, 25). It is possible that heterotrophic bacteria mediate the oxidation of $SO_3^{2-}{}_{(aq)}$ to $SO_4^{2-}{}_{(aq)}$ indirectly, and that the $\delta^{18}O$ of the $SO_4^{2-}{}_{(aq)}$ produced is similar to that of $SO_4^{2-}{}_{(aq)}$ formed by abiotic oxidation pathways (as above; Figure 2).

Oxidation of Thiosulfate. There is apparently no information on the $\delta^{18}O$ values of $SO_4^{2-}{}_{(aq)}$ formed by strictly abiotic oxidation of thiosulfate ($S_2O_3^{2-}{}_{(aq)}$). Kelly (18) suggested that the dominant microbial oxidation pathways do not involve polythionates and thiosulfate, and that these are merely side reaction by-products. In contrast, Jorgensen (26) argues that $S_2O_3^{2-}{}_{(aq)}$ is an important intermediate of S cycling reactions in marine sediments. Suzuki et al. (27) outline other evidence for microbially mediated reactions that produce or consume $S_2O_3^{2-}{}_{(aq)}$ and other intermediate valence compounds (e.g. polythionates).

Based on the model of Schedel and Trüper (28) and the summary of Kelly (18), most evidence suggests that microbial oxidation of $S_2O_3^{2-}{}_{(aq)}$ proceeds stepwise via $SO_3^{2-}{}_{(aq)}$. Rather than the theoretically expected value of 25 % (Equations 5, 6), Santer (29) found that 22 % of $SO_4^{2-}{}_{(aq)}$ oxygen was derived from ^{18}O-labelled phosphate following oxidation of $S_2O_3^{2-}{}_{(aq)}$ to $SO_4^{2-}{}_{(aq)}$ by *T. thioparus*. As above, enzyme-catalyzed phosphate-water oxygen exchange may have occurred, thus lowering the observed ^{18}O-label in $SO_4^{2-}{}_{(aq)}$ below 25 %.

Krouse (30; Series V) found no exchange of ^{18}O between H_2O and ^{18}O-enriched $O_{2\ (aq)}$ during experimental enzyme-catalyzed (rhodanese) conversion of $S_2O_3^{2-}{}_{(aq)}$ plus CN^- to SCN^- plus $SO_3^{2-}{}_{(aq)}$. Because $SO_3^{2-}{}_{(aq)}$ rapidly exchanges oxygen with water, in this experiment O_2 oxygen was apparently not incorporated into $SO_3^{2-}{}_{(aq)}$ during or after its cleavage from $S_2O_3^{2-}{}_{(aq)}$.

Oxidation of Elemental S. Abiotic or microbial oxidation of solid elemental S ($S_{(s)}$) proceeds via stepwise reactions in which $SO_3^{2-}{}_{(aq)}$ is an intermediate, and other compounds such as thiosulfate or polythionates may form (27). Suzuki (31) found that during oxidation of $S_{(s)}$ by *Thiobacillus thiooxidans*, the $SO_4^{2-}{}_{(aq)}$ formed did not include any oxygen from ^{18}O-labelled O_2, even though $O_{2\ (aq)}$ was the electron acceptor during S oxidation. This observation indicates that either 1) the ^{18}O-labelled oxygen was transferred rapidly from reaction intermediates such as $SO_3^{2-}{}_{(aq)}$ or a glutathione derivative (31) to water, or 2) an enzymatic electron shuttle system ("transport chain") was utilized in which $O_{2\ (aq)}$ did not react directly with S. The electron shuttle mechanism is more likely the correct explanation, because these chemotrophs derive metabolic energy from S oxidation (18). The oxygen in $SO_4^{2-}{}_{(aq)}$ likely was derived from water (18) and/or APS (Equations 5, 6).

Mizutani and Rafter (32) oxidized $S_{(s)}$ with bacterial cultures of unknown identity in two waters of distinctly different isotopic compositions, and found that water contributed all of the $SO_4^{2-}{}_{(aq)}$ oxygen, which was virtually identical isotopically to the water oxygen ($\Delta^{18}O_{SO4-H2O} \approx 0$). Apparently all $SO_4^{2-}{}_{(aq)}$ oxygen was derived directly from water with an insignificant kinetic enrichment factor (ε_{H2O}). Reaction intermediates, such as $SO_3^{2-}{}_{(aq)}$ did not exchange oxygen with water significantly (see above). As in Suzuki's (31) experiment, probably an electron-transport system was operative in which the electron acceptor ($O_{2\ (aq)}$ or other) did not react directly with S.

Suzuki (31) found that during oxidation of $S_{(s)}$ by a cell-free *Thiobacillus thiooxidans* enzyme extract, a small amount of ^{18}O-labelled oxygen from $O_{2\ (aq)}$ was incorporated into $S_2O_3^{2-}{}_{(aq)}$. However, this minor labelling may be due to a "side reaction or artifactual functioning of the ... oxidizing enzyme" (33). The side reaction could be an abiotic oxidation pathway. This explanation may also apply to minor incorporation of ^{18}O from $O_{2\ (aq)}$ during oxidation of $S_{(s)}$ to $SO_3^{2-}{}_{(aq)}$ by *Sulfolobus brierleyi* (34), although the authors who reported these isotope data argued that they are evidence for an oxygenase enzyme.

Oxidation of Sulfides. Experimental isotope data on oxidation of sulfides (dissolved and mineral forms) compliment the numerous hydrochemical studies of this group of reactions. As in the case of $S_2O_3^{2-}{}_{(aq)}$ and $S_{(s)}$, oxidation of sulfides to $SO_4^{2-}{}_{(aq)}$ occurs stepwise, via intermediates including $S_{(s)}$ and $SO_3^{2-}{}_{(aq)}$.

Role of Fe^{3+} In Abiotic Oxidation of Pyrite. Studies of abiotic sulfide-oxidation mechanisms, conducted under sterile conditions, have been instructive. The mechanisms involved in abiotic oxidation of pyrite have received particular attention (35-39). Based on theoretical considerations (40) and experimental data (39), it appears that, regardless of pH conditions, $Fe^{3+}{}_{(aq)}$, rather than $O_{2\ (aq)}$, is the primary agent responsible for direct abiotic oxidation of pyrite S, and possibly for oxidation of intermediate valence S forms (e.g., $S_{(s)}$) on the surfaces of weathering pyrite crystals

(39). Dissolved O_2 reconverts $Fe^{2+}_{(aq)}$ to $Fe^{3+}_{(aq)}$ during ongoing pyritic S oxidation and also may oxidize directly intermediate valence sulfur oxyanions released to the ambient solution (39).

Recently, Reedy et al. (41) used Fourier transform infrared and laser Raman vibrational spectroscopy to examine the role of water and $O_{2\ (aq)}$ as contributors of oxygen to $SO_4^{2-}{}_{(aq)}$ during abiotic oxidation of pyrite, with and without Fe^{3+} added, at pH 1 and 7, and temperatures of 70 and 20°C. They were able to measure the relative abundance of three SO_4^{2-} isotopomers, $S^{18}O_4^{2-}$, $S^{16}O^{18}O_3^{2-}$, and $S^{16}O_2^{18}O_2^{2-}$ and found that ~ 90 to 100 % of the $SO_4^{2-}{}_{(aq)}$ oxygen was derived from water. The percentage of oxygen contributed by water increased when Fe^{3+} was added, suggesting that Fe^{3+} was the dominant oxidant of the pyrite and S-oxyanion intermediates. This isotope evidence supports the above interpretation of Moses and Herman (39), who concluded that Fe^{3+} is the direct agent of pyrite S oxidation based on experimental hydrochemistry.

$\delta^{18}O$ of Sulfate Formed by Abiotic Oxidation of Pyrite. In several sterile pyrite oxidation experiments, Taylor et al. (42) found that when Fe^{3+} was added, the $\delta^{18}O$ of $SO_4^{2-}{}_{(aq)}$ formed was quite similar to that of ambient water (Figure 3). They concluded that the agent oxidizing pyrite to $SO_4^{2-}{}_{(aq)}$ was mainly Fe^{3+} rather than $O_{2\ (aq)}$. Data from Qureshi (43) on aqueous abiotic oxidation of pyrite (bacterial inhibitors added) indicate a wider range of $\delta^{18}O_{SO4}$ values for a given $\delta^{18}O_{H2O}$ (Figure 3). A regression of the data from Qureshi's submersed experiments at pH = 2 (43), in which $\delta^{18}O_{H2O}$ was varied, gives the following equation: $\delta^{18}O_{SO4} = 0.62(\delta^{18}O_{H2O}) + 5.6$. The slope of this equation (i.e. m in Equation 3) indicates that water contributed 62 % of the $SO_4^{2-}{}_{(aq)}$ oxygen.

A plot summarizing $\delta^{18}O_{H2O}$ versus $\delta^{18}O_{SO4}$ for abiotic oxidation of pyrite is shown on Figure 3. Note that all data lie within Area B. The upper boundary of this area, where $\delta^{18}O_{SO4} = 0.62(\delta^{18}O_{H2O}) + 9$, is based on the slope of the regression of the submersed abiotic sulfide oxidation experiments of Qureshi (43; submersed pH 2 data shown as open circles in Figure 3), and an empirically chosen intercept value to include all experimental data. The lower boundary, where $\delta^{18}O_{SO4} = \delta^{18}O_{H2O}$ is the inferred lower limit for all $SO_4^{2-}{}_{(aq)}$ oxygen derived from water, excluding the range R shown in Figure 1.

Figure 3 indicates that $SO_4^{2-}{}_{(aq)}$ formed during abiotic oxidation of pyrite (Area B) is, on average, ^{18}O-depleted relative to that formed experimentally by abiotic oxidation of $SO_3^{2-}{}_{(aq)}$ (Area A, derived from Figure 2). Typically, 60 to 75 % of the oxygen in $SO_4^{2-}{}_{(aq)}$ formed by oxidation of $SO_3^{2-}{}_{(aq)}$ is contributed by the water (10). Some of the $SO_4^{2-}{}_{(aq)}$ formed by pyrite oxidation in Qureshi's (43) experiments (pH = 2) has similar water contribution (62 %). This similarity may be coincidental, but it suggests that during abiotic oxidation of pyrite, $\delta^{18}O_{SO4}$ is sometimes controlled by the last subreaction step: oxidation of $SO_3^{2-}{}_{(aq)}$ to $SO_4^{2-}{}_{(aq)}$. Perhaps direct $O_{2\ (aq)}$ oxidation of $SO_3^{2-}{}_{(aq)}$ was the dominant process in Qureshi's experiments at pH 2. The fact that $SO_4^{2-}{}_{(aq)}$ formed by sulfide oxidation (Area B) is generally ^{18}O-depleted in comparison to that formed by $SO_3^{2-}{}_{(aq)}$ oxidation (Area A) may, in part, reflect different degrees of oxygen exchange. During sulfide oxidation, the reaction S oxyanion intermediates (e.g. $SO_3^{2-}{}_{(aq)}$) may have undergone partial or negligible oxygen exchange with water, whereas the $SO_3^{2-}{}_{(aq)}$ in Holt's experiments had equilibrated entirely with water (Figure 1; further discussion of ε_{H2O} below). Consideration of the data for sulfite-water oxygen exchange (9) suggests that $SO_3^{2-}{}_{(aq)}$ produced during stepwise oxidation of pyrite to $SO_4^{2-}{}_{(aq)}$ in some of the experiments

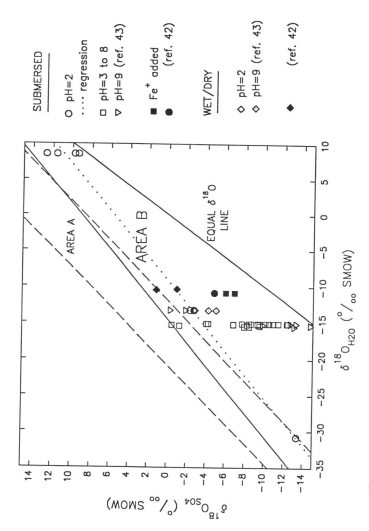

Figure 3. Experimental data for $\delta^{18}O_{SO4}$ as a function of $\delta^{18}O_{H2O}$ during abiotic oxidation of pyrite. Regression lines are based on data using a range of ^{18}O-enriched waters (30, 50). The upper boundary of Area B, $\delta^{18}O_{SO4} = 0.62(\delta^{18}O_{H2O}) + 9$, has a slope based on the regression of the Qureshi data (43) for submersed experiments at pH = 2 (open circles); its intercept was chosen to include all plotted data. The lower boundary of Area B, $\delta^{18}O_{SO4} = \delta^{18}O_{H2O}$, is the theoretical lower limit, assuming $\varepsilon_{H2O} = 0$ (Limit I in Figure 1).

of Taylor et al. (42) and Qureshi (43) had a very short half-life (<< 1 min). Qureshi's (43) abiotic oxidation of pyrite experiments included a range of pH conditions, from 2 to 9. Qureshi observed that, as pH increased, $\delta^{18}O_{SO4}$ tended to become more negative, and closer to ambient $\delta^{18}O_{H2O}$. He concluded that the major control of $\delta^{18}O_{SO4}$ is the variation in dominance of different oxidation reaction pathways as a function of pH. Qureshi suggested that at higher pH, more thiosulfate and polythionates form and incorporate more water oxygen through hydrolysis, and sulfite-water exchange also becomes more important as $SO_3^{2-}{}_{(aq)}$ persists. However, the data of Holt et al. (10) indicate that equilibration of $SO_3^{2-}{}_{(aq)}$ with water leads to large values of ϵ_{H2O} (+8 to +10 ‰; Table I), which would tend to make $\delta^{18}O_{SO4}$ more positive (further from ambient $\delta^{18}O_{H2O}$) under Qureshi's experimental conditions. Further, as noted above, the rate of sulfite-water oxygen exchange decreases dramatically as pH increases (9). Thus, the trend toward more negative $\delta^{18}O_{SO4}$ values with higher pH observed by Qureshi (43) may be due to <u>less</u> sulfite-water oxygen exchange.

Table I. Estimates of kinetic ^{18}O enrichment factors (ϵ) for uptake of oxygen from H_2O and O_2 during experimental oxidation of S compounds to $SO_4^{2-}{}_{(aq)}$ at ~ 25°C (see Equation 3)

ϵ_{H2O}	ϵ_{O2}	Source of data
9.7[1,2]	0	Holt et al. (10) stoichiometric oxidation of $SO_3^{2-}{}_{(aq)}$ by $O_{2\,(aq)}$
8[1]	--[3]	Holt et al. (10) anaerobic oxidation of $SO_3^{2-}{}_{(aq)}$ by $Br_{2\,(aq)}$
0[1]	--[4]	Mizutani & Rafter (31) oxidation of $S_{(s)}$
0	-8.7	Lloyd (14) oxidation of $Na_2S_{(aq)}$ by $O_{2\,(aq)}$
	-4.3	Taylor et al. (42) abiotic oxidation of pyrite
+3.5	-11.4	Taylor et al. (42) microbial oxidation of pyrite
+2.6	--[4]	Van Everdingen et al. (44); minimum based on field data from ref. 45

[1]Not reported by authors listed at right, but based on their data.
[2]Assuming atmospheric O_2 ($\delta^{18}O = +23$) contributed ¼ of oxygen.
[3]Not applicable.
[4]Not available.

Role of Bacteria in Sulfide Oxidation. Various microbes can mediate the oxidation of sulfide, directly or indirectly (22, p. 523-528). Some bacteria attach themselves to sulfide-mineral surfaces (46) and employ an enzymatic electron-transport mechanism to oxidize the S atoms directly (47). Others oxidize mineral sulfide indirectly by generating $Fe^{3+}_{(aq)}$ from $Fe^{2+}_{(aq)}$, and the $Fe^{3+}_{(aq)}$ in turn oxidizes the sulfide.

In principle, microbes may mediate one or more step(s) of a "serial" sulfide oxidation process (e.g., sulfide to S^0 step), while abiotic reactions may dominate other steps (e.g., $SO_3^{2-}_{(aq)}$ to $SO_4^{2-}_{(aq)}$). It is also probable that concurrent ("parallel") abiotic and microbiological mechanisms, competing at the same oxidation step, are common in natural environments (42).

Using a comparison of concentrations in sterile and non-sterile wastewater batch tests, Wilmot et al. (48) estimated that 12 to 56 % of dissolved-sulfide oxidation was microbially mediated, and that the balance occured via abiotic oxidation. Others have concluded that *Thiobacillus ferrooxidans* may greatly enhance the rate of pyrite oxidation, under low pH (~ 3 to 5) conditions (42, 49, 50), but may have minimal effect on pyrite oxidation rates at higher pH values (e.g., pH ~ 6; 50). Taylor et al. (42) observed a 25 % to 500 % increase in pyrite oxidation rate due to presence of *T. ferrooxidans*, in comparison with sterile tests at pH ~ 2. Under vadose conditions and unspecified pH, Kleinmann and Crerar (49) observed a 10 to 400 % increase in the rate of pyrite oxidation in coal due to presence of *T. ferrooxidans* relative to sterile controls, based on measurements of acid production. The latter study concluded that the role of the bacteria increased with greater frequency of wetting events.

$\delta^{18}O$ of Sulfate Formed by Oxidation of Sulfides with Bacterial Isolates.
Holding water isotope composition constant, Taylor et al. (42) tested *T. ferrooxidans* oxidation under wet/dry and submerged conditions. Qureshi (43) carried out similar experiments, varying $\delta^{18}O_{H2O}$, pH, and using cultures of either *T. ferrooxidans* or *T. thiooxidans*, or a mixture of both. Using flasks inoculated with *T. ferroxidans*, Krouse (30) found that water contributed 74 % of the oxygen to $SO_4^{2-}_{(aq)}$ during pentlandite oxidation, and 80 to 90 % of $SO_4^{2-}_{(aq)}$ oxygen during pyrite oxidation. Other studies of oxidation of pyrite, sphalerite and pentlandite, mixtures, or ore samples using *T. ferroxidans* by Krouse (51) and Gould et al. (52) indicate ~ 55 to 75 % of $SO_4^{2-}_{(aq)}$ oxygen was contributed by water.

The data (or linear regressions of data) from the pyrite oxidation experiments using *Thiobacillus* are plotted on Figure 4. The large majority lie within the same field as for abiotic oxidation (Area B, derived from Figure 3). However, under wet/dry conditions (sediment was allowed to air-dry between leaching events), some $SO_4^{2-}_{(aq)}$ formed was slightly enriched in ^{18}O relative to Area B (Figure 4). Perhaps the capillary sediment water became enriched in ^{18}O due to evaporation. This effect, in turn, may have caused ^{18}O enrichment in $SO_4^{2-}_{(aq)}$ that formed during drying, even though the bulk leaching solution did not shift isotopically during the experiments (42).

More than 95 % of the data for oxidation of other individual sulfide minerals or mixtures of two sulfides (pentlandite, sphalerite, chalcopyrite), and sulfide-bearing ore samples in the presence of *Thiobacillus*, and water with $\delta^{18}O$ ~ -11 ‰ (51, 52) also plot within Area B derived from Figure 3 (not shown). Apparently there is no significant effect of sulfide mineralogy on $\delta^{18}O_{SO4}$.

The most negative $\delta^{18}O_{SO4}$ values in some of the *Thiobacillus* experiments (lower portion of Area B) may indicate that the dominant oxidizing agent was Fe^{3+} and/or an enzymatic electron-transport mechanism was involved, as discussed above. Similar to the case for abiotic oxidation of pyrite, Qureshi (43) observed more negative $\delta^{18}O_{SO4}$ values as the pH of his *Thiobacillus* experiments increased. As outlined above, this

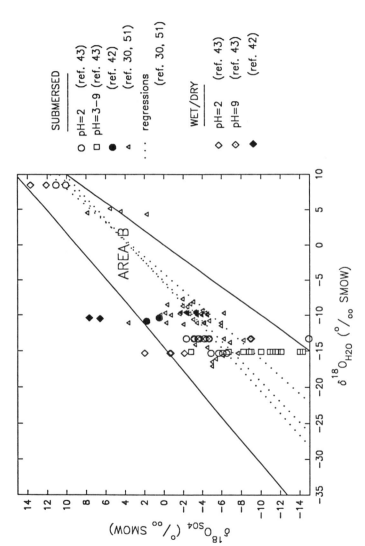

Figure 4. Experimental data for $\delta^{18}O_{SO_4}$ as a function of $\delta^{18}O_{H_2O}$ during "bacterial" (inoculated with *Thiobacillus*) oxidation of pyrite or pentlandite. The significance of parallel abiotic oxidation in these experiments is uncertain. The few data that plot outside Area B (from Fig 3) are from wet-dry experiments and may have been affected by evaporative enrichment of $\delta^{18}O_{H_2O}$.

trend might be due to a decrease in sulfite-water oxygen exchange at the higher pH conditions (i.e., minimum $\Delta^{18}O_{SO4-H2O}$).

Based on a comparison of Figures 3 and 4, it appears that $SO_4^{2-}{}_{(aq)}$ produced via bacterial or abiotic reaction pathways has virtually indistinguishable $\delta^{18}O$ values. However, as discussed above, concurrent parallel or serial microbial and abiotic oxidation mechanisms occurred in the "microbial" oxidation experiments. Abiotic oxidation mechanisms may have dominated in some of these experiments, based on sterile vs. nonsterile oxidation rate comparisons. Thus Figures 3 and 4 do not provide a conclusive comparison of $\delta^{18}O$ in $SO_4^{2-}{}_{(aq)}$ formed via bacterial versus abiotic oxidation mechanisms (42).

$\delta^{18}O$ of Sulfate in Other Sulfide Oxidation Experiments. Figure 5 shows data from various other non-sterile oxidation experiments with sulfides (pyrite, Na_2S, glacial till), in which the role of bacteria is unknown. By varying water isotope composition independently, various investigators found that water generally contributes 60 to 75 % of the $SO_4^{2-}{}_{(aq)}$ oxygen during oxidation of Na_2S (14) or pyrite (2, 53). The large majority of points fall within Area B, the area based on abiotic pyrite oxidation experiments (Figure 3). Exceptions are data from the 31-day experiment of van Everdingen and Krouse (2), where water contributed around 40 % of the oxygen, in contrast to ~ 65 % after 547 days. For the shorter term experiment, background $SO_4^{2-}{}_{(aq)}$ may have been a problem.

Under vadose conditions, Qureshi (43) observed a trend towards more negative $\delta^{18}O_{SO4}$ values with increased P_{O2}. As for the similar trend with increased pH noted above, Qureshi attributed this result to variations in oxidation pathways and changing importance of intermediate valence S compounds with changing P_{O2}. However, it is possible that the more negative $\delta^{18}O_{SO4}$ values with increasing P_{O2} are due to less sulfite-water exchange of oxygen, due to a shorter half-life of $SO_3^{2-}{}_{(aq)}$ at higher P_{O2}. This trend requires further investigation.

Figure 5 also includes data for experimental oxidation of glacial till (54, 55), in which pyrite is the dominant pre-oxidation S component. The till $SO_4^{2-}{}_{(aq)}$ data plot in Area B, the proposed field for abiotic pyrite oxidation.

A comparison of Figure 5 with Figures 3 and 4 supports the hypothesis that the $\delta^{18}O$ of $SO_4^{2-}{}_{(aq)}$ that forms by oxidation of sulfides tends to fall into Area B irrespective of the sulfide mineralogy or the oxidation mechanism.

Oxidation of Reduced Organic S. Microorganisms are known to oxidize solid or dissolved organic sulfides, sulfonates, and other organic forms of reduced S to $SO_4^{2-}{}_{(aq)}$ in a step-wise manner. Abiotic reactions may also produce $SO_4^{2-}{}_{(aq)}$ from these compounds. In some cases, reduced organic S may be released from its organic moeity prior to oxidation, and thus can be considered to follow the same oxidation pathways as sulfide, $S_{(s)}$ and $SO_3^{2-}{}_{(aq)}$ described above.

Very little information exists on the oxygen isotopic composition of $SO_4^{2-}{}_{(aq)}$ produced by oxidation of reduced organic S, or the enzymatic or abiotic mechanisms. Exceptions are studies of $SO_4^{2-}{}_{(aq)}$ in human urine, which is largely produced by oxidation of S-containing amino acids. Krouse (56) and Katzenberg and Krouse (57) demonstrated that the majority of the oxygen in the urine $SO_4^{2-}{}_{(aq)}$ comes from ambient body water. As for inorganic sulfides, the $\delta^{18}O_{SO4}$ versus $\delta^{18}O_{H2O}$ data (not shown) for human urine tend to plot within Area B derived from Figure 3.

Based on $\delta^{18}O_{SO4}$ and $\delta^{18}O_{H2O}$ data from litter leachate experiments and soil samples (58; not shown), the $SO_4^{2-}{}_{(aq)}$ formed largely from organic S generally has an

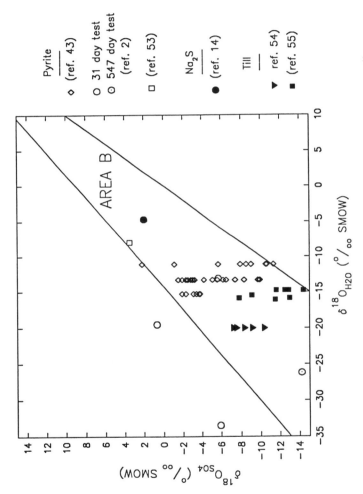

Figure 5. A plot of $\delta^{18}O_{SO4}$ as a function of $\delta^{18}O_{H2O}$ from various non-sterile oxidation experiments. The relative importance of microbial mediated and abiotic reactions are unknown. Data from a 31-day experiment (2) that plot outside Area B may have been affected by SO_4^{2-} contamination of pyrite.

oxygen isotope composition similar to that predicted for oxidation of $SO_3^{2-}{}_{(aq)}$ (Area A of Figure 2) or sulfide (Area B of Figure 3). However, the experiments with ^{18}O-enriched water indicate 21 % contribution by water, considerably lower than that of the sulfite and sulfide oxidation experiments. These data may be biased by memory effects and/or S-stress (loss of S during leaching), which may have induced fission of organic sulfate (discussed below), rather than oxidation of reduced organic S (58).

Variations of Isotope Enrichment Factors During S Oxidation. Based on Equation 3, $SO_4^{2-}{}_{(aq)}$ oxygen depends not only on the relative amounts of water and $O_{2\ (aq)}$ oxygen, but also the enrichment factors ε_{H2O} and ε_{O2}, as well as $\delta^{18}O_{O2}$ and $\delta^{18}O_{H2O}$. Based on previous studies (10, 14, 42), the inferred ε_{O2} can vary significantly (0 to -11.4), is different for abiotic and microbial pathways, and will become negligible when $O_{2\ (aq)}$ is limiting (Table I). $\delta^{18}O_{O2}$ is relatively constant for atmospheric oxygen (+23 ‰), but becomes more positive during biological respiration in semi-closed systems due to preferential consumption of ^{16}O-enriched oxygen by microorganisms (42, 59).

Values of ε_{H2O} can be determined directly when all $SO_4^{2-}{}_{(aq)}$ oxygen is contributed by water (10, 31), but this factor is a complex combination of kinetic fractionation and equilibrium water exchange during formation of $SO_4^{2-}{}_{(aq)}$ from reduced S compounds. As noted above, equilibration with water at 22°C leads to a $\delta^{18}O$ enrichment in SO_2 of ~ 24 ‰ relative to water (10). For $SO_4^{2-}{}_{(aq)}$ in isotopic equilibrium with water, data from Holt et al. (10) indicate $\varepsilon_{H2O} \approx +10$ for reaction of stoichoimetric amounts of $SO_3^{2-}{}_{(aq)}$ and $O_{2\ (aq)}$ at 24°C, and $\varepsilon_{H2O} \approx +8$ ‰ during $SO_3^{2-}{}_{(aq)}$ oxidation by bromine at an unspecified T. This enrichment is likely T-dependent as noted above (16). Lloyd (14) and Taylor et al. (42) calculated ε_{H2O} values of 0 to +3.5 during sulfide oxidation, based on Equation 3. Based on the data of Mizutani and Rafter (32) $\varepsilon_{H2O} = 0$ during rapid oxidation of $S_{(s)}$. These results (Table I) indicate that sulfite-water exchange is limited during rapid step-wise oxidation of reduced S compounds, and that in some cases the $SO_4^{2-}{}_{(aq)}$ acquires the same isotopic composition as the water. Thus, the difference between $\delta^{18}O_{SO4}$ and $\delta^{18}O_{H2O}$ ($\Delta^{18}O_{SO4-H2O}$) may be controlled largely by variation in ε_{H2O} as a function of oxidation rate and extent of sulfite-water exchange (pH-dependent), rather than the relative contributions of water and $O_{2\ (aq)}$, as some previous investigators have assumed.

Unless $\delta^{18}O_{H2O}$ is varied experimentally, the use of Equation 3 for the determination of the % water contribution to $SO_4^{2-}{}_{(aq)}$ oxygen is of limited value. This is due to the wide variation in values of ε_{H2O}, $\delta^{18}O_{O2}$, and ε_{O2}, and the potential role of other oxygen sources such as phosphate or $H_2O_{2\ (aq)}$. Thus, hypothetical calculations of water versus $O_{2\ (aq)}$ contributions of oxygen to $SO_4^{2-}{}_{(aq)}$ in environmental samples (e.g. 42, 60) should be viewed as qualitative, first approximations.

Other Factors Controlling the $\delta^{18}O$ of Environmental Sulfate.

After formation, the $\delta^{18}O$ of $SO_4^{2-}{}_{(aq)}$ may be modified by various processes. Isotope fractionation during diffusion of $SO_4^{2-}{}_{(aq)}$ probably has a negligible effect on environmental $\delta^{18}O_{SO4}$. Precipitation of sulfate minerals (e.g. gypsum) may deplete the $\delta^{18}O$ of $SO_4^{2-}{}_{(aq)}$ due to fractionation of oxygen between dissolved and mineral sulfate on the order of 2 ‰ (61). However, rapid redox cycling (62), or a lack of mixing between $SO_4^{2-}{}_{(aq)}$ in an evaporating basin and the open ocean (63) may negate such $\delta^{18}O_{SO4}$ shifts.

Sorption-desorption does not shift $\delta^{18}O_{SO4}$ significantly in oxic marine sediments (64) or in upland forest soils (65). Substantial $\delta^{18}O$ shifts in $SO_4^{2-}{}_{(aq)}$ added to organic-rich lake (66) and ocean (64) sediments are likely due to $SO_4^{2-}{}_{(aq)}$ reduction (65; discussion below).

There are three non-oxidative processes that can play a significant role in altering $\delta^{18}O_{SO4}$: exchange of oxygen between $SO_4^{2-}{}_{(aq)}$ and water, hydrolysis of organic sulfates to form $SO_4^{2-}{}_{(aq)}$, and rate-dependent fractionation of $SO_4^{2-}{}_{(aq)}$ oxygen during microbial $SO_4^{2-}{}_{(aq)}$ reduction.

Exchange of Oxygen Between Sulfate and Water. Lloyd (61) estimated the time required for 97% equilibration of ocean-water and sulfate oxygen exchange (pH 8.2, 4°C) to be on the order of 250,000 yr, based in part on laboratory experiments at pH 7 and 25°C. In contrast, Zak et al. (67) reported $\delta^{18}O_{SO4}$ versus $\delta^{18}O_{H2O}$ data from porewaters of deep marine sediments that showed no sign of sulfate-water oxygen exchange and inferred exchange $t^{\frac{1}{2}} > 5 \times 10^9$ yr. Further, Chiba and Sakai (68) extrapolated their laboratory data for conditions of 100 to 300°C and pH 2 to 7 and inferred that the $t^{\frac{1}{2}}$ for ocean water-sulfate oxygen exchange rate is ~ 10^9 yr. The rates reported by Chiba and Sakai are in general agreement with those extrapolated from the earlier study of Hoering and Kennedy (69), who reported experimental water-sulfate oxygen exchange at low pH (≤ 0) and 10, 25 and 100°C.

Based on the latter three studies, it appears that in most Earth-surface and shallow groundwater environments, the rate of oxygen exchange between chemically inactive $SO_4^{2-}{}_{(aq)}$ and ambient water is extremely slow ($t^{\frac{1}{2}} = 10^5$ to 10^9 y or more). In contrast, the mean age of $SO_4^{2-}{}_{(aq)}$ ions is relatively short due to biologically mediated redox S cycling (58, 62). Thus, sulfate-water oxygen exchange is insignificant in most Earth-surface and near-surface environments.

The rate of sulfate-water oxygen exchange increases with increase in temperature or decrease in pH (68, 69). The pH-dependency of the rate is actually controlled by exchange between either $H_2SO_4^0{}_{(aq)}$ or $HSO_4^-{}_{(aq)}$ and H_2O, which become more abundant at lower pH, rather than direct exchange between $SO_4^{2-}{}_{(aq)}$ and H_2O (68, 69). In environments with elevated temperatures and/or low pH (68), sulfate-water oxygen exchange may have a significant effect on $\delta^{18}O_{SO4}$. In extreme cases, $\delta^{18}O_{SO4}$ may equilibrate with $\delta^{18}O_{H2O}$ and all previous factors that influenced $\delta^{18}O_{SO4}$, as considered above, become negligible.

Sulfate-water oxygen exchange is important in hydrothermal systems (61, 68-71). For example, at pH 6 and 200°C the $t^{\frac{1}{2}}$ for sulfate-water oxygen exchange is ~ 12 yr; at pH = 7 and 300°C the $t^{\frac{1}{2}}$ is around 2 weeks (68). At equilibrium, $\delta^{18}O$ of $HSO_4^-{}_{(aq)}$ and/or $SO_4^{2-}{}_{(aq)}$ tends to be ^{18}O-enriched relative to $\delta^{18}O_{H2O}$ by ~ 5 $^o/_{oo}$ at 300°C and ~ 20 $^o/_{oo}$ at 100°C (61, 70). Plots of $\delta^{18}O_{SO4}$ versus $\delta^{18}O_{H2O}$ in hydrothermal waters have been used to infer temperatures at depth in these systems (70-73). In a similar way, $\delta^{18}O_{SO4}$ and $\delta^{18}O_{OH}$ in alunite [$KAl_3(SO_4)_2(OH)_6$] crystals that formed in hydrothermal environments can be used to infer paleotemperatures, assuming that both OH and SO_4 in alunite were in isotopic equilibrium with the hydrothermal H_2O (73).

Some acid mine drainage has pH < 0 (74). Unfortunately, data are not available for exchange rates and isotope enrichment between water and sulfate under Earth-surface acid mine drainage conditions. Based on extrapolations of the experimental data of Hoering and Kennedy (69), and Chiba and Sakai (68), for lower pH or higher temperature conditions than acid mine drainage repectively, at 25°C the inferred exchange $t^{\frac{1}{2}}$ at pH 0 is ~ 1 to 30 yr, at pH 1 is ~ 25 to 100 yr, and at pH 3 is ~ 10^3 to

10^5 yr. This analysis indicates that the $\delta^{18}O$ of $SO_4^{2-}{}_{(aq)}$ in acid mine drainage with pH ≤ 1 may be partially or completely controlled by sulfate-water exchange. Similarly, in naturally acidic systems (e.g. Australian groundwaters with pH of 3 to 4: 75), the $\delta^{18}O_{SO4}$ may be affected by partial sulfate-water exchange. By extrapolation from hydrothermal experiments (61, 70), sulfate in oxygen isotope equilibrium with water at 25°C would probably be enriched in ^{18}O by ~ 30 to 35 $^o/_{oo}$ relative to the water.

Hydrolysis of Organic Sulfates to Form Aqueous Sulfate. Organic sulfates are compounds of the general formula $R-OSO_3^-$ and are produced by all major types of organisms, via enzyme-catalyzed incorporation of $SO_4^{2-}{}_{(aq)}$. Organic sulfates are abundant in natural environments such as soils and sediments (76), but these compounds have received little detailed study. Sulfatases are organism-secreted enzymes that liberate SO_4^{2-} from specific types of organic sulfates. Using $H_2^{18}O$ Spencer (77) found that various arylsulfatases (isolated from mammals and microorganisms) break the O-S bond, and thus the new $SO_4^{2-}{}_{(aq)}$ incorporates one ambient water oxygen. Later work (78-80) confirmed this mechanism. Some alkylsulfates are hydrolyzed by sulfatases via C-O fission, during which the original sulfate oxygen remains intact (80, 81). Very little is known about the relative abundances of arylsulfates and alkylsulfates and their respective sulfatases in natural environments.

If arylsulfates are abundant, their fission will produce $SO_4^{2-}{}_{(aq)}$ with 25% new water oxygen via O-S rupture, or will release $SO_3^{2-}{}_{(aq)}$ (82) that will follow the oxidation pathways discussed above. If, on the other hand, alkylsulfates are more important for biocycling than arylsulfates, then no corresponding oxygen isotope shift may occur. Based on ^{35}S experiments, organic sulfate cycling may be the dominant pathway between organic S and $SO_4^{2-}{}_{(aq)}$ in some soil horizons (83). As discussed above, the $\delta^{18}O_{SO4}$ data from litter experiments (58) also suggest that organic sulfate cycling may be the dominant source of $SO_4^{2-}{}_{(aq)}$ in this soil horizon.

Effects During Sulfate Reduction. In anaerobic environments, microbially mediated reduction of $SO_4^{2-}{}_{(aq)}$ to sulfide can lead to substantial enrichments of both ^{34}S and ^{18}O in residual $SO_4^{2-}{}_{(aq)}$ (84, 85). This produces $\delta^{18}O_{SO4}$ versus $\delta^{18}O_{H2O}$ data (not shown) that generally plot above Area B derived from Figure 3. Mizutani and Rafter (84) suggested that this heavy isotope enrichment is a kinetic effect, due to preferential reduction of isotopically light APS (Equation 5) and/or $SO_3^{2-}{}_{(aq)}$, in the following reversible sulfate-reduction pathway:

$$SO_4^{2-} \Leftrightarrow APS \Leftrightarrow SO_3^{2-} \Leftrightarrow H_2S$$

In batch experiments, as $SO_4^{2-}{}_{(aq)}$ reduction nears completion and $\delta^{34}S_{SO4}$ continues to become more positive, $\delta^{18}O_{SO4}$ approaches a constant value (84, 85). These $\delta^{18}O_{SO4}$ data indicate that water oxygen is somehow incorporated indirectly into $SO_4^{2-}{}_{(aq)}$ during ongoing $SO_4^{2-}{}_{(aq)}$ reduction. Mizutani and Rafter (84) and Fritz et al. (85) suggested that oxygen exchange between water and APS and/or sulfite may affect the $\delta^{18}O_{SO4}$, because each step in the sulfate reduction pathway, as outlined above, is reversible. Experiments using ^{35}S-labelled H_2S (86, 87) indicate that simultaneous $SO_4^{2-}{}_{(aq)}$ reduction and oxidation of H_2S are likely common occurences in anoxic sediments. Mizutani and Rafter (84) also suggested that oxygen isotope equilibrium between $SO_4^{2-}{}_{(aq)}$, $SO_3^{2-}{}_{(aq)}$ and water might occur via reversible interconversion with APS, and rapid sulfite-water oxygen exchange.

By analogy with $SO_4^{2-}{}_{(aq)}$ (68) and organic sulfates (e.g. 80) which do not exchange oxygen with water significantly under most Earth-surface conditions, Van Stempvoort (88) argued that direct APS-water oxygen exchange is probably negligible. Further, he reasoned that $SO_4^{2-}{}_{(aq)}$ and $SO_3^{2-}{}_{(aq)}$ cannot reach oxygen isotope equilibrium under such conditions. During oxidation of $SO_3^{2-}{}_{(aq)}$ to $SO_4^{2-}{}_{(aq)}$, 25% of $SO_4^{2-}{}_{(aq)}$ oxygen is obtained from the neighboring phosphoryl group in APS (Equations 5, 6), and this oxygen is entirely independent of $SO_3^{2-}{}_{(aq)}$. Conversion between $SO_4^{2-}{}_{(aq)}$, APS, and $SO_3^{2-}{}_{(aq)}$ are kinetically controlled processes, which will not lead to isotope equilibrium.

Because exchange of oxygen between $SO_3^{2-}{}_{(aq)}$ and H_2O is rapid (9, 10), this mechanism could enable the indirect incorporation of water oxygen into $SO_4^{2-}{}_{(aq)}$ (84). It is also likely that when sulfide formed by sulfate reduction is reoxidized to $SO_3^{2-}{}_{(aq)}$ it will incorporate water oxygen via hydrolysis, which could be passed on to $SO_4^{2-}{}_{(aq)}$ (88).

Thus, it appears that $\delta^{18}O_{SO4}$ during $SO_4^{2-}{}_{(aq)}$ reduction is controlled by an interplay of 3 or 4 mechanisms:

1) preferential reduction of isotopically light APS-sulfate to sulfite, and/or sulfite to sulfide,

2) kinetic effects during re-oxidation of sulfite to APS, in which one $SO_4^{2-}{}_{(aq)}$ oxygen is derived from the APS phosphate group,

3) sulfite-water oxygen exchange,

4) incorporation of oxygen from water and/or other sources during reoxidation of sulfide.

These combined effects can explain the isotope trend during ongoing $SO_4^{2-}{}_{(aq)}$ reduction, in which there is an interplay of progressive ^{18}O-enrichment and partial water control of residual $SO_4^{2-}{}_{(aq)}$ oxygen (84, 85).

Comparison of Field Data to Laboratory Findings

The above sections discuss: 1) experimental data that illustrate the factors that influence $\delta^{18}O_{SO4}$ during formation, and 2) other factors that can alter $\delta^{18}O_{SO4}$ of aqueous solutions by removal or addition of $SO_4^{2-}{}_{(aq)}$, or isotope exchange. Clearly, gaps exist between the above experimental conditions and natural environments. A key difference is that each experiment focussed on a single process controlling $\delta^{18}O_{SO4}$, whereas in many natural environments $SO_4^{2-}{}_{(aq)}$ is generally derived from several sources. Natural mixtures may involve dissolution of mineral sulfate, weathering of mineral sulfide, atmospheric emissions, and other sources. In some cases, SO_4^{2-} may be rapidly introduced into water of isotopic composition different from that in which it formed. Analysis of mixing should consider both $\delta^{18}O_{SO4}$ and $\delta^{34}S_{SO4}$ values of potential sources, as well as $SO_4^{2-}{}_{(aq)}$ concentrations. Some of the mathematical relationships between these parameters for sulfate mixtures are discussed by Krouse and Tabatabai (89), Krouse (56) and Staniaszek (90). For example, linear plots of $\delta^{34}S$ versus $\delta^{18}O$ suggest mixing of $SO_4^{2-}{}_{(aq)}$ from two major sources, and triangular distributions suggest three or more dominant sources of $SO_4^{2-}{}_{(aq)}$.

Some of the above laboratory experiments were conducted using a wide variation in $\delta^{18}O_{H2O}$ (ranges of several hundred to thousand ‰). In contrast, field data have a relatively small range in $\delta^{18}O_{H2O}$ (< 40 ‰), which can vary either spatially or temporally (e.g. by season).

In the following sections, plots of $\delta^{18}O_{SO4}$ versus $\delta^{18}O_{H2O}$ from various Earth-surface and near-surface environments are compared to the Areas A or B derived above from laboratory S oxidation experiments. This comparison will demonstrate whether or not ongoing natural S oxidation processes have a controlling influence on the $\delta^{18}O$ of environmental $SO_4^{2-}{}_{(aq)}$. For each type of environment considered, the data from several field sites are plotted on a single figure. The data distributions are discussed in general terms, relative to Areas A and B, and then outliers are given further attention.

$\delta^{18}O$ in Atmospheric Sulfate. In the continental atmosphere, precipitation $SO_4^{2-}{}_{(aq)}$ and aerosol sulfate (particulate) are similar isotopically (91, 92). Nearly all atmospheric SO_4^{2-} data shown in Figure 6 plot above Area A, which is the field for $SO_3^{2-}{}_{(aq)}$ oxidation at $\sim 25°C$ (Figure 2). The ^{18}O-enriched atmospheric SO_4^{2-} may include some "primary" SO_4^{2-} that formed by oxidation of SO_2 at several hundred °C in stacks or exhaust (17). This ^{18}O-enriched SO_4^{2-} apparently mixes with "secondary" SO_4^{2-} that forms by oxidation of $SO_3^{2-}{}_{(aq)}$ in the atmosphere and has an isotopic composition represented by Area A.

Most atmospheric $\delta^{18}O_{SO4}$ data correspond to continental sites in the Northern Hemisphere, and generally plot along a trend represented as RAWSO (Regression of Atmospheric Water versus SO4 Oxygen) in Figure 6 (regression of all data shown except the Crossfield, Alberta anomaly discussed below). The slope of this regression (~ 0.5), in comparison to that of Area A (~ 0.75) suggests mixing of approximately 2/3 "secondary" SO_4^{2-} with approximately 1/3 "primary" SO_4^{2-}. In Illinois (91), the seasonal variations of $\delta^{18}O_{SO4}$ in rain or snow are in phase with the seasonal trends of $\delta^{18}O_{H2O}$ in rain or snow, indicating that relatively rapid atmospheric hydrolysis of SO_2 does play a significant role.

There are notable exceptions to the RAWSO trend:

1. The Bermuda data (92) lie entirely within the Area A, well below the RAWSO, suggesting that primary, ^{18}O-enriched SO_4^{2-} is a minimal component here. When $\delta^{34}S$ data are considered, the bulk of this secondary SO_4^{2-} may have originated as SO_2 from distant continental sources (92). No significant seawater spray contribution is evident.

2. Of snow SO_4^{2-} collected near a sour gas plant near Crossfield, Alberta (92), one sample is a remarkable anomaly; it plots below Area A, the expected range for $SO_3^{2-}{}_{(aq)}$ oxidation. This unusually ^{18}O-depleted SO_4^{2-} was generated as a consequence of snow falling through an industrial SO_2 plume at a temperature of $\sim -25°C$. However, the exact mechanism(s) that produced this isotopically light SO_4^{2-} is unknown.

At Pisa and Venice, Italy, ^{18}O-rich precipitation $SO_4^{2-}{}_{(aq)}$ (not shown) is also affected by oxidation of anthropogenic SO_2 (93, 94), and, similar to Bermuda, these coastal sites have little seawater spray influence.

$\delta^{18}O$ in Throughfall and Stemflow Sulfate. At forested sites in eastern Canada, the $\delta^{18}O_{SO4}$ of throughfall (precipitation water that has passed through the forest canopy) is similar, but slightly depleted in ^{18}O on average, in comparison to precipitation $\delta^{18}O_{SO4}$ (82, 95). Most throughfall $\delta^{18}O_{SO4}$ data (not shown) plot above Area A, which is based on SO_2 oxidation experiments. However, the $\delta^{18}O_{SO4}$ of stemflow (precipitation water that flows down stems of trees) is depleted, on average by 2 to 3 $^o/_{oo}$ relative to precipitation $\delta^{18}O_{SO4}$ (95), and the majority of the data plot within Area A (not shown). These data indicate that most throughfall and some stemflow $SO_4^{2-}{}_{(aq)}$ is of precipitation and aerosol derivation, and is mixed with some ^{18}O-depleted SO_4^{2-},

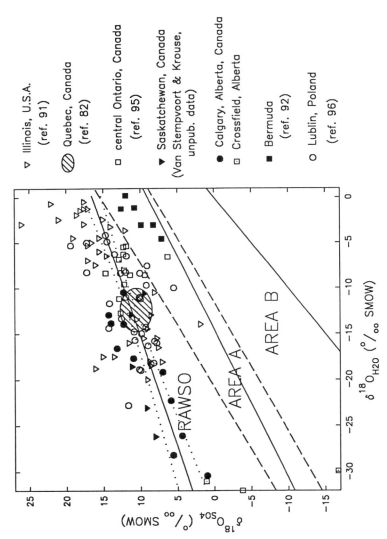

Figure 6. Relationship of $\delta^{18}O_{SO_4}$ as a function of $\delta^{18}O_{H_2O}$ for atmospheric samples (rain and snow) from various sites. The line labelled RAWSO (Regression of Atmospheric Water versus Sulfate Oxygen) is the regression of all data, excluding the anomalous sample from Crossfield, Alberta in lower left (see text).

probably from oxidized plant-organic S or SO_2 that was deposited from the atmosphere to the vegetation surfaces (95).

$\delta^{18}O$ in Soil Sulfate. The available data indicate that, in general, soil SO_4^{2-} is depleted in ^{18}O significantly relative to precipitation and aerosol SO_4^{2-} at the same site (58, 82, 97, 98). For soil seepage collected from litter-humus and B horizons by lysimeters at 2 forested sites in eastern Canada (58, 82), the majority of data plot within Area A (not shown), which is the general field for $SO_3^{2-}{}_{(aq)}$ oxidation in isotopic equilibrium with water. There is considerable range in the data, and some samples plot either slightly above or below Area A. This range suggests that soil SO_4^{2-} is a mixture of 2 endmembers: a) residual atmospheric SO_4^{2-} that is ^{18}O-rich relative to Area A, and b) $SO_4^{2-}{}_{(aq)}$ formed in the soil by oxidation of organic S, whose composition is ^{18}O-depleted relative to Area A and is perhaps described by Area B (sulfide oxidation). However, it is likely that the $\delta^{18}O_{H2O}$ of some of the seepage, which is similar to that of meteoric input, is not representative of the capillary water present in the uppermost soil during oxidation of organic S to $SO_4^{2-}{}_{(aq)}$ (58). Between precipitation events soil capillary water can be strongly fractionated isotopically by evaporation (99), even though the local recharge water may not have an evaporative shift (100). This evaporation effect probably explains some of the ^{18}O-rich soil $\delta^{18}O_{SO4}$ values (58).

As well, some soil O_2 may be enriched in ^{18}O relative to atmospheric O_2 due to respiration (59); this process also could enrich soil SO_4^{2-} in ^{18}O, if its oxygen is partially derived from ambient $O_2{}_{(aq)}$. Dissimilatory sulfate reduction can enrich $SO_4^{2-}{}_{(aq)}$ in ^{18}O, as discussed above, but is not considered an important mechanism in unsaturated soils and therefore cannot be used to explain the ^{18}O-enriched $SO_4^{2-}{}_{(aq)}$ values.

Dowuona (101) and Miller (102) studied sulfate in saline soils in western prairies of Canada, and found that, in surface, post-glacial evaporite crusts and some other crystals within the soil profile, the SO_4^{2-} had anomalously high $\delta^{18}O$ values (~ 0 to +17 ‰). Unfortunately, corresponding $\delta^{18}O_{H2O}$ values are not available. The cause of this strong ^{18}O-enrichment of SO_4^{2-} in these saline soils is unknown, but may be related to evaporative ^{18}O-enrichment of soil water during oxidative weathering of sulfides.

$\delta^{18}O$ in Groundwater Sulfate. At several sites in the prairies of Western Canada, the majority of groundwater samples collected from shallow (< 50 m) wells plot within Area B, or slightly below it (Figure 7). This is not surprising because other evidence indicates that the $SO_4^{2-}{}_{(aq)}$ in shallow groundwater of the prairies is derived largely from pyrite oxidation (55). Note that some of the Alberta samples (103) plot near or below the lower boundary of Area B (Figure 7). These $\delta^{18}O_{SO4}$ values are virtually identical to ambient $\delta^{18}O_{H2O}$. Based on laboratory findings discussed above, the latter $SO_4^{2-}{}_{(aq)}$ may have formed largely by microbial process(es) that involved an electron shuttle process, or oxidation by $Fe^{3+}{}_{(aq)}$, with minimal sulfite-water oxygen exchange.

The data points from Saskatchewan and Alberta that plot above Area B correspond to relatively deep wells at each site, with intake zones located within the unweathered zone. The $\delta^{18}O_{SO4}$ data, combined with $\delta^{34}S_{SO4}$ and other data, indicate that these deep samples were affected by $SO_4^{2-}{}_{(aq)}$ reduction.

The samples from Winnipeg, Manitoba that plot above Area B may have also been affected by $SO_4^{2-}{}_{(aq)}$ reduction. However, Mkumba (104) concluded that the

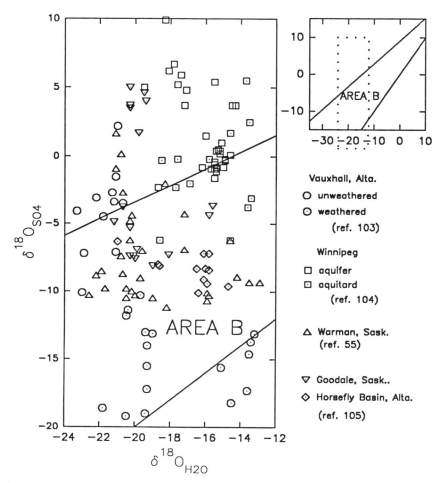

Figure 7. Relationship of $\delta^{18}O_{SO4}$ as a function of $\delta^{18}O_{H2O}$ for groundwater sampled in the prairie region of western Canada. Diagram in upper right indicates the area on left relative to larger area of Fig. 3.

^{18}O-enrichment is mainly due to dissolution of ancient marine evaporite mineral sulfate, which occurs in shallow bedrock in the Winnipeg area.

Figure 8 illustrates a plot $\delta^{18}O_{SO4}$ versus $\delta^{18}O_{H2O}$ of groundwater, mainly from southern Ontario, Canada, but also from Wisconsin and Italy. In this figure, the majority of the data plot above Area B. As an exception, the data from Wisconsin plot entirely within Area B, consistent with the hypothesis that the $SO_4^{2-}{}_{(aq)}$ was formed mainly by sulfide oxidation (60).

The data from landfills in southern Ontario (Figure 8) have been affected by $SO_4^{2-}{}_{(aq)}$ reduction (106). At other sites in southern Ontario, there is a data trend from near the center of Area B to a region above Area B (104, 107). The Kitchener range, like that of Winnipeg in Figure 7, has been affected by dissolution of ^{18}O-enriched, Paleozoic evaporite sulfate (104). Other examples of evaporite dissolution affecting $\delta^{18}O$ values of groundwater $SO_4^{2-}{}_{(aq)}$ are described by Gilkenson et al. (108; Illinois, U.S.A.) and Clark (109; Persian Gulf).

Wassenaaar's data (107; Figure 8) show two clusters of $\delta^{18}O_{SO4}$ values. At one site (Rodney), the $SO_4^{2-}{}_{(aq)}$ probably formed by oxidation of organic S and/or mineral sulfides, and the data fall within Area B. Wassenaar (107) attributed a trend towards slightly heavier $\delta^{18}O_{SO4}$ (by ~ 3 $^o/_{oo}$) in the deepest samples at this site to $SO_4^{2-}{}_{(aq)}$ reduction. At the other site (Alliston), the ^{18}O-enriched $SO_4^{2-}{}_{(aq)}$ (above Area B) has probably retained an atmospheric $\delta^{18}O_{SO4}$ signature (107). Wassenaar argued that sulfate reduction is not an important factor at Alliston because of high $O_{2\ (aq)}$ (~ 8 to 10 mg/L).

The data from Abbott (110) shown on Figure 8 are somewhat puzzling; Abbott inferred a pyrite oxidation source of the $SO_4^{2-}{}_{(aq)}$, but virtually all the data lie above Area B. Perhaps the $\delta^{18}O$ values were affected by $SO_4^{2-}{}_{(aq)}$ reduction.

Groundwater sampled in central Italy contains ^{18}O-enriched $SO_4^{2-}{}_{(aq)}$ (Figure 8) that has probably been affected by $SO_4^{2-}{}_{(aq)}$ reduction (111).

$\delta^{18}O$ in Sulfate from Springs.

Both hot and cold springs from various sites in the mountainous Cordilleran Region of western Canada (44, 90, 112) show a $\delta^{18}O$ data trend ranging from the center of Area B to a zone well above Area B. For the sites he investigated, Staniaszek (90) inferred mixing of SO_4^{2-} from two major sources: marine-derived bedrock evaporite sulfate (^{18}O-enriched endmember) and oxidized sulfide ($\delta^{18}O_{SO4}$ plotting near center of Area B). For some springs, $SO_4^{2-}{}_{(aq)}$ reduction may also have been important.

Data from hot springs in central Italy that plot well above Area B (Figure 9) are also apparently controlled by dissolution of marine-derived evaporite sulfate (113). In contrast, $SO_4^{2-}{}_{(aq)}$ from cold springs in northwest Italy that plots within or just above Area B is apparently derived mainly from oxidized sulfide (51).

Some $SO_4^{2-}{}_{(aq)}$ in hot springs may have exchanged oxygen with water at depth in these hydrothermal systems (112), as discussed above. The experiments of Chiba and Sakai (68) indicate that the half-life of sulfate-water oxygen exchange at 250-300°C in neutral-pH hydrothermal systems is around 2 weeks to 4 years. The latter authors argue that re-equilibration of $SO_4^{2-}{}_{(aq)}$ and water oxygen isotopes at lowered T during ascent of geothermal waters can be neglected.

$\delta^{18}O$ of Sulfate in Runoff.

The majority of $SO_4^{2-}{}_{(aq)}$ in surface stream samples plot above Area B (Figure 10). In some cases, the $SO_4^{2-}{}_{(aq)}$ in runoff is apparently enriched in ^{18}O relative to Area B because of dissimilatory $SO_4^{2-}{}_{(aq)}$ reduction; this includes $SO_4^{2-}{}_{(aq)}$ in streams from Ontario, Canada (88), and in central Italy (downstream

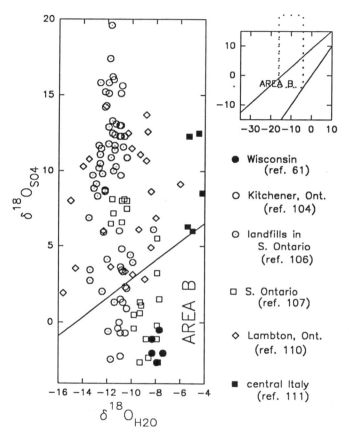

Figure 8. Relationship of $\delta^{18}O_{SO4}$ as a function of $\delta^{18}O_{H2O}$ for groundwater sampled in Ontario, Canada, Wisconsin, U. S. A. and central Italy. Diagram in upper right indicates the area on left relative to larger area of Fig. 3.

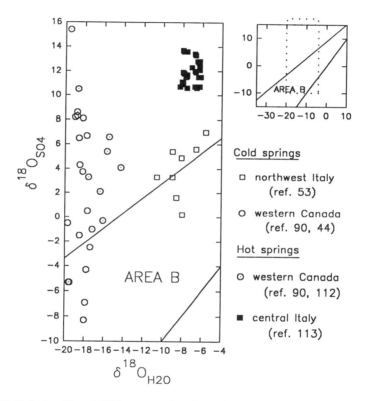

Figure 9. Relationship of $\delta^{18}O_{SO_4}$ as a function of $\delta^{18}O_{H2O}$ for hot and cold springs. Diagram in upper right indicates the area on left relative to larger area of Fig. 3.

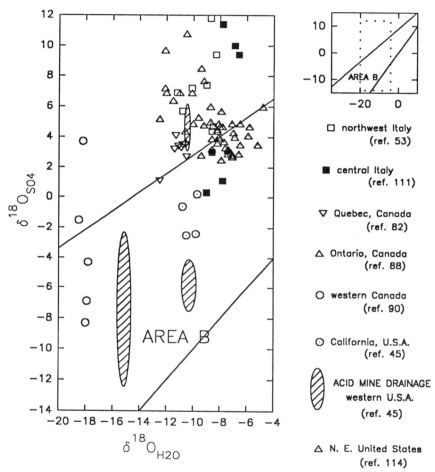

Figure 10. Relationship of $\delta^{18}O_{SO4}$ as a function of $\delta^{18}O_{H2O}$ for surface runoff. Diagram in upper right indicates the area on left relative to larger area of Fig. 3.

samples: 111). Elsewhere, runoff $SO_4^{2-}{}_{(aq)}$ is probably ^{18}O-enriched due to dissolution of evaporite sulfate (northwest Italy and some data from western Canada - Figure 10).

In contrast, in acid mine drainage from western U. S. A. and runoff in California, central Italy (upstream samples), and at some locations in western Canada, $\delta^{18}O_{SO4}$ versus $\delta^{18}O_{H2O}$ data plot in Area B (Figure 10). In these examples, the $SO_4^{2-}{}_{(aq)}$ in runoff has formed mainly by oxidation of sulfide minerals (45, 90, 111).

The $\delta^{18}O_{SO4}$ versus $\delta^{18}O_{H2O}$ in streams from northeastern U. S. A. (114) and Quebec (82) plot near the upper boundary of Area B, and probably were influenced by precursor soil or groundwater $\delta^{18}O_{SO4}$.

$\delta^{18}O$ of Sulfate in Lakes, Oceans, Seas. For the majority of lake samples, $\delta^{18}O_{SO4}$ versus $\delta^{18}O_{H2O}$ plots above Area B (Figure 11). Sulfate reduction causes ^{18}O-enrichment in some lakes, including those studied in Ontario, Canada (88, 115), Arctic Canada (116) and central Italy (117). Presumably, $SO_4^{2-}{}_{(aq)}$ reduction also explains some of the ^{18}O-enrichment of $SO_4^{2-}{}_{(aq)}$ in other lakes plotted on Figure 11, including Lake Vanda, Antarctica (118), and the saline lakes (119; authors' unpublished data).

Based on $\delta^{34}S_{SO4}$ and $\delta^{18}O_{SO4}$ values and concentrations of $SO_4^{2-}{}_{(aq)}$ in lakes of western Canada, Staniaszek (90) inferred mixing of relatively ^{18}O-depleted SO_4^{2-} from oxidized sulfides with ^{18}O-enriched SO_4^{2-} from marine bedrock evaporites (Figure 11).

The controls of $\delta^{18}O$ of marine $SO_4^{2-}{}_{(aq)}$ have come under considerable attention in the last 25 years since the pioneering efforts of Lloyd (14) and Longinelli and Craig (119). Lloyd (14) suggested that biological redox cycling in the ocean is the major control of ocean $\delta^{18}O_{SO4}$. The relatively constant $\delta^{18}O_{SO4}$ (~ +9.5) in ocean water ($\delta^{18}O_{H2O} \approx 0.0$) plots near the upper boundary of Area B (Fig. 11), supporting Lloyd's idea that S oxidation may be a dominant factor. Further support is given by the work of Pierre (62) who found that the enrichment of $\delta^{18}O$ of $SO_4^{2-}{}_{(aq)}$ in marine salt pans in southern France closely followed a parallel ^{18}O-enrichment in water during evaporation. If redox cycling does control $\delta^{18}O$ of ocean $SO_4^{2-}{}_{(aq)}$, then the $\delta^{18}O$ of ancient evaporite sulfate units may represent a paleorecord of localized redox and evaporitic enrichment conditions, rather than recording a secular trend in the $\delta^{18}O$ of "global" ocean $SO_4^{2-}{}_{(aq)}$ over geologic time, as modelled by Holser et al. (120) and Claypool et al. (121).

Depending on the rate of S redox cycling in the ocean environment, other factors that may affect the ocean $\delta^{18}O_{SO4}$ include the $\delta^{18}O$ of river input $SO_4^{2-}{}_{(aq)}$ (120), and exchange of $SO_4^{2-}{}_{(aq)}$ and H_2O oxygen in mid-ocean hydrothermal systems (122, 123). Isotope fractionation during precipitation of evaporite sulfate minerals (14, 120) may not influence $SO_4^{2-}{}_{(aq)}$ in the open ocean significantly because of the limited return-flow of water from the marginal marine evaporative basins (63).

The slightly ^{18}O-depleted $SO_4^{2-}{}_{(aq)}$ in bottom waters of the Red Sea (119) relative to mean ocean $SO_4^{2-}{}_{(aq)}$ is possibly a local effect of an influx of hydrothermal water, and/or enhanced S oxidation (perhaps of hydrothermal H_2S). Slightly ^{18}O-depleted $SO_4^{2-}{}_{(aq)}$ in near-surface polar ocean waters sampled during winter is possibly affected by high rates of biological S uptake (124). The cause of slightly ^{18}O-depleted ocean $SO_4^{2-}{}_{(aq)}$ in polar regions at intermediate depths or depths < 2800 m (124) has not been explained.

Alpers et al. (75) found ^{18}O-enriched $SO_4^{2-}{}_{(aq)}$ in acid-hypersaline lakes in Australia. The $\delta^{18}O$ of $SO_4^{2-}{}_{(aq)}$ in these lakes is possibly affected by $SO_4^{2-}{}_{(aq)}$ reduction. Alternatively, the ^{18}O-enrichment may be due to partial sulfate-groundwater exchange of oxygen, given the low pH (3 to 4), relatively long residence

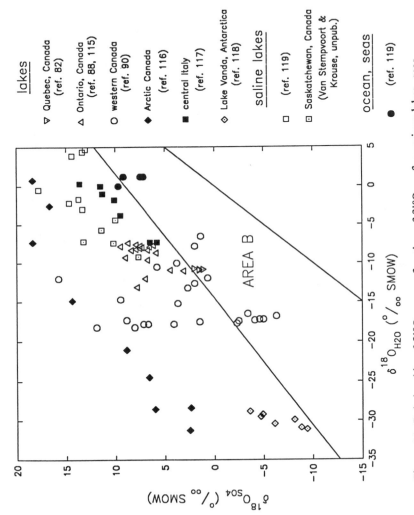

Figure 11. Relationship of $\delta^{18}O_{SO_4}$ as a function of $\delta^{18}O_{H_2O}$ for various lakes, seas, and oceans.

time of groundwater in the vicinity of the lakes, and the uncertainty in the pH-dependent rate of sulfate-water oxygen exchange (75).

Summary: Interpretation of $\delta^{18}O$ in Sulfate in Various Environments.

Based on presently available information, five general rules for interpreting $\delta^{18}O$ of environmental $SO_4^{2-}{}_{(aq)}$ can be applied:

1) Sorption, precipitation, diffusion and sulfate-water oxygen exchange probably have negligible effects on the $\delta^{18}O$ of $SO_4^{2-}{}_{(aq)}$ in most Earth-surface and near-surface environments.

2) Redox S cycling, and S oxidation mechanisms in particular, are of primary importance in controlling the $\delta^{18}O$ of environmental $SO_4^{2-}{}_{(aq)}$.

3) During aqueous abiotic or biological oxidation of S compounds, water is the source of 50-100 % of the $SO_4^{2-}{}_{(aq)}$ oxygen.

4) Mixing of SO_4^{2-} from various sources must be considered for all environments.

5) Sulfate concentrations and $\delta^{34}S$ values are important additional data for interpreting mixing relationships and identifying sources.

In addition, we offer the following general observations, which should be regarded as working hypotheses that may require modification as additional data become available:

6) Abiotic oxidation of $SO_3^{2-}{}_{(aq)}$ in the atmosphere produces "secondary" SO_4^{2-} with an isotopic composition described by Area A in Figure 2, based on the experiments of Holt et al. (10).

7) This secondary SO_4^{2-} will generally mix in the atmosphere with ^{18}O-enriched SO_4^{2-}, relative to Area A, which includes "primary" SO_4^{2-} that forms by high temperature anthropogenic S oxidation, and sea spray $SO_4^{2-}{}_{(aq)}$.

8) Abiotic or bacterially mediated oxidation of mineral sulfides in Earth-surface and near-surface environments will produce $SO_4^{2-}{}_{(aq)}$ that falls approximately into Area B in Figures 3 to 5. The type of sulfide mineral(s) does not seem to affect $\delta^{18}O_{SO4}$ significantly.

9) Aqueous sulfate that plots in the lower portion of Area B may have formed largely as a product of oxidation of reduced S by Fe^{3+} and/or mediated by chemolithotrophic bacteria (e.g. thiobacilli), utilizing an electron-transport-chain mechanism. In such processes the $\delta^{18}O_{SO4}$ becomes similar or virtually identical to the ambient $\delta^{18}O_{H2O}$ (32).

10) Based on the work of Qureshi (43), higher pH or P_{O2} will tend to produce relatively negative $\delta^{18}O_{SO4}$ values (lower portion of Area B).

11) If evaporite sulfate that precipitated in ^{18}O-enriched water (e.g. marginal marine) is dissolved in more ^{18}O-depleted water (e.g. meteoric-derived continental), the $\delta^{18}O_{SO4}$ versus $\delta^{18}O_{H2O}$ data may plot above Area B.

12) Microbial sulfate reduction, which enriches residual $SO_4^{2-}{}_{(aq)}$ in ^{18}O, can also produce data that lie above Area B.

Topics that deserve special attention in future studies include:

a) further experimental examination of the effects of various oxidation reaction rates, P_{O2} and pH conditions on $\delta^{18}O_{SO4}$,

b) experimental data on equilibrium fractionation and kinetics of water-sulfate exchange of oxygen under low pH, low T conditions typical of acid mine drainage,

c) the influence of water evaporation on the $\delta^{18}O_{SO4}$ in soils, and

d) the importance of biologically mediated cycling between $SO_4^{2-}{}_{(aq)}$ and organic sulfate forms as a factor affecting the $\delta^{18}O$ of $SO_4^{2-}{}_{(aq)}$.

Acknowledgments

This manuscript is based in part on an earlier review contained in the Ph.D. thesis by D. Van Stempvoort (ref. 88). The Stable Isotope Laboratory at the University of Calgary is supported by the Natural Sciences and Engineering Research Council of Canada. We thank C. Alpers, B. Taylor and R. Stoffregen for constructive review comments.

Literature Cited

1. Rees, C. E.; Holt, B. D. In Stable Isotopes in the Assessment of Natural and Anthropogenic Sulphur in the Environment; Krouse, H. R.; Grinenko, V. A., Eds.; Wiley & Sons: New York, NY, 1991, pp. 43-64.
2. van Everdingen, R. O.; Krouse, H. R. In Mine Drainage and Surface Mine Reclamation; U.S. Bureau of Mines: Washington, D.C., 1988, Vol. 1, pp. 147-156.
3. Toran, L.; Harris, R. F. Geochim. Cosmochim. Acta 1989, 53, 2341-2348.
4. Hoffman, M. R.; Calvert, J. G. Chemical Transformation Modules for Eulerian Acid Deposition Models; National Center for Atmospheric Research: Boulder, CO, 1985, Vol. II.
5. Seinfeld, J. H. Atmospheric Chemistry and Physics of Air Pollution. Wiley-Interscience: New York, NY, 1986.
6. Lee, Y. N.; Schwartz, S. E. Proc., Fourth Ann. Conf. Precipitation Scavenging, Dry Deposition and Resuspension, Santa Monica, Calif., 1982.
7. Lettl, A. Folia Microbiol. 1982, 27, 147-149.
8. Wainwright, M.; Johnson, J. Plant Soil 1980, 54, 299-305.
9. Betts, R. H.; Voss, R. H. Can. J. Chem. 1970, 48, 2035-2041.
10. Holt, B. D.; Kumar, R.;Cunningham, P. T. Atmos. Environ. 1981, 15, 557-566.
11. Holt, B. D.; Cunningham, P. T.; Engelkemeir, A. G.; Graczyk, D. G.; Kumar, R. Atmos. Environ. 1983, 17, 625-632.
12. Krouse, H. R.; Gould, W. D.; McCready, R. G. L.; Rajan, S. Earth Planet. Sci. Lett. 1991, 107, 90-94.
13. Chang, S. G.; Littlejohn, D.; Hu, K. Y. Science 1987, 237, 756-758.
14. Lloyd, R. M. Science 1967, 156, 1228-1231.
15. Holt, B. D.; Kumar, R. Am. Chem. Soc., Div. of Fuel Chem. 1985, 30, 207-211.
16. Caron, F.; Tessier, A.; Kramer, J. R.; Schwarcz, H. P.; Rees, C. E. Appl. Geochem. 1986, 1, 601-606.
17. Holt, B. D.; Kumar, R. Atmos. Environ. 1984, 18, 2089-2094
18. Kelly, D. P. Phil. Trans. R. Soc. Lond., B, 1982, 298, 499-528.
19. Peck, H. D.; Stulberg, M. P. J Biol. Chem. 1962, 237, 1648-1652.
20. Peck, H. D. Bact. Rev. 1962, 26, 67-94.
21. Boyer, P. D. In Current Topics in Bioenergetics; Sanadi, D. R., Ed.; Academic Press: New York, NY, 1967, Vol. 2, pp. 99-149.
22. Ehrlich, H. L. Geomicrobiology. Marcel Dekker Inc.: New York, NY, 1990; 2nd ed.
23. Germida, J. J.; Wainwright, M.; Gupta, V. V. S. R. In Soil Biochemistry; Stotzky, G.; Bollag, J.-M., Eds.; Marcel Dekker: New York, NY, 1992, pp. 1-53.
24. Wainwright, M. Advances in Agron. 1984, 37, 349-396.
25. Germida, J. J.; Lawrence, J. R.; Gupta, V. V. S. R. In Proc. Sulphur-84, an Int. Conf. Sponsored by SUDIC, Calgary, Canada, 1984, pp. 703-710.
26. Jorgensen, B. B. Science 1990, 249, 152-154.

27. Suzuki, I.; Chan, C. W.; Takeuchi, T. L. In this volume.
28. Schedel, M.; Trüper, H. G. Arch. Microbiol. 1980, 124, 205-210.
29. Santer, M. Biochem. Biophys. Res. Commun. 1959, 1, 9-12.
30. Krouse, H. R. Oxygen Isotope Composition Produced during Bacterial Oxidaton of Metal Sulphides: Dependence upon O_2 Pressure and Oxygen Isotope Compositions of H_2O and O_2. Report, Energy, Mines and Resources Canada, Ottawa, Canada, Contract File 55SS.23440-8-9161, 1989.
31. Suzuki, I. Biochim. Biophys. Acta 1965, 110, 97-101.
32. Mizutani, Y.; Rafter, T. A. New Zealand J. Sci. 1969, 12, 60-68.
33. Kelly, D. P. Microbiol. Sci. 1985, 2, 105-109.
34. Emmel, T.; Sand, W.; Konig, W. A.; Bock, E. J. Gen. Microbiol. 1986, 132, 3415-3420.
35. Lowson, R. T. Chem. Rev. 1982, 22, 55-69.
36. Nordstrom, D. K. In Acid Sulfate Weathering; Kittrick J. A. et al., Eds.; Soil Sci. Soc. Amer.: Madison, Wis., 1982, pp. 37-56.
37. Goldhaber, M. B. Am. J. Sci. 1983, 283, 193-217.
38. Moses, C. O.; Nordstrom, D. K.; Herman, J. S.; Mills, A. L. Geochim. Cosmochim. Acta 1987, 51, 1561-1571.
39. Moses, C. O.; Herman, J. S. Geochim. Cosmochim. Acta 1991, 55, 471-482.
40. Luther, G. W. III. Geochim. Cosmochim. Acta 1987, 51, 3193-3199.
41. Reedy, B. J.; Beattie, J. K.; Lowson, R. T. Geochim. Cosmochim. Acta 1991, 55, 1609-1614.
42. Taylor, B. E.; Wheeler, M. C.; Nordstrom, D. K. Geochim. Cosmochim. Acta 1984, 48, 2669-2678.
43. Qureshi, R. M. The Isotopic Composition of Aqueous Sulfate (A laboratory investigation). Ph.D. thesis, U. of Waterloo, Waterloo, Ont., Canada, 1986.
44. van Everdingen, R. O.; Shakur, M. A.; Michel, F. A. Can. J. Earth Sci. 1985, 22, 1689-1695.
45. Taylor, B. E.; Wheeler, M. C.; Nordstrom, D. K. Nature 1984, 308, 539-541.
46. Murr, L. E.; Berry, V. K. Hydrometallurgy 1976, 2, 11-24.
47. Arkesteyn, G.J.M.W. Antonie van Leeuwenhoek 1979, 45, 423-435.
48. Wilmot, P. D.; Cadee, K.; Katinic, J. J.; Kavanaugh B. V. J. Water Pollut. Control Fed. 1988, 60, 1264-1270.
49. Kleinmann, R.L.P.; Crerar, D. A. Geomicrob. J. 1979, 1, 373-388.
50. Arkesteyn, G.J.M.W.; Plant & Soil 1980, 54, 119-134.
51. Krouse, H. R. Oxygen and Sulphur Isotope Fractionation During Bacterial Oxidation of Metal Sulphides. Report prepared for Energy, Mines and Resources Canada, Ottawa, Canada, Contract File 38ST.23440-6-9197, 1988.
52. Gould, W. D.; McCready, R. G. L.; Rajan, S.; Krouse, H. R. In Biohydrometallurgy 1989, Proc. Int. Symp. Biohydrometallurgy, Jackson Hole, Wyo.; Salley, J.; McCready, R. G. L.; Wichlacz, P, Eds.; CANMET SP89-10, pp. 81-92, 1990.
53. Schwarcz, H. P.; Cortecci, G. Chem. Geol. 1974, 13, 285-294.
54. Hendry, M. J. Origin of Groundwater Sulfate in a Fractured Till in an Area of Southern Alberta, Canada. Ph.D. thesis, Univ. of Waterloo, Waterloo, Ont., Canada, 1984.
55. Van Stempvoort, D. R.; Hendry, M. J.; Schoenau, J. J.; Krouse, H. R. Sources and Dynamics of Sulfur in Weathered Till, Western Glaciated Plains of North America (Chem. Geol., in press)

56. Krouse, H. R. In Studies on Sulphur Isotope Variations in Nature; International Atomic Energy Agency: Vienna, Austria, 1987, pp. 19-29.
57. Katzenberg, M. A.; Krouse, H. R. Can. Soc. Forens. Sci. J. 1989, 22, 7-19.
58. Van Stempvoort, D. R.; Fritz, P.; Reardon, E. J. Appl. Geochem. 1992, 7, 159-175.
59. Lane, G. A.; Dole, M. Science 1956, 123, 574-576.
60. Toran, L. J. Contam. Hydrol. 1987, 2, 1-29.
61. Lloyd, R. M. J. Geophys. Res. 1968, 73, 6099-6110.
62. Pierre, C. Chem. Geol. 1985, 53, 191-196.
63. Thode, H. G.; Monster, J. In Fluids in Subsurface Environments; Young A.; Galley, J. E. Eds.; Amer. Assoc. Petrol. Geol. Memoir; 1965, Vol. 4, pp. 367-377.
64. Cortecci, G. In Stable Isotopes in Earth Sciences; Robinson, B. W., Ed.; New Zeal. Dept. Sci. & Indust. Research Bulletin, 1978, Vol.220, pp. 49-52.
65. Van Stempvoort, D. R.; Reardon, E. J.; Fritz, P. Geochim. Cosmochim. Acta 1990, 54, 2817-2826.
66. Nriagu, J. O. Earth Planet. Sci. Let. 1974, 22, 366-370.
67. Zak, I.; Sakai, H.; Kaplan, I. R. In Isotope Marine Chemistry; Miyake, Y., Ed.; Uchida Rokakuho, Tokyo, Japan, 1980, p. 339-373.
68. Chiba, H.; Sakai, H. Geochim. Cosmochim. Acta 1985, 49, 993-1000.
69. Hoering, T. C.; Kennedy, J. W. J. Am. Chem. Soc. 1957, 79, 56-60.
70. Mizutani, Y.; Rafter, T. A. New Zealand J. Sci. 1969, 12, 54-59.
71. Mizutani, Y. Geochem. J. 1972, 6, 67-73.
72. Truesdell, A. H.; Hulston, J. R. In Handbook of Environmental Isotope Geochemistry; Fritz, P.; Fontes, J. C., Eds.; Elsevier: New York, NY, 1980, Vol. 1, pp. 179-226.
73. Rye, R. O; Bethke, P. M.; Wasserman, M. D. Econ. Geol. 1992, 87, 225-262.
74. Alpers, C. N.; Nordstrom, D. K. In 1991 MEND Proceedings, 2nd Int'l Conf. Abatement of Acidic Drainage, Montreal, Quebec, Canada, Tome 2, 1991, pp 321-335.
75. Alpers, C. N.; Rye; R. O.; Nordstrom, D. K.; White, L. D.; King, B.-S. Chem. Geol. 1992, 96, 203-226.
76. Houghton, C.; Rose, F. A. Appl. Environ. Microbiol. 1976, 31, 969-976.
77. Spencer, B. Biochem. J. 1958, 69, 155-159.
78. Sampson, E. J.; Vergara, E. V.; Federer, J. M.; Funk, M. O.; Benkovic, S. J. Arch. Biochem. Biophys. 1975, 169, 372-383.
79. Dodgson, K. S.; Rose, F. A. In Metabolic Pathways, 3rd Ed.; Greenberg, D. M., Ed; Academic Press: New York, NY, 1975, Vol. 7, pp. 359-431.
80. Cloves, J. M.; Dodgson, K. S.; Games, D. E.; Shaw, D. J.; White, G. F. Biochem. J. 1977, 167, 843-846.
81. Bartholomew, B.; Dodgson, K. S.; Matcham, G. W. J.; Shaw, D. J.; White, G. F. Biochem. J., 1977, 165, 575-580.
82. Gelineau, M.; Carignon, R.; Tessier, A. Appl. Geochem. 1989, 4, 195-201.
83. David, M. B.; Schindler, S. C.; Mitchell, M. J.; Strick, J. E. Soil Biol. Biochem. 1983, 15, 671-677.
84. Mizutani, Y.; Rafter, T. A. Geochem. J. 1973, 6, 183-191.
85. Fritz, P.; Basharmal, M.; Drimmie, R. J.; Ibsen, J.; Qureshi, R. M. Chem. Geol. (Isot. Geosci. Sect.) 1989, 79, 99-105.
86. Trudinger, P. A.; Chambers, L. A. Geochim. Cosmochim. Acta 1973, 37, 1775-1778.
87. Elsgaard, L.; Jorgensen, B. B. Geochim. Cosmochim. Acta 1992, 56, 2417-2424.

88. Van Stempvoort, D. R. The Use of Stable Isotope Techniques to Investigate the Sulfur Cycle in Upland Forests of Central and Southern Ontario. Ph.D. thesis, U. of Waterloo, Waterloo, Ont., Canada, 1989.
89. Krouse, H. R.; Tabatabai, M. A. In Sulfur in Agriculture; Tabatabai, M. A., Ed; Amer. Soc. Agron., Crop Sci. Soc. Amer., Soil Sci. Soc. Amer., Madison, Wis., 1986, pp. 160-205.
90. Staniaszek, P. Isotope Composition of Environmental Sulphate. Ph.D. thesis, U. of Calgary, Calgary, Alta., Canada, 1992.
91. Holt, B. D.; Cunningham, P. T.; Kumar, R. Int. J. Environ. Anal. Chem. 1979, 6, 43-53.
92. Norman, A.-L. Atmospheric Sulphur in Bermuda and Alberta. M.Sc. thesis, U. of Calgary, Calgary, Alta., Canada, 1991.
93. Cortecci, G.; Longinelli, A. Earth Planet. Sci. Lett. 1970, 8, 36-40.
94. Longinelli, A.; Bartelloni, M. Water Soil Air Pollut. 1978, 10, 335-341.
95. Van Stempvoort, D. R.; Wills, J. J.; Fritz, P. Water Air Soil Poll. 1991, 60, 55-82.
96. Trembàczowski, A. Nordic Hydrol. 1991, 22, 49-66.
97. Feenstra, S. The Isotopic Evolution of Sulfate in a Shallow Groundwater Flow System on the Canadian Shield. M.Sc. thesis, U. of Waterloo, Waterloo, Ont., Canada, 1980.
98. Mayer, B.; Fritz, P.; Li, G.; Fischer, M.; Rehfuess, K.-E.; Krouse, H. R. In Proc. Int. Symp. Use of Stable Isotopes in Plant Nutrition, Soil Fertility and Environmental Studies, IAEA, Vienna, Austria, 1991, pp. 581-591.
99. Allison, G. B.; Barnes, C. J.; Hughes, M. W.; Leaney, F. W. J. In Proc. Int. Symp. Isotope Hydrology in Water Resources Development; International Atomic Energy Agency: Vienna, Austria, 1983, pp. 105-123.
100. Fontes, J. C. In Handbook of Environmental Isotope Geochemistry; Fritz, P; Fontes, J. C., Eds.; Elsevier: New York, NY, 1980; Vol. 1, p. 75-140.
101. Dowuona, G. N.-N. Origin, Mineralogy and Geochemistry of Salts in Saskatchewan Soils. Ph.D. thesis, U. of Saskatchewan, Saskatoon, Sask., Canada, 1989.
102. Miller, J. J. The Origin of Dryland Salinity near Nobleford, Alberta. Ph.D. thesis, U. of Alberta, Edmonton, Alta., Canada, 1989.
103. Hendry, M. J.; Krouse, H. R.; Shakur, M. A. Water Resour. Res. 1989, 25, 567-572.
104. Mkumba, J. T. K. $\delta^{34}S$ and $\delta^{18}O$ Variations in Aqueous Sulfates in Groundwater Systems of Winnipeg and Kitchener-Waterloo. M.Sc. thesis, U. of Waterloo, Waterloo, Ont., Canada, 1983.
105. Van Stempvoort, D. R.; Schoenau, J. J. Investigation of Sources and Dynamics of Sulphate in Saline Soils, Prairies of Western Canada. Workshop Proc., Sulphur Transformations in Soil Ecosystems, National Hydrology Research Institute, Saskatoon, Sask., Canada, Nov. 5-7, 1992 (in press).
106. Basharmal, M. The Isotopic Compostion of Sulphur Compounds in Landfills. M.Sc. thesis, U. of Waterloo, Waterloo, Ont., Canada, 1985.
107. Wassenaar, L. Geochemistry, Isotopic Composition, Origin, and Role of Dissolved Organic Carbon Fractions in Groundwater Systems. Ph.D. thesis, U. of Waterloo, Waterloo, Ont., Canada, 1990.
108. Gilkenson, R. H.; Perry, E. C., Jr.; Cartwright, K. U. of Illinois at Urbana-Champaign, Water Resources Center, Res. Report, 1981, Vol. 165.
109. Clark, I. D. Groundwater Resources in the Sultanate of Oman: Origin, Circulation Times, Recharge Processes and Paleoclimatology: Isotopic and

Geochemical Approaches. Doctorate thesis, l'Université Paris XI, Paris, France, 1988.
110. Abbott, D. E. The Origin of Sulphate and the Isotopic Geochemistry of Sulphate-rich Shallow Groundwater in the St. Clair Clay Plain, Southwestern Ontario. M.Sc. thesis, U. of Waterloo, Waterloo, Ont., Canada, 1987.
111. Longinelli, A.; Cortecci, G. Earth Planet. Sci. Lett. 1970, 7, 376-380.
112. Clark, I. D. Isotope Hydrogeology and Geothermometry of the Mount Meager Geothermal Area. M.Sc. thesis, U. of Waterloo, Waterloo, Ont., Canada, 1980.
113. Longinelli, A. Earth Planet. Sci. Lett. 1968, 4, 206-210.
114. Holt, B. D.; Kumar, R. Water Air Soil Pollut. 1986, 31, 175-186.
115. Fritz, P.; Fries, C. A.; Drimmie, R. J. Isotopic Investigation of the Sulphur Cycle in Selected Dorset Watersheds. Report, National Water Research Institute, Burlington, Ont., Canada, Project 509-09, 1986, 32 p.
116. Jeffries, M.O.; Krouse, H. R.; Shakur, M. A.; Harris, S. A. Can. J. Earth Sci. 1984, 21, 1008-1617
117. Cortecci, G. Geochim. Cosmochim. Acta 1973, 37, p. 1531-1542.
118. Rafter, T. A.; Mizutani, Y. New Zealand J. Sci. 1967, 10, 816-840.
119. Longinelli, A.; Craig, H. Science 1967, 156, 56-59.
120. Holser, W. T.; Kaplan, I. R., Sakai, H.; Zak, I. Chem. Geol. 1979, 25, 1-17.
121. Claypool, G. E.; Holser, W. T.; Kaplan, I. R.; Sakai, H.; Zak, I. Chem. Geol. 1980, 28, 199-260.
122. Muehlenbachs, K.; Clayton, R. N. J. Geophys. Res. 1976, 81, 4365-4369.
123. Holland, H. D. The Chemical Evolution of the Atmosphere and Oceans. Princeton Univ. Press: Princeton, NJ, 1984.
124. Leone, G.; Ricchiuto, T. E.; Longinelli, A. In Studies on Sulphur Isotope Variations in Nature, IAEA; Vienna, Austria, 1987, pp. 5-14.

RECEIVED October 20, 1993

Chapter 30

Sulfur- and Oxygen-Isotope Geochemistry of Acid Mine Drainage in the Western United States
Field and Experimental Studies Revisited

B. E. Taylor[1] and Mark C. Wheeler[2]

[1]Geological Survey of Canada, 601 Booth Street, Ottawa, Ontario K1A 0E8, Canada
[2]Groundworks Environmental, Inc., 1022 North Second Street, San Jose, CA 95112

Sulfate in acid mine drainage formed from the oxidation of pyrite and other sulfides in ore bodies and/or pyritic rocks from the West Shasta district and Leviathan Mine (California) and the Argo Tunnel and Alpine Gulch (Colorado) was analyzed for its oxygen- and sulfur-isotope composition. Except for Alpine Gulch, the $\delta^{34}S$ values of the sulfate and source pyrite were identical, whereas the $\delta^{18}O_{SO4}$ values were variable (‰, V-SMOW): -7.4 to 7.0, West Shasta district; -8.5 to -3.5, Leviathan Mine area; -11.5 to -6.7, Argo Tunnel; and -3.9 to -1.3, Alpine Gulch. The $\delta^{18}O$ value of dissolved sulfate in acid mine drainage is principally controlled by the local surface waters: mean $\delta^{18}O_{H2O} = -10.1$, West Shasta district; -15.0, Leviathan Mine; -15.0, Argo Tunnel; -14.9, Alpine Gulch. Most isotope compositions of sulfate formed in acid mine drainage under dominantly saturated conditions plot on a $\delta^{18}O_{SO4}$ vs. $\delta^{18}O_{H2O}$ diagram in a region bounded by a "Zero Fractionation Trend" (i.e., $\Delta^{18}O_{SO4-H2O} = 0$) and an experimental "*T. ferrooxidans.* Trend". In flooded environments, variations in $\delta^{18}O_{SO4}$ may largely be explained by isotope exchange between a short-lived sulfite intermediate and water, plus isotope fractionations accompanying microbially mediated oxidation (*T. ferrooxidans* and *T. thiooxidans* were present in mine effluents). Stoichiometric isotope balance would suggest that, in this case, sulfate contains at least 50-80% water-derived oxygen. Elevated $\delta^{18}O_{SO4}$ values were found in West Shasta portal effluents sampled after the wet season. $\delta^{18}O_{SO4}$ values higher than those of the "*T. ferro.* Trend" characterize Richmond and Lawson portal drainage. $\Delta^{18}O_{SO4-H2O}$ values are similar to those from wet/dry experiments. We interpret these results to indicate dissolution of ferrous- and ferric-sulfate minerals with high $\delta^{18}O_{SO4}$ acquired by isotope exchange between dissolved sulfate (and/or short-lived intermediate sulfur species) and very low pH (<0) waters. ^{18}O-enrichment of drip water by evaporation may have also been indirectly responsible for the hypothesized high $\delta^{18}O_{SO4}$ values of the precipitated minerals. Anomalously high $\delta^{18}O_{SO4}$ values, where metal sulfide oxidation is the only source of sulfate, may indicate the presence of high acidity in the flow path of acid waters.

This study reports the stable isotope composition of sulfate in acidic drainage produced during the oxidation of pyrite-rich ore bodies in California and Colorado, and describes the use of these data to characterize the environments of oxidation (or, hydrologic setting) and some aspects of the oxidation processes. The individual steps in the oxidation of pyrite are complex and not entirely agreed upon. The isotope properties of sulfate measured in this study reflect a number of contributing processes. Our conclusions are therefore rather general and many aspects remain to be tested. One of the more exciting is that, in some cases, subsurface, very low pH conditions may be indicated by the oxygen-isotope composition of dissolved sulfate. Previous suggestions (1-2) that microbial contributions to oxidation processes can also contribute to the oxygen-isotope enrichment of sulfate in restricted hydrologic environments (e.g., O_2-depleted waters in flooded mines) are largely substantiated by newer experimental work (3), but we reinterpret some of the extreme ^{18}O enrichments previously reported (1), and attribute them to abiologic processes. Following a review of previous work, we describe the geologic and hydrologic settings of four study areas, summarize isotope data obtained from each site, and then discuss possible interpretations of the data based on available experimental data on isotope effects during pyrite oxidation.

Previous Work

Abiologic Sulfide Oxidation. Numerous studies of the rates of pyrite oxidation under various conditions have been conducted in the last two decades. Of particular interest here are those studies conducted under conditions similar to environments producing acid drainage in the absence of neutralizing agents, and/or those which addressed the effects of different oxidants, mechanisms and pathways of sulfide mineral oxidation, and the role of intermediate sulfur species. Two unidirectional reactions commonly used to describe the overall oxidation process of pyrite are (4-6):

$$FeS_2 + 7/2\ O_{2(aq)} + H_2O \rightarrow Fe^{2+} + 2\ SO_4^{2-} + 2\ H^+ \quad (1)$$

$$FeS_2 + 14\ Fe^{3+} + 8\ H_2O \rightarrow 15\ Fe^{2+} + 2\ SO_4^{2-} + 16\ H^+ \quad (2)$$

Reaction (2) is rate-limited under acid conditions by the following reaction (7):

$$Fe^{2+} + 1/4\ O_{2(aq)} + H^+ \rightarrow Fe^{3+} + 1/2\ H_2O \quad (3)$$

which is greatly accelerated in the presence of the iron-oxidizing bacterium *Thiobacillus ferrooxidans* (8).

Moses et al. (9) measured oxidation rates which indicated that Fe^{3+} was the direct oxidant of pyrite under both aerobic and anaerobic conditions, at rates which varied only slightly over the pH range 2-9. Control of the oxidation rate of pyrite at pH=2 by Fe^{3+} was also supported by experiments of McKibben and Barnes (10). In natural environments, these rates are strongly influenced by the formation of hydroxide coatings on pyrite (11), or by adsorption of Fe^{2+} (12), as well as by the presence of other metals which act readily as electron acceptors (e.g., Mn^{+4}; 13).

The initial mechanisms and subsequent oxidation steps leading to sulfate following the stoichiometry of reactions (1) and (2) are not completely understood. Numerous schemes have been devised (e.g., 9,12,14-15), and different mechanisms may operate more efficiently under different conditions. From voltametric measurements, Biegler and Swift (16) suggested that the initial oxidation mechanism probably changes above pH 3-4. Elemental sulfur may be involved as an intermediate reaction product above pH 4.6 (17). The appearance and increase in abundance of dissolved sulfur intermediates (sulfite, SO_3^{2-}; thiosulfate, $S_2O_3^{2-}$; and polythionates, $S_nO_6^{2-}$) with increasing pH (9, 13) and dissolved oxygen (3) suggest increased complexity of the oxidation processes. At low pH, intermediate species comprise but a trace amount of the total dissolved sulfur, detectable only in vigorously stirred solutions and not at all under anoxic conditions (9). This implies complete oxidation within the mineral-solution boundary layer. Hiskey and Schlitt (18) suggested adsorption of oxygen as the initial abiologic, rate-limiting step, and many of the above-cited studies generally agree that the adsorbed oxygen is water-derived, whereas Moses and Herman (12) concluded that dissolved oxygen is not involved in direct attack on pyrite at neutral pH.

Microbial Mediation of Oxidation. The biochemical literature on microbial mediation of sulfide oxidation is extensive; a lucid review is given by Kelly (19). Only a few points are noted here. The oxidation of pyrite can be mediated by Thiobacilli which oxidize ferrous iron (*T. ferrooxidans*; 20) and sulfur (e.g., *T. ferrooxidans*; *T. thiooxidans*; 19). *T. ferrooxidans* obtains energy for growth by the oxidation of Fe^{2+} to Fe^{3+} while at the same time consuming molecular oxygen and increasing the rate of reaction (3) by up to six orders of magnitude relative to the abiological rate (21). The rate of *T. ferrooxidans*-enhanced pyrite oxidation depends not so much on the cell density, but rather on the availability of Fe^{2+}-binding sites to promote reaction (3) by reduction of O_2 inside the cells. Inhibition of the oxidation rate of pyrite can occur because of competition among the cells for these sites (22), the presence of Fe-hydroxide coatings, soluble Fe^{3+}, or direct attachment to pyrite surfaces (e.g., 23).

T. *ferrooxidans* also derives energy from the enzymatic oxidation of sulfur and sulfur species including sulfite and thiosulfate (24, and references therein). Two enzymatic pathways of elemental sulfur oxidation suggested by Vestal and Lundgren are: (1) $S^o \rightarrow SO_3^{2-} \rightarrow SO_4^{2-}$, and (2) $S^o \rightarrow S_2O_3^{2-} \rightarrow S_4O_6^{2-}$ then possibly to other polythionates. A similar enzymatic system also operates in *T. thiooxidans*, where the sulfur oxidase requires molecular oxygen as the electron acceptor in order to produce thiosulfate from elemental sulfur (25).

Isotope Studies. The pioneering work of Lloyd (26-27) on the oxygen-isotope composition of sulfate in geologic systems, and its very slow isotope exchange with surface waters, suggested that the isotope composition of sulfate might be retained in nature and provide information on the origin of the sulfate. An important aspect is the potential for determination of the relative contributions of two isotopically distinct sources of oxygen: air (O_2) and water.

Prior to the mid-1970's, and with the exception of Lloyd's work (27), experimental isotope studies primarily focussed on the use of ^{18}O in tracer experiments which monitored the transfer of ^{18}O from enriched water, oxygen gas, or bacterial energy sources. These studies indicated that (1) in the abiologic oxidation of pyrite, the end-product sulfate contains dominantly water-derived oxygen (28); (2) despite consumption of O_2 by *T. thiooxidans* during oxidation of elemental sulfur, water was the principal source of oxygen in the product thiosulfate or sulfate (25); and (3) sulfate incorporated oxygen from phosphate during enzymatic oxidation of sulfite and thiosulfate by *T. thioparus* (29-30).

Schwarcz and Cortecci (33) were the first to document a correlation between $\delta^{18}O_{SO4}$ and $\delta^{18}O_{H2O}$ in pyrite oxidation experiments in near-neutral water. They suggested that this correlation is likely due to the rapid exchange of oxygen isotopes between sulfite and water. Taylor et al. (2) determined the kinetic oxygen-isotope fractionations for the incorporation of water- and air-derived oxygen during biologic and abiologic oxidation of pyrite under acid conditons. Krouse et al. (35) demonstrated the incorporation of ^{18}O-labelled O_2 into sulfate and water during microbially mediated oxidation. The $\delta^{18}O_{SO4}$ values were higher than those resulting from abiologic oxidation, and were also higher when sulfite oxidation occurred in the presence of an oxidase enzyme. Experiments of Gould et al. (34) reconfirmed the enrichment of ^{18}O in sulfate formed during bacterially mediated sulfide oxidation, and the correlation of $\delta^{18}O_{SO4}$ with $\delta^{18}O_{H2O}$. Qureshi's (3) experimental studies of pyrite oxidation over a range of pH, including conditions of acid mine drainage, emphasized the potential importance of oxygen-isotope fractionation among sulfur intermediate species.

Few oxygen-isotope studies of acid mine drainage are available. Smejkal (31) published isotope data on sulfate in acid drainages from Bohemia from which he inferred that water was the dominant source of oxygen. Taylor et al. (1) documented a broad correlation between $\delta^{18}O_{SO4}$ and $\delta^{18}O_{H2O}$ in acid mine drainage, and suggested variable contributions of atmospheric and water-derived oxygen to account for local variation in $\delta^{18}O_{SO4}$. This concept was applied to describe the origin of dissolved sulfate in acid discharges in northwest Canada by van Everdingen et al. (32). Admittedly, determination of the relative contributions of atmospheric and water-derived oxygen does not define the electron transfer pathway (cf., 15). However, evaluation of $\delta^{18}O_{SO4}$ in acid mine drainage based on the hydrologic environment, potential mineral-water reactions and available experimental data can reveal certain aspects of the oxidation process and environment.

Geologic and Hydrologic Setting of Sample Sites

The four sample sites selected for this study offer variations of geologic and hydrologic setting, in addition to differences in the isotope composition of local waters. Except for Alpine Gulch, Colorado, where pyrite oxidation appears to take place in open, relatively exposed bogs, the hydrologic environments can be represented by those shown schematically in Figure 1. Sample collection (1981-82) and some details of the hydrologic setting discussed in this paper largely predate subsequent remedial action for the abatement of acid waters in each of the mining areas (as noted below).

West Shasta District, California. The West Shasta copper-zinc mining district, ca. 15 km northwest of Redding in Shasta County, California, comprises an area 13 km long by 3 km wide containing numerous volcanogenic massive sulfide type deposits (36; Figure 2). The deposits contain from a few million to as many as 25 million tons of sulfide (36), composed dominantly of pyrite (FeS_2), with subordinate chalcopyrite ($CuFeS_2$), and sphalerite ($(Zn,Fe)S$), with minor gold and silver. Aspects of the mineralogy, geology, ore deposits and stable isotope geochemistry have been previously reported (36-40). Rugged topography, faults and fractures and extensive but intermittent mining activity during 1879-1963 have exposed the deposits to oxidation, producing severe acid mine drainage. In addition to the ore deposits, the Balaklala Rhyolite and interbedded and underlying andesitic basalt flows of the Copley Greenstone each contain up to a few percent of pyrite. Also, piles of waste rock from mining, and tailings from sulfide processing at Iron Mountain Mine which have washed into Boulder and Slickrock Creeks (Figure 2), enhance the production of acid waters. Middle Devonian (Kennett Fm.) and Mississippian (Bragdon Fm.) siltstone, shale, limestone and conglomerate, and Recent surficial deposits which unconformably overlie the volcanic rocks provide little to no natural abatement of the acid drainage. Three mining areas (Iron Mountain Mine; Balaklala Mine; and the Mammoth Mine) were selected for detailed study.

Iron Mountain Mine. The principal mined ore bodies include the Old Mine, No. 8, Brick Flat, Richmond-Complex, and Hornet. The latter three consitute portions of an originally contiguous ore body displaced along the Scott and Camden faults (36). The hydrology of these ore bodies is variable, depending upon previous mining activity as well as geologic characteristics. Open-pit mining of the Old Mine and the Brick Flat deposits exposed remnants of these ore bodies to oxidation at the surface, whereas

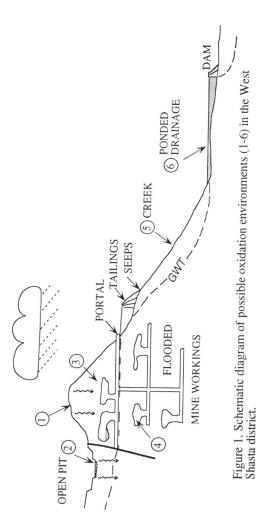

Figure 1. Schematic diagram of possible oxidation environments (1-6) in the West Shasta district.

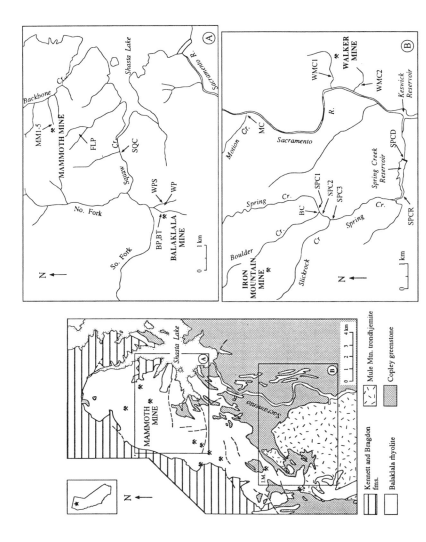

Figure 2. General geologic map (36) of the West Shasta district, California, and location of sampling sites; I.M., Iron Mountain Mine; other abbreviations as in Table I.

underground mining created stopes and haulage ways which constrain the flow and pooling of ground water. The actual exposure to, and length of contact, with ground water of the sulfide ore bodies remains unknown in detail.

Acidic water, pooled in the Brick Flat open pit, drains into Slickrock Canyon and percolates downward along the major faults (and presumably other minor faults and fractures as well), into the underground workings of the Richmond-Complex and Hornet ore bodies, and discharges from the Richmond and Lawson portals. The residence time of water entering the underground mines is difficult to establish, and likely quite variable. The time lag following a rainstorm before increased portal discharge is noted can be on the order of several days to weeks (41-42). Landslides during the mid-1950's covered the portals of the Old Mine and No.8 ore bodies, restricting drainage which now occurs as acid seeps. The absence of known major faults in the Old Mine and No. 8 ore bodies suggests a longer residence time for ground waters in these mines, and the pooling of acidic waters in contact with massive sulfide. Subsequent lining of the Brick Flat pit has virtually eliminated leakage into the underground workings, whereas capping of caved areas above the Richmond Mine has apparently not stopped the inflow of water because of the permeability afforded by many open fractures and other mine workings (C. Alpers, written commun., 1993).

The Iron Mountain Mine area contributes acidic drainage primarily to Boulder, Slickrock, and Spring Creeks. The combined drainage from the Richmond and Lawson portals, which has a pH of 1.0 (41) is first fed to a copper cementation/recovery plant along stainless steel flumes, and then enters Boulder Creek, where it contributes at least 90% of the dissolved sulfate in the creek (from data in 41).

Spring Creek debris dam, built in 1963, prevents Iron Mountain Mine mining debris from collecting in the tail race of the power plant, and offers some control on release of acidic effluent into Keswick Reservoir. Previous releases of acidic waters from the dam have caused fish kills in the reservoir, as have discharges from Spring Creek during natural runoff (41). In 1983, the Iron Mountain Mine was listed as one of 60 sites to be cleaned up with the so-called "Superfund" established by the U.S. Environmental Protection Agency (43). An emergency lime neutralization plant operated by ICI Americas, from December to March/April each year since 1990, presently processes ca. 140 gallons of acid water/minute (C. Alpers, written commun., 1993).

Balaklala Mine. Two ore bodies, the Windy Camp and Weil, represent segments of a once-continuous ore body displaced by fault movement. Groundwater penetrates the mine primarily along faults and fractures in the locally pyritized Balaklala Rhyolite, and drains from the mine via Tunnel 11 and the Weil Tunnel. Effluent from Tunnel 11 is diluted during the wet season by flow of surface waters into a shaft at higher elevations. Effluent from both tunnels had discharged directly into the headwaters of Squaw Creek which flows into Shasta Lake. Silver King Mines (then owner of the mine), installed a steel-reinforced concrete plug in the Weil Tunnel, some 80 meters from the portal, during November 1981. Flow from a pressure-relieving drain pipe in the plug was subsequently stopped March 4, 1982. Water pooling in the workings had apparently not leaked through other portals or seeps in significant amounts by the end of our sampling (1983).

Mammoth Mine. Numerous tabular massive sulfide bodies comprise the Mammoth Mine. Penetration of surface waters to the mine workings is primarily via faults, fractures, and primary permeability (e.g, lithologic contacts) in the Balaklala Rhyolite. Acid mine waters drained freely from the Main Tunnel into Backbone Creek, and then into Shasta Lake, until sealed by Sharon Steel Corp. (then owner of the mine) in December 1981. Discharge from the Friday Lowden Tunnel, situated at a lower elevation than the Mammoth Mine Tunnel, increased notably between April, 1981 and March, 1982, suggesting that water pooled behind the seal in the Main Tunnel of the Mammoth Mine had begun to flow into the Friday Lowden workings.

Leviathan Mine, California. The Leviathan Mine, located in the Monitor district, Alpine County, California (Figure 3), was mined intermittently for chalcopyrite and sulfur by underground methods from the late 1800's to the early 1900's, and from 1953 to 1962 for sulfur by open-pit methods (44). Hydrothermally altered Tertiary rhyolite and dacite tuffs and andesitic intrusions comprise a faulted lens of mineralized rock some 700 meters long and 30 meters thick containing as much as 35% of uniformly distributed elemental sulfur; veins of sulfur up to 0.5 meter thick also occur (44). Primary sulfide minerals include small amounts of marcasite, pyrite, and chalcopyrite, in addition to secondary copper- and iron-sulfate minerals. Both the sulfur and sulfides may be attributed to deposition from magmatic-hydrothermal gases associated with the emplacement of volcanic magmas (44). An overlying agglomerate, similar to the Pliocene Mehrten Fm. of the western Sierra Nevada foothills, and a younger tuff are unmineralized (44-45).

Acidic waters from tailings in an open pit and from a tunnel portal drain into Leviathan Creek, and then into the East Fork of the Carson River in Nevada. The State Water Resources Board estimated contributions of acid water to Leviathan Creek in 1971 from the mine area as follows: tailings pile seepage (58%), direct runoff from dumps (21%), tunnel seepage (19%), and surface drainage from the open pit (2%) (44). Additional hydrologic data were collected by the U.S. Geological Survey (46). The samples in our study were collected prior to remediation efforts carried out in 1984.

Argo Tunnel, Colorado. The Argo Tunnel, located in Gilpin and Clear Creek Counties, Colorado (Figure 4), drains mines which exploited veins in the Central City mining district for gold, silver, uranium and base metals. The 6.7 km-long tunnel was cut through Precambrian gneisses, granites and pegmatites between 1893 and 1907 in order to drain ground water from mine workings. Sulfide minerals in 26 veins intersected by the Argo Tunnel (47) include pyrite, chalcopyrite, tennantite ($Cu_3AsS_{3.25}$), enargite (Cu_3AsS_4), sphalerite, and galena (48).

Ground-water flow in the bedrock is controlled primarily by fractures and veins. Wentz (47) found that the acid waters draining from the Argo Tunnel in the period January 1976 to March 1977 had a relatively constant pH of 3.0 and a temperature of 16 °C. The Argo Tunnel is one of several sources of acid mine drainage which renders Clear Creek unsuitable for aquatic life or agricultural or other usage.

Alpine Gulch, Colorado. Two natural bogs in Alpine Gulch, 6 km southwest of Lake City in the western San Juan Mountains, Colorado (Figure 4) are sites of natural sulfide oxidation and acid water formation. Pyrite in altered Picayune quartz latite oxidizes to form sulfate-rich acid waters which drain into the East Fork of Alpine Gulch. The bogs do not create a serious water quality problem because the acid waters are greatly diluted by Alpine Gulch, but provide an example of a natural, acid-producing environment uncomplicated by mining activity.

Methodology

Collection and Storage of Water Samples. Water samples from mine portals, open-pit drainage, and sites downstream were filtered (49) and stored in polyethylene bottles for analysis of dissolved sulfate, and in 125 ml glass bottles with polyseal caps for hydrogen and oxygen-isotope analysis of the water (50). Sequential gravimetric and isotope analysis of sulfate during short- (37-day) and long-term (61- and 96-day) storage of filtered and unfiltered waters from Weil, Balaklala, and Friday Lowden portals, and from Boulder and Squaw Creeks indicated no significant differences (i.e., variation beyond 2σ) between aliquots of either filtered or unfiltered sample from the same site. Variation of the mean sulfate concentration ((2σ/mean)*100%), determined on 5 to 23 aliquots, was <1.7 % for high sulfate waters (>9,000 mg/l) and < 5.5% for waters with less sulfate. Samples with very low sulfate contents (e.g., Squaw Creek, mean = 13

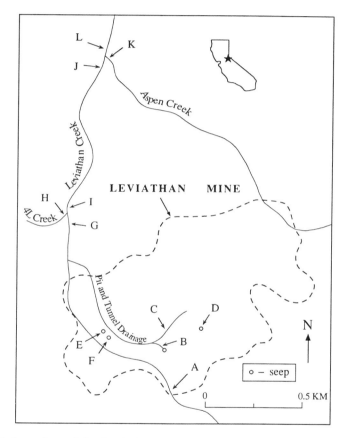

Figure 3. Location of the Leviathan Mine area, California, and sampling sites (upper case letters).

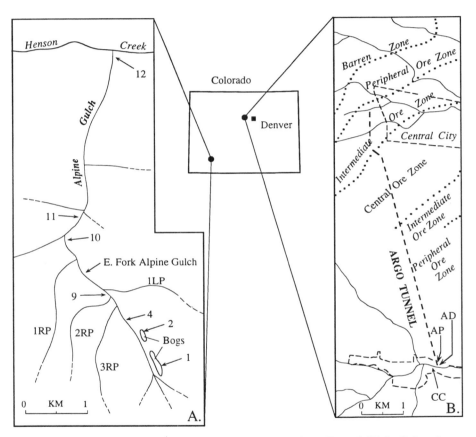

Figure 4. Location of the Alpine Gulch area (A) and Argo Tunnel (B) in Colorado, and sampling sites in each locality; numbered in (A), lettered in (B). Ore zone designations (B; 48), abbreviations as in Table I.

mg/l), had higher associated analytical uncertainties (31%). Additional details of collection and extraction procedures are described by Wheeler (50).

Gravimetric and Isotope Analysis. Sulfate (HSO_4^-, SO_4^{2-}) was quantitatively removed from an aliquot of each sample by precipitation as $BaSO_4$ (1-2), which was dried and homogenized by grinding prior to weighing and analysis. Periodic analysis of aliquots of a synthetic ferrous sulfate solution with a mean SO_4 concentration of 20,877 mg/l (1σ = 405.9; n = 11) indicated a maximum uncertainty of 3.9% on sulfate concentration; natural samples yielded a lower uncertainty. Determination of sulfite (27) suggested that it usually consituted ≤ 7% of the total sulfur in acidic, high-sulfate waters (cf. 9, 13).

The isotope compositions of dissolved sulfate, sulfide minerals, elemental sulfur and water (1-2) are reported in the usual δ-notation (i.e., ‰, or permil; 51) relative to V-SMOW (for oxygen and hydrogen), and to CDT (for sulfur). A maximum uncertainty (1σ) for $\delta^{18}O_{SO_4}$ of 0.2‰ was determined by replicate analysis of a synthetic ferrous sulfate solution and a natural sulfate. Replicate analyses of a standard $BaSO_4$ (mean $\delta^{34}S$ = +1.28) yielded a 1σ uncertainty of 0.05‰; the uncertainty in $\delta^{34}S$ was slightly lower for sulfide minerals.

Results

The results of chemical, physical and isotope determinations on acidic and neutral drainages at each of the four sites are presented in Table I, along with abbreviations for specific sample sites. General characteristics and relationships among these and other data are presented below.

Sulfate, pH and Temperature. The range of pH measured during this study was 1.6 (Weil Portal; WP) to 8.9 (Mountaineer Creek; N, Leviathan Mine area). Sulfate concentrations varied from < 1.0 mg/l (Walker Mine Creek 1; WMC1) to 30,800 mg/l (Richmond Portal; RP). Nordstrom (41) reported a pH of 0.9 and sulfate concentration of 43,050 mg/l from the combined effluent of the Richmond and Lawson tunnels. Values of pH < -2.0 have been measured underground in the Richmond mine, in drip water adjacent to oxidized ore by Alpers and Nordstrom (52), and their calculations suggest that pH values <1.0 probably characterize the active sites of sulfide mineral oxidation.

The general negative correlation between sulfate concentration and pH in Figure 5 illustrates the mixing of very acid, sulfate-rich water generated at the site of oxidation of pyrite and other, less abundant sulfides, with dilute, neutral waters. For discussion purposes, we arbitrarily categorize the waters with a pH <6.0 and a concentration of sulfate >100 mg/l as acid-sulfate waters, and will refer to waters with pH >6.0 and sulfate concentrations generally <100 mg/l as dilute neutral waters.

Dilution from storm runoff during the wet season (winter and spring months) was often indicated by a decrease in dissolved sulfate relative to samples collected during the dry seasons (50). These effects were more apparent for downstream collection sites than for samples collected at mine portals. Variation in the Zn/Cu ratio in effluent from the Richmond portal and Lawson Tunnel, among other factors, led Alpers et al. (42) to suggest that the composition of the effluent from the Richmond-Complex and Hornet Mines is influenced by formation and dissolution of Cu- and Zn-bearing Fe-sulfate minerals, in particular, melanterite ([Fe^{II}, Zn, Cu]$SO_4 \cdot 7H_2O$; 53). The differences in overall sulfate concentration between sites corresponds to the combined effects of differences in oxidation rates, nature of oxidation environments contributing to the drainage, volume of sulfide exposed to oxidation, and dilution by surface waters.

Temperatures of acid waters in the West Shasta district measured during June 1981 varied from 12.0 to 18.0, irrespective of sulfate concentration (50). The highest

Table I. Isotopic, chemical, and physical data for samples

Sample/Date Site	F/UF[a]	pH	T (°C)	SO$_4$ (mg/l)	$\delta^{18}O_{SO4}$ (‰)	$\delta^{18}O_{H2O}$ (‰)	$\Delta^{18}O$[b] (‰)	$\delta^{34}S_{SO4}$ (‰)
West Shasta district, California								
Weil Portal (WP)								
*[c] 11/1/8	F			11500	-6.6	-9.6	3.0	4.6
* do.	U			11400	-6.4	-9.6	3.2	
* 4/3/81	F			7860	-5.0	-10.6	5.6	4.8
* 6/23/81	F	1.6	18	11200	-6.5	-10.1	3.6	4.5
3/3/82	F			9330	-4.9	-10.0	5.1	4.7
do.	F			9310	-4.7	-10.0	5.3	4.8
do.	F			9270	-5.2	-10.0	4.8	4.8
* do.	U			9240	-6.2	-10.0	3.8	
3/4/82	F			9200	-5.7	-9.7	4.0	4.6
* do.	F			9370	-5.4	-9.7	4.3	4.8
do.	F			9350	-4.8	-9.7	4.9	
do.	U			9320	-5.3	-9.7	4.4	5.0
do.	U			9250	-5.7	-9.7	4.0	
do.	U			9380	-5.7	-9.7	4.0	4.6
Weil Portal, stream (WPS)								
* 4/3/81	F			6	0.4	-10.4	10.8	
6/23/81		6.2	14					
Balaklala Portal (BP)								
11/1/80	F			1600	-5.8	-9.9	4.1	4.8
do.	U			1600	-5.7	-9.9	4.2	
4/3/81	F			490	-1.1	-9.9	8.8	4.8
6/23/8	F	2.8	12	680	-3.6	-10.6	7.0	4.9
3/3/82	F			240	-5.5	-10.6	5.1	4.9
* do.	F			220	-5.3	-10.6	5.3	
do.	U			220	-5.4	-10.6	5.2	4.8
do.	U			230	-5.6	-10.6	5.0	
* 3/4/82	F			220	-6.7	-9.8	3.1	4.8
* do.	U			220	-6.7	-9.8	3.1	
Balaklala Portal, tailings (BPTLS)								
6/23/81	F		18	700	-4.5	-10.9	6.4	5.0
Squaw Creek (SQC)								
* 4/3/81	F			50	-5.8	-10.9	5.1	
do.	F			50	-1.0	-10.9	9.9	4.7
do.	U			40	-4.8	-10.9	6.1	4.5
6/24/81	F	3.2	17	140	-5.5	-10.6	5.1	4.8
3/4/82	F			20	-4.3	-9.7	5.4	4.4
* do.	U			10	-4.0	-9.7	5.7	
* do.	U			10	-3.2	-9.7	6.5	4.3
Friday Lowden Portal (FLP)								
4/3/81	F			190	-1.6	-10.0	8.4	2.1
* 6/24/81	F	6.3	13	180	-6.0	-10.0	4.0	1.6
3/4/82	F			670	-5.7	-9.7	4.0	4.5
do.	F			670	-5.6	-9.7	4.1	
* do.	U			670	-7.4	-9.7	2.3	4.5
Mammoth Mine (MMx)								
-1 4/3/81	F			890	-4.3	-9.8	5.5	4.3
6/24/81	F	2.9	12	830	-5.8	-10.2	4.4	4.1
* do.	U			850	-7.0	-10.2	3.2	

Table I. Continued

Sample/Date Site	F/UF[a]	pH	T (°C)	SO_4 (mg/l)	$\delta^{18}O_{SO4}$ (‰)	$\delta^{18}O_{H2O}$ (‰)	$\Delta^{18}O$[b] (‰)	$\delta^{34}S_{SO4}$ (‰)
do.	U			830	-5.3	-10.2	4.9	4.3
-2 do.	F	3.0	12	850	-5.2	-10.2	5.0	4.3
-3 do.	F	3.0	13	850	-5.8	-10.2	4.4	4.3
-4* do.	F			860	-6.9	-10.2	3.3	3.2
-5 do.	F	2.9	15	830	-4.0	-9.8	5.8	4.1
Richmond Portal (RP)								
* 12/14/82	U			30800	3.0	-10.2	13.2	
Lawson Portal (LP)								
* 12/14/82	U			18700	2.7	-10.2	12.9	
Old Mine #8 (OM/8)								
12/14/82	U			9450	-4.9	-10.6	5.7	
Boulder Creek (BC)								
4/2/81	F			3440	7.0	-10.1	17.1	
do.				3480	6.5	-10.1	16.6	4.5
Spring Creek (SPCx)								
-1 4/2/81	F			10	-1.2	-10.3	9.1	
-2 do.	F			810	6.8	-10.8	17.6	4.4
-3* do.	F			840	4.7	-10.2	14.9	4.4
Slickrock Creek (SKC)								
4/2/81	F			640	-5.9	-9.9	4.0	4.9
Spring Creek Reservoir (SPCR)								
4/1/81	F			190	3.2	-10.0	13.2	4.5
Spring Creek Dam (SPCD)								
* 11/1/80	F			2680	3.8	-10.0	13.8	4.7
* 4/1/81	F			260	5.4	-10.7	16.1	4.9
Motion Creek (MC)								
4/2/81	F			10	-1.2	-9.4	8.2	5.0
Walker Mine, creek (WMC1)								
* 4/2/81	F			10	1.8	-8.9	10.7	4.3
Leviathan Mine, California								
B* 10/9/81	F			5180	-3.5	-15.1	11.6	-17.5
B* 6/16/82	F	1.8		7370	-4.6	-15.4	10.8	-17.6
B* 10/6/82	F	2.3	12	5600	-4.9	-15.4	10.5	
C* 6/16/82	F	1.9		10600	-5.3	-14.4	9.1	-17.6
E* 10/6/82	F	3.3		2880	-6.3	-14.6	8.3	
J 6/16/82	F	3.3		660	-5.2	-14.7	9.5	-17.1
K* 6/16/82	F	8.0		280	-8.5	-15.5	7.0	-14.5
L 6/16/82	F	3.7		550	-5.4	-14.7	9.3	-16.3
S* 6/14/82	F	5.5		190	-5.2	-14.9	9.7	-15.1
Argo Tunnel, Colorado								
AP* 6/26/81	F	2.6	16	2260	-10.8	-15.0	4.2	-0.7
AD* do.	F	2.7	19	2270	-11.5	-15.0	3.5	-0.8
CC* do.	F			10	-6.7	-15.0	8.3	
Alpine Gulch, Colorado								
1* 9/20/82	F	3.4	7	160	-1.3	-15.0	13.7	-4.2
2* do.	F	4.1	7	300	-2.0	-15.1	13.1	-4.4
4* do.	F	6.9	7	50	-3.9	-15.0	11.1	
MIX2*do.	F	6.6	7	40	-3.8	-15.0	11.2	

[a] F=filtered, UF=unfiltered; [b] $\Delta^{18}O = \Delta^{18}O_{SO4-H2O} = (\delta^{18}O_{SO4}) - (\delta^{18}O_{H2O})$
[c] * $\delta^{18}O_{SO4}$ and $\delta^{18}O_{H2O}$ previously published in Figure 2, reference (1)

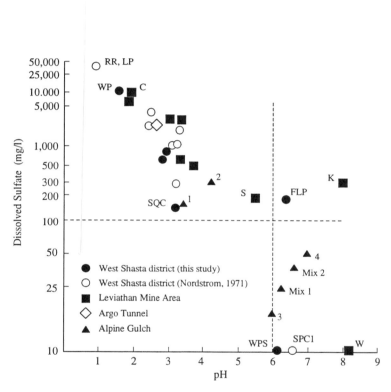

Figure 5. Plot of sulfate concentrations vs. pH for waters of representative sites in the indicated field areas. Abbreviations as in Table I.

temperature (18.0 °C) measured did correspond with the highest sulfate concentration (WP site), as expected for the exothermic nature of pyrite oxidation. Temperatures as high as 47 °C have been recently measured underground, near to the actual sites of oxidation in the Richmond Tunnel (52). However, a simple correlation at other sites was not found, probably because of cooling underground along flow paths, mixing along the flow paths, and heating of waters on the surface.

Sulfur Isotopes. No discernible sulfur-isotope fractionation was found between dissolved sulfate and the mean $\delta^{34}S$ values found in previous studies for pyrite (the dominant sulfide mineral) in deposits in the West Shasta district. Figures 6a and 6b compare histograms for sulfur-isotope compositions of sulfides and sulfate from the West Shasta District. The mean $\delta^{34}S$ of pyrite from the West Shast district is +4.25 ‰ (1σ=1.64; n=154) compared to a mean $\delta^{34}S$ for dissolved sulfate of +4.47 ‰ (1σ=0.64; n=42). For the Leviathan Mine area, mean $\delta^{34}S$ = -15.27 ‰ for whole-rock sulfur (1σ=1.64; n=7), and for elemental sulfur, mean $\delta^{34}S$ = -12.2 ‰ (1σ=0.46; n=3) compared to a mean $\delta^{34}S$ = -16.53 ‰ (1σ=1.28; n=7) for dissolved sulfate. The lack of sulfur-isotope fractionation between sulfides and the sulfate produced during oxidation has been previously demonstrated both experimentally (e.g., 2) and in field studies (e.g., 54), and is consistent with complete, layer-by-layer sulfide reaction. Thus, the isotope composition of the sulfate in the acidic waters (Table I) serves as measure of the average sulfur-isotope composition of the oxidizing ore bodies. At the Leviathan Mine, sulfide is indicated to be oxidizing in preference to elemental sulfur (Figure 6b).

Precipitate from the banks of acidic drainages in the West Shasta district yielded $\delta^{34}S$ values of 4.1 ‰ (Weil Portal) and 3.8 and 5.0 ‰ (Spring Creek Dam). The mineralogical composition of the precipitated material was not determined, but probably comprised a mixture of complex compounds which may have included hydrated ferrous oxides (e.g., ferrihydrite; 55-56), ferric hydrosulfates or feroxyhite (56-57), jarosite (58), and goethite. Murad et al. (59) indicate that schwertmannite (ferric hydroxy-sulfate; a.k.a. MDM, mine drainage material) is a ubiquitous precipitate from waters with pH 2-5. It is not known whether the sulfur extracted from the precipitate was mineralogically bound, or simply sulfate that was adsorbed (e.g., by goethite). However, the lack of detectable sulfur isotope fractionation between dissolved sulfate and the sulfur-isotope composition of the analyzed precipitates is consistent with the occurrence of the sulfur as sulfate (cf. $\Delta^{34}S$ GYPSUM-DISSOLVED SULFATE = 1.65 ‰; 60-61), and further indicates the lack of significant sulfur reduction in the overall process of sulfide mineral oxidation in the West Shasta district.

Oxygen Isotopes. The oxygen-isotope composition of the various mine and surface waters sampled in this study differed between areas by ca. 5‰ (mean $\delta^{18}O_{H2O}$: West Shasta district, -10.1‰; Leviathan Mine, -15.0‰; Argo Tunnel, -15.0‰; and Alpine Gulch, -14.9‰), but varied < 2.5 ‰ in $\delta^{18}O$ within each area (Table I). Only at the Leviathan Mine area did acid and neutral waters differ significantly in mean $\delta^{18}O_{H2O}$, and even there by only 0.5‰. This suggests incomplete mixing of surface and ground waters. The maximum seasonal variation in $\delta^{18}O_{H2O}$ at any site in the West Shasta district was only 1.2‰ (Squaw Creek; 50).

In contrast to the nearly constant $\delta^{18}O_{H2O}$ at any one site, a large range was found for $\delta^{18}O_{SO4}$ (Figure 7). In the West Shasta district, $\delta^{18}O_{SO4}$ varied from -7.4 to +7.0‰; from -8.5 to -3.5‰ in the Leviathan Mine area; from -11.5 to -6.7‰ at the Argo Tunnel; and -3.9 to -1.3‰ at Alpine Gulch. Generally speaking, dissolved sulfate with the lowest $\delta^{18}O_{SO4}$ was found in the acid portal effluents draining from underground workings. However, in the West Shasta district, acid portal effluents contained dissolved sulfate with the highest values of $\delta^{18}O_{SO4}$. The permil oxygen-isotope

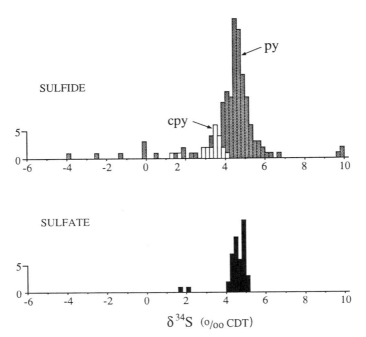

Figure 6a. Histogram plots of the sulfur-isotope compositions of pyrite (py) and chalcopyrite (cpy) from massive sulfide ore bodies in the West Shasta district (37, 39, 40, 50), and of sulfate in acid drainage (50).

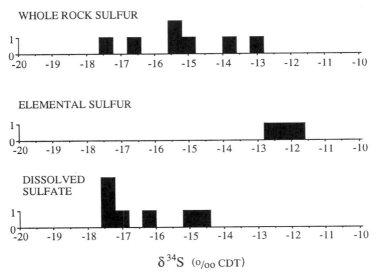

Figure 6b. Histogram plots of the sulfur-isotope compositions of whole-rock sulfur (pyrite with elemental sulfur), elemental sulfur, and dissolved sulfate from acid drainage in the Leviathan Mine area (50).

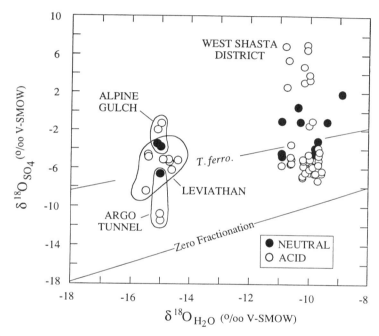

Figure 7. Plot of $\delta^{18}O_{SO_4}$ vs. $\delta^{18}O_{H_2O}$ for sulfate from acid and neutral waters in the West Shasta district, Leviathan Mine area, Argo Tunnel, and Alpine Gulch. Lines for the "Zero Fractionation Trend" (1) for sterile oxidation, and "*T. ferro.* Trend" (34) are shown for comparison.

fractionation ($\Delta^{18}O_{SO4-H2O} = \delta^{18}O_{SO4}-\delta^{18}O_{H2O}$) between dissolved sulfate and water in any one area reflects primarily the variation of $\delta^{18}O_{SO4}$. We emphasize the value of $\Delta^{18}O_{SO4-H2O}$ in later discussion as it is a principal quantity by which data from all areas can be compared, and with which isotope processes may be associated.

Microbial Population. During our study both *T. ferrooxidans* and *T. thiooxidans* were found in portal effluent and associated contaminated creeks, and generally comprised 10-15% of the total microbial population (A. Mills and C. Moses, Univ. Virginia, pers. commun., 1983). There was no direct correlation between the sulfate concentration and the cell count, however. The cell counts (cells/ml) of *T. ferrooxidans* generally ranged from 3.1 to 7.2 x10^3, and from 3.2 to 14.1 x 10^3 for *T. thiooxidans*. Effluent from the Weil portal was exceptional, containing 57.8 x 10^3 (*T. ferrooxidans*) and 50.1 x 10^3 (*T. thiooxidans*).

Discussion

A goal at the outset of this study was to ascertain whether there exists a correlation between the isotope properties of sulfate and the highly acidic hydrogeochemical environments formed from the oxidation of pyrite and other sulfides. The following discussion emphasizes sulfate formation in highly acidic environments, extends our previously published accounts of experimental and preliminary field isotope data (1-2), and incorporates information from recent experimental data (3,34-35) on the isotope systematics of sulfate. We focus primarily on the relationship between the oxygen-isotope composition of sulfate and water as a means of distinguishing different field environments and processes.

Variation of $\Delta^{18}O_{SO4-H2O}$ with Environment. The range in $\Delta^{18}O_{SO4-H2O}$ for each of the areas studied is: 2.3 to 17.6‰ in the West Shasta district; 7.0 to 11.6‰ in the Leviathan Mine area; 3.5 to 8.3‰ at the Argo Tunnel; and 11.1 to 13.7‰ at Alpine Gulch (Table I). The pronounced range in $\Delta^{18}O_{SO4-H2O}$ for the West Shasta represents the largest variation yet recorded from a single isotope and mineralogic source of sulfide sulfur. Only two other analyses of sulfide-derived sulfate from acidic discharge and a warm spring in the Yukon and Northwest Territories, respectively, indicate such large values for $\Delta^{18}O_{SO4-H2O}$ (both =18.2‰; 32). Waters with the lowest values of $\Delta^{18}O_{SO4-H2O}$ measured in this study (2.3 to 8.8‰; most: 3.0 to 5.6‰; Table I) are those high-sulfate effluents from underground mine workings sampled at mine portals. In the West Shasta district, these include effluents from the Weil, Balaklala, Friday Lowden, Mammoth, and Old Mine No. 8 (Iron Mountain area) portals, plus Slickrock and Squaw Creeks which are dominated by these effluents. Also included are the Argo Tunnel (Colorado) effluent and one sample of seepage (E, Table I; $\Delta^{18}O_{SO4-H2O}$ = 8.3‰) from underground workings at the Leviathan Mine.

We infer that the environment of oxidation sampled by portal effluent corresponds to that shown schematically as (3) and/or (4) in Figure 1. We presume that below the level of the portal (environment 4; Figure 1) the workings are flooded, whereas workings in environment (3) may be partially flooded because of dams or other obstructions, and the wall rocks are probably largely saturated. Low values of $\Delta^{18}O_{SO4-H2O}$ correspond generally to the range of experimental fractionations (2-3) measured under submersed, sterile conditions analogous to environment (4).

The correspondence implies a similar oxidation environment and/or process in both nature and the laboratory. By analogy, the natural environments of pyrite oxidation characterized by values of $\Delta^{18}O_{SO4-H2O}$ on the order of a few permil are those in which the oxidation proceeds relatively slowly in O_2-depleted waters, probably without significant catalysis by oxidizing thiobacilli (i.e., environment 4; Figure 1). It is unlikely that subsequent isotope exchange has significantly altered the $\delta^{18}O_{SO4}$ of mine waters in environment (4) because: (1) the slow rate of oxygen-isotope exchange (discussed later) between dissolved sulfate and water expected for the temperature and pH of the portal effluents; (2) lack of correlation between $\Delta^{18}O_{SO4-H2O}$ and pH (as low as 2); and (3) the tendency for isotope exchange to increase $\Delta^{18}O_{SO4-H2O}$. Therefore, the correspondence between field and experimental data remains despite the contrast in duration of the experiments (order of days) and storage of acid mine waters (probably weeks to years).

Values of $\Delta^{18}O_{SO4-H2O}$ larger than about 8‰ appear to be associated with environments more likely to be aerated, including neutral waters (Figure 7) and high-sulfate acid waters sampled from creeks and selected mine portals, and ultra-low pH waters discussed later. The Richmond and Lawson portal effluents ($\Delta^{18}O_{SO4-H2O}$ = 13.2 and 12.9‰, respectively), and sites downstream from these portals (Boulder Creek, Spring Creek, Spring Creek Reservoir, and Spring Creek Dam: $\Delta^{18}O_{SO4-H2O}$ = 9.1 to 17.6‰) contrast markedly with other mine effluents in the district. These portal effluents appear to be associated with hydrologic settings characterized by higher flow rates due to the influx of surface waters. At the time of sampling, effluent from the Richmond portal was, at least in part, drainage from the Brick Flat open-pit. Chemical and hydrographic evidence for the influx of surface waters to the Hornet ore body accessed by the Lawson portal was presented by Alpers et al. (42). Effluent from the portal of Tunnel 5 (sample B; Table I) at the Leviathan Mine ($\Delta^{18}O_{SO4-H2O}$ = 10.5-11.6‰) poses a similar contrast to the low $\Delta^{18}O_{SO4-H2O}$ values found for many portal drainages in the West Shasta district. For other sites (creeks, open-pits, etc.), exposure to the air is indicated, such as at the Leviathan Mine, where samples from such environments had a large range values of $\Delta^{18}O_{SO4-H2O}$ (8.3 to 11.6‰). The larger values of $\Delta^{18}O_{SO4-H2O}$ found in wet/dry relative to submersed type experiments (2-3) suggest partially saturated to unsaturated environments (e.g., environments 1-3, 5-6 in Figure 1; 1).

Environment (3) in Figure 1 is difficult to characterize exactly, and, except for the recent work of Alpers and Nordstrom (52), has not been studied in detail from a geochemical standpoint. The mine environment must grade through a vadose zone to an unsaturated zone ((1) in Figure 1), but the state of saturation probably varies greatly over short distances in excavated areas. Dissolved oxygen (DO) in infiltrating waters is normally expected to be consumed within several tens of centimeters below the surface, but mine workings can provide access to O_2 (air) and also to the rapid, localized influx of surface waters. The content of DO could vary considerably, from nil to saturation (cf. 62). Such an environment as (3) in Figure 1 is thought to characterize the Argo Tunnel, and, to some degree, the underground workings of the Leviathan Mine as well.

The wet season in the West Shasta district brings an influx of aerated surface waters into the underground workings by numerous natural pathways (faults, fractures) and man-made openings (manways, draw points, stopes, etc.), and increased flow from portals occurs. The increase in portal flow may have a variable time-lag, from days to weeks (42). Seasonal variation of $\Delta^{18}O_{SO4-H2O}$ shown in Figure 8 for the West Shasta study area indicates that larger values of $\Delta^{18}O_{SO4-H2O}$ during late winter and early spring months follow higher seasonal precipitation and discharge. The reason for this apparent isotope response to the seasonal perturbation of the underground hydrologic environment is probably linked to one or more factors such as increased DO, increased pH, influx of iron- and sulfur-oxidizing bacteria and nutrients, dissolution of previously precipitated minerals, and displacement of stored acid waters. We explore the possible isotope effects in a later section.

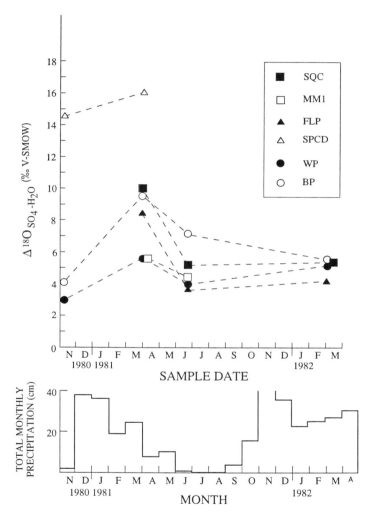

Figure 8. Plot of temporal variation in $\Delta^{18}O_{SO4-H2O}$ at selected sites in the West Shasta district (abbreviations as in Table I), compared with average monthly precipitation (42) at the Shasta Dam.

Figure 9 illustrates the lack of a simple relationship between dissolved SO_4 concentration vs. $\Delta^{18}O_{SO4-H2O}$, but amplifies the previous discussion. Two groups of high-sulfate samples in the West Shasta district, one with low values of $\Delta^{18}O_{SO4-H2O}$, and the other with higher values are evident; each shows dilution. Comparison of Figures 5 and 9 also indicates a lack of correlation between pH and $\Delta^{18}O_{SO4-H2O}$. To the extent that values of $\Delta^{18}O_{SO4-H2O}$ may reflect environments and/or processes of pyrite oxidation, the generalized trends from experiments (2-3) at pH of 2-3 for O_2-depleted submersed conditions, and for wet/dry conditions are shown for comparison. Sources of oxygen and possible mechanistic controls on $\delta^{18}O_{SO4}$ are discussed in following sections.

Oxidation of pyrite in shallow bogs in Alpine Gulch is accompanied by a positive correlation between SO_4 concentration and $\Delta^{18}O_{SO4-H2O}$. In comparison with the workings of the mining areas sampled, oxidation processes at Alpine Gulch may have greater access to free oxygen. However, oxygen-isotope enrichment during bacterial sulfate reduction (such as by *Desulfovibrio desulfuricans*) in O_2-depleted waters could also yield high values of $\delta^{18}O_{SO4}$ (e.g., 63). Whereas sulfur-isotope data for the West Shasta and Leviathan areas indicate complete oxidation, the negative values of $\delta^{34}S_{SO4}$ (where values of 0 - 5‰ were expected) may suggest a more complicated explanation of apparent ^{18}O enrichment of dissolved sulfate at Alpine Gulch.

Sources of Oxygen in Sulfate: the Stoichiometric Isotope-Balance Model.
The significance of the oxygen-isotope composition of sulfate produced by sulfide mineral oxidation has been the subject of some discussion in the last few years (1-2, 15, 31, 64-66). Taylor et al. (1-2) suggested that the proportions of oxygen in sulfate from acid mine drainage contributed from water and DO could be modelled in terms of equations (1) and (2), with water-derived oxygen varying from 12.5% to 100%. Similarly, following Lloyd (27), the isotope composition of sulfate can be expressed (64):

$$\delta^{18}O_{SO4} = X(\delta^{18}O_w + \varepsilon_w) + (1-X)[0.875 (\delta^{18}O_{O2} + \varepsilon_{O2}) + 0.125(\delta^{18}O_w + \varepsilon_w)] \quad (4)$$

where X is the fraction of H_2O-derived oxygen, and ε_w and ε_{O2} are the isotope enrichment factors for incorporation of oxygen from water and atmospheric oxygen, respectivley. Equation 4, which we refer to as the stoichiometric isotope-balance model, describes a "permissable" field of isotope compositions of sulfate formed from the oxidation of sulfide minerals, bounded by the conditions of 12.5% and 100% water-derived oxygen.

The application of equation (4) assumes implicitly, however, that the oxygen-isotope fractionations involved are constant and that isotope re-equilibration does not occur. This also means that participation of the oxygen sources is considered to be by a fully molecular path in which atoms of oxygen are transferred. However, oxygen may also be involved in an electrochemical pathway, involving only electron transfer. Similarly, microbial pathways of oxidation may involve molecular transfer of oxygen and/or electron transfer. The stoichiometric isotope balance model is valid only for sulfate formed from oxidation of sulfide sulfur, as described by equations (1) and (2).

Alternatively, contributions of water-derived and atmospheric oxygen in acid-drainage sulfate can be estimated (following 27) from:

$$\delta^{18}O_{SO4} = X(\delta^{18}O_w + \varepsilon_w) + (1-X)(\delta^{18}O_{O2} + \varepsilon_{O2}) \quad (5)$$

which we refer to as the general isotope-balance model. Figures 10 and 11 each show an additional 0% water-oxygen curve, plotted for an appropriate value of ε_{O2},

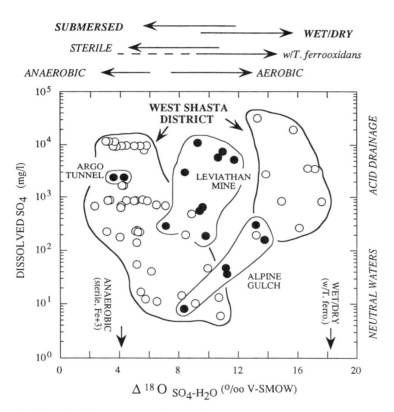

Figure 9. Plot of sulfate concentration vs. $\Delta^{18}O_{SO4\text{-}H2O}$ for neutral (typically <100 mg SO_4^{2-}/l) and acid waters in the West Shasta district (open circles), and Leviathan Mine, Argo Tunnel, and Alpine Gulch areas (filled circles). Tendencies in $\Delta^{18}O_{SO4\text{-}H2O}$ are indicated for experimental environmental conditions (2).

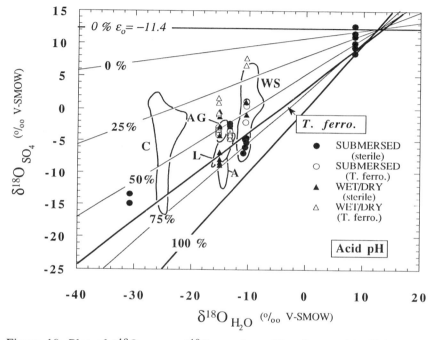

Figure 10. Plot of $\delta^{18}O_{SO4}$ vs. $\delta^{18}O_{H2O}$ for sulfate from acid pH in pyrite oxidation experiments (2-3) compared with isopleths for percent water-derived oxygen (from equation (4) in text) and "*T. ferro.* Trend" (34). Horizontal isopleth "0%, ε_o= -11.4" represents sulfate formed from atmospheric O_2. Fields and sources of data for sulfate from acid mine drainage are: C, Canada, 32; L, Leviathan Mine; A, Argo Tunnel; AG, Alpine Gulch; WS, West Shasta district; 50). Isopleth for 100% represents a "Zero Fractionation Trend"; see text for discussion. Circles: sterile experiments; triangles: experiments with *T. ferrooxidans* or *T. ferrooxidans* + *T. thiooxidans*.

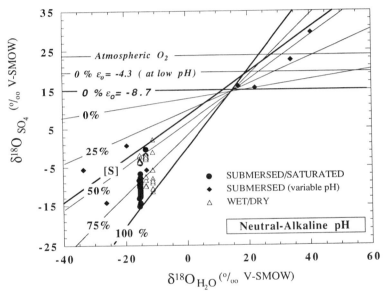

Figure 11. Plot of $\delta^{18}O_{SO4}$ vs. $\delta^{18}O_{H2O}$ for sulfate oxidation of pyrite at neutral to alkaline pH compared with: isopleths for water-derived oxygen (from equation (4) in text) and a sulfite trend, [S] (35). Horizontal isopleths represent sulfate formed from atmospheric O_2 (67) with $\varepsilon_o=0$, $\varepsilon_o=-4.3$ (2), and $\varepsilon_o=-8.7$ (27). Filled circles (3) and diamonds (66): submersed experiments; open triangles (3): wet/dry experiments.

representing the contribution of water-derived oxygen based on equation (5), and expanding slightly the "permissable" field for sulfate noted above.

The general isotope-balance model provides a slightly higher estimate for water-derived oxygen than does the stoichiometric isotope-balance model. From equation (5), the contributions of water-derived oxygen are: West Shasta district, 24-90%; Leviathan Mine, 58-75%; Argo Tunnel, 70-87%; and Alpine Gulch, 50-61%. A visual inspection of the plotted data in Figures 5 and 10 indicates that sulfate from some samples in the West Shasta district and from Argo Tunnel have the highest apparent contributions of water-oxygen, Leviathan Mine have intermediate contributions, and sulfate in bog water from Alpine Gulch have the least. The distribution of the data in Figure 10 (and also for western Bohemia, 31; not plotted) compare closely with available experimental data for pyrite oxidation under acid conditions. Together, the field and experimental data are consistent with dominant oxidation (greater than 50%) by ferric iron (equation (2)). The very large, positive values of $\delta^{18}O_{SO4}$ (i.e., high $\Delta^{18}O_{SO4-H2O}$) are discussed separately in a later section.

The oxygen-isotope fractionations used in equations 4 and 5 are essentially kinetic. The extent to which these fractionations remain relatively constant, or change with environmental factors, is unknown, and we acknowledge that this as a principal uncertainty. For the present, we use those fractionations derived from our earlier experiments (2) and from Lloyd (27). Because of the potential descriptive value of the stoichiometric or general-isotope balance models for sulfate in acid mine drainage, we evaluate the enrichment factors for different pH environments of oxidation.

Acid Environments. For the fractionation of DO ($\delta^{18}O_{O2}$=+23.8‰; 67) during oxidation in sterile environments, ε_{O2}= -4.3‰ and, in the presence of *T. ferrooxidans*, ε_{O2}= -11.4‰ (2). The value of ε_{O2}= -11.4‰ (which is probably generally applicable in nature) appears to be consistent with experiments by Qureshi (3; see Figure 10) and Gould et al. (34) over a large range in $\delta^{18}O_{H2O}$. A best-fit equation to the data in (34) for experiments in the presence of *T. ferrooxidans* (labelled "T. ferro." on Figure 10), passes through the intersection points of curves for 0% and 100% water-oxygen from the stoichiometric isotope-balance model.

The agreement between our value of ε_{O2}= -11.4‰ and the results of Gould et al. (34) is striking. Especially so, because the ε_{O2} measured by Taylor et al. (2) reflects the effect of isotope fractionation during both microbial respiration and incorporation of oxygen (DO) into sulfate (which is possibly closer to -4.3‰). Comparison with data for mixed microbe experiments (3) indicates no significant species-dependency of ε_{O2}.

A value of ε_W=4.0‰ was initially chosen (1) based on results for submersed experiments and available data for nearly all of the low $\Delta^{18}O_{SO4-H2O}$ acid drainage samples. However, a value of ε_W = 4.0‰ causes a number of data for submersed, sterile experiments (see Figure 10) to fall below the 100% water-oxygen curve, i.e., outside of the permissable field. The average $\Delta^{18}O_{SO4-H2O}$ for all available sterile, submersed experiments (2-3) pH = 2 to 3 is 5.5‰. For the sake of compatibility with the range of all experimental data, we set ε_W = 0‰.

A test of the stoichiometric isotope-balance for acid conditions (pH<3) is that sulfate formed in air-saturated water with $\delta^{18}O_{H2O}$ = +12.0‰ (which is not a naturally attained composition for surface water) should also have $\delta^{18}O_{SO4}$= +12.0‰ (cf., 64-66). Available experimental data for acid conditions (pH<3) are consistent with this prediction.

Neutral-to-Alkaline Environments. For neutral-to-alkaline conditions (Figure 11), a value of ε_{O2}= -8.7‰ (27) provides a better fit to available experimental data than does ε_{O2}= -11.4‰ for acid conditions (2). The equation for $\delta^{18}O_{SO4}$ derived from the oxidation of sulfite over a large range in $\delta^{18}O_{H2O}$ at pH = 8.5 (labelled "sulfite" in Figure 11; 35) passes close to the intersection of the 0% and 100% water oxygen curves. The uncertainty on this equation would encompass both the intersection and the 50% isopleth.

Available experimental data plotted in Figure 11 appear to be consistent with the intersection of the 100% and 0% boundaries (i.e., with our choice of ε_{O2} and ε_W) in the stoichiometric isotope-balance for neutral-to-alkaline environments ($\delta^{18}O_{H2O}=\delta^{18}O_{SO4}$ =15 ‰; cf. 66). However, we recognize that the constraints are not especially stringent. For example, if we chose ε_{O2}= -4.3 ‰, the "sulfite" trend in Figure 11 would pass through the intersection of recalculated 100% and 0% curves, as may be surmised from inspection. In this case, only a single point, a 547-day experiment of van Everdingen and Krouse (66), for which there was an unspecified evaporation effect on $\delta^{18}O_{H2O}$, would fall outside of the permissable fields. Nevertheless, available experimental and field data on the oxidation of pyrite (± other sulfides) are consistent with the formulation we use for the stoichiometric isotope balance.

Slopes to best-fit equations of the type

$$\delta^{18}O_{SO4} = m\, \delta^{18}O_{H2O} + b \tag{6}$$

for the experimental data of Gould et al. (34; $m_{ave.}$= 0.68, $b_{ave.}$= 3.65) on oxidation of pyrite under acid conditions (pH = 2), those of Krouse et al. (35; m=0.56 ±0.02, b=8.3±4.8) for oxidation of aqueous sulfite at pH = 8.5, and those of Schwarcz and Cortecci (33; m=0.6), are very similar. Only the intercept, b, appears to differ between experiments with and without microbes and under different pH. The close agreement between the three different sets of experiments and the stoichiometric isotope-balance models for acid and neutral-to-alkaline conditions suggests that our choices for ε_{O2} and ε_W are generally correct, and that a common step (i.e., sulfite-water exchange of oxygen isotopes) occurs in both sterile and microbially mediated oxidation processes.

Stoichiometric Contributions and Environments of Oxidation. Taken at face value, Figure 10 suggests that, in terms of oxygen, equation (2) contributes on the order of 50-80% to the overall oxidation of pyrite under submersed, sterile conditions, and roughly 20-65% of the oxidation process under wet/dry (acid) conditions. Overall, the impact of *T. ferrooxidans* or mixtures of *T. ferrooxidans* and *T. thiooxidans* on $\delta^{18}O_{SO4}$ is intermediate, and discernable only for submersed conditions. Under these conditions, $\delta^{18}O_{SO4}$ is larger during microbially mediated oxidation than during abiologic oxidation, and data for *T. ferrooxidans*-mediated oxidation plot on or above the *T. ferrooxidans* trend in Figure 11.

Fields for available data on dissolved sulfate in acid mine drainage are shown on Figure 10 for comparison with the experimental data discussed above. Acidic drainage that can reasonably be expected to have concentrations of oxygen near atmospheric saturation (i.e., all but waters from most flooded mines) should plot on or above the "*T. ferro.*" trend in Figure 10. This trend can be viewed as a "biological lower limit" in aerated waters where *T. ferrooxidans* plays an active role in oxidation (i.e., ca. pH 2-4). Data for acid waters plotting well above this trend may reflect other processes, as discussed later.

Under neutral-to-alkaline pH, from which the lowest values of $\Delta^{18}O_{SO4-H2O}$ (as low as 0) result, sulfate formed under submersed and wet/dry experiments cannot be distinguished by oxygen-isotope data. The apparent increase in % of equation (2) under neutral-to-alkaline conditions relative to similar experiments under acid conditions corresponds roughly with the marked increase in the rate of oxidation of Fe^{2+} by DO at higher pH, surpassing that of reaction (1) by orders of magnitude (6, 9, 12). Reaction (2) is then favored.

Isotope Signatures and Oxidation Processes. Although larger values of $\Delta^{18}O_{SO4-H2O}$ appear to correspond with oxidation environments expected to have higher DO, and with results from wet/dry experiments (2-3), Qureshi's finding (3), that $\Delta^{18}O_{SO4-H2O}$ decreased as pO$_2$ increased in partially saturated experiments, contradicts a simple "availability of DO" explanation. How well, then, do these stoichiometric percentages approximate the actual reactions? This question cannot be answered in its entirety, because isotope data and knowledge on the intermediate sulfur oxyanions, on all important steps, and on the extent of possible isotope exchange are lacking. Fossing and Jorgensen (68) used ^{35}S to follow transfer of sulfur between small "reservoirs" of intdermediate species, simultaneous oxidation and reduction processes, and disproportionation of thiosulfate in estuarine sediments under generally oxidizing conditions. Sulfur-reducing pathways may also permit ^{18}O exchange (63, 69). More such studies are needed using ^{18}O as a label. Based on available data, we briefly discuss below some of the factors which appear to be important.

Abiologic Oxidation. Experimental studies provide abundant evidence that surface adsorption/desorption processes are important controls on the overall rate of pyritic sulfur oxidation under acid conditions (e.g., 10, 12, 18, 70). The lack of sulfur oxyanions in solution, except when vigorously stirred, indicates rapid, complete oxidation occuring at the pyrite surface. The sulfur-isotope data are consistent with this view, because isotope differences between source sulfur and end-product sulfate occur only when a significant quantity of sulfur is partitioned among several reservoirs of intermediate species, as happens at higher pH (e.g., 3).

The effect of sulfur oxyanions on the final oxygen-isotope composition of sulfate is difficult to ascertain. The relative abundances of sulfite, thiosulfate and tetrathionate increase with pH (3, 12-13). Qureshi (3) found that in carbonate-buffered experiments under controlled pO$_2$, that $\Delta^{18}O_{SO4-H2O}$ decreased as the ratio of sulfate to other aqueous sulfur species increased with pO$_2$. In "incompletely oxidized" solutions (i.e., those with detectable intermediate sulfur species) the range and magnitude of $\Delta^{18}O_{SO4-H2O}$ increased with decreasing pH. This could imply that $\delta^{18}O_{SnOy}$ > $\delta^{18}O_{SO4}$. Such a relationship is (possibly because of the complexity of the molecule involved) opposite to that found following isotope equilibrium tendencies, for which $^{18}O/^{16}O$ ratios would increase with the oxidation state of sulfur, and suggests partial oxygen-isotope equilibration between water and the intermediates. An increase in $\Delta^{18}O_{SO4-H2O}$ due to oxygen-isotope exchange between <u>sulfate</u> and water in these experiments (2-3) is unlikely for their temperature (ca. 30 °C), pH of 2 to 3 and short duration (several days to two weeks), based on present knowledge of isotope exchange rates (discussed later).

Thiosulfate, plus lesser amounts of other polythionates, comprise <10% of the total sulfur (as sulfate) under acid conditions (pH<4; 3, 9-10). The Wachenroder reaction is thought to govern the homogeneous oxidation of these species such that (10):

$$S_nO_6^{2-} + S_2O_3^{2-} \Leftrightarrow S_{n+1}O_6^{2-} + SO_3^{2-} \tag{7}$$

goes to the left at pH>7 and to the right at pH<7. Qureshi (3) found that thiosulfate was the principal sulfur oxyanion at low pH, and this should decompose (71-72) according to:

$$8\ S_2O_3^{2-} + H^+ \rightarrow S_8 + 7\ SO_3^{2-} + HSO_3^- \tag{8}$$

The rate of isotope exchange between sulfite and water has been well documented using ^{18}O tracer techniques (73). In 0.1 M HCl (pH = 1.1), the average proportion of water oxygen in the resultant sulfate was 58% (74), which is the same as found in previously cited experiments using natural abundance levels of ^{18}O. Halpern and Taube (75) reported that oxidation of sulfite by O_2 yielded sulfate with 56% of its oxygen from derived from O_2. The availability of DO at this step can potentially have some isotope effect on $\delta^{18}O_{SO4}$. Although there are evidently a number of possible side reactions, once formed, sulfite exchanges oxygen isotopes with water relatively rapidly at low-to-neutral pH, exerting a dominant control on $\delta^{18}O_{SO4}$. This control is evidently not affected by the presence of sulfite-oxidizing enzyme (35).

Since sulfite-water isotope exchange should affect $\delta^{18}O_{SO4}$ in neutral-to-acidic environments, the isotope composition of water at the pyrite/solution interface is a main factor. The variability of $\delta^{18}O_{SO4}$ shown in Figures 10 and 11, may also be additionally influenced by factors governing the first steps of pyrite oxidation which take place on the surface of the pyrite, prior to desorption of the sulfur oxyanion(s). These factors may be both environmental (e.g., evaporative enrichment in ^{18}O of H_2O in the pyrite/solution boundary layer; isotope exchange between SO_4 and H_2O; change in reaction mechanism), or biologic (e.g, isotopically variable enzymatic pathways of oxidation).

Other factors which are not well understood require investigation, such as the possible effect of sulfide surface area on $\Delta^{18}O_{SO4-H2O}$. In carbonate-buffered experiments, $\Delta^{18}O_{SO4-H2O}$ can apparently also vary with grain size: larger values associated with smaller grain sizes, irrespective of reaction rates (3). This may indicate that availability of oxidation sites plays a part in competing reaction steps having unequal isotope effects.

Biologically Mediated Oxidation. *T. ferrooxidans*-mediated oxidation of pyrite can produce ^{18}O-enriched sulfate relative to abiologic oxidation. Despite some overlap in range, average $\Delta^{18}O_{SO4-H2O}$ values for submersed, sterile experiments at pH<3 under low-to-negligible pO_2 are lower than for similar experiments with *T. ferrooxidans*; the same relationship is seen for wet/dry environments (Table II). This is further demonstrated by experiments of Gould et al. (34). Absolute differences between the different sets of data (2-3) may be due to factors such as mechanical stirring and bubble aeration which could aid the mass transport of oxygen into the pyrite-solution boundary layer, or to biologic factors not yet identified.

Table II. Average experimental oxygen isotope fractionations (‰) between dissolved sulfate and water ($\Delta^{18}O_{SO4-H2O}$) during pyrite oxidation at pH < 3

Experiment	Sterile	with *T. ferro.*	Reference
Submersed	4.9	8.2	(2)
	6.8	10.5	(3)
Wet/Dry	7.7	10.5	(2)
	10.6	11.7	(3)

Because a variable increase in $\Delta^{18}O_{SO4-H2O}$ was observed in microbe-mediated experiments of both submersed and wet/dry types, a component of biologically induced ^{18}O enrichment is probably indicated. Taylor et al. (2) suggested that some DO was incorporated into sulfate. Krouse et al. (35) proved that this is so from submersed experiments at pH 2.5 under an atmosphere of highly ^{18}O-enriched oxygen, in which

T. ferrooxidans-mediated oxidation produced higher $\delta^{18}O_{SO4}$ than in abiologic experiments. Suzuki (25) documented the transfer of some ^{18}O-labelled DO to thiosulfate during enzymatic oxidation of elemental sulfur to thiosulfate by *T. thiooxidans*. The overall increase in $\Delta^{18}O_{SO4-H2O}$ probably also reflects the isotope effects of the sulfite-to-sulfate step, which, as Krouse et al. (35) showed, can include oxidation by DO. They also pointed out that this transfer cannot be observed in the experiments using natural abundances of oxygen because of an insufficiently large contrast of $\delta^{18}O_{O2}$ and $\delta^{18}O_{H2O}$.

One fourth of the sulfate oxygen was shown by Santer (29) and Peck and Stulberg (30) using ^{18}O as a label to have been derived from the energy substrate during enzymatic oxidation of thiosulfate. Sulfate can also serve as such a substrate, and is involved in the stabilization of an aqueous Fe^{2+} complex (76). Utilization of suspected high $\delta^{18}O$ sulfate minerals precipitated in nearby very low pH environments (discussed later), could perhaps provide a local high-^{18}O source of oxygen for such enzymatically facilitated oxygen-isotope exchange between sulfate and water. Probably the more common source of relatively ^{18}O-enriched oxygen is DO adsorbed by Thiobacilli ($\delta^{18}O \approx 11.3‰$; 2). Additional tracer studies are needed, especially to compare the different potential enzymatic oxidation pathways.

High-$\delta^{18}O$ Sulfates in the West Shasta District. The anomalously large excursions in $\delta^{18}O_{SO4}$ (up to 10 ‰ above the *T. ferrooxidans* trend in Figure 7) for sulfate in drainage from the Richmond and Lawson portals (and, to a lesser degree from the Friday Lowden portal) require explanation by mechanisms other than those previously described. We propose that dissolution of ^{18}O-rich secondary sulfate minerals, formed as a consequence of processes discussed below, may explain the high $\delta^{18}O_{SO4}$ in acid effluents of some mines. This proposition, and the efficacy of the mechanisms discussed below, can readily be tested in future studies by the isotope analysis of drip water (O, H), atmospheric oxygen (O_2), and precipitated sulfate minerals (O in SO_4) in mine workings.

Hypothesis 1: Thin-film Effects During Wet/Dry Conditions. In this hypothesis, higher values of $\delta^{18}O_{SO4}$ may occur as a result of an increase in $\delta^{18}O_{(H2O+O2)}$ in the pyrite-solution boundary layer. Mechanisms may include evaporation, microbial respiration, and absorption of atmospheric oxygen. The volume of liquid affected is envisioned to be primarily that which coats the exposed sulfide surfaces in the mine and held in near-surface fractures in the exposed, or excavated workings. Therefore, we refer to these possible processes as "thin-film" effects because similar processes would not be effective in larger bodies of water.

The isotope effect of evaporation on $\delta^{18}O_{SO4}$ depends on the extent of evaporation prior to pyrite oxidation. For example, Pierre (77) measured an increase of ca. 2‰ in $\delta^{18}O_{H2O}$ in free water up to the point at which gypsum precipitated, and a total ^{18}O enrichment of about 3‰ in water by the time halite saturation was reached. Larger ^{18}O enrichments of H_2O (e.g., 12-14‰) can result from evaporation to near dryness in sand-packed columns (e.g., 78), and much larger enrichments (up to 75‰) are possible at high evaporation rates (R. Krouse, written commun., 1993). Both the occurrence of abundant Fe-sulfates in the Richmond Mine and the ultra-low pH drip waters are compatible with a marked degree of evaporation (42). In spite of high temperatures (20-50 °C), humidity levels as high as 76% in stopes in the Richmond Mine (52; Nordstrom and Alpers, written commun., 1990) would tend to reduce the magnitude of possible ^{18}O enrichments (78), however. Formation of sulfate from water enriched in ^{18}O by evaporation could account for high values of $\delta^{18}O_{SO4}$. Rapid isotope exchange between a sulfite intermediate and water could also be involved as a step in this process.

Incorporation of atmospheric oxygen might be promoted by its reduction at very low pH (see equation 3), increasing the $\delta^{18}O_{H_2O}$ of the "thin-film" solution. The incorporation of atmospheric oxygen in a small volume of water in this way would promote the formation of high $\delta^{18}O_{SO_4}$. By simple mass balance, addition of 26% of atmospheric O_2 to the water could account for the observed increase in $\delta^{18}O$ of sulfate formed via equation (2). Oxygen contents (ca. 19%; C. Alpers, oral commun., 1992) below atmospheric concentration in the underground workings of the Richmond mine attest to the consumption of oxygen by abiologic and/or microbially mediated oxidation processes. Isotope exchange between the sulfate salts formed on mined surfaces and atmospheric oxygen as the salts dehydrate, dissolve and reprecipitate, or recrystallize under changing humidity conditions might also result in an increase in $\delta^{18}O$ of these minerals.

Microbial respiration could enrich residual DO in a thin film of water in ^{18}O (2). However, the very high acidity of drip water found in the underground workings (pH< 0: 52-53) is far removed from the pH range of 2.3 to 2.5 for optimum metabolic activity (79). Similarly, isotope fractionation during consumption of dissolved oxygen by either abiologic or microbial oxidation during infiltration will enrich the residual oxygen in ^{18}O, based on fractionation factors (2).

Hypothesis 2. Oxygen-Isotope Exchange Between SO_4 and H_2O.
The rate of oxygen-isotope exchange between SO_4 and H_2O at low temperatures (<100 °C), is poorly quantified at pH<3. Extrapolation of data from Chiba and Sakai (80; cf. 27, 72) suggests near 50% exchange equilibrium might be possible within about 4 months at pH = 0. Fifty percent exchange would result in $\delta^{18}O_{SO_4}$ values of ca. +3‰ for mine water in the West Shasta district, and about 65% isotope exchange is required to explain the highest $\delta^{18}O_{SO_4}$ values. However, for the sampled portal effluents, which have a pH of 1 to 3 (Table I), attainment of 50% isotope exchange equilibrium between sulfate and water would require more than 100 years (27, 80-81). Therefore, any significant isotope equilibration between dissolved sulfate and water must occur in the "thin-film" of drip water on, or in cracks just beneath, excavated surfaces in the mines. Conditions in the Richmond mine (ca. 50 °C, pH<0 in drip water; 53) would promote hypothesis 2 in addition to hypothesis 1. We emphasize, however, that extrapolation of isotope exchange rates to low pH environments is risky, and isotope exchange under these acid conditions needs to be investigated.

Hypothesis 3. Storage and Release of High $\delta^{18}O_{SO_4}$ in Sulfate Minerals. Sulfate minerals with high values of $\delta^{18}O_{SO_4}$ are hypothesized to precipitate in very acidic (pH<1) waters, and to contribute aqueous sulfate with high values of $\delta^{18}O$ upon dissolution. High values of $\delta^{18}O_{SO_4}$ are evidently not characteristic of all low-pH, high-sulfate portal effluents analyzed (e.g., Weil Portal; Table I), and therefore the flushing of sequestered high-$\delta^{18}O_{SO_4}$ waters is a less likely, alternative explanation for the high-$\delta^{18}O_{SO_4}$ portal effluents.

Influx of dilute, higher pH waters during wet periods tends to dissolve low pH-stable minerals such as jarosite, melanterite, and copiapite (e.g., 6, 52). Correspondence between seasonal influx of fresh water and variation in Zn/Cu ratios in effluent from the Richmond and Lawson portals suggests control by solution and precipitation of salts such as melanterite ([Fe^{II},Zn,Cu]$SO_4 \cdot 7H_2O$; 42, 53). During leaching experiments of pyrite-bearing shales, Sullivan et al. (82) found that the activities of sulfate were controlled by a melanterite below pH=3, and by Fe(OH)SO_4 between pH 3 to 6. At very low pH, they suggested sulfate activity may be controlled by jarosite ([K,Na]Fe$_3$(SO_4)$_2$(OH)$_6$). Field studies (83-84) on storm runoff from coal mine areas also indicate control on the chemical composition of acid drainage by dissolution of sulfate salts.

The ease with which low-pH-stable sulfate salts may grow even in "normal" environments is illustrated by Wiese et al. (85) who report the growth of melanterite (and other related species) on or near the polished surfaces of sulfide mineral specimens stored under the ambient concentrations of oxygen and humidity of a museum. The suite of minerals present was thought to change with time and relative humidity. These minerals may explain part of the oxygen-isotope enrichment noted in wet/dry experiments, and when pyrite surfaces are not scrupulously cleaned immediately before experiments (e.g, expt. 1 in 2).

The effect of dissolution of secondary sulfate minerals on $\Delta^{18}O_{SO_4-H_2O}$ depends on the their oxygen-isotope composition. Experimental studies (86-87) have demonstrated that, at isotope equilibrium, sulfate minerals concentrate ^{18}O relative to H_2O at all temperatures studied. Extrapolation of data for jarosite-water exchange equilibrium (87) to 30 °C predicts $\delta^{18}O_{SO_4}$ of jarosite (as a proxy for exisiting Fe-sulfate minerals) in the Richmond Tunnel of 31.5‰, which is close to dissolved sulfate-water equilibrium (27, 72). For comparison, sulfate in red and yellow ochre crust deposits analyzed by Shakur (in 32) is, in one case, enriched in ^{18}O ca. 2 to 4‰ relative to dissolved sulfate in the associated acid waters (pH 2.4 to 3.7), and, in another case, depleted in ^{18}O by 1.4‰. Thus, production of high $\delta^{18}O$ sulfate minerals in the Richmond mine requires mechanism 1 and/or 2, above, to produce high $\delta^{18}O_{SO_4}$ values prior to Fe-sulfate mineral precipitation.

Conclusions

The oxygen-isotope composition of dissolved sulfate formed in acidic water from the oxidation of pyrite in nature will vary directly with that of the water in which occurs. This is largely due to the isotope exchange between sulfite and water. For oxidation occurring largely under low-pO_2, submersed conditions, isotope data for sulfate and water will typically plot near or between two general trends on a $\delta^{18}O_{SO_4}$ vs. $\delta^{18}O_{H_2O}$ diagram: a "Zero Fractionation" trend for sterile oxidation, and a "*T. ferrooxidans*" trend, which shows a tendency for ^{18}O-enrichment of sulfate (relative to sterile conditions) during microbially mediated oxidation. Sulfates plotting in a field defined by these trends contain oxygen derived largely (50-80%) from water. Whereas such contributions are consistent with experimentally determined kinetics of pyrite oxidation under sterile conditions, isotope exchange between sulfite and water presently adds a degree of uncertainty to a detailed interpretation by possibly obscuring the isotope record.

Large values of $\Delta^{18}O_{SO_4-H_2O}$ in acid drainage in the West Shasta district, i.e., those plotting well-above the "*T. ferrooxidans*" trend, correspond with the results of experiments under wet/dry, partially saturated conditions. These can likely be explained as due to the dissolution of ^{18}O-rich ferrous sulfate minerals like melanterite, which can form in very acid waters, or even on sulfide surfaces exposed to humid air. The postulated ^{18}O-rich nature of these sulfates can be explained by partial isotope equilibration between sulfate and water at very low pH and relatively high temperatures (e.g., 40-50 °C). Evaporation of drip water will cause enrichment in ^{18}O (and reduce the pH), increasing $\delta^{18}O_{SO_4}$, perhaps involving isotope exchange between intermediary sulfite and water. In addition, isotope exchange between Fe-salts and moist air during solution/redeposition or recrystallization are other potential mechanisms of ^{18}O enrichment. Detection of very high $\Delta^{18}O_{SO_4-H_2O}$ values in acid drainage may, where the oxidation of metal sulfides is the overwhelming source of sulfate, be an indicator of very low pH conditions along the flow path.

Acknowledgments

This research was initially supported by National Science Foundation Grant EAR 79-11144 while BET and MCW were at the Univ. of California (Davis). The results of field studies reported in this paper were derived in part from the M.Sc. thesis of MCW at UCD. D. K. Nordstrom provided samples from the Leviathan Mine area and the Argo Tunnel, and D. Murphy supplied samples from Alpine Gulch. D. Watanabe, and R. Lancaster and K. Nguyen assisted in collection of literature and preparation of final figures, respectively. C. Alpers, D. Boyle, R. Krouse and two anonymous reviewers are thanked for their thoughtful reviews.

Literature Cited

1. Taylor, B. E.; Wheeler, M. C.; Nordstrom, D. K. Nature 1984, 308, 538-541.
2. Taylor, B. E.; Wheeler, M. C.; Nordstrom, D. K. Geochim. Cosmochim. Acta 1984, 48, 2669-2678.
3. Qureshi, R. M. Ph.D. Thesis, Univ. of Waterloo, Waterloo, Ont., 1986.
4. Garrels, R. M.; Thompson, M.E. Amer. J. Sci. 1960, 258, 57-63.
5. Smith, E. E.; Svanks, K.; Shumate, K. S. Sulfide to sulfate reaction studies, 2nd. Symp. Coal Mine Drainage Res., Pittsburgh, Pa., 12-34.
6. Nordstrom, D. K. 1982, In Acid Sulfate Weathering: Pedogeochemistry and Relationship to Manipulation of Soil Materials; Kittrick, J. A.; Fanning, D. S.; Hossner, L. R., Eds.; Soil Sci. Soc., America Press: Madison, WI.; pp. 37-56.
7. Singer, P. C.; Stumm, W. Science 1970, 167, 1121-1123.
8. Ehrlich, H. L. Geomicrobiology, Marcel Dekkar, Inc., New York, 1981, 393 pp.
9. Moses, C.O.; Nordstrom, D.K.; Herman, J.S.; Mills, A.L. Geochim. Cosmochim. Acta 1987, 51, 1561-1571.
10. McKibben, M. A.; Barnes, H. L. Geochim. Cosmochim. Acta 1986, 50, 1509-1520.
11. Nicholson, R. V.; Gillham, R. W.; Reardon, E. J. Geochim. Cosmochim. Acta 1990, 54, 395-402.
12. Moses, C. O.; Herman, J. S. Geochim. Cosmochim. Acta 1991, 55, 471-482.
13. Aller, R. C.; Rude, P. D. Geochim. Cosmochim. Acta 1988, 52, 751-765.
14. Goldhaber, M. B. Amer. J. Sci. 1983, 283, 193-217.
15. Toran, L.; Harris, R. F. Geochim. Cosmochim. Acta 1989, 53, 2341-2348.
16. Biegler, T.; Swift, D. A. Electrochim. Acta 1979, 24, 415-420.
17. Hamilton, I. C.; Woods, R. J. Electroanal. Chem. 1981, 118, 327-343.
18. Hiskey, J. B.; Schlitt, W. J. In Interfacing Technologies in Solution Mining; Schlitt, W. J.; Hiskey, J. B., Eds.; Proc. Second SME-SPE Int. Soln. Mining Symp.; Denver, Colo.; 1982, 55-74.
19. Kelly, D. P. Phil. Trans. R. Soc. Lond. 1982, B 298, 499-528.
20. Ingledew, W. J. Biotechnol. Bioeng. Symp. 1986, 16, 23-33.
21. Lacey, E.T.; Lawson, F. Biotech. Bioeng. 1970, 12, 29-50.
22. Suzuki, I.; Lizama, H. M.; Tackaberry, P. D., Appl Environ. Microbiol. 1989, 55, 1117-1121.
23. Wakao, N.; Mishina, M.; Sakurai, Y.; Shiota, H., J. Gen. Appl. Microbiol. 1984, 30, 63-77.
24. Vestal, J. R.; Lundgren, D. G., Can. J. Biochem. 1971, 49, 1125-1130.
25. Suzuki, I., Biochim. Biophys. Acta 1965, 104, 359-371.
26. Lloyd, R. M., Science 1967, 156, 1228-1231.
27. Lloyd, R. M., Jour. Geophys. Res. 1968, 73, 6099-6110.
28. Bailey, L. K.; Peters, E., Can. Metal. Quart. 1976, 15, 333-344.
29. Santer, M., Biochem. Biophys. Res. Commun. 1959, 1, 9-12.
30. Peck, H. D.; Stulberg, M. P., J. Biol. Chem. 1962, 237, 1648-1652.
31. Smejkal, V., Isotope Hydrology 1978; IAEA, Vienna, 1978; Vol. 1, 83-98.

32. van Everdingen, R. O.; Shakur, M. A.; Michel, F. A. Can. J. Earth Sci. 1985, 22, 1689-1695.
33. Schwarcz, H. P.; and Cortecci, G. Chem. Geol. 1974, 13, 285-294.
34. Gould, W. D.; McCready, R. G. L.; Rajan, S.; Krouse, H. R. In Biohydrometallurgy 1989; Salley, J.; McCready, R. G. L.; Wichlacz, P., Eds.; CANMET SP89-101990; Ottawa, 1990; pp. 81-92.
35. Krouse, H. R.; Gould, W. D.; McCready, R. G. L.; Rajan, S. Earth Planet. Sci. Lett., 1991, 107, 90-94.
36. Kinkel, A. R., Jr.; Hall, W. E.; Albers, J. P., U.S. Geol. Survey Prof. Paper 285; 1956; 156 pp.
37. Taylor, B. E..; South, B. C. Econ. Geol. 1985, 80, 2149-2163.
38. Reed, M. H. Econ. Geol. 1984, 79, 1299-1318.
39. Casey, W. H.; Taylor, B. E., Econ. Geol. 1982, 77, 38-49.
40. South, B. C.; Taylor, B. E. Econ. Geol. 1985, 80, 2177-2195.
41. Nordstrom, D. K. Ph.D. Thesis, Stanford Univ., Palo Alto, Calif., 1977.
42. Alpers, C. N.; Nordstrom, D. K.; Burchard, J. M., U. S. Geol. Survey Water-Resour. Invest. Rept. 91-4160, 1992, 173 pp.
43. San Francisco Chronicle; July 27, 1983; p. 4.
44. Evans, J. R., In Mines and Mineral Resources of Alpine County, California; Clark, W. B., Ed.; Calif. Div. Mines and Geol. County Rept. 8; Sacramento, Calif., 1977, 35-39.
45. Herbst, C. M.; Sciacca, J. E. Calif. State Water Res. Control Board, 1982; 22 pp.
46. Hammermeister, D. P.; Walmsley, S. J., U. S. Geol. Survey Open-File Rept. 85-160; 1985, 160 pp.
47. Wentz, D. A., Review Report for Water and Related Land Resources Management Study, Metropolitan Denver and South Platte River and Tributaries, Colorado, Wyoming, and Nebraska; U.S. Army Corps of Eng.: Omaha, NE., Vol. 5, App.F, 61 pp.
48. Simms, P. K.; Drake, A. A., Jr.; Tooker, E. W., U. S. Geol. Survey Prof. Paper 359, 1963, 231 pp.
49. Kennedy, V. C.; Jenne, E. A.; Burchard, J. M., U.S. Geol. Survey Open-File Rept. 76-126, 1976, 12 pp.
50. Wheeler, M. C. M.Sc. Thesis, Univ. of California, Davis, Calif., 1984.
51. O'Neil, J. R., In Stable Isotopes in High Temperature Geological Processes, Reviews in Min., Vol. 16; Valley, J. W., Taylor, H.P., Jr., O'Neil, J. R., Eds.; Mineral. Soc. America, Chelsea, MI., 1986, pp. 561-570.
52. Alpers, C. N.; Nordstrom, D. K., In Proc. Sec. Int. Conf. on the Abatement of Acidic Drainage, MEND (Mine Environ. Neutral Drainage); Montreal, Canada; 1991, Vol. 2, 321-342.
53. Alpers, C. N.; Nordstrom, D. K.; Thompson, J. M., In this volume.
54. Field, C. W., Econ. Geol. 1966, 61, 1428-1435.
55. Ferris, F. G.; Tazaki, K.; Fyfe, W. S., Chem. Geol. 1989, 74, 321-330.
56. Brady, K. S.; Bigham, J. M.; Jaynes, W. F.; Logan, T. J., Clays and Clay Minerals. 1986, 34, 266-274.
57. Lazaroff, N.; Sigal, W.; Wasserman, A., Appl. Environ. Microbiol. 1982, 43, 924-938.
58. Alpers, C. N.; Nordstrom, D. K.; Ball, J. W., Sci. Geol. Bull. 1989, 42, 281-298.
59. Murad, E.; Schwertmann, U.; Bigham, J. M.; and Carlson, L., In this volume.
60. van Stempvoort, D. R.; Reardon, E. J.; and Fritz, P., Geochim. Cosmochim. Acta 1990, 54, 2817-2826.
61. Thode, H. G.; Monster, J., Am. Assoc. Pet. Geol. 1965, Mem. No. 4, 367-377.
62. Bennett, J. W.; Harries, J. R.; Ritchie, A. I., U. S. Bur. Mines Info. Circ. 9183, 1988; 104-108.

63. Fritz, P.; Basharmal, G. M.; Drimmie, R. J.; Ibsen, J.; Qureshi, R. M., Chem. Geol. 1989, 79, 99-105.
64. van Everdingen, R. O.; Krouse, H. R., Nature, 1985, 315, 395-396.
65. van Everdingen, R. O.; Krouse, H. R., Extended Abstracts, Fifth Int. Symp. Water-Rock Interaction; Int. Assoc. of Geochem. Cosmochem.: Reykjavik, Iceland, 1986; 663-666.
66. van Everdingen, R. O.; Krouse, H. R., U.S. Bureau of Mines Info. Circ. 9183, 1988, 147-156.
67. Horibe, Y.; Shigehara, K.; Takakuwa, Y., J. Geophys. Res. 1973, 78, 2625-2629.
68. Fossing, H.; Jorgensen, B. B., Geochim. Cosmochim. Acta 1990, 54, 2731-2742.
69. Mizutani, Y.; and Rafter, T. A., N. Z. Jour. Science 1969, 12, 54-59.
70. Wiersma, C. L.; Rimstidt, J. D., Geochim. Cosmochim. Acta 1984, 48, 85-92.
71. Davis, R. E., J. Amer. Chem. Soc. 1958, 80, 3565-3569.
72. Meyer, B.; Peter, L; Spitzer, K., Inorg. Chem. 1977, 16, 27-33.
73. Betts, R. H.; Voss, R. H., Can. J. Chem. 1970, 48, 2035-2041.
74. Halpern, J.; Taube, H., J. Am Chem. Soc. 1952, 74, 375-380.
75. Halpern, J.; Taube, H., J. Am Chem. Soc. 1952, 74, 380-382..
76. Lazaroff, N., Science 1983, 222, 1331-1334.
77. Pierre, C., Chem. Geol. 1985, 53, 191-196.
78. Allison, G. B., J. Hydrology 1982, 55, 163-169.
79. Torma, A. E., In Advances in Biochemical Engineering; Ghose, T. K.; Fiechter, A.; Blakebrough, N., Springer-Verlag: New York, N.Y., Vol. 6; 1-37.
80. Chiba, H.; Sakai, H., Geochim. Cosmochim. Acta 1985, 49, 993-1000.
81. Hoering, T. C.; Kennedy, J. W., J. Am. Chem. Soc. 1957, 79, 56-60.
82. Sullivan, P. J.; Yelton, J. L.; Reddy, K. J., Environ. Geol. Water Sci. 1988, 11, 289-295.
83. Bayless, E. R., ACS National Meetings Book of Abstracts; Amer. Chem. Soc., 1992, Vol. 1, GEOC 79.
84. Cravotta, C. A., III, In this volume.
85. Wiese, R. G., Jr.; Powell, M. A.; Fyfe, W. S., Chem. Geol. 1987, 63, 29-38.
86. Stoffregren, R. E.; Rye, R. O.; Wasserman, M. D., Geochim. Cosmochim. Acta, in press.
87. Stoffregren, R. E.; Rye, R. O., ACS National Meetings Book of Abstracts; Amer. Chem. Soc., 1992, Vol. 1, GEOC 86.

RECEIVED October 20, 1993

SUPERGENE OXIDATION AND ENRICHMENT OF SULFIDE ORE DEPOSITS

Chapter 31

Applications of Mass-Balance Calculations to Weathered Sulfide Mine Tailings

Edward C. Appleyard and David W. Blowes

Department of Earth Sciences, University of Waterloo, Waterloo, Ontario N2L 3G1, Canada

The oxidation of sulfide minerals in mine tailings impoundments releases SO_4, Fe(II) and other metals to tailings pore-waters. The subsequent precipitation of secondary oxide and hydroxide minerals can decrease dissolved metals through accompanying precipitation, coprecipitation or adsorption reactions. Quantitative assessments of the fluxes of metals in tailings systems resulting from weathering require determinations of the changes in total solid phase metal which are independent of apparent changes caused by mass, and concomitant volume and density changes resulting from sulfide oxidation and hydroxide reprecipitation.

Mass changes during weathering were calculated for samples collected through the weathered zones in the high-sulfide Heath Steele Zn-Pb-Cu tailings impoundment, New Brunswick, and the low-sulfide Delnite Au tailings impoundment, Ontario. The results of these calculations show that the degree of metal mobility varies according to the duration of weathering, the sulfide content and the buffering capacity of the tailings. Results of the mass-balance calculations complement the results of pore-water chemical analyses.

Sulfide-oxidation reactions and subsequent acid-neutralization reactions occurring in inactive mill tailings result in the depletion of sulfide minerals, carbonate minerals and some aluminosilicate minerals with the concomitant accumulation of metal-hydroxide and sulfate minerals. To assess the potential effects of a tailings impoundment on water quality it is necessary to distinguish the amounts of all the constituents that remain within the tailings from those that have been displaced out of the tailings impoundment.

Recently, numerical models that combine geochemical reactions with solute-transport mechanisms have been used to describe the movement of dissolved constituents in tailings impoundments and in adjacent aquifers (1,2). To apply these models accurately it is necessary to constrain model simulations as closely as possible using field observations of dissolved and solid masses of chemical components.

Oxidation of sulfide minerals and precipitation of hydroxide and sulfate minerals typically result in changes in the density of tailings as well as in the volume occupied by the solid materials. Thus, changes in mass occur during weathering that may produce erroneous indications of elemental mobilities if raw chemical analyses are relied upon, i.e. it is usually incorrect to compare 100 g of the altered material with 100 g of the starting material. Chemical data corrected for weathering-related changes in mass provide a way of determining elemental mobilities independent of apparent variations due to mass-change effects.

This paper describes the application of mass-balance corrections to mine tailings from two mineralogically dissimilar impoundments, the high-sulfide Heath Steele Zn-Pb-Cu tailings impoundment, New Brunswick, and the low-sulfide Delnite Au tailings impoundment, Ontario (Figure 1). To apply this approach successfully a group of elements must be identified with effectively immobile characteristics under the conditions of weathering. Immobile elements provide a reference for evaluating changes in mass resulting from chemical weathering and permit the quantification of elemental fluxes independent from calculations based on pore-water analyses. In addition, these studies reveal the influences of different mineralogical compositions of tailings on the mobilities of elements. In our experience, this is the first application of a mass-balance technique to mine tailings.

Mass-Balance Calculations

Quantitative estimates of elemental fluxes (gains and losses) which occur during alteration of any solid geological materials (e.g. soil, rock, ore etc.), can only be determined if the ratio of the *after-alteration mass* to the *starting mass* is known. This ratio, F_M, is generally unknown and must be determined to proceed. A method of determining the value of F_M for each sample of altered material is provided through the application of Gresens' General Alteration equation ($\underline{3}$).

There is a family of equations, one for each element, which express the relationships between the *before-alteration* and the *after-alteration* compositions of any solid material and the resultant mass-change. The general form of these equations can be represented as follows:

$$\Delta\chi_\eta = \left(\frac{m^a}{m^p} \times \chi_\eta^a\right) - \chi_\eta^p \qquad (1)$$

where $\Delta\chi_\eta$ = change in mass of the element η, m^a/m^p = mass-change ratio for the whole sample (F_M), χ_η^a = concentration of element η in the *altered* sample, and χ_η^p = concentration of element η in the *precursor (unaltered)* material. Because the value of m^a/m^p is initially unknown, an alternative form of the equation substituting the product of density ρ, and volume v, for mass m, is more commonly used:

$$\Delta\chi_\eta = \left(\frac{v^a}{v^p} \times \frac{\rho^a}{\rho^p} \times \chi_\eta^a\right) - \chi_\eta^p \qquad (2)$$

where v^a/v^p = the volume-change ratio (F_V) and ρ^a/ρ^p = the density-change ratio (F_D). In all cases, p refers to the starting material which should be unaltered or altered as little as possible, and a refers to the altered material obtained from a precursor of identical composition.

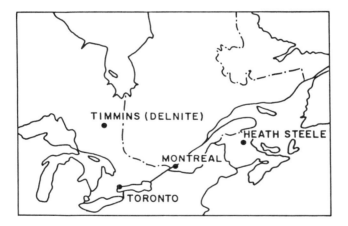

Figure 1. Location map for Heath Steele and Delnite tailings impoundments.

If an element has been *immobile*, i.e. mass was neither gained nor lost during alteration, the sample volume-change that is required for this constraint can be obtained by setting $\Delta \chi_n = 0$ for that element. The volume-change value so obtained is referred to as the *zero-mass-change volume factor* (F_V°) and is derived as follows:

$$F_V^\circ = \frac{\rho^p \times \chi_\eta^p}{\rho^a \times \chi_\eta^a} \qquad (3)$$

This quantity is the volume-change ratio for the altered sample required if there has been zero-mass-change in the given element. If a group of mutually immobile elements was present during the alteration, these elements can be recognized by their closely similar (ideally identical) F_V° values in the same sample. Thus, if clusters of F_V° values for the same group of elements are repeatedly found in a set of related samples, these elements can be identified as *immobile* and the mean of F_V° values of the immobile elements for each sample fixes the volume-change ratio for that sample.

When more than one precursor sample is available, the number of estimates of the true volume-change factor of the sample is equal to the product of the number of parent analyses and the number of immobile elements. These values should have a log-normal distribution but it is common to encounter some outlying values far above or below the main concentration. This phenomenon is usually caused by inhomogeneities within the population of precursor samples or, occasionally, by analytical errors. In the current study, a conventional statistical practice of dealing with outlying values was adopted. Values of F_V° beyond ±1 standard deviation of the log-transformed mean of the complete set were excluded before determining the final mean value and assigning it as the volume-change factor of that sample. An example of zero-mass-change volume factors (F_V°) of immobile elements for a weathered sample from site OW8 at the Heath Steele tailings impoundment is given in Table I.

Table I. Zero-Mass-Change Volume Factors for 6 Immobile Elements in Sample OW8-3 (depth 0.25 m) *

Compared with	OW8-8	OW8-7	OW8-6
Ti	0.881	0.863	0.829
Sc	0.901	0.882	*1.10*
Zr	0.948	0.886	0.770
La	0.816	0.834	0.888
Tb	*0.633*	--	0.751
Yb	0.674	0.660	0.895

* Zero-mass-change factors have been calculated using equation 3. Values lying outside the limits of the log normalized mean ±1 standard deviation are declared to be outlying and excluded; these values are in italics. The log normalized mean of the remaining 13 values is the best estimate of the actual relative volume change resulting from weathering, i.e. $F_V = 0.86$, a 14% volume loss.

Mobile elements, on the other hand, will have higher or lower F_V° values depending on the abundance of the element and the magnitude and sign of its flux. These calculations are, however, spurious because, for these elements, the initial assumption of immobility is false. Once the volume-change ratio for each sample has been established, the respective Gresens equations have only one unknown and the fluxes of all elements, i.e. their $\Delta \chi_n$ values, can be calculated. Results may be expressed for convenience as *net compositions*, i.e. $\chi_n^p + \Delta \chi_n$, and are given in this form in Table II and in Figures 2, 4, 5 and 7. Net compositions reflect the absolute amount of each element remaining after an arbitrary volume, chosen in this study to be 100 cm^3 (for major elements) or 1 m^3 (for trace elements), of the precursor material was altered. Pore-water geochemical profiles for three of these sites are illustrated in Figures 3, 6 and 8.

In both the Heath Steele and Delnite sample sites, a zone of weathering overlies tailings showing no mineralogical evidence of weathering or alteration. Samples from the unweathered section are therefore assumed to be chemically unaltered and to have experienced *no* change in mass since deposition. For these *unweathered* samples, therefore, $F_M = 1.0$ and is so indicated in Figures 2, 4, 5 and 7.

Description of the Study Areas

The Heath Steele tailings are located in northeastern New Brunswick (Figure 1), 50 km from the town of Newcastle. The tailings are contained in two impoundments, an older impoundment, covering *ca.* 10 ha, which operated from 1957 to 1965, and a new impoundment covering *ca.* 200 ha, which operated from 1965 to 1982 and from 1989 to the present. Core samples were collected from two locations on the old impoundment (holes OW3 and OW8) to a depth of *ca.* 5 m, and to a depth of *ca.* 3 m at one location on the new impoundment (hole NW10).

The tailings contain approximately 85 wt% sulfide minerals and 3 wt% carbonate minerals. Boorman and Watson (4) conducted a detailed study of the tailings and discerned the geochemical zones within the old tailings impoundment to be, from the surface downwards, an Oxidation Zone, a Hardpan Layer, and a Reduction Zone overlying unweathered tailings. Our sampling, conducted in 1986–1987, confirmed the presence of the same three alteration zones in the old impoundment.

The Delnite Au tailings are located 5 km from Timmins, Ontario (Figure 1). The tailings were deposited continuously from 1937 to 1964, and are contained in a single elevated tailings impoundment, *ca.* 25 ha in area. Core samples were collected from site D5 on the Delnite tailings. The water table at this location was at a depth of 2.7 m when the samples were collected. No hardpan layer was detected in the section. The tailings contain *ca.* 5 wt% sulfide minerals, primarily pyrite and pyrrhotite, with lesser amounts of arsenopyrite. The tailings also contain about 20 wt% of carbonate minerals, primarily dolomite and siderite. A revegetation program, initiated in 1971, established a self-sustaining cover of grass and small trees on the tailings surface.

Methods

Mass-balance calculations in this study were conducted using the SOMA computing package (5). The procedures utilized are outlined in greater detail by Taylor and Appleyard (6) and Appleyard (7).

Samples for the whole-rock analyses were collected in 7.6 cm (3 in.) thin-walled aluminum tubing. Core samples were cut into 10 cm intervals and lightly ground.

Particle density measurements of all samples were made prior to analysis using a Beckman Model 930 Air Comparison Pycnometer. Total carbonate was determined using the technique of Barker and Chatten (8) and total sulfur concentrations were determined using a LECO induction furnace. Sample splits were analyzed by Activation Laboratories Ltd., Ancaster, Ontario. The major elements, Si, Ti, Al, Fe_{total}, Mn, Mg, Ca, Na, K, and P and the trace elements Ba, Sr and Y were analyzed by inductively coupled argon plasma atomic emission spectrometry (ICP-AES) following a Li-metaborate fusion. Copper, Pb, Zn, Ni and V were measured by ICP-AES following a HNO_3-HCl-$HClO_4$-HF digestion. Silver, As, Au, Mo, Sb and Se were determined by ICP-AES following digestion with solvent extraction (9). Rubidium, Zr, and Nb were analyzed by X-ray fluorescence analysis on a pressed powder pellet. Cobalt, Cr, Cs, Hf, Sc, Ta, Th, U, and the rare earth elements La, Ce, Nd, Sm, Eu, Tb, Yb and La were determined by instrumental neutron activation analysis and B was measured using the prompt gamma modification.

Results

Application of Mass-Balance Studies to Heath Steele Tailings.

General. Immobile elements identified within the weathered zones of the Heath Steele tailings impoundments include Ti, Sc and Zr at sites OW3, OW8 and NW10. In addition, V and Hf were immobile at sites OW3 and NW10. Rare-earth element data for site OW8 indicate that La, Yb and Lu were generally immobile.

Compositional variability within unweathered sections presents a problem in establishing the compositions of the precursors of the weathered samples. The procedure adopted was to use the mean composition of tailings immediately underlying the weathered sections. For site OW3, the precursor composition was taken as the mean of five samples collected over the depth range 90 to 180 cm. For site OW8, two samples over the depth range 90 to 120 cm and for site NW10, four samples over the depth range 60 to 100 cm served as the precursor compositions. Uncertainties in establishing precursor compositions for weathered samples form the greatest limitations on the credibility of calculated mass-changes.

Mass-change ratios ($F_M = m^a/m^p$) calculated for weathered samples range from 0.765 (23.5% mass loss) to 1.23 (23% mass gain). The Oxidation Zone in the old tailings is marked by prominent mass loss, signifying net leaching, while the Hardpan Layer and the top of the Reduction Zone are marked by mass gains presumably associated with precipitation, coprecipitation and adsorption.

The best-developed profiles are found at sites OW3 and OW8 in the old tailings impoundment. Patterns in the new impoundment (hole NW10) are marked by a near-surface zone of mass loss corresponding to leaching, but a subjacent hardpan layer was not detected either megascopically or geochemically.

Heath Steele: Old Tailings. The effects of sulfide alteration are most evident in the older tailings impoundment (Figures 2,4), most clearly at location OW8. At this location (Figure 2, Table II), the mass-change ratio, F_M, shows an abrupt increase at the hardpan layer, with decreases in mass immediately above and below. The similar distribution

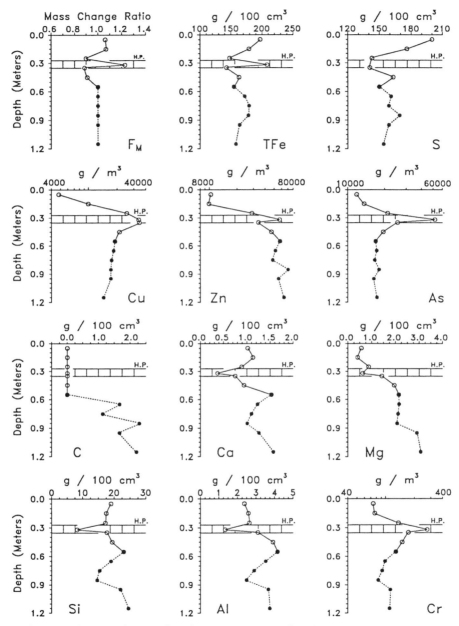

Figure 2. Geochemical profiles for mass change ratios (F_M) and selected element concentrations for site OW8 from the Heath Steele "old" tailings impoundment. *Unweathered* samples are indicated in solid circles and dashed line. *Weathered* samples located near the tailings surface are marked by open circles and solid line. The compositions of the *weathered* samples include calculated mass changes and are given as "net compositions" (see text). H.P. = Hardpan Layer. The water table was at a depth of 2.5 m at the time of sampling.

Table II. Average Compositions of Weathered
and Unweathered Tailings from Heath Steele Hole OW8 *

Zone	Upper Oxidation	Lower Oxidation	Hard Pan	Reduction	Unweathered
Depth	0–20 cm	20–40 cm	32 cm	40–60 cm	60–120 cm
N	2	2	1	2	5
S.G.	4.43	4.42	4.40	4.28	4.32
F_M	1.07	0.893	1.23	0.938	1.00
Si	18.3	17.5	8.22	21.2	19.1
Ti	0.21	0.19	0.19	0.20	0.21
Al	2.48	2.87	1.35	4.04	3.26
ΣFe^{2+}	188	146	208	159	170
Mn	0.09	0.12	0.12	0.42	0.75
Mg	0.49	1.15	0.63	2.05	2.46
Ca	1.08	0.82	0.37	1.24	1.23
Na	0.16	0.11	0.08	0.14	0.12
K	0.29	0.50	0.22	0.40	0.53
P	0.20	0.18	0.77	0.15	0.17
C^{4+}	0.00	0.00	0.00	0.00	1.79
S^{2-}	188	142	--	157	160
Sc	3.75	4.84	5.19	5.23	4.94
V	79.0	109.	166.	112.	96.6
Cr	75.5	153.	275.	138	98.7
Co	2140	1850	2800	2590	2790
Ni	33.4	199.	25.9	41.8	37.6
Cu	6900	29400	33700	20100	16900
Zn	10200	31200	57100	51900	57300
As	12700	23300	51900	18200	17100
Se	221	162	223	162	150
Rb	54.8	18.8	25.9	20.8	21.6
Sr	89.6	54.7	25.9	62.6	89.5
Y	18.9	18.4	31.1	20.5	28.0
Zr	85.1	90.4	88.2	85.6	97.4
Mo	14.2	28.2	5.19	31.3	15.4
Ag	161	88.6	130	115	96.7
Sb	756	564	934	542	518
Cs	0.471	0.377	0.052	0.418	0.432
Ba	2680	1450	1070	1570	1610
La	41.1	31.9	32.7	34.3	33.6
Ce	60.0	56.7	51.9	58.9	54.8
Nd	26.0	26.9	7.78	24.5	15.8
Sm	8.02	8.86	3.68	8.53	9.10
Eu	6.37	6.97	8.77	9.29	12.6
Tb	1.02	1.30	0.259	1.29	0.852
Yb	1.53	1.88	0.208	1.94	1.62
Lu	0.259	0.299	0.026	0.332	0.173
Hf	4.00	3.69	4.67	3.38	3.10
Au	3.37	2.01	2.99	2.42	2.47
Th	6.34	8.81	13.0	6.64	5.67
U	1.18	6.62	1.30	8.97	3.60

* Values for *weathered* samples (first four columns) are "net compositions", i.e. corrected for mass changes during weathering (see text); values for *unweathered* samples are raw analytical results. Mean values are arithmetic means for major constituents (Si to S) and logarithmic means for trace components (Sc to U). ΣFe^{2+} is Fe_{total} expressed as Fe^{2+}. Units are g/100 cm^3 of the precursor for major constituents and g/m^3 of the precursor for trace constituents. Column 5 is the assumed precursor composition.

of Fe_{total} probably accounts for most of the F_M change. Pore-waters contain >60 g/l dissolved Fe in the shallow tailings water and >100 g/l dissolved SO_4^{2-} (Figure 3; 10). Pyrrhotite is depleted near the tailings surface (11). The sulfophile elements, Zn, Cu and As are also depleted near the tailings surface as a result of sulfide-oxidation reactions. Shallow pore-waters contain dissolved concentrations >6 g/l Zn, >100 mg/l Cu, (Figure 3) and >0.4 g/l As, indicating that oxidation of sphalerite, chalcopyrite, tetrahedrite-tennantite and arsenopyrite has released some of the solid-phase concentrations of these elements to the tailings pore-water. (See Figure 4.)

Copper and As in the hardpan layer are enriched in the solid phases compared to unweathered tailings below the hardpan layer (Figure 2, Table II). This secondary solid-phase enrichment is accompanied by decreases in the dissolved concentrations of Cu and As. High concentrations of covellite (CuS) and an As-bearing secondary oxidation product occur at this depth (11). Zinc is depleted at the tailings surface but is enriched in the hardpan layer where it is incorporated into melanterite, which occurs as a cementing mineral. The behavior of Cr is similar to that of Cu and As. Chromium may have been derived from the milling reagents or from the oxidation of magnetite. Geochemical speciation calculations conducted using the computer code MINTEQA2 suggest that Cr may accumulate near the depth of the hardpan layer as $Cr(OH)_3$.

Acid-neutralizing, carbonate-dissolution reactions have consumed the carbonate content of the upper 50 cm of the old tailings at location OW8. The pore-water pH in this zone increases from <0.8 at the tailings surface to >4.5 near the depth of the hardpan (Figure 3). These pH-buffering reactions are probably the cause of the depletion of solid-phase Ca and Mg near the tailings surface. The concentration of Ca is further depleted near the hardpan layer, possibly through aluminosilicate dissolution, and enriched 15–20 cm below the hardpan layer as the result of gypsum precipitation.

Aluminosilicate dissolution is suggested by the depletion of Al relative to Si at the tailings surface. High concentrations of dissolved Al (up to 1,030 mg/l) and Si (up to 722 mg/l H_4SiO_4) are present in the oxidation zone above the hardpan layer. MINTEQA2 calculations indicate that these water samples show supersaturation with respect to amorphous silica. Similar degrees of supersaturation have been observed at other oxidized mine-tailings impoundments where aluminosilicate dissolution has been inferred (13–15). These dissolved Al and Si concentrations decrease abruptly at the depth of the hardpan layer. Slight solid-phase enrichment of Al occurs immediately below the hardpan layer, possibly because of the formation of aluminum hydroxide or aluminum hydroxy-sulfate minerals. Mineralogical study has confirmed the presence of aluminum sulfate, but not aluminum hydroxide precipitates.

Heath Steele: New Tailings. The composition of the new tailings is variable, as indicated by the elemental profiles for the unweathered samples in NW10 (solid dots in Figure 5). These fluctuations are indicative of irregularities in the composition of the ore feed to the mill and/or in the efficiency of the beneficiation processes. In spite of this variability, the mass-change-corrected whole-rock data reveal some distinct trends after calculation of net composition values for weathered samples.

The mass-change ratio indicates that a loss of mass has occurred near the tailings surface. This change is the result of losses of Fe_{total}, S, Cu, Zn, C, Ca, Mg, Al, Si and K. The losses of S, and the metals Fe, Cu and Zn, probably result from sulfide-oxidation

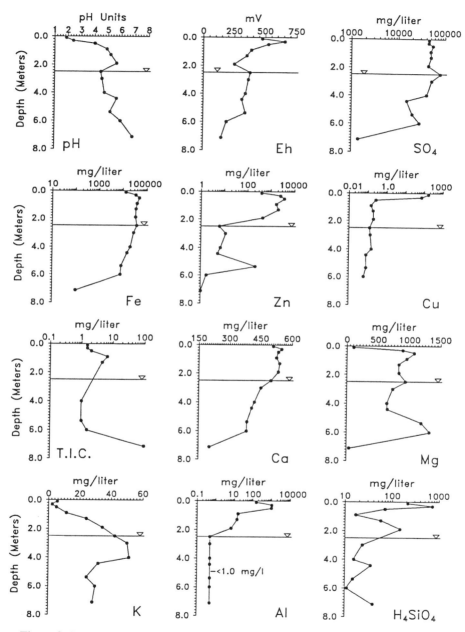

Figure 3. Pore-water geochemical profiles for pH, Eh and selected elements for site OW8 from the Heath Steele "old" tailings impoundment. Sampling was carried out in June, 1986. T.I.C. = total inorganic carbon. ▽ = water-table at time of sampling.

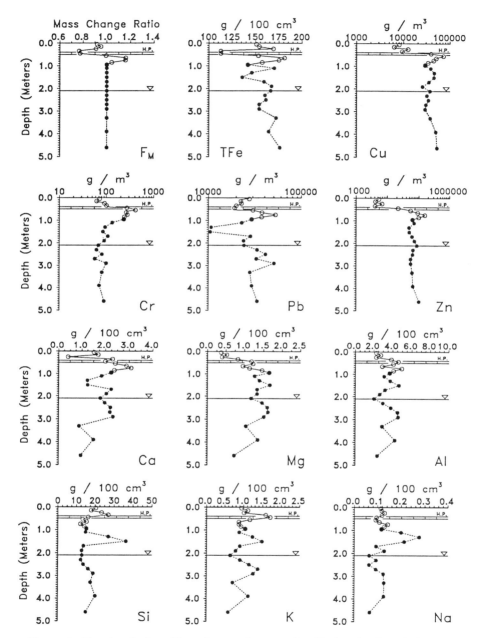

Figure 4. Geochemical profiles of mass change ratios (F_M) and selected element concentrations for site OW3 from the Heath Steele "old" tailings impoundment. ▽ = water-table at time of sampling. • and dashed line = unweathered samples; ○ and solid line = weathered samples. H.P = Hardpan Layer.

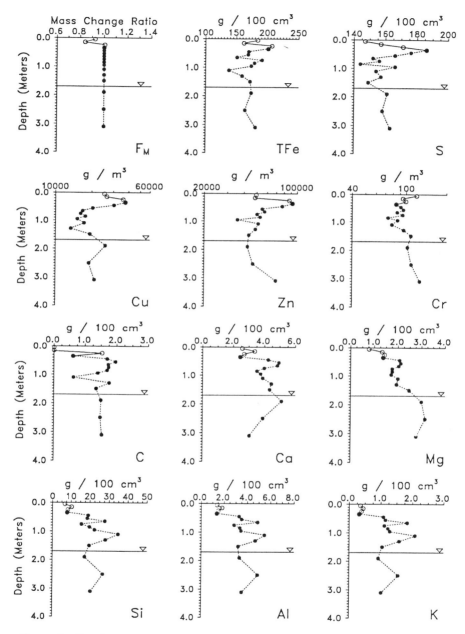

Figure 5. Geochemical profiles of mass change ratios (F_M) and selected element concentrations for site NW10 from the Heath Steele "new" tailings impoundment. ▽ = water-table at time of sampling. • and dashed line = unweathered samples; ○ and solid line = weathered samples.

reactions occurring in the shallow tailings. Of these elements, S, Cu and Zn show an enrichment 30-40 cm below the tailings surface.

The depletions of C, Ca and Mg near the tailings surface probably result from the dissolution of calcite, dolomite and other carbonates through pH-buffering reactions. The apparent depletion of Al, Si and K from the tailings surface may represent aluminosilicate dissolution, but may also represent unknown changes in the precursor composition of the tailings (i.e. lower initial aluminosilicate content than that in the assumed precursor).

Variations in pore-water composition with depth are illustrated in Figure 6. The brief period of sulfide oxidation in the new impoundment is reflected in the tailings pore-water geochemistry. High concentrations of dissolved Fe and SO_4 are limited to the shallow vadose zone. The maximum concentrations of Fe (10,200 mg/l) and SO_4 (24,800 mg/l) are observed in the shallowest two samples (10 and 30 cm depth), near the depth of active sulfide oxidation. The pore-water pH in the shallow vadose zone is >5.5 indicating that H^+ produced by sulfide oxidation has been consumed by reaction with carbonate minerals. The aqueous concentrations of Mg are high in the shallow vadose zone, also probably reflecting dissolution of dolomite. The concentrations of Ca and H_4SiO_4 are relatively constant throughout the profile and geochemical calculations suggest that these concentrations are limited by the solubilities of gypsum and amorphous SiO_2. High concentrations of Zn (200 mg/l) and Ni (2 mg/l) are restricted to the shallowest sampling points (not shown on Figure 6), also reflecting the brief duration of sulfide oxidation.

Results of Mass-Balance Studies for Delnite Tailings. Immobile elements identified in the Delnite weathered-zone tailings include Ti, Al, Sc, V, Zr, Yb and Lu. Primary variances of some of these elements in the unweathered tailings are relatively high so the estimates of mass-change ratios and fluxes of individual elements are less precise than in the case of the Heath Steele tailings.

The precursor composition used for the Delnite site was the mean composition of three unweathered samples collected from the depth range 90 to 140 cm. Mass-change ratios (Figure 7) range from 0.988 (1.2% mass loss) to 1.32 (32% mass gain) but because there is no mineralogical indication of the presence of a hardpan layer the latter value may represent an artificial increase due to a lower initial immobile-element content of the tailings at this depth relative to the average precursor values. Despite the uncertainty associated with this sample, a subtle leaching effect just below the tailings surface is suggested by decreases in the mobile elements S and As in a manner that parallels the behavior of these elements at Heath Steele. Similarly, loss of Ca, Mg and C in the upper 50 cm of the hole is probably associated with the dissolution of carbonate minerals. Such reactions are also indicated by high gas-phase CO_2 concentrations, >10 vol% (12).

The Delnite Tailings Area: Results. The effects of sulfide oxidation on tailings geochemistry (Figure 7) are superimposed on probable compositional variations in the tailings as they were deposited.

The mass-change ratio shows a distinct increase at 65 cm. As noted before, this increase may not be due to post-depositional mass transfer, because it is reflected by increases in the masses of Si, Al and Na which are suggestive of increased amounts of aluminosilicate minerals in the mill feed. Mineralogical study identified the first occurrence of talc at this approximate depth (16). The addition of talc to the tailings

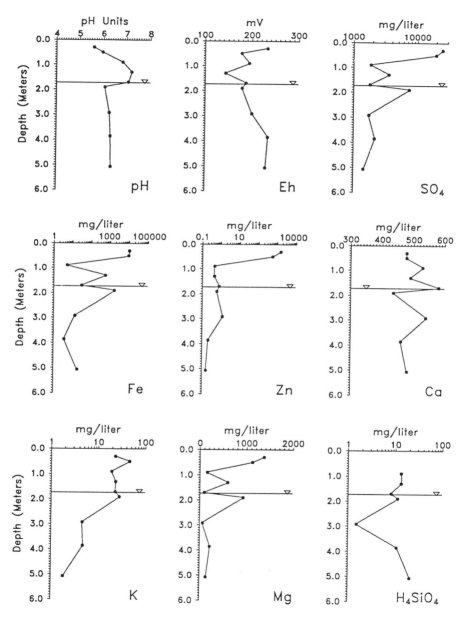

Figure 6. Pore-water geochemical profiles for pH, Eh and selected elements for site NW10 from the Heath Steele "new" tailings impoundment. Sampling was carried out in June, 1986. ▽ = water-table at time of sampling.

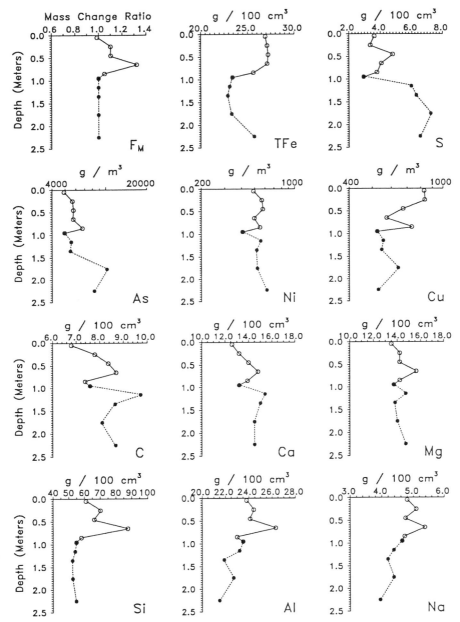

Figure 7. Geochemical profiles for mass change ratios (F_M) and selected element concentrations for site D5 from the Delnite Au tailings impoundment. The water table was at a depth of 2.7 m at the time of sampling. • and dashed line = unweathered samples; o and solid line = weathered samples.

would not explain the enrichment of Al but it might be responsible for the peak in Mg at that depth.

The effects of sulfide oxidation are most evident in the depletion of S and As near the tailings surface. Unlike Heath Steele, pH in the weathered zone at Delnite has remained near neutral, ranging from 6.5 to 8.5 (Figure 8). Mineralogical study indicates that the sulfide minerals in the vadose zone have been extensively oxidized and are replaced by Fe(III)-oxyhydroxide minerals, predominantly goethite (16). Fe_{total} remains relatively constant throughout the weathered zone suggesting that Fe released through sulfide-oxidation reactions is reprecipitated locally. Dissolved Fe concentrations in pore-waters in the shallow tailings are consistently <50 mg/l.

The formation of geothite is favored by the high-pH conditions present at Delnite resulting from the pH-buffering effect of the abundant carbonate minerals. Although such reactions have consumed much of the initial carbonate content of the shallow tailings (Figure 7), >6 g/100 cm^3 of the C from carbonate remains in the sulfide-depleted zone of the tailings. This large mass of carbonate will probably be sufficient to prevent the extensive acidification observed at Heath Steele from occurring at the Delnite site.

Lower solid-phase Fe_{total} levels deeper in the profile are attributed to primary variations in the mill feed. Sulfate-rich waters move downward through the tailings with infiltrating precipitation/recharge water. Throughout the entire depth of the tailings sulfate is removed from the pore-water by precipitation of gypsum, which is abundant as a secondary mineral. Equilibrium with respect to gypsum maintains pore-water SO_4 concentrations between 1,500 and 3,500 mg/l.

Dissolved As concentrations are in excess of 30 mg/l although they are low near the tailings surface where solid-phase As has been depleted. The maximum pore-water As concentrations occur 1 to 2.5 m below the tailings surface and decline sharply at 3.5 m suggesting that As is removed at a depth of *ca.* 3.5 m. Mineralogical examination of the tailings solids (16) indicates that As accumulates in goethite rims surrounding the oxidized sulfide grains. Goethite rims surrounding grains of pyrite, pyrrhotite and arsenopyrite contain similar concentrations of As irrespective of the composition of the host grain, suggesting that As accumulation is through a secondary adsorption or coprecipitation process (16).

The metals Cu and Ni, as well as Cr, Zn and Au (not illustrated in Figure 7), do not display relative changes of the same magnitude as at Heath Steele. Copper, in contrast to its usual behavior, shows a mass increase at the tailings surface suggesting increased Cu discharge in the original tailings. Nickel is relatively constant throughout the depth of sampling. Chromium, Zn and Au show some evidence of depletion near the tailings surface with enrichment at greater depths. The low mobility of these metals at Delnite, compared with Heath Steele, is consistent with the high-pH conditions that result from the large quantities of carbonate minerals present at Delnite.

Discussion

The mineral compositions of the Delnite and Heath Steele tailings are distinctly different, as shown by the relative abundances of individual elements and the apparent mobilities of these elements. The Heath Steele tailings contain *ca.* 85 wt% sulfide minerals, with high concentrations of S, Fe, Cu, Zn and Pb. Depletion of S, Cu, Zn and Pb from near

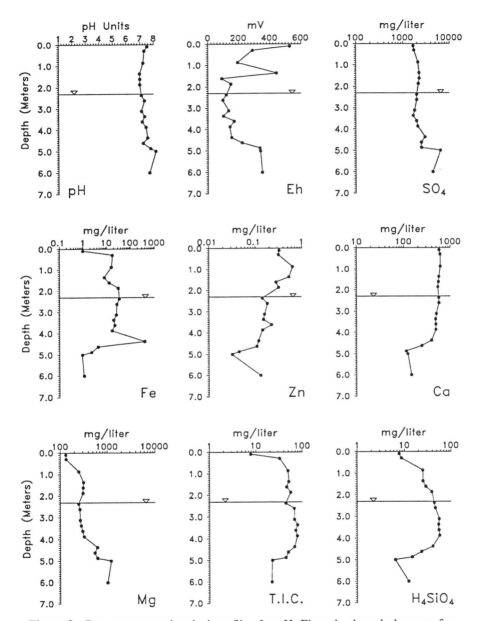

Figure 8. Pore-water geochemical profiles for pH, Eh and selected elements for site D5 from the Delnite tailings impoundment. Sampling was carried out in June, 1986. ▽ = water-table at time of sampling. T.I.C. = total inorganic carbon.

the tailings surface and enrichment at depth is evident. The elements present in the pH-buffering carbonate minerals, i.e. C, Ca and Mg, are also depleted near the tailings surface.

The Delnite tailings on the other hand contain only 5 wt% sulfide minerals and as much as 20 wt% carbonates. Compositional changes due to weathering at Delnite are subtle and easily masked by primary variations in the composition of the tailings. The ore minerals at the Delnite mine are found in veins which cross-cut a variety of rock-types in different ore bodies, leading to greater compositional variability in the tailings than would be anticipated in massive sulfide ore deposits such as Heath Steele.

Compositional changes at Delnite suggest that S, As and possibly Zn have been mobilized close to the surface through sulfide oxidation reactions and displaced downward through the tailings. By contrast, Fe, Cu and Ni show no significant mobility. The generation of H^+ at Heath Steele depleted the solid-phase carbonate content and resulted in very low pH conditions near the tailings surface. Under these low-pH conditions there is a tendency for dissolved metals to remain in solution. The abundance of carbonates at Delnite have maintained near-neutral pH values near the tailings surface which favor the precipitation of secondary Fe-bearing oxyhydroxides, principally goethite. The precipitation of geothite has stabilized Fe within the shallow tailings, resulting in much less metal mobility than is evident at Heath Steele.

Conclusions

In this exploratory study, mass-balance calculations were employed to investigate the weathering systematics of sulfide-bearing mine tailings. It was possible to identify a group of elements that are sufficiently immobile under the conditions of weathering to estimate mass-change ratios and to calculate the probable associated elemental fluxes in a manner believed to be more accurate than would be obtained using raw data. The most immobile elements were the first transition-series elements Ti, Sc and V, the high field-strength elements Zr and Hf, and the rare earth elements La (in some samples), Yb and Lu.

Difficulties in establishing correct precursor compositions for individual weathered samples present the greatest limitations on the credibility of the results. Additional studies of the compositional systematics of the tailings are needed to improve altered rock/precursor matching and thereby to improve the accuracy of the calculated elemental fluxes.

Large differences in element mobilities (e.g. metals, S, As), in different tailings impoundments can be ascribed to initial tailings characteristics such as the amount of acid-generating sulfides on one hand and the quantity of pH-buffering carbonates on the other.

Whole-rock analyses of tailings, when assessed for mass changes during weathering, provide estimates of the mobilities of elements independent of those predicted from pore-water geochemistry or geochemical calculations and can be used to constrain fluid mass-balance calculations. This source of independent data will be useful in future studies directed at the development of numerical models describing the physical and chemical processes controlling the environmental effects of mine wastes.

Lithogeochemical data can provide a time-dependent record of cumulative element fluxes over the duration of exposure of tailings impoundments that can not be obtained from pore-water studies without repeated sampling dating from the inception of the impoundment. Inasmuch as this study was reconnaissance in nature, integrated elemental flux rates over the periods of weathering were not estimated, but it would be feasible to do so if sampling were undertaken on a more extensive and systematic basis. Geochemical mass-balance calculations, therefore, provide a useful tool when used in association with pore-water and mineralogical studies of weathered tailings.

Acknowledgments

Funding for the project was provided by Noranda Minerals Incorporated. We thank C.J. Ptacek, T.A. Al, C. Hanton-Fong, T. Barrett and an anonymous ACS reviewer for critically and constructively reviewing the manuscript. C.J. Coggans and K.R. Waybrant assisted with sample preparation and analysis.

Literature Cited

1. Walter, A.L. M.Sc. thesis, University of Waterloo, ON; 1992.
2. Yeh, G.T.; Tripathi, V.S. Water Resour. Res. 1991, 27, 3075–3094.
3. Gresens, R.L. Chem. Geol. 1967, 2, 47–65.
4. Boorman, R.S.; Watson, D.M. Can. Inst. Mining Metal. Bull. 1976, 69, 86–96.
5. Appleyard, E.C. SOMA — A Package of Fortran Programs for Calculating Mass Exchange in Metasomatic and Altered Rocks (PC/DOS Compatible Version); Department of Earth Sciences, University of Waterloo: Waterloo, ON, Canada, 1991, 38p.
6. Taylor, G.F.; Appleyard, E.C. Jour. Geochem. Expl. 1983, 18, 87–110.
7. Appleyard, E.C. In Modern Exploration Techniques; Beck, L.S.; Harper, C.T., Eds.; Saskatchewan Geological Society: Regina, SK, 1990, Special Publication 10, p.50–60.
8. Barker, J.; Chatten, S. Chem. Geol. 1982, 36, 317–323.
9. Clark, J.R.; Viets, J.G. Anal. Chem. 1981, 53, 61–65.
10. Ptacek, C.J.; Blowes, D.W. In this volume, 1993.
11. Jambor, J.L.; Blowes, D.W. CANMET Division Report MSL 89–137(IR); Energy, Mines and Resources: Ottawa, ON, 1991.
12. Blowes, D.W. Ph.D. thesis, University of Waterloo, ON; 1990.
13. Dubrovsky, N.M. Ph.D. thesis, University of Waterloo, ON; 1986.
14. Coggans, C.J. M.Sc. thesis, University of Waterloo, ON; 1992.
15. Johnson, R.H. M.Sc. thesis, University of Waterloo, ON; 1993.
16. Jambor, J.L.; Blowes, D.W. Proc.: Second Internat. Conf. on Abatement of Acidic Drainage; MEND (Mine Environment Neutral Drainage); Montréal, QC, 1991, Vol. 4, 173–198.

RECEIVED August 10, 1993

Chapter 32

Oxidation of Massive Sulfide Deposits in the Bathurst Mining Camp, New Brunswick
Natural Analogues for Acid Drainage in Temperate Climates

D. R. Boyle

Geological Survey of Canada, 601 Booth Street, Ottawa, Ontario K1A 0E8, Canada

Mass-balance techniques have been applied to gossan zones overlying massive sulfide deposits in the Bathurst camp of New Brunswick to determine the relative order of mobility of toxic elements during prolonged oxidation of this type of mineralization under temperate climatic conditions. The oxidation sequence of sulfide minerals and the relative mobilities of various metals out of these oxidized zones are remarkably similar to the oxidation-mobility characteristics observed for massive sulfide tailings impoundments in Canadian temperate climates. The oxidation rates for various types of massive sulfide tailings deposits are greater than those of natural consolidated massive sulfide ore bodies by a factor of 200 or more.

The safe environmental disposal of sulfide tailings from massive sulfide mining operations is fast becoming a major economic factor in determining the profitability of mining operations. The capital, operating, remediation and monitoring costs of sulfide tailings disposal are escalating in accordance with new environmental standards (1). This economic pressure makes research on the processes controlling oxidation of sulfide tailings deposits, and the mobility of toxic elements from them, of utmost importance. Knowledge of the factors leading to the formation of supergene deposits, and the elemental distributions and enrichment-depletion trends within them, can be useful in trying to understand the long-term behavior of tailings deposits in near-surface oxidizing environments.

Detailed studies of supergene ore bodies formed over massive sulfide ores in many countries have demonstrated that many of these deposits are significantly enriched in Au and Ag in the oxide gossan zone, and in a variety of metals in the transitional oxide and supergene sulfide zones (2-10). Environmentally toxic elements, such as Pb, As, Hg and Sb are generally retained in the supergene zone as relatively stable hydrated sulfates, hydrated oxides or carbonates (4). As with sulfide-bearing tailings and waste rock piles, however, these elements may be released to the environment after mining and waste storage.

This paper describes the geology and geochemistry of supergene zones developed over massive sulfide Zn-Pb-Cu ore bodies in the Bathurst mining camp of northern New Brunswick, and discusses the relationship of these findings to tailings management. The present research is part of an on-going detailed study by the author of pre-and post-glacial weathering events in the Bathurst camp.

Supergene Mineral Deposits of the Bathurst Camp, New Brunswick

The Bathurst mining camp contains 37 massive sulfide (Zn-Pb-Cu-Ag) deposits with defined tonnages and grades, and approximately 100 smaller occurrences (11). Of the 37 main deposits, 10 have been mined. The largest is the Brunswick Mining and Smelting No. 12 ore body (161 million metric tonnes grading 8.83 wt% Zn, 3.55 wt% Pb, 0.31 wt% Cu and 98.6 g Ag/t; W.M. Luff, pers. comm., 1992).

The primary ore deposits of the Bathurst mining camp consist of fine-grained, layered stratiform lenses composed of pyrite, sphalerite, galena and chalcopyrite with lesser amounts of arsenopyrite, pyrrhotite, marcasite, tetrahedrite and Pb-Sb-As-Ag sulfosalts (Table I). Sulfides constitute more than 90% of the ore mined. Gangue minerals are mainly quartz, iron-bearing carbonates and chlorite-sericite micas. The deposits are hosted predominantly by Ordovician fine-grained metasedimentary rocks and a few are in felsic volcanic rocks (11).

Eight of the massive sulfide deposits of the Bathurst camp have well developed overlying supergene zones which formed prior to the last glaciation (Figure 1). Tonnages for these gossans vary from 20,000 (Half Mile Lake) to just under 2 million tonnes (Murray Brook), and three have been mined for Au and Ag (11).

The stratigraphy and mineralogy shown in Figure 2 and Table I, respectively, are typical of the precious- metal-bearing supergene deposits of the Bathurst camp. Acid solutions generated in the zone of sulfide oxidation have given rise to a Ferruginized Wallrock unit characterized by oxidation of Fe-bearing minerals and precipitation of Fe oxides from the acid solutions. These zones extend up to 30 m into the wallrocks beyond which up to 200 m of the surrounding rocks commonly have been strongly bleached, kaolinized and/or chloritized by acidic solutions moving out of the sulfide-oxidation zone.

Compared to the Massive-Sulfide-Zone Gossan, the Ferruginized Wallrock unit has excellent preservation of structural and stratigraphic features, much lower Au and Ag contents, and elevated Al, Mg, and K concentrations. Where the footwall zone of disseminated sulfides is exposed at surface through folding or structural dislocation, it is strongly oxidized to an Fe oxide-silica rock (Disseminated-Zone Gossan) showing excellent preservation of stratigraphic features but much lower Al, Mg and K levels than in the Ferruginized Wallrock Unit.

The Massive-Sulfide-Zone Gossan constitutes the main body of economic mineralization and consists mainly of massive goethite showing highly vesicular, cellular-boxwork and pseudomorphic-replacement textural features of completely oxidized massive sulfide. The Massive-Sulfide-Zone Gossan is composed of goethite, quartz (primary anhedral), secondary amorphous silica, a number of hydroxy-sulfate and oxide minerals containing combinations of K-Fe-Pb-As-Sb-Ag (beudantite, plumbojarosite, argentojarosite, jarosite, bindheimite, scorodite), trace amounts of cinnabar, woodhouseite and barite, and cassiterite of primary origin (12; Table I). Native Ag and Bi occur in the Massive-Sulfide-Zone Gossan and, Au, although not identified directly, is also assumed to be in the native form (12). Pyrite in the gossan is only present in trace amounts at the contact of the gossan and supergene sulfide zone.

Most of the gangue component in the Massive-Sulfide-Zone Gossan is silica, present as anhedral quartz or amorphous silica. During oxidation, the silicate minerals, and to a much lesser extent primary quartz, are dissolved by acidic solutions to form cation complexes and silicic acid. The silicic acid precipitates amorphous silica intimately mixed with goethite lower in the oxidizing

Table I. Mineralogy of Primary and Supergene Ore Zones, Massive Sulfide Deposits, Bathurst Camp, New Brunswick (after 13 and author's data)

GOSSAN ZONE

Goethite - $FeO(OH)$ (botryoidal, massive, reniform)
Silica - primary euhedral quartz
    - secondary silica (amorphous)
K-Fe-Pb-Sb-As-(Ag) -- Hydroxy-sulfates
    Beudantite - $PbFe_3(AsO_4)(SO_4)(OH)_6$
    Plumbojarosite - $PbFe_6(SO_4)_4(OH)_{12}$
Argentojarosite - $AgFe_3(SO_4)_2(OH)_6$
Jarosite - $KFe_3(SO_4)_2(OH)_6$
Bindheimite - $Pb_2Sb_2O_6(O,OH)$
Scorodite - $FeAsO_4 \cdot 2H_2O$
Woodhouseite - $CaAl_3(PO_4)(SO)_4(OH)_6$
Native - Au, Ag, Bi
Cassiterite - SnO_2
Cinnabar - HgS
Barite - $BaSO_4$
Pyrite - FeS_2 (rare to absent)

SUPERGENE SULFIDE ZONE

Covellite - CuS
Chalcocite - Cu_2S
Digenite - Cu_9S_5
Enargite - Cu_3AsS_4
Luzonite - Cu_3AsS_4
Beaverite
  - $Pb(Cu,Fe,Al)_3(SO_4)_2(OH)_6$
Meta-aluminite
  - $Al_2(SO_4)(OH)_4 \cdot 5H_2O$

Acanthite - Ag_2S
Anglesite - $PbSO_4$
Scorodite - $FeAsO_4 \cdot 2H_2O$
Goethite - $FeO(OH)$ (minor)
Litharge - PbO
Hinsdalite
  - $PbAl_3(PO_4)(SO_4)(OH)_6$
Brochantite
  - $Cu_4(SO_4)(OH)_6$

PRIMARY ORE

Pyrite - FeS_2
Marcasite - FeS_2
Arsenopyrite - FeAsS
Sphalerite - ZnS
Chalcopyrite - $CuFeS_2$
Galena - PbS
Tetrahedrite Group
  - $(Cu,Ag,Fe,Zn)_{12}(Sb,As)_4S_{13}$
Pyrrhotite - $Fe_{1-x}S$
Cassiterite - SnO_2
Stannite - Cu_2FeSnS_4
Pb-Sb-As-Ag Sulfosalts

Quartz - SiO_2
Ferroan Dolomite
  - $Ca(Fe,Mg,Mn)(CO_3)_2$
Siderite - $FeCO_3$
Calcite - $CaCO_3$
Magnetite - $(Fe,Mg)Fe_2O_4$
Chlorite
  - $(Mg,Fe,Al)_6AlSi_3O_{10}(OH)_8$
Sericite
  - $KAl_2(AlSi_3O_{10})(OH)_2$

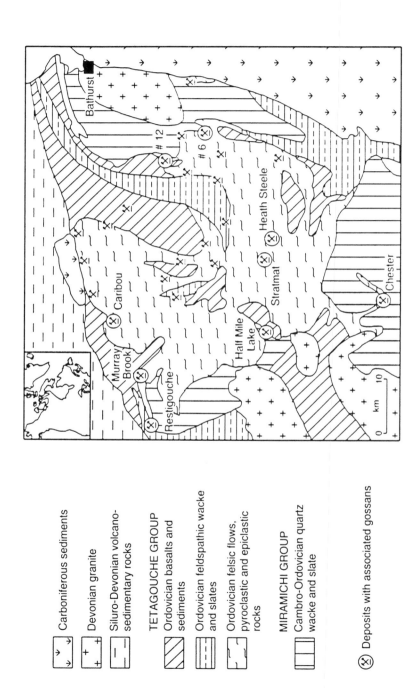

Figure 1. Geology of Bathurst Mining camp showing locations of massive sulfide mineral deposits with associated supergene ore zones.

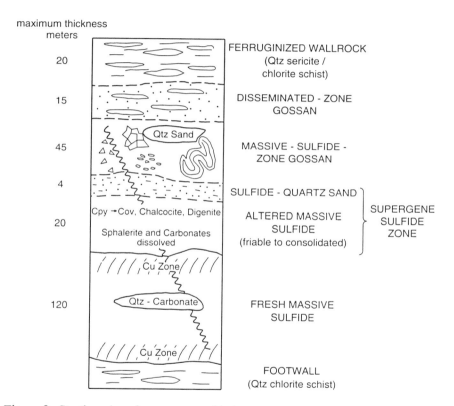

Figure 2. Stratigraphy of supergene oxidation zones over massive sulfide deposits in the Bathurst Mining camp.

profile; most of the primary quartz remains behind as residual anhedral grains in goethite. Primary carbonates (siderite, ferroan dolomite, calcite) are completely dissolved by the acid solutions, often leaving small lenticular cavities in the gossans. There are no secondary carbonates, halides, or metal silicates formed in any of the supergene zones.

The sulfides beneath the Massive-Sulfide-Zone Gossan show oxidation and replacement generally to depths up to 25 m. Alteration is dominated by almost total dissolution of carbonate-gangue phases, dissolution of sphalerite, alteration of chalcopyrite to supergene Cu sulfides (covellite, chalcocite, digenite), and partial oxidation of galena and arsenopyrite to anglesite and scorodite, respectively. Minor amounts of luzonite, brochantite, beaverite, hinsdalite, litharge and meta-aluminite occur in the supergene sulfide zone of some deposits (13). The supergene sulfide zone can be divided into two units: a) an upper unit at the sulfide-gossan contact composed of a highly friable sulfide-quartz sand up to 4 m thick, which is analogous in composition (pyrite, minor galena-sphalerite-quartz) and texture to the massive sulfide tailings produced at the Bathurst mines; and b) a lower unit characterized by incipient alteration of the sulfide and gangue.

The sequence of sulfide alteration at the Murray Brook deposit, shown in Figure 3, is typical for deposits in the Bathurst camp. Sphalerite is readily attacked, but no replacement phase has been identified; the process seems to be simply one of oxidative dissolution. Next, the tetrahedrite group of minerals (volumetrically minor in deposits) are replaced by supergene acanthite which is commonly the source of Ag for the formation of argentojarosite and native Ag. The tetrahedrite group may also supply significant amounts of Sb and As to form bindheimite and beudantite, respectively. Chalcopyrite is replaced by chalcocite-covellite-digenite with subsequent production of dissolved Cu^{2+}. Although Cu enrichment in the supergene sulfide zone of oxidized sulfide deposits in many mining camps is often strong (2,5,8-10,14) to incipient (13), only a few of the Bathurst deposits (e.g. Caribou, Heath Steele) have supergene Cu enrichment zones. Galena and arsenopyrite oxidize to produce anglesite and scorodite, respectively. Galena alteration eventually leads to the formation of Pb-bearing hydrated sulfates and oxides, whereas the arsenopyrite-scorodite alteration process commonly yields goethite, occasionally as a pseudomorph after scorodite. The As from the oxidation of arsenopyrite moves downward with the oxidizing solutions, eventually combining with Pb, Fe and SO_4 to form beudantite. Pyrite is the most persistent sulfide in these deposits and can be found in traces within the gossan at the sulfide-gossan contact and also as pristine grains in the sulfide-quartz sand unit and underlying oxidized sulfide zone.

Pyrrhotite is not included in the alteration sequence shown in Figure 3 because it is generally absent from these deposits (e.g. Murray Brook, Restigouche) or is present only locally at volumes less than 15% (e.g. Devils Elbow, No. 6, Orvan Brook). Where present, pyrrhotite greatly disrupts the alteration sequence shown in Figure 3. In the Pb-Zn and pyrite zones the presence of pyrrhotite greatly accelerates the oxidation of sphalerite-galena and pyrite. In the Bathurst camp this process is well defined in the Half Mile Lake deposit. Presumably, the ability of pyrrhotite to oxidize very rapidly increases the local concentration of ferric ion in pore solutions, thus accelerating the oxidation of nearby sulfide grains. Recent experimental data show that the abiotic oxidation rate of pyrrhotite is approximately 10 times greater than that of pyrite under the same experimental conditions (15).

Enrichment and Depletion of Elements During Formation of Supergene Zones

Accurate determination of elemental enrichments and depletions during supergene weathering of sulfide ore bodies is essential to an understanding of the factors controlling element mobilization and precipitation. During oxidation of massive

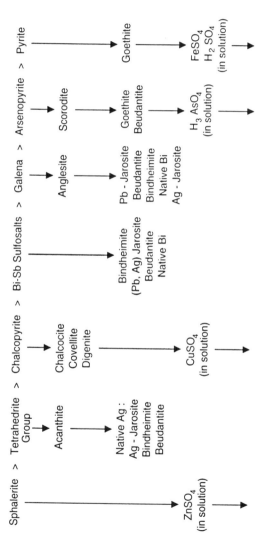

Figure 3. Alteration sequence for oxidation of sulfides in the massive sulfide deposits of the Bathurst Mining camp.

sulfide deposits to form gossans and supergene sulfide-enrichment zones, there is an important change in mass. Elemental concentrations in the gossan must, therefore, be corrected for mass and volume changes before interpretations of enrichment-depletion trends can be made. For the Bathurst area gossans, two methods of mass balancing have been employed: isovolumetric corrections based on specific gravity and porosity, and the "conservative element" approach.

In the Bathurst gossans, primary structural and sedimentary features such as sulfide layers, quartz veins and quartz-carbonate lenses are still well preserved as ghosted textures. Evidence of contemporaneous or post-gossan deformational features is lacking. There is, therefore, a good isovolumetric relationship between the Massive-Sulfide-Zone Gossan and its parent massive sulfide body (i.e. unit volumes equal). For the Murray Brook deposit, the average specific gravity of the gossan (3.29 g/cm^3 on 123 determinations) and massive sulfides (4.42 g/cm^3 on 45 determinations), and the average porosity of the gossan (21% on 448 determinations by image analysis) can be used in the following formula to obtain a volumetric mass-balance-adjustment factor (VMBAF) of 0.59:

$$\text{VMBAF} = [(\text{specific gravity gossan})-(\text{specific gravity gossan} \times \text{porosity gossan})] / \text{specific gravity primary ore}.$$
$$= (3.29)-(3.29 \times 0.21) / 4.42 = 0.59. \quad (1)$$

In the Massive-Sulfide-Zone Gossan at Murray Brook, Sn is present as cassiterite grains (<30 μm) in goethite. The cassiterite shows no signs of chemical attack even when present as 2-5 μm grains. Other than traces of stannite in the primary ore, Sn is present solely as cassiterite. In the gossan and primary ore the mean Sn concentrations are 447 ppm (n=147) and 251 ppm (n=64), respectively. This gives a "conservative element" mass balance adjustment factor (CMBAF) of 0.56 using the formula:

$$\text{CMBAF} = \text{Sn in primary ore (ppm)} / \text{Sn in gossan ore (ppm)}$$
$$= 251/447 = 0.56. \quad (2)$$

MBAF results for these two methods compare very well. The MBAF must be applied to elemental concentrations in the gossan in order to obtain elemental enrichment-depletion factors (EDF) and percentage losses or gains. The EDF values are obtained by the following formula:

$$\text{EDF} = (\text{C gossan} \times \text{VMBAF}) / \text{C massive sulfide} \quad (3)$$

where C is the elemental concentration. EDF values greater than one indicate enrichment whereas those less than one denote depletion.

Enrichments and depletions of various metals and sulfur during the formation of the Murray Brook gossan are given in Table II. These data show that, with complete oxidation of massive sulfides, practically all of the Zn, Cu, Cd and S and a large proportion of the Fe and Al are lost to the oxidizing fluids. The enrichments observed for Au, Pb, As, Sb, and Bi have a more complex interpretation. The values for the enrichment-depletion factor suggest that variable amounts of these elements have been mobilized downward through the gossan stratigraphy during weathering and have been concentrated in various mineral forms in the lower portions of the gossan. The upper portions of the gossan have been constantly eroded by physico-chemical processes during a long period of weathering. Although elements such as Pb, As, Sb and Bi are enriched in the gossan, are they also removed from the system as a whole? If Au, which displays the greatest enrichment in the oxidized zone, is considered to be conservative within the supergene system (i.e., minimal net loss of Au to solutions leaving the

Table II. Enrichment-depletion factors (EDF) and percentage gains or losses of various base metals and sulfur during the formation of the Murray Brook supergene deposit

ELEMENT	PRIMARY ORE	MASSIVE-SULFIDE ZONE GOSSAN	EDF	% GAIN/LOSS
S(%)	36.30	0.48	0.01	-99
Cd(ppm)	6.9	0.3	0.03	-98
Zn(ppm)	3715	245	0.04	-96
Cu(ppm)	5620	1000	0.10	-90
Fe(%)	33.10	28.20	0.50	-50
Al(%)	0.23	0.20	0.52	-48
Hg(ppm)	34	36	0.64	-38
Ag(ppm)	28.2	39.0	0.82	-18
Bi(ppm)	36	78	1.28	+28
Sb(ppm)	380	1000	1.55	+55
As(ppm)	3160	9120	1.70	+70
Pb(ppm)	1515	6165	2.40	+140
Au(ppb)	160	1096	3.97	+295

$$\text{EDF} = \frac{(\text{conc. gossan} \times 0.59)}{\text{conc. massive sulfide}}$$

EDF < 1.0 = depletion
    = 1.0 = conservative
    > 1.0 = enrichment

oxidizing zone), then all elements with enrichment-depletion factors (EDF) less than that for Au must show some mobility out of the oxidizing system. If data in Table II are recalculated assuming Au is not lost from the system as a whole, then Pb, As and Hg show losses from the system of 40, 67 and 84%, respectively. These losses would have occurred over the entire period of oxidation which, from preliminary paleomagnetic results (D.T.A. Symons and D.R. Boyle, unpublished data), occurred over at least a one-million-year period. As demonstrated later, massive sulfide tailings deposits can be expected, on a volume-for-volume basis, to oxidize much more quickly than *in situ* massive sulfide ore deposits.

From Table II, the relative order of decreasing mobility of elements out of this oxidizing massive sulfide system is S,Cd,Zn > Cu >> Fe,Al > Hg > Ag > Bi > Sb > As >> Pb >> Au. This sequence, derived from data on complete oxidation of sulfides, does not address the issue of temporal mobility of elements under temperate climatic conditions. This problem can be examined by interpreting the mobility series in light of information given in the mineralogical alteration sequence shown in Figure 3. For example, Zn and Cu show similar overall mobilities (system losses) for a completely oxidized system (Table II). Sphalerite alteration, however, does not lead to formation of intermediate secondary sulfides, whereas Cu-bearing minerals alter through a series of secondary sulfides exhibiting higher oxidation states. Zinc, therefore, is mobilized out of the system at high concentrations over a short period of time, but Cu will be mobilized more slowly at lower concentrations over a longer period of time. A similar relationship between Cu and Ni has been described by Thornber (7) for the Kambalda Ni sulfide deposits of Western Australia. The temporal difference between the mobilities of Cu and Zn, as shown by high Zn/Cu ratios in waters associated with oxidizing sulfide deposits characterized by Cu grades equal to or exceeding those of Zn, has been noted in many studies on geochemical dispersion of elements around undisturbed sulfide deposits (16-22) and in groundwaters flowing from old mine workings (23). Another factor which may affect the relative release rates of Zn and Cu is the different kinetic rates of oxidation of sphalerite and chalcopyrite, the two most common primary mineral sources of these elements (24,25). With regard to seepage from massive sulfide tailings impoundments in temperate climates, Cu, therefore, can be expected to be more of a long-term problem than Zn. This differential mobility' and the greater toxicity of Cu to aquatic organisms (maximum effluent discharge levels of 4 ppb Cu and 30 ppb Zn for waters of moderate hardness, (26)) should have a strong influence, both on engineering and on chemical remediation methods for sulfide tailings. Whether Pb and As will display greater or lesser long-term release problems than Cu for tailings impoundments requires more research on the relative stabilities of the secondary Pb and As minerals that may form in tailings impoundments. A monitoring program of the surface and ground waters associated with the Murray Brook tailings from the milling of the gossan (D.R. Boyle, and N. Smith, unpub. data) shows no measurable releases of Pb and As after two years, indicating that the Pb and As minerals formed in the Bathurst gossans are relatively stable in the present-day temperate climate.

The Bathurst area ores are fine-grained, which has resulted in significant losses of sphalerite, galena and chalcopyrite to the tailings. Tailings are, therefore, equivalent to finely ground massive sulfide deposits with grades and mineralogy very similar to lower grade ore bodies in the camp (e.g. Murray Brook deposit). By analogy, the sulfide alteration sequence outlined in Figure 3 and the elemental enrichment-depletion trend shown in Table II may apply during weathering of tailings, provided process intervention such as heavy liming does not occur. Research on the oxidation of the Heath Steele tailings impoundment (27,28,29) shows extensive oxidation of both sphalerite (with no replacement phases) and chalcopyrite (with some covellite replacement). Galena is oxidized much more slowly because of shielding by anglesite coatings, and pyrite displays the slowest oxidation rate with many relatively fresh pyrite grains occurring in the oxidation zone. This observed oxidation sequence for the Heath Steele tailings is similar to that shown in Figure 2 for the massive sulfide ore deposits of the Bathurst camp. With regard to enrichment-depletion trends, Boorman and Watson (27) showed severe depletions of Zn and Cu and slight enrichments in Pb in the oxidized zone of the Heath Steele tailings relative to the underlying fresh sulfides. Their data, however, were not mass balanced. For a single profile in these tailings, Blowes et al. (29) showed mass-balanced data in which Cu and Zn are depleted by up to 20% in the oxidation zone; both elements also show less depletion with depth, although Zn is more depleted overall than Cu. These depletions are much lower than those observed for the Murray Brook gossan (Zn = 96%; Cu = 90%; Table II). Lead is only enriched by about 5% in the oxidation zone of the Heath Steele tailings impoundment (27), whereas enrichment is 140% in the Murray Brook ore deposit. That oxidation of the Heath Steele tailings is still quite immature is indicated by the common presence of relatively fresh pyrite, chalcopyrite, galena, and sphalerite in the oxidation zone and the presence of most of the oxidized iron as Fe(II) oxyhydroxides and sulfates (29). Eventually the Fe (II) minerals will convert to Fe (III) oxides.

Gangue-induced pH-buffering reactions, which may have an effect on mobility and concentrations of elements in the saturated zone of tailings deposits (30,31), are minimal in the Bathurst camp due to a low gangue content (10-30%; mostly quartz with small amounts of carbonates and aluminosilicates). Thus, if left to oxidize without interventions such as heavy liming or engineered covers containing neutralizing agents, the Bathurst area tailings may form fairly rich Au-Ag oxide deposits, especially since 70% of the Au and about 35% of the Ag in these deposit (e.g. No. 12) end up in the tailings impoundments. Blowes et al. (29) have noted slight enrichments of Au and Ag in the as-yet-immature, oxidation zone at Heath Steele tailings.

Recent Oxidation of Massive Sulfide Deposits in the Bathurst Camp

The Bathurst camp supergene deposits were formed largely before onset of glaciation some 40,000 years ago and have undergone post-glacial oxidation. Evidence accumulated to date on climatic indicators, such as mineralogy (lack of complex metal halide, -carbonate, and -silicate minerals indicative of arid and semi-arid weathering conditions) and on-going paleomagnetic research (D.T.A. Symons and D.R. Boyle, unpub. data), indicates that these deposits underwent pre-glacial oxidation under warm temperate climatic conditions similar to the summer climate of the region today. Isostatic uplift and glacial erosion-deposition has since elevated the deposits above their pre-glacial hydrological regimes.

Under present oxidizing conditions, the Murray Brook gossan is considered here to be analogous to what we might expect from the weathering of tailings impoundments in northern temperate climates. Gossan depth at Murray Brook is 40 m, below which is an extensive sulfide-alteration zone of up to 25 m. The most striking feature of this altered-sulfide zone is the formation of a highly friable sulfide-quartz sand unit below the gossan (present in all the oxidized deposits of the camp). The composition and texture of this unit is remarkably similar to that of massive sulfide tailings produced in the camp.

The present-day water table in the Murray Brook deposit is slightly above the contact between the gossan and sulfide-quartz sand unit, and there is a capillary fringe of about 1.5 m. The sulfide-quartz sand unit is the major aquifer for lateral groundwater movement out of the deposit. Because the water table is above the gossan-sulfide contact the diffusion of atmospheric O_2 into the sulfide zone is considered to be minor compared to O_2-CO_2 input from groundwater flow.

Although electrochemical potentials have not been measured, the electrochemical acceleration of sulfide oxidation is also considered to be operative in this type of oxidizing environment (4,6,7). Groundwaters in the sulfide sand unit are acidic (pH 3.0-4.0) and base flow from the Murray Brook deposit has given rise to strong Zn, Cu, As, and Pb enrichments in the stream sediments of Gossan Creek down stream from the deposit (25-, 20-, 15- and 10-fold increases above stream sediment baseline concentrations, respectively). Acid extractable concentrations of Fe, Zn and Cu in these sediments are as high as 8.0, 0.3 and 0.05 wt%, respectively. However, Gossan Creek and metal-laden groundwaters entering its head have pH values between 6.5 and 8.5, indicating that acid neutralization by the wallrocks occurs along the flow path but considerable metal mobility still exists.

Oxidation of Massive Sulfide Tailings

As with massive sulfide ore systems, the rate of oxidation of massive sulfide tailings will depend on the following: a) position of sulfide body in hydrologic regime, b) diffusion rate of atmospheric O_2 into sulfide body, c) flow characteristics of O_2-CO_2-bearing groundwaters through sulfides (degree of O_2-CO_2 saturation), d) strength of electrochemical processes within sulfide body, e) porosity and permeability (primary and secondary), f) types and abundances of sulfides, g) types and abundances of gangue minerals, h) grain size distributions (reactive surface area), i) degree and types of microbiological activity, j) climate, k) ambient and. internal temperatures, l) nature and composition of surrounding wallrocks or overburden sediments, and m) geometry and structure of sulfide body.

Research on abatement of acid mine drainage from tailings impoundments has concentrated mainly on trying to eliminate atmospheric O_2 from the sulfides. However, research on natural sulfide oxidation systems demonstrates that factors such as movement of O_2-CO_2- saturated groundwaters into the sulfide body, the strength of an overall electrochemical oxidation cell within the sulfide body, (4,6,7) and possibly the degree of microbiological activity above and below the water table (32,33,34) are also important. Some sulfide ore bodies undergoing near-surface oxidation display extensive alteration of sulfides to depths of up to 500 m below the water table (4,6,35). Clearly this oxidation cannot be controlled by molecular diffusion of atmospheric O_2.

Sound field or laboratory data on the oxidation rates of high sulfide-bearing tailings under open atmospheric conditions are scarce. Estimates of sulfide oxidation rates for certain conditions, however, can be obtained from the works of Boorman and Watson (27) and Blowes and Jambor (30). For the Heath Steele mine tailings in the Bathurst camp, which are composed mainly of pyrite and small amounts of sphalerite, galena, chalcopyrite, arsenopyrite, and pyrrhotite, Blowes et al. (29) describe a 50 cm-thick zone of extensive oxidation that formed over approximately 20 years after the end of tailings deposition.

Oxidation at the Waite Amulet tailings impoundment in the Rouyn district of Quebec, which is composed of 25-30% sulfide minerals (pyrite, pyrrhotite, sphalerite, and chalcopyrite) and magnetite, has been studied by Blowes and Jambor (30). This impoundment was covered shortly after closure with a limestone-vegetation cover, and in approximately 26 years complete oxidation of sulfides has occurred to depths of 30-50 cm. Using the oxidation-rate model of Davis and Ritchie (36), Blowes and Jambor (30) determined that complete oxidation of the Waite Amulet tailings down to the present water table (approx. 300 cm) would take 150-640 years.

The above types of estimates, whether based on field observations, empirical evidence, or deterministic modelling, do not account for the rate of oxidation of sulfide bodies below the water table under non-steady- state groundwater conditions and electrochemical potentials. These are the conditions which are considered to contribute significantly to the oxidation of many types of natural sulfide deposits to depths in excess of 500 m (4,6,37,38) and to the considerable mobility of elements under less oxidizing conditions. In this respect the massive sulfide ore bodies of the Chibougamau camp in Quebec (Copper Rand and Henderson deposits) are excellent examples (35). These deposits display alteration down to 400 m with almost complete conversion of primary sulfides to secondary sulfides and minor oxides. The high secondary porosities of these zones indicate that considerable mobility of metals and sulfur has occurred both within and out of the ore zones. Other examples of deep oxidation of sulfide ore bodies well below existing water tables have been described (2,39,40).

Although not linear with time, the oxidation rates of sulfides above the water table at the two tailings impoundments described are orders of magnitude greater than those for massive sulfide deposits exposed at surface in Canadian climates since the last period of glaciation (8,000-10,000 years ago). Most exposed massive sulfide bodies examined by the author in Canada, many of them in the mining camps where the above-mentioned tailings are located, rarely display more than 1 m of intensive oxidation since glaciation. This includes massive and vein sulfide deposits that have been scraped clean by glaciers and which have been exposed to atmosphere for 8,000 to 10,000 years. Similar observations have been noted for the temperate climatic regions of Scandinavia (41) and Russia (22), thereby suggesting a maximum post-glacial rate of massive sulfide oxidation of only 1 cm per 100 years. The oxidation rate for tailings is therefore about 200 times greater than for consolidated massive sulfide bodies (using an average of 2 cm/year for tailings). The differences can be attributed largely to increases in porosity and reactive surface area of the tailing sulfides, although such factors as increased microbiological activity and stronger electrochemical oxidation processes in tailings have not been evaluated. Many other factors in these two environments, such as hydrologic setting, presence of neutralizing gangue minerals, sulfide mineralogy, and climate would seem to be equivalent.

An important factor to consider in the oxidation of sulfide deposits below the water table is the introduction and flow characteristics of O_2-CO_2- saturated groundwaters within the sulfide body (ore deposit or tailings). This point has been emphasized previously in the section on recent oxidation of massive sulfide ore bodies. Recent research (28,29,42,43) on oxidation of tailings and groundwater movement within and through tailings impoundments indicates that, for tailings remediation methods to be effective, greater emphasis on controlling the flow of groundwaters will be as important as controlling the ingress of atmospheric O_2. In many cases, simply cutting off access to atmospheric O_2 by using systems such as engineered covers may only prolong the acid drainage problem at sub-toxic levels over a much longer period of time if the movement and oxidative powers of groundwaters in the system are not remediated as well.

Although considerable laboratory and metallurgical research has been carried out on bacterial behavior in the oxidation-reduction process involving sulfides (33,34,44), very little field research has been done on this important oxidation control. In particular, the effects of climate (ambient and internal temperatures), groundwater composition, and changes in the types and relative abundances of the sulfides on lithoautotrophic bacterial reactions have not been addressed. In at least two of the abandoned sulfide tailings impoundments in eastern Canada (Heath Steele and Waite Amulet) bacterial activity has been identified (27,45). The magnitudes of these activities and the factors controlling microbial oxidation in these environments have, however, not been addressed. It has been noted by researchers that bacterial activity below the water table is greatly reduced compared to that in the vadose zone (32 and references therein). What is not known, however, is whether the reduced activity within saturated zones of high-sulfide environments is still sufficient to cause significant oxidation of sulfides and hence mobility of toxic metals. Lack of *in situ* research on this subject has been compounded by our inability, as yet, to recognize microbial oxidation textures in continental supergene sulfide deposits, although considerable data has now been generated on the microbial oxidation of sea floor sulfide deposits (46-48).

Conclusions

Supergene zones derived from the massive sulfide deposits of the Bathurst camp are both pre- and post-glacial in origin. Oxidation in both cases occurred under temperate climatic conditions and has led to: a) total conversion of sulfides to goethite and stable hydrated oxides and sulfates of Fe, Pb, As, and to a lesser extent Sb, Bi and Ag in the gossan zone; b) development of considerable secondary porosity (15-30%); c) strong enrichment of Au (295%) in the gossans; d) considerable loss of S (-99%), Zn (-96%), and Cu (-90%) to the oxidizing fluids; and e) retention of Pb (+140%) and As (+70%) in the gossans. The oxidation sequence of sulfide minerals and the relative mobilities of various metals out of these oxidized zones are remarkably similar to the oxidation-mobility characteristics observed for massive sulfide tailings impoundments in temperate Canadian climates. Oxidation rates for massive sulfide tailings deposits are greater than those of natural consolidated massive sulfide ore bodies by a factor of 200 or more. Most of this difference is thought to be due to the much higher porosities and effective surface areas of sulfides in the tailings.

Studies of natural sulfide oxidation processes in the Bathurst camp and elsewhere would indicate that for long-term abatement of surface and subsurface effluents from massive sulfide tailings a better understanding is needed of a) groundwater mass transfer-reaction processes in massive sulfide tailings impoundments and their surrounding overburden environments, b) the *in situ* processes that control microbiological activity of sulfide oxidizing bacteria, and c) the factors that control the establishment and functioning of electrochemical corrosive cells within sulfide bodies. Greater knowledge of these processes will be important to a better understanding of the factors controlling the mobility of toxic elements out of sulfide zones and the development of remediation methods for lowering toxic metal concentrations to regulatory levels over both the short and long term.

Acknowledgments

I would like to thank N. Smith and K. McNeil of Nova Gold Inc. for supplying valuable data on site operations and greatly assisting the author in carrying out geological and environmental studies at the Murray Brook mine. W.M. Luff of Brunswick Mining and Smelting Ltd. is also thanked for supplying valuable technical and geological information on the #12 mine. Reviews by J.L. Jambor of Energy, Mines and Resources Canada and two external reviewers greatly aided clarity of the text.

Literature Cited

1. Svela, O. The Northern Miner Magazine 1992, 7 (4), 19-20.
2. Guilbert, J.M.; Park, C.F. Jr., The geology of ore deposits; 1986, W.H. Freeman and Co.:, New York, NY, Chapter Seventeen, pp774-831.
3. Nickel, E.H.; Ross, J.R.; Thornber, M.R. Economic Geology 1974, 69, 93-107.
4. Blain, C.F.; Andrew, R.L. Minerals Science Engineering 1977 9, (3), 119-150.

5. Boyle, R.W. Geol. Surv. Canada Bulletin 280 1979, 584 p.
6. Thornber, M,R, Chemical Geology 1975, 15, (1), 1-14.
7. Thornber, M.R. Chemical Geology 1975, 15, (2), 117-144.
8. Taylor, G.F. J. Geochem. Explor. 1983, 18 (2), 87-130.
9. Taylor, G.F. J. Geochem. Explor. 1984 22 (1-3), 351 -352.
10. Scott, K.M.; Taylor, G.F. J. Geochem. Explor. 1987, 27 (1-2), 103-124.
11. McCutcheon, S.R. Exploration and Mining Geology Journal 1992, 1, No. 2, 105-120.
12. Boyle, D.R. In Source, Transport and Deposition of Metals, M. Pagel and J.L. Leroy (eds), Balkema Publ.: Rotterdam 1991, pp647-652.
13. Jambor, J.L. Canadian Centre for Mineral and Energy Technology. 1978, Report 78-14, 26 p.
14. Alpers, C.N.; Brimhall, G.H., Econ. Geol. 1989, 84, 229-255.
15. Nicholson, R.V.; Scharer, J.M. In this volume.
16. Boyle, R.W.; Tupper, W.M.; Lynch, J.; Friedrich, G.; Ziauddin, M.; Shafiqullah, M.; Carter, M.; Bygrave, K. Geol. Surv. Canada Paper 65-42, 49 p.
17. Chernyayev, A.M.; Chernyayev, L.E. Geochemistry Inter. 1962, 10, 1030-1041.
18. Kovalev, V.F.; Kozlov, A.V.; Koval'chuk, A.I. Geochemistry Inter. 1961, 7, 638-646.
19. Naumov, V.N.; Pachadzhanov, D.N.; Burichenko, T.I. Geochemistry Inter. 1972, 9 (1), 129-134.
20. Hoag, R.B.; Webber, G.R. J. Geochemical Explor. 1976, 5 (1), 39-57.
21. Andrews-Jones, D.A. Colorado School of Mines Mineral Industry Bull. 1968, 11 (6), 1-31.
22. Smirnov, S.S. Oxidation of Sulfide Deposits. Akademiya Nauk SSSR, 1951. Geological Survey of Canada Translation No. 6740, 334p.
23. Alpers, C.N.; Nordstrom, D.K.; Thompson, J.M.; Lund, M. In this volume.
24. Rimstidt et al. In this volume.
25. Alpers, C.N.; Nordstrom, D.K. Geological Society of America Abstracts with Programs, 1989, 21 (6), A102.
26. Inland Waters Directorate (Canada). Canadian Water Quality Guidelines. Inland Waters Directorate, Water Quality Branch Publ., Ottawa, 1987, Section 3.10.
27. Boorman, R.S.; Watson, D.M. Canadian Institute of Mining and Metallurgy Bulletin 1976 69, (772), 86-96.
28. Blowes, D.W.; Cherry, J.A.; Reardon, E.J.; Jambor, J.L. Applied Geochemistry 1993, (in press).
29. Blowes, D.W.; Jambor, J.L.; Appleyard, E.C.; Reardon, E.J. and Cherry, J.A. 1992, Explor. Mining Geol., 1 (3), 251-264.
30. Blowes, D.W.; Jambor, J.L. Applied Geochemistry 1990, 5, 327-346.
31. Dubrovsky, N.M.; Cherry, J.A.; Reardon, E.J.; Vivyurka, A.J. Canadian Geotechnical Journal 1984 22, 110-128.
32. Nordstrom,D.K. In: Acid Sulfate Weathering; Editors Kittrick, J.A., Fanning, D.S., Hossner, L.R., Kral, D.M.; Hawkins, S. Soil Science of Amer. Spec. Publ. 1982, 37-61.
33. Lundgren, D.G.; Silver, M. Ann. Rev. Microbiol. 1980, 34, 263-283.
34. Trudinger, P.A. Minerals Sci. Engng. 1971, 3, (4), 13-15.
35. Allard, G.O. Geological Association of Canada Program and Abstracts, 74th Annual Meeting, St. John's, Newfoundland 1974, p. 1.
36. Davis, G.B.; Ritchie, A.I.M. Applied Mathematical Modelling 1986, 10, 314-322.
37. Thornber, M.R.; Wildman, J.E. Chemical Geology 1979, 24, 97-110.

38. Sveshnikov, G.B.; Ryss, Yu-S. Geochemistry International 1964, 3, 198-204.
39. Jensen, M.L.; Bateman, A.M. Economic Mineral Deposits 1981, John Wiley and Sons, New York, 593p.
40. Boyle, R.W.; Daas, A.S. Canadian Mineralogist, 1971. 11, 358-390.
41. Schneiderhohn, H.; Ramdohr, P. Lehrbuch der Ermikroskopie Verlag von Gebruder Borntraeger: Berlin, 1931, 168 p.
42. Dubrovsky, N.M.; Morin, K.A.; Cherry, J.A.; Smyth, D.J.A. Canadian J. Water Poll. Res. 1984, 19, 55-89.
43. Morin, K.A.; Cherry, J.A.; Dave, N.K.; Lim, T.P.; Vivyurka, A.J. J. Contam. Hydrol. 1988, 2, 271-303.
44. Suzuki, I; Chan, C.W.; Takeuchi, T.L. In this volume.
45. Dave, N.K.; Lim, T.P.; Siwik, R.S.; Blackport, R. In Natl. Sympos. on Mining, Hydrology, Sedimentology and Reclamation, Univ. of Kentucky, Lexington, Kentucky, 1986, pp13-19.
46. Juniper, S.K.; Fouguet, Y. Canadian Mineralogist 1988, 26, 859-869.
47. Alt, J.C. EOS, Trans. Amer. Geophys. Union 1986, 18, 526.
48. Jannasch, H.W.; Wirsen, C.O. Appl. Environ. Microb. 1981, 41, 528-538.

RECEIVED September 2, 1993

Chapter 33

Thiosulfate Complexing of Platinum Group Elements

Implications for Supergene Geochemistry

Elizabeth Y. Anthony[1] and Peter A. Williams[2]

[1]Department of Geological Sciences, University of Texas at El Paso, El Paso, TX 79968
[2]Department of Chemistry, University of Western Sydney, Nepean, P.O. Box 10, Kingswood, New South Wales 2747, Australia

> Studies of placers and profiles above weathering sulfide deposits demonstrate that platinum-group elements (PGE) are mobile during weathering processes. Given that chloride complexes of these elements are stable only at high Eh, low pH and high salinity, other ligands may be involved. Leaching experiments of Pt and Pd native metals have shown that solutions of thiosulfate ($S_2O_3^{2-}$) are strikingly effective in dissolution of the PGE. Dissolution rates are independent of pH in the range of 6 to 9. ^{195}Pt nuclear magnetic resonance studies indicate the formation of a single aqueous species of Pt(II) with thiosulfate, $Pt(S_2O_3)_4^{6-}$, for solutions with ratios of $S_2O_3^{2-}$:Pt from 2:1 to 10:1, when $PtCl_4^{2-}$ is reacted with $S_2O_3^{2-}$. However, this complex is not thermodynamically stable, and S-bridged oligomers are ultimately formed. Such is also the case with Pd(II). Thiosulfate complexing appears likely to be important during the weathering of PGE-bearing sulfide ores.

The platinum group elements (PGE) have traditionally been considered to be inert during weathering. This concept results from the very limited solubility of the elements under most conditions. In common with gold, chloride complexing is only effective at low pH, high Eh, and high concentrations of chloride (1).

There is, however, a growing body of evidence that the PGE are mobile in aqueous solutions, both at high temperature during the generation of protore (1), and also in the supergene environment. Evidence for supergene mobility is found in the textures of nuggets from placers and in laterites (2, 3) and also in the dispersion of PGE in weathered profiles above sulfide deposits. Figure 1 summarizes some available data concerning such weathered profiles. Most of the elements show complex patterns of enrichment and depletion relative to the more immobile elements Rh and Ir, indicating that residual enrichment is not solely responsible. Further, the patterns of enrichment often diverge; for example, Pt as sperrylite is enriched and

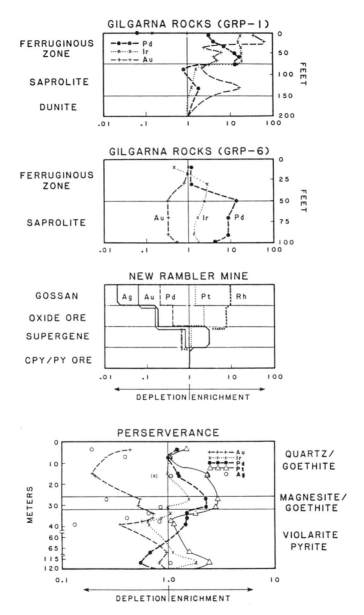

Figure 1. Patterns of enrichment and depletion of selected elements in weathering profiles. Enrichment factors are concentrations relative to bedrock or protore. Data from Gilgarna rocks (4), the New Rambler Mine, Wyoming (5), the Stillwater complex (6), and the Perseverance deposit, W. Australia (7) are illustrated. No corrections for density or volume changes during weathering have been applied.

Pd as tellurides depleted in the New Rambler Mine gossan/protore, suggesting that the supergene geochemistry of the various related elements and minerals is not uniform. Aqueous complex coordination chemistry in conjunction with the refractory nature of some of these minerals is ultimately responsible for the dispersion patterns. Colloidal and detrital transport phenomena may also be significant.

Given the limited conditions for the efficacy of Cl⁻ complexing, it is natural to seek alternative ligands. Thiosulfate has been suggested as being significant (8). Though it is a metastable species formed during the oxidation of sulfide to sulfate, many studies have documented its key role in the oxidation of pyrite under a variety of conditions (9-11). Recent studies have demonstrated its importance as a natural ligand for Au and Ag in groundwaters (12, 13). Indeed, thiosulfate is a key intermediate in natural sulfur systems. It needs to be taken into account as a ligand for a number of metal ions under appropriate conditions for its formation. We present here the results of a series of dissolution experiments designed to test the effectiveness of thiosulfate complexing of the PGE with respect to geochemical mobility. We report also the results of ^{195}Pt NMR studies undertaken to determine the distribution of species in aqueous solutions containing Pt(II) and $S_2O_3^{2-}$.

Experimental methods

Dissolution Experiments. Metal blacks used in dissolution experiments were precipitated by reduction from dissolved chloride salts, resulting in fine-grained powders which were stored under water. Two different methods for precipitation were used, that of Willstätter (14) for Pt and Pd, and that of Brown (15) for Rh, Ir and Ru. Electron microscope measurements showed that the dense metal blacks were aggregates of particles greater than 10 microns in size. The metals were reacted in 500 cm³ Ehrlenmeyer flasks with $Na_2S_2O_3$ solutions of varying concentrations. Carbonate buffer was used in the pH range 8 to 9, and phosphate buffer in the pH range 6 to 8. $Na_2S_2O_3$-free solutions were tested to assess effects of buffers on PGE solubility. In all cases, metal concentrations were below detection limits (< 0.2 ppm). For some experiments Cl⁻ was added (as dissolved NaCl) to determine whether it enhanced solubility. The flasks were covered with an oxygen-permeable membrane and submersed in a thermostatted water bath at 25°C.

Aliquots were removed at intervals and pH measured to verify that the buffer was effective. Although the pH increased gradually, the final pH was within half a unit of initial pH except for experiments 6 and 13 which changed from 8.5 to 9.3 and 8.5 to 9.7 respectively. The aliquots were filtered using Whatman GF/F fiberglass filters (0.4 micron) prior to analysis, to remove colloid-size solids. Analyses by ICP-AES incorporated standards prepared with thiosulfate of appropriate concentrations. Accuracy is estimated to be within 10 percent relative.

^{195}Pt NMR Experiments. Nuclear magnetic resonance (NMR) spectroscopy takes advantage of spin, which is one of the quantum properties of nuclei. When a nucleus with non-zero spin interacts with a magnetic field, it increases or decreases its energy by quantized frequencies which vary as a function of the electron density in the vicinity of the nucleus. Because absolute values of these frequencies are difficult to obtain, they are reported as shifts (δ, in ppm of change) relative to a useful standard. A more detailed discussion of NMR spectroscopy and its application to geological questions may be found in Kirkpatrick (16). In addition, there are numerous textbooks and reference books for NMR spectroscopy in the chemical literature. ^{195}Pt NMR spectra for our experiments were recorded at 21°C using a Bruker WM 360 spectrometer operating at 77.29 MHz and a Jeol FX90 operating at 19.21 MHz and 27°C employing K_2PtCl_4 as an internal standard (δ 0 ppm).

Results

Leaching Experiments. Table I and Figure 2 illustrate results of the dissolution experiments for Pt and Pd. The most significant results were the high concentrations in solution of the PGE, especially Pd. The maximum solubilities observed were 111 ppm after 75 days for Pd and 23 ppm after 46 days for Pt. The solutions were colorless at the time of mixing of the Pd metal and the thiosulfate, and they developed a a clear lemon-yellow color over a period of days. The intensity of the color increased each day and was directly correlated with the measured abundance of Pd. The capacity of the solutions to continue to dissolve Pd diminished with time, and eventually a plateau was reached. The approach to the plateau was accompanied by development of a more intense brown color. Finally the solutions became opaque and in some cases brown solids settled out leaving a colorless solution. Pd in these colorless solutions was below detection limits. The solids have not yet been characterized. It appears from this behavior that the diminished capacity to dissolve Pd was associated with slow destabilization of the aqueous Pd thiosulfate complex. The rate of change from the lemon-yellow aqueous complex to the precipation of solids increased with decreasing pH and decreasing concentrations of thiosulfate. Both factors have the same effect on the rate of decomposition of free thiosulfate (17, 18).

Aliquots removed from the reaction flasks retained their lemon- yellow color much longer, depending on pH. After 8 months' storage, the aliquots with pH of 10 still retained their yellow colour, while those with pH of 8.5 had turned brown and opaque, perhaps due to the formation of colloidal phases. It seems that breakdown of the aqueous complex was catalyzed by both metal and solid products in the reaction mixture; these factors have also been shown to enhance the breakdown of free thiosulfate (17).

The experiments demonstrate that the initial rates of dissolution are essentially independent of pH in the range 5.9 to 9. Experiments 7 and 8 used the same batch of metal and the same type of buffer, and differ only in pH. They have quite similar dissolution rates for the first 30 days. Experiments 3 and 6 differ from the previous two in the batch of metal used; they show higher rates of dissolution. The solution at pH 10, which was self-buffered, shows a slightly smaller capacity for dissolution than the solution at pH 8.5 (Fig. 2).

Similar dissolution rates for solutions 0.01 M in $S_2O_3^{2-}$ but with various pH values were found. All were run with the same buffer and the same batch of metal, pH ranging from 6 to 8.5. After 10 days, however, the lowest pH solution levelled off in concentration quite rapidly, followed by the solution at pH 7. The solution at pH 8.5 continued to dissolve metal until the experiments were discontinued. No enhanced solubility of Pd in the presence of added Cl$^-$ under these conditions was observed.

Table I. Pd concentration (ppm) from dissolution experiments

Time, days	Experiment number								
	3	6	7	8	11	12	13	15	18
1			2	4	10	6	6		14
2			5	9				12	
3					19	22	17	16	
4	22	36			22	26	22	21	20
6					27	31	26	27	25
8			23	25	32	38	33		
14	43	70	29	28	33	46	44	36	32
17		73							
20	46	76	32	32			54	50	
23					32		56	53	41
26	54	81	35	37					
31			37	42					
32	62	97							
37	67	99							
40			41	49					
46	69	102							56
60			42	57				75	
66	84	106							
75	84	111							
$[S_2O_3^{2-}]/M$ [a]	0.1	0.1	0.1	0.1	0.05	0.05	0.05	0.05	0.01
Metal [b]	1	1	2	2	2	2	2	2	2
pH	10	8.5	8.2	9.0	5.9	6.9	8.5	8.7	8.9

[a] Initial concentration. [b] Particular batch of metal "black".

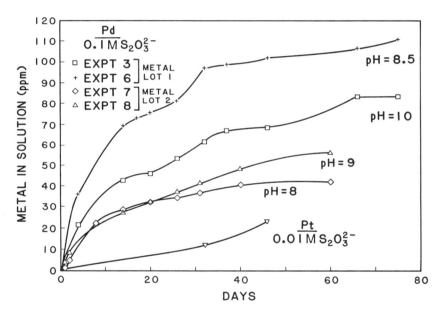

Figure 2. Extent of dissolution of Pt and Pd for various concentrations of $S_2O_3^{2-}$.

The rates of dissolution of the PGE metals vary inversely with the concentration of thiosulfate; there is a tendency for the rates to increase with decreasing concentration. The concentration of Pd after 24 hours was 4 ppm for 0.1 M $S_2O_3^{2-}$, 6 ppm for 0.05 M, 14 ppm for 0.01 M, and 20 ppm for 0.001 M. This kind of behavior is expected if the ligand chemisorbs at the metal, thus inhibiting diffusion of oxygen to the surface.

Preliminary experiments were carried out with Os, Ru and Rh. These elements exist predominantly in oxidation state (III) in supergene environments, and prefer octahedral coordination (19). Mixed $S_2O_3^{2-}$ and Cl^- ligand complexes may be more important than for Pt(II) and Pd(II), which prefer square planar coordination (19), but little is known of the coordination chemistry of these elements with $S_2O_3^{2-}$. Ru and Rh were reacted with both 0.01 M $S_2O_3^{2-}$ and mixed 0.01 M $S_2O_3^{2-}$ plus 0.01 M Cl^- solutions. Ir was reacted only with the latter solution.

All three elements were detected in solution after 28 days; Ru in the absence of chloride: 1.5 ppm; Ru in thiosulfate and chloride: 2.5 ppm; Rh: 1 ppm in the chloride-free case and 1.4 ppm in the mixed ligand solution. The solution containing Ir had changed to a pale lemon-yellow color, and had a concentration of 200 ppm dissolved Ir. These experiments demonstrate the potential importance of thiosulfate and mixed thiosulfate-chloride complexing of all PGE. Simple Os complexes of thiosulfate have also been reported recently (20), but this metal has not been tested in batch leach runs.

^{195}Pt NMR Experiments. Ryabchikov (21, 22) proposed four complexes of Pt (or Pd) with thiosulfate alone: cis- and trans-$Pt(S_2O_3)_2^{2-}$, $Pt(S_2O_3)_3^{4-}$ and $Pt(S_2O_3)_4^{6-}$. The mixed ligand species $Pt(S_2O_3)Cl_2^{2-}$ was also claimed to have been isolated. In order to ascertain which, if any, of these species might be present in the leach solutions reported above, we have carried out ^{195}Pt NMR experiments on reaction solutions containing $PtCl_4^{2-}$ and $S_2O_3^{2-}$. Isolation of the $Pt(S_2O_3)_3^{4-}$ ion as its $Co(NH_3)_6^{3+}$ salt in pure form has enabled unequivocal assignment of the ^{195}Pt NMR resonance for this species (-2103 ppm). At all ratios of Pt(II):$S_2O_3^{2-}$, for periods of days, only three resonances are detected in ^{195}Pt NMR spectra of aqueous solutions: that of the starting material, $PtCl_4^{2-}$, that of the tetrakis species $Pt(S_2O_3)_4^{6-}$ and that of a transient species (-1227 ppm) which is almost certainly trans-$[Pt(S_2O_3)_2Cl_2]^{4-}$. Solutions deficient in ligand react to yield $Pt(S_2O_3)_4^{6-}$, and excess $PtCl_4^{2-}$ remains as such. Thus it seems certain that the redox measurements and the stability constant reported for the tetrakis complex (20) are reliable. No other species, such as described by Ryabchikov (21, 22), was detected in solution under comparable conditions to synthetic cases.

Nevertheless, at an M:L ratio of 1:2, solid, lemon-yellow $K_2[Pt(S_2O_3)_2]\cdot 3H_2O$ (from K_2PtCl_4) precipitates after some time. This species, which contains the trans-$Pt(S_2O_3)_2^{2-}$ ion, undergoes a solid-state reaction to yield an isomeric yellow-ochre compound of the same stoichiometry, a salt of the S-bridged coordination polymer.

No tetrakis complex appears to exist for the analogous Pd(II) system; only the related bis species can be isolated from solution, although it is apparent that S-bridged oligomers are present in solution. It should be noted that $Pt(S_2O_3)_4^{6-}$ is not thermodynamically stable. Prolonged reaction times at all M:L ratios give oligomers as well. These results, taken together, provide a further explanation of the redox results for the Pd(II) and Pt(II) systems reported previously (23). It should be noted that although the concentrations of Pt in the NMR experiments were high (typically 0.5 M), the reaction to form the oligomers is not dependent upon this condition; the reaction is irreversible. In this connection we have observed that the polymeric yellow-ochre compound $K_2[Pt(S_2O_3)_2] \cdot 3H_2O$ is not soluble in concentrated aqueous solutions of thiosulfate.

We note, however, that the various thiosulfate complexes are very stable towards acid. For example, $Pt(S_2O_3)_4^{6-}$ is not decomposed by concentrated HCl and by aqua regia only with difficulty. Thus, while free $S_2O_3^{2-}$ is a fairly reactive metastable intermediate, Pt(II) and Pd(II) complexes of thiosulfate may persist for appreciable times. Coordination to the metal has the effect of a protecting group as far as the ligand is concerned. This fact has profound implications for the transport of the elements in the supergene zone when PGE-bearing sulfide ores are oxidized, in that the complexes are sufficiently stable to persist for a significant time under ambient conditions.

Discussion

The experiments reported above demonstrate that thiosulfate is very effective in promoting dissolution of the platinum group elements via the oxidation of the native elements in the presence of O_2. The observations confirm predicted efficacy based on the thermodynamic stability of PGE thiosulfate complexes ([8]) and show that kinetic parameters for the reactions are favorable for the dissolution of large amounts of PGE even for short reaction times. Experiments involving the dissolution of the metals themselves are especially pertinent. Of course, extension of these observations to natural conditions would depend on the likelihood or not of the occurrence of native Pd and Pt in oxidizing PGE ores. It is well established ([24]) that binary primary minerals such as PtS, $PtAs_2$ and their congeners, including the other PGE, are oxidized to the native metals in the first instance. This seems at first somewhat paradoxical, but the reaction, for example, of PtS to Pt plus SO_4^{2-} represents in fact an overall oxidation. Sperrylite altered to native platinum has been observed in oxidized Australian ores (N. Gray and P.A. Williams, unpublished results). The remaining question of interest is under what near-surface conditions thiosulfate complexing should be encountered.

Numerous studies have demonstrated the importance of thiosulfate as an intermediate in the oxidation of sulfides; however, its longevity is compromised by low pH, high temperature, bacteria and other catalysts. Moses et al. ([10, 11]) report its detection at pH 3.9 and greater although they find that it decomposes quickly at a pH less than 5. Goldhaber ([9]) reports detection of thiosulfate at pH 7 during oxidation of pyrite, although it is not dominant among the sulfur species below pH 8. One of the few kinetic studies in alkaline conditions is that of Pryor ([25]). He reports an observed first-order rate constant for the decomposition of free thiosulfate in H_2O of approximately 10^{-4} sec^{-1} for pH 6 to 7 and 270°C, but the reaction is slow at room temperature. Studies of the temperature dependence of the decomposition rate are few ([26, 27]), although it has been demonstrated that 98 percent of original thiosulfate will persist after 200 days in sterile, stoppered vessels kept at room temperature ([17]).

The general limits for Eh and pH in weathered profiles are well-known ([28, 29]). These limits have direct applicability to weathering profiles over PGE deposits.

Both Mann (30) and Thornber and co-workers (31, 32) have emphasized the importance of the goethite buffer in the weathering of nickel sulfide deposits. They have characterized specific conditions for incipient weathering in the "depth" and "transitional" environments and find two different regimes. The first is close to the goethite equilibrium boundary with more reduced species and is characteristic of the weathering of disseminated ore. Attendant higher pH is attributed to buffering by silicates and carbonates in the matrix. These conditions are appropriate for formation of PGE thiosulfate complexes, in that thiosulfate itself can be present in considerable amounts.

Considering the remarkable chemical metastability of Pt(II) and Pd(II) complexes of $S_2O_3^{2-}$, it seems probable that such species play a role in the mobilization of the PGE in weathering sulfides akin to that now recognized for Ag and Au (12, 30). Depletion of Pd in the near-surface environment in the Perseverance deposit (7) and in the oxide zone of the New Rambler mine (5) may be interpreted in terms of thiosulfate chemistry. Supergene enrichment of Pt and Pd in the Perseverence deposit may in turn be viewed as being analogous to the behavior, for example, of copper during the oxidation and secondary enrichment of base-metal sulfides. In addition, the relative reactivities of Pd and Pt leading to removal of Pd in placer deposits (2) may arise from (inter alia) faster oxidation of Pd metal in the presence of $S_2O_3^{2-}$ as compared to Pt.

Finally, it is worth commenting on the possibility that thiosulfate complexing of PGE metals might be significant for the development of low-temperature (< 200°C) hydrothermal Pd-Au-carbonate veins which have arisen from relatively oxidized, high pH solutions. Particular examples include deposits at Hope's Nose, Devon, U.K. (33) and at Goodsprings, Nevada, U.S.A. (34). Limestone hosts in these environments would provide conditions conducive to the stabilization of thiosulfate. However, the exact role of thiosulfate complexing of the PGE group in such settings remains to be fully assessed.

Acknowledgments

E.Y.A. wishes to thank the Leverhulme Trust for a USA/Commonwealth post-doctoral fellowship. We wish to thank Dr M.J.E. Hewlins of the Department of Chemistry, University of Wales College of Cardiff for assistance with collection of the NMR spectra. Johnson-Matthey provided precious metal chemicals for the laboratory work. Special thanks are given to John Anthony for his helpful discussions throughout the course of the study. Referees' comments on the manuscript have been greatly appreciated.

Literature Cited

1. Mountain, B.W.; Wood, S.A. Econ. Geol., 1988, 83, 492.
2. Stumpfl, E.F.; Tarkian, M. Econ. Geol., 1976, 71, 1451.
3. Bowles, J.F.W. Econ. Geol., 1986, 81, 1278.
4. Travis, G.A.; Keays, R.R.; Davison, R.M. Econ. Geol., 1976, 71, 1229.
5. McCallum, M.E.; Loucks, R.R.; Carlson, R.R.; Cooley E.F.; Doerge, T.A. Econ. Geol., 1976, 71, 1429.
6. Fuchs, W.A.; Rose, A.W. Econ. Geol., 1974, 69, 332.

7. McGoldrick, P.J.; Keays, R.R. Econ. Geol., 1981, 76, 1752.
8. Plimer, I.R.; Williams, P.A. In Geoplatinum 87; Prichard, H.M.; Potts, P.J.; Bowles, J.F.W.; Cribb, S.J., Eds.; Elsevier: London, 1988; pp 83-92.
9. Goldhaber, M.B. Amer. Jour. Sci., 1983, 283, 193.
10. Moses, C.O.; Herman, J.S. Geochim. Cosmochim. Acta, 1991, 55, 471.
11. Moses, C.O.; Nordstrom, D.K.; Herman, J.S.; Mills, A.L. Geochim. Cosmochim. Acta, 1987, 51, 1561.
12. Webster, J.G. Geochim. Cosmochim. Acta, 1986, 50, 1837.
13. Benedetti, M.; Boulegue, J. Geochim. Cosmochim. Acta, 1991, 55, 1539.
14. Willstätter, R.; Waldschmidt-Leitz, E. Berichte d. D. chem. Gesellschaft, 1921, 54, 113.
15. Brown, H.C.; Brown, C.A. J. Am. Chem. Soc., 1962, 84, 1493.
16. Kirkpatrick, R.J. In Spectroscopic methods in mineralogy and geology; Hawthorne, F.C., Ed.; Mineral. Soc. Am., Reviews in Mineralogy, 1988, 18, 341.
17. Kilpatrick, M., Jr.; Kilpatrick, M. L. J. Am. Chem. Soc., 1923, 45, 2132.
18. Skoog, D.A.; West, D.M. Fundamentals of Analytical Chemistry; Saunders College Publishing: Philadelphia, 1982.
19. Wilkinson, G.; Gillard, R.D.; McCleverty, J.A. Comprehensive co-ordination chemistry; Pergamon Press: New York, 1987.
20. Edwards, C.F.; Griffith, W.P.; Williams, D.J. J. Chem. Soc., Chem. Comm., 1990, 1523.
21. Ryabchikov, D.I. Comptes Rendus (Doklady) Acad. Sci. URSS, 1943, 41, 208.
22. Ryabchikov, D.I. Akad. Nauk SSSR (Izvestiya Sectora Platiny), 1948, 21, 74.
23. Hancock, R.D.; Finkelstein, N.P.; Evers, A. J. Inorg. Nucl. Chem., 1977, 39, 1031.
24. Westland, A.D. In Platinum Group Elements: Mineralogy, Geology, Recovery; Cabri, L.J. Ed.; Canadian Institute of Mining and Metallurgy Special Volume, 1981, 23, 5-18.
25. Pryor, W.A., J. Am. Chem. Soc., 1960, 82, 4794.
26. Giggenbach, W. Inorg. Chem., 1974, 13, 1730.
27. Murray, R.C., Jr.; Cubicciotti, D. J. Electrochem. Soc., 1983, 130, 866.
28. Baas Becking, L.G.M.; Kaplan, I.R.; Moore, D. Jour. Geol., 1960, 68, 243.
29. Sato, M. Econ. Geol., 1960, 55, 928.
30. Mann, A.W. Econ. Geol., 1984, 79, 38.
31. Thornber, M.R. Chem. Geol., 1975, 15, 117.
32. Thornber, M.R.; Allchurch, P.D.; Nickel, E.H. Econ. Geol., 1981, 76, 1764.
33. Stanley, C.J.; Criddle, A.J.; Lloyd, D. Min. Mag., 1990, 54, 485.
34. Mertie, J.B., Jr. Economic Geology of the Platinum Metals. Prof. Paper U.S. Geol. Survey, 1969, 630, 120 pp.

RECEIVED April 12, 1993

Remediation and Prevention of the Environmental Effects of Sulfide Oxidation

Chapter 34

Suppression of Pyrite Oxidation Rate by Phosphate Addition

Xiao Huang[1,3] and V. P. Evangelou[1-3]

[1]Department of Agronomy and [2]Agricultural Experiment Station, University of Kentucky, N-122 Agricultural Science Center North, Lexington, KY 40546-0091
[3]Pittsburgh Research Center, U.S. Bureau of Mines, Pittsburgh, PA 15236

Pyrite, commonly found in various ore deposits, produces highly acidic drainage water when exposed to the atmosphere. Current acidic drainage remediation technologies are not long lasting or cost effective. In this study we demonstrate that a ferric phosphate coating can form on pyrite surfaces when contacted with a solution of KH_2PO_4 and H_2O_2. This ferric phosphate coating was shown to inhibit pyrite oxidation.

Pyrite is a mineral commonly found in coals or other ore deposits formed in chemically reduced environments (1). Mining operations expose the overburden or ore containing pyrite to the atmosphere. Pyrite is also a waste product of ore processing plants. As a consequence of pyrite oxidation, the drainage from mining sites becomes highly acidic and enriched with sulfate, iron, manganese, and sometimes many other heavy metals (2). This acidic drainage finds its way into streams and lakes, causing a severe environmental pollution problem. In view of these facts, pyrite oxidation mechanisms and possible controls are of great interest to mining engineers, chemical engineers, and environmental scientists.

Research on pyrite oxidation and its control has focused mainly on preventing oxidizing components from coming in contact with pyrite (3). For instance, Fe-oxidizing bacteria are considered to be responsible for the rapid oxidation rate of pyrite (4). This leads to the development and use of bactericides and slow-release bactericidal formulations (5). Other methods include sealing and insulation. The effectiveness of these techniques is site-specific and high cost precludes their use.

Inspired by the phosphating technology widely used to treat the surface of steel for rust-proofing purposes (6,7), we conceived that a $FePO_4$ coating could also be established on pyritic surfaces by treating them with a solution of phosphate and hydrogen peroxide. We hypothesized that H_2O_2 can be used to oxidize pyritic surfaces, generating Fe^{3+}, so that insoluble $FePO_4$ would form directly on the pyrite surfaces. Thus, at the expense of a certain fraction of pyrite, pyrite oxidation can be prevented. In this study, we examined the potential of forming a $FePO_4$ coating on pyritic surfaces and then evaluated the influence of this coating on the kinetics of pyrite oxidation, employing a solution of H_2O_2.

Methods

Iron sulfide was separated from a shale by density separation using 97% tetrabromoethane (density of 2.97 g mL^{-1}). The sulfide separate was washed with 4 M hydrofluoric acid (HF) and distilled water. The iron sulfide was then dried in a vacuumed desiccator and several properties relevant to this study were determined. X-ray diffraction was employed to establish that iron sulfide obtained from the shale was pyrite. It had a specific surface area of 7.15 m^2 g^{-1} and contained 76% pyrite, as determined by dissolution of the sample with 30% H_2O_2. The impurity was expected to be iron oxyhydroxide, which was produced due to the exposure to air, and hydrolysis of Fe^{2+} during washing with distilled water. We removed the iron oxyhydroxide impurity by leaching the pyrite sample with 0.1 M HCl before conducting oxidation experiments.

Leaching-oxidation experiments were conducted at 40°C, employing a porous bed-reactor system. This porous bed-reactor system consisted of a chromatographic column (threaded chromaflex borosilicate glass column with acrylic water jacket and 20 micrometer polyethylene bed support). Fifty milligrams of pyrite were suspended in 500 milligrams of sand that had passed through a 140 mesh sieve. The mixture was placed on top of a styrene filter that was located on the bottom of the chromatographic column. An oxidizing solution was passed through the porous bed-reactor, at a constant flow rate of 0.5 mL min^{-1}, employing a peristaltic pump. A water jacket was used to maintain a constant temperature. Aliquots were collected at certain time intervals using a Buchler Alpha 200 fraction collector. Sulfate (SO_4) was determined turbidometrically and iron was determined with atomic adsorption spectrophotometry.

Results and Discussion

We conducted three leaching experiments with the following solutions: 1) 0.5% H_2O_2 in 0.2 M NaCl at pH 4; 2) 0.5% H_2O_2 and 0.013 M EDTA in 0.2 M NaCl at pH 4; 3) 0.5% H_2O_2 and 0.02 M KH_2PO_4 in 0.1 M NaCl at pH 4.

Experiment 1 was designed to test maximum pyrite oxidation potential. Experiment 2 was designed to represent the situation where the influence of Fe^{3+} on kinetics of pyrite oxidation was eliminated through complexation of Fe^{3+} by EDTA; and experiment 3 was aimed at examining the potential of creating phosphate coatings and observing the influence of these coatings on the kinetics of pyrite oxidation.

The oxidation of pyrite by hydrogen peroxide can be represented schematically

$$FeS_2 + H_2O_2 \rightarrow Fe^{3+} + SO_4^{2-} + H^+$$

$$FeS_2 \left(\begin{array}{c} \uparrow \\ \\ \downarrow \end{array} \right) H_2O_2 \qquad (1)$$

$$Fe^{2+} + SO_4^{2-} + H^+$$

Based on the mechanism proposed in equation 1, the pyrite oxidation rate is the sum of the direct oxidation of S_2^{2-} by H_2O_2 and the oxidation of S_2^{2-} by Fe^{3+}. Note that because the oxidation of Fe^{2+} by H_2O_2 is a rapid reaction, it is expected that at any time t, all the Fe released from pyrite oxidation is in the form of Fe^{3+}. Based also on the above, the rate law of oxidation can be expressed as:

$$-\frac{dM}{dt} = (k_1[H_2O_2] + k_2[Fe^{3+}])\, S \qquad (2)$$

where M represents the number of moles of pyrite remaining in the system and S represents the surface area of pyrite at time t; k_1 and k_2 denote rate constants; $[H_2O_2]$ and $[Fe^{3+}]$ represent concentrations of H_2O_2 and Fe^{3+} at time t. Examination of the pyrite sample by a scanning electron microscope revealed that the particles were relatively homogeneous in size. Thus, we assumed that surface area (S) at any time t was proportional to the number of moles (M) of pyrite remaining in the system during oxidation ([8]) and can be described by:

$$S = K\,[M] \qquad (3)$$

where K is a constant. Substituting equation 3 into equation 2 gives:

$$-\frac{dM}{dt} = (k_1[H_2O_2] + k_2[Fe^{3+}])\, K\,[M] \qquad (4)$$

By moving M to the left-hand side and integrating with respect to M, equation 4 can be rearranged to:

$$-\frac{d(\ln M)}{dt} = (Kk_1[H_2O_2] + Kk_2[Fe^{3+}]) \qquad (5)$$

According to equation 5, if kinetic data of pyrite oxidation (SO_4 release) obtained by employing a constant concentration of H_2O_2 were plotted as a first-order reaction, ln (M/M_o) vs. t ($M=M_o$ at t=0) will give a curvilinear function. In leaching oxidation experiments, the concentration of Fe^{3+} was expected to decrease with time. Thus, a plot of ln (M/M_o) vs. t was expected to concave up (progressively less negative values of the slopes with increasing time). When oxidation of S_2^{2-} by Fe^{3+}

was inhibited, a ln (M/M_o) vs. t plot was expected to be a straight line as $k_2[Fe^{3+}]$ approaches zero.

We hypothesized the following: if phosphate, introduced into the leaching-oxidation system, reacted with all Fe^{3+} to form a discrete phase of $FePO_4$, the ln (M/M_o) vs. t plot would produce a straight line. Any further suppression of pyrite oxidation by phosphate beyond what was described by precipitation of Fe^{3+} as a discrete phase of $FePO_4$, would be attributed to the formation of a $FePO_4$ coating established on the surface of pyrite.

In this study, we used EDTA to complex Fe^{3+} by forming Fe-EDTA complexes, thereby preventing oxidation of pyrite by Fe^{3+} and the formation of $Fe(OH)_3$, which might also coat pyrite particles. Peck (9) indicated that the standard redox potential of Fe^{3+}/Fe^{2+} could be lowered from +0.77 to 0 by EDTA. Thus, we expected that pyrite oxidation by H_2O_2 in the presence of EDTA represented a situation where the observed rate of pyrite oxidation was solely due to the direct oxidation of S_2^{2-} by H_2O_2. The kinetic oxidation data from this treatment were expected to give a ln (M/M_o) vs. t plot with a single slope.

Figure 1 shows that pyrite oxidation by H_2O_2 was very rapid in the first 500 minutes and became slow from 500 to 1000 minutes. The first-order plot (ln (M/M_o) vs. t) shown in Figure 2 is a concave curve, demonstrating the influence of Fe^{3+} on the oxidation of pyrite. In the presence of EDTA, the oxidation of pyrite was suppressed. The first-order plot of the data representing the EDTA treatment shown in Figure 2 exhibits a straight line. This straight line indicates that the suppression was due to the prevention of the direct oxidation of S_2^{2-} by Fe^{3+} (Figure 3).

As shown in Figure 1, phosphate suppressed pyrite oxidation to a much greater extent than EDTA. The slope of the ln (M/M_o) vs. t plot, representing the phosphate treatment is almost parallel to that representing the EDTA treatment in the initial 300 minutes of the oxidation process. This observation indicates that, over this period, phosphate played the same role as EDTA, i.e. to remove Fe^{3+} and to inhibit the direct oxidation of S_2^{2-} by Fe^{3+}. After 300 minutes, the rate of oxidation of pyrite in the presence of phosphate dropped rapidly, as evidenced by the change in slope of the plot around 300 minutes (Figures 1 and 2). The plot representing the phosphate treatment deviated from the straight line obtained with the data representing the EDTA treatment (Figure 2). This deviation resulted from the faster decrease in active surface area of pyrite or from an increase in surface coating coverage of pyrite. As shown in Figure 3, almost all Fe^{3+} produced during oxidation was precipitated by phosphate. These results strongly suggest that $FePO_4$ was coating the pyrite surfaces, rather than simply precipitating as a discrete phase.

To obtain direct evidence of a $FePO_4$ surface coating, we separated residual pyrite particles from the sand-pyrite mixture, after oxidation and phosphatation were terminated, and examined the surfaces of the pyrite particles by scanning electron microscopy (SEM) and element-specific X-ray energy-dispersive analysis from the SEM (e.g. Figure 5). Figure 4A shows the morphology of residual pyrite particles after oxidation with 0.5% H_2O_2. Pyrite particles were coated with a thin layer of presumably amorphous iron (III) hydroxide. As shown in Figure 3, slightly less than

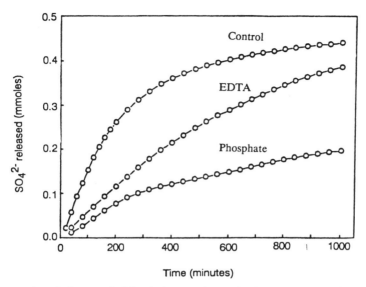

Figure 1. Release of SO_4 during pyrite oxidation under the following conditions: (1) 0.5% H_2O_2; (2) 0.5% H_2O_2 in the presence of 0.013 M EDTA; and (3) 0.5% H_2O_2 in the presence of 0.02 M KH_2PO_4.

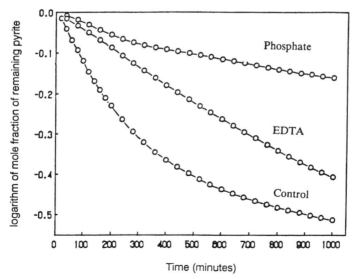

Figure 2. First-order plots of pyrite oxidation under the following conditions: (1) 0.5% H_2O_2; (2) 0.5% H_2O_2 in the presence of 0.013 M EDTA; and (3) 0.5% H_2O_2 in the presence of 0.02 M KH_2PO_4.

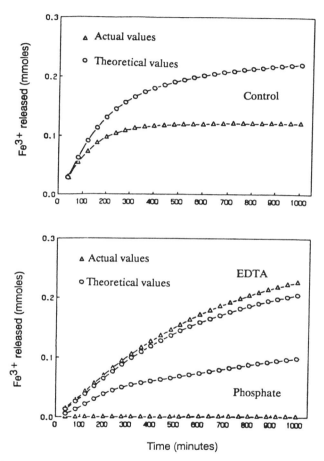

Figure 3. Release of Fe during pyrite oxidation under the following conditions: (1) 0.5% H_2O_2; (2) 0.5% H_2O_2 in the presence of 0.013 M EDTA; and (3) 0.5% H_2O_2 in the presence of 0.02 M KH_2PO_4 (actual values are those determined by measuring Fe in the leachate; theoretical values are those calculated with SO_4^{2-} data according to the stoichiometry).

half of the Fe produced during oxidation was not released to solution and presumably ended up as Fe(OH)$_3$. Comparison of the kinetic data from experiments 1 and 2 (Figure 1) suggests that the iron(III) hydroxide formed during oxidation did not inhibit pyrite oxidation. Actually, it might have accelerated pyrite oxidation due to its function as a reservoir of Fe^{3+}, because the localized high acidity on the surface of pyrite during oxidation would prevent the deposition of Fe(OH)$_3$. Thus, most of the Fe(OH)$_3$ would have formed as a discrete phase.

Figure 4B shows the residual pyrite particles displayed a morphology typical of framboidal pyrite (15). These particles consist of small pyrite crystals with easily identified octahedrons. The surfaces of these particles were free of coatings. The holes observed on the octahedrons indicate the locations where oxidation took place. The absence of any coating on the surfaces of pyrite shown in Figure 4B is probably due to removal of all Fe released by pyrite oxidation as Fe-EDTA complexes (Figure 3), which prevents the formation of iron hydroxide precipitates.

Figure 4C shows the morphology of pyrite particles oxidized in the presence of phosphate. In this photograph the surfaces of the framboidal pyrite particles are heavily coated. We conducted X-ray scanning of these coated pyrite particles to examine the distribution of Fe, S, and P. The distribution of P was similar to the distribution of Fe and S but the intensity of Fe was much higher than that of S and P (Figure 5). The higher density of Fe is due to the presence of FePO$_4$ and FeS$_2$, both of which contain Fe. The above results further suggest that the differences in oxidation rates between the EDTA and phosphate treatments resulted from the formation of an FePO$_4$ coating.

In order to further understand the chemical properties of the iron phosphate coating, we repeated the leaching experiment with a set of columns of pure pyrite and solutions containing 0.147 M H$_2$O$_2$ and 0.01 M KH$_2$PO$_4$. At the end of this experiment, each column was leached with 50 mL of 2 M HCl. The leachate was analyzed for iron and phosphate. We found that the mole ratio of iron to phosphate was 1.0, indicating that the coating was most likely amorphous FePO$_4$.

Conclusions

Pyrite particles can be coated with a protective coating by treatment with a mixed solution of KH$_2$PO$_4$ and H$_2$O$_2$, at the expense of a certain fraction of pyrite. The data in Figure 6 show that phosphate coatings established with a mixed solution of 10^{-3} M KH$_2$PO$_4$ and 0.5% H$_2$O$_2$ consumed 20% of the pyrite sample. We believed that this consumed fraction can be decreased by decreasing the concentration of H$_2$O$_2$. To test the stability of the coatings, we exposed pyrite with coatings to 0.5% H$_2$O$_2$ in the absence and presence of phosphate. As shown in Figure 6 (curve B), the pyrite sample with phosphate coating gradually decomposed, suggesting the partial collapse of the coating with time. The collapse of the coating resulted from the dissolution of FePO$_4$ by the strong acid produced by oxidation of exposed pyritic surfaces. However, when compared with the oxidation of pyrite with no coating (curve A), the suppression of oxidation due to the coating was significant. Moreover, FePO$_4$ dissolution can be inhibited in the presence of phosphate

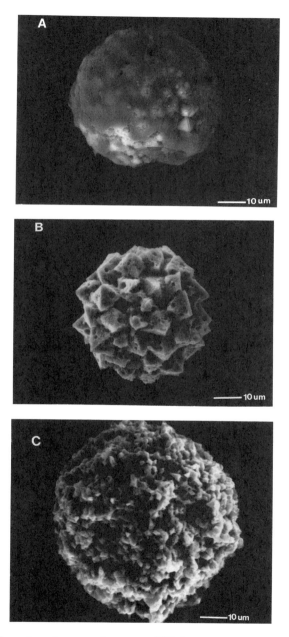

Figure 4. Scanning electron microscope (SEM) photos of residual framboidal pyrite particles after oxidation experiments: (A) pyrite particle oxidized with 0.5% H_2O_2; (B) and (C) pyrite particles oxidized with 0.5% H_2O_2 in the presence of 0.013 M EDTA. *Continued on next page.*

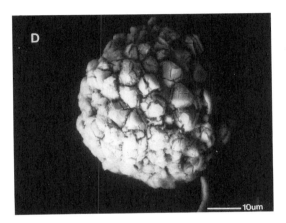

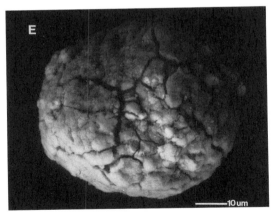

Figure 4. Continued. Scanning electron microscope (SEM) photos of residual framboidal pyrite particles after oxidation experiments: (D) and (E) pyrite particles oxidized with 0.5% H_2O_2 in the presence of 0.02 M KH_2PO_4.

34. HUANG & EVANGELOU *Suppression of Pyrite Oxidation Rate* 571

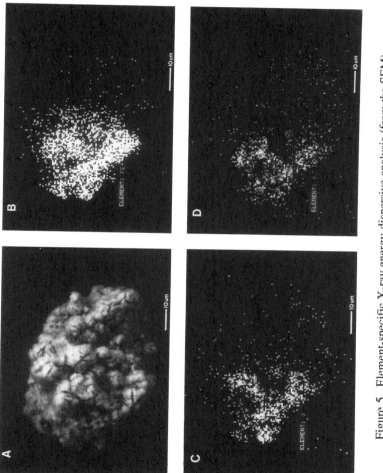

Figure 5. Element-specific X-ray energy-dispersive analysis (from the SEM) of pyrite particles oxidized in the presence of phosphate, showing the distribution of Fe, S, and P: (A) SEM microphoto of the particle examined; B) distribution of Fe; (C) distribution of S; and (D) distribution of P.

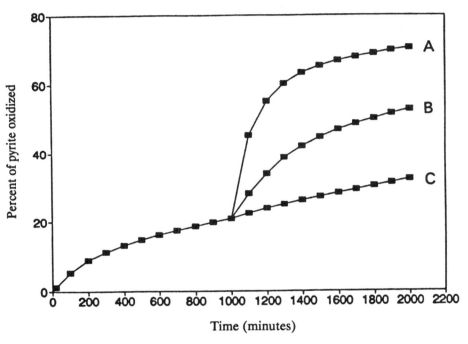

Figure 6. Pyrite oxidation as a function of time. Before 1000 min represents coating with a mixed solution of 10^{-3} M KH_2PO_4 and 0.1 % H_2O_2. After 1000 min, curve A represents pyrite oxidation with coating removed by leaching with 50 mL of 2 M HCl and exposed to 0.25% H_2O_2; curve B represents the oxidation of $FePO_4$-coated pyrite exposed to 0.25% H_2O_2 in the absence of phosphate solution; and curve C represents oxidation of $FePO_4$-coated pyrite exposed to 0.25% H_2O_2 in the presence of 10^{-4} M KH_2PO_4.

concentration as low as 10^{-4} M (curve C in Figure 6), which decreased the solubility of iron phosphate (10). Thus, when pyritic ores are to be exposed to strong oxidizers, the stability of phosphate coating will need to be strengthened by a small concentration of phosphate in solution.

In summary, pyritic surfaces can be coated with $FePO_4$, which can prevent pyrite oxidation. However, when coated pyrite is exposed to strong oxidizers, the coating can be stabilized by maintaining 10^{-4} M KH_2PO_4 in solution.

Acknowledgments

The investigation in the paper (91-3-179) is in connection with a project of the Agric. Exp. Stan. of the University of Kentucky and is published with the approval of the director. Also, the authors wish to thank Mr. R.W. Hammack, and Dr. R. Kleinmann of the U.S. Department of the Interior, Bureau of Mines, Pittsburgh, PA, and Ms. Patricia M. Erikson of the U.S. EPA, Cincinnati, OH, for the enlightening discussions on pyrite oxidation, and Ms. Libby Reed for typing and proofing the manuscript.

Literature Cited

1. Lowson, R.T. Chem. Rev. 1982, 82, 461-493.
2. Krothe, N.C.; Edkins, J.E.; Schubert, J.P. In Proceedings of 1980 Symposium on Surface Hydrology, Sedimentology, and Reclamation; University of Kentucky: Lexington, KY, 1980; pp. 455-564.
3. Singer, P.C.; Stumm, W. Science 1970, 167, 1121-1123.
4. Temple, K.L.; Delchamps, E.W. Appl. Microbio. 1953, 1, 255-258.
5. Kleinmann, R.L.P. In Proceedings of 1980 Symposium on Surface Hydrology, Sedimentology, and Reclamation; University of Kentucky: Lexington, KY, 1980; pp. 333-337.
7. Philips, D. Plating and Surface Finishing 1990, 77, 31-35.
8. Gorecki, G. Metal Finishing 1988, 86, 15-16.
9. Peck, H.P.J. An. Rev. Microbiol. 1968, 22, 489-518.
10. Lindsay, W. Chemical Equilibrium in Soils; J. Wiley and Sons: New York, NY, 1979.

RECEIVED October 13, 1993

Chapter 35

Iron Sulfide Oxidation

Impact on Chemistry of Leachates from Natural and Pyrolyzed Organic-Rich Shales

Thomas L. Robl

Center for Applied Energy Research, University of Kentucky, 3572 Iron Works Pike, Lexington, KY 40511

Retorted and raw Devonian oil shales were placed in large field lysimeters and allowed to weather for a period of 1300 days. The shales were composed of iron sulfides and silicates (quartz, illite, chlorite and kaolinite). Retorting induced the dehydroxylation and decrystallization of kaolinite and illite and converted part of the pyrite to pyrrhotite. The leachates generated in the field lysimeters were highly acidic (pH 2.5 to 3.5) with high concentrations of SO_4^{2-}, Mg and Fe (1,000 to 10,000 ppm); Al, Zn, Na, Ca, K (100 to 1,000 ppm); Ni and Mn (10 to 100 ppm). The concentrations of most of the elements in the raw-shale leachates were a function of the rate of acid generation. The matrix of the retorted shale was more reactive than the raw shale. Elemental release was transport controlled, with the exception of Ca, which appeared to be near equilibrium with respect to gypsum.

Organic-rich Devonian black shale occurs throughout much of the mid-western United States. In addition to serving as a major source of oil and gas, certain stratigraphic intervals are sufficiently high in kerogen to be considered oil shales (1). A significant research effort was initiated in the early 1980's to study the economic feasibility of developing this resource, with environmental research as a component. A focus of concern was the acid-generating and elemental-release characteristics of retorted and raw shale.

An opportunity to examine the weathering characteristics of these shales occurred in 1983 when a pilot-plant study of Devonian oil shale was undertaken by a consortium, led by the Southern Pacific Petroleum Company, in response to a solicitation by the Synthetic Fuels Corporation. The pilot plant was the Dravo traveling-grate facility located in Cleveland, Ohio. This plant consists of a circular grate retort 8.2 m² in area, capable of retorting up to 250 tonnes/day of raw shale (2-4). The test utilized 1,000 tonnes of 25 mm x 6 mm sized shale, two-thirds of which was mined from the Cleveland Member of the Ohio shale and one-third from the overlying Sunbury Shale. The shale was mined in Montgomery County, Kentucky. The spent shale along with the <6 mm size fraction of raw unprocessed material was trucked to Kentucky to a field-research station located near the mine site.

Composition of Study Materials

In addition to organic matter, the shales used in the study consist of silicates (principally clays and quartz) and sulfides (primarily pyrite, FeS_2 and pyrrhotite, $Fe_{1-x}S$). The small amount of carbonates in this shale (Table I) are present as calcite, $CaCO_3$. The clays include illite, a potassium-deficient 2:1 mica (i.e. two silica tetrahedral layers and one aluminum tetrahedral layer); chlorite (2:1 clay inter-layered with a brucite, $Mg(OH)_2$, sheet); and kaolinite, a 1:1 clay. X-ray diffraction (XRD) was used in combination with heat and chemical treatment to determine the mineralogic composition of the materials (5,6). The silicate fraction of the shale consists of approximately 20% quartz, 60% illite, 11% chlorite and 9% kaolinite (7). Mixed-layer clays are, at most, a small component (i.e. <4%).

Processing Affects. Retorting has a major effect on the shale mineralogy and weathering behavior. A major change observed in the spent shale was the loss of the kaolinite peak on an XRD scan (Figure 1). Kaolinite undergoes dehydroxylation, beginning at ~450°C, which destroys its crystallinity. Changes in the illite component of the spent shale included the loss of most of the sharp 1.0 nm 001 peak as well as the 0.5 nm 002 peak. The crystallinity of the illite seemed to be progressively destroyed as the temperature increased. For comparison, a similar shale retorted in Petrobras's PETROSIX retort, an indirectly heated technology which operates at lower temperature than Dravo's, is also presented in Figure 1. Thus, much of the inorganic matrix of the shale must exist in something of a glassy, or at least partially decrystallized, state. Quartz and chlorite did not appear to be strongly affected. The iron sulfides were also affected by retorting. Estimates from XRD data indicated that about half of the pyrite (FeS_2) was converted to pyrrhotite ($Fe_{1-x}S$).

Table I. Analysis of Raw and Retorted Oil Shale

	C-organic	C-mineral	H	N	S	Ash		
Raw Shale	11.4	0.53	1.38	0.82	3.35	80.5		
Retorted Shale	6.7	0.41	0.20	0.34	2.38	90.2		
	SiO_2	Al_2O_3	TiO_2	Fe_2O_3	CaO	MgO	K_2O	Na_2O
Ash Analysis Raw Shale	66.59	15.19	0.80	7.03	0.77	1.34	4.09	0.42

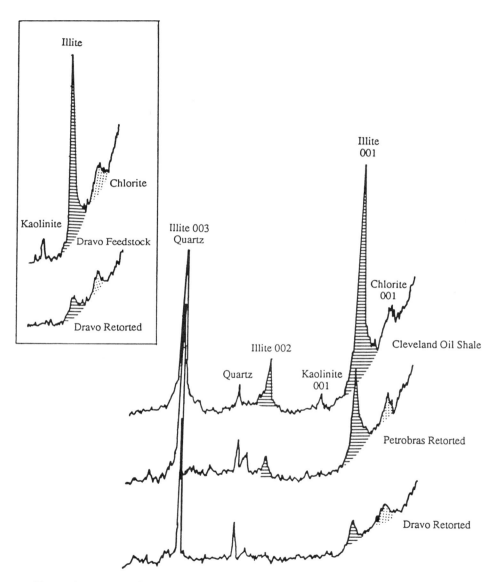

Figure 1. X-ray diffraction patterns of silt (>0.2 μm) and clay (<0.2 μm, insert) size fractions of oil shale samples. Samples are magnesium saturated and glycerated.

Experimental Apparatus, Design and Procedures

The weathering characteristics of the shales were studied in eight concrete field lysimeters. These lysimeters were constructed in two clusters of four, each concentric to a central leachate-collection chamber (Figure 2). The design and construction of the units are discussed elsewhere (8).

Four types of materials were used in the lysimeter fills: retorted oil shale, raw oil shale, raw oil shale-retorted oil shale mixtures, and soil and overburden materials. The overburden material was a plastic clay from the Nancy Member of the Borden Formation, which overlies the oil shale. The soil was a local silty clay.

The spent oil shale (SS) from the pilot runs, was placed in thicknesses of 3, 2.4, 1.8 and 1.2 m in the four chambers numbered SS-3, SS-2.4, SS-1.8 and SS-1.2, respectively (Figure 2). A mixture (MS) of approximately 80% spent oil shale and 20% <6 mm size fraction raw oil shale was placed to a 1.84 m thickness in the lysimeters numbered MS-1.8O and MS-1.8M. A 1.8 m thickness of the <6 mm size fraction raw shale (RS) was placed in lysimeter RS-1.8. The final lysimeter, SO-3, consisted of 2.1 m of soil and 0.9 m of overburden material. In filling, an average dry density of 1.30 g/cm^3 was achieved for retorted shale, 1.71 g/cm^3 for the mixed material and 2.19 g/cm^3 for the raw shale. All lysimeters were covered with soil, overburden or mixtures of the two.

Water sample collection began in July 1984 and continued until May 1988, a period of more than 1300 days. In general, the samples were collected once per week, except during periods of high precipitation when the collection frequency was increased to twice per week. In general, the bottom lysimeter tubes provided samples more consistently, whereas shallower tubes flowed only during periods of high precipitation.

Total sample volume from the individual lysimeters was measured. This data and the void-space volume were used to calculate the total number of pore-volume changes. This ranged from a low of ≈ 1 for lysimeter SS-3 to a high of ≈ 7 for RS-1.8, with the rest of the lysimeters having a value ≈ 2. These values are estimates, as seepage around transport-tube portholes, sample-bottle overflow, and compaction of materials all resulted in low estimates of pore-change values, by factors as great as 30 to 50%.

The amount of flow through the spent-shale chambers was related to the thickness and nature of the soil and clay overburden covering the oil-shale materials. The smallest total sample volumes were from the MS-1.8M lysimeter (4,000 l), which had 1.2 m of mixed soil and overburden, and the SS-1.2 lysimeter (6,900 l), which had a total of 1.8 m of layered soil and overburden. The spent shale covered with thinner soil and overburden, i.e. 1.2 m (SS-1.8, 8,600 l) and 0.6 m (SS-2.4 10,000 l), had proportionately more flow.

Analytical Protocol. Standard methods and procedures were used for the analysis of conductivity, pH, sulfate, chloride, nitrate, ammonium, As, B, Ca, Cr, Cu, Fe, K, Mn, Mg, Na, Ni, Pb, Se and Zn. Aluminum was analyzed later but is missing from the early data. Most elemental concentrations were above the detection limits for the techniques employed (inductively coupled plasma/direct coupled plasma, atomic absorption, electrode, titration and gravimetry). Some elements such as As and Pb,

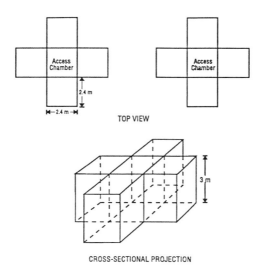
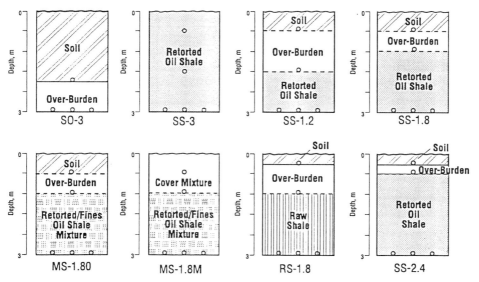

Figure 2. Schematic diagram of lysimeter design (upper) and lysimeter fills (lower).

were present in low concentrations and rapidly fell below the detection limits. The following list gives typical values for the worst-case detection limits determined during the study (in ppm): As (0.5), B (0.03), Cd (0.01), Cu (0.1), Cr (0.1), Fe (0.03), Ni (0.06), Mo (0.2), Mn (0.05), Pb (0.5) and Se (0.4). These values were considerably lower, by factors of 3 to 5, during times when the instruments were in peak adjustment.

Leachate Chemistry

Data were compiled on a project-year basis. The project years spanned the interval from September 1 to August 31, with the exception of the first year, which started on July 1 (Day 1) 1984 and the last year which ended on May 31, 1988. During the study over 3,800 samples were collected and 76,000 chemical determinations made. Thus, an adequate presentation of all the data from the study is difficult. This discussion focuses on the leachates collected from the lowest transport tubes, for those lysimeters filled with either raw or retorted shale. Leachates from the mixed material generally fell between these two end-members in composition. A partial summary of the data, compiled on a project-year basis, is presented in Table II. The "mean" in the table is the average of the concentrations of all the 3 m transport tube samples collected for the year. Because of the large number of samples, this average and the volume-weighted average are essentially identical, i.e. within 2% relative concentration. Table III provides the means, ranges and standard deviations of the leachate chemistry for two lysimeters (RS-1.8 and SS-3) by project year.

Elemental-Concentration Ranges and Averages. Although elemental concentrations varied widely in the leachates, they can be ranked into concentration ranges. Those elements which were generally present in the shale leachates in concentrations of 1 ppm or less included Cr, Cu, Mo, Pb, As and Se. Those which were present in concentrations of 1 to 10 ppm included Cd and B. Those present from 10 to 100 ppm in concentration included K, Mn and Ni, and those from 100 to 1,000 ppm included Zn, Na, Ca and Al. Elements present in concentrations >1,000 ppm included sulfur (as sulfate) and Mg. Some elements, including Fe and Mg, spanned several ranges.

The highest concentrations for most elements were found in the leachates from the raw shale during the first year of the study. Sulfate concentrations of 66,000 ppm, and Fe and Mg values of 12,500 and 3,640 ppm, respectively, were measured, along with Zn, Al, Ni, Mn, Cu and Cd concentrations of 7,750, 1,200, 380, 550, 120 and 35 ppm, respectively. The mean concentrations of the retorted shale were considerably lower, at least initially. The mean concentration values from the mixed-material leachates fell between the retorted- and raw-shale materials.

With a few exceptions, the mean elemental concentrations in the leachates decreased during the study, with the greatest change between years 1 and 2. The leachates from the unretorted or raw shale (Lysimeter RS-1.8) changed the most. For example, mean sulfate concentrations in the leachates for the RS-1.8 samples decreased from 30,000 ppm (1984/85) to 9,500 ppm the second year (1985/86), to 4,700 ppm the third year (1986/87) and to 3,580 ppm the fourth year (1987/88). The decline in mean-elemental concentrations in the leachates from the retorted oil shales was not as great. Mean sulfate concentrations decreased in the SS-1.8 lysimeter from

Table II. Data Summary of Leachate Chemistry (1984-1988)

	SO-3 Soil-Overburden Control				RS-1.8 Raw Shale			
	84/85	85/86	86/87	87/88	84/85	85/86	86/87	87/88
Samples	134	136	192	86	153	145	193	115
pH	5.60	6.33	6.39	6.53	2.66	2.65	3.21	2.93
Cond	0.25	0.19	0.20	0.23	20.34	8.34	5.29	4.49
Cr	ns	ns	ns	ns	3.15	0.91	ns	ns
Cu	ns	ns	ns	ns	11.69	0.44	ns	0.04
Mo	ns	ns	ns	ns	2.25	0.40	ns	ns
Pb	ns	ns	ns	ns	2.08	ns	ns	ns
As	ns	ns	ns	ns	1.38	ns	ns	ns
Se	ns	ns	ns	ns	1.94	ns	ns	ns
Cd	ns	ns	ns	ns	10.2	0.6	0.2	0.1
B	0.1	ns	ns	ns	1.8	1.3	1.3	1.0
Mn	ns	ns	ns	ns	140	54	22	13
Ni	0.1	0.1	ns	ns	155	49	16	5
Fe	0.3	ns	ns	ns	5665	2434	1166	750
Zn	1	5	0.1	ns	913	197	37	4
K	3	4	5	5	1	5	9	10
Na	15	19	16	13	16	33	28	24
Ca	8	19	24	24	398	436	443	469
Al	nd	nd	ns	nd	457	58	6	ns
Mg	1	1	1	1	1841	654	270	172
SO_4	37	35	35	37	29870	9484	4698	3580

	SS-3 Retorted Shale				SS-2.4 Retorted Shale			
	84/85	85/86	86/87	87/88	84/85	85/86	86/87	87/88
Samples	94	88	102	78	66	116	119	61
pH	2.96	3.02	3.50	3.37	3.13	3.14	3.42	3.22
Cond	8.77	8.55	8.22	7.49	8.98	5.90	4.84	5.26
Cr	0.33	0.46	ns	ns	0.31	0.23	ns	ns
Cu	0.39	0.27	0.31	ns	0.37	0.29	0.19	0.16
Mo	0.93	0.53	ns	ns	0.78	ns	ns	ns
Pb	1.17	0.70	ns	ns	1.06	ns	ns	ns
As	0.49	ns	ns	ns	0.42	ns	ns	ns
Se	0.69	0.60	ns	ns	0.56	ns	ns	ns
Cd	3.6	3.1	2.3	1.7	4.3	1.9	0.9	0.7
B	5.3	6.5	6.5	6.2	6.5	5.5	4.5	4.3
Mn	69	64	52	36	72	38	20	18
Ni	80	69	53	37	78	38	19	16
Fe	150	211	142	143	63	68	71	146
Zn	415	294	175	149	502	162	77	67
K	155	129	81	64	145	100	75	71
Na	373	371	273	219	451	219	118	105
Ca	455	394	402	430	448	454	439	426
Al	341	549	710	737	291	371	282	388
Mg	775	785	696	529	799	372	254	251
SO_4	8567	9782	9312	8372	8412	5858	4664	5241

ns: not significant, i.e. more than 50% of samples were below the detection limit

Table II. Continued

	SS-1.8 Retorted Shale				SS-1.2 Retorted Shale			
	84/85	85/86	86/87	87/88	84/85	85/86	86/87	87/88
Samples	48	86	168	99	71	127	108	53
pH	2.92	2.97	3.39	3.22	2.91	2.95	3.38	3.10
Cond	6.49	5.17	4.13	4.09	6.65	5.79	4.04	4.47
Cr	ns	ns	ns	ns	0.38	0.34	ns	ns
Cu	0.07	0.10	0.15	ns	0.17	0.13	0.10	0.13
Mo	0.59	ns	ns	ns	1.07	0.72	ns	ns
Pb	1.02	ns	ns	ns	1.35	0.96	ns	ns
As	0.42	ns	ns	ns	0.59	ns	ns	ns
Se	0.74	ns	ns	ns	0.82	ns	ns	ns
Cd	0.9	0.5	0.4	0.4	1.6	1.0	0.5	0.4
B	5.1	3.6	2.8	2.5	5.2	5.0	3.1	3.0
Mn	44	30	17	15	51	36	13	13
Ni	39	23	14	10	52	35	13	11
Fe	531	315	135	173	389	379	73	110
Zn	146	83	54	46	214	132	60	52
K	113	96	81	69	90	86	75	74
Na	209	145	92	76	243	194	96	92
Ca	448	440	414	411	449	428	422	423
Al	nd	nd	148	174	nd	318	183	251
Mg	432	289	210	181	470	346	172	177
SO_4	5704	4363	3491	3475	5623	4999	3532	3931

	MS-1.8S Mixed Shale				MS-1.8M Mixed Shale			
	84/85	85/86	86/87	87/88	84/85	85/86	86/87	87/88
Samples	65	75	81	84	56	39	59	66
pH	2.97	2.98	3.49	3.24	2.89	2.93	3.22	3.12
Cond	10.00	7.16	5.49	5.51	6.63	5.10	4.65	4.97
Cr	0.62	0.50	ns	ns	0.59	0.22	ns	ns
Cu	0.05	0.12	ns	0.11	0.06	0.03	ns	0.04
Mo	1.91	0.46	ns	ns	0.67	0.25	ns	ns
Pb	1.92	ns	ns	ns	0.83	ns	ns	ns
As	0.64	ns	ns	ns	ns	ns	ns	ns
Se	1.06	ns	ns	ns	0.63	ns	ns	ns
Cd	3.5	1.1	0.6	0.6	0.7	0.2	0.2	0.2
B	5.0	4.5	4.0	3.6	4.1	2.9	2.6	2.5
Mn	102	61	39	34	44	29	29	29
Ni	100	51	31	27	33	13	17	16
Fe	803	459	233	217	899	498	194	283
Zn	459	188	120	115	121	64	66	71
K	70	59	51	48	72	68	71	69
Na	315	249	167	154	144	122	112	114
Ca	436	388	418	432	437	391	415	435
Al	155	134	152	235	17	22	69	87
Mg	1104	664	426	396	436	282	327	334
SO_4	10408	6864	4918	5005	5828	4041	3873	4140

Table III. Summary of Leachate Chemistry for the SS-3 Lysimeter

	SS-3 84/85 #Samples 66				SS-3 85/86 #Samples 116			
	Mean	Std	Max-Min	%>dl	Mean	Std	Max-Min	%>dl
pH	3.13	6	3.53-2.68		3.14	7	3.72-2.36	
Cond	8.98	17	13.1-6.74		5.90	12	7.37-4.55	
Cr	0.31	45	0.68-0.2		0.23	59	0.99-<dl	66
Cu	0.37	68	1.14-0.1		0.29	66	0.90-.04	95
Mo	0.78	48	1.74-0.1		ns	ns	0.57-<dl	49
Pb	1.06	44	2.00-<dl	77	ns	ns	2.20-<dl	43
As	0.42	40	0.74-<dl	56	ns	ns	1.20-<dl	38
Se	0.56	40	1.08-<dl	77	ns	ns	1.10-<dl	41
Cd	4.3	38	8.2-0.7		1.9	42	4.0-0.6	
B	6.5	20	9.3-2.7		5.5	21	8.8-3.6	
Mn	72	30	137-36		38	42	79-18	
Ni	78	25	138-42		38	36	72-19	
Fe	63	85	260-7		68	77	200-5	
Zn	502	34	877-188		162	42	460-47	
K	145	20	205-83		100	25	210-55	
Na	451	29	832-295		219	28	348-70	
Ca	448	22	897-335		454	11	554-220	
Al	291	43	458-24	61	371	13	502-272	20
Mg	799	24	1210-504		372	31	604-125	
SO_4	8412	20	12428-5060		5858	18	7893-3309	
	SS-3 86/87 #Samples 119				SS-3 87/88 #Samples 61			
	Mean	Std	Max-Min	%>dl	Mean	Std	Max-Min	%>dl
pH	3.42	7	3.98-2.73		3.22	7	3.67-2.7	
Cond	4.84	12	6.07-3.35		5.26	10	6.27-4.01	
Cr	ns	ns	0.30-<dl	40	ns	ns	<dl-<dl	0
Cu	0.19	79	0.70-.01		0.16	133	1.22-.02	
Mo	ns	ns	0.40-<dl	10	ns	ns	0.42-<dl	10
Pb	ns	ns	0.80-<dl	4	ns	ns	0.40-<dl	3
As	ns	ns	0.50-<dl	3	ns	ns	<dl-<dl	0
Se	ns	ns	0.80-<dl	10	ns	ns	0.63-<dl	15
Cd	0.9	50	2.0-0.1		0.7	51	1.4-0.2	
B	4.5	23	7.9-2.2		4.3	16	5.7-2.6	
Mn	20	29	35-9		18	17	23-10	
Ni	19	28	31-8		16	18	23-9	
Fe	71	77	250-3		146	92	520-7	
Zn	77	27	130-29		67	17	92-39	
K	75	17	111-51		71	18	97-55	
Na	118	29	280-39		105	16	135-65	
Ca	439	12	880-360		426	9	520-370	
Al	282	19	410-200	34	388	42	635-168	
Mg	254	31	630-110		251	23	385-160	
SO_4	4664	19	6724-2881		5241	17	7284-3745	

Table III. Continued

	RS-2.4 84/85 #Samples 153				RS-2.4 85/86 #Samples 145			
	Mean	Std	Max-Min	%>dl	Mean	Std	Max-Min	%>dl
pH	2.66	7	3.28-2.1		2.65	9	3.88-2.16	
Cond	20.34	38	38.5-8.22		8.34	20	12.24-4.69	
Cr	3.15	166	36.00-<dl	99	0.91	106	3.60-<dl	60
Cu	11.69	199	123.00-0.0		0.44	123	2.38-<dl	82
Mo	2.25	63	6.24-<dl	71	0.40	41	0.74-<dl	56
Pb	2.08	63	4.98-<dl	61	ns	ns	1.40-<dl	43
As	1.38	81	7.15-<dl	64	ns	ns	0.70-<dl	29
Se	1.94	59	4.35-<dl	69	ns	ns	1.60-<dl	44
Cd	10.2	87	35.3-0.6		0.6	71	2.1-0.1	
B	1.8	82	8.4-<dl	97	1.3	30	2.4-0.3	
Mn	140	48	550-44		54	38	116-26	
Ni	155	44	384-45		49	42	109-20	
Fe	5665	42	12500-2380		2434	30	4900-1340	
Zn	913	46	1750-191		197	49	445-56	
K	1	88	5-0		5	43	11-0	
Na	16	45	36-1		33	29	88-19	
Ca	398	14	736-239		436	11	740-360	
Al	457	62	1200-63		58	89	251-3	
Mg	1841	41	3640-622		654	37	1400-262	
SO_4	29870	54	66794-7329		9484	30	17509-4617	

	RS-2.4 86/87 #Samples 193				RS-2.4 87/88 #Samples 115			
	Mean	Std	Max-Min	%>dl	Mean	Std	Max-Min	%>dl
pH	3.21	14	5.43-2.09		2.93	12	3.95-2.21	
Cond	5.29	18	7.85-3.55		4.49	15	6.08-3.25	
Cr	ns	ns	0.90-<dl	30	ns	ns	<dl-<dl	0
Cu	ns	ns	1.00-<dl	48	0.04	96	0.22-<dl	69
Mo	ns	ns	0.44-<dl	7	ns	ns	0.44-<dl	8
Pb	ns	ns	0.50-<dl	3	ns	ns	0.42-<dl	5
As	ns	ns	0.70-<dl	8	ns	ns	0.72-<dl	4
Se	ns	ns	0.70-<dl	10	ns	ns	0.80-<dl	18
Cd	0.2	53	0.8-0.04		0.1	46	0.3-0.03	
B	1.3	30	2.3-0.6		1.0	43	2.0-0.1	
Mn	22	40	49-9		13	34	24-6	
Ni	16	53	39-4		5	66	11-1	
Fe	1166	30	2100-540		750	34	1520-221	
Zn	37	95	230-3		4	83	12-1	
K	9	29	14-1		10	28	16-3	
Na	28	32	70-16		24	26	65-17	
Ca	443	9	600-220		469	8	600-415	
Al	6	122	36-0.3		ns	ns	15-<dl	
Mg	270	39	750-120		172	36	341-70	
SO_4	4698	25	7551-2667		3580	21	5300-2407	

5,700 ppm the first year to 3,470 ppm the fourth year. This pattern was similar for most other elements as well. Mean concentrations for the fourth year decreased to 20 to 40% of their first year levels in the retorted-shale leachates. For the raw-shale leachates this decline was to a level of 2 to 10% of the first year concentration (Table II). Thus, by the end of the study, many elements, including Zn, Ni, Mn and Cd, had higher concentrations in the retorted-shale leachates than in the raw-shale leachates. This change is partly due to the higher intensity of leaching (i.e. larger sample volume/pore volume) in the raw-shale lysimeter.

The SS-3, SS-2.4, SS-1.8 and SS-1.2 lysimeters had thicknesses of 3, 2.4, 1.8, and 1.2 m of retorted shale, respectively, and calculated pore displacements (i.e. number of sample volumes/pore volume) of 0.9, 1.7, 1.9 and 2.3. A comparison of these leachates provided some insight into the effects of leachate residence times on composition. The elemental concentrations of the SS-3 leachates were higher than the others throughout the study. Samples from the SS-2.4 lysimeter generally ranked second in concentration. However, a simple proportionality is not found. For example, the SS-2.4 lysimeter had twice the thickness of shale and half of the computed pore-volume change of SS-1.2. The mean elemental concentrations for the fourth year of the study in the SS-2.4 leachates were only about 25% higher than SS-1.2, and the range of concentrations observed over the year largely overlapped. A simple breakthrough curve for the elemental-concentration changes is not indicated.

Exceptions to the general trends in elemental concentrations were found for several elements. Sodium and K concentrations were higher in the leachates from the spent shale than in leachates from the raw shale. Concentrations as high as 238 ppm K and 832 ppm Na were measured in the retorted-shale leachates during the first year of the study, compared to maximums of 5 ppm K and 38 ppm Na in the raw-shale leachates. Sodium and K concentrations both declined with time in the retorted-shale leachates, but K increased in concentration in the raw-shale leachates. Aluminum, initially higher in the raw-shale leachates, rapidly decreased in concentration with time to below the detection limit by the fourth year. Conversely, Al increased in concentration in the retorted-shale leachates.

Calcium concentrations remained relatively constant throughout the study in all the leachates and had approximately the same average concentration (~430 ppm), with the exception of the SO-3 lysimeter. Calcium also had less variability (lowest standard deviation) of concentration of any of the elements.

Elemental-Release Patterns. The relative cationic compositions of the raw- and spent-shale leachates were also found to differ considerably. The principal cation in the raw-shale leachate was Fe. Mg and Al also were important constituents during the first year, but by the end of the study, Al concentration had greatly decreased. In the retorted-shale leachate, Mg was the predominant cation for the first year along with significant Al, K and Na. By the end of the study, Al concentration had increased and was the principal cation.

Plots of elemental concentration versus time for leachates show contrasting patterns of elemental release (Figure 3). Elemental concentrations in the raw-shale leachates were generally the highest within the first 140 to 170 days of the study, i.e. between mid-November and mid-December 1984, and rapidly declined to lower

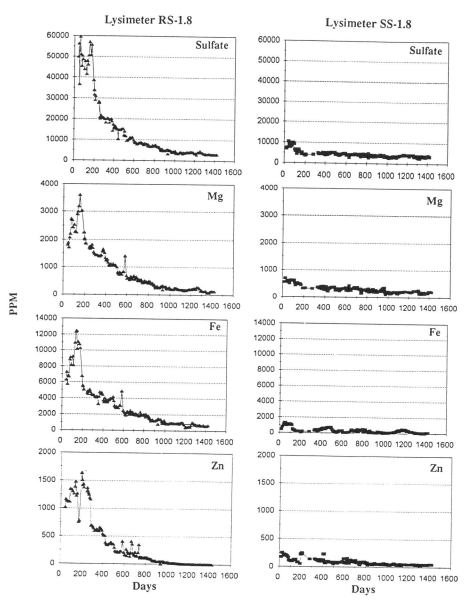

Figure 3. Comparison of elemental-release patterns between raw- (lysimeter RS-1.8) and retorted-shale leachates (lysimeter SS-1.8).

concentrations by the end of the study. Many of the elements, including Mg, Fe, Zn, Ni and Cd, exhibited very similar trends.

Elemental concentrations in the retorted-shale leachates generally reached the highest values somewhat later than the raw shale, i.e. late November 1984 to late March 1985 (days 170 to 270), and decreased in concentration at a lower rate. In contrast to the raw-shale leachate, the elemental-release pattern for the retorted-shale leachates had more scatter and did not show similar trends for many elements (Figure 3). For example, Fe in the retorted-shale leachates showed periodic maximums reaching peak concentrations during the early fall of the year, i.e. September and October. Iron also showed the highest degree of variability in concentration.

Factors Controlling Leachate Chemistry

Pyrite Oxidation. The low pH and high concentrations of dissolved ions present in the shale leachates result from the oxidation of iron sulfides and the hydrolysis of dissolved ferric iron. The important reactions in the oxidation of iron disulfide are expressed in the following equations:

$$FeS_2(s) + 7/2 O_2 + H_2O \rightarrow Fe^{2+} + 2SO_4^{2-} + 2H^+ \tag{1}$$

$$2Fe^{2+} + 1/2 O_2 + 2H^+ \rightleftharpoons 2Fe^{3+} + H_2O \tag{2}$$

$$Fe^{3+} + 3H_2O \rightleftharpoons Fe(OH)_3(s) + 3H^+ \tag{3}$$

Raw-Shale Leachate Chemistry. The chemistry of the raw-shale leachates can be largely explained by the above reactions. The oxidation of iron sulfides in the shale results in the formation of ferrous iron, sulfate and 2 hydrogen ions, with Fe^{2+} and SO_4^{2-} generated in molar proportion of 1:2. When these molar proportions were compared in the raw-shale leachates they were found to be reasonably close, with mean values of 0.7:2, 0.9:2, 0.9:2 and 0.7:2 calculated on an individual sample basis for the four project years, respectively. Estimated Fe using sulfate values and the stoichiometry of equation 1 were similar to measured Fe (Figure 4), with the exception of some of the very high sulfate concentrations. That iron was present in a proportion of less than 1:2 to sulfate can be explained by the oxidation of Fe^{2+} to Fe^{3+} and precipitation of ferric hydroxide (equations 2 and 3). The samples from the RS-1.8 lysimeter were frequently turbid and had a brownish-yellow filtrate, indicating the presence of precipitated iron oxides and hydroxides.

Magnesium was present in the shale primarily in the brucite interlayer of the chlorite clay. The high concentration of Mg^{2+} in the leachates was due to acid attack on brucite:

$$Mg(OH)_2(s) + 2H^+ \rightleftharpoons Mg^{2+} + 2H_2O \tag{4}$$

The chemistry of the raw-shale leachates can be defined by simple chemical reactions, dominated by the hydrolysis and oxidation of pyrite and resulting in the generation of acid. The concentration of trace metals, such as Mn, Ni, Zn and Cr, is a function of the degree of acid generation. Strong correlations were found between sulfate concentration and most of the other elements measured. For example, the correlation coefficient between sulfate concentration and Al, Cd, Fe, Ni, Mg and Zn concentrations were 0.96, 0.96, 0.93, 0.95, 0.96. and 0.91, respectively. Early in the study, acid generation was sufficient to attack the octahedral layers of the clays and dissolve high concentrations of Al. Other less soluble elements, such as Cr were also mobilized. The low concentration of K in these leachates is somewhat unexpected. The higher K concentrations in the retorted-shale leachates illustrate the effect of thermal disruption of the clay.

Retorted-Shale Leachate Chemistry. The leachate chemistry of the spent shale strongly contrasts with that of the raw shale. The correlation between sulfate concentration and Al, Cd, Fe, Ni, Mg and Zn concentrations was weak or absent, with coefficients of 0.35, 0.49, 0.02, 0.50, 0.69. and 0.22, respectively, for the SS-3 Leachates. The mineralogic changes induced by retorting provide explanation for some of the major differences in leachate chemistry. For example, the disruption of illite must be a major factor in the significantly higher K concentrations in the retorted-shale leachates, because K largely resides in the illite interlayer in the shale. The disruption of the octahedral layers in the illite, and particularly the dehydroxylation of kaolinite, results in minerals which are much more susceptible to acid attack and Al dissolution.

As in the case of the raw shale, the high sulfate concentrations and low leachate pH values must be due to the oxidation of the iron sulfides. However, the concentration of iron was low compared to that of sulfate. For example, the mean $Fe:SO_4$ molar ratios for the SS-3 leachates were 0.06:2, 0.08:2, 0.05:2 and 0.06:2, respectively, for the four successive years of the study, much less than the values of almost 1:2 in the raw-shale leachates.

Magnesium and Al were the most abundant elements in the retorted-shale leachates. These two elements displayed inverse elemental-release patterns, with Mg concentration decreased as Al concentration increased (Figure 5). The sum of their concentrations is present in a stoichiometric proportion of 1:2 with sulfate (Figure 6).

Aluminum and Fe displayed a product-reactant relationship (Figure 7). Low concentrations of Al clearly corresponded to high Fe concentrations and low sulfate concentrations. Reactions of Fe^{2+} with the spent shale (equations 5 and 6) must occur, in combination with oxidation of Fe^{2+} to Fe^{3+}, acid dissolution, and hydrolysis reactions (equations 7 and 8).

$$2Fe^{2+} + Al_2O_3(s) + 2H_2O + 1/2 O_2 + 2H^+ \rightleftharpoons 2Fe(OH)_3(s) + 2Al^{3+} \tag{5}$$

$$2Fe^{2+} + 3Mg(OH)_2(s) + 2H^+ + 1/2 O_2 \rightleftharpoons 2Fe(OH)_3(s) + 3Mg^{2+} + H_2O \tag{6}$$

$$6H^+ + Al_2O_3(s) \rightleftharpoons 2Al^{3+} + 3H_2O \tag{7}$$

$$Al^{3+} + H_2O \rightleftharpoons AlOH^{2+} + H^+ \tag{8}$$

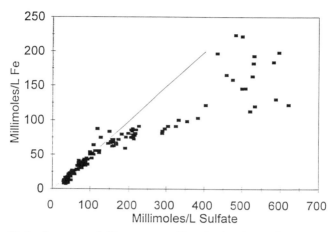

Figure 4. Plot of measured Fe versus sulfate in leachates from the raw shale (lysimeter RS-1.8). The line indicates the stoichiometric composition for pyrite oxidation [equation 1].

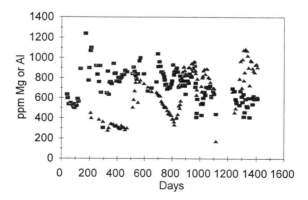

Figure 5. Mg (■) and Al (▲) concentrations in retorted-shale leachates (lysimeter SS-3).

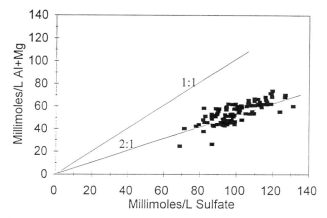

Figure 6. Concentration of Mg plus Al versus sulfate for retorted-shale leachates (lysimeter SS-3).

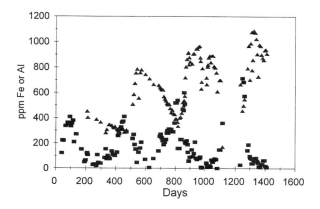

Figure 7. Concentration of Fe (■) and Al (▲) in retorted-shale leachates (lysimeter SS-3).

Some additional insight in the product-reactant relationship of Al and Fe can be gained from examining the short-term changes in the leachate chemistry. October 1986 was dry, and by the end of the month, sample volume was very low. During the week of November 4 to 11 (days 856 to 863) severe storms resulted in more than 150 mm of precipitation, effectively flushing the lysimeters. In the November 10 sample, collected immediately after the precipitation, Fe concentration increased by a factor of three and rapidly decreased in subsequent collections. Aluminum concentration steadily increased in the samples, doubling during the period October 13 to December 3. Sulfate also increased in concentration, reaching its maximum in the same sample as Al. The samples collected immediately after the precipitation event represent the rapid displacement of waters which were present at levels above field capacity, and moving downward. Later, water which was wetting the shale particles was displaced. These waters were higher in sulfate and Al and represent solutions with longer residence times than the initial flow after the precipitation.

This effect is further illustrated by comparing the variation in Fe concentration with the volume of the samples collected. The highest concentrations of Fe were present during, and immediately after, periods of very low sample volume, when the lysimeters were barely flowing. As the lysimeters were drained by gravity, the samples collected must represent water present in excess of field capacity. These sample waters had shorter residence times than those still present in the lysimeter wetting the shale particles (i.e. water held by the specific retention of the shale). Thus, the variation in Fe concentration appears to be a function of residence time, with longer residence time correlating to lower Fe and higher Al concentrations.

Oxygen also has a role in the relationship of Fe and Al. The relative proportion of Fe to Al concentration can be related to the thickness of soils and overburden materials covering the spent shale. The ratio of mean Al to Fe concentration for the last four years of the study averaged 2.3, 2.4, 2.7 and 5 for the SS-1.2, SS-1.8, SS-2.4 and SS-3 lysimeters, respectively (2.4, 1.2, 0.6, and 0 m of cover). The thicker cover would restrict aeration, retarding the oxidation of Fe^{2+} and the generation of hydronium ion, and possibly the reaction of Fe with the shale.

The chemistry of the leachate-spent shale interactions is more complex than expressed above. Adsorption-desorption phenomena have an impact, and ion pairing effects play a role. Aluminum is not entirely present as the trivalent ion; a large degree of ion pairing is expected. The U.S.G.S. geochemical computer model WATEQ was used to calculate ionic species distribution in the leachates (9). Using the mean values for the SS-3 leachates for the final year of the project, approximately 18% of Al was calculated to be present as Al^{3+}, 42% as $AlSO_4^+$, and 40% as $Al(SO_4)_2^-$. $AlOH^{2+}$, $Al(OH)_2^+$, and $Al(OH)_4^-$ combined were calculated to comprise less than 0.1% of the aluminum bearing ions, suggesting that acid generation from Al hydrolysis was insignificant for the conditions of this study. More than 80% of the Fe present was calculated to be as $FeSO_4^+$.

Mineral Equilibria and Calcium Concentration. In general, the chemistry of the spent-shale leachates did not show strong evidence of mineral-equilibria control. The variations in elemental concentrations were too large to be provided by solution-precipitation reactions. During the time of observation, the chemistry of the leachates appeared to be a function of transport control, i.e. how fast water, gases and ions are moved into and out of the system.

One exception was the behavior of Ca. It was found in very similar concentrations in all of the shale leachates, regardless of shale type or thickness, and did not change in concentration to any large degree with respect to time. This pattern suggests that its concentration was controlled by equilibrium or near-equilibrium with respect to a solid phase. The high concentration of sulfate suggests that the most likely phase is gypsum ($CaSO_4 \cdot 2H_2O$) or anhydrite ($CaSO_4$). Using the free energy values of Garrels and Christ (10), an equilibrium solubility product was calculated for these minerals:

$$\gamma_{Ca^{2+}}(Ca^{2+})\gamma_{SO_4^{2-}}(SO_4^{2-}) = 10^{-4.4} \text{ anhydrite, } 10^{-4.6} \text{ gypsum} \quad (9)$$

where γ_X denotes the activity coefficient and (X) the concentration of species X. Again using WATEQ and the mean concentrations from the SS-3 leachates for the fourth year of the study, an ion activity product of $10^{-4.58}$ was calculated for Ca^{2+} and SO_4^{2-}, corrected for ion pairing. This value is essentially identical to the equilibrium solubility product, indicating that solution-precipitation reactions with gypsum were likely responsible for the narrow range of variability and concentration found for Ca in the leachates.

Summary

The leachates generated in the field lysimeters were highly acidic with pH values varying from approximately 2.5 to 3.5 during the study. The leachates contained high concentrations of dissolved elements, with sulfate, Mg and Fe present at concentrations of tens of thousands or thousands of ppm.

The intensity of acid generation and elemental concentrations in the leachates were highest in the first year and decreased significantly during the period of the study. The acidity of the leachates was caused by the oxidation of iron sulfides and the resultant generation of H^+, Fe(II) and SO_4^{2-}, followed by oxidation of Fe^{2+} to Fe^{3+}, hydrolysis of Fe^{3+}, and the further generation of hydrogen ions.

In the raw-shale leachates, iron, magnesium, and sulfate were present in the highest concentrations. The mobilization of most of these elements was, for the most part, a function of the intensity of acid generation.

The decrystallized clay minerals contained in the retorted shale were more reactive than those in the raw shale, providing leachates with higher Al and K concentrations. Aluminum and Fe concentrations varied inversely, suggesting a product-reactant relationship.

The overall leachate chemistry was probably transport-controlled, i.e. determined by kinetics of solution reactions and material transport. Little evidence, with the exception of Ca and SO_4, was found for mineral-equilibrium control.

Acknowledgements

This project required the help and cooperation of many individuals and organizations. The support of George Lloyd and John Gannon of Southern Pacific Petroleum was critical. David Meyers, the owner of the field-station property, Frank Walker of Walker Construction, Moshe Shirav of the Geological Survey of Israel, Richard Barnhisel of the UK Department of Agronomy, Robert Meade and David Allen of the UK Highway Transportation Research, were all important in the project development. The project was supported by funding from the Kentucky Energy Cabinet and the U.S. Department of Energy, and these organizations are gratefully acknowledged. I would like to thank Carl Roosmagi, the technical officer at U.S. DOE. I would in particular like to thank numerous current and former colleagues at the CAER who supported this work, including: David Koppenaal, Lance Barron, Jill Obley, Jerry Kung, Bill Schram and Bob Kruspe, Gerald Thomas, Bill Jones and Karen Cisler.

Literature Cited

1. Robl, T.L.; Bland, A.E.; Koppenaal, D.W.; Barron, L.S. In Geochemistry and Chemistry of Oil Shale; ACS Symp. Series No. 230; American Chemical Society: Washington, DC, 1983; pp. 159-180.
2. Forbes, F.; Kinsey, F.W. In Proc. 1981 East. Oil Shale Symp.; Inst. for Mining and Minerals Res.: Univ. of Kentucky, Lexington, KY, 1981; pp. 235-240.
3. Forbes, F.; Kinsey F.W.; Colaianni, L.J. In Proc. 1983 East. Oil Shale Symp.; Inst. for Mining and Minerals Res.: Univ. of Kentucky, Lexington, KY, 1983; pp. 337-344.
4. Lloyd, G.A. In Proc. 1983 East. Oil Shale Symp.; Inst. for Mining and Minerals Res.: Univ. of Kentucky, Lexington, KY, 1983; pp. 11-15.
5. Rich, C.I.; Barnhisel, R.I. Minerals and the Soil Environment; Soil Sci. Soc. Am.: Madison, WI, 1977; pp. 797-808.
6. Beavers, A.H.; Jones, R.L. Soil Sci. Soc. Amer. Proc. 1966, 30, 126-128.
7. Robl, T.L.; Barnhisel, R.I.; Rubel, A.M. In Proc. 1989 East. Oil Shale Symp.; Inst. for Mining and Minerals Res.: Univ. of Kentucky, Lexington, KY, 1989; pp. 212-218.
8. Robl, T.L.; Kruspe, R.; Koppenaal, D.W. In Proc. of the First Ann. Oil Shale/Tar Sands Contractors Meet.; U.S. Dept. of Energy: DOE/METC-85/6026, 1985; pp. 333-339.
9. Runnels, D.D.; Linberg, R.D. J. of Geoch. Explor. 1981, 17, 37-50.
10. Garrels, R.M.; Christ, C.L. Solutions, Minerals and Equilibria; Harper and Row: New York, NY, 1965; 450 pp.

RECEIVED August 10, 1993

Chapter 36

Oxidation of Sulfide Minerals Present in Duluth Complex Rock
A Laboratory Study

Kim A. Lapakko and David A. Antonson

Division of Minerals, Minnesota Department of Natural Resources, 1525 Third Avenue East, Hibbing, MN 55746–1461

> The average rate of sulfate release (mol (g rock)$^{-1}$ s^{-1}) from sixteen 75-g Duluth Complex rock samples (0.053 < d < 0.149 mm) was described by $[d(SO_4^{2-})/dt]_{ave} = (5.97 \times 10^{-13}) S^{0.984}$ (n = 32, r^2 = 0.801), where S is the solid-phase sulfur content in percent. The sulfate-release rate also increased as drainage pH decreased below 4. Drainage pH decreased with increased sulfur content and experimental duration. After 150 weeks the minimum drainage pH from samples containing 0.18-0.40% S was 6.1, while that from samples containing 0.41-0.71% S ranged from 4.8 to 5.3. Minimum drainage pH values from samples containing 1.12-1.64% S ranged from 4.3 to 4.9 (69 weeks), while those from samples containing 2.06 and 3.12% S were 4.3 and 3.5, respectively (78 weeks).

The dissolution of rocks and their component minerals has become a topic of interest for mine-waste management. Such dissolution determines the quality of drainage generated by pit walls, waste rock, and tailings. Of particular concern is the generation of acidic drainage by abandoned mine wastes. By predicting mine-waste drainage quality prior to the inception of mining, plans for mineral-resource development and mine-waste management can be developed to minimize adverse environmental impacts.

The Duluth Complex in northeastern Minnesota is a large copper and nickel resource ([1]), and contains elevated levels of platinum group elements ([2]). The Minnesota Department of Natural Resources (MDNR), Division of Minerals was aware of the possibility of mineral resource development in this formation. The Division was also cognizant that predicting the quality of drainage generated over period of decades and centuries by abandoned mine wastes is a relatively new and complex science. Consequently a program was developed to examine the quality of drainage generated by potential Duluth Complex mining wastes.

The major water-quality concern regarding mine waste is the generation of acidic drainage. The extent of acid release is dependent on the balance of acid-producing and acid-consuming mineral-dissolution reactions. Acid is produced as a result of the oxidation of iron-sulfide minerals present in mine waste. Some or all of the acid produced may be neutralized by dissolution of minerals present in the host rock.

Pyrrhotite ($Fe_{1-x}S$) is the predominant iron sulfide in the Duluth Complex, occurring in both monoclinic and hexagonal forms (3), and the presence of troilite (FeS) has also been reported (4). Other sulfides present, in decreasing order of abundance, are chalcopyrite, cubanite, and pentlandite, with minor amounts of bornite, sphalerite, and pyrite (5).

The complete oxidation of one mole of iron sulfide (either FeS or $Fe_{1-x}S$) releases two moles of acid and one mole of sulfate, as indicated by reaction 1.

$$Fe_{1-x}S(s) + [(5-3x)/2]H_2O + [(9-3x)/4]O_2(g) \rightarrow$$

$$(1-x)Fe(OH)_3(s) + 2H^+(aq) + SO_4^{2-}(aq) \quad (1)$$

The acid production is the net result of the oxidation of ferrous iron and the subsequent precipitation of ferric iron and the oxidation of elemental sulfur present. The ferric hydroxide may react further to form lepidocrocite, for example (6).

The most effective minerals for neutralizing acid are calcium carbonate and magnesium carbonate. The effectiveness of such neutralization was observed with calcite occurring naturally in the Duluth Complex (7) and with limestone added as a mitigative measure (8). However, the carbonate-mineral content of the Duluth Complex is typically very low.

The dominant host-rock minerals in the Duluth Complex are plagioclase, olivine, and pyroxenes. Dissolution of these minerals neutralizes acid as indicated by reactions 2 and 3 (9, 10).

$$CaAl_2Si_2O_8(s) + 2H^+(aq) + H_2O \rightarrow Ca^{2+}(aq) + Al_2Si_2O_5(OH)_4(s) \quad (2)$$

$$Mg_2SiO_4(s) + 4H^+(aq) \rightarrow 2Mg^{2+}(aq) + H_4SiO_4(aq) \quad (3)$$

The acid neutralization by these minerals will be rapid during the initial phase of mineral dissolution, when hydrogen ions are rapidly and reversibly exchanged with alkali ions on the mineral surface (11, 12). The rate of dissolution and consequent acid neutralization decreases, eventually becoming linear with respect to time (12, 13).

The objective of this program is to correlate laboratory drainage quality with rock composition and to compare this correlation with field results. An earlier dissolution experiment was conducted on 10 drill-core samples and a test-shaft sample from the Duluth Complex. The quality of drainage generated over a 17-week period was related to the rock chemistry, mineralogy, and surface area, and compared with field data (7).

The experiment presently in progress is examining the dissolution of 16 Duluth Complex samples collected from blast holes at the Dunka mine. The quality of drainage generated for periods of 69 to 150 weeks, considerably longer than the earlier experiment, is described in this paper. The data presented focus on the variation of pH and sulfide-oxidation rate as a function of solid-phase sulfur content.

Methods

Materials. The samples were collected from blast holes at the Dunka mine near Babbitt in northeastern Minnesota. Particle size was reduced either by hand, using a bucking maul, or mechanically with a pulverizor. The samples were sieved and particles with diameters from 0.053 to 0.149 mm (-100/+270 mesh) were retained for experimental use. Samples with sulfur contents of 1.12, 1.16, 1.40, 1.44, and 1.64 percent were wet-sieved while the remaining samples were dry-sieved. The maximum diameter was selected to ensure liberation of the sulfide minerals present (7).

The sulfur contents of the 16 samples, as determined with a LECO furnace, ranged from 0.18 to 3.12 percent. Metal content of the solids was determined using acid digestion (14) and subsequent analysis with a Perkin Elmer 603 atomic absorption spectrophotometer. The trace-metal content of the samples was fairly uniform, with trace-metal content decreasing in the order Cu > Ni > Zn > Co (Table I).

Procedures. Samples (75 g), run in duplicate, were placed into the upper segment, or reactor, of a two-stage filter unit. The solids were placed on a glass-fiber filter which rested on a perforated plastic plate near the bottom of the reactor. To each reactor, 200 mL of distilled-deionized water was added, allowed to remain in contact with the solids for four to seven minutes, and then filtered through a 0.45-μm filter on top of the lower stage of the filter unit. This rinsing was repeated three times at the inception of the experiment to remove oxidation products which accumulated between the time of sample crushing and the beginning of the experiment. The solids were subsequently rinsed weekly with a single 200-mL volume.

Reaction Conditions. Between rinses the solids were retained in the reactors to oxidize. The reactors were stored in individual cubicles which formed a rectangular matrix within a topless housing with a perforated base. A thermostatically controlled heating pad was placed beneath the housing to control temperature. The housing was stored in a small room equipped with an automatic humidifier and dehumidifier, to maintain a stable range of humidity. Temperature and relative humidity were monitored two to three times a week, and the average weekly values were determined. Variations in temperature and humidity did occur, due largely to seasonal variations in these parameters (Figure 1).

The experiment began on 14 February 1989 with reactors 1-24. Reactors 1-20 are yet in progress and data from the first 150 weeks are presented. Reactors 21-24 were terminated after 78 weeks, and reactors 29-38 were started at week 81 (4

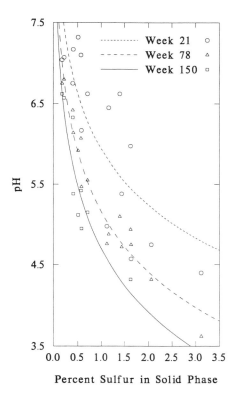

Figure 1. Temperature and relative humidity versus time.

September 1990). The last set of reactors is yet in progress and pH data for the first 78 weeks and sulfate data for 69 weeks are presented.

Analysis. The volume of filtered effluent rinse water, or drainage, was determined by weighing the drainage. Samples were then filtered and analyzed on site to determine pH, alkalinity, acidity, and specific conductance. Samples were also taken for subsequent determination of metal and sulfate concentrations. Samples taken for metal analyses were acidified with 0.2 mL AR Select nitric acid (Mallinckrodt) per 50 mL sample. An Orion SA 72 pH meter, with a Ross combination pH electrode (8165), was used for pH analysis and a Myron L conductivity meter was used to determine specific conductance. Alkalinity and acidity were analyzed using standard titration techniques (15). Sulfate was analyzed using an HF Scientific DRT-100 nephelometer for the barium sulfate turbidimetric method (15). Metals were analyzed with a Perkin Elmer 603 atomic absorption spectrophotometer.

Calculation of Sulfate-Release Rates. The mass of sulfate released was calculated as the product of the observed sulfate concentration and the volume of drainage. Although all drainage samples were not analyzed for sulfate, specific conductance was analyzed weekly. Missing sulfate values were estimated based on regression analyses of the measured sulfate concentrations vs specific conductance observed for each of the sixteen solids ($40 \leq n \leq 123$). The r^2 values for 13 of the solids ranged from 0.56 to 0.95, with a median value of 0.84. The r^2 values for the solids containing 0.18, 0.22, and 0.51 percent sulfur were 0.45, 0.22, and 0.41, respectively.

For most of the reactors the sulfate release in the initial weeks was inconsistent with that observed over the experiment as a whole. This initial period was ignored in the calculation of all sulfate-release rates. The average release rate was calculated as the total sulfate release during this modified period of record divided by the number of weeks in the modified period. To determine the variation in release rates during the experiment, cumulative sulfate release was plotted as a function of time for each reactor. Periods of linear release were selected based on visual examination of the graphs produced, and the release rate for each period was determined by linear regression.

Results and Discussion

The agreement between duplicate samples for all three rates was generally quite good, with the difference from the mean value for the two reactors typically less than ten percent. Using the chemical data from Table I, it was estimated that 0.9 grams of pyrrhotite were present in the sample containing 0.71 percent sulfur. Assuming all sulfate release from this sample was due to pyrrhotite oxidation yields an oxidation rate of about 2.4×10^{-11} mol s^{-1} g^{-1} pyrrhotite. Assuming the relationship between the pyrrhotite particle size and specific surface area is the same as that between quartz sand particle size and specific surface area (16) allows conversion of this pyrrhotite oxidation rate to 8.0×10^{-10} mol m^{-2} s^{-1}. Assuming equal oxidation rates for all metal-sulfide minerals present, and that release of one mole of sulfate indicates the oxidation of one mole of metal-sulfide mineral, yields a metal-sulfide oxidation rate of 4.9×10^{-10} mol m^{-2} s^{-1}. This rate is reasonably consistent with the

Table I. Chemical Composition of Dunka Blast Samples Subjected to Dissolution Experiments

S wt%	Cu wt%	Ni wt%	Co wt%	Zn wt%	S w/ FeS[a] wt%
0.18	0.190	0.055	0.011	0.018	0.039
0.22	0.180	0.062	0.010	0.019	0.081
0.40	0.072	0.032	0.005	0.032	0.328
0.41	0.105	0.047	0.009	0.021	0.316
0.51	0.179	0.038	0.009	0.027	0.381
0.57	0.179	0.041	0.011	0.027	0.438
0.58	0.208	0.055	0.018	0.025	0.423
0.71	0.190	0.059	0.018	0.021	0.562
1.12	0.124	0.042	0.009	0.041	1.010
1.16[b]	0.163	0.063	0.012	0.022	1.026
1.16[b]	0.183	0.066	0.011	0.022	1.015
1.40[b]	0.290	0.087	0.013	0.024	1.187
1.40[b]	0.318	0.097	0.015	0.024	1.167
1.44	0.174	0.052	0.011	0.026	1.305
1.63	0.181	0.059	0.015	0.026	1.486
1.64	0.333	0.084	0.013	0.031	1.404
2.06	0.187	0.058	0.009	0.037	1.911
3.12	0.201	0.050	0.007	0.035	2.970

[a] Calculated as the sulfur not bound by Cu, Ni, Co, or Zn assuming all trace metals were bound by sulfur in a 1:1 molar ratio of sulfur to metal.
[b] Duplicate analyses for metals.

overall metal-sulfide rate for Duluth Complex rock of 2.6 x 10^{-10} mol m^{-2} s^{-1} previously reported for batch reactor experiments (17).

Few rates are reported for the oxidation of pyrrhotite. The pyrrhotite-oxidation rate determined in the present study is roughly an order of magnitude lower than the 1 x 10^{-8} mol m^{-2} s^{-1} oxidation rate at pH 6 reported by Nicholson and Scharer (18) for pyrrhotite particles of similar size, and similar to that previously reported for pyrite (18). The similarity with reported rates for pyrite oxidation with those for the mixed sulfides in the present study suggests that the rate-controlling step may be similar. Examination of this hypothesis was beyond the scope of the present study. Previous work on the oxidation of sulfide minerals present in the Duluth Complex was consistent with a surface-reaction mechanism in which the rate limiting step was the rearrangement of molecules and/or electrons at sites at which oxygen was adsorbed to sulfide minerals (17).

The studies on pyrrhotite and pyrite previously mentioned (18) were conducted with relatively pure minerals, and the rates in the present study are possibly influenced by the complex mineral assemblage present. For example, Koch (19) reported that the presence of copper in a sulfide such as chalcopyrite stabilizes the ferric iron present. It is also possible that copper released from chalcopyrite may have participated in an exchange reaction with iron present in pyrrhotite and, consequently, inhibited its oxidation. Similar reactions have been reported in which metals in solution exchange with sulfide-bound metals of higher solubility (20). In contrast, if galvanic interactions of chalcopyrite and pyrrhotite contributed significantly to pyrrhotite oxidation (21), lower than normal oxidation rates might be expected for these samples.

Dependence of Sulfate-Release Rate on Sulfur Content. A first order of dependence on the sulfide-mineral surface area has been reported for the oxidation of pyrrhotite (18) and mackinawite (6) by oxygen. A similar dependence has been reported for the oxidation of the mixed sulfides present in the Duluth Complex (7, 17). The sulfur content of the solids in the present study was assumed to be proportional to the sulfide-mineral surface area available for reaction because (1) the sulfide minerals were liberated in the size fraction selected for the experiment and (2) the specific surface area of the sulfide minerals was assumed to be constant among sulfide minerals of the same particle size. The weekly release rates calculated were approximately first order with respect to solid-phase sulfur content, as indicated by equations 4, 5, and 6.

$[d(SO_4^{2-})/dt]_{min} = (4.13 \times 10^{-13}) S^{0.948}$, n = 32, r^2 = 0.833 (4)

$[d(SO_4^{2-})/dt]_{ave} = (5.97 \times 10^{-13}) S^{0.984}$, n = 32, r^2 = 0.801 (5)

$[d(SO_4^{2-})/dt]_{max} = (8.74 \times 10^{-13}) S^{0.954}$, n = 32, r^2 = 0.669 (6)

where $d(SO_4^{2-})/dt$ = rate of sulfate release in mol (g rock)$^{-1}$ s^{-1} and
S = solid-phase sulfur content in percent.

These equations were plotted on a single graph along with the observed average sulfate-release rates (Figure 2). Since the dependence on sulfur content was close to first order, linear regression analysis was conducted on the data and yielded the following equations.

$$[d(SO_4^{2-})/dt]_{min} = 4.96 \times 10^{-13} \, S - 5.39 \times 10^{-14}, \, n = 32, \, r^2 = 0.833 \qquad (7)$$

$$[d(SO_4^{2-})/dt]_{ave} = 1.17 \times 10^{-12} \, S - 4.73 \times 10^{-13}, \, n = 32, \, r^2 = 0.722 \qquad (8)$$

$$[d(SO_4^{2-})/dt]_{max} = 2.41 \times 10^{-12} \, S - 1.26 \times 10^{-12}, \, n = 32, \, r^2 = 0.661 \qquad (9)$$

The ratio of the maximum rate to the minimum rate was between 1.3 and 2.8 for 75 percent of the reactors. The sulfate release from the sample containing 1.12 percent sulfur was fairly constant, and the ratio was below the typical range (Table II). The variation in sulfate release from sample containing 3.12 percent sulfur was highest, with maximum to minimum rate ratios of 5.8 and 6.8 for the duplicate samples. The maximum to minimum rate ratios were generally higher for the samples with the 150 week period of record than for samples with shorter records, as indicated by median ratios of 2.35 and 1.65, respectively.

Examination of Figure 2 indicates that the sulfate-release rates for solids containing 0.40, 1.12, and 3.12 percent sulfur were higher than suggested by the relationship between rate and sulfur content for the samples as a whole. The rates for the 3.12 percent sample were most deviant from the remaining data, and the deviation can be attributed to increased sulfide oxidation at low pH.

The accelerated rates observed for the 0.40 and 1.12 percent S samples may be due to greater available sulfide-surface areas and/or chemical or mineralogical differences for these samples. First, the sulfide-surface area of these samples may be higher due to subtle differences in sulfide grain size and/or "roughness" of the mineral surfaces. Second, there may be sulfide minerals present in these two samples which oxidize more rapidly than the sulfides typically present in the samples examined. Third, these two samples have the lowest copper content of the samples examined. This difference, and/or its implication of a difference in chalcopyrite content, may influence the sulfide-oxidation rate for these samples. As previously discussed, it is possible that under certain conditions copper has an inhibitive effect on iron-sulfide oxidation.

Dependence of Sulfate-Release Rate on pH. The dependence of the rate of sulfide-mineral oxidation by oxygen on pH is reported to be slight. Nicholson and Scharer (18) found little variation in the abiotic pyrrhotite oxidation rate over the pH range of 2 to 6, reporting "maximum differences in oxidation rates were about a factor of 2 or within 50% of the mean rate at a specified temperature." Nelson (6) reported that the rate of mackinawite oxidation by oxygen at pH 6.5 was five times that at pH 9.0. For the mixture of sulfides present in Duluth Complex rock the sulfide-oxidation rate was reported as proportional to $[H^+]^{0.2}$ over the pH range of 5 to 8 (17).

Research on pyrite oxidation, however, indicates that as "pH decreases to 4.5, ferric iron becomes more soluble and begins to act as an oxidizing agent" (22). As pH further decreases, bacterial oxidation of ferrous iron becomes the rate limiting

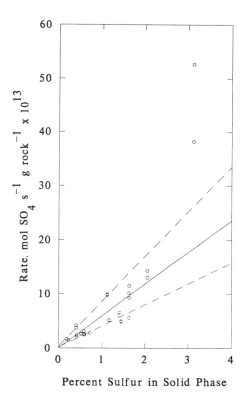

Figure 2. Average sulfate-release rate as a function of solid phase sulfur content. The lines are regressions for minimum, average, and maximum rates (equations 4-6): $[d(SO_4^{2-})/dt]_{min} = (4.13 \times 10^{-13}) \, S^{0.948}$; $[d(SO_4^{2-})/dt]_{ave} = (5.97 \times 10^{-13}) \, S^{0.984}$; $[d(SO_4^{2-})/dt]_{max} = (8.74 \times 10^{-13}) \, S^{0.954}$.

Table II. Summary of Sulfate Release Rates and Minimum Drainage pH

Sulfur %	Min pH s.u.	Rate, mol SO$_4$ s^{-1} g rock^{-1} x 10^{13}			Rate, mol SO$_4$ s^{-1} mol S^{-1} x 10^9			Time weeks
		Min	Avg	Max	Min	Avg	Max	
0.18	6.20	1.23	1.67	2.29	2.20	2.99	4.08	150
0.18	6.25	1.17	1.72	3.26	2.08	3.06	5.81	150
0.22	6.40	1.10	1.43	2.29	1.61	2.09	3.34	150
0.22	6.33	1.12	1.45	2.77	1.64	2.12	4.05	150
0.40	6.16	2.38	4.16	7.39	1.91	3.34	5.94	150
0.40	6.10	1.98	3.72	6.84	1.59	2.99	5.50	150
0.41	5.30	1.36	2.38	4.25	1.07	1.86	3.33	150
0.41	5.20	1.36	2.07	3.10	1.07	1.62	2.43	150
0.51	5.08	1.61	2.60	4.11	1.01	1.64	2.59	150
0.51	4.98	1.69	2.66	4.09	1.07	1.68	2.58	150
0.57	5.25	2.33	3.08	3.54	1.31	1.74	2.00	150
0.57	5.35	2.40	3.15	3.67	1.35	1.77	2.07	150
0.58	4.85	2.05	2.60	3.65	1.13	1.44	2.02	150
0.58	4.82	1.83	2.46	4.03	1.01	1.37	2.23	150
0.71	4.98	2.20	2.82	3.94	1.00	1.27	1.78	150
0.71	5.05	1.63	2.82	4.88	0.74	1.27	2.21	150
1.12	4.62	9.46	9.77	10.10	2.71	2.80	2.90	69
1.12	4.75	9.66	9.90	10.05	2.77	2.84	2.88	69
1.16	4.69	3.96	5.13	6.73	1.10	1.42	1.87	69
1.16	4.60	4.05	5.17	6.62	1.12	1.43	1.83	69
1.40	4.78	4.53	6.09	7.88	1.04	1.40	1.81	69
1.40	4.92	4.84	6.53	7.50	1.11	1.50	1.72	69
1.44	4.59	3.85	4.86	5.61	0.86	1.09	1.25	69
1.44	4.41	3.87	4.97	5.63	0.86	1.11	1.26	69
1.63	4.95	4.58	6.05	6.62	0.90	1.19	1.31	69[a]
1.63	4.75	6.80	6.80	6.80	1.34	1.34	1.34	69[a]
1.63	4.37	3.98	5.59	6.62	0.79	1.10	1.31	150
1.63	3.90	6.71	10.10	18.22	1.32	1.99	3.59	150
1.64	4.42	7.59	9.26	9.86	1.49	1.81	1.93	69
1.64	4.32	9.50	11.48	19.12	1.86	2.25	3.75	69
2.06	4.20	10.54	14.23	28.23	1.64	2.22	4.40	78
2.06	4.30	10.93	12.96	21.63	1.71	2.02	3.37	78
3.12	3.70	15.88	38.24	91.48	1.64	3.94	9.42	78
3.12	3.50	15.91	52.65	104.32	1.64	5.42	10.75	78

[a] Sixty-nine week data included for 1.63% S sample for comparison with samples of similar sulfur content.

step in the oxidation of pyrite by ferric iron (23), which is the only significant oxidant in this pH range (22, 23, 24). Consequently, under these low-pH conditions the reaction is independent of sulfide-mineral surface area (24).

The large variation observed for the rate of sulfate release from the sample containing 3.12 percent sulfur suggests that the rate was affected by the low-pH conditions created by rapid iron-sulfide oxidation. The pH of drainage from this sample decreased steadily from around 6.0 to about 3.5. As pH decreased below 4.0, the sulfate-release rate increased rapidly. The rates observed in the pH range of 3.5 to 4.05 were roughly six to seven times those observed in the pH range of 5.35 to 6.1. This acceleration was probably due to increased biologically mediated ferric-iron oxidation of the sulfide minerals as pH decreased below 4.0. Abiotic oxidation may have also increased, although the extent of increase was most likely slight relative to the increase in biological oxidation (22). Because the decreased pH conditions which accelerated sulfide oxidation were produced as a result of elevated iron-sulfide oxidation, the variation in sulfate release with sulfur content is influenced by a form of autocatalysis.

Dependence of Sulfate-Release Rate on Temperature and Relative Humidity. Analysis of the data indicates, qualitatively, that variations in sulfide-oxidation rates were related to changes in temperature and relative humidity in the reaction environment, as well as changes in pH within the reactor. Temperature and/or relative humidity may have influenced the rate of sulfide-mineral oxidation in a manner similar to that reported for pyrite oxidation (26). Inspection of the rates reported by Nicholson and Scharer (18) indicates that the pyrrhotite oxidation rate at pH 6 roughly quadrupled with a temperature increase of 10°C. Nelson (6) reported that the rate of mackinawite oxidation was approximately doubled by a temperature increase of 8 - 10°C. Recent studies indicate that the effect of relative humidity on pyrite oxidation is dependent on the type of pyrite oxidized (27).

The observed temperature and relative humidity during periods of maximum and minimum sulfate release qualitatively indicate that sulfide-oxidation rates were affected by temperature and, perhaps, relative humidity. The qualitative assessment of these effects considered 23 reactors for which the ratio of maximum to minimum sulfate-release rate exceeded 1.5. It excluded the sample containing 3.12 percent sulfur. For 20 of the 23 cases, maximum rates occurred during times of elevated temperature, specifically weeks 6 to 28, 65 to 81, and 117 to 134. The relative humidity was above average in the middle period and near average in the others. Likewise, for 20 of the 23 cases, minimum rates occurred during periods of low temperature: weeks 36 to 58, 86 to 109, and 136 to 150. The relative humidity during the first two periods was near average and below average during the third (Figure 1).

Dependence of Drainage pH on Sulfur Content. Assuming iron-sulfide oxidation was the major contributor of sulfate, the rate of acid production also increased linearly with solid-phase sulfur content. The observed drainage pH data support this hypothesis. The pH of drainage from the 16 samples was observed generally to decrease as solid-phase sulfur content and time of dissolution increased (Figure 3,

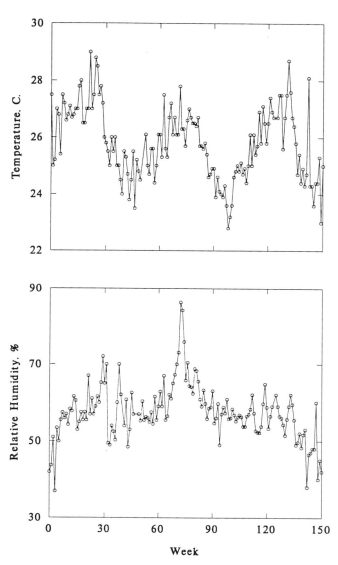

Figure 3. Drainage pH decreased as sulfur content and time increased.

Table II). The decrease in drainage pH over time can be attributed to the depletion of the more reactive neutralizing components.

The samples were divided into four classes based on sulfur content and the minimum pH of drainage from the samples (Table II). This classification was complicated by the variable periods of record for the samples and was based partly on the "best guess" extrapolation of drainage pH over time. As the experiment progresses, and as other samples are examined, the classification is likely to be revised. The terms describing the drainage pH ranges (circumneutral to strongly acidic) were used to simplify data presentation. They were selected as descriptions relative to the experimental data generated, and not with respect to mine waste drainage in general. The gross rates of sulfate release (not normalized for sulfur content) associated with the various groups are also presented. The rates were not normalized in order to provide a measure of the relative magnitudes of acid production for the four classes.

Samples containing 0.18 to 0.40 percent sulfur generated minimum drainage pH values of 6.10 to 6.40 over the 150 week period of record. The pH of drainage from these solids meets the typical water-quality standard of pH 6.0, thus these samples present the lowest potential for generating acidic drainage. The rates of sulfide oxidation were the lowest observed, averaging approximately 1.5×10^{-13} mol (g rock)$^{-1}$ s^{-1}, and were consistent with the general trends observed for the solids. Apparently the dissolution of silicate minerals present in these samples (for example reactions 2, 3) neutralized the acid resulting from iron-sulfide oxidation.

Samples containing 0.41 to 0.71 percent sulfur produced "mildly acidic" drainage, with minimum drainage pH values of 4.82 to 5.25 over a period of 150 weeks (Figure 3). The average sulfide-oxidation rate for these samples was in the range of 2.2×10^{-13} to 3.3×10^{-13} mol (g rock)$^{-1}$ s^{-1}, with maximum rates roughly 50 percent higher. Although mitigative measures might be required for such drainages to meet pH standards, the extent of mitigation required would be relatively low.

Samples containing 1.12 to 1.64 percent sulfur produced "moderately acidic" drainage, with minimum drainage pH values of 4.32 to 4.95 over a period of 69 weeks. Although the minimum drainage pH range for these samples intersects that for the Group 2 samples, the period of record for the Group 3 samples is considerably shorter. Based on the observed temporal variation, the pH of drainage from these samples would be expected to decrease as dissolution continues (Figure 3). Indeed, the period of record for reactors containing the 1.63 percent sulfur sample is 150 weeks, and the minimum drainage pH values for this period are 4.37 and 3.90. It is likely that other samples in this group will exhibit similar decreases in drainage pH over time.

The average sulfide-oxidation rates from these samples ranged from about 4.4×10^{-13} to 11×10^{-13} mol (g rock)$^{-1}$ s^{-1}, indicating a higher rate of acid production than samples from the first two groups. The oxidation rate for the 1.12 percent sulfur sample was anomalously high. Unlike most samples, the rate of sulfate release from this sample was also relatively constant. This suggests that the sulfide-mineral composition of this sample may be slightly different from that of the samples in general. Mineralogical analyses are planned to elucidate these phenomena.

The samples containing 2.06 and 3.12 percent sulfur produced "strongly acidic" drainage, with minimum pH values of 4.20 and 3.50, respectively, over a period of 78 weeks (Figure 3). The average sulfide-oxidation rates for these samples were roughly 13×10^{-13} and 46×10^{-13} mol (g rock)$^{-1}$ s^{-1}. With the observed pH variation, accelerated sulfide oxidation would be expected at the lower pH conditions observed for the higher sulfur content. Based on samples examined to date, Duluth Complex mine wastes with similar sulfur contents would require the most rigorous mitigation to meet water-quality standards for pH.

Summary

The release of sulfate from Duluth Complex rock samples was observed to be first order with respect to solid-phase sulfur content. The majority of the sulfate released was assumed to be due to the oxidation of iron sulfide, mainly pyrrhotite, which is the dominant sulfide in the Complex. An observed decrease in drainage pH with sulfur content was consistent with this hypothesis.

Based on the observed drainage quality, the samples were divided into four ranges of sulfur content. Distinct ranges for minimum drainage pH and sulfide-oxidation rate were observed for each group. Continued testing of the present samples, dissolution tests on additional well-characterized Duluth Complex samples, and field verification will be necessary to evaluate the validity of the results presented. Trace metal release will also be examined.

Additional data analysis is planned to quantify the effects of pH, temperature, and relative humidity on the observed sulfide-oxidation rate. Drainage samples will be analyzed for calcium and magnesium to obtain insight into dissolution reactions which contribute to acid neutralization in various pH ranges. Furthermore, the rock samples examined will be subjected to more extensive physical, chemical, and mineralogical analysis. The rates of sulfide oxidation and acid neutralization will be compared with other reported values, and factors which potentially control dissolution will be addressed.

Acknowledgments

Bill Conger of LTV Steel Mining Company was responsible for particle size reduction and sample sieving. Kate Willis conducted the majority of the laboratory experiments and Jean Matthew analyzed samples for sulfate. Anne Jagunich and Jason Perala were responsible for data input and Jim Porter managed data output. The U. S. Bureau of Mines provided funding to support data compilation and analysis as part of a cooperative project on mine waste dissolution.

Literature Cited

1. Kingston, G. A.; Carillo, F. V.; Gray, J.J.; Mallory, P. U.S. Bur. Mines IC 8469, 1970, 57 pp.
2. Anonymous. Skillings' Mining Review, 1992, 81, 5.
3. Foose, M.; Weiblen, P. In Geology and Metallogeny of Copper Deposits, Freidrich, G. H., Ed., Springer-Verlag: Berlin, 1986, pp. 8-24.
4. Boucher, M. L. U.S. Bur. Mines RI 8084, 1975, 55 pp.

5. Stevenson, R. J.; Kreisman, P. J.; Sather, N. P. In Minnesota Regional Copper-Nickel Study, Chap. 1. MN Environ. Quality Board: St. Paul, MN, 1979; Vol. 3, 70 pp.
6. Nelson, M. Ph.D. Thesis, Stanford University, Palo Alto, CA, 1978.
7. Lapakko, K. A. In Proc. 1988 Mine Drainage and Surface Mine Reclamation Conference, U.S. Bur. Mines IC9183, 1, 1988, pp. 180-190.
8. Lapakko, K. A.; Antonson, D. A. In Proc. Second International Conference on the Abatement of Acidic Drainage, MEND (Mine Environment Neutral Drainage), Montreal, QC, 1991, pp. 343-358.
9. Busenberg, E. Geochim. Cosmochim. Acta 1978, 42, 1679-1686.
10. Hem, J. D. U.S. Geol. Survey Water-Supply Paper 1473, 1970, 363 pp.
11. Holdren, G. R.; Berner, R. A. Geochim. Cosmochim. Acta 1979, 43, 1161-1171.
12. Busenberg, E.; Clemency, C. Geochim. Cosmochim. Acta 1976, 40, 41-49.
13. White, A. F.; Classen, H. C. In Chemical Modeling in Aqueous Systems, Jenne, E. A., Ed., ACS Symposium Series No. 93, American Chemical Society: Washington, DC, 1979, pp. 447-473.
14. U.S. Bur. Mines. U.S. Bur. Mines RI 8480, 1980; 14 pp.
15. American Public Health Association; American Water Works Association; Water Environment Federation. Standard Methods for the Examination of Water and Wastewater, 18th edition, American Public Health Association: Washington DC, 1992.
16. Parks, G. A. In Mineral-Water Interface Geochemistry, Rev. Mineral. 1990, 23, 133-175.
17. Lapakko, K. A. M.S. Thesis. University of Minnesota, Minneapolis, MN, 1980.
18. Nicholson, R. V.; Scharer, J. M. In this volume.
19. Koch, D. F. In Modern Aspects of Electrochemistry, Bockris, J. O.; Conway, B. E., Eds.; Plenum Press: New York, New York, 1975, Vol. 10, pp. 211-237.
20. Gaudin, A. M.; Fuerstenau, D. W.; Mao, G. W. Trans. AIME, 1959, 214, 430-432.
21. Ahonen, L.; Hiltunen, P.; Tuovinen, O. H. In Fundamental and Applied Biohydrometallurgy; Lawrence, R. W.; Branion, R. M. R.; Ebner, H. G. Eds., Elsevier: Amsterdam, 1986; pp. 13-22.
22. Nordstrom, D. K. Acid Sulfate Weathering, Soil Sci. Soc. Am.: Madison, WI. 1982, pp. 37-56.
23. Singer, P. C.; Stumm, W. Science 1970, 167, 1121-1123.
24. Kleinmann, R. L. P.; Crerar, D. A.; Pacelli, R. R. Mining Eng. 1981, 300-305.
25. Stumm, W.; Morgan, J. J. Aquatic Chemistry, Wiley: New York, 1981, 780 pp.
26. Morth, A. H.; Smith, E. E. In preprints from ACS Division of Fuel Chemistry Symposium on Fossil Fuels and Environmental Pollution; American Chemical Society: Washington, DC, 1966, Vol. 10 (1), pp. 83-92.
27. Borek, S. L. In this volume.

RECEIVED October 20, 1993

Chapter 37

Chemical Predictive Modeling of Acid Mine Drainage from Metallic Sulfide-Bearing Waste Rock

W. W. White III and T. H. Jeffers

Salt Lake City Research Center, U.S. Bureau of Mines, Salt Lake City, UT 84108

> The Bureau of Mines is developing a geochemical predictive model for acid mine drainage (AMD) from waste rock associated with metal mining. The model will identify AMD potential during property exploration and development, and will facilitate preplanning of waste-rock handling. This paper presents results to date on waste-rock characterization, and static and kinetic tests for three selected samples. Sulfide content of the samples was 1.5, 3.5, and 13.0 %, mainly as a mix of euhedral/subhedral and framboidal pyrite. Although acid-base accounting classified all three samples as potential acid producers, effluent pH from the 1.5 % sulfide sample was neutral to slightly basic during 51 weeks of accelerated weathering. Samples containing 3.5 and 13.0 % sulfide continuously produced acidic effluent during the same 51-week period. Sulfate release rates resulting from sulfide oxidation increased with solid-phase sulfide content and decreasing pH. Extrapolated calcium and magnesium release-rate data from the 1.5 % sulfide sample suggest that it should develop AMD after 110 to 130 weeks of laboratory accelerated weathering.

Acid mine drainage (AMD) is contaminated effluent that results from the oxidation of iron-sulfide minerals exposed to air and water. AMD emanating from base- and precious-metal mining operations is one of the most difficult environmental problems facing the mining and mineral industries.

A significant amount of the AMD draining into the Nation's rivers and streams, particularly in the western states, is released from waste rock. Waste rock generated during mining activities was formerly placed in large piles with little or no provision for preventing AMD. By contrast, modern operations attempt to construct and manage waste dumps to minimize AMD formation. However, this task is hindered by the complex rock types encountered and the lack of simple, inexpensive test techniques to predict if a specific rock type will eventually produce acid. Once AMD generation begins, the process is extremely difficult to reverse, and long-term treatment is required.

Current static and kinetic tests for determining AMD potential are largely based on eastern United States coal-related research. As a result, a debate exists as to which tests are appropriate for metal mines. As part of Canada's Mine Environment Neutral Drainage (MEND) program, the Canada Centre for Mineral and Energy Technology (CANMET) recently sponsored an evaluation of several test methods commonly used or proposed for use in predicting AMD from metals mines (1). These techniques include Acid-Base Accounting (ABA) (2), the B. C. Research Initial (BCRI) and Confirmation Tests (3), Humidity-Cell Tests (2), Alkaline Production Potential vs. Sulfur Ratio (4), Hydrogen Peroxide Test (5), Net Acid Production Test (6) (Lawrence, R. W.; Jaffe, S.; Boughton, L. M., Coastech Research, West Vancouver, BC, unpublished data.), Modified Confirmation Test (Lawrence, R. W.; Sadeghnobari, A., Coastech Research, West Vancouver, BC, unpublished data.), Shake Flasks Tests (7), and the Soxhlet Extraction Test (8-9). Although the CANMET report presented comments and recommendations regarding the effectiveness of these predictive tests (1), there was no general agreement on which test or series of tests will best determine potential to generate metals mine AMD (10-12). Additionally, comparison and correlation of test-generated data with field data has been minimal (Ferguson, K. D., Environment Canada, North Vancouver, BC, personal communication, 1991), (Lawrence, R. W., Coastech Research, North Vancouver, BC, personal communication, 1991), (1, 10, 13-16).

Accelerated-weathering humidity-cell tests and ABA are predictive-test techniques commonly used by U.S. and Canadian governments and industry to assess waste-rock AMD potential. The U.S. Bureau of Mines (Bureau) has selected these two test techniques to provide initial experimental data for use in its predictive-modelling project. The intent of the Bureau's research is not to model waste-rock dumps, but rather to model acidic-drainage potential for discrete waste-rock types based on drill core and bulk samples obtained during property exploration and development. Identification of the tendency for a discrete waste-rock type to develop AMD will enable companies to plan waste-rock treatment and placement during the feasibility stage of property development. This is especially appropriate for the western states because several state and Federal land managing agencies such as the Bureau of Land Management and the U.S. Forest Service require companies to assess the AMD potential of future operations as part of the permitting process.

For this study, acidic drainage potential of a waste-rock sample is defined as its tendency to produce AMD over a specific period of time, under the established conditions characteristic of accelerated-weathering humidity-cell tests. This potential is quantified by using cation and anion release rates derived from weekly humidity-cell effluent. Years of accelerated weathering required to deplete each sample of its neutralizing and acid-producing potentials (NP and AP, respectively) are estimated using these release rates.

The Bureau is conducting AMD modelling research through four activities: (1) collecting waste-rock samples from discrete rock types during exploration and development phases of an operation, (2) performing mineralogical and chemical characterization of each sample, (3) subjecting samples to predictive tests (ABA and humidity-cell accelerated weathering), and (4) using data generated from sample characterization and predictive tests as input to the developing model. The objective of the preliminary model is to predict the potential for a waste-rock sample to produce AMD beyond the 20 weeks of humidity-cell test duration common in commercial testing. Lapakko (17) demonstrated that some samples require at least 50 to 80 weeks of laboratory accelerated weathering before NP is depleted and actual acid production occurs. The preliminary model (1) calculates specific cation and anion concentrations contained in weekly humidity-cell effluent and (2) extrapolates the predicted weekly effluent compositions beyond 20 weeks. The model has been calibrated by comparing modelled effluent composition with experimental humidity-cell effluent data from 51 weeks of Bureau tests on discrete waste-rock type samples.

The predictive model development is a cooperative effort between the Bureau and the University of Utah Chemical Engineering Department. Static and kinetic laboratory tests are being conducted by the Bureau, and the resulting experimental data are used by the University to calibrate the developing mathematical model.

Methods and Materials

Sample Collection and Preparation. The Bureau collected samples of drill core or cuttings intervals that were identified as waste and segregated by rock type. Where possible, waste-rock core and cuttings from exploration and development phases were matched with bulk samples from subsequent mining operations. Samples known to contain a high concentration of sulfide or excessive neutralizing potential were initially selected to provide baseline data for the developing model.

Forty-six samples from six western-U.S. metals mines have been collected. Of these six mines, three have identified AMD, two have potential to develop AMD, and one is reported to contain neutralization potential that far exceeds acid-producing potential. Three samples (1-B, 1-C, and 1-D) from one of the mines experiencing AMD have been selected for discussion in this paper because they are each comprised of different rock types, and contain 3.5, 13.0, and 1.5 % sulfide, respectively. The three different rock types represented by these samples match rock types that comprise more than 200 million tons of waste rock. Sample 1-B is from blast-hole drill cuttings, and samples 1-C and 1-D are bulk samples.

Bulk samples and drill core samples were crushed to 100 % passing 1/4 inch, which is the recommended feed size for accelerated-weathering humidity-cell tests (18). A size-fraction analysis was performed on the crushed samples and 11 fractions were produced and used for both chemical and mineral characterization. Part of each sample was subsequently pulverized to 80 % passing 150 mesh for chemical characterization and static testing.

Mineralogy. Selected size fractions (-10/+20, -48/+65, and -200 mesh) from the three samples were examined with optical and scanning electron microscopy (SEM) and analyzed by X-ray diffraction (XRD). These examinations were conducted to (1) identify metallic-sulfide mineral species and their grain morphologies, (2) determine size ranges and liberation sizes of sulfide-mineral grains, and (3) identify and quantify the proportions of minerals known to provide acid-neutralizing potential. The surface area of each sample was calculated by using the average particle diameter of each of the eleven fractions that represent the minus 1/4-inch sample. The percent distribution and liberation of pyrite in each fraction were also included in the calculation of surface area.

Solid-Phase Sulfate-Salt Storage, Hot-Water Leach Determination. Sulfate-salt storage occurs when interstitial water evaporates during dry periods and precipitates previously-solubilized sulfate on waste-rock fragments. Source of the sulfate is natural pyrite/pyrrhotite oxidation that occurred prior to sampling. Quantification of salt storage is important as it causes sulfate-release rates to be overestimated during the first 3 to 6 weeks of kinetic accelerated-weathering tests. The sulfate-salt storage for each waste-rock sample was determined by a hot-water leach of the minus 1/4-inch fraction, followed by analysis of the resulting leachate for sulfate. The sulfate concentration is expressed as milligrams per kilogram of solid sample and compared to cumulative sulfate released during the first five weeks of humidity-cell effluent.

Acid-Base Accounting. The goal of acid-base accounting (ABA) is to predict the net-neutralization potential of a mine-waste sample. ABA is the measure of balance between acid-producing potential (AP) and neutralizing potential (NP). The AP and NP results can be expressed as a difference (NP - AP = Net NP or NNP), or as a fraction (NP/AP), and are most useful when used together. To calculate AP, percent sulfur as sulfide is multiplied by a factor of 31.25 (19). Units of AP and NP are expressed as parts per thousand of calcium-carbonate equivalent (i.e., tons $CaCO_3/10^3$ tons of waste rock).

Neutralization potential determination is a critical component of ABA; if NP is overestimated, a waste sample could be misclassified as non-acid producing. Several investigators (1, 10, 17) suggest that various methods currently being used to determine NP may overestimate a waste rock's actual field NP capacity. To reduce the potential overestimation of NP for samples 1-B, 1-C, and 1-D, two different NP-determination methods were used on each of the samples. The Sobek method (2) was performed on the minus 1/4-inch and 80 % passing 150 mesh fractions from each sample, and a modified version of the B.C. Research Initial (BCRI) method (12) was performed on duplicate minus 150-mesh fractions. Selection of the different screen fractions was based on the following: (1) waste-rock charge size for the Bureau's humidity cells is 100 % passing 1/4 inch; and (2) 80 % passing 150 mesh is midway between the minus 60-mesh size used in the Sobek method (2), and the variable size ranges (-100, -325, and -400 mesh) described for the BCRI method (3, 10, 12). Resulting NP data from each method were then compared. If good agreement existed between the NP values determined by the two different methods, then the NP value was considered representative of the actual field NP capacity.

Sobek Method. NP determination was accomplished by leaching 2.0 g of waste-rock sample with 20 ml of 0.1 N HCl. The resulting slurry was brought up to 120-ml volume with de-ionized (D.I.) water, and titrated with 0.1 N NaOH to a pH endpoint of 8.4. Measured acid consumption was converted to calcium-carbonate equivalents (2, 18).

Modified B.C. Research Initial (BCRI) Method. The modified BCRI method consists of titrating a waste-rock slurry comprised of 10 g solid and 100 ml D.I. water with 1 N sulfuric acid to endpoints of pH 6, 5, and 3.5. The resulting volumes of acid for the respective endpoints are then converted to parts per thousand calcium carbonate equivalent (NP). The NP obtained at pH 6 is termed the "effective" NP, or the calcium carbonate equivalent available in the sample to maintain the pH above 6 (12).

Bureau Humidity-Cell Array. A single humidity-cell array is comprised of sixteen individual cylindrical cells, each 20.32 cm long, with an inside diameter (I.D.) of 10.16 cm (Figure 1). Waste-rock charges for each cell are comprised of 1,000 g of sample crushed to 100 % passing 1/4 inch. The test protocol is comprised of weekly leach cycles that include 3 days of dry air and 3 days of wet air pumped up through the sample, followed by a drip-trickle leach with 500 ml of de-ionized water on day 7 (18). Duration of the leach is approximately 2 hours. Bureau tests are continued beyond the 20-week duration currently used in commercial practice; the rationale is to evaluate calculated rates of NP depletion and subsequent acid production from samples whose effluents may appear benign at the end of 20 weeks of testing. The following data are collected weekly from each humidity cell: (1) pH, Eh, and conductivity readings, (2) CO_2 concentrations, (3) alkalinity and acidity values, (4) aqueous Ca, Fe, Mg, SO_4^{2-}, and selected heavy metals concentrations, (5) wet and dry airflow rates, temperatures, and relative humidities, (6) cell weights before and after the drip-trickle leach, and (7) cell weights before the dry-air portion of each weekly cycle.

Geochemical Predictive Model

Continuous-Stirred-Tank Reactor (CSTR) Model. The University of Utah Chemical and Fuels Engineering Department, under contract with the Bureau, has designed a preliminary model that simulates the chemical changes which occur in the weekly interstitial water remaining in each humidity cell (Trujillo, E. M.; Leonora, S.; Lin, C.; Department of Chemical Engineering, University of Utah, Salt Lake City, UT, personal and written communication, 1992). The interstitial water is defined as the water retained by the waste-rock charge at the end of each weekly 500 ml drip-trickle leach. The model assumes that the drip-trickle leach dilutes and flushes most of the previous week's remaining interstitial water from the humidity cell. The non-recoverable part of the drip-trickle leach becomes the interstitial water for the following week. The weekly chemical

Figure 1. A single humidity-cell array comprised of 16 cells.

changes in interstitial water are mathematically described by simulating the contents of each loaded humidity cell as a continuous-stirred-tank reactor (CSTR). Assumptions made in the simulation are: (1) solids and liquids are well mixed in each CSTR, and represent a pseudo-homogeneous mixture; and (2) except for the component oxygen, only one phase is represented.

The 3-day dry-air portion of each weekly cycle is modelled by a decreasing-volume CSTR which represents the evaporation of interstitial water. The 3-day wet-air portion of each weekly cycle is simulated by a constant-volume CSTR; this CSTR represents the assumption that the interstitial-water volume remains constant due to the 100 % relative humidity of wet air. The drip-trickle leach which marks the end of each weekly cycle is represented by a unit comprised of 20 CSTRs in series; the output of one serves as input to another until the simulated drip-trickle leachant has traversed all 20 CSTRs. Each component in the interstitial water from the 20 CSTRs in series is averaged and becomes the interstitial input to the dry-air CSTR and a new week's cycle begins. The predicted effluent composition from the last CSTR in series is integrated over time to obtain the predicted composition in the leachate. Predicted leachate composition is then compared to that obtained experimentally.

Contributing Chemical Reactions. Seven preliminary chemical reactions (1-7) have been selected to initially describe each cell's solid- and aqueous-phase chemistry. These chemical equations represent the following:

- Inorganic and biologic oxidation and depletion of pyrite (1,3);
- Growth of oxidizing bacteria on the sulfide minerals (7);
- Production of significant cations and anions such as sulfate, ferrous, and ferric ions (1,2,3,6,7);
- Production of hydrogen ion (1,6,7); and
- Acid-consuming reactions with carbonate minerals such as calcite (4,5).

$$FeS_2 + \frac{7}{2}O_2 + H_2O \xrightarrow{k_1 + bacteria} Fe^{2+} + 2SO_4^{2-} + 2H^+ \quad (1)$$

$$Fe^{2+} + \frac{1}{4}O_2 + H^+ \xrightarrow{k_2} Fe^{3+} + \frac{1}{2}H_2O \quad (2)$$

$$FeS_2 + 2Fe^{3+} \xrightarrow{k_4} 3Fe^{2+} + 2S^o \quad (3)$$

$$CaCO_3 + H^+ \xrightarrow{k_5} Ca^{2+} + HCO_3^- \quad (4)$$

$$CaCO_3 + 2H^+ \xrightarrow{k_6} Ca^{2+} + H_2CO_3 \quad (5)$$

$$S^o + \frac{3}{2}O_2 + H_2O \xrightarrow{k_9} 2H^+ + SO_4^{2-} \quad (6)$$

$$\text{Bacteria-growth/death equation} \quad (7)$$

Differential Equations. From the seven chemical reactions, 10 chemical species were identified as important components of the reactions. These species are FeS_2 (pyrite), SO_4^{2-}, S^o, Fe^{2+}, Fe^{3+}, O_2, H^+, Ca^{2+}, $CaCO_3$ (calcite), and bacteria (designated as the concentration [X] in the following equations). A mass-balance has been derived for each of the species and is comprised of four functions that affect the rate of change in species concentration during each weekly cycle of the accelerated-weathering humidity-cell test. The first two functions describe rate of change due to abiotic oxidation of pyrite and elemental sulfur; the third function represents the rate of change due to bacterial catalysis of pyrite oxidation; and the fourth function describes rate of change due to interstitial water-volume changes caused by the dry-air portion of the weekly cycle. The bacterial-catalysis expression is included in the differential equations for pyrite, sulfate, ferric ion, oxygen, and hydrogen-ion species. While the model simultaneously solves for rates of change in each species concentration with respect to time through each weekly cycle, volume functions drop out of the equation periodically during each cycle as conditions change. An illustration of this periodic presence or absence can be shown in the following two equations which describe rate of change in sulfate concentrations during the dry-air and wet-air portions of each weekly cycle.

Dry Air:

$$\frac{d[SO_4^{2-}]}{dt} = 2k_1[FeS_2][O_2]^{\frac{7}{2}} + k_9[S^o][O_2]^{\frac{3}{2}} + \left(\frac{2\mu_{max}[X][H^+]}{k_{M7a}+[H^+]}\right)\left(\frac{[FeS_2]}{k_{M7b}+[FeS_2]}\right)$$

$$- \left(\frac{v_o}{V_o+v_o t}\right)[SO_4^{2-}] \quad (8)$$

Wet Air:

$$\frac{d[SO_4^{2-}]}{dt} = 2k_1[FeS_2][O_2]^{\frac{7}{2}} + k_9[S^\circ][O_2]^{\frac{3}{2}} + \left(\frac{2\mu_{max}[X][H^+]}{k_{M7a}+[H^+]}\right)\left(\frac{[FeS_2]}{k_{M7b}+[FeS_2]}\right) \quad (9)$$

where:

$2k_1[FeS_2][O_2]^{\frac{7}{2}}$ = rate of change in sulfate concentration due to abiotic oxidation of pyrite,

$k_9[S^\circ][O_2]^{\frac{3}{2}}$ = rate of change in sulfate concentration due to abiotic oxidation of elemental sulfur,

$\left(\frac{2\mu_{max}[X][H^+]}{k_{M7a}+[H^+]}\right)\left(\frac{[FeS_2]}{k_{M7b}+[FeS_2]}\right)$ = rate of change in sulfate concentration due to bacteria-catalyzed pyrite oxidation, and

μ_{max} = specific growth rate of bacteria;

$\left(\frac{v_o}{V_o + v_o t}\right)[SO_4^{2-}]$ = rate of change due to interstitial water volume changes during dry-air segments of the weekly cycle, and

V_o = interstitial water volume after 500 ml drip-trickle leach, and

v_o = interstitial water volume lost due to evaporation (i.e., -1.03 ml/h during the dry-air portion of the weekly cycle).

Note that the expression $\left(\frac{v_o}{V_o + v_o t}\right)[SO_4^{2-}]$ drops out in equation (9) because of the assumption that the interstitial-water volume remains constant during the wet-air period.

Results and Discussion

Mineralogy. Mineral-characterization studies indicate that the sulfur present as sulfide in the three samples occurs mainly as pyrite (FeS_2); in addition to pyrite, a trace of pyrrhotite (Fe_5S_6 to $Fe_{16}S_{17}$) was also found in sample 1-D. Three different pyrite-crystal morphologies are represented in the three waste-rock samples: euhedral (mineral grain bounded by crystal faces), subhedral (mineral grain bounded by partially-developed crystal faces), and framboidal (microscopic aggregate comprised of spheroidal pyrite clusters). Sample 1-B contains euhedral pyrite; sample 1-C is dominated by framboidal pyrite; and sample 1-D contains a mix of subhedral and framboidal pyrite. These morphologic characteristics are significant because framboidal pyrite has much greater specific surface area available for oxidation and acid formation than do euhedral and subhedral pyrite of the same size. Relative differences in calculated surface areas exhibited by the three samples are consistent with their contained pyrite morphology: 6.5 X 10^{-3} m^2/g for euhedral, 1.3 X 10^{-2} m^2/g for framboidal, and 1.2 X 10^{-2} m^2/g for subhedral/framboidal morphologies. Variation of surface area as a function of particle size related to waste-rock mineral constituents is the subject of ongoing Bureau research.

Very few occurrences of carbonate and non-acid producing sulfate minerals (e.g., barite and gypsum) were noted in samples 1-B, 1-C, and 1-D. Sample 1-D had the only confirmed observation of a carbonate mineral (ankerite). All three samples contained traces of barite, but neither gypsum nor jarosite was detected.

Pyrite liberation is minimal (0 to 10 %) for all samples in the -10/+20-mesh fraction. The euhedral and subhedral/framboidal samples (1-B, 1-D) are both 70 % liberated at -48/+65-mesh, and at least 90 % liberated at -200 mesh. The framboidal sample (1-C) is poorly liberated at -48/+65-mesh and only 60 % liberated at -200 mesh. Although only 60 % liberated compared to the other two samples, sample 1-C is comprised of 3.7 to 8.7 times more sulfide, and exhibits a surface area that is an order of magnitude greater than that of sample 1-B. Significant differences in iron sulfide mineralogy, morphology, and liberation among the three samples are presented in Table I.

Table I. Summary of Morphology, Percent Liberation by Size Fraction, and Percent Sulfide Contained in Dominant Iron Sulfide Minerals Comprising Samples 1-B, 1-C, and 1-D

Sample	Sulfur as Sulfide (%)	Dominant Sulfide Mineral	Pyrite Morphology (-48/+65 fraction)	Percent Liberation by Size Fraction (mesh)		
				-10/+20 (%)	-48/+65 (%)	-200 (%)
1-B ...	3.5	pyrite	euhedral	0	70	95
1-C ...	13	pyrite	framboidal ..	0	5 - 10	60
1-D ...	1.5	pyrite/tr pyrrhotite ..	subhedral ...	10	70	90

Because only one confirmed occurrence of carbonate mineral was found during mineral characterization (ankerite in sample 1-D), no statement on carbonate liberation can be made at this time. The single occurrence of ankerite was encapsulated in a fragment from the -48/+65-mesh screen fraction and measured -100/+150 mesh.

Acid-Base Accounting. Preliminary NP determinations by the Bureau using the Sobek method showed that NP for a given sample increased as particle size decreased. The Sobek-determined NP values for the -150-mesh fraction were 3 to 7.5 times as great as those obtained from the -1/4-inch fraction (Table II).

Table II. Comparison of Waste-Rock NP Obtained from Sobek Method Applied to Two Different Size Fractions

Sample	AP[1]	NP[1]		NNP[1]		NP/AP	
		-1/4 inch	-150 mesh	-1/4 inch	-150 mesh	-1/4 inch	-150 mesh
1-B ..	109.4	0	0	-109.4	-109.4	0	0
1-C ..	406.8	2.4	18.1	-404.4	-388.7	0	.04
1-D ..	46.9	3.3	10.2	-43.6	-36.7	.07	.21

[1]Expressed as tons $CaCO_3$ equiv/10^3 tons waste rock

The increase of NP with decreasing particle size can result in overestimation of NP because reaction kinetics change with particle-size and attendant surface-area differences. For example, acidic conditions typical of the Sobek method may result in more complete dissolution of minerals (i.e., silicates) not usually solubilized under natural weathering conditions (12). This problem is the subject of additional Bureau research and results will be presented at a later date.

Comparison of NP Results from Two Different Methods. The Sobek and BCRI NP determination methods were applied to samples 1-B, 1-C, and 1-D and results for two different size fractions were compared to determine if any meaningful correlation existed between methods. If results between the two methods differed markedly, the more conservative (lower) value was used in calculations to estimate the number of years required for NP depletion and total sulfide oxidation. This is considered a temporary solution until more test work on NP determination is completed.

Both test methods determined NP to be zero or near zero for sample 1-B; however, NP for sample 1-C was 18.1 and 0 when Sobek and BCRI methods were applied to its -150-mesh fraction. The zero NP was selected for sample 1-C. While good agreement existed between Sobek and BCRI derived NP for sample 1-D's -150-mesh fraction, NP for its -1/4-inch fraction was 1/3 of the -150-mesh value. Therefore, the conservative 3.3 NP value was selected for sample 1-D.

Comparison of Bureau Sample ABA Results with Established ABA Criteria. Criteria currently followed in British Columbia, Canada, classify samples with NP/AP values less than 1 and NNP values less than -20 (tons $CaCO_3$ equivalent/10^3 tons waste rock) as high-probability acid producers (13). According to these criteria, both screen sizes for all three samples (1-B, 1-C, and 1-D) are rated as acid producers by the Sobek method. The -150-mesh material from all three samples is also rated as acid producing by the BCRI method.

Humidity-Cell Tests. Rates of pyrite oxidation and carbonate-mineral dissolution were determined from 51 weeks of humidity-cell-effluent chemistry. Sulfate-release rates were used to approximate pyrite-oxidation rates. Similarly, calcium-plus-magnesium-release rates were used to approximate carbonate-dissolution rates so that time to NP depletion could be estimated. Observed concentrations of sulfate, calcium, and magnesium were relatively high during the first 3 to 5 weeks of testing compared to those observed throughout the remaining 46 to 48 weeks (Figures 2b, 3b, and 4b). These elevated concentrations were attributed to release of salts that formed during oxidation prior to collection. Consequently, release rates were calculated for the period from weeks 5 through 51.

Temporal Variation of Effluent pH. Weekly humidity-cell effluents from samples 1-B and 1-C were acidic (pH 3.2 and 4.4, respectively) at week 5, and effluent pH values continued to decline steadily to 2.3 and 2.2, respectively, at week 51 (Figures 2, 3). Sample 1-D remained consistently neutral through week 51 (Figure 4). Elapsed time required for sample effluents to reach pH thresholds of 3.0 and 2.5 were 14 and 34 weeks, respectively for sample 1-B, and 23 and 34 weeks, respectively for sample 1-C. These pH values were selected because they coincide with marked increases in observed sulfate-release rates for the samples.

Sulfate Release Rates. Effluent from samples 1-B and 1-C exhibited four distinct release rates during 51 weeks of testing (Figures 2a and 3a). Sample 1-D had two release rates for the same test period (Figure 4a). The first release rate for each sample represents sulfate-salt storage being flushed from the samples during the initial 5 weeks of testing. The remaining three release rates exhibited by samples 1-B and 1-C represent effluent transitions from one pH threshold to another (pH $\geq$ 3.0, 2.5 < pH $\leq$ 3.0, and pH $\leq$ 2.5) during the course of the test; these rates are summarized in Table III. Sample 1-D exhibited a constant sulfate release rate of 23 mg/kg/wk from week 5 to 51.

Table III. Summary of SO_4 Release Rates for Samples 1-B, 1-C, and 1-D (mg/kg/wk)

Sample	1-B	1-C	1-D
SO_4 storage (wk 0 to wk 5)	230	813	67
To pH 3.0 (from wk 5)	162	423	NA
From pH 3.0 to pH 2.5	272	854	NA
From pH 2.5 (to wk 51)	380	1,836	NA

NA Not Applicable

Sulfate-Salt Storage. The hot-water leach test on samples 1-B, 1-C, and 1-D yielded 1,150, 3,870, and 310 mg of sulfate per kg of waste-rock, respectively.

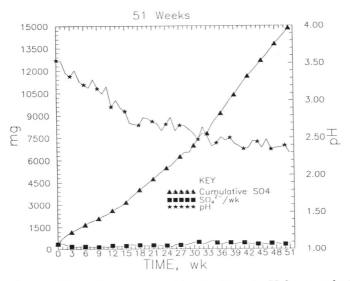

Figure 2a. Sulfate-release rate and leach-effluent pH for sample 1-B.

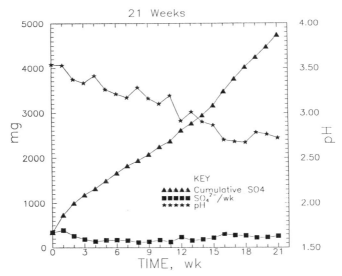

Figure 2b. First inflection point on cumulative sulfate plot (week 2) represents sulfate-salt storage release for 1-B.

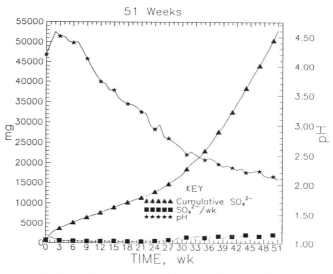

Figure 3a. Sulfate-release rate and leach-effluent pH for sample 1-C.

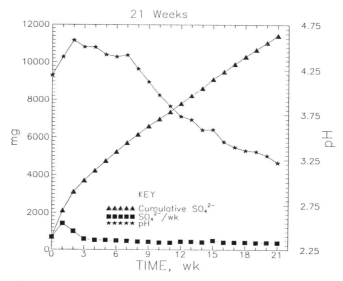

Figure 3b. First inflection point on cumulative sulfate plot (week 2) represents sulfate-salt storage release for 1-C.

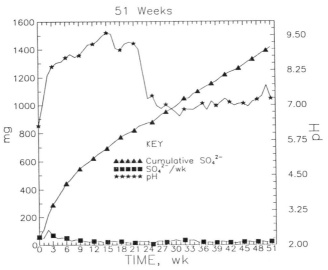

Figure 4a. Sulfate-release rate and leach-effluent pH for sample 1-D; marked drop in pH at week 24 was due to a change in the filter media from Pyrex wool to Teflon wool.

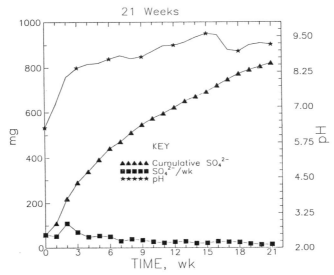

Figure 4b. First inflection point on cumulative sulfate plot (week 3) represents sulfate-salt storage release for 1-D.

These values are in reasonable agreement with the sulfate released from 1-B, 1-C, and 1-D during the first 3 weeks of humidity-cell testing (1,185, 3,710, and 290 mg). They also correspond with the first inflection point on the cumulative sulfate plots for samples 1-B, 1-C, and 1-D (Figures 2b, 3b, and 4b). This inflection point marks the change in release rate from flushing of sulfate-salt storage to sulfate release due to pyrite oxidation.

Sulfide Content Normalization. A marked difference in release rates was observed between samples 1-B and 1-C in Table III. Release rates increased as sulfide content increased. To better differentiate possible differences between samples, release rates were normalized by dividing by percent sulfide contained in each sample (3.5 and 13.0 %, respectively); results are presented in Table IV.

Table IV. Normalized Sulfate-Release Rates (Sulfate Release Rate Divided by Percent Sulfide) for Samples 1-B and 1-C (mg/kg/wk/%sulfide)

Sample	1-B	1-C
To pH 3.0 (from wk 5)	46	33
From pH 3.0 to pH 2.5	78	66
From pH 2.5 (to wk 51)	106	141

The normalized rates above pH 2.5 were slightly higher for sample 1-B than for 1-C. However, in view of significant mineralogical and chemical differences between the two samples, the rates are actually quite similar. This observation suggests that the net effect of differences in pyrite morphology and degree of liberation on sulfate-release rates from these two samples was minimal.

Variation with pH. A significant increase in sulfate-release rates for samples 1-B and 1-C was observed when respective effluent pH values dropped below 3 at weeks 14 and 23 (Figures 2a and 3a). This decrease corresponds to a 68 % increase in the average weekly sulfate release rate beginning at week 14 for sample 1-B and more than a 100 % increase beginning at week 23 for sample 1-C. When pH decreased below 2.5 (coincidentally beginning at week 34 for both samples 1-B and 1-C), weekly sulfate-release rates increased an additional 35 and 115 %, respectively.

This increase in sulfate-release rates at and below pH 3 has been observed by others. Kleinmann and others (20) have demonstrated that the actual pyrite oxidation process occurs in a three-stage sequence dependent upon the activity of *Thiobacillus ferrooxidans* and solution pH and Eh. Initially, pyrite is abiotically oxidized by oxygen. However, as pH decreases to 4.5, bacterial catalysis of oxidation begins to influence the reaction. As ferric iron becomes more soluble at pH 4.5, it begins to act as an oxidizing agent, and below pH 3.0 ferric iron is the only important oxidizer of pyrite. Ferrous to ferric oxidation is considered the rate-controlling step for development of AMD, and the iron-oxidizing bacterium, *T. ferrooxidans*, is the catalyst for this reaction. *T. ferrooxidans* is known to increase the ferrous to ferric oxidation rate by five to six orders of magnitude and is the mechanism believed responsible for the rapid, self-perpetuating process of pyrite oxidation as effluent pH drops below 4.5 (21).

To facilitate comparison with other studies, average weekly sulfate-release rates were converted to pyrite oxidation rates as moles pyrite per m^2 per second. Samples 1-B, 1-C, and 1-D exhibited average oxidation rates of 3.83×10^{-10}, 6.84×10^{-10}, and 1.68×10^{-11} moles per m^2 per second, respectively, for weeks 5 through 51. These rates (with the exception of sample 1-D) are in reasonable agreement with an average pyrite oxidation rate of 5.58×10^{-10} moles per m^2 per second determined by Scharer and others (Scharer, J. M., University of Waterloo, Waterloo, Ontario, personal communication, 1993) from batch-reactor studies conducted on samples of pure pyrite. The oxidation rate for sample 1-D may more closely approach Scharer's experimental rate when 1-D starts producing acidic effluent.

Estimation of AP and NP Depletion. Calcium, magnesium and sulfate-release rates were used to estimate time of accelerated weathering required to deplete each sample of its acid-producing potential (AP) and neutralizing potential (NP). Release rates were determined for the periods of weeks 5 through 21 and 5 through 51, and are shown in Table V.

Table V. Average Weekly Sulfate Release Rates After 21 and 51 Weeks of Accelerated Weathering and Calculated Time for Total Sulfide Oxidation

Sample	Total Sulfide as Sulfate (mg/kg)	Sulfate Release Rate (mg/kg/wk)		Time for Total Sulfide Oxidation (years)	
		wks 5-21	wks 5-51	Rate from wks 5-21	Rate from wks 5-51
1-B	105,000	203	290	10	7
1-C	390,000	418	1,048	18	7
1-D	45,900	27	23	33	39

Contained sulfide in each sample is assumed to oxidize completely to sulfate. Sulfate-release rates calculated through 21 weeks were compared with those calculated through 51 weeks. This comparison was made to demonstrate that the common commercial practice of terminating humidity-cell tests at 20 weeks may result in underestimation of release rates and consequent overestimation of the number of years required for AP depletion. Note that by week 51, release rates had increased for samples 1-B and 1-C by 43 and 150 %, respectively, over rates exhibited at week 21. This difference resulted in significant decreases in calculated years to AP depletion for 1-B and 1-C (30 to 61 pct less, respectively).

Calculation of years to NP depletion is calculated by dividing the sample's NP by the sum of calcium and magnesium release rates. Samples 1-B and 1-C had zero NP values and therefore zero years to NP depletion. This observation was confirmed by the steady, weekly acidic effluent emanating from both samples throughout the 51 weeks of humidity-cell testing. Sample 1-D had an NP of 3.3 parts per thousand, or 33 millimoles calcium carbonate equivalent; average calcium plus magnesium release rates for the periods 5 through 21 and 5 through 51 weeks were 0.3 and 0.25 millimoles per week respectively. Calculation based on these figures indicate that 110 to 130 weeks of laboratory accelerated weathering would be required to deplete the NP of sample 1-D (Table V). By contrast, acid release will continue for 33 to 39 years. These data suggest that effluent from sample 1-D should become acidic after 2 to 2.5 years of accelerated humidity-cell weathering.

It should be emphasized that these time estimates for acid- and neutralization-potential depletion refer only to the individual 1 kg samples subjected to accelerated weathering in the humidity cells. Although no correlation between humidity-cell release rates and release rates exhibited by actual waste-rock dumps is currently available, comparison of individual sample release rates provides a relative gauge of the potential for each waste-rock sample to produce AMD.

Geochemical Predictive Model. Preliminary fit of model-generated data to corresponding experimental data from 51 weeks of humidity-cell testing is encouraging. Figures 5a through 5d are examples of current output from the developing model. Figure 5a compares model-predicted sulfate concentration with actual sulfate concentration contained in weekly humidity-cell effluent from sample 1-C. Fairly good agreement between predicted and experimental concentrations from sample 1-C have also been observed for ferrous and ferric iron and pH (Figures 5b, 5c, and 5d).

Best fit of model-generated curves to experimental data was obtained as follows: (1) the magnitude of each rate constant was initially established by using trial and error curve-fitting techniques, and (2) sensitivity analysis was then performed to assess the importance of each rate constant on the modelled output.

The next phase in model calibration is to have the model predict the number of weeks required for NP depletion of sample 1-D. The model-generated estimate could then be compared with the time range of 110 to 130 weeks that was projected from estimated release rates determined from humidity-cell-effluent analyses. The anticipated utility of the humidity-cell model is that it will project humidity-cell-effluent concentrations beyond the 20-week period common to commercial tests.

Once the model has been calibrated with experimental data from a variety of waste-rock types, the next phase of modelling will be to apply principles learned from the humidity-cell model to the development of a model designed to describe effluent concentrations from waste-rock test piles.

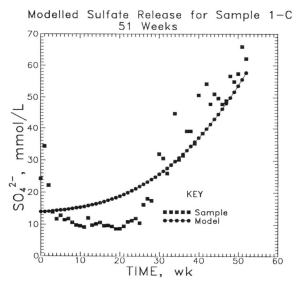

Figure 5a. Model-generated curve for weekly sulfate concentration, compared with 51 weeks of effluent data from sample 1-C.

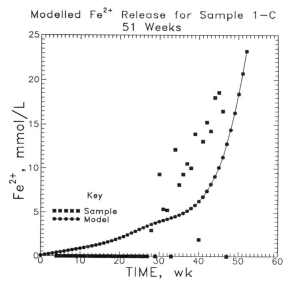

Figure 5b. Model-generated curve for weekly ferrous-iron concentration, compared with 51 weeks of effluent data from sample 1-C.

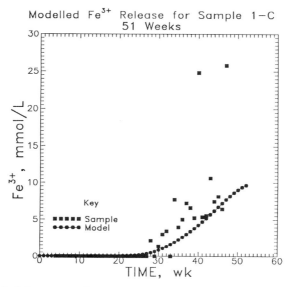

Figure 5c. Model-generated curve for weekly ferric-iron concentration, compared with 51 weeks of effluent data from sample 1-C.

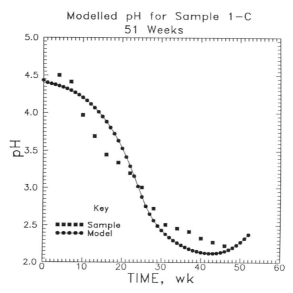

Figure 5d. Model-generated curve for weekly pH, compared with 51 weeks of effluent data from sample 1-C.

Summary and Conclusions

Three waste-rock samples were subjected to characterization and predictive testing to generate data for a developing acid-mine-drainage-predictive model. Although static testing acid-base accounting classified all three samples as acid producers, only two of the three samples subjected to accelerated-weathering (humidity-cell) tests have produced acid effluent. Samples 1-B and 1-C have produced acid since the first leach and have become progressively more acidic. They are significant because they collectively represent rock types similar to those contained in 40 million tons of waste rock capable of generating acid equivalent to at least 5.5 million tons of calcium carbonate. Although the effluent from the third sample (1-D) has ranged from alkaline to neutral throughout 51 weeks of humidity-cell testing, combined static and kinetic test results indicate that it has as much as 39 years of acid-producing potential, but less than 3 years of neutralizing potential. On the basis of monitored calcium and magnesium release rates compared with sulfate release rates, onset of acid production from sample 1-D is predicted to occur after 110 to 130 weeks.

If kinetic testing of sample 1-D had been terminated at 20 weeks, as has been common practice, the sample could have been mis-classified as non-acid producing. This is of considerable concern because sample 1-D is predicted to become acid-generating after 130 weeks, and is made up of the same rock type that comprises 200 million tons of waste rock. The 200 million tons of rock could potentially generate an amount of sulfuric acid equivalent to at least 9 million tons of calcium carbonate. Should a highly reactive waste rock similar to sample 1-C be mixed with 200 million tons of waste rock similar to sample 1-D, the end result could be the accelerated production of acid. Because AMD generation can result in surface- and ground-water contamination that requires expensive water-treatment techniques and involves potential liability in perpetuity, accurate prediction of acid-producing potential through chemical and mineralogical characterization and geochemical modelling is essential.

Preliminary fit of model-generated data to 51 weeks of corresponding experimental data for sample 1-C has been encouraging. Fairly good agreement between predicted and experimental concentrations was achieved for pH, ferrous and ferric iron, and sulfate ion. The ultimate goal of this modelling approach is to enable the mine operator to use short-term laboratory static- and kinetic-test data along with field measurements to make long-term predictions about rates of AP and NP depletion and consequent acid-generation behavior.

Acknowledgments

Design and initial operating parameters for the "modified" humidity cells were generously supplied by Richard W. Lawrence, Lawrence-Marchant LTD, Vancouver, B.C. Keith Ferguson of Environment Canada, Vancouver, B.C., and Kim Lapakko, Minnesota Department of Natural Resources, St. Paul, MN, provided numerous references and technical advice. William Schafer, President of Schafer and Associates, Bozeman, MT, provided a copy of his modified acid-base accounting protocol. Lawrence J Froisland, Bureau of Mines, Salt Lake

City, UT, generously gave of his time and talents in the initial humidity-cell assembly and testing. Bureau colleagues from Pittsburgh, Reno, and Spokane Research Centers contributed helpful suggestions and discussions to benefit the developing research. The authors especially thank the participating mining companies who provided the waste-rock samples; without their participation, this study would not be possible.

Literature Cited

1. Canadian Centre for Mineral and Energy Technology. Investigation of Prediction Techniques for Acid Mine Drainage; Mine Environment Neutral Drainage (MEND) project 1.16.1a; Coastech Research: West Vancouver, BC, Canada, 1989; 61 pp.
2. Sobek, A. A.; Schuller, W. A.; Freeman, J. R.; Smith, R. M. U.S. EPA Report; EPA 600/2-78-054, 1978; 203 pp.
3. Duncan, D. W.; Bruynesteyn, A. Met. Soc. AIME preprint 1979, A-79-29, 10 pp.
4. Carruccio, F. T.; Geidel, G.; Pelletier, M. J. Energy (Amer. Soc. Civil Engineers) 1981, 107, (EY1).
5. Finkelman, R. B.; Giffin, D. E. Recreation and Revegetation Research 1986, 5, 521-534.
6. Albright, R. In Proceedings, Acid Mine Drainage Workshop; Environment Canada and Transport Canada; Minister of Supply and Services, Canada: Halifax, NS, Canada, 1987, 146-162.
7. Scharer, J. M.; Garga, V.; Smith, R.; Halbert, B. E. Use of Steady-State Models for Assessing Acid Generation in Pyritic Mine Tailings. In Proceedings, 2nd International Conference on Abatement Acidic Drainage: Montreal, QC, Canada, 1991, Tome 2, 211-229.
8. Singleton, G. A.; Lavkulich, L. M. Soil Sci. 1978, 42, 984-986.
9. Sullivan, P. J.; Sobek, A. A. Minerals and the Environment 1982, 4, (1), 9-16.
10. Canadian Centre for Mineral and Energy Technology. Acid Rock Drainage Prediction Manual: A Manual of Chemical Evaluation Procedures for the Prediction of Acid Generation from Mine Wastes; MEND project 1.16.1b; Coastech Research: North Vancouver, BC, Canada, 1990; 132 pp.
11. Bradham, W. S.; Caruccio, F. T. In Proceedings, Second International Conference on the Abatement of Acidic Drainage. MEND: Ottawa, ON, Canada, 1991; Tome 1, 157-173.
12. Lapakko, K. Evaluation of Tests for Predicting Mine Waste Drainage pH: Draft Report to the Western Governors' Association; 1992, Vol. 1, 171 pp.
13. Ferguson, K. D.; Morin, K. A. In Proceedings, Second International Conference on the Abatement of Acidic Drainage; MEND: Ottawa, ON, Canada, 1991; 83-106.
14. British Columbia Acid Mine Drainage Task Force. Draft Acid Rock Drainage Technical Guide; Can. Dept. Energy, Mines, and Resources: Vancouver, BC, Canada, 1989; Vol. 1, 260 pp.

15. Lawrence, R. W.; Ritcey, G. M.; Poling, G. W.; Marchant, P. B. Strategies for the Prediction of Acid Mine Drainage. In <u>Proceedings, 13th annual Mine Reclamation Symposium</u>; British Columbia Technical and Research Committee on Reclamation, Ministry of Energy, Mines and Petroleum Resources: Victoria, BC, Canada, 1989; 16 pp.
16. Ferguson, K. D.; Erickson, P. M. In <u>Proceedings, Acid Mine Drainage Seminar/Workshop</u>; Environment Canada and Transport Canada; Minister of Supply and Services, Canada: Halifax, NS, Canada, 1987; 215-244.
17. Lapakko, K. In <u>Proceedings, Mining and Mineral Processing Wastes</u>; Doyle, F. M., Ed.; (Proc., Western Regional Symposium on Mining and Mineral Processing Wastes, Berkeley, CA), Soc. Min. Eng. AIME: Littleton, CO, 1990, 81-86.
18. Lawrence, R. W. In <u>Mining and Mineral Processing Wastes</u>; Doyle, F. M., Ed.; (Proc., Western Regional Symposium on Mining and Mineral Processing Wastes, Berkeley, CA), Soc. Min. Eng. AIME: Littleton, CO, 1990, 115-121.
19. Erickson, P. M.; Hedin, R. S. In <u>Proceedings, Mine Drainage and Surface Mine Reclamation</u>; U.S. Bureau of Mines; Info. Circ. 9183; 1988, 11-19.
20. Kleinmann, R. L. P.; Crerar, D. A.; Pacelli, R. R. <u>Min. Eng.</u> (NY), 1981, <u>33</u>.
21. Nordstrom, D. K. In <u>Proceedings, Acid Sulfate Weathering</u>; Kittrick, J. A.; Fanning, D. S.; Hossner, L. R., Eds.; Soil Sci. Soc. Am. Spec. Pub. 10, 1982; 37-56.

RECEIVED September 23, 1993

Chapter 38

Composition of Interstitial Gases in Wood Chips Deposited on Reactive Mine Tailings
Consequences for Their Use as an Oxygen Barrier

N. Tassé, M. D. Germain, and M. Bergeron

Institut National de la Recherche Scientifique—Géoressources, P.O. Box 7500, Sainte-Foy, Quebec G1V 4C7, Canada

> Interstitial gases were measured within a 1 m thick wood-chip cover on top of base-metal mine tailings in northwestern Québec, Canada, to evaluate the efficiency of the cover as an oxygen interceptor. The oxygen concentration decreases sharply to less than 5 vol. % below 50 cm, as O_2 is consumed by organic-matter oxidation. At these depths, CO_2 and CH_4 concentrations increase to >20 and >10 vol. %, respectively, as a result of oxidation and methanogenesis. Profiles measured three times within a month show that gas generation is sensitive to surface temperature. The calculated rates of biomass removal are so high that they jeopardize the long-term effectiveness of such a cover. Moreover, the potential for methylation of heavy metals, as a consequence of methanogenesis, must be considered.

Covers for reactive mine tailings are intended to isolate sulfide minerals from oxygen, with the goal of minimizing the generation of acid water and the release of heavy metals. Forest by-products, such as wood chips, are an attractive material for use in close-out designs. The rate of oxygen diffusion through these materials should be slow as O_2 is consumed by the oxidation of the wood particles. These industrial by-products are inexpensive and abundant in some mining areas. The potential use of wood wastes as covers for sulfide tailings was explored by Reardon and Poscente ([1]), who investigated pore-gas composition, temperature, dissolved organic carbon and acidity in four sawmill-waste deposits. One conclusion of their study is that the observed rate of oxidative degradation in a wood pile is so large that the use of wood chips in a close-out design must be carefully planned, to provide adequate long-term O_2 consumption. In the lab, Reardon and Moddle ([2]) tested the impact of organic covers containing peat, a more refractory substance, on uranium-mill tailings. They found little effect on the chemical composition of the effluents.

This paper reports temperature and pore-gas composition data collected in a wood-chip cover laid down over several tens of hectares of reactive mine tailings in northwestern Québec. The prospective use of wood chips for prevention and remediation of acid drainage from mine tailings is discussed, followed by an examination of possible interactions between minerals and degrading organic compounds.

Study Site

The East Sullivan mine-waste deposit (Figure 1) contains 15 Mt of reactive tailings accumulated between 1949 and 1966. The impoundment forms a plateau extending over 136 ha, rising 4 to 5 m above the surrounding terrain. Since 1984, the site has been used as a dump for wood residues, sludges from a municipal wastewater-treatment plant, and sewage from septic tanks (Paquet, A., unpublished report, Ministère de l'Environnement du Québec, 1989). The wood residues overlay more than 25 % of the surface of the impoundment with an average thickness of 2-3 m, reaching 6 m in some areas. The deposition rate on the site is about 110 000 t of wood residue per year. Residues are composed of fine chips of aspen (33 vol. %), resinous trunks (3 vol. %) and bark (33 vol. %), presswood fragments (21 vol. %), and coarser fragments (10 vol. %) of presswood, aspen and resinous trunks (Sherbrooke University, Dept. of Civil Engineering, unpublished report, 1988). Occasionally, metallic remnants are found.

The studied location is about 30 m from the southern margin of the wood-waste pile (Figure 1). At this location, the thickness of the wood chips is 1 m. The average chip size is less than 10 mm.

Methodology

Wood-Chip Sampling. Cores of wood chips were collected twice, in August and October 1991, in aluminum tubes (7.6 cm inside diameter), once by slam hammer, and once by hydraulic pressure. The sample tubes were sealed with polystyrene plugs wrapped in a plastic or polyethylene film, and preserved in a refrigerator until water content and porosity were measured. All measurements were performed on 10-cm-long sub-samples. Water content, defined here as the free water extracted by a centrifuge, was calculated by the difference in weight before and after centrifugation, assuming a water density of 1 g/cm^3. The rotation speed was adjusted to maintain the centripetal pressure slightly lower than the capillary pressure in the wood particles. The required centrifugation time was determined by cumulative tests over a period of 2.5 hrs., with readings at every 30 min. Centrifugation was done on all samples at 1 000 rpm for a period of 2 hrs.

The effective porosity was calculated by the weight difference between the samples after their centrifugation and after their saturation with water. Volume lost by compaction at time of sampling was added. The pores within the wood material itself were not considered in the calculation of the effective porosity, as the water mobility within the organic matter is significantly less then that of the free water between the wood chips.

Gas Sampling. The device used for pore-gas sampling was a narrow-diameter stainless-steel pipe with a perforated tip at the lower end, connected to two plastic syringes via a three-way glass stopcock (Figure 2). One of the syringes was used to purge atmospheric gas from the pipe before sampling. The other syringe was used to sample the interstitial gas. Silicone ensured the integrity of the apparatus at its weakest junction, *i.e.* at the tip of the sampling syringe.

In the field, the stainless-steel pipe was driven to the required sampling depth (10 cm intervals) with a sledge hammer. A 1-cm-diameter pipe was used to increase the lateral support of the stainless-steel stem. The stopcock and the two syringes were then installed, and the atmospheric gas present in the steel pipe and in the sampling tubes was removed by creating a vacuum with the purge syringe for 30 s. The clamp between the stopcock and the purge syringe (Figure 2) was closed, and the gases in the vicinity of the perforated tip were collected by applying a suction with the sampling syringe for 1 min. The stopper next to the sampling syringe was then closed. The sampling syringe was removed and stored for later analysis. About

38. TASSE ET AL.   *Composition of Interstitial Gases in Wood Chips*

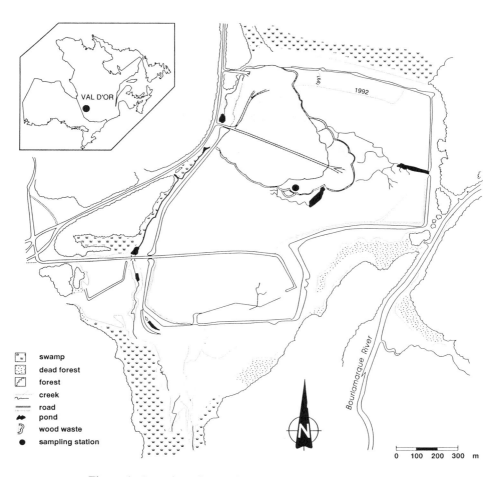

Figure 1. Location of sampling site in northwestern Québec.

50 ml of gas was sampled near the waste surface. Closer to the water table, the sample volume was reduced to avoid the migration of water toward the perforated tip.

Three series of analysis were conducted, on September 23, October 17 and October 25, 1991. The gases were analyzed the day of collection on a Varian 3300 gas chromatograph equipped with a thermal conductivity detector. A standard was analyzed at the beginning and at the end of each series of analyses. Detected gases included oxygen, methane, carbon dioxide and nitrogen. Total analytical concentrations ranged from 95.9 to 104.1 vol. %.

Temperature Measurement. Four thermocouples were installed at depths of 40, 80, 120 and 150 cm below the surface of the wood-chip cover in 1991. Two additional thermocouples were installed at depths of 100 and 180 cm in 1992 and the temperature measured over a few months to document long-term variations. For the installation of each thermocouple, a hole was pierced down to the desired depth, less 2 cm, with a 1-cm-diameter rod fitted with a conical tip at its lower end. The thermocouple wire was then introduced in the hole, and driven down an extra 2 cm. The holes were then filled with silica sand. The readings of the thermocouples were made using a Kane-May 457XP instrument.

Results

Water Content and Porosity. The wood-chip samples were significantly compacted regardless of the sampling technique, *i.e.* slam hammer or hydraulic pressure. Compaction of 40 % was observed in the samples collected in August. It is more difficult to estimate the extent of compaction for the October samples, because a branch jammed inside the sampling tube (Table I).

The compaction observed within each August sample was uniformly distributed for the porosity and water-content calculations. Most compaction probably results from a reduction of the air-filled porosity and is likely variable through the wood-chip cover, as the water-filled porosity increases downward. Consequently, calculated water contents are minimum estimates. Keeping in mind that the figures are crude estimates, the porosity values were found to decrease from 70 to 75 % near the surface to about 40 % at the lower end of the wood-chip cover (Figure 3). The free-water content increased from close to zero near the surface to 8 % at the base of the cover.

Gas. Results are summarized in Table II and Figure 4. Sampling could not be done below 90 cm depth, because the sampling tube became filled with water. The pore-gas profiles show essentially the same general trend for the three sampling campaigns. The oxygen concentration decreased from the atmospheric mean value (21 vol. %) to between 1 and 4 vol. % below 50 cm. The decrease in O_2 concentration is associated with an increase in CO_2 concentration. The concentration of CO_2 increased from below detection at the surface of the cover to up to 50 vol. % at 90 cm depth, during September. Methane was first detected at a depth of 40 cm, at concentrations varying between 0.2 and 1.1 vol. %. These concentrations increased to 10 to 15 vol. % at 80 cm.

Temperature. The three temperature profiles measured when pore-gases were sampled show an increase in temperature with depth in the wood-chip cover (Figure 5). These trends are in agreement with the fall and winter trends observed during the long-term temperature survey done at the sampling site the following year. The temperature increase is probably due to direct or bacterially mediated oxidation reactions and breakdown of wood components. The profiles of October 17 and 25 show lower temperatures compared to that of September 23.

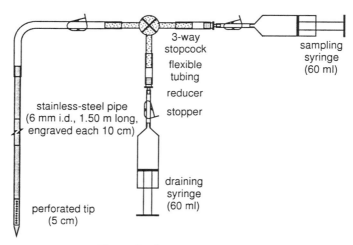

Figure 2. Gas sampling device.

Table I. Porosity and Water and Gas Contents of Wood Chips

Sampling Date (1991)	Sample Number	Sample Length (cm)	Pre-compaction Length (cm)	Depth Interval (cm)	Porosity(Vol%)		
					Total	Water-filled	Gas-filled
Aug. 27	A-1-1	7.5	17.5	0-17.5	76.9	1.4	75.5
Aug. 27	A-1-2	10.0	23.3	17.5-40.8	68.2	2.0	66.2
Aug. 27	A-1-3	10.0	23.3	40.8-64.1	69.8	4.6	65.2
Aug. 27	A-1-4	6.8	15.9	64.1-80.0	70.4	4.0	66.5
Aug. 27	A-2-1	7.5	8.1	80.0-88.1	56.4	8.1	48.3
Aug. 27	A-2-2	11.0	11.9	88.1-100.0	38.5	6.3	32.2
Oct. 23	O-1-1	7.5	17.5	0-17.5	73.3	4.8	68.6
Oct. 23	O-1-2	8.0	18.6	17.5-36.1	73.0	2.9	70.1
Oct. 23	O-1-3	6.0	14.0	36.1-50.1	---	branch	---
Oct. 23	O-2-1	10.7	17.4	50.1-67.5	57.7	1.9	55.8
Oct. 23	O-2-2	11.7	19.0	67.5-86.5	49.3	1.8	47.5
Oct. 23	O-2-3	8.4	13.6	86.5-100.1	62.2	3.0	59.2

Data used for calculation of pre-compaction lengths:
A-1 driven 80 cm into wood chips; A-2 driven 60 cm into wood chips and tailings
O-1 driven 50 cm into wood chips; O-2 driven 50 cm into wood chips

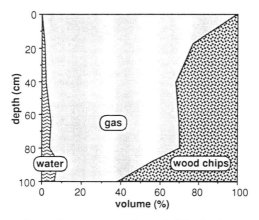

Figure 3. Proportions of gas, water and wood chips in the wood-chip cover.

Table II. Composition of Interstitial Gases in Wood Chips (Vol. %)

Sampling Date (1991)	Depth (cm)	Gas content* (%)	Gas composition				
			O_2 (%)	CO_2 (%)	CH_4 (%)	N_2 (%)	TOTAL (%)
Sep. 23	10	75.48	16.10	8.20	0.00	79.80	104.10
Sep. 23	20	73.14	9.50	14.70	0.00	77.60	101.80
Sep. 23	30	66.18	5.50	21.50	0.00	77.10	104.10
Sep. 23	40	66.18	2.70	25.00	1.10	73.00	101.80
Sep. 23	50	65.30	1.80	28.50	2.30	70.20	102.80
Sep. 23	60	65.22	1.30	30.00	2.70	69.10	103.10
Sep. 23	70	65.94	1.20	30.10	5.80	65.20	102.30
Sep. 23	80	66.45	- - - - - - contamination - - - - - -				
Sep. 23	90	45.20	1.50	50.80	15.00	32.90	100.20
Sep. 23	100	32.23	- - - - - - contamination - - - - - -				
Oct. 17	10	75.48	17.39	4.18	0.00	76.75	98.32
Oct. 17	20	73.14	14.68	7.04	0.00	77.71	99.43
Oct. 17	30	66.18	10.19	11.01	0.00	76.25	97.45
Oct. 17	40	66.18	6.01	16.03	0.18	75.54	97.76
Oct. 17	50	65.30	3.15	20.92	1.29	72.69	98.05
Oct. 17	60	65.22	2.92	20.28	1.57	71.14	95.91
Oct. 17	70	65.94	4.36	18.99	15.00	58.72	97.50
Oct. 25	10	75.48	13.44	7.17	0.00	78.66	99.27
Oct. 25	20	73.14	11.51	11.53	0.00	76.50	99.54
Oct. 25	30	66.18	9.00	16.04	0.00	76.26	101.30
Oct. 25	40	66.18	7.41	17.32	0.31	75.26	100.30
Oct. 25	50	65.30	5.30	20.45	0.79	74.77	101.31
Oct. 25	60	65.22	3.96	22.19	1.39	73.49	101.03
Oct. 25	70	65.94	4.49	21.45	1.92	73.16	101.02
Oct. 25	80	66.45	3.93	20.70	11.00	64.50	100.13

* calculated for 10 cm intervals from August data of Table I.

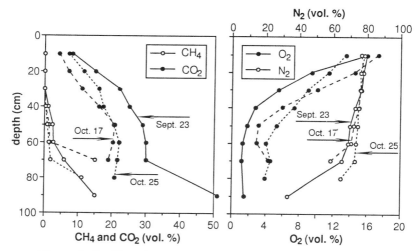

Figure 4. Profiles of gas concentrations in the wood-chip cover.

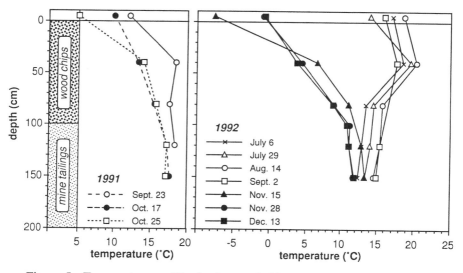

Figure 5. Temperature profiles in the wood-chip cover and underlying mine tailings at the time of pore-gas sampling (1991) and during a long term survey (1992).

September Sampling *vs*. October Sampling. The temperature variations over a month have consequences on the pore-gas profiles observed. According to the Arrhenius equation, the rate of a reaction varies with respect to temperature; a decrease in temperature is associated with a decrease in the reaction rate. According to this theory, a lower consumption of O_2, and, thus, a lower production of CO_2 and CH_4 should be observed in October (cold) than in September (warm). Our O_2 and CO_2 concentrations are in agreement with this prediction. For instance, in October, the O_2 concentration (9 and 10.2 vol. %) is twice the September concentration (5.5 vol. %) at 30 cm (Table II; Figure 4). The CO_2 concentrations are also lower, decreasing from 21.5 vol. % in September to 11.0 and 16.0 vol. % in October. Methane concentrations show smaller variations. The zone of methanogenesis, which occurs deeper in the wood-waste cover, is isolated by the insulating properties of the overlying wood chips, and probably did not cool off as rapidly as the overlying material. As a result, the rate of release of CH_4 probably did not decrease as rapidly as the rate of O_2 consumption in the shallow cover.

Discussion

The decrease in oxygen content, and the generation of heat, carbon dioxide and methane, are well known results of wood-waste degradation (*e.g.* 1, 3, 4). However, the use of wood chips as a cover over mine tailings requires a closer examination of eventual interactions between degrading minerals and organic compounds.

Theory. Froelich *et al.* (5) have presented a model describing the reaction sequence of degrading organic matter in a mineral matrix from a completely aerobic system to a progressively anaerobic one. In this model, organic matter is oxidized by the oxidant for which the reaction has the lowest value of Gibbs free energy ($\Delta G°$) per mole of organic carbon. As this oxidant is exhausted, the oxidation continues using the next oxidant for which the reaction has the lowest value of $\Delta G°$. This process continues until either all oxidants or all organic matter are consumed. More specifically, the oxidation reactions sequence of an organic matter, assuming it has the Redfield's composition (*i.e.* $(CH_2O)_{106}(NH_3)_{16}(H_3PO_4)$), is the following:

Oxidation by O_2 ($\Delta G° = -3190$ kJ per mole of glucose):

$$(CH_2O)_{106}(NH_3)_{16}(H_3PO_4) + 138\ O_2$$
$$\rightarrow 106\ CO_2 + 16\ HNO_3 + H_3PO_4 + 122\ H_2O \quad (1)$$

Oxidation by MnO_2 ($\Delta G° = -3090$ kJ per mole of glucose, depending on mineral species):

$$(CH_2O)_{106}(NH_3)_{16}(H_3PO_4) + 236\ MnO_2 + 472\ H^+$$
$$\rightarrow 236\ Mn^{2+} + 106\ CO_2 + 8\ N_2 + H_3PO_4 + 366\ H_2O \quad (2)$$

Oxidation by HNO_3 ($\Delta G° = -2750$ kJ per mole of glucose):

$$(CH_2O)_{106}(NH_3)_{16}(H_3PO_4) + 94.4\ HNO_3$$
$$\rightarrow 106\ CO_2 + 55.2\ N_2 + H_3PO_4 + 177.2\ H_2O \quad (3)$$

Oxidation by Fe_2O_3 ($\Delta G° = -1410$ kJ per mole of glucose):

$$(CH_2O)_{106}(NH_3)_{16}(H_3PO_4) + 212\ Fe_2O_3 + 848\ H^+$$
$$\rightarrow 424\ Fe^{2+} + 106\ CO_2 + 16\ NH_3 + H_3PO_4 + 530\ H_2O \quad (4)$$

Oxidation by SO_4^{2-} ($\Delta G° = -380$ kJ per mole of glucose):

$$(CH_2O)_{106}(NH_3)_{16}(H_3PO_4) + 53\ SO_4^{2-}$$
$$\longrightarrow 106\ CO_2 + 16NH_3 + 53\ S^{2-} + H_3PO_4 + 106\ H_2O \quad (5)$$

Methanogenesis ($\Delta G° = -350$ kJ per mole of glucose):

$$(CH_2O)_{106}(NH_3)_{16}(H_3PO_4) \longrightarrow 53\ CO_2 + 53\ CH_4 + 16\ NH_3 + H_3PO_4 \quad (6)$$

Methanogenesis implies predominantly the fermentation of methylated substrata such as acetate, methanol, methylated amines, etc. Once these labile methanogenic precursors have been exhausted, methanogenesis may continue by shifting to carbonate reduction (6, 7).

Application to the Wood Chips. The previous sequence of oxidation-reduction reactions has commonly been observed in marine sediment environments. It also applies to the present system.

The oxidation by atmospheric O_2 diffusing into the wood chips is the first oxidation reaction to occur (Equation 1). The decrease with depth in the O_2 concentration and the increase in CO_2 (Figure 4) result from an aerobic oxidation of the wood chips that proceeds until the complete consumption of O_2.

Because the wood chips probably lack MnO_2, the next oxidant is HNO_3, which is a minor product of the oxidation of organic matter by O_2 (Equation 1), and which is reduced as described by Equation 3. An increase in N_2 resulting from the reduction of nitrates could indicate the occurrence of this reaction. However, this change is unlikely to be detected, given the small amounts of N_2 released compared to the large concentration of atmospheric N_2 diffusing into the wood chips. The CO_2 production in Equation 3 is added to the large quantities already produced by the other oxidation reactions.

A priori, the wood-chip cover does not contain an appreciable amount of Fe_2O_3 and SO_4^{2-}. Therefore, the next likely reaction is methanogenesis (Equation 6). Significant amounts of CH_4 are observed. Methanogenesis is a terminal step in organic matter degradation and is not depth-limited, because fermentation is a dissimilation process that is independent of the availability of an oxidant. Accordingly, the organic source does not need to be restricted to the wood chips. It can also include the dissolved organic carbon (DOC) derived from other degradation processes. Methanogenesis releases CO_2 without O_2 consumption as the disproportionation of organic molecules occurs, e.g.

$$CH_3COOH \longrightarrow CH_4 + CO_2 \quad (7)$$

contrary to carbonate reduction which consumes CO_2 and releases only CH_4:

$$CO_2 + 4H_2 \longrightarrow CH_4 + 2H_2O \quad (8)$$

The production of CO_2 from fermentation may thus increase the CO_2 concentration, relative to the O_2 loss, well above the levels predicted by the stoichiometry of the aerobic oxidation reaction alone. Consumption of CO_2 by carbonate reduction is less likely, since easily metabolizable DOC is formed at concentrations up to 0.45 M in comparable degrading wood chips (1). The onset of methanogenesis implies an anoxic medium. Oxygen was measured at concentrations between 1 and 4 vol. % in the lower tens of centimeters of the cover, where it coexists with a few percent CH_4. These small amounts of oxygen probably result from atmospheric contamination.

The anoxic zone likely occurs due to the high water saturation of the wood chips which limits the diffusion of O_2. The eventual development of anoxia and methanogenesis will thus strongly depend not only on the grain size and permeability of the wood chips, but also on the hydraulic conductivity of the underlying tailings. Close to the sampling station, where wood-waste thickness reached 4 to 6 m, the water table was only 2-3 m below the surface.

No detectable amounts of CH_4 could be found at depths less than 30 cm. This observation probably reflects the fact that the upward migration of CH_4 is counteracted by the oxidation of methane by bacteria (*e.g.* 8):

$$CH_4 + 2\ O_2 \longrightarrow CO_2 + 2\ H_2O \qquad (9)$$

Gas Transfer and Mass Transport Calculations. The stoichiometry of oxidative degradation of organic matter predicts that one molecule of CO_2 will be released for each molecule of O_2 consumed (Equation 1). This stoichiometry explains why the concentration profiles of these gases in wood-chip piles are mirror images, when only that type of reaction takes place (1). In the wood-chip cover studied, O_2 decreases with depth as a result of its consumption by organic matter and methane oxidation, but proportions of CO_2 exceed 25 vol. % and even 50 vol. % at depths greater than 50 cm in the September data (Figure 4). These values are much greater than the equivalent amounts of O_2 in air (21 vol. %). The presence of CH_4 suggests that the excess of CO_2 is the result of methanogenesis, because fermentation releases one molecule of CO_2 for each molecule of CH_4 (Equations 6 and 7). This relationship is not, however, as clear for the two data sets obtained in October (Figure 4). Distortions of the "mirror image" that could be misinterpreted as an excess of CO_2 can arise from dissimilar diffusion coefficients, as a result of molecular weight differences.

The amounts of CO_2 derived from processes other than oxidative degradation can be evaluated by comparing O_2 and CO_2 concentration gradients. At steady state, fluxes of O_2 consumed and of CO_2 produced by oxidation of organic matter should be equal, so that the relation D_{CO_2} (dc_{CO_2}/dx) = D_{O_2} (dc_{O_2}/dx), where D is the diffusion coefficient and dc/dx the concentration gradient in Fick's first law, implies D_{CO_2}/D_{O_2} = (dc_{O_2}/dx)/(dc_{CO_2}/dx). The ratio of the diffusion rate of CO_2 relative to O_2 will be given by $(m_{O_2}/m_{CO_2})^{1/2}$, where m is the molecular weight. It follows from the ratio that D_{CO_2}/D_{O_2} should equal 0.85. Accordingly, the stoichiometric consumption of O_2 and release of CO_2 should give a concentration gradient ratio of 0.85. Concentration gradients calculated between 10 and 40 cm (Figure 4) are given in Table III. The 0.77 and 0.61 ratios for the September 23 and October 25 data suggest that the CO_2 gradients were greater by 10 and 28 %, respectively, than those expected from an oxidative process only. However, the 0.95 ratio for the concentration gradients of October 17 suggests a gradient lower than expected for CO_2. It is possible that the system was not at steady state at sampling time. One way to assess this possibility is to examine the CO_2 behavior through time.

Both O_2 and CO_2 concentration gradients decreased from September to the end of October (Table III), as expected from the decreasing temperature and thus the consequent decreasing rates of oxidative degradation (Figure 5). In terms of flux, O_2 consumption at the end of October decreased to about 48 % of its September value, whereas CO_2 production shows a smaller decrease, at 60 %. This difference is probably related to the seasonal temperature decrease which will have a greater impact on oxidative reactions that occur at shallower depths, than on methanogenesis, which occurs under anaerobic conditions at greater depths. The large CO_2 excess for October 25 profiles, as inferred from the concentration gradient ratios of O_2 and CO_2 (0.61 *versus* a theoretical 0.85), supports this speculation. The rate of methanogenesis seems to be less affected by the temperature decrease, and the CO_2 released by fermentation increases relative to the

Table III. Oxygen and Carbon Dioxide Gradients in Wood Chips, and Ratios for Evaluation of Excess CO_2 from Methanogenesis

Sampling Date (1991)	dc_{O_2}/dx	dc_{CO_2}/dx	$(dc_{O_2}/dx)/(dc_{CO_2}/dx)$	Deviation (%) from ideal 0.85 ratio
Sept. 23	1.91E-07	2.48E-07	0.77	9.6
Oct. 17	1.72E-07	1.81E-07	0.95	-11.8
Oct. 25	9.15E-08	1.49E-07	0.61	28.0

CO_2 derived from oxidative degradation. The "anomalous" $(dc_{O_2}/dx)/(dc_{CO_2}/dx)$ of October 17 may represent a transitory value in the adjustment of pore-gas profiles and reaction rates to colder temperatures. Discrepancies can arise because the oxidative reactions and methanogenesis implied in gas generation and consumption take place at different depths that are cooling at different rates.

Mass transport of reaction products derived from the degradation of organic matter occurs as gas transport of CO_2, and aqueous transport of dissolved inorganic carbon (DIC) and dissolved organic carbon (DOC). Reardon and Poscente (1) provide an approximation of the mass budget in a wood pile undergoing oxidative degradation that is similar to ours. They calculate that the removal of wood chips as CO_2, DIC and DOC correspond to 13, 0.04 and 0.2 cm a^{-1}, respectively. That is, 98.2, 0.3 and 1.5 mass % of the oxidized carbon will be lost as gas, DIC and DOC. Our concentration gradients for CO_2 vary between 1.5×10^{-7} and 2.5×10^{-7} moles cm^{-3} cm^{-1} (Table III) and are similar to Reardon and Poscente's gradient (2×10^{-7} moles cm^{-3} cm^{-1}). Using a diffusion coefficient of 0.035 cm^2 s^{-1} for CO_2 measured in similar wood chips (1), the CO_2 flux varies between 5.3×10^{-9} and 8.7×10^{-9} moles cm^{-2} s^{-1}. These fluxes include the assumed approximation of the diffusion coefficient. In the field, the diffusion coefficient will vary on a yearly basis as the moisture content varies, possibly to a very low value if an ice sheet forms over the wood-chip cover during the winter months (1). Moreover, the amounts of gases released to the pore spaces will also vary, according to the temperature and reaction rates, especially for near-surface oxidative processes (Figure 5). Assuming that the measured gradients and the inferred diffusion coefficients are good approximations of what occurs over 6 months (probably more for methanogenesis) of effective biodegradation per year, we estimate an annual carbon flux of 84 to 137 mmoles cm^{-2}. This flux accounts for carbon loss from both aerobic and anaerobic processes, as the carbon mobilized by methanogenesis is added directly as CO_2 (Equation 7) or as oxidized CH_4 (Equation 9) to the CO_2 released by oxidative reactions.

The loss of carbon to the groundwater can be evaluated by calculating the amounts of CH_4 and CO_2 dissolved in pore water. At 15°C and under 1 atm pressure, the solubility of methane in H_2O is 0.00260 gCH_4 / 100 gH_2O. Assuming a partial pressure of 0.15 atm, the amount of dissolved CH_4 will be 0.24 mmoles/l. At the same temperature, the pK describing the equilibrium between $CO_2(g)$ and $H_2CO_3^0$ is 1.34. Assuming a partial pressure of 0.22 atm and acidic pore water, *e.g.* pH < 4.5 (1), so that no other inorganic carbonate species need to be considered, total DIC will amount to 9.7 mmoles/l. Assuming a rate of water infiltration of 40 cm a^{-1}, the annual flux of dissolved gas to the water table would be about 0.40 mmoles cm^{-2}, that is 200 to 350 times less than the carbon lost to the atmosphere as gas. DOC was not measured, but carbon lost this way should be minor, relative to that as gas, as discussed by Reardon and Poscente (1).

Impact of a Wood-Chip Cover on the Underlying Mine Tailings. The combined effects of O_2 consumption by aerobic oxidation and the rising of the water table limit the migration of $O_2(g)$. Anaerobic conditions, indicated by methanogenesis, prevail in the base of the 1 m thick cover. Consequently, pyrite oxidation will be limited by the shortage of the primary oxidant, O_2. This limitation should inhibit the genesis of groundwater acidity within the tailings.

Wood-chip covers are also proposed to restore old mine tailings which were exposed to periods of oxidation of varying lengths and intensities. Some organic matter derived from the degradation of the wood-chip cover can migrate to the tailings as soluble organic compounds and interact with primary and secondary minerals, and dissolved metals.

The first layer of inorganic material met by the organic compounds dissolved in water will be mainly composed of iron oxy-hydroxides. The Gibbs free energy of

Equation 4, describing the oxidation of organic matter by Fe_2O_3, predicts organic-carbon oxidation and iron-oxide reduction. This reaction releases four moles of Fe^{2+} to solution, for each mole of organic carbon consumed. This reaction will proceed until one of the two constituents, organic matter or iron oxy-hydroxide, is exhausted.

Two scenarios, with very different consequences, may occur depending on the relative availability and reactivity of the organic compounds and iron oxy-hydroxides. In the first scenario, if the amount of ferric iron is small and if dissolved organic complexes are able to leave the oxy-hydroxide-rich zone, the organic-carbon compounds may be oxidized by the reduction of SO_4^{2-}, a compound present in large quantities in mine-tailings waters. Production of S^{2-} will occur, according to Equation 5. At least part of the total dissolved Fe^{2+} and other dissolved metals should then precipitate as sulfides (FeS, FeS_2, PbS, CuS, ZnS). In the second scenario, if the organic material is entirely oxidized in the oxy-hydroxide rich-zone, Fe^{2+} produced by the reduction of Fe_2O_3 may remain in solution. Reduced iron can precipitate as ferrous carbonate (*e.g.* siderite) or sulfate (*e.g.* melanterite) minerals if saturation with respect to these compounds is attained. Ferrous iron can also migrate outside the mine-tailings impoundment and precipitate, according to the equation:

$$2 Fe^{2+} + 1/2\ O_2 + 5\ H_2O \longrightarrow 2\ Fe(OH)_3 + 4\ H^+ \tag{10}$$

As a consequence of this mechanism, acidity may be generated in the near surface zone around the mine-tailings impoundment. Therefore, the wood-chip cover overlying old impoundments will be more efficient where the layer of Fe_2O_3 is thin.

These reduction-oxidation reactions are only part of the organic-inorganic interactions that can occur in the tailings pore-water. Other reactions can have severe environmental impacts. For instance, methylation can occur as a result of methanogenesis in the presence of high concentrations of dissolved metals, with the possibility of releasing highly toxic complexes such as methyl mercury, methyl arsenic and lead (9). Organic molecules can also form stable soluble complexes with heavy metals, increasing the mobility of these contaminants and their dispersion into the environment.

Summary and Conclusions

The available data that most closely relates to our study concern wood-chip piles studied by Reardon and Poscente (1). The geometry is, however, far different, with wood chips piled in mounds about 3 m high and 15-30 m in diameter in their study, whereas our 1 m cover merges into a 2-3 m thick continuous sheet extending over several tens of hectares. Oxic and oxygen-depleted profiles were recognized in their wood piles, with methane in only one instance, in trace amounts. The oxygen loss over 50 cm of their oxygen-depleted profiles compares well with our data, but methane is a major component that occupies several percent of the gas-filled porosity. The fine grain size of the wood chips, the wide lateral extent of the cover and the presence of a low-permeability substratum combined to increase the water content of the wood chips, allowing the establishment of such extreme reducing conditions that the terminal step in organic-matter degradation, methanogenesis, occurs. Fermentation of the dissolved organic compounds released by the oxidation of the adjacent biomass probably explains why large amounts of methane were measured.

Oxidation of wood chips is an important process if this material is planned to be used in the closure of pyritic mine tailings, not only because of their long-term stability as an oxygen interceptor, but also because of other environmental reasons.

The degradation of a large biomass of wood chips can generate organic leachates that can add to the complexity of a problem which is already complex. In addition, soluble organic compounds derived from direct oxidation by atmospheric oxygen, and dissolved in the pore water, can interact in undesirable ways with the dissolved and solid inorganic components. For example, dissolved organic compounds can participate in the reduction of iron oxy-hydroxides formed prior to the emplacement of the cover. In that case, soluble reduced iron may be released to groundwater and may add to the acid-generation problem if precipitation of the ferric forms occurs, after oxidation, at the periphery of the tailings impoundment. Dissolved organic molecules can also be involved in methanogenesis, with the risk of heavy-metal methylation; they can also form stable soluble complexes with metals, increasing the mobility of these contaminants and their release to the environment.

One approach to the problem, although not eliminating it, would be to use wood chips in closure designs in which high levels of water saturation are maintained. This approach would reduce the rate of oxidation, and thus the amounts of organic compounds transferred to the pore water. This approach would not prevent reductive dissolution, methylation or complexation problems, but could reduce these problems to an extent that could be handled by the natural environment. Thick accumulations of wood chips directly exposed to the atmosphere should be avoided, because oxidation rates are so high that their long-term use as an oxygen interceptor is compromised. The efficiency of such a barrier is low if the chip size is large. In addition, the dissolved organic compounds transferred to the pore waters can have unpredictable effects on the environment.

Acknowledgments

This study was funded by the Mine Environmental Neutral Drainage (MEND) group of the Centre de recherches minérales (Ministère de l'Énergie et des Ressources du Québec). Our gratitude is expressed to André Paquet, Centre de recherches minérales, to whom we owe our involvement in the problems of organic covers. Ms. M.-A. Cimon and K. Oravec are thanked for their help with the collection of field data. Comments by anonymous reviewers, D.W. Blowes and L. Corriveau greatly helped to improve the manuscript.

Literature Cited

1. Reardon, E.J.; Poscente, P.J. Reclam. Reveg. Res. 1984, 3, 109-128.
2. Reardon, E.J.; Moddle, P.M. Uranium 1985, 2, 83-110.
3. Hajny, G.J. Tappi. 1966, 49, 97A-105A.
4. Springer, E.L.; Hajny, G.J. Tappi. 1970, 53, 85-86.
5. Froelich, P.N.; Kinkhammer, G.P.; Bender, M.L.; Luedtke, N.A.; Heath, G.R.; Cullen, D.; Dauphin, P.; Hammond, D.; Hartman, B.; Maynard, V. Geochim. Cosmochim. Acta. 1979, 43, 1075-1090.
6. Whiticar, M.J.; Faber, E.; Schoell, M. Geochim. Cosmochim. Acta. 1986, 50, 693-709.
7. Whiticar, M.J. Org. Geochem. 1989, 16, 531-547.
8. Coleman, D.D.; Risatti, J.B.; Schoell, M. Geochim. Cosmochim. Acta. 1981, 45, 1033-1037.
9. Fergusson, J.E. The Heavy Elements: Chemistry, Environmental Impact and Health Effects; Pergamon Press: New York, NY, 1991; 614 p.

RECEIVED April 5, 1993

Chapter 39

Field Research on Thermal Anomalies Indicating Sulfide-Oxidation Reactions in Mine Spoil

Weixing Guo and Richard R. Parizek

Department of Geosciences, Pennsylvania State University, University Park, PA 16802

> Field investigations were conducted at two reclaimed surface coal mining sites in western Pennsylvania. A number of deep holes that penetrated to the mine floor and shallow holes 0.3 to 0.6 m deep were installed on transects containing the deep holes. Temperature within the mine spoil and at ground surface was measured during the last three years. Spoil temperature was 1.5 to 2.2°C higher than the background temperature, which was measured in a deep hole outside mine spoil, over the field investigation period. Thermal anomalies were detected near buried coal refuse in mine spoil, which was confirmed by using geophysical methods and test drilling. Application of sewage sludge showed an effect in suppressing spoil temperature. The intensity of thermal anomalies was found to decrease with time at an average rate -0.1°C/year. This may be the result of gradual depletion of pyrite in mine spoil.

Acid mine drainage (AMD), a well known water pollution problem in some coal and metal mining districts, results from the oxidation of sulfide minerals. Through a series of reactions in which pyrite is oxidized, hydrogen ions and sulfate are produced. Highly acidic mine waters can leach heavy metals from spoil into surface and subsurface waters. Soil erosion also can be excessive where acidic waters have come into contact with and killed vegetation.

Field study of spoil temperature is of importance in the practice of mined-land reclamation. Temperature can be measured comparatively easily and accurately in the field using simple equipment. Minor heat differences resulting from sulfide-oxidation reactions can be detected because mine spoil is a poor medium for heat transfer. The perturbation of mine spoil temperature due to the exothermic pyrite oxidation reactions has been known for a long time (1,2).

The ability to determine the location of chemically active pyrite-concentrated materials is very desirable in the practice of mined-land reclamation. Financial limitations often do not allow one to treat all disturbed areas in an attempt to abate acid mine drainage or to slow its production. Geophysical methods have been applied to locate the sites of coal refuse, a highly concentrated pyritic material (3,4). However, it is more important to locate the chemically active coal refuse, because the presence of pyritic material does not necessarily generate acid mine drainage unless other conditions are favorable. Thermal anomaly investigation has the advantage in providing the information on the locations of active pyrite-concentrated zones in mine spoil, which is a primary concern in land reclamation and acid mine drainage control.

Long-term spoil-temperature monitoring may reveal the intensity changes of pyrite-oxidation reactions and AMD generation from mine spoil. The effect of abatement measures can be evaluated by monitoring the intensity changes of the thermal anomalies in mine spoil. Spoil temperature should decrease with time as the result of pyrite depletion, even if no abatement measures are taken. Spoil temperature should be further suppressed if AMD generation is under control. Long-term temperature observation is required to test this hypothesis, but no such published efforts are known. Most field temperature observations reportedly lasted for only one year.

Non-uniform temperature distribution in spoil may also induce air circulation which supplies oxygen for pyrite oxidation. The amount of oxygen supplied through convection may be much greater than that supplied by diffusion together with that dissolved in infiltrating water. Field investigation of the distribution of thermal anomalies in spoil will help us understand the mechanisms of oxygen transport in mine spoil.

The primary objectives of this research were to locate hot spots in mine spoil and to characterize the sulfide-oxidation intensity changes with time using thermal-surveying methods. Hot spots are referred to as the locations where the oxidation reactions are taking place at a significant rate and where temperature is higher than the normal or background temperature due to the heat released from pyrite-oxidation reactions. Both near-surface and deep temperature surveys were conducted to determine if air, surface or near-surface surveys could be used for this purpose and if so, during what season of the year deeper temperature anomalies might be detected at or near land surface.

Methods and Materials

Two sites were selected for this study (Figure 1). Site 1 is located near Clearfield, Pennsylvania. It was mined by surface methods from 1968 to 1982. The lower Kittanning coals were mined to a depth of nearly 33 m. An older portion of the mine, mainly in the west of the site, was abandoned 20 years ago. This part of spoil was not treated with sewage sludge and was covered by only sparse evergreen trees (no sludge area). The central part of disturbed land was treated with sludge in 1986 (old sludge area). In the summers of 1988 and 1989, sewage sludge, mixed with wood chips and lime, was spread over the east part of the mine site (new sludge area). These sludge-treated areas were covered by dense grass. Several seeps were identified as the drainage from the site. The water quality of the seepage remains poor long after all mining activity has ceased (Table I).

Table I. Seep-water Chemistry at Site 1 (February, 1990) [Data from C. Cravotta, U.S. Geological Survey, written commun.; all data in mg/L except pH (std. units)]

pH (lab)	SO_4^{2-}	Ca^{2+}	Mg^{2+}	Cu^{2+}	Fe^{2+}	Fe (total)	Alkalinity (total)
3.4	673	99.3	93.5	0.093	0.47	0.58	0.0

During the summer of 1989, 15 deep holes with a 15.24 cm open diameter, were drilled (Figure 2). One piezometer (1.27 cm internal diameter (ID), polyvinyl chloride (PVC) pipe) was installed in each deep hole screened at the bottom 0.33 m. Piezometers provided access for temperature probes and were used to determine the presence of saturated spoil and the elevation of the water table. The size of the piezometers was chosen to limit the heat convection through the pipes.

One additional deep hole, the Control Hole, was drilled about 600 m outside of the mining-disturbed area. It was drilled through the undisturbed coal-bearing formations to a depth of 26.4 m, below the lower Kittanning coal seams that were mined. The Control Hole contained a PVC piezometer 10.16 cm in diameter with a 3.05 m long screen at its bottom. The background temperature was measured in this Control Hole.

The site contains 130 shallow holes (Figure 2). In each shallow hole, two short PVC pipes (1.27 cm ID) were installed, 0.3 and 0.6 m in length respectively. The bottom ends of these short pipes were left open. Shallow holes were aligned along five transect lines distributed over the site. The spacing between two shallow holes was approximately 15 m. The effects of thermal anomalies on the temperature near the ground surface were investigated by measuring the temperature in these shallow holes.

Fine sand, bentonite and cuttings were used to backfill the boreholes. All of the deep and shallow holes were capped to prevent heat convection through the pipes and the entry of rain or snow.

Site 2 is located in Clarion, Pennsylvania. Prior to 1968, the lower and upper Clarion and the lower Kittanning coals were mined from a portion of the site. During 1972 and 1973, the site was mined more extensively for the upper and lower Clarion coals. In this site, 20 deep holes and 140 shallow holes were established in the same manner as described above (5).

Digital thermometers, together with PSI-400 temperature probes were used for quick temperature measurement. The precision of these thermometers is ± 0.15°C and the minimum reading unit is 0.05°C. Digital thermometers also helped to reduce the delay time between inserting the sensors and taking the readings.

Spoil temperature at shallow depths is strongly affected by diurnal changes of ambient temperature. Shallow-hole temperatures were measured each day in the very early morning. Three groups of students worked at the same time to minimize the time difference, which may cause serious errors in shallow temperature measurements.

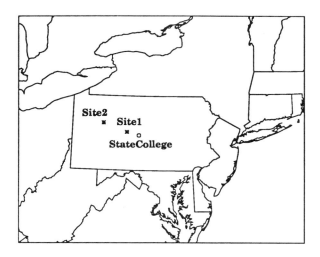

Figure 1. Location of the field sites.

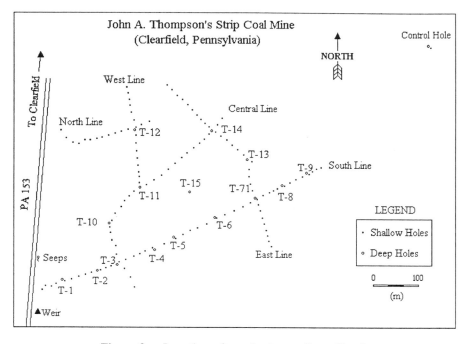

Figure 2. Location of monitoring wells at Site 1.

Results and Discussion

The temperature of spoil reflects its energy balance. The input energy includes solar radiation, geothermal conduction, chemical reactions and decomposition of organic materials, as well as the energy transferred by groundwater and spoil gas. Solar radiation and chemical reactions are the principal energy sources while the contributions of the other heat sources are relatively small under the conditions investigated. Heat radiation from spoil surface to the atmosphere is the major process dissipating energy. Some losses also are associated with mine-water seeps that drain the spoil.

Seasonal Variation and Amplitude Attenuation. The temperature data measured in spoil revealed that spoil temperature is chiefly controlled by the solar radiation at the ground surface. The monthly average temperature at the ground surface, observed at Philipsburg Weather Station, 15 km from Site 1, from September, 1989 to May, 1992 is shown in Figure 3. The temperature data observed in selected deep holes show strong seasonal variations (Figure 4), which have a similar pattern to that shown in Figure 3. These data plots look like sine or cosine curves similar to those observed in a natural soil profile. Data presented in Figure 4 imply that solar radiation was the most important energy source in the spoil thermal regime. Heat from other sources was not strong enough to distort these sine-like curves. The amplitude of seasonal temperature variation tended to decrease with increasing depth, as observed in natural soil profiles not influenced by *in-situ* heat production (6,7).

Thermal Anomalies Induced from Oxidation Reactions. Thermal-surveying data clearly show that spoil temperature was higher than the background temperature, which was measured in the Control Hole. Figure 5a shows the temperature values at various depths measured in both mine spoil and in the Control Hole. A similar phenomenon was also observed at Site 2 (Figure 5b). These diagrams are only selected examples that show this general tendency. The average temperature difference between the spoil holes and the Control Hole was about 1.5 to 2.2°C, and varied with time, depth and location within the mine spoil.

Based upon field measurements over the period of observation, thermal anomalies were identified near T-3 at Site 1, TO-2, TO-7, TO-9 and TO-19 at Site 2. Spoil temperature at these locations was substantially higher than the background temperature.

All of the thermal anomalies are not generated by pyrite oxidation reactions. Thermal anomalies observed in mine spoil may be induced by a number of causes. To locate active pyrite oxidation reaction zones, the thermal anomalies caused by processes other than pyrite oxidation reactions must be excluded.

(i) Heterogeneity of spoil physical properties: Various spoil and bedrock materials have different physical properties, such as thermal diffusivity, water content, etc. Spoil with lower thermal-conductivity values may cause temporary local thermal anomalies due to a time lag effect. If there is no heat source in the spoil, the temperature difference between this part of spoil and its surrounding areas should change with the seasons. When the entire spoil begins to warm up, the temperature within this part will be lower than its surrounding temperature and *vice versa*. An appearance of

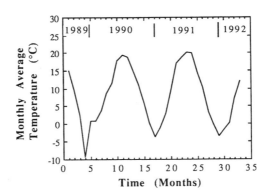

Figure 3. Monthly average temperature (September, 1989 to May, 1992) recorded at Philipsburg Weather Station, PA.

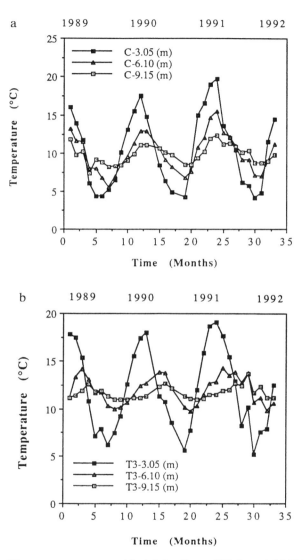

Figure 4. Temperature measured in (a) the Control Hole and (b) T-3 at depths 3.05 m, 6.10 m and 9.15 m. The first month is September, 1989.

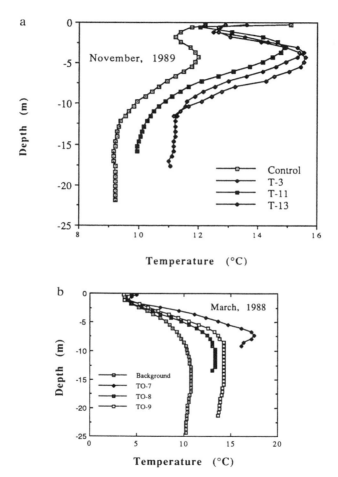

Figure 5. Temperature profiles observed at (a) Site 1 and (b) Site 2.

thermal anomalies all year long, as observed at both sites, indicates that these thermal anomalies were not caused by the heterogeneity of spoil physical properties.

(ii) Groundwater flow: Many studies have proven that groundwater may transfer heat, causing temperature differences within aquifers (8). However, this is not the principal cause of temperature variations noted in this study. The groundwater temperature was observed to be almost uniform in the region. The water table was detected at T-10 only, which is located in a topographic low area, while all the other deep holes drilled in the spoil at Site 1 were dry during the entire investigation period. Groundwater temperatures, measured in T-10 and the Control Hole, were relatively stable and lower than the spoil temperature all the time.

(iii) Heat production from oxidation reactions: In spite of the fact that reactive pyrite is contained in the undisturbed rock sequence penetrated by the Control Hole, the oxidation reactions would not start or would proceed only very slowly because of the lack of oxygen. In the spoil, oxidation reactions release a significant amount of heat, 374.74 kJ for each mole of pyrite oxidized. This heat production can cause the spoil temperature to be higher than that of the unmined material. In addition, the intensity of heat production varies with location in spoil, in part because of the uneven distribution of reactive pyrite within the spoil.

To ensure that thermal anomalies were consistent with coal refuse sites, geophysical methods were applied at Site 2. Results of geophysical surveys are consistent with the existence of highly reactive pyrite-concentrated coal refuse. Apparent conductivity surveying data showed high conductivity values appearing nearby TO-2, TO-7, TO-9 and TO-19 at Site 2 (4). These same areas were determined to be hot spots, with temperature differences of 2.8 to 8.0°C. A magnetometer survey was conducted in 1986 to locate more precisely the buried coal refuse (9). Areas near these hot spots were confirmed as sites of buried coal refuse by test drilling. Six samples of the buried coal refuse ranged in total sulfur content from 1.5% to 5.9% by weight (5).

Based on the evidence collected from our field investigation, it is clear that pyrite-oxidation reactions generate detectable thermal anomalies and these anomalies can be used to locate active pyrite-concentrated regions in mine spoil. In general, the intensities of thermal anomalies observed in this study were relatively low, compared to the temperatures reported in the literature. In most cases, spoil temperatures were in the range of 8.5°C to 13.0°C at both sites. Temperatures in excess of 80°C have been observed in low grade copper dumps (10).

The best time to detect hot spots is in the spring, when the surface temperature disturbance reaches a minimum. During the rest of the year, it is not easy to distinguish these thermal anomalies from solar heating unless much stronger anomalies are present.

Shallow Temperature Interpretation. Thermal anomalies at ground surface should reflect variations in the intensity of pyrite oxidation in the spoil as well as the seasonal changes in air temperature, solar radiation, infiltration and evaporation of spoil water, surface slope and orientation, shade and other factors. Because these shallow holes were buried only 0.3 and 0.6 m deep, they were more likely to be influenced by short-term weather and environmental changes. Without the disturbance of the heat from pyrite oxidation, temperatures at 0.3 m should be higher than those at 0.6 m during warm seasons. When the temperature disturbances induced by pyrite oxidation reactions are great enough, reversed thermal gradients result. Lower temperatures at 0.3 m during warm seasons, will be very useful in locating reactive hot spots. Shallow temperatures near a hot spot are also expected to be higher than those of the surrounding area.

Thermal anomalies were detected by measuring the near-ground surface temperature. Figure 6 shows the temperature taken along South Line at Site 1 in May, 1990. Two "temperature humps" were found. One thermal anomaly, between hole numbers 5 and 15 along South Line, is located near T-3, in which strong thermal anomalies were observed. Another temperature hump was detected between hole numbers 30 and 40 in the areas treated with sludge in 1988 and 1989. This hump was probably due to the dark color of sewage sludge material, which absorbed more heat from solar radiation.

It is interesting to note the effect of sludge treatment on the ground-surface temperature. Recently, sewage sludge has been widely used in mine-spoil abatement (11,12). Sewage sludge may increase the pH, moisture and organic-matter contents, and cause a reduction in heavy-metal concentrations. The general effects of such treatment have been to reduce temperature, to help in revegetation, and to slow down the sulfide-oxidation reactions by depleting part of the oxygen available in spoil.

From the field data obtained from treated and non-treated portions of the mine, it was found that the temperatures of shallow holes in untreated areas generally were higher than those in recently sludge-treated areas. The lowest shallow-hole temperatures were found in the older sludge treatment area (Figure 7).

The reversed temperature gradients, that may indicate *in situ* heating effects, were not detected. This observation implies that the intensity of the thermal anomalies produced by sulfide-oxidation reactions was not strong enough to reverse the normal gradient directions between 0.3 m and 0.6 m depths during warm seasons.

Intensity Changes of Thermal Anomaly with Time. Thermal-anomaly surveys are useful not only in locating acid-producing sites within mine spoil, but also in studying the intensity changes of pyrite-oxidation reactions. Thermal anomalies are induced by the heat generated from pyrite-oxidation reactions. The intensity of these anomalies should decay with time, because the amount of pyrite in spoil decreases with time. The difference between the spoil temperature and the background temperature should decline.

Because the spoil temperature is strongly influenced by seasonal changes and air temperature fluctuations, the decay of spoil temperature due to the intensity change of pyrite-oxidation reactions is easily smeared. Under some circumstances, the spoil temperature may show an increase instead of a decline. To distinguish the spoil temperature changes caused by changes in pyrite-oxidation reactions from those caused by seasonal variations, the magnitude of the difference between spoil and the background temperature versus time should be considered. In order to reduce the effect of air-temperature fluctuations and the heterogeneity of spoil thermal properties, the average values of temperature data were used in interpretations.

Positive temperature departure from normal values were recorded between June, 1990 and February, 1992 (Figure 8). This observation indicates that the weather in this area became warmer. The background temperature increased at a rate of 0.44°C

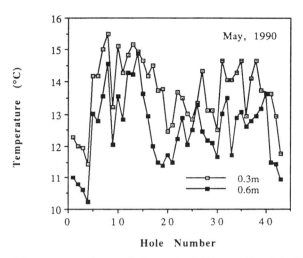

Figure 6. Temperature observed along South Line at Site 1 in May, 1990.

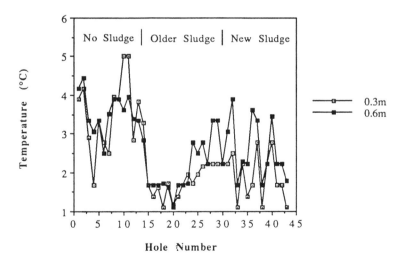

Figure 7. Shallow temperature observed in different spoil areas in March, 1990. "No sludge" refers to the area not being treated with sewage sludge; "old sludge" refers to the area treated with sewage sludge in 1986; "new sludge" refers to the area treated with sewage sludge in 1988 and 1989.

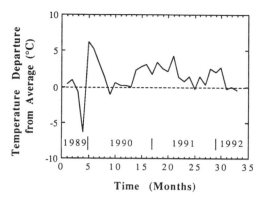

Figure 8. Monthly temperature departure (September, 1990 to May, 1992) based on the average values from 1931 to 1991 recorded at Philipsburg Weather Station, PA.

per year from 1989 to 1992, which is consistent with the air temperature changes. The spoil temperature at T-3, where hotspots were located, decreased at a rate of 0.08°C per year during the same period.

Conclusions

Acid mine drainage is the result of sulfide-oxidation reactions in which pyritic materials are oxidized and acid is released. Spoil temperature perturbed by the exothermic pyrite-oxidation reactions can be used to locate hot spots within a mine spoil, and to study the intensity changes of sulfide-oxidation reactions under field conditions.

Field research was conducted at two surface mining sites in western Pennsylvania. Thermal surveying methods were successfully used to locate the "hot spots" in mine spoil which varied from less than 10 m to nearly 30 m in thickness. Spoil temperature was found to be 1.5°C to 2.2°C higher than the background temperature over the period of field investigation. This temperature difference varied with time and space.

Our field temperature data showed that the spoil was cooling off slowly many years after mining ceased, at a rate of approximately -0.1°C/year. This temperature die-off trend most likely reflects the slowing down of the sulfide-oxidation reactions in mine spoil due to the depletion of pyrite.

This temperature-monitoring method is economical, efficient and accurate, and a sensitive indicator of sulfide-oxidation reaction rates and sites of these reactions. Only active, concentrated zones of sulfide oxidation can be identified by measuring the thermal anomalies both at and near ground surface and along boreholes drilled in mine spoil. This technique can be used to obtain information on the spatial distribution of hot spots within the spoil to identify where special, additional abatement steps should be taken. This application may be very important when budgets are limited and AMD generation persists following initial attempts of mine reclamation. Real-time temperature monitoring of reclaimed mines is likely to provide more sensitive indications of the early benefits, or lack thereof, of reclamation and acid-abatement measures than soilwater- and groundwater-quality monitoring methods.

Acknowledgments

Financial support for the early stage of this study was provided from the U.S. Bureau of Mines through grant No. G1125132-4261. Continuation of temperature monitoring at Site 1 was supported by Federal funds from the U.S. Bureau of Mines and the National Mine Land Reclamation Center under cooperative agreement CO 388026.

Literature Cited

1. Harries, J.; Ritchie, A. Water, Air and Soil Pollution 1981, 15, 405-423.
2. Jaynes, D.B.; Rogowski, A.S.; Pionke, H.B. Water Res. Res. 1984, 20, 233-250.
3. Williams, J.H.; Henke, J.R.; Pattison, J.R.; Parizek, R.R.; Hornberger, R.J. Hydrogeology and Water Quality at a Surface Coal Mine in Clarion County, Pennsylvania; Penn. State Univ., Coal Research Section, 1987.
4. Schueck, J.H. Mine drainage and surface mine reclamation, U.S. Bureau of Mines Information Circular 9183, 1988, pp. 117-130.
5. Fielder, D. Unpublished M.S.Thesis, Penn State Univ., 1989.
6. Hillel, D. Introduction to Soil Physics; Academic Press: New York, 1982; pp. 155-175.
7. Ciolkosz, E.R.; Cronce, R.; Cunningham, R.; Petersen, G. Soil Science 1985, 193, 232-238.
8. Wang, J.; Xiong, L. In Hydrogeological Regimes and Their Subsurface Thermal Effects; Geophy. Monograph 47; American Geophysical Union: Washington, DC, 1989; pp. 87-99.
9. Ladwig, K.J. Ground Water Monitoring Rev. 1983, 3, 46-51.
10. Beck, J.V. Biotech. Bioeng. 1967, 9, 487-490.
11. Plass, W.T. J. of Soil and Water Conservation 1978, 29, 119-121.
12. Wittwer, R.; Carpenter, S.; Graves, D. In Symposium on Surface Mining Hydrology and Reclamation; Univ. of Kentucky: KY, 1980; pp. 193-197.

RECEIVED March 26, 1993

INDEXES

Author Index

Ahonen, Lasse, 79
Alpers, Charles N., 324
Anthony, Elizabeth Y., 551
Antonson, David A., 593
Appleyard, Edward C., 516
Bencala, Kenneth E., 224
Bergeron, M., 365,631
Bhatti, Tariq M., 90
Bhupathiraju, Vishveshk, 68
Bigham, Jerry M., 90,190
Blowes, David W., 172,516
Borek, Sandra L., 31
Boyle, D. R., 535
Broshears, Robert E., 224
Calmano, W., 298
Carlson, L., 190
Cathles, L. M., 123
Chan, C. W., 60
Chermak, John A., 2
Cravotta, C. A. III, 345
Eggleston, Carrick M., 201
Evangelou, V. P., 562
Förstner, U., 298
Fuge, Ron, 261
Gagen, Patrick M., 2
Germain, M. D., 365,631
Guo, Weixing, 645
Halbert, Bruce, 132
Hammack, Richard W., 431
Hochella, Michael F., Jr., 201
Hong, J., 298
Huang, Xiao, 562
Jang, Joon H., 45
Jeffers, T. H., 608
Jeon, Yongseog, 412
Kimball, Briant A., 224
Krouse, H. R., 446
Kwong, Y. T. J., 382
Lapakko, Kim A., 593
Lichtner, Peter C., 153

Manowitz, Bernard, 412
McInerney, Michael J., 68
McKnight, Diane M., 224
Millero, Frank J., 393
Montgomery, Anne D., 68
Moore, Johnnie N., 276
Morse, John W., 289
Murad, E., 190
Nicholson, Ronald V., 14,132
Nimick, David A., 276
Nordstrom, D. Kirk, 244,324
Parizek, Richard R., 645
Pearce, Fiona M., 261
Pearce, Nicholas J. G., 261
Perkins, William T., 261
Ptacek, C. J., 172
Rimstidt, J. Donald, 2
Ritchie, A. I. M., 108
Robl, Thomas L., 574
Scharer, Jeno M., 14,132
Schwertmann, U., 190
Smith, Kathleen S., 244
Snodgrass, William J., 132
Sublette, K. L., 68
Suzuki, Isamu, 60
Takeuchi, T. L., 60
Tassé, N., 365,631
Taylor, B. E., 481
Thompson, J. Michael, 324
Tuovinen, Olli H., 79,90
Vairavamurthy, Appathurai, 412
Van Stempvoort, D. R., 382,446
Vuorinen, Antti, 90
Wadsworth, Milton E., 45
Webster, Jenny G., 244
Wheeler, Mark C., 481
White, W. W. III, 608
Williams, Peter A., 551
Zhang, Jia-Zhong, 393
Zhou, Weiqing, 412

Affiliation Index

Australian Nuclear Science and Technology Organization, 108
Brookhaven National Laboratory, 412
Cornell University, 123
Geological Survey of Canada, 481,535
Geological Survey of Finland, 79
Groundworks Environmental, Inc., 481
Institut National de la Recherche Scientifique–Géoressources, 365,631
Institute of Environmental Health and Forensic Science, 244
McMaster University, 132
Minnesota Department of Natural Resources, 593
National Hydrology Research Institute, 382,446
National Water Research Institute, 172
O.H.M. Corporation, 2
Ohio State University, 79,90,190
Pennsylvania State University, 645
SENES Consultants Ltd., 132
Swiss Federal Institutes of Technology, 201
Technical University of Hamburg-Harburg, 298
Technische Universität München, 190
Texas A&M University, 289
U.S. Bureau of Mines, 31,431,562,608
U.S. Geological Survey, 224,244,276,324,345
Universität Bern, 2,153
University of Calgary, 446
University of Helsinki, 90,190
University of Kentucky, 562,574
University of Manitoba, 60
University of Miami, 393
University of Montana, 276
University of Oklahoma, 68
University of Texas at El Paso, 551
University of Tulsa, 68
University of Utah, 45
University of Wales, 261
University of Waterloo, 14,132,172,516
University of Western Sydney, 551
Virginia Polytechnic Institute and State University, 2,201

Subject Index

A

Abiologic oxidation, evidence from isotope signatures, 507–508
Abiologic oxidation of sulfide, studies, 482–483
Abiotic oxidation of pyrite
  $\delta^{18}O$ of sulfate formed, 454–456t
  role of Fe^{3+}, 453–454
Abiotic oxidation of sulfite, $\delta^{18}O$ of sulfate formation, 449–451
Absorption edge, description, 414
Acid–base accounting
  advantages and disadvantages, 432
  description, 432,611
  objective, 611
Acid environments, sources of oxygen in sulfate, 503f,505
Acid generation, assessment in pyritic tailings using computer program, 132–150

Acid mine drainage
  application of heap leaching models, 128,130f,131
  definition, 608
  dissolved components, 325
  Duluth Complex rock
    acid production and neutralization, 594
    analytical procedures, 594–597
    chemical composition of samples subjected to dissolution, 597,598t
    drainage pH vs. sulfur content, 602–606
    future work, 606
    metal sulfide oxidation rate, 597,599
    pyrrhotite oxidation rate, 599
    reasons for study, 593
    sulfate release rate calculation procedure, 597
    sulfate release rate vs. pH, 600,603
    sulfate release rate vs. relative humidity, 603

Acid mine drainage—*Continued*
Duluth Complex rock—*Continued*
   sulfate release rate vs. sulfur content, 599–602*t*
   sulfate release rate vs. temperature, 603
East Sullivan, Quebec
   acid neutralization processes, 374,376
   CO_2 and O_2 concentration profiles, 369,371*f*
   experimental description, 365
   geochemical calculation procedure, 368
   grain size characteristics, 368,369*t*
   hydraulic conductivity, 369,370*f*
   location, 365,366*f*
   pore gas analysis, 367,369
   pore-water analysis, 365,367–368,372
   precipitation of secondary minerals, 376–378
   sulfide oxidation, 372,374,375*f*
   tailing solids analysis, 368,372,373*f*
environmental problem, 608
formation, 345
geochemistry, 329,332*t*,333
industrial pollution, source, 431
Iron Mountain, California, seasonal variations of Zn/Cu ratios, 324–343
Leviathan and Bryant Creeks, transport and natural attenuation of Cu, Zn, As, and Fe, 244–259
metallic sulfide bearing waste rock
   chemical predictive modeling, 609–627
   static and kinetic tests for determination of potential, 609
new mines, requirements, 431–432
pyritic tailings, importance of control, 133
release from waste rock, 608–609
sulfide mineral oxidation, 645
sulfur and oxygen isotope geochemistry in western United States, 481–511
Wales
   chemistry of acid lake water, 264,267–269
   chemistry of mine water, 264,266*t*
   experimental procedure, 264
   history, 261–262
   ochre chemistry, 270
   pH vs. water type, 269
   sampling procedure, 264
   study sites, 262–265*f*
   water chemistry, 269–270

Acid mine drainage—*Continued*
Wales—*Continued*
   water–ochre chemistry relationship, 270–273*t*
Acid neutralization, description, 594
Acid neutralization processes, acid mine tailings in East Sullivan, Quebec, 374,376
Acid production, description, 594
Acid rock drainage
   attenuation in natural wetland system, 383–391
   environmental problem, 382
   passive solutions, 382–383
   treatment techniques, 382
Acid sulfate waters, mineralogical characteristics of poorly crystalline precipitates of Fe^{2+} oxidation, 190–199
Acidic drainage potential of waste rock sample
   definition, 609
   quantification using release rates, 609
Acidic ground water, geochemical evolution at reclaimed surface coal mine in Pennsylvania, 345–362
Acidic solutions, reaction rates of sulfide minerals with Fe^{3+}, 2–12
Acidity, secondary iron sulfate minerals as sources, 346–347
Acidophilic thiobacilli, 90–91
Adenosine phosphosulfate reductase, role in inorganic sulfur compound oxidation, 62
Adularia, role in geochemical evolution of acidic ground water, 345–362
Air convection, control of leaching of copper-bearing mine waste, 124
Albite, role in geochemical evolution of acidic ground water, 345–362
Alkaline-earth carbonates, evolved-gas analysis, 440,442–443*f*
Alpine Gulch, Colorado, geology and hydrology, 488
Aluminum
   concentration in leachates from organic-rich Devonian black shale, 574–591
   concentration in massive sulfide deposits in Bathurst Mining Camp, New Brunswick, 536–548
   concentration in weathered sulfide mine tailings, 516–533

INDEX

Aluminum—*Continued*
  levels in acid mine drainage in Wales, 261–273
  pH modification effect on transport, 224–241
  role in geochemical evolution of acidic ground water, 345–362
  transport during sulfide oxidation, 237–240t
Anglesite, role in acid mine tailings, 377
Anoxic environments, hydrogen sulfide oxidation, 403,405–408
Anoxic sediments from Elbe River, heavy metal mobilization and scavenging after resuspension, 298–319
Antimony, concentration in massive sulfide deposits in Bathurst Mining Camp, New Brunswick, 536–548
Apparent rate of consumption of Fe^{3+}, calculation, 5
Argo Tunnel, Colorado, geology and hydrology, 488,490f
Arsenic
  attenuation rate from acid mine drainage, 255,256t
  chemical modeling, 256,257t
  concentration in massive sulfide deposits in Bathurst Mining Camp, New Brunswick, 536–548
  concentration in sulfidic flood plain sediments, 282,284f,285
  concentration in weathered sulfide mine tailings, 516–533
  coprecipitation with pyrite in anoxic sediments, 290
  levels in acid mine drainage in Wales, 261–273
  release via authigenic pyrite oxidation in resuspended sediments, 289–297
  site of removal from solution, 253–255
Arsenopyrite
  concentration in massive sulfide deposits in Bathurst Mining Camp, New Brunswick, 536–548
  levels in acid mine drainage in Wales, 261–273
  reaction rates with Fe^{3+} in acidic solutions, 2–12
  reactivity in acid rock drainage in natural wetland system, 383
Arsenopyrite oxidation, reaction, 3

Atmospheric sulfate, $\delta^{18}O$, 465,466f
Atomic structure, PbS{100} surfaces, 201–221
Attenuation
  Cu, Zn, As, and Fe in acid mine drainage of Leviathan and Bryant Creeks, 244–259
  definition, 245
Attenuation of acid rock drainage in natural wetland system
  analytical procedure, 385
  aqueous and sulfur isotope geochemistry of surface water, 385,386t
  experimental objectives, 383
  future work, 391
  general setting, 385
  mechanisms, 388–391
  pore-water geochemistry of sediments, 386–389f
  sampling procedure, 385
  site location and description, 383,384f
  sulfide minerals, reactivity, 383
Authigenic pyrite oxidation in resuspended sediments, toxic metal release, 289–297

B

Bacteria, role in sulfide oxidation, 456–457
Bacterial isolates, $\delta^{18}O$ of sulfate formed by sulfide oxidation, 457–459
Balaklala Mine, geology and hydrology, 487
Barite
  concentration in massive sulfide deposits in Bathurst Mining Camp, New Brunswick, 536–548
  levels in acid mine drainage in Wales, 261–273
Bathurst Mining Camp, New Brunswick, oxidation of massive sulfide deposits, 535–548
Bioleaching of complex sulfide ore
  chalcopyrite identification, 84,85f
  covellite identification, 82,83f
  experimental procedure, 81–82
  H_2S evolution, 82
  pentlandite identification, 82,85f
  pH effect, 84,86,87f
  previous studies, 80
  solubility effect, 84,86,87f
  sphalerite identification, 82,83f

Bioleaching of complex sulfide ore—
Continued
  theoretical distribution of aqueous species of ferric iron, 82,83*f*
Biologically mediated oxidation, evidence from isotope signatures, 508*t*,509
Biota in aquatic environments, factors controlling availability of toxic metals, 289
Bismuth sulfosalts, levels in acid mine drainage in Wales, 261–273
Black Sea, hydrogen sulfide oxidation, 403,405*f*,406
Bornite, concentration in Duluth Complex rock, 594–606
Bryant Creek, transport and natural attenuation of Cu, Zn, As, and Fe in acid mine drainage, 244–259

C

Cadmium
  Cd^{2+}, H_2S oxidation rate effect, 395–396
  concentration in sulfidic flood plain sediments, 282,283*f*
  levels in acid mine drainage in Wales, 261–273
  mobilization and scavenging in resuspended anoxic sediments from Elbe River, 298–319
  precipitation from acid mine tailings, 376
Calcite
  concentration in massive sulfide deposits in Bathurst Mining Camp, New Brunswick, 536–548
  evolved-gas analysis, 440,442–443*f*
  role in acid mine tailings, 378
Calcite dissolution, role in acid neutralization, 374,376
Calcium
  concentration in leachates from organic-rich Devonian black shale, 574–591
  concentration in weathered sulfide mine tailings, 516–533
  role in geochemical evolution of acidic ground water, 345–362
Canada Centre for Mineral and Energy Technology, studies on modeling of acid generation in mine tailings, 133
Carbon, concentration in weathered sulfide mine tailings, 516–533

Carbonate, evolved-gas analysis, 440,442–443*f*
Cariaco Trench, hydrogen sulfide oxidation, 408
Cassiterite, concentration in massive sulfide deposits in Bathurst Mining Camp, New Brunswick, 536–548
Chalcedony, role in geochemical evolution of acidic ground water, 345–362
Chalcocite, concentration in massive sulfide deposits in Bathurst Mining Camp, New Brunswick, 536–548
Chalcopyrite
  bioleaching, 79–87
  concentration in Duluth Complex rock, 594–606
  concentration in massive sulfide deposits in Bathurst Mining Camp, New Brunswick, 536–548
  hydrothermal enrichment, kinetics, 45–57
  levels in acid mine drainage in Wales, 261–273
  reaction rates with Fe^{3+} in acidic solutions, 2–12
  reactivity in acid rock drainage in natural wetland system, 383
Chalcopyrite oxidation, reaction, 3
Change in mass of element, calculation, 517
Chemical attenuation rate, definition, 245
Chemical predictive modeling of acid mine drainage from metallic sulfide bearing waste rock
  acid–base accounting, 617,618*t*
  acid–base accounting calculation procedure, 611–612
  acid-producing potential and neutralizing potential depletion, estimation, 624*t*,625
  calibration of model, 625
  contributing chemical reactions, 614–615
  differential equations, 615–616
  effluent pH vs. time, 619–622*f*
  humidity cell tests, 619–624
  mineralogy, 616,617*t*
  mineralogy determination procedure, 611
  model description, 612,614
  objectives, 609–610
  predicted vs. experimental data, 625–627*f*
  sample collection and preparation, 610
  solid-phase sulfate salt storage, hot water leach determination procedure, 611

INDEX

Chemical predictive modeling of acid mine drainage from metallic sulfide bearing waste rock—*Continued*
  sulfate release rate vs. pH, 620–624
  sulfate release rate vs. time, 619t–622f
  sulfate salt storage vs. time, 619–623
  sulfide content normalization, 623t
Chemisorption–oxidation reactions, PbS{100} surfaces, 201–221
Chemistry, sulfidic flood plain sediments from Upper Clark Fork Valley, Montana, 282–286
Chesapeake Bay, hydrogen sulfide oxidation, 406,408
Chromium, coprecipitation with pyrite in anoxic sediments, 290
Cinnabar, concentration in massive sulfide deposits in Bathurst Mining Camp, New Brunswick, 536–548
CO_2, evolution, 433
Cobalt
  Co^{2+}, H_2S oxidation rate effect, 395–396
  precipitation from acid mine tailings, 376
  solubilization during microbiological oxidation of pyrrhotite and pyrite, 90–103
Computer program, acid generation assessment in pyritic tailings, 132–150
Continuous stirred tank reactor model
  contributing chemical reactions, 614–615
  description, 612
  differential equations, 615–616
Copentlandite, bioleaching, 79–87
Copiapite, role in geochemical evolution of acidic ground water, 345–362
Copper
  attenuation rate from acid mine drainage, 255,256t
  chemical modeling, 257t,258
  concentration in acid rock drainage in natural wetland system, 382–391
  concentration in massive sulfide deposits in Bathurst Mining Camp, New Brunswick, 536–548
  concentration in sulfidic flood plain sediments, 282–285
  concentration in weathered sulfide mine tailings, 516–533
  content in acid mine tailings, 377

Copper—*Continued*
  Cu^{2+}
    H_2S oxidation product distribution effect, 397,399f
    H_2S oxidation rate effect, 395–396
  levels in acid mine drainage in Wales, 261–273
  mobilization and scavenging in resuspended anoxic sediments from Elbe River, 298–319
  pH modification effect on transport, 224–241
  precipitation from acid mine tailings, 376
  release via authigenic pyrite oxidation in resuspended sediments, 289–297
  site of removal from solution, 254f,255
  solubilization during microbiological oxidation of pyrrhotite and pyrite, 90–103
  transport during sulfide oxidation, 237,239f,240t
  See also Zn/Cu ratios in acid mine water from Iron Mountain, California
Copper-bearing mine waste, leaching model, 123–131
Copper enrichment, reactions, 45–46
Copper leaching, procedure, 123–124
Coquimbite, role in geochemical evolution of acidic ground water, 345–362
Covellite, concentration in massive sulfide deposits in Bathurst Mining Camp, New Brunswick, 536–548
Covers for reactive mine tailings, 631
Cubanite, concentration in Duluth Complex rock, 594–606

D

$\delta^{18}O$ in sulfate
  atmospheric sulfate, 465,466f
  comparison of field data to laboratory findings, 464–475
  elemental S oxidation, 453
  experimental objectives, 446
  ground-water sulfate, 467–470f
  influencing factors, 447,448f
  isotope fractionation effect, 461
  organic sulfate hydrolysis effect, 463
  reduced organic S oxidation, 459,461
  rules for interpretation in various environments, 475

$\delta^{18}O$ in sulfate—*Continued*
  soil sulfate, 467
  sorption–desorption effect, 462
  sulfate from springs, 469,471f
  sulfate in runoff, 469,472f
  sulfate mineral precipitation effect, 461
  sulfate reduction effect, 463–464
  sulfate–water oxygen exchange effect, 462–463
  sulfide oxidation, 453–460
  sulfite oxidation, 447,449–452
  thiosulfate oxidation, 452–453
  throughfall and stemflow sulfate, 465,467
  variations of isotopic enrichment factors during S oxidation, 461
Degree of pyritization, definition, 291–292
Degree of trace metal pyritization, definition, 292
Devonian black shale, *See* Organic-rich Devonian black shale
Digenite, concentration in massive sulfide deposits in Bathurst Mining Camp, New Brunswick, 536–548
Disposal of sulfide tailings, effect on profitability of mining operations, 535
Dissolution of rocks and component minerals, interest for mine waste management, 593
Dissolved ferric iron, role in chemical and biological leaching systems, 79
Dissolved oxygen, role of transport below water table in supergene enrichment, 154–155
Dolomite, role in acid mine tailings, 378
Duluth Complex Rock, sulfide mineral oxidation, 593–606

E

East Sullivan, Quebec, acid mine tailings, 365–378
Effective gaseous oxygen diffusion coefficient, calculation using Reactive Tailings Assessment Program, 137
Effective rate constants, definition, 158
Elbe River
  contamination of sediments, 298
  mobilization and scavenging of heavy metals after resuspension of anoxic sediments, 298–319

Electronic structure, PbS{100} surfaces, 201–221
Elemental S oxidation, $\delta^{18}O$ of sulfate formation, 453
Enthalpy transport module of Reactive Tailings Assessment Program, temperature calculation, 143
Ethylenediaminetetraacetic acid, pyrrhotite oxidation kinetics effect, 23,24t
Evolved-gas account
  carbonate determination, 440,442–443f
  CO_2 and SO_2 evolution monitoring, 433
  experimental procedure, 432–434
  future work, 444
  gases monitored, 434
  instrument, 433,435f
  predictive capability of acid–base accounting, 433
  pyrite determination, 434–441
  temperature of evolution vs. gas type, 433

F

$FeCO_3$, *See* Siderite
Feldspar, alteration during microbiological oxidation of black-schist ore, 98,103
FeOOH, H_2S oxidation product distribution effect, 397,399f
Ferric-bearing sulfate minerals, 333
Ferric ion(s), role in geochemical evolution of acidic ground water, 345–362
Ferric ion content of effluent, calculation, 4–5
Ferric sulfate, role in geochemical evolution of acidic ground water, 345–362
Ferrihydrite
  characteristics, 191
  Mössbauer spectra, 192,195f
  role in geochemical evolution of acidic ground water, 345–362
  X-ray diffraction diagram, 191–193f
Ferroan dolomite, concentration in massive sulfide deposits in Bathurst Mining Camp, New Brunswick, 536–548
Ferrous sulfate products, formation from pyrite oxidation, 35,37–38f
FeS_2, *See* Pyrite
$FeSO_4$, levels in mine tailing impoundments, 172–187

INDEX

Field research, thermal anomalies indicating sulfide-oxidation reactions in mine spoil, 645–656
Fine-grained terrigenous sediments, oxic to anoxic transition near sediment–water interface, 289–290
Finger dumps, 124
Forest byproducts, use as covers for reactive mine tailings, 631
Framvaren Fjord, hydrogen sulfide oxidation, 406,407f

G

Galena
  concentration in massive sulfide deposits in Bathurst Mining Camp, New Brunswick, 536–548
  levels in acid mine drainage in Wales, 261–273
  reaction rates with Fe^{3+} in acidic solutions, 2–12
  See also PbS{100} surfaces and chemisorption–oxidation reactions
Galena oxidation, reaction, 3
Gases, sour, treatment by microbial oxidation of sulfides, 68–77
General isotope-balance model, 501,505
Geochemical evolution of acidic ground water at reclaimed surface coal mine in Pennsylvania
  comparison to mass-balance models, 360–361
  experimental description, 345–346
  future work, 362
  geology, 347,349f,350
  hydrochemical data collection and evaluation procedure, 350–353t
  hydrochemical trends, 354,355f
  hydrology, 350
  inverse modeling, 354,356–360
  location and extent of oxidation reactions in subsurface, 353
  mine location, 347,348f
  saturation index, 351,353t
  secondary iron sulfate minerals as sources of acidity, 346–347
Geochemical model
  additions for field applications, 183,185,186f
  description, 178–180t,182t

Gibbsite, role in geochemical evolution of acidic ground water, 345–362
Goethite
  characteristics, 194,197
  concentration in massive sulfide deposits in Bathurst Mining Camp, New Brunswick, 536–548
  Mössbauer spectra, 197,198f
  X-ray diffraction diagram, 193f,197
Gold
  concentration in massive sulfide deposits in Bathurst Mining Camp, New Brunswick, 536–548
  concentration in weathered sulfide mine tailings, 516–533
Granite host rock
  supergene enrichment and weathering, 153–169
  time–space continuum formulation of supergene enrichment and weathering, 164,166–167f
Ground-water sulfate, $\delta^{18}O$, 467–470f
Gypsum, role in acid mine tailings, 375f,377

H

Hamburg harbor
  dredging, 298–299
  oxidation of sediments after dredging, 299
Heap leaching models, application to acid mine drainage, 128,130f,131
Heavy metals in resuspended anoxic sediments from Elbe River
  analytical procedure, 300–301
  Cd release, 302–304t
  characteristics of sediment, 300t
  Cl^- effect on release, 312t,313
  Cu release, 304t–306
  data treatment, 301–302
  laboratory simulation experimental procedure, 300
  net metal release at worst situation, 318,319f
  Pb release, 304t,305f
  previous studies, 299
  pyrite oxidation effect on release, 309,312,314f
  readsorption–coprecipitation, 306t
  release and scavenging process model, 316–319f

Heavy metals in resuspended anoxic sediments from Elbe River—*Continued*
  results from other studies, 306,308–309
  sampling site, 300,303f
  scavenging from aqueous phase, 313–316
  sediment sampling procedure, 300
  sources of released metals, 309–314
  sulfate concentration effect on release, 309t–311f
  Zn release, 304t,306,307f
Hematite, formation from pyrite oxidation, 35,39,40–41f
High-$\delta^{18}O$ sulfates in West Shasta District
  hypothesis of oxygen isotope exchange between SO_4 and H_2O, 510
  hypothesis of storage and release of high-$\delta^{18}O_{SO_4}$ in sulfate minerals, 510–511
  hypothesis of thin-film effects during wet–dry conditions, 509–510
Humidity effect on pyrite oxidation
  experimental procedure, 32,33t
  future work, 44
  identification of weathering products, 33,35–41
  isomer shift results, 33,34t
  previous studies, 31–32
  quadrupole splitting results, 33,34t
  weathering vs. pyrite sample vs. humidity, 39,42–43f
Hydrogen sulfide
  formation in natural waters, 393
  microbial oxidation by *Thiobacillus denitrificans* for treatment of sour water and gases, 68–77
  production in hydrothermal systems, 393–394
  production in pore water, 393
  removal and disposal by microbial process, 68–69
  removal via oxidation, 394
Hydrogen sulfide oxidation
  anoxic environments, 403,405–408
  Black Sea, 403,405f,406
  Cariaco Trench, 408
  Chesapeake Bay, 406,408
  Framvaren Fjord, 406,407f
  intermediate S^{4+}, 397,400
  kinetics model, 400–404f
  measurement methods, 413
  metal effect on rate, 395–396,398f

Hydrogen sulfide oxidation—*Continued*
  overall rate, 394–395
  previous studies, 413
  products, 396–399f
Hydrogen sulfide oxidation product determination by sulfur K-edge X-ray absorption near-edge structure spectroscopy
  air exposure effect, 422,424f
  beam-line setup, 420,422,423f
  experimental procedure, 413–414,420,422
  medium effect, 422,426t
  Ni^{2+} effect, 422,427–428f
  principles of X-ray absorption spectroscopy, 414–417f
  quantitative analytical approach, 416,418,419f
  self-absorption correction, 418,419f
  standard mixture analysis, 418,420,421t
  sulfur standards used for fitting, 422,425f
Hydrothermal enrichment of chalcopyrite
  equipment, 46–47,49f
  experimental procedure, 47–48
  mechanism, 52,53f
  oxygen feed rate effect, 48–51f
  particle size effect, 50,51f
  photomicrograph and X-ray examination, 50,52
  rate equation, 52,54–57
  steps, 48
  temperature effect, 50,53f
Hydrothermal systems, H_2S production, 393–394
Hydroxysulfate and oxide minerals, concentration in massive sulfide deposits in Bathurst Mining Camp, New Brunswick, 536–548

I

Immobile elements, description, 519
Inactive mill tailing(s)
  models describing movement of dissolved constituents, 516
  water quality effect, 516
Inactive mine tailing impoundments, siderite effect on pore-water chemistry, 172–187
Initial inventory module of Reactive Tailings Assessment Program
  effective gaseous oxygen diffusion coefficient calculation, 137

INDEX 669

Initial inventory module of Reactive Tailings Assessment Program—*Continued*
 particle size distribution calculation, 137,138*f*
Inorganic sulfur compound oxidation by thiobacilli
 enzymes involved, 62–63
 mechanism, 60–62,64–65
 polythionates, 64
 reactions, 63–64
Instream pH modification effect on sulfide oxidation product transport
 Al, 237,238*f*,240*t*
 analytical methods, 229
 calculated activities vs. pH, 229,231*f*,232
 Cu, 237,239*f*,240*t*
 experimental description, 225,227
 Fe, 237,239*f*–241
 geochemical reactions, 232–233
 geochemical simulation, 234,235*t*
 initial effects, 229,230*f*
 mass transfer for initial chemical reactions, 233*t*,234
 modification experimental procedure, 227,228*f*
 sampling procedure, 227,229
 sampling sites, location, 225,226*f*
 transport vs. pH, 235–237
Interstitial gas composition in wood chips deposited on reactive mine tailings
 composition, 634,636*t*,637*f*
 degradation of organic matter in mineral matrix, 638–639
 experimental description, 631
 gas sampling, 632–635*f,t*
 gas transfer, 640–642
 mass transport, 642
 oxidation–reduction reactions, application to wood chips, 639–640
 porosity of samples, 634–636*f*
 study site, 632,633*f*
 temperature measurement procedure, 634
 temperature vs. depth in cover, 634,637*f*
 temperature vs. reaction rate, 638
 water content of samples, 634,635*t*
 wood chip cover, impact on underlying mine tailings, 642–643
 wood chip sampling, 632

Inverse modeling of geochemical evolution of acidic ground water
 constraints, 356,357*t*
 description, 354
 flow from sludge-treated spoil to underlying bedrock, 359
 flow from untreated spoil to underlying bedrock, 356–358*t*
 waters from bedrock below untreated and sludge-treated areas mixed in deep bedrock below sludge-treated area, 359–360
 waters from untreated spoil and unmined coal mixed beneath sludge-treated spoil, 357,359
Iron
 attenuation rate from acid mine drainage, 255,256*t*
 chemical modeling, 256,257*t*
 concentration in acid rock drainage in natural wetland system, 382–391
 concentration in leachates from organic-rich Devonian black shale, 574–591
 concentration in sulfidic flood plain sediments, 282–285
 concentration in weathered sulfide mine tailings, 516–533
Fe^{2+}
 content in acid mine tailings, 365–378
 H_2S oxidation product distribution effect, 397,399*f*
 H_2S oxidation rate effect, 395–396
Fe^{2+} oxidation, mineralogical characteristics of poorly crystalline precipitates in acid sulfate waters, 190–199
Fe^{3+}
 H_2S oxidation product distribution effect, 397,399*f*
 H_2S oxidation rate effect, 395–396
 reaction rates with sulfide minerals in acidic solutions, 2–12
 role in abiotic oxidation of pyrite, 453–454
Fe^{3+} concentration, determination of effect on reaction rate, 6
 levels in acid mine drainage in Wales, 261–273
 pH modification effect on transport, 224–241
 release via authigenic pyrite oxidation in resuspended sediments, 289–297

Iron—*Continued*
role in geochemical evolution of acidic ground water, 345–362
role in sulfide mineral oxidation, 2–3
site of removal from solution, 253–255
transport during sulfide oxidation, 237,239f–241
Iron Mountain, California, seasonal variations of Zn/Cu ratios in acid mine water, 324–343
Iron Mountain Mine, geology and hydrology, 484,487
Iron sulfate minerals, secondary, sources of acidity, 346–347
Iron sulfide oxidation, *See* Pyrite oxidation
Iron transformation, column bioleaching of complex sulfide ore, 79–87
Isomer shift, description, 32
Isotope, studies, 483–484
Isotope fractionation effect, $\delta^{18}O$ in sulfate, 461
Isotope geochemistry of acid mine drainage in western United States
abiologic oxidation, 507–508
biologically mediated oxidation, 508t,509
collection and storage of water samples, 488,491
experimental description, 482
geologic and hydrologic setting of sample sites, 484–490
gravimetric analytical procedure, 491
high-$\delta^{18}O$ sulfates in West Shasta District, 509–511
isotope analytical procedure, 491
isotopic, chemical, and physical data, 491–493t
microbial population, 498
oxygen isotope(s), 495,497f,498
oxygen isotope fractionation between sulfate and water vs. environment, 498–502
pH values, 491,494f
previous studies
abiologic sulfide oxidation, 482–483
isotopes, 453–454
microbial mediation of oxidation, 453
sources of oxygen in sulfate, 501,503–507
sulfate concentration, 491,494f
sulfur isotopes, 495,496f
temperatures of acid waters, 491,495

J

Jarosite
role in acid mine tailings, 375f,377
role in geochemical evolution of acidic ground water, 345–362

K

Kaolinite, role in geochemical evolution of acidic ground water, 345–362
Kinetics
hydrothermal enrichment of chalcopyrite, 45–57
pyrrhotite oxidation, 14–28
Kinetics module of Reactive Tailings Assessment Program, description for sulfide oxidation, 139–141t
Kinetics of H_2S oxidation in natural waters
Black Sea, 403,405f,406
Cariaco Trench, 408
Chesapeake Bay, 406,408
Framvaren Fjord, 406,407f
metal effect on rate, 395–396,398f
model, 400–403,404f
overall rate, 394–395
oxidation of intermediate S^{4+}, 397,400
oxidation products, 396–399f

L

Lakes, sulfate, $\delta^{18}O$, 473–475
Leachate chemistry of organic-rich Devonian black shale, pyrite oxidation effect, 574–591
Leached rims, development on waste fragments in dumps, 125
Leaching of copper-bearing mine waste
air convection as control, 124
climate effect, 128,129f
information obtained from modeling, 127–128
leach rate control using leached rims, 125
models, 125–128,130f,131
Lead
concentration in massive sulfide deposits in Bathurst Mining Camp, New Brunswick, 536–548
concentration in sulfidic flood plain sediments, 282,284f,285

INDEX

Lead—*Continued*
  concentration in weathered sulfide mine tailings, 516–533
  content in acid mine tailings, 377
  levels in acid mine drainage in Wales, 261–273
  mobilization and scavenging in resuspended anoxic sediments from Elbe River, 298–319
  Pb^{2+}
    H_2S oxidation product distribution effect, 397,399f
    H_2S oxidation rate effect, 395–396
  precipitation from acid mine tailings, 376
Leviathan Creek, transport and natural attenuation of Cu, Zn, As, and Fe in acid mine drainage, 244–259
Leviathan Mine, California, geology and hydrology, 488,489f
Limestone host rock
  supergene enrichment and weathering, 153–169
  time–space continuum formulation of supergene enrichment and weathering, 164,167–169
Lower tailings, description, 279t
Lutetium, concentration in weathered sulfide mine tailings, 516–533

M

Magnesium
  concentration in leachates from organic-rich Devonian black shale, 574–591
  concentration in massive sulfide deposits in Bathurst Mining Camp, New Brunswick, 536–548
  concentration in weathered sulfide mine tailings, 516–533
  role in geochemical evolution of acidic ground water, 345–362
Magnetic splitting, description, 32
Mammoth Mine, geology and hydrology, 487
Manganese
  concentration in leachates from organic-rich Devonian black shale, 574–591
  concentration in sulfidic flood plain sediments, 282–285
  levels in acid mine drainage in Wales, 261–273

Manganese—*Continued*
  Mn^{2+}
    H_2S oxidation product distribution effect, 397,399f
    H_2S oxidation rate effect, 395–396
    role in geochemical evolution of acidic ground water, 345–362
Marcasite
  concentration in massive sulfide deposits in Bathurst Mining Camp, New Brunswick, 536–548
  evolved-gas analysis, 437,440,441f
  reactivity in acid rock drainage in natural wetland system, 383
Mass-balance calculations on weathered sulfide mine tailings
  comparison of Heath Steele and Delnite tailings, 531,533
  element mass calculation, change, 517
  experimental procedure, 516,520–521
  geochemical profiles
    Delnite tailings, 528,530f–532f
    Heath Steele, new tailings, 524,527f–529f
    Heath Steele, old tailings, 521–526
  Heath Steele tailings, procedure, 521
  study area descriptions, 518f,520
  tailing impoundments, location, 517,518f
  theory, 517–519
  zero-mass-change volume factor calculation, 519t
Mechanisms that govern pollutant generation from pyritic wastes, rates, 108–122
Mediation of oxidation, microbial, studies, 483
Melanterite
  role in geochemical evolution of acidic ground water, 345–362
  role in metal-rich sulfuric acid solution formation, 325
  seasonal variations of Zn/Cu ratios, 325–343
  structure, 325
  Zn/Cu partitioning, 335–342
Mercury
  concentration in massive sulfide deposits in Bathurst Mining Camp, New Brunswick, 536–548
  coprecipitation with pyrite in anoxic sediments, 290

Mercury—*Continued*
 release via authigenic pyrite oxidation in resuspended sediments, 289–297
Metal-rich sulfuric acid solutions, factors affecting formation, 324–325
Metallic sulfide bearing waste rock, acid mine drainage, 609–627
Metalliferous mining operations
 cessation of operation, 261–262
 history, 261
Mica, alteration during microbiological oxidation of black-schist ore, 98,103
Microbial mediation of oxidation, studies, 483
Microbial oxidation of sulfides by *Thiobacillus denitrificans* for treatment of sour water and gases
 behavior under upset conditions, 70–71
 coculture of*Thiobacillus denitrificans* with floc-forming heterotrophs, 71–72
 control of H_2S production by sulfate-reducing bacteria, 73–75
 field test to control H_2S production, 75–77t
 septic operation effect, 71
 sulfide-tolerant strains of *Thiobacillus denitrificans*, 72
 Thiobacillus denitrificans, growth on H_2S, 69–70t
 treatment of sour water, 73
Microbial oxidation of sulfite, $\delta^{18}O$ of sulfate formation, 451–452
Microbial process, 68–69
Microbiological oxidation of pyrrhotite and pyrite, alteration of mica and feldspar, 90–103
Mine drainage minerals
 characteristics, 192
 definition, 190
 IR spectra of arsenate treatment results, 194,196f
 Mössbauer spectra, 192,194,195f
 X-ray diffraction diagrams, 192,193f
Mine spoil, thermal anomalies indicating sulfide-oxidation reactions, 645–656
Mine tailing(s), weathered sulfide, mass-balance calculations, 516–533
Mine tailing impoundments, inactive, siderite effect on pore-water chemistry, 172–187

Mine waste
 Cu bearing, leaching model, 123–131
 environmental problem of pyrite oxidation, 108
Mineralogical changes during bacterial leaching of black-schist ore material
 abiotic condition effect, 94,96f
 Al concentration effect, 98,102f
 Ca concentration effect, 98,102f
 dissolution of metals, 94t,95f
 elemental S formation, 98,99–100f
 experimental procedure, 91–92
 Fe concentration effect, 98,101f
 graphite dissolution, 103
 inoculated system effect, 94,97f
 jarosite formation, 98–101f
 K concentration effect, 98,101f
 Mg concentration effect, 98,102f,103
 pH vs. oxidative dissolution, 94,95f
 redox potentials vs. oxidative dissolution, 94,95f
 Si concentration effect, 98,102f
 X-ray diffractogram of untreated ore, 92,93f
Mineralogical characteristics of poorly crystalline precipitates of Fe^{2+} oxidation in acid sulfate waters
 experimental procedure, 191
 ferrihydrite, 191–193f,195f
 geothite, 194,197,198f
 mine drainage mineral, 192,194–196f
 natural precipitates, 197,199
 synthetic precipitates, 199
MnO_2, H_2S oxidation product distribution effect, 397,399f
Mobile elements, description, 519
Modeling, industrial-scale leaching of copper-bearing mine waste, 123–131
Modified British Columbia research initial method, neutralization potential determination, 612
Molybdenum
 coprecipitation with pyrite in anoxic sediments, 290
 levels in acid mine drainage in Wales, 261–273
 release via authigenic pyrite oxidation in resuspended sediments, 289–297
Mössbauer spectroscopy, analysis of pyrite oxidation products, 32–44

INDEX

N

Na$_2$CO$_3$, sulfide-oxidation product transport effect, 224–241
Natrojarosite, role in geochemical evolution of acidic ground water, 345–362
Natural semiconducting materials, 201
Natural waters, kinetics of H$_2$S oxidation, 393–410
Natural wetland system, attenuation of acid rock drainage, 382
Neutral-to-alkaline environments, sources of oxygen in sulfate, 504f–506
Neutralization potential
  determination problems, 611–612
  humidity cell array, 612,613f
  modified British Columbia research, initial method for determination, 612
  Sobek method for determination, 612
Nickel
  concentration in leachates from organic-rich Devonian black shale, 574–591
  concentration in sulfidic flood plain sediments, 282,283f
  concentration in weathered sulfide mine tailings, 516–533
Ni^{2+}
  H$_2$S oxidation rate effect, 395–396
  role in H$_2$S oxidation, 412–428
  precipitation from acid mine tailings, 376
  solubilization during microbiological oxidation of pyrrhotite and pyrite, 90–103

O

Oceans, sulfate, δ^{18}O, 473–475
Ochre precipitation, water chemistry effect for acid mine drainage, 261–273
Olivine, dissolution in Duluth Complex rock, 594
Ore deposits, sulfide bearing, time–space continuum formulation of supergene enrichment and weathering, 153–169
Organic-rich Devonian black shale
  source of kerogen, 574
  weathering, 574–591
Organic sulfate hydrolysis effect, δ^{18}O in sulfate, 463

Overbank tailings
  description, 279t,280
  distribution, 280t–282
Overburden analysis, description, 432
Oxidation
  inorganic sulfur compounds by thiobacilli, 60–65
  microbial, See Microbial oxidation of sulfides by *Thiobacillus denitrificans* for treatment of sour water and gases
  stoichiometric contributions and environments, 506–507
Oxidation kinetics, pyrrhotite, See Pyrrhotite oxidation kinetics
Oxidation of H$_2$S, kinetics in natural waters, 393–410
Oxidation of massive sulfide deposits, Bathurst Mining Camp, New Brunswick, 535–548
Oxidation of sulfide minerals, 2
Oxidative–dissolution reactions of sulfide minerals, production or consumption of acid, 79–80
Oxygen barrier, use of wood chips, 631–643
Oxygen feed rate, chalcopyrite hydrothermal enrichment effect, 48–51f
Oxygen isotopes, geochemistry of acid mine drainage in western United States, 481–511
Oxygen transport module of Reactive Tailings Assessment Program, 142–143

P

Particle size, chalcopyrite hydrothermal enrichment effect, 50,51f
Particle size distribution, calculation using Reactive Tailings Assessment Program, 137,138f
PbS{100} surfaces and chemisorption–oxidation reactions
  electronic structure of clean surfaces, 202–205f
  experimental procedure, 201–202
  fresh surfaces, reaction with air and oxygen, 203,206f–208
  kinetics, 219
  mechanism constraints, 220
  [110] parallel reaction borders, 219–221f

PbS{100} surfaces and chemisorption–oxidation reactions—*Continud*
  real-time scanning tunneling microscopy, 212,214–219
  static scanning tunneling microscopy, 208–213
Pennsylvania, geochemical evolution of acidic ground water at reclaimed surface coal mine, 345–362
Pentlandite, concentration in Duluth Complex rock, 594–606
pH
  pyrrhotite oxidation kinetics effect, 20,21*f*,24
  sulfate release rate of acid mine drainage effect, 600,603
  sulfur content of acid mine drainage effect, 602–606
  supergene enrichment effect, 155
Phosphate addition, suppression of pyrite oxidation rate, 562–573
Plagioclase, dissolution in Duluth Complex rock, 594
Platinum group elements
  mobility in aqueous solution, 551
  thiosulfate complexing, 553–559
  weathering profiles, 551–553
Pollutant generation from pyritic wastes, rates of governing mechanisms, 108–122
Polythionates, oxidation reactions, 64
Poorly crystalline precipitates of Fe^{2+} oxidation in acid sulfate waters, mineralogical characteristics, 190–199
Poorly crystalline products of sulfate oxidation, identification, 190–191
Pore water, H_2S production, 393
Pore-water chemistry of inactive mine tailing impoundments, siderite effects, 172–187
Porphyry Cu deposits, concept of secondary enrichment, 45
Potassium
  concentration in leachates from organic-rich Devonian black shale, 574–591
  concentration in massive sulfide deposits in Bathurst Mining Camp, New Brunswick, 536–548
  concentration in weathered sulfide mine tailings, 516–533
  role in geochemical evolution of acidic ground water, 345–362

Precipitation of secondary minerals, acid mine tailings in East Sullivan, Quebec, 376–378
Premining flood plain deposits, 279*t*,280
Pyrite
  abiotic oxidation, 453–456*t*
  acid mine drainage from metallic sulfide bearing waste rock, 609–627
  acid mine tailings, 372,374,375*f*
  alteration of mica and feldspar associated with microbiological oxidation, 90–103
  biological leaching systems, 91
  concentration in Duluth Complex rock, 594–606
  concentration in massive sulfide deposits in Bathurst Mining Camp, New Brunswick, 536–548
  coprecipitation with toxic metals in anoxic sediments, 290
  evolved-gas analysis, 434–441
  fractions, 290
  geochemical evolution of acidic ground water, 345–362
  levels in acid mine drainage in Wales, 261–273
  reaction, 31
  reactivity in acid rock drainage in natural wetland system, 383
  reasons for interest, 562
  sources, 562
Pyrite oxidation
  analytical methods, 562
  humidity effect, 31–44
  intrinsic rate, 109,110*f*
  leachate chemistry of organic-rich Devonian black shale, 574–591
  studies, 14–15,173,175
Pyrite oxidation in mine wastes, 108–109
Pyrite oxidation in resuspended sediments, authigenic, toxic metal release, 289–297
Pyrite oxidation rate suppression by phosphate addition
  direct evidence for $FePO_4$2 surface coating, 565,568–571
  experimental procedure, 562–563
  Fe release during oxidation, 565,567*f*
  first-order plots of oxidation, 565,566*f*
  oxidation reaction, 563–564
  oxidation vs. time, 570,572*f*,573

INDEX

Pyrite oxidation rate suppression by phosphate addition—*Continued*
  rate law of oxidation, 564–565
  release of SO_4 during oxidation, 565,566*f*
Pyritic tailings, computer program to assess acid generation, 132–150
Pyritic wastes, rates of mechanisms that govern pollutant generation, 108–122
Pyroxenes, dissolution in Duluth Complex rock, 594
Pyrrhotite
  acid mine drainage from metallic sulfide bearing waste rock, 609–627
  acid mine tailings, 377
  alteration of mica and feldspar associated with microbiological oxidation, 90–103
  biological leaching systems, 79–87,91
  chemistry, 15
  concentration in Duluth Complex rock, 594–606
  concentration in massive sulfide deposits in Bathurst Mining Camp, New Brunswick, 536–548
  reactivity in acid rock drainage in natural wetland system, 383
Pyrrhotite oxidation
  ferric iron reaction, 15–16
  knowledge, 14–15
  oxygen reaction, 15
  present work, 17
  previous work, 16–17
Pyrrhotite oxidation kinetics
  comparison with pyrite oxidation kinetics, 27,28*t*
  ethylenediaminetetraacetic acid effect, 23,24*t*
  experimental procedure, 17–20
  future work, 27–28
  iron concentration vs. time, 20,21*f*
  pH effect, 20,21*f*,24
  pyrite–pyrrhotite mixtures, 24,25*f*,27
  reactor design, 18,19*f*
  sulfate concentration vs. time, 20,21*f*
  surface area effect, 20,23,24*f*
  surface/iron ratios, 23*t*,26–27
  temperature effect, 20,22–25

Q

Quadrupole splitting, description, 32

Quartz
  concentration in massive sulfide deposits in Bathurst Mining Camp, New Brunswick, 536–548
  role in biological leaching systems, 91

R

Random simulation module of Reactive Tailings Assessment Program, 149
Rates of mechanisms that govern pollutant generation from pyritic wastes
  future work, 121
  intercomparison of oxidation rates, 109,111*t*,112
  intrinsic oxidation rate, 109,110*f*
  nomenclature, 121–122
  pollutant transit times, 117,119
  rehabilitation of environment, 119–120
  shrinking core model, 117
  simple constant rate model, 112–116
  simple homogeneous model, 116,118*t*
  time scales, 117,118*t*
Reaction rate for standard system, calculation, 5
Reaction rates of sulfide minerals with Fe^{3+} in acidic solutions
  Arrhenius plot for Fe^{3+}–arsenopyrite reaction, 6,9*f*
  chemical reaction mechanism, 11
  experimental procedure, 3–5
  rate data, 6,8*t*
  rate data for chalcopyrite experiments, 6,7*f*
  rate laws based on best fit of experimental data, 6,9*t*
  reaction order, 10–11
  reaction processes, 6,10
  reaction rate vs. Fe^{3+} concentration, 11,12*f*
  reaction rate vs. mineral, 11,12*t*
  scanning electron microscopic results, 6,10
Reactive Tailings Assessment Program
  development, 133
  enthalpy transport module, 143
  initial inventory module, 137,138*f*
  kinetics module, 139–141*t*
  model simulations, 144–150
  modules and subroutines, 135–137
  objectives, 133
  oxygen transport module, 142–143

Reactive Tailings Assessment Program—*Continued*
  physical concept, 133–135
  random simulation module, 149,150f
  reactive sulfide minerals in tailings and common oxidation products, 133,135t
  solute transport module, 143
  sulfide-oxidation module, 140–142
Reclaimed surface coal mine in Pennsylvania, geochemical evolution of acidic ground water, 345–362
Reduced organic S oxidation, $\delta^{18}O$ of sulfate formation, 459,461
Reduced tailings, description, 279t,280
Relative humidity, sulfate release rate from acid mine drainage effect, 603
Release and scavenging process model for heavy metals
  net metal release at worst situation, 318,319f
  release stage, 316,317f
  scavenging stage, 317f,318
  schematic representation, 316,317f
  steady-state stage, 317f–319f
  transition stage, 316–318
Release of metals from sediment, calculation, 301
Resuspended sediments, toxic metal release via authigenic pyrite oxidation, 289–297
Reworked tailings
  description, 279t,280
  distribution, 280t–282
Rhodanese, role in inorganic sulfur compound oxidation, 62
Rhodochrosite, evolved-gas analysis, 440,442–443f
Romerite, role in geochemical evolution of acidic ground water, 345–362
Runoff, sulfate, $\delta^{18}O$, 469,472f,473

S

Sandstone host rock
  supergene enrichment and weathering, 153–169
  time–space continuum formulation of supergene enrichment and weathering, 162–165f
Saturation index
  calculation, 185,187
  description, 351

Scandium, concentration in weathered sulfide mine tailings, 516–533
Scanning tunneling microscopy, analysis of PbS{100} surfaces and chemisorption–oxidation reactions, 201–221
Scavenging concentration, calculation, 301–302
Seas, sulfate, $\delta^{18}O$, 473–475
Seasonal variations of Zn/Cu ratios, acid mine water from Iron Mountain, California, 324–343
Secondary iron sulfate minerals, sources of acidity, 346–347
Secondary sulfate minerals, 333,334t
Self-neutralization, limit in acid mine tailings, 365–378
Shrinking core model, oxidation rate for pollutant generation from pyritic wastes, 117
Shrinking radius kinetics, concept, 140–142
Siderite
  concentration in massive sulfide deposits in Bathurst Mining Camp, New Brunswick, 536–548
  evolved-gas analysis, 440,442–443f
Siderite effect on pore-water chemistry of inactive mine tailing impoundments
  additions to geochemical model for field application, 183,185,186f
  dissolution, 175
  Eh-buffering mechanisms, 176
  experimental description, 173
  field procedure, 178
  field study results, 183–187
  geochemical model description, 178–180t,182t
  geochemical modeling of tailings water, 176–177
  laboratory procedure, 177
  laboratory study results, 178–183
  pH-buffering mechanisms, 175–176
  pyrite oxidation, 173,175
  saturation index calculation, 185,187
  solubility product calculation, 181–183
Silicate minerals, role in biological leaching systems, 91
Silicon
  concentration in weathered sulfide mine tailings, 516–533
  role in geochemical evolution of acidic ground water, 345–362

INDEX

Silver, concentration in massive sulfide deposits in Bathurst Mining Camp, New Brunswick, 536–548

Simple constant rate model, oxidation rate for pollutant generation from pyritic wastes, 112–116

Simple homogeneous model, oxidation rate for pollutant generation from pyritic wastes, 116,118t

SO_2, evolution, 433

SO_4
concentration in acid rock drainage in natural wetland system, 382–391
concentration in weathered sulfide mine tailings, 516–533
levels in acid mine drainage in Wales, 261–273

SO_4^{2-}
concentration in leachates from organic-rich Devonian black shale, 574–591
content in acid mine tailings, 365–378
pH modification effect on transport, 224–241

Sobek method, neutralization potential determination, 612

Sodium
concentration in leachates from organic-rich Devonian black shale, 574–591
concentration in weathered sulfide mine tailings, 516–533
role in geochemical evolution of acidic ground water, 345–362

Soil sulfate, $\delta^{18}O$, 467

Solid-phase alteration, column bioleaching of complex sulfide ore, 79–87

Solubility products, calculation, 181–183

Solute transport module of Reactive Tailings Assessment Program, sulfide-oxidation determination, 143

Sorption–desorption effect, $\delta^{18}O$ in sulfate, 462

Sour water and gases, treatment by microbial oxidation of sulfides, 68–77

Sphalerite
bioleaching, 79–87
concentration in Duluth Complex rock, 594–606
concentration in massive sulfide deposits in Bathurst Mining Camp, New Brunswick, 536–548

Sphalerite—*Continued*
levels in acid mine drainage in Wales, 261–273
reaction rates with Fe^{3+} in acidic solutions, 2–12
role in acid mine tailings, 377

Sphalerite oxidation, reaction, 3

Spoil temperature
importance of field studies, 645
information obtained from measurement, 646

Springs, sulfate, $\delta^{18}O$, 469,471f

Stemflow sulfate, $\delta^{18}O$, 465,467

Stoichiometric isotope-balance model, description, 501,505

Stratigraphy, sulfidic flood plain sediments from Upper Clark Fork Valley, Montana, 277,279t,280

Sulfate
factors affecting concentration and isotopic composition, 446
key component of global biogeochemical S cycle, 446
lakes, $\delta^{18}O$, 473,474f,475
oceans, $\delta^{18}O$, 473,474f,475
runoff, $\delta^{18}O$, 469,472f,473
seas, $\delta^{18}O$, 473,474f,475
springs, $\delta^{18}O$, 469,471f

Sulfate mineral precipitation effect, $\delta^{18}O$ in sulfate, 461

Sulfate reduction effect, $\delta^{18}O$ in sulfate, 463–464

Sulfate–water oxygen exchange effect, $\delta^{18}O$ in sulfate, 462–463

Sulfide(s), microbial oxidation by *Thiobacillus denitrificans* for treatment of sour water and gases, 68–77

Sulfide-bearing ore deposits, time–space continuum formulation of supergene enrichment and weathering, 153–169

Sulfide-bearing tailing oxidation, source of heavy metal contamination, 365

Sulfide deposits, oxidation in Bathurst Camp, New Brunswick, 535–548

Sulfide mine tailings, weathered, mass-balance calculations, 516–533

Sulfide mineral(s)
dissolution reactions, 80
oxidation in Duluth Complex rock, 593–606
oxidation reactions, 79–80

Sulfide mineral(s)—*Continued*
  reaction rates with Fe^{3+} in acidic
    solutions, 2–12
Sulfide mineral oxidation reactions
  mechanism study techniques, 201
  mine tailing impoundments, 172–174f
  siderite formation, 173
  surface studies, 201
Sulfide oxidation
  abiologic, studies, 482–483
  acid mine tailings in East Sullivan,
    Quebec, 372,374,375f
  $\delta^{18}O$ of sulfate formed by abiotic
    oxidation of pyrite, 454–456t
  $\delta^{18}O$ of sulfate formed by sulfide oxidation
    with bacterial isolates, 457–459
  $\delta^{18}O$ of sulfate formed in other sulfide
    oxidation experiments, 459,460f
  product identification, 190
  role of bacteria, 456–457
  role of Fe^{3+} in abiotic oxidation of
    pyrite, 453–454
Sulfide oxidation from mines, 153
Sulfide-oxidation module of Reactive
  Tailings Assessment Program, 140–142
Sulfide-oxidation products in surface water
  factors affecting transport and fate, 224–225
  field studies of transport and fate, 225
  pH modification effect on transport, 224–241
Sulfide-oxidation reactions, thermal
  anomalies in mine spoil, 645–656
Sulfide tailing deposit oxidation,
  importance of control, 535
Sulfidic flood plain sediments from Upper
  Clark Fork Valley, Montana
  analytical methods, 277
  chemistry, 282–286
  distribution of tailings, 280t–282
  downstream movement, 276–277
  experimental description, 277
  origination, 276
  sampling procedure, 277
  sources of trace elements, 286–287
  stratigraphy, 277,279t,280
  study site, 277,278f
Sulfite oxidase, role in inorganic sulfur
  compound oxidation, 62
Sulfite oxidation
  $\delta^{18}O$ of sulfate formed by abiotic
    oxidation, 449–451

Sulfite oxidation—*Continued*
  $\delta^{18}O$ of sulfate formed by microbial
    oxidation, 451–452
  oxygen isotope exchange between sulfite
    and water, 449
  reaction rates, 447,449
Sulfosalts, concentration in massive
  sulfide deposits in Bathurst Mining
  Camp, New Brunswick, 536–548
Sulfur
  concentration in weathered sulfide mine
    tailings, 516–533
  role in geochemical evolution of acidic
    ground water, 345–362
Sulfur compounds, inorganic, oxidation by
  thiobacilli, 60–65
Sulfur cycles, 412–413
Sulfur isotopes, geochemistry of acid mine
  drainage in western United States,
  481–511
Sulfur K-edge X-ray absorption near-edge
  structure spectroscopy, H_2S
  oxidation product determination, 412–428
Sulfur transferase, role in inorganic
  sulfur compound oxidation, 62
Super concentrates, preparation, 46
Supergene enrichment and weathering of
  sulfide-bearing ore deposits
  continuum representation of mass
    transport and reaction, 155–158
  effective rate constants, 158,160,161t
  experimental objective, 154
  factors controlling sulfide-oxidation
    rate, 154–155
  granite host rock, 164,166–167f
  limestone host rock, 164,167–169
  minerals, 158,159t
  modal compositions, 158,161t
  model calculations, 158–169
  pH vs. depth, 160–162
  sandstone host rock, 162–165f
  structure of weathering profile, 154
  time–space continuum formulation,
    153–169
Supergene mineral deposits of Bathurst
  Camp, New Brunswick
  alteration sequence for sulfide
    oxidation, 540,541f
  depletion of elements during
    formation, 540,543t–545

INDEX

Supergene mineral deposits of Bathurst Camp, New Brunswick—*Continued*
  enrichment of elements during formation, 540,542–544
  experiment description, 535–536
  factors affecting oxidation rate, 546–548
  geology of camp, 536,538f
  mineralogy, 536,537t
  recent oxidation, 545–546
  stratigraphy of oxidation zones, 536,539f,540
Supergene weathering, 153–154
Suppression, pyrite oxidation rate by phosphate addition, 562–573
Surface area, pyrrhotite oxidation kinetics effect, 20,23,25f
Synchrotron-radiation-based X-ray absorption near-edge structure spectroscopy
  advantages, 413
  hydrogen sulfide oxidation product determination, 412–428

T

Temperature
  chalcopyrite hydrothermal enrichment effect, 50,53f
  pyrrhotite oxidation kinetics effect, 20,22–25
  sulfate release rate from acid mine drainage effect, 603
Tetrahedrite, concentration in massive sulfide deposits in Bathurst Mining Camp, New Brunswick, 536–548
Thallium, levels in acid mine drainage in Wales, 261–273
Thermal anomalies indicating sulfide oxidation reactions in mine spoil
  amplitude attenuation, 649
  evidence, 649,653
  experimental procedure, 646–647
  field sites, location, 646,648f
  intensity changes of thermal anomaly with time, 654,656f
  monitoring wells, location, 647,648f
  seasonal variation, 649,650–651f
  seep water chemistry, 646,647t
  shallow temperature interpretation, 653–655f
  temperatures vs. depth, 649,652

Thiobacilli, oxidation of inorganic sulfur compounds, 60–65
Thiobacillus denitrificans
  description, 69
  microbial oxidation of sulfides for treatment of sour water and gases, 68–77
Thiobacillus ferrooxidans, reactions, 63
Thiobacillus versutus, reactions, 63–64
Thiosulfate complexing of platinum group elements
  complex identification, 557–558
  dissolution experimental procedure, 553
  dissolution results, 554–557
  experimental description, 553
  formation conditions, 559
  pH effect, 558
  ^{195}Pt NMR experimental procedure, 553–554
  temperature effect, 558–559
Thiosulfate oxidation, $\delta^{18}O$ of sulfate formation, 452–453
Thiosulfate-oxidizing enzyme, role in inorganic sulfur compound oxidation, 62
Throughfall sulfate, $\delta^{18}O$, 465,467
Time scales, pollutant generation from pyritic wastes, 117,118t
Tin, levels in acid mine drainage in Wales, 261–273
Titanium, concentration in weathered sulfide mine tailings, 516–533
Total concentration of net release of metals, calculation, 302
Toxic metal(s), factors controlling availability to biota in aquatic environments, 289
Toxic metal release via authigenic pyrite oxidation in resuspended sediments
  analytical procedure, 291
  extent of release vs. extent of oxidation, 292,296f
  initial degree of pyritization vs. percent oxidation, 292,295f
  metal extraction procedure, 291
  oxidation kinetics experimental procedure, 291
  oxidative release, 292,294t
  pyritization of metals, 291–293f
  sample collection, 291
  sediment age effect, 292,297t
  study area, 290

Trace metal attenuation in acid mine
  drainage of Leviathan and Bryant Creeks
  analytical and physical data, 248,250–251t
  analytical techniques, 245,248
  attenuation rates, 255,256f
  chemical modeling, 256–258
  computational methods, 248,252
  flow rate determination, 248,249f
  influencing factors, 258–259
  locality map of features and sampling sites, 244–247f
  modifications of original data, 259
  previous studies, 245
  sampling techniques, 248
  stream description, 252
  trace metal removal from solution, 252–254f
Transit times, pollutant generation from pyritic wastes, 117,119
Transport, Cu, Zn, As, and Fe in acid mine drainage of Leviathan and Bryant Creeks, 244–259
Transport of sulfide-oxidation products
  field studies, 225
  influencing factors, 224–225
  pH modification effect, 225–241
Troilite, concentration in Duluth Complex rock, 594–606

U

Upper Clark Fork Valley, Montana, stratigraphy and chemistry of sulfidic flood plain sediments, 276–287
Upper tailings, description, 279t

V

Vanadium, concentration in weathered sulfide mine tailings, 516–533

W

Wales
  acid mine drainage, 266–273
  history of metalliferous mining, 261–262
Wastes from nonferrous metal mines, quantities of iron sulfide minerals, 14

Water
  natural, kinetics of H_2S oxidation, 393–410
  sour, treatment by microbial oxidation of sulfides, 68–77
Water chemistry of acid mine drainage, ochre precipitation effect, 261–273
Weathered sulfide mine tailings, mass-balance calculations, 516–533
Weathering of organic-rich Devonian black shale
  analytical procedure, 577,579
  apparatus, 577,578f
  calcium concentration vs. leachate chemistry, 591
  composition of materials, 575t
  elemental concentration ranges and averages, 579,584
  elemental release patterns, 584–586
  experimental procedure, 577
  leachate chemistry vs. lysimeter, 579,582–584
  leachate chemistry vs. project year, 579–581t,584
  mineral equilibria vs. leachate chemistry, 590–591
  pyrite oxidation, 586
  raw shale leachate chemistry vs. pyrite oxidation, 587–588f
  retorted shale leachate chemistry vs. pyrite oxidation, 587–590
  retorting effect on mineralogy and weathering behavior, 575,576f
  study site, 574–575
West Shasta District, California, geology and hydrology, 484–487
Western United States, sulfur and oxygen isotope geochemistry in acid mine drainage, 481–511
Wetland system, natural, attenuation of acid rock drainage, 382
White lines, description, 414
Wood chips, composition of interstitial gases, 631–643
Wood wastes, potential use as covers for sulfide tailings, 631
Woodhouseite, concentration in massive sulfide deposits in Bathurst Mining Camp, New Brunswick, 536–548

INDEX

X

X-ray absorption spectroscopy
  elemental sulfur, 414,415f
  gaseous argon, 414,415f
  principles, 414,415f
  sulfur speciation and edge shift, 414,416,417f

Y

Ytterbium, concentration in weathered sulfide mine tailings, 516–533

Z

Zero-mass-change volume factor
  calculation, 519
  values for immobile elements, 519t
Zinc
  attenuation rate from acid mine drainage, 255,256t
  chemical modeling, 257t,258
  concentration in acid rock drainage in natural wetland system, 382–391
  concentration in leachates from organic-rich Devonian black shale, 574–591
  concentration in sulfidic flood plain sediments, 282,283f,284f,285
  concentration in weathered sulfide mine tailings, 516–533
  content in acid mine tailings, 377

Zinc—*Continued*
  levels in acid mine drainage in Wales, 261–273
  mobilization and scavenging in resuspended anoxic sediments from Elbe River, 298–319
  precipitation from acid mine tailings, 376
  site of removal from solution, 254f,255
  solubilization during microbiological oxidation of pyrrhotite and pyrite, 90–103
Zn/Cu partitioning between melanterite and water
  analytical procedure, 335–336
  heating–cooling experimental procedure, 336–340
  sample collection, 335
  temperature dependence, 336–338,341–342
Zn/Cu ratios in acid mine water from Iron Mountain, California
  acid mine drainage, 329
  analytical data for elements in water samples, 329,332t,333
  composition of sulfide deposits, 328
  deposit sites, 325,327f,328
  flow rates vs. portals, 329–333
  future work, 342–343
  mine location, 325,326f
  rainfall vs. flow rates, 329,331f,333
  secondary sulfate minerals, 333,334t
  Zn/Cu partitioning between melanterite and water, 335–342
Zirconium, concentration in weathered sulfide mine tailings, 516–533

Production: Meg Marshall
Indexing: Deborah H. Steiner
Acquisition: Rhonda Bitterli

Printed and bound by Maple Press, York, PA

RETURN TO ➡

CHEMISTRY LIBRARY
100 Hildebrand Hall • 642-3753

LOAN PERIOD 1	2	3
4	5	6 1 MONTH

ALL BOOKS MAY BE RECALLED AFTER 7 DAYS
~~Renewable by telephone~~

DUE AS STAMPED BELOW

MAY 2 2 1999	SENT ON ILL	
~~DEC 1 6 1999~~	AUG 1 6 2013	
AUG 2 2 2000	U.C. BERKELEY	
DEC 1 4 2000		
AUG 1 6 2002		
MAY 2 0 2005		
AUG 1 7		
~~NOV 0 7~~		
~~DEC 1 1~~		
~~SEP 1 6~~		

FORM NO. DD5

UNIVERSITY OF CALIFORNIA, BERKELEY
BERKELEY, CA 94720-6000